Exploring the Unknown

ISBN 978-0-16-081381-8

For sale by the Superintendent of Documents, U.S. Government Printing Office
Internet: bookstore.gpo.gov   Phone: toll free (866) 512-1800;   DC area (202) 512-1800
Fax: (202) 512-2104 Mail: Stop IDCC, Washington, DC 20402-0001

ISBN 978-0-16-081381-8

Exploring the Unknown

Selected Documents in the History of the U.S. Civil Space Program

Volume VII
Human Spaceflight: Projects Mercury, Gemini, and Apollo

Edited by John M. Logsdon with Roger D. Launius

The NASA History Series

National Aeronautics and Space Administration
NASA History Division
Office of External Relations
Washington, DC   2008

NASA SP-2008-4407

Library of Congress Cataloging-in-Publication Data

Exploring the Unknown: Selected Documents in the History of the U.S. Civil Space Program/ John M. Logsdon, editor ...[et al.]
p. cm.—(The NASA history series) (NASA SP: 4407)

Includes bibliographical references and indexes.
Contents: v. 7. Human Spaceflight: Mercury, Gemini, and Apollo Programs
1. Astronautics—United States—History.   I. Logsdon, John M., 1937–
             II. Series   III. Series   V. Series: NASA SP: 4407.
TL789.8.U5E87   1999   96-9066
387.8'0973-dc20   CIP

Dedicated to the Pioneers of Human Spaceflight:
George Low, Robert Gilruth, and the members of the Space Task Group

Contents

Chapter Two

Documents

Acknowledgments

This volume is the seventh in a series that had its origins almost two decades ago. The individuals involved in initiating the series and producing the initial six volumes have been acknowledged in those volumes [Volume I—*Organizing for Exploration* (1995); Volume II—*External Relationships* (1996); Volume III—*Using Space* (1998); Volume IV—*Accessing Space* (1999); Volume V—*Exploring the Cosmos* (2001); Volume VI—*Space and Earth Science* (2004)]. Those acknowledgments will not be repeated here.

We owe thanks to the individuals and organizations that have searched their files for potentially useful materials, and for the staffs at various archives and collections who have helped us locate documents, especially Shelley Kelly at the University of Houston Clear Lake Library. Graduate students Chirag Vyas, Eric Dickinson, Daphne Dador, Angela Peura, and Audrey Schaffer provided essential assistance in the preparation of the volume.

My thanks go to all those mentioned above, and again to those who helped get this effort started and who have been involved along the way.

John M. Logsdon
George Washington University

* * * * * * * *

Numerous people at NASA associated with historical study, technical information, and the mechanics of publishing helped in myriad ways in the preparation of this documentary history. In the NASA History Division, Stephen J. Garber oversaw much of the editorial and production work. Nadine J. Andreassen provided key administrative support for this project. Intern Matthew Barrow capably researched and wrote the entries for the biographical appendix and Amelia Lancaster assisted with the final production. Archivists Jane Odom, Colin Fries, and John Hargenrader also helped in a number of ways. In addition, the staffs of the NASA Headquarters Library, the Scientific and Technical Information Program, and the NASA Document Services Center provided assistance in locating and preparing for publication the documentary materials in this work.

On the production end in the NASA Headquarters Communications Support Services Center, Stacie Dapoz oversaw the careful copyediting of this volume. Shelley Kilmer-Gaul laid out the book and designed the dust jacket. Gail Carter-Kane and Cindy Miller assisted in the overall process. Printing specialists Hanta Ralay and Tun Hla expertly oversaw this critical final stage.

Thanks are due to all these fine professionals.

Steven J. Dick
NASA Chief Historian

Introduction

One of the most important developments of the twentieth century has been the movement of humanity into space with machines and people. The underpinnings of that movement—why it took the shape it did; which individuals and organizations were involved; what factors drove a particular choice of scientific objectives and technologies to be used; and the political, economic, managerial, and international contexts in which the events of the Space Age unfolded—are all important ingredients of this epoch transition from an Earthbound to a spacefaring people. This desire to understand the development of spaceflight in the United States sparked this documentary history series.

The extension of human activity into outer space has been accompanied by a high degree of self-awareness of its historical significance. Few large-scale activities have been as extensively chronicled so closely to the time they actually occurred. Many of those who were directly involved were quite conscious that they were making history, and they kept full records of their activities. Because most of the activity in outer space was carried out under government sponsorship, it was accompanied by the documentary record required of public institutions, and there has been a spate of official and privately written histories of most major aspects of space achievement to date. When top leaders considered what course of action to pursue in space, their deliberations and decisions often were carefully put on the record. There is, accordingly, no lack of material for those who aspire to understand the origins and evolution of U.S. space policies and programs.

This reality forms the rationale for this series. Precisely because there is so much historical material available on space matters, the National Aeronautics and Space Administration (NASA) decided in 1988 that it would be extremely useful to have easily available to scholars and the interested public a selective collection of many of the seminal documents related to the evolution of the U.S. civilian space program. While recognizing that much space activity has taken place under the sponsorship of the Department of Defense and other national security organizations, the U.S. private sector, and in other countries around the world, NASA felt that there would be lasting value in a collection of documentary material primarily focused on the evolution of the U.S. government's civilian space program, most of which has been carried out since 1958 under the Agency's auspices. As a result, the NASA History Division contracted with the Space Policy Institute of George Washington University's Elliott School of International Affairs to prepare such a collection. This is the seventh volume in the documentary history series; one additional volume containing documents and introductory essays related to post-Apollo human spaceflight will follow.

The documents collected during this research project were assembled from a diverse number of both public and private sources. A major repository of primary source materials relative to the history of the civil space program is the NASA Historical Reference Collection of the NASA History Division located at the Agency's Washington headquarters. Project assistants combed this collection for the "cream" of the wealth of material housed there. Indeed, one purpose of this series from the start was to capture some of the highlights of the holdings at Headquarters. Historical materials housed at the other NASA installations, at institutions of higher learning, and Presidential libraries were other sources of documents considered for inclusion, as were papers in the archives of individuals and firms involved in opening up space for exploration.

Copies of the documents included in this volume in their original form will be deposited in the NASA Historical Reference Collection. Another complete set of project materials is located at the Space Policy Institute at George Washington University. These materials in their original forms are available for use by researchers seeking additional information about the evolution of the U.S. civil space program, or wishing to consult the documents reprinted herein in their original form.

The documents selected for inclusion in this volume are presented in two chapters: one covering the Mercury and Gemini projects and another covering Project Apollo.

Volume I in this series covered the antecedents to the U.S. space program, and the origins and evolution of U.S. space policy and of NASA as an institution. Volume II dealt with the relations between the civilian space program of the United States and the space activities of other countries; the relations between the U.S. civilian and national security space and military efforts; and NASA's relations with industry and academic institutions. Volume III provided documents on satellite communications, remote sensing, and the economics of space applications. Volume IV covered various forms of space transportation. Volume V covered the origins of NASA's space science program and its efforts in solar system exploration and astrophysics and astronomy. Volume VI covered space and Earth science. As noted above, one more future volume will cover post-Apollo human spaceflight.

Each section in the present volume is introduced by an overview essay. In the main, these essays are intended to introduce and complement the documents in the section and to place them in a chronological and substantive context. Each essay contains references to the documents in the section it introduces, and also contains references to documents in other volumes in this series. These introductory essays are the responsibility of their individual authors, and the views and conclusions contained therein do not necessarily represent the opinions of either George Washington University or NASA.

The documents included in each section were chosen by the project team in concert with the essay writer from those assembled by the research staff for the overall project. The contents of this volume emphasize primary documents or long-out-of-print essays or articles and material from the private recollections of important actors in shaping space affairs. The contents of this volume thus do not comprise in themselves a comprehensive historical account; they must be supplemented by other sources, those both already available and to become available in the future. The documents included in each section are arranged chronologically, with the exception that closely related documents are grouped together. Each document is assigned its own number in terms of the section in which it is placed. Thus, the first document in the second section of this volume is designated "Document II-l." Each document or group of related documents is accompanied by a headnote setting out its context and providing a background narrative. These headnotes also provide specific information about people and events discussed. We have avoided the inclusion of explanatory notes in the documents themselves and have confined such material to the headnotes.

The editorial method we adopted for dealing with these documents seeks to preserve spelling, grammar, paragraphing, and use of language as in the original. We have sometimes changed punctuation where it enhances readability. We have used the designation [not included, or omitted] to note where sections of a document have not been included in this publication, and we have avoided including words and phrases that had been deleted in the original document unless they contribute to an understanding of what was going on in the mind of the writer in making the record. Marginal notations on the original documents are inserted into the text of the documents in brackets, each clearly marked as a marginal comment. Except insofar as illustrations and figures are necessary to understanding the text, those items have been omitted from this printed version. Page numbers in the original document are noted in brackets internal to the document text. Copies of all documents in their original form, however, are available for research by any interested person at the NASA History Division or the Space Policy Institute of George Washington University.

We recognize that there are certain to be quite significant documents left out of this compilation. No two individuals would totally agree on all documents to be included from the many we collected, and surely we have not been totally successful in locating all relevant records. As a result, this documentary history can raise an immediate question from its users: why were some documents included while others of seemingly equal importance were omitted? There can never be a fully satisfactory answer to this question. Our own criteria for choosing particular documents and omitting others rested on three interrelated factors:

- Is the document the best available, most expressive, most representative reflection of a particular event or development important to the evolution of the space program?

- Is the document not easily accessible except in one or a few locations, or is it included (for example, in published compilations of presidential statements) in reference sources that are widely available and thus not a candidate for inclusion in this collection?

- Is the document protected by copyright, security classification, or some other form of proprietary right and thus unavailable for publication?

As general editor of this volume, I was ultimately responsible for the decisions about which documents to include and for the accuracy of the headnotes accompanying them. It has been an occasionally frustrating but consistently exciting experience to be involved with this undertaking; I and my associates hope that those who consult it in the future find our efforts worthwhile.

John M. Logsdon
Director
Space Policy Institute
Elliott School of International Affairs
George Washington University

Biographies of Volume VII Editors

Roger D. Launius is a member of the space history department of the Smithsonian Institution's National Air and Space Museum and is the former NASA Chief Historian. He has produced many books and articles on aerospace history, including *Innovation and the Development of Flight* (Texas A&M University Press, 1999); *NASA & the Exploration of Space* (Stewart, Tabori, & Chang, 1998); *Frontiers of Space Exploration* (Greenwood Press, 1998); *Organizing for the Use of Space: Historical Perspectives on a Persistent Issue* (Univelt, Inc., AAS History Series, Volume 18, 1995), editor; *NASA: A History of the U.S. Civil Space Program* (Krieger Publishing Co., 1994); *History of Rocketry and Astronautics: Proceedings of the Fifteenth and Sixteenth History Symposia of the International Academy of Astronautics* (Univelt, Inc., AAS History Series, Volume 11, 1994), editor; *Apollo: A Retrospective Analysis* (Monographs in Aerospace History, Vol. 3, 1994); and *Apollo 11 at Twenty-Five* (electronic picture book issued on computer disk by the Space Telescope Science Institute, Baltimore, MD, 1994).

John M. Logsdon is Director of the Space Policy Institute of George Washington University's Elliott School of International Affairs, where he is also Professor Emeritus of Political Science and International Affairs. He holds a B.S. in physics from Xavier University and a Ph.D. in political science from New York University. He has been at George Washington University since 1970, and previously taught at The Catholic University of America. He is also a faculty member of the International Space University. He is the author and editor of numerous books and articles on space policy and space history. He is an elected member of the International Academy of Astronautics and a former member of the board of The Planetary Society. He is a member of the NASA Advisory Council and served during 2003 on the Columbia Accident Investigation Board. Dr. Logsdon has lectured and spoken to a wide variety of audiences at professional meetings, colleges and universities, international conferences, and other settings, and has testified before Congress on numerous occasions. He is frequently consulted by the electronic and print media for his views on various space issues. He has twice been a Fellow at the Woodrow Wilson International Center for Scholars and was the first holder of the Chair in Space History of the National Air and Space Museum. He is a Fellow of the American Association for the Advancement of Science and the American Institute of Aeronautics and Astronautics.

Chapter 1

First Steps into Space: Projects Mercury and Gemini

by Roger D. Launius

Introduction

Humanity has dreamed of traveling into space for centuries, but in the twentieth century, scientific and technical capabilities converged with this dream for the first time. From the work of Robert H. Goddard through the heroic era of spaceflight into the 1960s, the modern age of rocketry signaled a beginning that would eventually lead to human flights beyond Earth to the Moon.[1] All of these enthusiasts believed humanity would soon explore and eventually colonize the solar system. And many of them worked relentlessly to make that belief a reality. They successfully convinced a large majority of Americans of spaceflight's possibility. Through their constant public relations efforts during the decade following World War II, they engineered a sea change in perceptions, as most Americans went from skepticism about the probabilities of spaceflight to an acceptance of it as a near-term reality.[2]

This is apparent in the public opinion polls of the era. In December 1949, Gallup pollsters found that only 15 percent of Americans believed humans would reach the Moon within 50 years, while 15 percent had no opinion, and a whopping 70 percent believed that it would not happen within that time. In October 1957, at the same time as the launching of Sputnik I, only 25 percent believed that it would take longer than 25 years for humanity to reach the Moon, while 41 percent believed firmly that it would happen within 25 years, and 34 percent were not sure. An important shift in perceptions had taken place, and it was largely the result of well-known advances in rocket technology coupled with a public relations campaign that emphasized the real possibilities of spaceflight.[3]

Indeed, by the end of World War II, all the technical assessments suggested that it was only a matter of a few years before the United States would be able

---

1. Robert H. Goddard. "R. H. Goddard's Diary," 16–17 March 1926 in Esther C. Goddard, ed., and G. Edward Pendray, assoc. ed., *The Papers of Robert H. Goddard* (New York: McGraw-Hill Book Co., 1970), 2: pp. 580–581; Milton Lehman, *This High Man* (New York: Farrar, Straus, 1963), pp. 140–144; David A. Clary, *Rocket Man: Robert H. Goddard and the Birth of the Space Age* (New York: Hyperian, 2003), pp. 120–122.

2. This is the core argument of Howard E. McCurdy, *Space and the American Imagination* (Washington, DC: Smithsonian Institution Press, 1997).

3. George H. Gallup, *The Gallup Poll: Public Opinion, 1935–1971* (New York: Random House, 1972), 1: pp. 875, 1152.

to place a satellite in orbit around Earth and, ultimately, to place a human in a capsule for orbital activities. In 1946, for instance, the forerunner of the Rand Corporation completed an engineering analysis of an Earth satellite vehicle for the Army Air Forces, finding important military support functions possible ranging from weather forecasting to secure global communications to strategic reconnaissance.[4] Later, military analysts thought there might be a role for piloted military missions in space, and that, along with the exploration imperative, drove efforts to make human spaceflight a reality. By the middle part of the 1950s, the spaceflight advocacy community was actively advocating, as later ensconced in the NASA long-range plan of 1959, "the manned exploration of the Moon and nearby planets." They called for the "first launching in a program leading to manned circumlunar flight and to a permanent near-Earth space station" that would make a human mission to the Moon possible.[5]

The von Braun Paradigm

All of the prospective futures for the near term contemplated by spaceflight pioneers ended with a human expedition to Mars. Without question, the most powerful vision of spaceflight since the early 1950s has been that articulated by Wernher von Braun, one of the most important rocket developers and champions of space exploration during the period between the 1930s and the 1970s. Working for the German Army between 1934 and 1945, von Braun led the technical effort to develop the V-2, the first ballistic missile, and deliberately surrendered to the Americans at the close of World War II because he said he desired to work for a rich and benevolent uncle, in this case Uncle Sam. For 15 years after World War II, von Braun worked with the U.S. Army in the development of ballistic missiles. Von Braun became one of the most prominent spokesmen of space exploration in the U.S. in the 1950s. In 1952 he gained note as a participant in an important symposium dedicated to the subject and he gained notoriety among the public in the fall of 1952 with a series of articles in *Collier's*, a popular weekly periodical of the era. He also became a household name following his appearance on three Disney television shows dedicated to space exploration in the mid-1950s.[6] Indeed, no one became more significant as an advocate for space

4. Douglas Aircraft Company, Inc., "Preliminary Design of an Experimental World-Circling Spaceship," Report No. SM-11827, 2 May 1946. Folder 18674, NASA Historical Reference Collection, NASA History Division, NASA Headquarters, Washington, DC.

5. Office of Program Planning and Evaluation, "The Long Range Plan of the National Aeronautics and Space Administration," 16 December 1959, document III-2 in James M. Logsdon, gen. ed., with Linda J. Lear, Janelle Warren Findley, Ray A. Williamson, Dwayne A. Day, *Exploring the Unknown: Selected Documents in the History of the U.S. Civil Space Program, Volume I, Organizing for Exploration* (Washington, DC: NASA Special Publication 4407, 1995), pp. 403–407.

6. See Erik Bergaust, *Wernher von Braun* (Washington, DC: National Space Institute, 1976); Ernst Stuhlinger, Frederick I. Ordway, III, *Wernher von Braun: Crusader for Space*, 2 vols. (Malabar, FL: Krieger Publishing Co., 1994). See Michael J. Neufeld, *Wernher von Braun: Dreamer of Space, Engineer of War* (New York: Alfred A. Knopf, 2007). Also see the *Collier's* series of articles conveniently reprinted in Cornelius Ryan, ed., *Across the Space Frontier* (New York: Viking Press, 1952); and Cornelius Ryan, ed., *Conquest of the Moon* (New York: Viking Press, 1953). The three Disney programs have recently been

exploration in the first part of the Space Age than von Braun, whose ideas influenced millions and charted the course of space exploration in the U.S. Central to von Braun's ideas was the human exploration of space; there was virtually no room in his vision for robotic spaceflight.

From the 1950s on, this German émigré called for an integrated space exploration plan centered on human movement beyond this planet and involving these basic steps accomplished in this order:

1. Earth orbital satellites to learn about the requirements for space technology that must operate in a hostile environment (initially soft-pedaled by von Braun but later embraced in such missions as Explorer 1).
2. Earth orbital flights by humans to determine whether or not it was possible to explore and settle other places.
3. A reusable spacecraft for travel to and from Earth orbit, thereby extending the principles of atmospheric flight into space and making routine space operations.
4. A permanently inhabited space station as a place both to observe Earth and from which to launch future expeditions. This would serve as the base camp at the bottom of the mountain or the fort in the wilderness from which exploring parties would depart.
5. Human exploration of the Moon with the intention of creating Moon bases and eventually permanent colonies.
6. Human expeditions to Mars, eventually colonizing the planet.

This has become known over time as the von Braun paradigm for the human colonization of the solar system. This approach would lead, von Braun believed, in the establishment of a new and ultimately perfect human society elsewhere in the solar system.

This integrated plan has cast a long shadow over American efforts in space for over 50 years. It conjured powerful images of people venturing into the unknown to make a perfect society free from the boundaries found on Earth. As such, it represented a coherent and compelling definition of American ideals in space. In many respects, von Braun's vision of space exploration has served as the model for U.S. efforts in space through the end of the 20th century.[7] His vision was constrained by the time in which he lived, for without a coherent vision of the rise of electronics, he failed to perceive the role of robotic explorers. As John H. Gibbons, Assistant to the President for Science and Technology during the Clinton administration, said in 1995:

---

released in DVD as *Tomorrow Land: Disney in Space and Beyond* (Burbank, CA: Buena Vista Home Entertainment, 2004).

    7. Dwayne A. Day, "The Von Braun Paradigm," *Space Times: Magazine of the American Astronautical Society* 33 (November to December 1994): pp. 12–15; "Man Will Conquer Space Soon," *Collier's* (22 March 1952): pp. 23–76ff; Wernher von Braun, with Cornelius Ryan, "Can We Get to Mars?" *Collier's* (30 April 1954): pp. 22–28.

The von Braun paradigm—that humans were destined to physically explore the solar system—which he so eloquently described in *Collier's* magazine in the early 1950's was bold, but his vision was highly constrained by the technology of his day. For von Braun, humans were the most powerful and flexible exploration tool that he could imagine. Today we have within our grasp technologies that will fundamentally redefine the exploration paradigm. We have the ability to put our minds where our feet can never go. We will soon be able to take ourselves—in a virtual way—anywhere from the interior of a molecule to the planets circling a nearby star—and there exclaim, "Look honey, I shrunk the Universe!"[8]

Most important, von Braun's integrated approach to space exploration was ensconced in the NASA long-range plan of 1959, and, with the exception of a jump from human orbital flights to a lunar (Apollo) mission driven by political concerns, the history of spaceflight has followed this paradigm consistently. Following the Apollo missions, NASA returned to the building of winged reusable spacecraft (the Space Shuttle), and a space station (*Freedom*/International Space Station) and, in 2004, embarked on human lunar and Mars expeditions. This adherence to the paradigm is either a testament to the amazing vision of Wernher von Braun or to a lack of imagination by NASA leaders, but the best guess suggests that it lies somewhere between the two.

The NACA and Spaceflight Research

During the latter part of World War II, leaders of the National Advisory Committee for Aeronautics (NACA), the predecessor to NASA, had become interested in the possibilities of high-speed guided missiles and the future of spaceflight. It created the Pilotless Aircraft Research Division (PARD), under the leadership of a young and promising engineer at the Langley Research Center in Hampton, Virginia, Robert R. Gilruth. In early 1945, NACA asked Congress for a supplemental appropriation to fund the activation of a unit to carry out this research, and a short time later the NACA opened the Auxiliary Flight Research Station (AFRS), which was later redesignated the name by which it gained fame, PARD, with Gilruth as Director.[9]

Established at Wallops Island as a test-launching facility of Langley on 4 July 1945, PARD launched its first test vehicle, a small two-stage, solid-fuel rocket to

---

8. John H. Gibbons, "The New Frontier: Space Science and Technology in the Next Millennium," Wernher von Braun Lecture, 22 March 1995, National Air and Space Museum, Smithsonian Institution, Washington, DC, available online at *http://clinton4.nara.gov/textonly/WH/EOP/OSTP/other/space.html*, accessed 2 October 2008.

9. James M. Grimwood, *Project Mercury: A Chronology* (Washington, DC: NASA SP-4001, 1963), Part 1A, p. 1; Joseph Adams Shortall, *A New Dimension: Wallops Island Flight Test Range, the First Fifteen Years* (Washington, DC: NASA Reference Publication [RP]-1028, 1978). At first, only part of the land on Wallops Island was purchased; the rest was leased. In 1949 NACA purchased the entire island.

check out the installation's instrumentation. Beyond a series of exploratory flight tests of rocket models, Gilruth's PARD advanced the knowledge of aerodynamics at transonic and, later, hypersonic speeds. They did so through exhaustive testing, which some at Langley considered excessive and overly expensive, launching at least 386 models between 1947 and 1949, leading to the publication of NACA's first technical report on rocketry, "Aerodynamic Problems of Guided Missiles," in 1947. From this, Gilruth and PARD filled in tremendous gaps in the knowledge of spaceflight. As historian James R. Hansen writes: "the early years of the rocket-model program at Wallops (1945–1951) showed that Langley was able to tackle an enormously difficult new field of research with innovation and imagination."[10]

The NACA leadership believed that human spaceflight could be achieved within a decade after 1952, and Gilruth served as an active promoter of the idea within the organization. He helped to engineer the creation of an interagency board to review "research on spaceflight and associated problems" with an end to advancing the cause of human spaceflight (I-1).[11] For example, while Gilruth was interested in orbiting an artificial satellite, it did not capture his imagination. As he recalled, "When you think about putting a man up there, that's a different thing. That's a lot more exciting. There are a lot of things you can do with men up in orbit."[12] This led to concerted efforts to develop the technology necessary to make it a reality. In 1952, for example, PARD started the development of multistage, hypersonic, solid-fuel rocket vehicles. These vehicles were used primarily in aerodynamic heating tests at first and were then directed toward a reentry physics research program. On 14 October 1954, the first American four-stage rocket was launched by PARD, and in August 1956 it launched a five-stage, solid-fuel rocket test vehicle, the world's first, that reached a speed of Mach 15.[13]

At the same time, H. Julian Allen at NACA's Ames Research Center began research on recovery of objects from orbit. In the early 1950s, he found that a blunt-nose body experienced less heating and dissipated it more quickly than a pointed body during the reentry; the pointed body was likely to burn up before reaching Earth's surface. Allen's work fundamentally shaped the course of spaceflight research and provided the basis for all successful reentry vehicles. It became the standard technology used in reconnaissance, warhead, and human reentry missions from the;1950s to the present. Based upon this research, in 1955 General Electric (GE) engineers began work on the Mark 2 reentry vehicle. While an overall success, GE adopted a heat-sink concept for the Mark 2 vehicle, whereby the heat

---

10. Robert R. Gilruth, "Aerodynamic Problems of Guided Missiles," NACA Report, draft, 19 May 1947, Gilruth Papers, Special Collections, Carol M. Newman Library, Virginia Polytechnic Institute and State University, Blacksburg, VA; James R. Hansen, *Spaceflight Revolution: NASA Langley Research Center from Sputnik to Apollo* (Washington, DC: NASA SP-4308, 1995), p. 270.

11. H. J. E. Reid, Director, NACA, to NACA, "Research on Spaceflight and Associated Problems," 5 August 1952. Folder 18674, NASA Historical Reference Collection, NASA History Division, NASA Headquarters, Washington, DC.

12. Third oral history interview of Robert R. Gilruth, by Linda Ezell, Howard Wolko, Martin Collins, National Air and Space Museum, Washington, DC, 30 June 1986, pp. 19, 44.

13. NASA Space Task Group to NASA Headquarters, 5 July 1960. Folder 18674, NASA Historical Reference Collection, NASA Hisotry Division, NASA Headquarters, Washington, DC; Eugene M. Emme, *Aeronautics and Astronautics: An American Chronology of Science and Technology in the Exploration of Space, 1915–1960* (Washington, DC: National Aeronautics and Space Administration, 1961), p. 76; House Rpt. 67, 87th Cong., 1st Sess., p. 27.

of reentry was conducted from the surface of the vehicle to a mass of material that could soak it up quickly. The key was to dissipate the heat away from the surface fast enough so that it did not melt. By 1956, Allen and other researchers had noticed that reinforced plastics had proven more resistant to heating than most other materials. They proposed coating the reentry vehicle with a material that absorbed heat, charred, and either flaked off or vaporized. As it did so, these "ablative" heatshields took away the absorbed heat (I-2).[14]

While Gilruth experimented with launch technology, and Allen worked on spacecraft recovery, both became very interested in the prospects for human spaceflight. They became aware of the *Collier's* series of articles on space, the first of which appeared on 22 March 1952. In it readers were asked by Wernher von Braun, "What Are We Waiting For?" and urged to support an aggressive space program.[15] Clearly the *Collier's* series helped to shape the perceptions of many at NACA that spaceflight was something that was no longer fantasy. Gilruth recalled of von Braun and his ideas: "I thought that was fascinating. He was way ahead of all of us guys . . . everybody was a space cadet in those days. I thought a space station was very interesting."[16]

In more than 12 years NACA made some significant strides in the development of the technology necessary to reach orbital flight above the atmosphere. Clearly, PARD held the lion's share of knowledge in NACA about the nascent field of astronautics. And it enjoyed renewed attention and funding once the Soviet Union launched the world's first satellite, Sputnik I, on 4 October 1957. "I can recall watching the sunlight reflect off of Sputnik as it passed over my home on the Chesapeake Bay in Virginia," Gilruth commented in 1972. "It put a new sense of value and urgency on things we had been doing. When one month later the dog Laika was placed in orbit in Sputnik II, I was sure that the Russians were planning for man-in-space."[17]

In the aftermath of the Sputnik crisis, NACA proceeded with efforts to advance human spaceflight even as plans were underway in 1958 to transform it into a new space agency. NACA engineers developed plans for a human space-

---

14. H. Julian Allen, NACA, to A. J. Eggers Jr., NACA, "Research Memorandum: A Study of the Motion and Aerodynamic Heating of Missiles Entering the Earth's Atmosphere at High Supersonic Speeds," 25 August 1953. Folder 18674, NASA Historical Reference Collection, NASA History Division, NASA Headquarters, Washington, DC; H. Julian Allen and Alfred J. Eggers Jr. "A Study of the Motion and Aerodynamic Heating of Ballistic Missiles Entering the Earth's Atmosphere at High Supersonic Speeds," NACA Technical Report 1381, *Forty-Fourth Annual Report of the NACA—1958* (Washington, DC: 1959), pp. 1125–1140; H. Julian Allen, "Hypersonic Flight and the Reentry Problem," *Journal of the Aeronautical Sciences* 25 (April 1958): pp. 217–230; Alfred J. Eggers Jr., "Performance of Long Range Hypervelocity Vehicles," *Jet Propulsion* 27 (November 1957): pp. 1147–1151; Loyd S. Swenson, James M. Grimwood, and Charles C. Alexander, *This New Ocean: A History of Project Mercury* (Washington, DC: NASA SP-4201, 1966), pp. 55–82; David K. Stump, *Titan II* (Fayetteville, AR: University of Arkansas Press, 2000), pp. 56–63.

15. "Man Will Conquer Space Soon" series, *Collier's*, 22 March 1952, pp. 23–76ff.

16. Robert Gilruth Oral History No. 6 by David DeVorkin and John Mauer, 2 March 1987, Glennan-Webb-Seamans Project, National Air and Space Museum.

17. NASA Press Release H00-127, "Dr. Robert Gilruth, an Architect of Manned Spaceflight, Dies," 17 August 2000. Folder 18674, NASA Historical Reference Collection, NASA History Division, NASA Headquarters, Washington, DC.

flight proposal during the spring of the year.[18] As a part of this effort they considered the best method for reaching space. At a series of meetings to discuss planning for human-in-space program approaches being developed by U.S. industry in January–February 1958, NACA officials found:

> Proposals fell into two rough categories: (a) a blunt-nose cone or near-spherical zero-lift high-drag vehicle of a ton to a ton-and-a-half weight, and (b) a hypersonic glider of the ROBO or Dyna-Soar type. The first category of vehicles used existing ICBM vehicles as boosters; the second used more complex and arbitrary multiplex arrangements of existing large-thrust rocket engines. A number of contractors looked at the zero-lift high-drag minimum weight vehicle as the obvious expedient for beating the Russians and the Army into space. Others, notably Bell, Northrup, and Republic Aviation, set this idea aside as a stunt and consequently these contractors stressed the more elaborate recoverable hypersonic glider vehicle as the practical approach to the problems of flight in space (I-3).[19]

By April 1958, NACA engineers had concluded that the first of these options should become the basis for NACA planning for an initial human spaceflight (I-4).[20]

It soon became obvious to all that an early opportunity to launch human spacecraft into orbit would require the development of blunt-body capsules launched on modified multistage intercontinental ballistic missiles (ICBMs). Robert Gilruth recalled one of these decisions:

> Because of its great simplicity, the non-lifting, ballistic-type of vehicle was the front runner of all proposed manned satellites, in my judgment. There were many variations of this and other concepts under study by both government and industry groups at that time. The choice involved considerations of weight, launch vehicle, reentry body design, and to be honest, gut feelings. Some people felt that man-in-space was only a stunt. The ballistic approach, in particular, was under fire since it was such a radical departure from the airplane. It was called by its opponents 'the man in the can,' and the pilot was termed only a 'medical specimen.' Others thought it was just too undignified a way to fly.[21]

---

18. Abe Silverstein, Associate Director, NACA, to Langley, "Review of Prospective Langley Report Entitled "Preliminary Study of a Manned Satellite" by Maxime A. Faget, Benjamine E. Garland, and James J. Buglia, 7 March 1958; Paul E. Purser, Aeronautical Research Engineer, NACA, Memorandum for Mr. Gilruth, "Langley Manned-Satellite Program," 11 April 1958. Folder 18674, NASA Historical Reference Collection, NASA History Division, NASA Headquarters, Washington, DC.

19. Adelbert O. Tischler, Head, Rocket Combustion Section, NACA, Memorandum for Associate Director, NACA, "Minimum Man-In-Space Proposals Presented at WADC, January 29, 1958 to February 1, 1958," 10 April 1958. Folder 18674, NASA Historical Reference Collection, NASA History Division, NASA Headquarters, Washington, DC.

20. Silverstein to NACA, "Review of Prospective Langley Report," 7 March 1958. Folder 18674, NASA Historical Reference Collection, NASA History Division, NASA Headquarters, Washington, DC.

21. Robert R. Gilruth, "Memoir: From Wallops Island to Mercury; 1945–1958," paper, Sixth

While initially criticized as an inelegant, impractical solution to the challenge of human spaceflight, the ballistic spacecraft concept gained momentum as NACA engineers, led by Maxime A. Faget, championed the approach. At a meeting on human spaceflight held at Ames on 18 March 1958, a NACA position emerged on this approach to human spaceflight, reflecting Faget's ideas.[22] By April 1958, NACA had completed several studies "on the general problems of manned-satellite vehicles," finding that they could build in the near term "a basic drag-reentry capsule" of approximately 2,000 pounds and sufficient volume for a passenger.[23]

In August 1958, Faget and his designers developed preliminary specifications that then went to industry, especially the McDonnell Aircraft Corporation, for a ballistic capsule. Faget and his colleagues emphasized the simplicity, if not the elegance, of a ballistic capsule for the effort:

> The ballistic reentry vehicle also has certain attractive operational aspects which should be mentioned. Since it follows a ballistic path there is a minimum requirement for autopilot, guidance, or control equipment. This condition not only results in a weight saving but also eliminates the hazard of malfunction. In order to return to Earth from orbit, the ballistic reentry vehicle must properly perform only one maneuver. This maneuver is the initiation of reentry by firing the retrograde rocket. Once this maneuver is completed (and from a safety standpoint alone it need not be done with a great deal of precision), the vehicle will enter Earth's atmosphere. The success of the reentry is then dependant only upon the inherent stability and structural integrity of the vehicle. These are things of a passive nature and should be thoroughly checked out prior to the first man-carrying flight. Against these advantages the disadvantage of large area landing by parachute with no corrective control during the reentry must be considered.[24]

The Mercury spacecraft that flew in 1961 to 1963 emerged from these early conceptual studies by NACA engineers (I-9).

International History of Astronautics Symposium, Vienna, Austria, 13 October 1972, pp. 31–32.

22. Swenson et al., *This New Ocean*, p. 86; James M. Grimwood, *Project Mercury: A Chronology* (Washington, DC: NASA SP-4001, 1963), p. 17; "How Mercury Capsule Design Evolved," *Aviation Week*, 21 September 1959, pp. 52–53, 55, and 57.

23. Paul E. Purser, Aeronautical Research Engineer, NACA, Memorandum for Mr. Gilruth, "Langley Manned-Satellite Program," 11 April 1958. Folder 18674, NASA Historical Reference Collection, NASA History Division, NASA Headquarters, Washington, DC.

24. Maxime A. Faget, Benjamine J. Garland, and James J. Buglia, Langley Aeronautical Laboratory, NACA, "Preliminary Studies of Manned Satellites," 11 August 1958. Folder 18674, NASA Historical Reference Collection, NASA History Division, NASA Headquarters, Washington, DC; Grimwood, *Project Mercury: A Chronology*, pp. 19–24; Gilruth, "Memoir: From Wallops Island to Mercury," pp. 34–37.

Man-in-Space Soonest

At the same time that NACA was pursuing its studies for a human spaceflight program, the U.S. Air Force (USAF) proposed the development of a piloted orbital spacecraft under the title of "Man-in-Space Soonest" (MISS).[25] Initially discussed before the launch of Sputnik I in October 1957, afterwards the Air Force invited Dr. Edward Teller and several other leading members of the scientific/technological elite to study the issue of human spaceflight and make recommendations for the future. Teller's group concluded that the Air Force could place a human in orbit within two years and urged that the department pursue this effort. Teller understood, however, that there was essentially no military reason for undertaking this mission and chose not to tie his recommendation to any specific rationale, falling back on a basic belief that the first nation to do so would accrue national prestige and advance, in a general manner, science and technology.[26] Soon after the new year, Lieutenant General Donald L. Putt, the USAF Deputy Chief of Staff for Development, informed NACA Director Hugh L. Dryden of the intention of the Air Force to aggressively pursue "a research vehicle program having as its objective the earliest possible manned orbital flight which will contribute substantially and essentially to follow-on scientific and military space systems." Putt asked Dryden to collaborate in this effort, but with NACA as a decidedly junior partner.[27] Dryden agreed; however, by the end of the summer he would find the newly created NASA leading the human spaceflight effort for the United States, with the Air Force being the junior player.[28]

Notwithstanding the lack of clear-cut military purpose, the Air Force pressed for MISS throughout the first part of 1958, clearly expecting to become the lead agency in any space program of the U.S. Specifically, it believed hypersonic space planes and lunar bases would serve national security needs in the coming decades well. To help make that a reality, it requested $133 million for the MISS program and secured approval for the effort from the Joint Chiefs of Staff.[29] Throughout this period, a series of disagreements between Air Force and NACA officials rankled both sides. The difficulties reverberated all the way to the White House, prompting a

---

25. The MISS program called for a four-phase capsule orbital process, which would first use instruments, to be followed by primates, then a pilot, with the final objective of landing humans on the Moon. See David N. Spires, *Beyond Horizons: A Half Century of Air Force Space Leadership* (Peterson Air Force Base, CO: Air Force Space Command, 1997), p. 75; Swenson et al., *This New Ocean*, pp. 33–97.

26. Swenson et al., *This New Ocean*, p. 73–74.

27. Lt. Gen. Donald L. Putt, USAF Deputy Chief of Staff, Development, to Hugh L. Dryden, NACA Director, 31 January 1958. Folder 18674, NASA Historical Reference Collection, NASA History Division, NASA Headquarters, Washington, DC.

28. NACA to USAF Deputy Chief of Staff, Development, "Transmittal of Copies of Proposed Memorandum of Understanding between Air Force and NACA for joint NACA-Air Force Project for a Recoverable Manned Satellite Test Vehicle," 11 April 1958. Folder 18674, NASA Historical Reference Collection, NASA History Division, NASA Headquarters, Washington, DC.

29. The breakdown for this budget was aircraft and missiles, $32M; support, $11.5M; construction, $2.5M; and research and development, $87M. See Memorandum for ARPA Director, "Air Force Man-in-Space Program," 19 March 1958. Folder 18674, NASA Historical Reference Collection, NASA History Division, NASA Headquarters, Washington, DC.

review of the roles of the two organizations (I-5, I-6, I-7).[30] The normally staid and proper Director of NACA, Hugh L. Dryden, complained in July 1958 to the President's science advisor, James R. Killian, of the lack of clarity on the role of the Air Force versus NACA. He asserted that "the current objective for a manned satellite program is the determination of man's basic capability in a space environment as a prelude to the human exploration of space and to possible military applications of manned satellites. Although it is clear that both the National Aeronautics and Space Administration and the Department of Defense should cooperate in the conduct of the program, I feel that the responsibility for and the direction of the program should rest with NASA." He urged that the President state a clear division between the two organizations on the human spaceflight mission (I-8).[31]

As historians David N. Spires and Rick W. Sturdevant have pointed out, the MISS program became derailed within the Department of Defense (DOD) at essentially the same time because of funding concerns and a lack of clear military mission:

> Throughout the spring and summer of 1958 the Air Force's Air Research and Development Command had mounted an aggressive campaign to have ARPA convince administration officials to approve its Man-in-Space-Soonest development plan. But ARPA [Advanced Research Projects Agency] balked at the high cost, technical challenges, and uncertainties surrounding the future direction of the civilian space agency.[32]

Dwight D. Eisenhower signed the National Aeronautics and Space Act of 1958 into law at the end of July and, during the next month, assigned the USAF's human spaceflight mission to NASA. Thereafter, the MISS program was folded into what became Project Mercury.[33] By early November 1958, the DOD had acceded to the President's desire that the human spaceflight program be a civil-

---

30. Maurice H. Stans, Director, Bureau of the Budget, Memorandum for the President, "Responsibility for "Space" Programs," 10 May 1958; Maxime A. Faget, NACA, Memorandum for Dr. Dryden, 5 June 1958; Clotaire Wood, Headquarters, NACA, Memorandum for files, "Tableing [*sic*] of Proposed Memorandum of Understanding Between Air Force and NACA For a Joint Project For a Recoverable Manned Satellite Test Vehicle," 20 May 1958, with attached Memorandum, "Principles for the Conduct by the NACA and the Air Force of a Joint Project for a Recoverable Manned Satellite Vehicle," 29 April 1958; Donald A. Quarles, Secretary of Defense, to Maurice H. Stans, Director, Bureau of the Budget, 1 April 1958. Folder 18674, NASA Historical Reference Collection, NASA History Division, NASA Headquarters, Washington, DC.

31. Hugh L. Dryden, Director, NACA, Memorandum for James R. Killian Jr., Special Assistant to the President for Science and Technology, "Manned Satellite Program," 19 July 1958. Folder 18674, NASA Historical Reference Collection, NASA History Division, NASA Headquarters, Washington, DC.

32. David N. Spires and Rick W. Sturdevant, "'. . . to the very limit of our ability . . .': Reflections on Forty Years of Civil-Military Partnership in Space Launch," in Roger D. Launius and Dennis R. Jenkins, eds., *To Reach the High Frontier: A History of U.S Launch Vehicles* (Lexington, KY: University Press of Kentucky, 2002), p. 475.

33. For an overall discussion of the early military human program see Dwayne A. Day, "Invitation to Struggle: The History of Civilian-Military Relations in Space," in John M. Logsdon, with Dwayne A. Day and Roger D. Launius, eds., *Exploring the Unknown: Selected Documents in the History of the U.S. Civil Space Program, Volume II, External Relationships* (Washington, DC: NASA SP-4407, 1996), 2: pp. 248–251.

ian effort under the management of NASA. For its part, NASA invited Air Force officials to appoint liaison personnel to the Mercury program office at Langley Research Center, and they did so.[34]

Beginning Project Mercury

Everyone recognized that time was of the essence in undertaking the human spaceflight project that NASA would now lead. Roy Johnson, director of ARPA for the DOD, noted in September 1958 that competition with the Soviet Union precluded taking a cautious approach to the human spaceflight initiative and advocated additional funding to ensure its timely completion. As he wrote to the Secretary of Defense and the NASA Administrator:

> I am troubled, however, with respect to one of the projects in which there is general agreement that it should be a joint undertaking. This is the so-called "Man-in-Space" project for which $10 million has been allocated to ARPA and $30 million to NASA. My concern over this project is due 1) to a firm conviction, backed by intelligence briefings, that the Soviets next spectacular effort in space will be to orbit a human, and 2) that the amount of $40 million for FY 1959 is woefully inadequate to compete with the Russian program. As you know our best estimates (based on some 12–15 plans) were $100 to $150 million for an optimum FY 1959 program.
>
> I am convinced that the military and psychological impact on the United States and its Allies of a successful Soviet man-in-space "first" program would be far reaching and of great consequence.
>
> Because of this deep conviction, I feel that no time should be lost in launching an aggressive Man-in-Space program and that we should be prepared if the situation warrants, to request supplemental appropriations of the Congress in January to pursue the program with the utmost urgency (I-10).[35]

Johnson agreed to transfer a series of space projects from ARPA to NASA but urged more timely progress on development of the space vehicle itself. Two weeks later, ARPA and NASA established protocols for cooperating in the aggressive development of the capsule that would be used in the human spaceflight program (I-11).[36]

---

34. Memorandum for Dr. Silverstein, "Assignment of Responsibility for ABMA Participation in NASA Manned Satellite Project," 12 November 1958; Abe Silverstein to Lt. Gen. Roscoe C. Wilson, USAF Deputy Chief of Staff, Development, 20 November 1958; Hugh L. Dryden, Deputy Administrator, NASA, Memorandum for Dr. Eugene Emme for NASA Historical Files, "The 'signed' Agreement of April 11, 1958, on a Recoverable Manned Satellite Test Vehicle," 8 September 1965. Folder 18674, NASA Historical Reference Collection, NASA History Division, NASA Headquarters, Washington, DC.

35. Roy W. Johnson, Director, ARPA, DOD, Memorandum for the Administrator, NASA, "Man-in-Space Program," 3 September 1958. Folder 18674, NASA Historical Reference Collection, NASA History Division, NASA Headquarters, Washington, DC.

36. Roy W. Johnson, Director, ARPA, DOD, Memorandum for the Administrator, NASA, "Man-in-Space Program," 19 September 1958, with attached Memorandum of Understanding,

To aid in the conduct of this program, ARPA and NASA created a panel for Manned Spaceflight, also referred to as the Joint Manned Satellite Panel, on 18 September 1958. Holding its first meeting on 24 September, the panel established goals and strategy for the program. Chaired by Robert Gilruth and including such NASA leaders as Max Faget and George Low, the panel focused on a wide range of technical requirements necessary to complete the effort. Under this panel's auspices, final specifications for the piloted capsule emerged in October 1958, as did procurement of both modified Redstone (for suborbital flights) and Atlas (for orbital missions) boosters (I-12, I-13, I-14).[37]

Just six days after the establishment of NASA on 1 October 1958, NASA Administrator T. Keith Glennan approved plans for a piloted satellite project to determine if human spaceflight was possible, and on 8 October he established the Space Task Group at Langley Research Center under Robert Gilruth. Thirty-five key staff members from Langley, some of whom had been working the military human spaceflight plan, were transferred to the new Space Task Group, as were 10 others from the Lewis Research Center near Cleveland, Ohio (I-15, I-16).[38] These 45 engineers formed the nucleus of the more than 1,000-person workforce that eventually took part in Project Mercury, so named on 26 November 1958 (I-17, I-18).[39] On 14 November, Gilruth requested the highest national priority procurement rating for Project Mercury, but that did not come until 27 April 1959 (I-23).[40] As Glennan recalled, "the philosophy of the project was to use known technologies, extending the state of the art as little as necessary, and relying on the unproven Atlas. As one looks back, it is clear that we did not know much about what we were doing. Yet the Mercury program was one of the best organized and managed of any I have been associated with."[41] Throughout

"Principles for the Conduct by NASA and ARPA of a Joint Program for a Manned Orbital Vehicle," 19 September 1958. Folder 18674, NASA Historical Reference Collection, NASA History Division, NASA Headquarters, Washington, DC.

37. Minutes of Meetings, Panel for Manned Spaceflight, 24 and 30 September, 1 October 1958; NASA, "Preliminary Specifications for Manned Satellite Capsule," October 1958; Paul E. Purser, Aeronautical Research Engineer, NASA, to Mr. R. R. Gilruth, NASA, "Procurement of Ballistic Missiles for use as Boosters in NASA Research Leading to Manned Spaceflight," 8 October 1958, with attached, "Letter of Intent to AOMC (ABMA), Draft of Technical Content," 8 October 1958. Folder 18674, all in NASA Historical Reference Collection, NASA History Division, NASA Headquarters, Washington, DC.

38. S. B. Batdorf, ARPA, Memorandum for File, "Presentation of MIS Program to Dr. Glennan," 14 October 1958; Robert R. Gilruth, Project Manager, NASA, Memorandum for Associate Director, NASA, "Space Task Group," 3 November 1958. Folder 18674, NASA Historical Reference Collection, NASA History Division, NASA Headquarters, Washington, DC.

39. Abe Silverstein, Director of Spaceflight Development, NASA, Memorandum for Administrator, NASA, "Code Name "Project Mercury" for Manned Satellite Project," 26 November 1958; George M. Low, NASA, Memorandum for Dr. Silverstein, NASA, "Change of Manned Satellite Project name from "Project Mercury" to "Project Astronaut," 12 December 1958. Folder 18674, both in NASA Historical Reference Collection, NASA History Division, NASA Headquarters, Washington, DC; Linda Ezell, NASA Historical Data Book: Volume II: Programs and Projects 1958–1968 (Washington, DC: NASA SP-4012, 1988), pp. 102, 139–140; James M. Grimwood, Project Mercury: A Chronology (Washington, DC: NASA SP-4001, 1963), pp. 31–32.

40. George M. Low, NASA, Memorandum for House Committee on Science and Astronautics, "Urgency of Project Mercury," 27 April 1959. Folder 18674, NASA Historical Reference Collection, NASA History Division, NASA Headquarters, Washington, DC.

41. T. Keith Glennan, The Birth of NASA: The Diary of T. Keith Glennan, J. D. Hunley, ed. (Washington, DC: NASA SP-4105, 1993), p. 13.

the fall of 1958, therefore, NASA leaders worked to press the Mercury program through to flight initially conceived as possible before the end of 1959 (I-19).[42]

The Role of the Mercury Seven Astronauts

As an important step in moving forward with Project Mercury, NASA selected and trained the astronaut corps.[43] Although NASA at first intended to hold an open competition for entry into the astronaut corps, over the 1958 Christmas holiday, President Dwight D. Eisenhower directed that the astronauts be selected from among the armed services' test pilot force. Indeed, NASA Administrator T. Keith Glennan visited the White House over Christmas of 1958. "When he came back to NASA," NASA Chief Historian Eugene Emme wrote in 1964, "Project Mercury was to possess classified aspects and the astronauts were to be military test pilots."[44] Although this had not been NASA leadership's first choice, this decision greatly simplified the selection procedure. The inherent riskiness of spaceflight, and the potential national security implications of the program, pointed toward the use of military personnel. It also narrowed and refined the candidate pool, giving NASA a reasonable starting point for selection. It also made good sense in that NASA envisioned the astronaut corps first as pilots operating experimental flying machines, and only later as working scientists.[45]

As historian Margaret Weitekamp has concluded in a recent study:

> From that military test flying experience, the jet pilots also mastered valuable skills that NASA wanted its astronauts to possess. Test pilots were accustomed to flying high-performance aircraft, detecting a problem, diagnosing the cause, and communicating that analysis to the engineers and mechanics clearly. In addition, they were used to military discipline, rank, and order. They would be able to take orders. Selecting military jet

---

42. George M. Low, Memorandum for Administrator, "Status of Manned Satellite Program," 23 November 1958; George M. Low, Program Chief, Manned Spaceflight, NASA, Memorandum for Administrator, NASA, "Status Report No. 1, Manned Satellite Project," 9 December 1958; Abe Silverstein, Director of Spaceflight Development, NASA, Memorandum for Administrator, NASA, "Schedule for Evaluation and Contractual Negotiations for Manned Satellite Capsule," 24 December 1958; Message from NASA to Commanding General, Army Ordnance Missile Command, 8 January 1959. Folder 18674, NASA Historical Reference Collection, NASA History Division, NASA Headquarters, Washington, DC.

43. See Allan C. Fisher Jr., "Exploring Tomorrow with the Space Agency," *National Geographic*, July 1960, pp. 48, 52–89; Kenneth F. Weaver, "Countdown for Space," *National Geographic*, May 1961, pp. 702–734.

44. George M. Low to NASA Administrator, "Pilot Selection for Project Mercury," 23 April 1959; Eugene M. Emme to Mae Link and James Grimwood, "Military Status of Mercury Astronauts," 23 March 1964. Folder 18674, NASA Historical Reference Collection, NASA History Division, NASA Headquarters, Washington, DC.

45. This was in striking contrast to the Soviet Union's cosmonauts, whom space program leaders believed were essentially passengers without complex tasks to perform. See Slava Gerovitch, "Trusting the Machine: The Technopolitics of Automation in the Soviet Space Program," paper presented at Society for History in the Federal Government annual meeting, 10 October 2003, copy in possession of author.

test pilots as their potential astronauts allowed NASA to choose from a cadre of highly motivated, technically skilled, and extremely disciplined pilots.[46]

In addition, since most NASA personnel in Project Mercury came out of the aeronautical research and development arena anyway, it represented almost no stretch on the Agency's part to accept test pilots as the first astronauts. (It also guaranteed, as Weitekamp notes, that all of the original astronauts would be male.) After all, NACA had been working with the likes of them for decades and knew and trusted their expertise. It also tapped into a highly disciplined and skilled group of individuals, most of whom were already aerospace engineers, who had long ago agreed to risk their lives in experimental vehicles.[47]

NASA pursued a rigorous process to select the eventual astronauts that became known as the Mercury Seven. The process involved record reviews, biomedical tests, psychological profiles, and a host of interviews.[48] In November 1958, aeromedical consultants working for the Space Task Group at Langley had worked out preliminary procedures for the selection of astronauts to pilot the Mercury spacecraft. They then advertised among military test pilots for candidates for astronauts, receiving a total of 508 applications (I-20).[49] They then screened the service records in January 1959 at the military personnel bureaus in Washington and found 110 men that met the minimum standards established for Mercury:

1. Age—less than 40
2. Height—less than 5'11"
3. Excellent physical condition
4. Bachelor's degree or equivalent
5. Graduate of test pilot school
6. 1,500 hours total flying time
7. Qualified jet pilot

This list of names included 5 Marines, 47 Navy aviators, and 58 Air Force pilots. Several Army pilots' records had been screened earlier, but none was a graduate of a test pilot school.[50] The selection process began while the possibility

---

46. Margaret A. Weitekamp, "The Right Stuff, The Wrong Sex: The Science, Culture, and Politics of the Lovelace Woman in Space Program, 1959–1963," Ph.D. Diss., Cornell University, 2001, p. 98. Dr. Weitekamp's dissertation has been published as *Right Stuff, Wrong Sex: America's First Women in Space Program* (Baltimore, MD: Johns Hopkins Press, 2004).

47. In some cases this was literally the case. The best example is Neil A. Armstrong, who worked with the NACA and NASA as a civilian research pilot on the X-15 program at its Flight Research Center in the Mojave Desert prior to selection for astronaut training in 1962. For an excellent account of flight research at NACA/NASA see Michael H. Gorn, *Expanding the Envelope: Flight Research at NACA and NASA* (Lexington, KY: University Press of Kentucky, 2001).

48. This process is well told in Swenson et al., *This New Ocean*, pp. 140–164.

49. "Invitation to Apply for Position of Research Astronaut-Candidate, NASA Project A, Announcement No. 1," 22 December 1958. Folder 18674, NASA Historical Reference Collection, NASA History Division, NASA Headquarters, Washington, DC.

50. See Swenson et al., *This New Ocean*, pp. 155–165; Joseph D. Atkinson, Jr. and Jay M. Shafritz, *The Real Stuff: A History of NASA's Astronaut Recruitment Program* (New York: Praeger Publishing, 1985), pp. 8–12.

of piloted Mercury/Redstone flights late in 1959 still existed, so time was a critical factor is the screening process.[51]

A grueling selection process began in January 1959. Headed by the Assistant Director of the Space Task Group, Charles J. Donlan, the evaluation committee divided the list of 110 arbitrarily into three groups and issued invitations for the first group of 35 to come to Washington at the beginning of February for briefings and interviews (I-22).[52] Donlan's team initially planned to select 12 astronauts, but as team member George M. Low reported:

> During the briefings and interviews it became apparent that the final number of pilots should be smaller than the twelve originally planned for. The high rate of interest in the project indicates that few, if any, of the men will drop out during the training program. It would, therefore, not be fair to the men to carry along some who would not be able to participate in the flight program. Consequently, a recommendation has been made to name only six finalists.[53]

Every one of the first 10 pilots interrogated on 2 February agreed to continue through the elimination process. The next week a second group of possible candidates arrived in Washington. The high rate of volunteering made it unnecessary to extend the invitations to the third group. By the first of March 1959, 32 pilots prepared to undergo a rigorous set of physical and mental examinations.

Thereafter each candidate went to the Lovelace Clinic in Albuquerque, New Mexico, to undergo individual medical evaluations. Phase four of the selection program involved passing an elaborate set of environmental studies, physical endurance tests, and psychiatric studies conducted at the Aeromedical Laboratory of the Wright Air Development Center, Dayton, Ohio. During March 1959 each of the candidates spent another week in pressure suit tests, acceleration tests, vibration tests, heat tests, and loud noise tests. Continuous psychiatric interviews, the necessity of living with two psychologists throughout the week, an extensive self-examination through a battery of 13 psychological tests for personality and motivation, and another dozen different tests on intellectual functions and special aptitudes—these were all part of the Dayton experience (I-29).[54]

---

51. Atkinson and Shafritz, *The Real Stuff*, pp. 18, 43–45.

52. George M. Low, Program Chief, Manned Spaceflight, NASA, Memorandum for Administrator, NASA, "Pilot Selection for Project Mercury," 23 April 1959. Folder 18674, NASA Historical Reference Collection, NASA History Division, NASA Headquarters, Washington, DC.

53. Quoted in Swenson et al., *This New Ocean*, p. 161.

54. Charles L. Wilson, Captain, USAF, WADC Technical Report 59-505, "Project Mercury Candidate Evaluation Program," December 1959. Folder 18674, NASA Historical Reference Collection, NASA History Division, NASA Headquarters, Washington, DC. Although depicted as comic relief in the film version of *The Right Stuff* (1982), the battery of physiological tests were the most sophisticated designed up to that point. On these examinations see W. Randall Lovelace II, "Duckings, Probings, Checks That Proved Fliers' Fitness," *Life*, 20 April 1959; Mae Mills Link, *Space Medicine in Project Mercury* (Washington, DC: NASA SP-4003, 1965); John A. Pitts, *The Human Factor: Biomedicine in the Manned Space Program to 1980* (Washington, DC: NASA SP-4213, 1985).

Finally, without conclusive results from these tests, late in March 1959 NASA's Space Task Group began phase five of the selection, narrowing the candidates to 18. Thereafter, final criteria for selecting the candidates reverted to the technical qualifications of the men and the technical requirements of the program, as judged by Charles Donlan and his team members. NASA finally decided to select seven. The seven men became heroes in the eyes of the American public almost immediately, in part due to a deal they made with *Life* magazine for exclusive rights to their stories, and without NASA quite realizing it, they became the personification of NASA to most Americans.[55]

NASA unveiled the Mercury Seven in the spring of 1959, a week before the cherry blossoms bloomed along the tidal basin in Washington, DC, drenching the city with spectacular spring colors. NASA chose to announce the first Americans who would have an opportunity to fly in space on 9 April 1959. Excitement bristled in Washington at the prospect of learning who those space travelers might be. Surely they were the best the nation had to offer, modern versions of medieval "knights of the round table" whose honor and virtue were beyond reproach. Certainly they carried on their shoulders all of the hopes and dreams and best wishes of a nation as they engaged in single combat the ominous specter of communism. The fundamental purpose of Project Mercury was to determine whether or not humans could survive the rigors of liftoff and orbit in the harsh environment of space. From this perspective, the astronauts were not comparable to earlier explorers who directed their own exploits. Comparisons between them and Christopher Columbus, Admiral Richard Byrd, and Sir Edmund Hillary left the astronauts standing in the shadows.[56]

NASA's makeshift Headquarters was abuzz with excitement. Employees had turned the largest room of the second floor of Dolly Madison House facing Lafayette Park near the White House, once a ballroom, into a hastily set-up press briefing room. Inadequate for the task, print and electronic media jammed into the room to see the first astronauts. One end of the room sported a stage complete with curtain and both NASA officials and the newly chosen astronauts waited behind it for the press conference to begin at 2:00 p.m. The other end had electrical cable strewn about the floor, banks of hot lights mounted to illuminate the stage, more than a few television cameras that would be carrying the event live, and movie cameras recording footage for later use. News photographers gathered at the foot of the stage and journalists of all stripes occupied seats in the

55. See Tom Wolfe, "The Last American Hero," in *The Kandy-Kolored Tangerine-Flake Streamline Baby* (New York: Farrar, Straus, and Giroux, 1965); Atkinson and Shafritz, *The Real Stuff*, pp. 8–12; James L. Kauffman, *Selling Outer Space: Kennedy, the Media, and Funding for Project Apollo, 1961–1963* (Tuscaloosa, AL: University of Alabama Press, 1994), pp. 68–92; Mark E. Byrnes, *Politics and Space: Image Making by NASA* (Westport, CT: Praeger, 1994), pp. 25–46.

56. On this dynamic, see Roger D. Launius, "Project Apollo in American Memory and Myth," in Stewart W. Johnson, Koon Meng Chua, Rodney G. Galloway, and Philip J. Richter, eds., *Space 2000: Proceedings of the Seventh International Conference and Exposition on Engineering, Construction, Operations, and Business in Space* (Reston, VA: American Society of Civil Engineers, 2000), pp. 1–13; Harvey Brooks, "Motivations for the Space Program: Past and Future," in Allan A. Needell, ed., *The First 25 Years in Space: A Symposium* (Washington, DC: Smithsonian Institution Press, 1983), pp. 3–26; Perry Miller, "The Responsibility of a Mind in a Civilization of Machines," *The American Scholar* 31 (Winter 1961–1962), pp. 51–69; Thomas Park Hughes, *American Genesis: A Century of Invention and Technological Enthusiasm, 1870–1970* (New York: Viking, 1989), p. 2.

gallery. Since the room was inadequate for the media, NASA employees brought in more chairs and tried to make the journalists as comfortable as possible in the cramped surroundings.[57]

Many of the Mercury Seven astronauts have recorded their recollections of this singular event and all expressed the same hesitation and dread that Glennan experienced. They also expressed irritation at the huge and unruly audience assembled for the press conference. Alan Shepard and Donald 'Deke' Slayton had a brief conversation as they sat down at the table behind the curtain and contemplated the event ahead:

> "Shepard," Deke leaned toward him. "I'm nervous as hell. You ever take part in something like this?"
>
> Alan grinned. "Naw." He raised an eyebrow. "Well, not really. Anyway, I hope it's over in a hurry."
>
> "Uh huh. Me, too," Deke said quickly.[58]

When the curtain went up NASA Public Affairs Officer par excellence Walter Bonney announced:

> Ladies and gentlemen, may I have your attention, please. The rules of this briefing are very simple. In about sixty seconds we will give you the announcement that you have been waiting for: the names of the seven volunteers who will become the Mercury astronaut team. Following the distribution of the kit—and this will be done as speedily as possible—those of you who have p.m. deadline problems had better dash for your phones. We will have about a ten- or twelve-minute break during which the gentlemen will be available for picture taking.[59]

Like a dam breaking, a sea of photographers moved forward and popped flashbulbs in the faces of the Mercury Seven astronauts. A buzz in the conference room rose to a roar as this photo shoot proceeded. Some of the journalists bolted for the door with the press kit to file their stories for the evening papers; others ogled the astronauts.

Fifteen minutes later Bonney brought the room to order and asked Keith Glennan to come out and formally introduce the astronauts. Glennan offered a brief welcome and added, "It is my pleasure to introduce to you—and I consider

---

57. "Press Conference, Mercury Astronaut Team," transcript of press conference, 9 April 1959. Folder 18674, NASA Historical Reference Collection, NASA History Division, NASA Headquarters, Washington, DC.

58. See Donald K. Slayton and Alan B. Shepard, *Moonshot: The Inside Story of America's Race to the Moon* (New York: Turner Publishing, Inc., 1994), p. 62; Roger D. Launius and Bertram Ulrich, *NASA and the Exploration of Space* (New York: Stewart, Tabori, and Chang, 1998), pp. 35–43; Donald K. "Deke" Slayton with Michael Cassutt, *Deke! U.S. Manned Space from Mercury to Shuttle* (New York: Forge Press, 1994), pp. 73–74.

59. "Press Conference, Mercury Astronaut Team," transcript of press conference, 9 April 1959. Folder 18674, NASA Historical Reference Collection, NASA History Division, NASA Headquarters, Washington, DC.; also quoted in Launius and Ulrich, *NASA and the Exploration of Space*, pp. 40–41.

it a very real honor, gentlemen—Malcolm S. Carpenter, Leroy G. Cooper, John H. Glenn, Jr., Virgil I. Grissom, Walter M. Schirra, Jr., Alan B. Shepard, Jr., and Donald K. Slayton . . . the nation's Mercury Astronauts!" These personable pilots faced the audience in civilian dress, and many people in the audience forgot that they were volunteer test subjects and military officers. Rather, they were a contingent of mature, middle-class Americans, average in build and visage, family men all, college-educated as engineers, possessing excellent health, and professionally committed to flying advanced aircraft.[60]

The reaction was nothing short of an eruption. Applause drowned out the rest of the NASA officials' remarks. Journalists rose to their feet in a standing ovation. Even the photographers crouched at the foot of the stage rose in acclamation of the Mercury Seven. A wave of excitement circulated through the press conference like no one at NASA had ever seen before. What was all of the excitement about?

The astronauts asked themselves the same question. Slayton nudged Shepard and whispered in his ear, "They're applauding us like we've already done something, like we were heroes or something." It was clear to all that Project Mercury, the astronauts themselves, and the American space exploration program were destined to be something extraordinary in the nation's history.[61]

The rest of the press conference was as exuberant as the introduction. At first the newly selected astronauts replied to the press corps' questions with military stiffness, but led by an effervescent and sentimental John Glenn, they soon warmed to the interviews. What really surprised the astronauts, however, was the nature of the questions most often asked. The reporters did not seem to care about their flying experience, although all had been military test pilots, many had combat experience and decorations for valor, and some held aircraft speed and endurance records. They did not seem to care about the details of NASA's plans for Project Mercury. What greatly interested them, however, were the personal lives of the astronauts. The media wanted to know if they believed in God and practiced any religion. They wanted to know if they were married and the names and ages and gender of their children, they wanted to know what their families thought about space exploration and their roles in it, and they wanted to know about their devotion to their country. God, country, family, and self, and the virtues inherent in each of them became the theme of the day.[62]

It was thus an odd press conference, with the reporters probing the characters of the pilots. But the motivation was never to dig up dirt on the astronauts, as has so often been the case with the media since, and was certainly something they could have profitably done with these men; instead, it was just the opposite. The reporters wanted confirmation that these seven men embodied the best virtues of the U.S. They wanted to demonstrate to their readers that the Mercury Seven strode Earth as latter-day saviors whose purity coupled with noble deeds

---

60. The Astronauts Themselves, *We Seven* (New York: Simon and Schuster, 1962); William Leavitt, "First American into Orbit," *Space Digest*, March 1959, pp. 62–65.

61. Slayton and Shepard, *Moonshot*, chapter 1.

62. "Space Voyagers Rarin' to Orbit," *Life*, 20 April 1959, p. 22

would purge this land of the evils of communism by besting the Soviet Union on the world stage. The astronauts did not disappoint.

John Glenn, perhaps intuitively or perhaps through sheer zest and innocence, picked up on the mood of the audience and delivered a ringing sermon on God, country, and family that sent the reporters rushing to their phones for rewrite. He described how Wilbur and Orville Wright had flipped a coin at Kitty Hawk in 1903 to see who would fly the first airplane and how far we had come in only a little more than 50 years. "I think we would be most remiss in our duty," he said, "if we didn't make the fullest use of our talents in volunteering for something that is as important as this is to our country and to the world in general right now. This can mean an awful lot to this country, of course." The other astronauts fell in behind Glenn and eloquently spoke of their sense of duty and destiny as the first Americans to fly in space. Near the end of the meeting, a reporter asked if they believed they would come back safely from space, and all raised their hands. Glenn raised both of his.[63]

The astronauts emerged as noble champions who would carry the nation's manifest destiny beyond its shores and into space. James Reston of the *New York Times* exulted in the astronaut team. He said he felt profoundly moved by the press conference, and even reading the transcript of it made one's heartbeat a little faster and step a little livelier. "What made them so exciting," he wrote, "was not that they said anything new but that they said all the old things with such fierce convictions. . . . They spoke of 'duty' and 'faith' and 'country' like Walt Whitman's pioneers. . . . This is a pretty cynical town, but nobody went away from these young men scoffing at their courage and idealism."[64]

These statements of values seem to have been totally in character for what was a remarkably homogeneous group. They all embraced a traditional lifestyle that reflected the highest ideals of the American culture. The astronauts also expressed similar feelings about the role of family members in their lives and the effect of the astronaut career on their spouses and children. In a recent study by sociologist Phyllis Johnson, analyzing several Apollo-era astronaut autobiographies, she found that the public nature of what the astronauts did meant that their family and work lives were essentially inseparable, often taking a toll on those involved in the relationship. She concluded:

> The data on these early astronauts need to be interpreted in light of the work-family views of the time: men were expected to keep their work and family lives compartmentalized. Their family life was not supposed to interfere with work life, but it was acceptable for work life to overlap into their family time. In high level professions, such as astronauts, the wife's support of his career was important; rather than 'my' career, it became 'our' career. The interaction between work and family is an

---

63. John Glenn with Nick Taylor, *John Glenn: A Memoir* (New York: Bantam Books, 1999); "Space Voyagers Rarin' to Orbit," *Life*, 20 April 1959, p. 22.

64. John H. Glenn, "A New Era: May God Grant Us the Wisdom and Guidance to Use It Wisely," *Vital Speeches of the Day*, 15 March 1962, pp. 324–326; Dora Jane Hamblin, "Applause, Tears and Laughter and the Emotions of a Long-ago Fourth of July," *Life*, 9 March 1962, p. 34; Launius and Ulrich, *NASA and the Exploration of Space*, p. 43.

important aspect of astronaut morale and performance, which has been neglected by researchers.[65]

The media, reflecting the desires of the American public, depicted the astronauts and their families at every opportunity. The insatiable nature of this desire for intimate details prompted NASA to construct boundaries that both protected the astronauts and projected specific images that reinforced the already present traditional and dominant structure of American society. NASA, for obvious reasons, wanted to portray an image of happily married astronauts, not extramarital scandals or divorce. Gordon Cooper, one of the Mercury Seven, recalled that public image was important to some inside NASA because "marital unhappiness could lead to a pilot making a wrong decision that might cost lives—his own and others."[66] That might have been part of it, but the Agency's leadership certainly wanted to ensure that the image of the astronaut as clean-cut, all-American boy did not tarnish.

Sometimes the astronauts caused NASA officials considerable grief, and they sometimes had to rule them with an authoritarian hand. More often, however, they were benevolent and patriarchal toward the astronauts. Often this had to do with what rules they needed to follow and the lack of well-understood guidelines for their ethical conduct. For example, when the Space Task Group moved to Houston in 1962, several local developers offered the astronauts free houses. This caused a furor that reached the White House and prompted the involvement of Vice President Lyndon B. Johnson. (In this case, the head of the Manned Spacecraft Center, Robert R. Gilruth, had to disallow an outright gift to the astronauts.)[67] Gilruth's boys also got into trouble over what they could and could not do to make additional money on the outside. NASA had facilitated the Mercury Seven to sell their stories to *Life* magazine. This had raised a furor, and NASA policies were changed thereafter, but in 1963, Forrest Moore complained to Johnson that the second group of astronauts was seeking to do essentially the same thing. Gilruth had to intervene and explain that any deals for "personal stories" would be worked through the NASA General Counsel and would only take place in a completely open and legal manner.[68] Gilruth also defended the astronauts to the NASA leadership when they accepted tickets to see the Houston Astros season opener baseball game in the new Astrodome in

65. Phyllis J. Johnson, "Work-Family Linkages in the Lives of Astronauts," presentation at 55th International Astronautical Congress, 6 October 2004, Vancouver, British Columbia, Canada.

66. Gordon Cooper and Barbara Henderson, *Leap of Faith: An Astronaut's Journey into the Unknown* (New York: HarperCollins, 2000), p. 26.

67. Edward Welsh to Lyndon B. Johnson, "Gift of Houses to Astronauts," 2 April 1962, VP Papers, LBJ Library, box 182, University of Texas, Austin, Texas; Robert Gilruth Oral History No. 6 by David DeVorkin and John Mauer, 2 March 1987, Glennan-Webb-Seamans Project, National Air and Space Museum, Washington, DC.

68. Robert C. Seamans, Jr., Associate Administrator, NASA, Memorandum for Robert R. Gilruth, Director, Manned Spaceflight, NASA, "Astronaut Activities," 30 August 1962. Folder 18674, NASA Historical Reference Collection, NASA History Division, NASA Headquarters, Washington, DC; LBJ to Forrest Moore, President, Rominger Advertising Agency, 14 June 1963, VP Papers, LBJ Library, box 237, University of Texas, Austin, Texas.

1965, although he reprimanded several for poor judgment. While he told his superiors that he saw no reason why the astronauts should not enjoy the experience, he ensured that this type of media problem did not repeat itself. He also privately chastised, but publicly defended, John Young over the famous corned beef sandwich episode during Gemini III. He took the licks for these actions from the NASA Administrator:

> If this were a military operation and this kind of flagrant disregard of responsibility and of orders were involved, would not at least a reprimand be put in the record? . . . The only way I know to run a tight ship is to run a tight ship, and I think it essential that you and your associates give the fullest *advance* consideration to these matters, rather than to have them come up in a form of public criticism which takes a great deal of time to answer and which make the job of all of us more difficult.[69]

None of this suggests that NASA officials let the astronauts run amuck. They tried to maintain order through more patriarchal means than military ones, but on occasion—as in the case of the Apollo 15 stamp cover sales by the crew—they could be enormously stern.[70] Gilruth later said he tried to keep issues in perspective. These men put their lives on the line and deserved some leniency when minor problems arose. After all, they rose to the challenge repeatedly in conducting Mercury, Gemini, and Apollo.

The bravery of the astronauts touched emotions deeply seated in the American experience of the 20th century. Even their close associates at NASA remained in awe of them. The astronauts put a very human face on the grandest technological endeavor in history and the myth of the virtuous, no-nonsense, able, and professional astronaut was born at that moment in 1959. In some respects it was a natural occurrence. The Mercury Seven were, in essence, each of us. None were either aristocratic in bearing or elitist in sentiment. They came from everywhere in the nation, excelled in the public schools, trained at their local state university, served their country in war and peace, married and tried to make lives for themselves and their families, and ultimately rose to their places on the basis of merit. They represented the best the country had to offer and, most importantly, they expressed at every opportunity the virtues ensconced in

---

69. James E. Webb to Robert R. Gilruth, 15 April 1965, James E. Webb Papers, Box 113, NASA-Astronaut Notes, Truman Library, Independence, Missouri.

70. On the stamp cover incident, in which the crew of Apollo 15 took collectibles with them to the Moon and then sold them after their return, see David Scott and Alexei Leonov, *Two Sides of the Moon: Our Story of the Cold War Space Race* (New York: Thomas Dunne Books, 2004), pp. 328–331. Jeff Dugdale, *Orbit* magazine, notes that "David Scott, the Commander, was famously dismissed from the Astronaut Corps on the first anniversary of his return from this mission as the Apollo 15 crew had smuggled 400 space covers with them. It was reported in newspapers in July 1972 that a West German stamp dealer had sold 100 of these at £570 each. Each of the three crew members had been expected to gain as much as £2,700 from the sale of covers. However they then declined to accept any money, acknowledging that their actions had been improper. Jim Irwin also resigned from the Astronaut Corps, and Worden was also moved out of the select group and made no more flights." See Jeff Dugdale, "Moonwalkers," available online at *http://www.asss.utvinternet.com/articles1/moonwalkers.htm*, accessed 11 October 2004.

the democratic principles of the republic. In many ways, the astronauts were the logical focal point of the space program because they were something that regular people could understand. Instead of mathematics, rockets, and acronyms, the astronauts served as an understandable entry point into a mysterious and elite world of science, technology, and exploration. In other words, the astronauts were the single most important element that made the space program something that resonated with the broader populace because of their (constructed to some degree) "everyman" status. They were not part of the technological elites that ran NASA, nor were they mechanical and alien like the machines they flew. They were quite aware of their status as national symbols and hoped to use that status to advance U.S. interests (I-28).[71]

The astronauts worked enormously hard to make Project Mercury a success, undergoing training far from their professional experience (I-21).[72] In December 1959, John Glenn described for a colleague some of the stress and strain of this effort:

> Following our selection in April, we were assigned to the Space Task Group, portion of NASA at Langley Field, and that is where we are based when not traveling. The way it has worked out, we have spent so much time on the road that Langley has amounted to a spot to come back to get clean skivvies and shirts and that's about all. We have had additional sessions at Wright Field in which we did heat chamber, pressure chamber, and centrifuge work and spent a couple of weeks this fall doing additional centrifuge work up at NADC, Johnsville, Pennsylvania. This was some program since we were running it in a lay-down position similar to that which we will use in the capsule later on and we got up to as high as 16 g's. That's a bitch in any attitude, lay-down or not (I-30).[73]

NASA kept the astronauts enormously busy training for future space missions. As Robert B. Voas of NASA's Space Task Group reported in May 1960: "The [training] program which has resulted from these considerations has allotted about one-half of the time to group activities and the other half to individually planned activities in each Astronaut's area of specialization" (I-31).[74]

When they were selected for Project Mercury in 1959, no one fully realized what would be the result of having highly skilled pilots involved in the effort.

---

71. Mercury Astronauts, Memorandum For [Mercury] Project Director, NASA, "Exchange of visits with Russian Astronauts," 21 October 1959. Folder 18674, NASA Historical Reference Collection, NASA History Division, NASA Headquarters, Washington, DC.

This is also the subject of Roger D. Launius, "Heroes in a Vacuum: The Apollo Astronaut as Cultural Icon," IAC-04-IAA.6.15.1.07, IAA History Session, International Astronautical Congress, Vancouver, British Columbia, Canada, 4–8 October 2004.

72. Among other things they undertook scientific training. See Dr. William S. Augerson, Human Factors Branch, NASA, Memorandum for Chief, Operations Division, NASA, "Scientific Training for Pilots of Project Mercury," 27 March 1959. Folder 18674, NASA Historical Reference Collection, NASA History Division, NASA Headquarters, Washington, DC.

73. John Glenn, Mercury Astronaut, NASA, to Lt. Commander Jim Stockdale, USN, 17 December 1959. Folder 18674, NASA Historical Reference Collection (Doc. VII-I-31), NASA History Division, NASA Headquarters, Washington, DC.

74. Robert B. Voas, NASA Space Task Group, "Project Mercury Astronaut Training Program,"

Originally they had been viewed as minor participants in the flights by engineers developing Project Mercury at NASA's Langley Research Center in the winter of 1958 to 1959. Numerous skirmishes took place between engineers and astronauts in the development of the Mercury capsule, the "man-rating" of the launch vehicle, and in determining the level of integration of the astronaut into the system. Donald K. Slayton, who early took the lead for the Mercury Seven and later officially headed the astronaut office, emphasized the criticality of astronauts not as passengers but as pilots. In a speech before the Society of Experimental Test Pilots in 1959, he said:

> Objections to the pilot [in space] range from the engineer, who semi-seriously notes that all problems of Mercury would be tremendously simplified if we didn't have to worry about the bloody astronaut, to the military man who wonders whether a college-trained chimpanzee or the village idiot might not do as well in space as an experienced test pilot . . . I hate to hear anyone contend that present day pilots have no place in the space age and that non-pilots can perform the space mission effectively. If this were true, the aircraft driver could count himself among the dinosaurs not too many years hence.
>
> Not only a pilot, but a highly trained experimental test pilot is desirable . . . as in any scientific endeavor the individual who can collect maximum valid data in minimum time under adverse circumstances is highly desirable. The one group of men highly trained and experienced in operating, observing, and analyzing airborne vehicles is the body of experimental test pilots represented here today. Selection of any one for initial spaceflights who is not qualified to be a member of this organization would be equivalent to selecting a new flying school graduate for the first flight on the B-70, as an example. Too much is involved and the expense is too great.[75]

Slayton's defense of the role of the Mercury astronauts has found expression in many places and circumstances since that time.

Notwithstanding arguments to the contrary from some quarters, officials overseeing Project Mercury always intended that the astronauts should have control over the spacecraft that they flew in. Making these devices safe enough for humans took longer and exposed more doubts than NASA had expected and the astronauts themselves aided immensely in moving this integration forward. As the official history of Mercury reported in 1966:

> During the curiously quiet first half of 1960, the flexibility of the Mercury astronaut complemented and speeded the symbiosis of man

---

26 May 1960; U.S. Naval School of Aviation Medicine, "Proposed Schedule, Project Mercury (NASA) Astronauts Training Program, 28 March-1 April 1960," 11 February 1960. Folder 18674, NASA Historical Reference Collection, NASA History Division, NASA Headquarters, Washington, DC.

75. Donald K. Slayton, speech, annual meeting, Society of Experimental Test Pilots, Los Angeles, CA, 9 October 1959. Folder 18674, NASA Historical Reference Collection, NASA History Division, NASA Headquarters, Washington, DC.

and missile, of astronaut and capsule. Technology, or hardware, and techniques, or procedures—sometimes called "software" by hardware engineers—both had to be developed. But because they were equally novel, reliability had to be built into the new tools before dexterity could be acquired in their use.[76]

From the beginning, therefore, Project Mercury managers accepted the integral role of astronauts in controlling the spacecraft.

Christopher C. Kraft, Jr., Chief Flight Director for Mercury, made the case that many in NASA wanted a "go slow" approach to astronaut integration because "at the beginning, the capabilities of Man were not known, so the systems had to be designed to function automatically. But with the addition of Man to the loop, this philosophy changed 180 degrees since primary success of the mission depended on man backing up automatic equipment that could fail."[77] Kraft and his colleagues came to realize that the astronauts served an exceptionally useful purpose for enhancing the chances of success with Project Mercury. As an example, when the astronauts first visited the McDonnell Aircraft Corporation facilities in May 1959 they reviewed progress of the capsule they would fly with a sense for the human factors that would be necessary to make it work. They came up with several requests for alterations—including an observation window, manual reentry thruster controls, and an escape hatch with explosive bolts—and based on their recommendations NASA and McDonnell engineers went to work to overcome their concerns.[78]

One incident concerning the astronauts' desire for changes to the Mercury capsule has entered the public consciousness as a representation of conflicts between the fliers and the engineers. One key alteration the astronauts pressed for was the addition of an observation window for navigational purposes. In the feature film, *The Right Stuff*, this incident is depicted as a nasty confrontation that required the astronauts to threaten to appeal directly to the public through the media for their changes to be adopted. Only in the face of perceived embarrassment would the NASA and McDonnell engineers back down.[79] This adversarial approach to astronaut involvement made for sparks on the screen, but it bore little resemblance to what actually took place. The design engineers working on the spacecraft were exceptionally concerned about weight, and glass thick enough to

76. Swenson et al., *This New Ocean*, p. 167.

77. Christopher C. Kraft, Jr., "A Review of Knowledge Acquired from the first Manned Satellite Program," MSC fact sheet No. 206, p. 1, Mercury Files, Special Collections, University of Houston-Clear Lake, Texas.

78. Minutes, "Mock-Up Review," 12–14 May 1959, with enclosure addressed to C. H. Zimmerman and George M. Low, 23 June 1959. Folder 18674, NASA Historical Reference Collection, NASA History Division, NASA Headquarters, Washington, DC.

79. *The Right Stuff* (Los Angeles: Warner Bros., 1983). Feature film directed by Philip Kaufman and produced by Irwin Winkler and Robert Chartoff, screenplay by Chartoff. A cast of unknown actors at the time depicted the development of aeronautics and astronautics from 1947 through the time of the Mercury program. Scott Glenn, cast as Alan Shepard, played the astronaut perfectly, and Ed Harris as John Glenn captured the essence of being an astronaut. A box office hit, the film also won an Academy Award for special effects.

withstand the harsh environments of launch, spaceflight, and reentry would weigh quite a lot. As Maxime A. Faget, designer of the Mercury spacecraft, remarked in an interview on 1 February 1991: "When we started off, we thought the Atlas could put about 2,000 pounds into orbit. So our design weight at the initiation of the program was 2,000 pounds. That was our goal. We had to build it at 2,000 pounds, and it was very challenging." To save weight Faget had only two portholes in the spacecraft and he thought that was good enough, but the astronauts pressed their point and got their navigation window. In the process of this and other changes, the Mercury capsule grew to a weight of about 2,700 pounds. Faget concluded, "Fortunately, as the Atlas was developed, we improved its performance, so it didn't have any trouble carrying the full weight. I think a great number of changes to the Mercury capsule would not have happened if the Atlas had not been improved." He added, "The astronauts were involved in the program decisions from the time they came on board. I think it was the right way to do it."[80]

Edward Jones made his point about human involvement even more succinctly in a paper delivered before the American Rocket Society in November 1959. He suggested that the astronaut was virtually necessary to the successful operation of Mercury missions. He commented:

> Serious discussions have advocated that man should be anesthetized or tranquillized or rendered passive in some other manner in order that he would not interfere with the operation of the vehicle. . . . As equipment becomes available, a more realistic approach evolves. It is now apparent with the Mercury capsule that man, beyond his scientific role, is an essential component who can add considerably to systems effectiveness when he is given adequate instruments, controls, and is trained. Thus an evolution has occurred . . . with increased emphasis now on the positive contribution the astronaut can make.[81]

The result of these efforts led to the development of a Mercury spacecraft that allowed considerable, but not total, control by the astronaut.

As Gordon Cooper recalled: "We weren't just mouthpieces or pilots milling around a hangar waiting to fly. We were involved in all aspects of the program, and there was a job for everybody." Of the Mercury Seven, Scott Carpenter took on communication and navigation, Alan Shepard handled worldwide tracking and capsule recovery, John Glenn worked on cockpit layout and design of the instrument panel in the spacecraft, Wally Schirra worked on spacesuits and life-support, Gus Grissom worked to develop automatic and manual control systems, Deke Slayton oversaw systems integration with the Mercury capsule and the Atlas

---

80. Interview with Maxime A. Faget and Alan B. Shepard, 1 February 1991, Hall of Science and Exploration, Academy of Achievement: A Museum of Living History, available online at *http://www.achievement.org/autodoc/page/she0int-2*, accessed 7 November 2004.

81. Edward R. Jones, "Man's Integration into the Mercury Capsule," paper presented at the 14th annual meeting of the American Rocket Society, Washington, DC, 16–19 November 1959, pp. 1–2. Folder 18674, NASA Historical Reference Collection, NASA History Division, NASA Headquarters, Washington, DC.

rocket, and Gordon Cooper served as liaison with the rocket team developing the launch systems.[82] When problems arose during MA-4, an unpiloted flight of the Mercury-Atlas system in September 1961, Robert Gilruth commented that had an astronaut been aboard he could have diagnosed and overcome the malfunctions of the automated system. That was why they were present, he asserted. In the end, Mercury as a system worked, but not without flaws, and the program successfully flew six humans in space between 5 May 1961 and 15 to 16 May 1963.[83]

Building the Mercury Capsule

The Mercury spacecraft flown by the first astronauts was the product of a genius incarnate in the form of a diminutive Cajun by the name of Dr. Maxime A. Faget, an engineering graduate of Louisiana State University and submarine officer in World War II. Working at the Langley Research Center in Hampton, Virginia, he was one of the most innovative and thoughtful engineers working on Mercury. While everyone thinking about spaceflight in the 1950s was obsessed with rocket planes, Faget realized that space was an entirely different environment and could effectively be accessed using an entirely different type of vehicle.[84]

During November and December 1958, the Space Task Group energetically pursued the development of the ballistic capsule flown by the astronauts. Faget became the chief designer of the Mercury spacecraft, and on 7 November 1958, held a briefing for 40 aerospace firms to explain the requirements for bidding on a NASA contract to build the capsule according to Faget's specifications. A week later, after 20 firms had indicated an interest, Faget's team mailed out requests for proposals. They received 11 proposals on 11 December and worked over the Christmas holidays to complete an evaluation. The Source Evaluation Board, convened under Faget's direction, recommended that the McDonnell Aircraft Corporation of St. Louis, Missouri, serve as the prime manufacturer for this system. The NASA leadership accepted this decision and announced the contract award on 9 January 1959. In the end NASA procured one dozen capsules at an estimated cost of $18.3 million—plus an award fee of $1.5 million— but the actual costs almost immediately spiraled upwards, causing considerable concern among senior government officials even as they made the funds avail-

82. Cooper with Henderson, *Leap of Faith*, pp. 20–22. See also Donald K. "Deke" Slayton with Michael Cassutt, *Deke! U.S. Manned Space from Mercury to the Shuttle* (New York: Forge, 1994), pp. 78–79.

83. John Catchpole, *Project Mercury: NASA's First Manned Space Programme* (Chichester, UK: Springer Praxis, 2001), p. 310. Before the astronauts flew, however, NASA launched primates into space to test the system. See George M. Low, Program Chief, Manned Spaceflight, NASA Memorandum for Mr. R. R. Gilruth, Director, Project Mercury, NASA, "Animal Payloads for Little Joe," 19 June 1959, with attached Memorandum from T.K.G to George M. Low, 15 June 1959; NASA, Information Guide for Animal Launches in Project Mercury, 23 July 1959. Folder 18674, NASA Historical Reference Collection, NASA History Division, NASA Headquarters, Washington, DC.

84. "Maxime A. Faget," biographical file, NASA Historical Reference Collection.

able to complete the effort (I-26).[85] This two-month procurement process, from start to contract award, deserves special notice as something of a speed record with respect to the convoluted manner in which the federal government buys everything from paperclips to nuclear powered aircraft carriers. In the end, the Mercury project would cost approximately $350 million for research and development as well as operations.[86]

McDonnell's Mercury team, under the leadership of John F. Yardley, immediately began wrestling with Faget's requirements. It had a good start on the capsule from work done the year before for the Air Force, but Yardley was unprepared for the difficulties encountered when actually building the spacecraft. First and most important, Yardley's team struggled with strict weight requirements so that the capsule could be launched atop the Atlas rocket. NASA's specifications for the capsule had been 2,000 pounds placed in orbit. McDonnell's bid had proposed a 2,400-pound spacecraft, plus or minus 25 percent. The minus side allowed a capsule of 1,800 pounds, perfect for the capability of the Atlas, but anything over 2,000 pounds could not be put into orbit by the envisioned launcher. A combination of paring the capsule design down to the lightest weight possible and increasing the thrust of the Atlas finally made successful launches in Project Mercury attainable, but it was a difficult task and the capability margins were always stretched. Everyone was keenly aware of this and other problems in building the spacecraft. Wernher von Braun wrote a friendly letter to Robert Gilruth about McDonnell's performance. "It has come to my attention that one of our ball carriers has his shoelaces untied and doesn't know it," he wrote. "If he trips and falls we may all lose the game and our astronaut his life. So I feel that I must pass along to you what has been brought to my attention, at the risk of making a few people sore" (I-27).[87] In response to such concerns, teams of NASA engineers swarmed over contractors in an effort to keep the program on track.

For the next year the NASA/McDonnell engineering team worked through the critical components of the spacecraft. They focused on the four major elements of any flying machine:
- Aerodynamics/stability and control
- Avionics/electronics
- Propulsion
- Materials

In addition, they had the critical area of human factors to oversee in the development of this entirely new type of spacecraft.[88] One of the McDonnell engi-

85. Cost had long been an issue, even before the McDonnell contract. See A. J. Goodpaster, Brigadier General, USA, Memorandum of Conference with the President, 29 September 1959. Folder 18674, NASA Historical Reference Collection, NASA History Division, NASA Headquarters, Washington, DC.

86. President's Science Advisory Committee, "Report of the Ad Hoc Panel on Man-in-Space," 14 November 1960 (final report 16 December 1960). Folder 18674, NASA Historical Reference Collection, NASA History Division, NASA Headquarters, Washington, DC.

87. Wernher von Braun, Director, Development Operations Division, NASA, to Robert R. Gilruth, Space Task Group, NASA, 9 October 1959. Folder 18674, NASA Historical Reference Collection, NASA History Division, NASA Headquarters, Washington, DC.

88. George M. Low, Program Chief, Manned Spaceflight, NASA, Memorandum for Administrator, NASA, "Status Report No. 1, Manned Satellite Project," 9 December 1958. Folder 18674, NASA

neering team's important decisions was to use a pure oxygen atmosphere at 5 psi. This atmosphere would become the standard for American spacecraft until the Space Shuttle, but it had a fundamental drawback as a fire hazard, something that proved fatal in the Apollo 1 accident of 1967.[89]

The Mercury capsule that emerged from this process stood 115 inches high with a tapering cylinder from 74 inches at its base so that it appeared to all as an upside-down ice cream cone. The pressurized cockpit for the pilot was the largest portion of the capsule, with most other systems packed throughout the cramped interior. Indeed, the astronaut had very little room for movement, being placed in an individually fitted contour seat for the duration of the flight. A smaller cylinder at the top housed other electronics as well as a parachute for recovery. Attitude control jets allowed the astronaut to orient the spacecraft during flight. An ablative heatshield with a ceramic coating affixed to the capsule's base would protect the spacecraft during reentry. Designed to adhere to strict weight restrictions and maximum strength, much of the spacecraft was titanium, but heat-resistant beryllium made up the upper cone of the vehicle since, other than the heatshield, it would suffer the greatest heat during reentry. Underneath the heatshield a retrorocket pack of three solid rocket motors served to slow the vehicle down and return it to Earth. Each motor produced 1,000 pounds of thrust for only about 10 seconds. The Mercury spacecraft also had 3 smaller posigrade rockets that produced 400 pounds of thrust each for a second, used for separating the capsule from its booster. Atop the capsule stood a launch escape tower with solid rocket motors producing 52,000 pounds of thrust that could shoot the capsule away from the rocket during an emergency on the launchpad or during ascent. The capsule proved a spare but serviceable space vehicle.[90]

Adapting Launch Vehicles

During Project Mercury two different boosters proved their mettle in sending astronauts into space. The first was the Redstone, built by Wernher von Braun's rocket team at the Army Ballistic Missile Agency (ABMA) in Huntsville, Alabama, as a ballistic missile and retrofitted for human flights.[91] NASA Administrator T. Keith Glennan materially aided this effort by securing the transfer of ABMA

Historical Reference Collection, NASA History Division, NASA Headquarters, Washington, DC.

89. This was discussed in "Report of the Ad Hoc Mercury Panel," 12 April 1961. Folder 18674, NASA Historical Reference Collection, NASA History Division, NASA Headquarters, Washington, DC.

The report stated: "The idea of using a single gas, O_2, atmosphere, in both the suit and capsule to simplify the system appears to be reasonable from an engineering standpoint if it meets the biomedical requirements. The environmental control system is capable of operating completely automatically if required and still provide redundancy in many areas against failure. In the automatic mode the only single point of failure without backup appears to be with the emergency oxygen rate valve. However, with man functioning in the system, this valve can be manually operated."

90. A description of the Mercury spacecraft may be found in Swenson et al., *This New Ocean*, pp. 223–262; Linda Ezell, *NASA Historical Data Book: Volume II*, pp. 134–139.

91. See Wernher von Braun, "The Redstone, Jupiter, and Juno," in Eugene M. Emme, ed., *The History of Rocket Technology* (Detroit, MI: Wayne State University Press, 1964), pp. 107–121.

to NASA thereby facilitating the tapping of expertise from the builders of the Redstone rocket.[92] In addition to a large number of other modifications, NASA engineers worked to lengthen the Redstone tanks and scrapped the original fuel, Hydyne, for alcohol. Hydyne proved too toxic and difficult to work with. In all, NASA's rocketeers made some 800 changes to the Redstone to prepare it for human spaceflight.[93]

Then there was the problem with the reliability of the Atlas rocket, envisioned as the launcher of choice for the Mercury orbital missions. A converted ICBM, the Atlas had been undergoing an on-again, off-again development since 1946. Canceled once and underfunded thereafter, the Air Force had been unable until the Sputnik crisis to secure sufficient resources to make serious progress on it. Because of this difficulty, its designers at the Convair Corp. had accepted, as a given, a 20 percent failure rate. In fact, the rate proved much higher in the early going. As 1959 began, seven out of eight launches had failed. Sometimes the Atlas blew up on the pad and sometimes it veered off course in flight only to be destroyed by the range safety officer. Instead of 80 percent reliability, still not acceptable for human flight, the Atlas had an 80 percent failure rate.[94] That would most assuredly not do with astronauts aboard. Robert Gilruth testified to Congress about this problem a few months after the creation of the Space Task Group. "The Atlas . . . has enough performance . . . and the guidance system is accurate enough, but there is the matter of reliability. You don't want to put a man in a device unless it has a very good chance of working every time." Gilruth urged time and money to test the hardware under actual flight conditions without people aboard. "Reliability is something that comes with practice," he said.

Ever so incrementally, Atlas project engineers improved the performance of the launch vehicle. They placed a fiberglass shield around the liquid oxygen tank to keep the engines from igniting it in a massive explosion, a rather spectacular failure that seemed to happen at least half the time. They changed out virtually every system on the vehicle, substituting tried and true technology wherever possible to minimize problems. They altered procedures and developed new telemetry to monitor the operations of the system. Most important, they developed an abort sensing system

---

92. T. Keith Glennan to the President, "Responsibilities and Organization for Certain Space Activities," 2 November 1959. Folder 18674, NASA Historical Reference Collection, NASA History Division, NASA Headquarters, Washington, DC.

93. Swenson et al., *This New Ocean*, p. 181; NASA Office of Congressional Relations, "Mercury-Redstone III," p. 5-1; William M. Bland, Jr., Space Task Group, Memorandum for Project Director, "Second Coordination Meeting Concerning Project Mercury with NASA, ABMA, and MAC Representatives Held 20 March 1959, at NASA, Space Task Group, Langley Field, Virginia," 27 March 1959; T. Keith Glennan, NASA Administrator, to Neil H. McElroy, Secretary of Defense, 14 July 1959. Folder 18674, NASA Historical Reference Collection, NASA History Division, NASA Headquarters, Washington, DC.

94. For able histories of the Atlas, see Dennis R. Jenkins, "Stage-and-a-Half: The Atlas Launch Vehicle," in Roger D. Launius and Dennis R. Jenkins, eds., *To Reach the High Frontier: A History of U.S. Launch Vehicles* (Lexington, KY: University Press of Kentucky, 2002), pp. 102–170; John Lonnquest, "The Face of Atlas: General Bernard Schriever and the Development of the Atlas Intercontinental Ballistic Missile, 1953-1960," Ph.D. Diss., Duke University, 1996; Davis Dyer, "Necessity is the Mother of Invention: Developing the ICBM, 1954–1958," *Business and Economic History* 22 (1993): pp. 194–209. Although dated, a useful early essay is Robert L. Perry, "The Atlas, Thor, Titan, and Minuteman," in Eugene M. Emme, ed., *History of Rocket Technology*, pp. 143–155.

(labeled ASS by everyone but the people involved in developing it) to monitor vehicle performance and to provide early escape of the Mercury capsule if necessary.[95]

Suborbital Flights

The first Mercury test flight took place on 21 August 1959, when a capsule carrying two rhesus monkeys was launched atop a cluster of Little Joe solid-fuel rockets (I-24). Other tests using both Redstone and Atlas boosters and carrying both chimpanzees and astronaut dummies soon followed (I-25). The first flight of a Mercury-Redstone combination took place on 21 November 1960 (Mercury-Redstone 1), but only with a "simulated man" in its capsule. It pointed out serious problems with the system. The rocket rose only 3.8 inches off the pad, and then it settled back on its fins. The parachutes deployed and fell to the launchpad while the capsule remained in place on the booster. The episode proved embarrassing, but NASA soon found that faulty grounding on electrical circuitry had caused a short in the system. They repaired the problem and the next test flight, Mercury-Redstone 1A, flown on 19 December 1960, went somewhat better but still experienced problems. The rocket boosted the capsule higher and at greater G forces than expected, pushing it some 20 miles downrange beyond the target area. This led to the 31 January 1961, Mercury-Redstone 2 launch with Ham the chimpanzee aboard on a 16-minute, 39-second flight. Again, the booster overperformed and carried him 42 miles higher and 124 miles further downrange than planned. In the process, Ham suffered about 17 g's going up and some 15 during reentry. NASA made one more test flight, on 24 March 1961, and this time the mission took place as planned.[96]

With these tests, NASA was prepared to move on to the piloted portion of the suborbital Mercury program. As preparations for this flight progressed throughout the spring, on 12 April 1961, the Soviet Union suddenly launched Yuri Gagarin into orbit, counting coup on the U.S. space effort one more time.[97] This spaceflight gave greater impetus to rescue national honor in the early launch of an astronaut in the U.S.'s Mercury program. Interestingly, the leaders of the program took extraordinary efforts to prepare for the release of public information about the mission. They kept the name of the astronaut assigned to fly the mission secret until only a short time before the scheduled launch.

Presidential science advisor Jerome B. Wiesner also expressed concern that the media should be prevented from making the flight "a Hollywood production, because it can jeopardize the success of the entire mission." Wiesner, concerned

95. "Report of the Ad Hoc Mercury Panel," 12 April 1961. Folder 18674, NASA Historical Reference Collection, NASA History Division, NASA Headquarters, Washington, DC.

96. Swenson et al., *This New Ocean*, pp. 293–318; Linda Ezell, *NASA Historical Data Book*, Volume II, pp. 134–143; NASA News Release, "Mercury Redstone Booster Development Test," 22 March 1961, NASA Historical Reference Collection (Doc. IV-I-24); Richard J. Wisniewski, for Warren J. North, Head, Manned Satellites, to Director of Spaceflight Programs, "Mercury Status as of March 2, 1961," 3 March 1961. Folder 18674, NASA Historical Reference Collection, NASA History Division, NASA Headquarters, Washington, DC.

97. Swenson et al., *This New Ocean*, pp. 332–335; Thomas A. Heppenheimer, *Countdown: A History of Spaceflight* (New York: John Wiley & Sons, Inc., 1997), pp. 189–192.

with NASA's preparations for the mission, chartered a panel of the Presidential Scientific Advisory Committee to conduct an independent review of the program; that panel gave a qualified endorsement to NASA's plans to launch the first U.S. astronaut (I-32, I-33, I-34).[98]

Alan Shepard made that first suborbital Mercury flight on 5 May 1961, in the process establishing that the U.S. could send an individual into space and return him to Earth. At 9:34 a.m., about 45 million Americans sat tensely before their television screens and watched a slim black-and-white Redstone booster, capped with a Mercury spacecraft containing Shepard, lift off its pad at Cape Canaveral and go roaring upward through blue sky toward black space. At 2.3 seconds after launch, Shepard's voice came through clearly to Mercury Control; minutes later the millions heard the historic transmission: "Ahh, Roger; lift-off and the clock is started. . . . Yes, sir, reading you loud and clear. This is *Freedom 7*. The fuel is go; 1.2 g; cabin at 14 psi; oxygen is go . . . *Freedom 7* is still go!" Reaching a speed of 5,146 miles per hour and an altitude of about 116.5 miles, well above the 62-mile international standard for the minimum altitude for spaceflight, Shepard's suborbital flight lasted only 15 minutes and 22 seconds and he was weightless only a third of that time. *Freedom 7* landed 302 miles downrange from the Cape Canaveral in the Atlantic Ocean (I-35). It was an enormously significant event for the U.S. The flight made Shepard a national hero, and his stoical persona and public countenance also served to solidify his stature among Americans as a role model. In the following months, how best to capitalize for propaganda purposes on the astronauts' experiences without distorting them became a matter of policy concern (I-37).[99]

NASA officials were euphoric in the aftermath of the Alan Shepard flight, and some even offered proposals for expansive follow-on missions such as a circumlunar flight using the Mercury hardware (I-36).[100] Those schemes went nowhere,

---

98. Abe Silverstein, Director of Spaceflight Programs, NASA, Memorandum for Administrator, NASA, "Astronaut Selection Procedure for Initial Mercury-Redstone Flights," 14 December 1960; Jerome B. Wiesner, The White House, Memorandum for Dr. Bundy, "Some Aspects of Project Mercury," 9 March 1961; James E. Webb, Administrator, NASA, to James C. Hagerty, Vice President, American Broadcasting Company, 1 June 1961; Abe Silverstein, Director of Spaceflight Programs, NASA, Memorandum for Administrator, "Use of a Television System in Manned Mercury-Atlas Orbital Flights," 6 September 1961; Wernher von Braun, Memorandum, "Sensitivity of Mercury Launching Dates," 3 March 1961. Folder 18674, NASA Historical Reference Collection, NASA History Division, NASA Headquarters, Washington, DC.

99. NASA MR-3 Technical Debriefing Team, "Debriefing," 5 May 1961; James E. Webb, Administrator, NASA, to James C. Hagerty, Vice President, American Broadcasting Company, 1 June 1961. Folder 18674, NASA Historical Reference Collection, NASA History Division, NASA Headquarters, Washington, DC; Alan B. Shepard, Jr., "A Pilot's Story," *National Geographic* 130 (September 1961): pp. 432–444; The Astronauts Themselves, *We Seven* (New York: Simon and Schuster, 1962), Shepard wrote three sections, "The Urge to Pioneer," pp. 65–69, "What to do Until the Ship Comes," pp. 164–167, and "A Range Around the World," pp. 285–299; Alan B. Shepard Jr., "The Astronaut's Story of the Thrust into Space," *Life*, 19 May 1961; "Shepard's Space Saga," *Naval Aviation News* 42 (June 1961): pp. 20–23; Roger D. Launius, "Alan B. Shepard Jr.," in *Research Guide to American Historical Biography* (Washington, DC: Beacham Publishing, 1991), 5: pp. 2742–2748.

100. Joachim P. Kuettner, Chief, Mercury-Redstone Project, NASA, to Dr. von Braun, 18 May 1961. Folder 18674, NASA Historical Reference Collection, NASA History Division, NASA Headquarters, Washington, DC.

and a second Mercury flight on 21 July 1961 proved less successful.[101] After landing the hatch blew off prematurely from the Mercury capsule, *Liberty Bell 7,* and it sank into the Atlantic Ocean before it could be recovered. As Grissom noted about the incident:

> I was just waiting for their call when all at once, the hatch went. I had the cap off and the safety pin out, but I don't think that I hit the button. The capsule was rocking around a little but there weren't any loose items in the capsule, so I don't see how I could have hit it, but possibly I did. I had my helmet unbuttoned and it wasn't a loud report. There wasn't any doubt in my mind as to what had happened. I looked out and saw nothing but blue sky and water starting to ship into the capsule. My first thought was to get out, and I did. As I got out, I saw the chopper was having trouble hooking onto the capsule. He was frantically fishing for the recovery loop. The recovery compartment was just out of the water at this time and I swam over to help him get his hook through the loop. I made sure I wasn't tangled anyplace in the capsule before swimming toward the capsule. Just as I reached the capsule, he hooked it and started lifting the capsule clear. He hauled the capsule away from me a little bit and didn't drop the horsecollar down. I was floating, shipping water all the time, swallowing some, and I thought one of the other helicopters would come in and get me. I guess I wasn't in the water very long but it seemed like an eternity to me. Then, when they did bring the other copter in, they had a rough time getting the horsecollar to me. They got in within about 20 feet and couldn't seem to get it any closer. When I got the horsecollar, I had a hard time getting it on, but I finally got into it. By this time, I was getting a little tired. Swimming in the suit is difficult, even though it does help keep you somewhat afloat. A few waves were breaking over my head and I was swallowing some water. They pulled me up inside and then told me they had lost the capsule (I-38).[102]

Some suspected that Grissom had panicked and prematurely blown the capsule's side hatch into the water—and a panicked Grissom is how most people routinely remember him today because of a graphic misrepresentation of the incident in the movie *The Right Stuff*—but he became a national hero because of that flight, and appropriately so.[103] Despite this problem, these suborbital flights proved valuable for NASA technicians who found ways to solve or work around

---

101. This mission is recorded in Swenson et al., *This New Ocean,* pp. 341–379.

102. MR-4 Technical Debriefing Team, Memorandum for Associate Director, NASA, "MR-4 Postflight Debriefing of Virgil I. Grissom," 21 July 1961, with attached, "Debriefing." Folder 18674, NASA Historical Reference Collection, NASA History Division, NASA Headquarters, Washington, DC.

103. Tom Wolfe, *The Right Stuff* (New York: Farrar, Straus, Giroux, 1979), pp. 280–296. See also William J. Perkinson, "Grissom's Flight: Questions," *Baltimore Sun,* 22 July 1961, which suggests that NASA had miscalculated in its rocketry and forced Grissom into the unenviable position of being outside the recovery area, thereby increasing the time it took to reach the spacecraft. Grissom had personally performed well, Perkinson noted.

literally thousands of obstacles to successful spaceflight. The success of these two missions led to the cancellation of any more Mercury-Redstone flights, although two more had been planned (I-39).[104]

Achieving Orbit

Even as these suborbital flights reached completion, NASA began final preparations for the orbital aspects of Project Mercury (I-40, I-41). In this phase, NASA planned to use a Mercury capsule capable of supporting a human in space not just for a few minutes, but eventually for as much as three days. As a launch vehicle for this Mercury capsule, NASA used the more powerful Atlas instead of the Redstone. But this decision was not without controversy. There were technical difficulties to be overcome in mating it to the Mercury capsule, to be sure, but most of the differences had been resolved by the first successful orbital flight of an unoccupied Mercury/Atlas combination in September 1961. On 29 November 1961, the final test flight took place, this time with the chimpanzee Enos occupying the capsule for a two-orbit ride before being successfully recovered in an Atlantic Ocean landing.[105]

Not until 20 February 1962, after several postponements, did NASA launch an astronaut on an orbital flight. After repeated delays, including a nationally televised 27 January 1962 scrub just 20 minutes before liftoff, John Glenn became the first American to circle Earth on 20 February, making three orbits in his *Friendship 7* Mercury spacecraft.[106] The flight had several difficulties, and Glenn proved the worth of a pilot in the spacecraft. During his first orbit, Glenn's spacecraft drifted out of proper orbit attitude, yawing to the right and not being corrected by the low-rate attitude thrusters. When it reached a 20-degree alteration, high-rate thrusters fired to correct the problem, but this was an inappropriate use of these thrusters. Glenn took control and manually corrected for the yaw throughout much of the remainder of the mission using the low-rate attitude control jets. It was an excellent object lesson in the advantage of having an astronaut step in to control the spacecraft in the event of a malfunction. Virtually every Mercury mission would require a similar type of action on the part of the astronaut, and with every demonstration, all those associated with the program became more comfortable with human/machine interaction. Even more significant, Glenn experienced a potentially disastrous event when he learned that

---

104. Richard J. Wisniewski, Memorandum for Administrator, "Mercury Redstone 4 Mission," 24 July 1961; Robert R. Gilruth, Director, Space Task Group, NASA, to Marshall, NASA, (attention: Dr. Wernher von Braun), "Termination of Mercury Redstone Program," 23 August 1961. Folder 18674, NASA Historical Reference Collection, NASA History Division, NASA Headquarters, Washington, DC; Swenson et al., *This New Ocean*, pp. 328–330; Linda Ezell, *NASA Historical Data Book, Volume II*, pp. 143–144.

105. Von Braun, "The Redstone, Jupiter, and Juno," in Eugene M. Emme, ed., *History of Rocket Technology*, pp. 107–122.

106. "Month's Delay for Glenn Seen," *Washington Star*, 31 January 1962; Art Woodstone, "Television's $1,000,000 (When & If) Manshoot; Lotsa Prestige & Intrigue," *Variety*, January 24, 1962; Swenson et al., *This New Ocean*, pp. 419–436.

on the back side of his *Friendship 7* Mercury pressure shell, a landing bag, programmed to inflate a few seconds before splashdown to help cushion the impact, had possibly inflated in orbit. The landing bag was located just inside the heatshield, an ablative material meant to burn off during reentry, and was held in place in part by a retropack of three rocket motors that would slow the capsule down and drop it from orbit. Because of this apparent problem, Glenn had to return to Earth after only three orbits, instead of a planned seven, and leave the retropack in place during his fiery reentry, hoping that it would hold the heatshield in place. It did, and Glenn returned safely to Earth (I-42, I-43).[107]

Glenn's flight provided a welcome increase in national pride, making up for at least some of the earlier Soviet successes. The public, more than celebrating the technological success, embraced Glenn as a personification of heroism and dignity. Hundreds of requests for personal appearances by Glenn poured into NASA Headquarters, and NASA learned much about the power of the astronauts to sway public opinion. The NASA leadership made Glenn available to speak at some events but more often substituted other astronauts and declined many other invitations. Among other engagements, Glenn did address a joint session of Congress and participated in several ticker-tape parades around the country. NASA discovered, in the process of this hoopla, a powerful public relations tool that it has employed ever since. It also discovered that there was a need to control the activities of the Mercury astronauts so that they did not become a source of political or public embarrassment (I-44).[108]

Three more successful Mercury flights took place during 1962 and 1963. Scott Carpenter made three orbits on 20 May 1962 (I-45), and on 3 October 1962, Wally Schirra flew six orbits. The capstone of Project Mercury came on the flight of Gordon Cooper, who circled Earth 22 times in 34 hours from 15 to 16 May 1963. The program had succeeded in accomplishing its purpose: to successfully orbit a human in space, explore aspects of tracking and control, and to learn about microgravity and other biomedical issues associated with spaceflight.[109]

As the Mercury program made strides toward enabling the U.S. to move on to a lunar landing, as promised by President John F. Kennedy in May 1961, the

---

107. Dr. Robert B. Voas, Training Officer, NASA, Memorandum for Astronauts, "Statements for Foreign Countries During Orbital Flights," 7 November 1961; Telegram, NASA—Manned Spacecraft Center, Port Canaveral, FL, to James A. Webb; "MA-6 Postlaunch Memorandum," 21 February 1962; R. B. Voas, NASA, Memorandum for Those Concerned, "MA-6 Pilot's Debriefing," 22 February 1962, with attached, John Glenn, NASA, "Brief Summary of MA-6 Orbital Flight," 20 February 1962; NASA Manned Spacecraft Center, "Postlaunch Memorandum Report for Mercury-Atlas No. 6 (MA 6)," 5 March 1962. Folder 18674, NASA Historical Reference Collection, NASA History Division, NASA Headquarters, Washington, DC; The Astronauts Themselves, *We Seven*, pp. 310–312; Roger D. Launius, *NASA: A History of the U.S. Civil Space Program* (Malabar, FL: Krieger Pub., Co., 2000 ed.), chapter 11.

108. Swenson et al., *This New Ocean*, pp. 422–436.

109. W. J. North, Senior Editor, Manned Spacecraft Center, NASA, "MA-7/18 Voice Communications and Pilot's Debriefing," 8 June 1962. Folder 18674, NASA Historical Reference Collection (Doc. VII-I-47), NASA History Division, NASA Headquarters, Washington, DC; NASA Manned Spacecraft Center, *Mercury Project Summary Including Results of the Fourth Manned Orbital Flight May 15 and 16, 1963* (Washington, DC: NASA SP-45, 1963); Swenson et al., *This New Ocean*, pp. 446–503.

human spaceflight program found itself in turmoil over the relocation of the Space Task Group from Langley Research Center to a new Manned Spacecraft Center in Houston, Texas. A decision taken in September 1961 as it became apparent that the scope, size, and support for human spaceflight necessitated an entirely separate center, the new human spaceflight center rested on land granted from Rice University.[110] Upon reaching Houston, the Space Task Group set to work not only settling into their new facility, but also in completing the design and development of their next projects. The center also became the home of NASA's astronauts and the site of mission control.[111] Within its first few months in Houston, said Robert Gilruth in June 1962, "the Manned Spacecraft Center has doubled in size; accomplished a major relocation of facilities and personnel; pushed ahead in two new major programs; and accomplished Project Mercury's design goal of manned orbital flights twice with highly gratifying results."[112]

The early astronauts were, in too many instances, rambunctious men, as many had recognized during the Mercury program. They roughhoused and drank and drove fast and got into sexual peccadilloes. Rumors swirled around several of the astronauts, especially Gus Grissom, whom NASA officials considered a consummate professional in the cockpit and an incorrigible adolescent whenever off-duty. Everyone laughed when Grissom said:

> There's a certain kind of small black fly that hatches in the spring around the space center south of Houston. Swarms of the bugs can splatter windshields, but their real distinction is that male and female catch each other in midair and fly along happily mated. Grissom told a *Life* magazine reporter that he envied those insects. "They do the two things I like best in life," he said, "flying and ****ing—and they do them at the same time." For years thereafter, the insects were known as Grissom Bugs to local residents.[113]

Several memoirs have recounted these and other anecdotes of the astronauts, many of which are the stuff of legend. It should come as no surprise to anyone that many astronauts had a wild, devil-may-care side to their personalities, the alter ego of the professional who faces danger and death in his or her daily work.[114]

Project Mercury had been formally established just after the birth of NASA in 1958 and completed in a little less than five years at a cost of $384 million. It

---

110. James C. Webb, NASA Administrator, Memorandum for the President, 14 September 1961. Folder 18674, NASA Historical Reference Collection, NASA History Division, NASA Headquarters, Washington, DC.

111. On the creation of this center see Henry C. Dethloff, *"Suddenly Tomorrow Came . . .": A History of the Johnson Space Center* (Washington, DC: NASA SP-4307, 1993).

112. *Space News Roundup* (Houston, TX), 11 July 1962.

113. Quoted in James Schefter, *The Race: The Uncensored Story of How America Beat Russia to the Moon* (Garden City, NY: Doubleday and Co., 1999), p. 72. LBJ also had great confidence in Grissom. See Lyndon B. Johnson to Gus Grissom, 21 July 1961, LBJ Papers, Vice Presidential Papers, box 116, January–July 1961, Space and Aero File, LBJ Library, University of Texas, Austin, Texas.

114. See Guenter Wendt and Russell Still, *The Unbroken Chain* (Burlington, Ontario, Canada: Apogee Books, an imprint of Collector's Guide Publishing Ltd., 2001).

may have been the best bargain ever in human spaceflight, in no small measure because its goals were so simple. Although lagging behind the original schedule, it had succeeded in proving the possibility of safe human space exploration and in demonstrating to the world U.S. technological competence during the Cold War rivalry with the Soviet Union. At the conclusion of the Mercury effort, Walter C. Williams noted that "in the period of about 45 months of activity, some 25 flights were made which was an activity of a major flight in something less than every 2 months." He then commented on what NASA learned in the context of completing Mercury:

> I think we learned . . . a lot about spacecraft technology and how a spacecraft should be built, what its systems should be, how they should perform, where the critical redundancies are that are required. I think we learned something about man-rating boosters, how to take a weapons system development and turn it into a manned transportation system. I think, in this area, we found primarily, in a nutshell, that this was a matter of providing a malfunction detection system or an abort system, and, also, we found very careful attention to detail as far as quality control was concerned. I think that some of the less obvious things we learned—we learned how to plan these missions and this takes a lot of detail work, because it's not only planning how it goes, but how it doesn't go, and the abort cases and the emergency cases always took a lot more effort than the planned missions. . . . We learned what is important in training crews for missions of this type. When the crew-training program was laid down, the program had to cover the entire gamut because we weren't quite sure exactly what these people needed to carry out the missions. I think we have a much better focus on this now. We learned how to control these flights in real time. This was a new concept on a worldwide basis. I think we learned, and when I say we, I'm talking of this as a National asset, not NASA alone, we learned how to operate the world network in real time and keep it up. And I think we learned a lot in how to manage development programs of this kind and to manage operations of this kind (I-47).[115]

As Christopher C. Kraft, senior flight controller, concluded, Mercury "changed quite a few concepts about space, added greatly to our knowledge of the universe around us, and demonstrated that Man has a proper role in exploring it. There are many unknowns that lie ahead, but we are reassured because we are confident in overcoming them by using Man's capabilities to the fullest" (I-48).[116]

---

115. Dr. Walter C. Williams, Deputy Director, NASA Manned Spacecraft Center, NASA, "Project Review," 3 October 1963. Folder 18674, NASA Historical Reference Collection, NASA History Division, NASA Headquarters, Washington, DC.

116. Christopher C. Kraft, Jr., "A Review of Knowledge Acquired from the First Manned Satellite Program." Folder 18674, NASA Historical Reference Collection, NASA History Division, NASA Headquarters, Washington, DC.

Bridging the Technology Gap: Project Gemini

Even as the Mercury program was underway and Apollo hardware was beginning development, NASA program managers recognized that there was a huge gap in the capability for human spaceflight between that acquired with Mercury and what would be required for a lunar landing. They closed most of the gap by experimenting and training on the ground, but some issues required experience in space. Several major areas immediately arose where this was the case. These included the following major mission requirements, as defined in the Gemini crew familiarization manual:

A. Accomplish 14-day Earth orbital flights, thus validating that humans could survive a journey to the Moon and back to Earth.
B. Demonstrate rendezvous and docking in Earth orbit.
C. Provide for controlled land landing as the primary recovery mode.
D. Develop simplified countdown techniques to aid rendezvous missions (lessens criticality of launch window).
E. Determine man's capabilities in space during extended missions (I-52).[117]

These major initiatives defined the Gemini program and its 10 human space-flight missions conducted in the 1965 to 1966 period.[118]

NASA conceived of Project Gemini first as a larger Mercury "Mark II" capsule, but soon it became a totally different vehicle. It could accommodate two astronauts for extended flights of more than two weeks. It pioneered the use of fuel cells instead of batteries to power the ship, and it incorporated a series of modifications to hardware. Its designers also toyed with the possibility of using a paraglider being developed at Langley Research Center for land landings instead of a "splashdown" in water and recovery by the Navy.[119] The whole system was to be powered by the newly developed Titan II launch vehicle, another ballistic missile developed for the Air Force. A central reason for this program was to perfect techniques for rendezvous and docking, so NASA appropriated from the military some Agena rocket upper stages and fitted them with docking adapters to serve as the targets for rendezvous operations.

---

117. NASA Flight Crew Operations Division, "Gemini Familiarization Package," 3 August 1962. Folder 18674, NASA Historical Reference Collection, NASA History Division, NASA Headquarters, Washington, DC.

118. The standard work on Project Gemini is Barton C. Hacker and James M. Grimwood, *On Shoulders of Titans: A History of Project Gemini* (Washington, DC: NASA SP-4203, 1977). See also David M. Harland, *How NASA Learned To Fly in Space: An Exciting Account of the Gemini Missions* (Burlington, ON, Canada: Apogee Books, 2004).

119. Barton C. Hacker, "The Idea of Rendezvous: From Space Station to Orbital Operations, in Space-Travel Thought, 1895–1951," *Technology and Culture* 15 (July 1974): pp. 373–388; Barton C. Hacker, "The Genesis of Project Apollo: The Idea of Rendezvous, 1929–1961," *Actes 10: Historic des techniques* (Paris: Congress of the History of Science, 1971), pp. 41–46; Barton C. Hacker and James M. Grimwood, *On Shoulders of Titans: A History of Project Gemini* (Washington, DC: NASA SP-4203, 1977), pp. 1–26.

The Gemini program emerged full-blown in October 1961 from a working group of NASA and McDonnell engineers. They developed a detailed project development plan that incorporated the following philosophy as central to the effort (I-49):

> In general, the philosophy used in the conception of this project is to make maximum use of available hardware, basically developed for other programs, modified to meet the needs of this project. In this way, requirements for hardware development and qualification are minimized and timely implementation of the project is assured.
>
> Another fundamental concept is that in the design of the spacecraft, all systems will be modularized and made independent of each other as much as possible. In this way, an evolutionary process of product improvement and mission adaptation may be implemented with a minimum of time and effort. Thus, it will be possible to use equipment of varying degrees of sophistication as it becomes available and as the mission requirements are tightened. It is important that a minimum of lead time can be obtained by making use of the latest hardware developments. This concept will make possible the attainment of mission and permits reasonable compromises to be made in the face of difficulties rather than excessive delays that otherwise might be required to meet the full objectives.
>
> This project will provide a versatile spacecraft/booster combination which will be capable of performing a variety of missions. It will be a fitting vehicle for conducting further experiments rather than be the object of experiments. For instance, the rendezvous techniques developed for the spacecraft might allow its use as a vehicle for resupply or inspection of orbiting laboratories or space stations, orbital rescue, personnel transfer, and spacecraft repair.[120]

It took only a little longer for the Gemini name to be attached to the program; by early January 1962 the new program received its official moniker, chosen because of its reference to classical mythology and the "twins," which D. Brainerd Holmes, NASA's Director of Manned Spaceflight, thought most appropriate for the two-person spacecraft. Associate Administrator Robert Seamans presented a bottle of Scotch whiskey to the first person to suggest Gemini as the project's name, engineer Al Nagy (I-50, I-51).[121]

The Gemini spacecraft was a marked improvement on the Mercury capsule. It was 19 feet long (5.8 meters), 10 feet (3 meters) in diameter, and weighed about

---

120. NASA Manned Spacecraft Center, "Project Development Plan for Rendezvous Development Utilizing the Mark II Two Man Spacecraft," 8 December 1961. Folder 18674, NASA Historical Reference Collection, NASA History Division, NASA Headquarters, Washington, DC.

121. Al Nagy, NASA, to George Low, NASA, 11 December 1961; D. Brainerd Holmes, NASA Director of Manned Spaceflight Programs, Memorandum for NASA Associate Administrator, "Naming Mercury-Mark II Project," 16 December 1961; D. B. Holmes, Director of Manned Spaceflight, NASA, Memorandum for the Associate Administrator, NASA, "Naming the Mercury Mark II Program," 2 January 1961. Folder 18674, NASA Historical Reference Collection, NASA History Division, NASA Headquarters, Washington, DC.

8,400 pounds (3,810 kilograms)—twice the weight of Mercury. But it had only 50 percent more cabin space for twice as many people, and was extremely cramped for the long-duration missions envisioned. Ejection seats replaced Mercury's escape rocket and more storage space was added for the longer Gemini flights. The long-duration missions also used fuel cells instead of batteries for generating electrical power, an enormously significant development in the methodology of generating power for the spacecraft.[122] An adapter module fitted to the rear of the capsule (and jettisoned before reentry) carried on-board oxygen, fuel, and other consumable supplies. Engineering changes, such as systems that could be removed and replaced easily, simplified maintenance. Since extra-vehicular activities (EVAs) were an essential part of these missions, the spacesuit became a crucial piece of equipment, the suit providing the only protection for astronauts in the extremely hostile environment of space.[123] By January 1964, NASA had developed a preliminary plan for one astronaut to conduct an EVA at some point during Gemini (I-53). To make EVAs possible, NASA redesigned the Gemini's mechanical hatch to permit astronauts to leave the spacecraft in orbit. As early as July 1964, Gemini Deputy Manager Kenneth Kleinknecht suggested that NASA might attempt an EVA during Gemini IV, but some were opposed to doing this on the second crewed mission of the program, and astronauts James McDivitt and Edward White, the primary crew for Gemini IV, had to lobby to make it a reality the next year. The demonstration of the EVA proved to be one of the huge successes, both from a public relations and a knowledge-advancement viewpoint, of the whole Gemini program.[124]

Problems with the Gemini program abounded from the start. The Titan II had longitudinal oscillations called the "pogo" effect because it resembled the behavior of a child on a pogo stick. Overcoming this problem required engineering imagination and long hours of overtime to stabilize fuel flow and maintain vehicle control. The fuel cells leaked and had to be redesigned, and the Agena reconfiguration also suffered costly delays. NASA engineers never did get the paraglider to work properly and eventually dropped it from the program in favor of a parachute system and ocean recovery, similar to the approach used for Mercury. All of these difficulties increased an estimated $350 million program cost to over $1 billion. The overruns were successfully justified by the Agency, however, as necessities to meet the Apollo landing commitment.[125]

---

122. Linda Carrette, K. Andreas Friedrich, and Ulrich Stimming, "Fuel Cells: Principles, Types, Fuels, and Applications," *ChemPhysChem* 1 (2000): pp. 162–193; Brian Cook, *An Introduction to Fuel Cells and Hydrogen Technology* (Vancouver, British Columbia, Canada: Heliocentris, April 2002), pp. 5–6; M. L. Perry and T. F. Fuller, "A Historical Perspective of Fuel Cell Technology in the 20th Century," *Journal of The Electrochemical Society*, 149 no. 71 (2002): S59-S67.

123. On EVAs and spacesuits, see David S.F. Portree and Robert C. Treviño, *Walking to Olympus: An EVA Chronology* (Washington, DC: Monographs in Aerospace History Series No. 7, 1997); David Shayler, *Walking in Space: Development of Space Walking Techniques* (Chicester, UK: Springer-Praxis, 2003); Gary L. Harris, *The Origins and Technology of the Advanced Extravehicular Space Suit* (San Diego, CA: Univelt, Inc., 2001).

124. Charles W. Mathews, Manager, Gemini Program, "Program Plan for Gemini Extravehicular Operation," 31 January 1964. Folder 18674, NASA Historical Reference Collection, NASA History Division, NASA Headquarters, Washington, DC.

125. James M. Grimwood and Ivan D. Ertal, "Project Gemini," *Southwestern Historical Quarterly*,

By the end of 1963, most of the difficulties with Gemini had been resolved, albeit at great expense, and the program was approaching its first test flights. As they took place, NASA officials considered the possibility of reconfiguring the Gemini spacecraft for a circumlunar mission in the 1966 time frame. With continued pressures from the Soviet Union, examining the possibility of an early circumlunar flight as a contingency for the future appeared appropriate. The initial review in the spring of 1964 showed promise and Edward Z. Gray, Director of NASA's Advanced Manned Missions Program, recommended: "I believe that a study should be initiated to more thoroughly investigate the Gemini circumlunar mode, utilizing the Saturn IB with a Centaur as the injection stage, in either a direct ascent or an Earth orbit rendezvous trajectory. . . . The purpose of such a study would be to more accurately determine the capability of each configuration, the key technical problems, relative costs, development schedules and key decisions points to provide a basis for possible contingency-type decisions in the 1965–66 time period" (I-54).[126]

Further study the next year yielded a decision not to pursue this option. Eldon Hall, Director of Gemini Systems Engineering, commented:

> I think the proposal is feasible, but not within the time and effort indicated. The equipment and mission are too marginal to absorb changes and additions that will be required without extensive redesign and testing. . . . I personally would prefer to see us advance our Earth orbital capability. With the same or fewer modifications to the spacecraft advocated in this proposal and additional Agena payloads, we could attain a significant lead in the design and operation of Earth-orbital space stations (I-55).[127]

In his typically convoluted "adminispeak" style, NASA Administrator James E. Webb communicated this perspective to Representative Olin E. Teague (D-Texas) in September 1965, adding, "I do not believe a decision not to make the substantial investment that would be required by a modified Gemini lunar fly-by will change the posture which our program has had for a number of years" (I-56).[128]

---

81 (January 1968): pp. 393–418; James M. Grimwood, Barton C. Hacker, and Peter J. Vorzimmer, *Project Gemini Technology and Operations* (Washington, DC: NASA SP-4002, 1969); Robert N. Lindley, "Discussing Gemini: A 'Flight' Interview with Robert Lindley of McDonnell," *Flight International*, 24 (March 1966): pp. 488–489.

126. Edward Z. Gray, Director, Advanced Manned Missions Program, Office of Manned Spaceflight, NASA, to Director, Gemini Program, NASA, "Gemini Lunar Mission Studies," 30 April 1964. Folder 18674, NASA Historical Reference Collection, NASA History Division, NASA Headquarters, Washington, DC.

127. Eldon. W Hall, Director, Gemini Systems Engineering, NASA, to Deputy Director, Gemini Program, NASA, "Circumlunar Missions," 29 June 1965. Folder 18674, NASA Historical Reference Collection, NASA History Division, NASA Headquarters, Washington, DC.

128. James E. Webb, NASA Administrator, to Olin E. Teague, Chairman, Subcommittee on NASA Oversight, Committee on Science and Astronautics, House of Representatives, 10 September 1965. Folder 18674, NASA Historical Reference Collection, NASA History Division, NASA Headquarters, Washington, DC.

At the same time, confident that Gemini's major technological challenges were being overcome, NASA moved out on mission planning for the human-piloted portion of the program. LeRoy E. Day, the Gemini Program's Deputy Director, outlined the missions in a 25 June 1964 memorandum:

Flights 4, 5, and 7 will provide experience in long duration orbital flight . . . Many measurements and experiments will be performed to assess the effects of orbital weightless flight on man and machine for periods up to 14 days—more than adequate for the Apollo lunar expedition. Among the medical experiments, for example; M-1, Cardiovascular Reflex, will determine the feasibility of using inflatable cuffs to prevent cardiovascular deterioration—evidence of which was noted in Project Mercury flights MA-8 and MA-9. . . . In addition to these experiments, we also plan to conduct extravehicular activity to evaluate man's performance outside the spacecraft.

With Flight No. 6, we will establish the feasibility of rendezvous and provide experience for the visual manual docking mode, which is common to both Gemini and Apollo . . . Whereas radar computer guidance will be the primary onboard mode for the terminal rendezvous phase of Flight No. 6; the radar optical and optical guidance modes will be primary for Flights 8 and 9 respectively.

By Flights 10 and 11, or earlier, we plan to flight test the feasibility of the LEM lunar orbit direct rendezvous mode in Earth orbit if possible. In this mode, the catch up or parking orbits are essentially by-passed and terminal rendezvous is initiated near first apogee. . .

For Flight No. 12, we plan to simulate LEM abort maneuvers; either abort from an equiperiod transfer orbit (I-57).[129]

Eldon Hall followed in July 1964 with another set of mission profiles that offered not only the already agreed-upon Gemini mission objectives, but also such proposals as tests of propellant transfer, rendezvous with an empty Apollo Command Module, rendezvous with a Lunar Module, using Gemini as a minimum space station, a joint NASA/Air Force Manned Orbiting Laboratory (MOL) using Gemini spacecraft, satellite recovery on-orbit, and a one-astronaut Gemini mission with a telescope mounted in the other seat of the spacecraft. Of course, these missions did not come to pass (I-58).[130]

---

129. L. E. Day, Deputy Director, Gemini Program, NASA, Memorandum for William C. Schneider, Deputy Director, Apollo Program, NASA, "Gemini Support of Apollo," 25 June 1964. Folder 18674, NASA Historical Reference Collection, NASA History Division, NASA Headquarters, Washington, DC.

130. John L. Hammersmith, Director, Gemini Systems Engineering, NASA, Memorandum for Deputy Director, Gemini Program, NASA, "List of Missions," 17 July 1964. Folder 18674, NASA Historical Reference Collection, NASA History Division, NASA Headquarters, Washington, DC.

Flying the Gemini Missions

Following two unoccupied orbital test flights, Gemini III, the first crew-carrying mission, took place on 23 March 1965; it was a three-orbit flight. (The mission was originally designated GT-3, for Gemini/Titan-3.) Mercury astronaut Gus Grissom commanded the mission, with John W. Young, a Naval aviator chosen as an astronaut in 1962, accompanying him. This mission proved to be a huge success for many reasons, serving "to flight qualify the crew-spacecraft combination as well as checkout the operational procedures." The system performed essentially as intended, although there were a few glitches in the technology that Mission Control and the astronauts aboard resolved satisfactorily. During this mission, as James Webb wrote to the President, "the two-man crew maneuvered their craft in orbit preparing the way for the rendezvous missions to follow. GT-3 also initiated the use of the Gemini spacecraft as an orbiting laboratory. Astronauts Grissom and Young also executed the first manned, controlled, lifting reentry" (I-66).[131]

Despite the success of Gemini III, or perhaps because of it, the White House became concerned about the possibility of losing a crew in Earth orbit during a future mission and questioned NASA and the DOD about plans for space rescue should they be stranded in orbit (I-59).[132] Both responded with analyses of the extremely low possibility of losing a crew because they were stranded in orbit, as well as by acknowledging the extremely risky nature of spaceflight. As Cyrus Vance told Bill Moyers, "It is possible we may strand an astronaut in orbit some day. It is very likely that astronauts will be killed, though stranding them is one of the less likely ways. The nation must expect such a loss of life in the space program. There have been several deaths already in our rocket development. We would be untruthful if we were to present any different image to our citizens" (I-60).[133] James Webb opined to President Lyndon B. Johnson, again in a masterpiece of indirect syntax, that:

> . . . in Gemini, we are building on all of the measures for safety that have come from our extensive experience in test flying and such advanced systems as the X-15—the measures which have also been instrumental in achieving our perfect record of astronaut safety thus far. The redundancy designed into the retro-system for return from orbit is optimized

---

131. James E. Webb, Administrator, NASA, Cabinet Report for the President, "Significance of GT-3, GT-4 Accomplishments," 17 June 1965. Folder 18674, NASA Historical Reference Collection, NASA History Division, NASA Headquarters, Washington, DC.

132. E.C. Welsh, National Aeronautics and Space Council, Executive Office of the President, Memorandum for the President, "Space Rescue," 21 May 1965. Folder 18674, NASA Historical Reference Collection, NASA History Division, NASA Headquarters, Washington, DC.

133. Bill Moyers, Special Assistant to the President, The White House, Memorandum for James Webb, Administrator, NASA, and Robert McNamara, Secretary of Defense, 29 May 1965, with attached; Joseph A. Califano, Jr., Special Assistant to the Secretary and Deputy Secretary of Defense, Memorandum for Mr. Valenti/Mr. Busby, Special Assistants to the President, 29 May 1965, with attached; Cyrus Vance, Office of the Secretary of Defense, Memorandum for Mr. Bill Moyers, The White House, "Comments on Need for Space Rescue," 29 May 1965. Folder 18674, NASA Historical Reference Collection, NASA History Division, NASA Headquarters, Washington, DC.

for crew safety. The orbital parameters of the next Gemini mission are planned so that the orbit will decay to reentry within 24 hours after the planned termination of the flight, should all other provisions for initiating the de-orbiting landing sequence fail. . . . It is our judgment that the knowledge needed to begin the design of such a space rescue system is not yet available, but will come from our present developmental and flight program. You may be assured, Mr. President, that we shall continue to give first priority to considerations of astronaut safety (I-61).[134]

NASA has tended to follow this approach to crew safety to the present, relying on the development of the best possible technologies and processes to ensure safety and reliability rather than some type of space rescue capability. It also developed procedures in dealing with the necessity of informing the public about possible accidents and loss of astronauts, should that eventuality occur (I-71).[135]

Also in the aftermath of the successful Gemini III mission, NASA began planning how to honor the astronauts after their flights. For the Mercury program there had been considerable pomp and circumstances, usually involving medals awarded by the President and ticker-tape parades. But Gemini was different, argued Julian Scheer, NASA's Director of Public Affairs. "We are now entering a new phase of our program," he wrote. "The image that is, perhaps, best for this nation is that of a nation with this capability, a nation that goes about its work in an orderly and well-planned manner. We will fly these flights as best we can and put these flyers right back into the flight schedule for a future mission" (I-62, I-63).[136] Because of this desire to "routinize" spaceflight and in the process downplay the heroism of the astronauts, except in truly exceptional circumstances, the aftermath of the Gemini missions was more restrained than in Project Mercury. The Gemini III crew did visit the White House and received medals from President Johnson. In the case of the Gemini IV crew, President Johnson came to Houston to congratulate them and NASA Administrator James Webb sent them, at the request of the President, to the Paris International Air Show, where they met Soviet Cosmonaut Yuri Gagarin. Later missions were less pronounced in their public relations hoopla.[137]

Based on the success of Gemini III, NASA accelerated plans to fly the next mission, a 66-revolution, 4-day mission that began on 3 June and ended on 7 June

---

134. James E. Webb, Administrator, NASA, Memorandum to the President, "Space Rescue," 2 June 1965. Folder 18674, NASA Historical Reference Collection, NASA History Division, NASA Headquarters, Washington, DC.

135. NASA, "Gemini Contingency Information Plan," 11 May 1966. Folder 18674, NASA Historical Reference Collection, NASA History Division, NASA Headquarters, Washington, DC.

136. Julian Scheer, Assistant Administrator for Public Affairs, NASA, Memorandum to Mr. Marvin Watson, The White House, 24 May 1965. Folder 18674, NASA Historical Reference Collection, NASA History Division, NASA Headquarters, Washington, DC.

137. Marvin Watson, The White House, Memorandum for the President, 25 May 1965. Folder 18674, NASA Historical Reference Collection, NASA History Division, NASA Headquarters, Washington, DC. On the "routinization" of power see Eric Hoffer, *The True Believer: Thoughts on the Nature of Mass Movements* (New York: Harper & Row, 1951), pp. 3–23, 137–155; Max Weber, "The Pure Types of Legitimate Authority," in *Max Weber on Charisma and Institution Building: Selected Papers*, S. N. Eisenstadt, ed. (Chicago: University of Chicago Press, 1968), p. 46.

1965 (I-67).[138] During that mission astronaut Edward H. White performed the first American EVA. During his 20 minutes outside Gemini IV, White remained connected to the spacecraft's life-support and communications systems by an "umbilical cord," and he used a hand-held jet thruster to maneuver in space. McDivitt remained inside the spacecraft during this event.[139] Although it turned out well, NASA leaders had debated intensely among themselves whether or not to allow the EVA on this mission. Those in favor emphasized the necessity of developing an EVA capability for the Apollo Moon landings and the necessity of haste because of the success of the Soviet efforts in space, including the first EVA by anyone, accomplished by Cosmonaut Alexey Leonov three months before the Gemini IV mission. Those opposed, who included NASA Deputy Administrator Hugh L. Dryden, argued that the EVA was premature, that it was risky, and that it looked like a direct response to Leonov's earlier spacewalk.

At a 24 May 1965 showdown at NASA Headquarters, Dryden raised the issue of "the element of risk to complete the 4-day Gemini flight because of EVA." The reply was that the added risk was simply having to depressurize the spacecraft, open the hatch, seal the hatch, and repressurize the spacecraft. This was not an insignificant set of concerns, Dryden countered. As the memorandum of the meeting recorded: "There was a strong feeling to ratify EVA for Gemini 4 in order to get the maximum out of the flight. There was unanimity in that EVA eventually would be carried out, but there was some reservation as to whether or not it was the best judgment to have EVA on Gemini 4 as a risk beyond that which has to be taken." Dryden, who was dying of cancer at the time and worked until his death on 2 December 1965, perhaps felt more keenly than others in the debate the weight of mortality and reflected this in his concern for the safety of the astronauts. No one could fault him for that concern, and everyone recognized the crew safety issue, but that had to be balanced against other factors that tipped the scales in favor of success. Calculating the risk and accepting the unknowns soon led NASA leaders to approve the EVA on Gemini IV. Since it turned out well, they looked like geniuses. Had it gone otherwise, they would have become scapegoats (I-64, I-65).[140] As James Webb wrote to the President: "It is significant that the first operational flight of Gemini, GT-4, has provided significant experience in each of the major mission areas of Gemini: long duration flight, rendezvous and docking, extra vehicular activity, and the conduct of experiments" (I-66).[141]

138. NASA Program Gemini Working Paper No. 5038, "GT-4 Flight Crew Debriefing Transcript," n.d. Folder 18674, NASA Historical Reference Collection, NASA History Division, NASA Headquarters, Washington, DC.

139. Reginald M. Machell, ed., *Summary of Gemini Extravehicular Activity* (Washington, DC: NASA SP-149, 1968).

140. Robert C. Seamans, Jr., Associate Administrator, NASA, to The Administrator, "Extra Vehicular Activity for Gemini IV," 24 May 1965; W. Vogel, Executive Officer, Memorandum for the Record, "Top Management Meeting on Gemini 4 Extra-Vehicular Activity," 8 June 1965. Folder 18674, NASA Historical Reference Collection, NASA History Division, NASA Headquarters, Washington, DC.

141. James E. Webb, Administrator, NASA, Cabinet Report for the President, "Significance of GT-3, GT-4 Accomplishments," 17 June 1965. Folder 18674, NASA Historical Reference Collection, NASA History Division, NASA Headquarters, Washington, DC.

Eight more Gemini missions followed through November 1966. Despite problems great and small encountered on virtually all of them, the program achieved its goals. This especially was the case in the development of rendezvous and docking procedures necessary for the successful accomplishment of the lunar landing commitment. For example, Buzz Aldrin, selected in the third group of NASA astronauts in 1963, had a unique impact in this area, given his Ph.D. in astronautics from the Massachusetts Institute of Technology. Aldrin had written his dissertation on orbital rendezvous and he applied this knowledge to solving one of the principal riddles of the Gemini program: how to accomplish rendezvous and docking of two spacecraft in Earth orbit.[142] Acquiring the nickname "Dr. Rendezvous" from his fellow astronauts, Aldrin worked more than the others to develop the orbital maneuvers essential to the program's success. During Project Gemini, Aldrin became one of the key figures working on the problem of spacecraft rendezvous and docking in Earth or lunar orbit. Without solutions to such problems, Apollo could not have been successfully completed. Rendezvous techniques remained largely in the realm of theory until Aldrin began to work on the problem. In 1963 and 1964, Aldrin worked hard to convince flight operations leaders that a concentric rendezvous would work. In his estimation, a target vehicle could be launched in a circular orbit with the rendezvousing spacecraft in a closer orbit to Earth. It would then take less time to circle the globe, he argued, and then catch up for rendezvous. Aldrin and others worked together to develop the trajectories and maneuvers that would allow the spacecraft to intercept a target vehicle.[143]

Moreover, Aldrin argued that a closed-loop concept that relied more on machines than on astronauts could easily spell failure. Ground controllers wanted to use radar and computers to guide the two spacecraft together from the ground, making rendezvous essentially automatic. Should either the equipment or procedures fail, however, the mission would be lost. Aldrin argued for the astronauts as active participants in the process, even more involved than taking action should the equipment malfunction.[144]

Systematically and laboriously, Aldrin worked to develop procedures and tools necessary to accomplish space rendezvous and docking. He was also central in devising the methods necessary to carry out the astronauts' EVA. That, too, was critical to the successful accomplishment of Apollo. Techniques he devised have been used on all space rendezvous and docking flights since. Aldrin also significantly improved operational techniques for astronautical navigation star displays for these missions. He and a critical ally, Dean F. Grimm from the Manned Spacecraft Center's (MSC) Flight Crew Support Division, convinced their supe-

---

142. See Edwin E. Aldrin Jr., "Line of Sight Guidance Techniques for Men in Orbital Rendezvous," Ph.D. Diss., Massachusetts Institute of Technology, 1964. Later Aldrin legally changed his first name to "Buzz."

143. Hacker and Grimwood, *On Shoulders of Titans*, pp. 266–268.

144. "Preflight Training Plan for Fourth Manned Gemini Flight Crew (GTA-6)," NASA Program Gemini working paper No. 5031, 23 August 1965. Folder 18674, NASA Historical Reference Collection, NASA History Division, NASA Headquarters, Washington, DC; Hacker and Grimwood, *On the Shoulders of Titans*, p. 267.

riors at MSC and McDonnell Aircraft to build a simulator to test this possibility. They explored how astronauts responded to various situations with maneuvers leading to target interception. Astronauts mastered procedures for overcoming the failure of any one piece of equipment, and soon convinced everyone that the astronaut as active participant was critical to successful rendezvous and docking of the Gemini systems.[145]

What emerged was a combination system that relied on automated systems to get the Gemini spacecraft close enough to the target vehicle so that the crew could complete the rendezvous and docking process using the control handles, observing the pilot displays, and observing the optical targets through windows in the spacecraft. At some point in the approach, typically at about 60 meters separation, the rendezvous radar could no longer give an accurate estimate of range because of the closeness of the target. Then, visual observations of the docking targets by the crew were heavily relied upon. This approach worked flawlessly throughout the Gemini program. In all, Gemini astronauts completed successful rendezvous and dockings on Gemini VIII in March 1996, Gemini X in July 1966, Gemini XI in September 1966, and Gemini XII in November 1996.[146]

The first test of rendezvous in space occurred on the twin flights of Gemini VI and VII in December 1965. Gemini VI was initially intended to rendezvous with an Agena target spacecraft, but when the Agena failed during launch the mission was hastily modified to rendezvous with a piloted spacecraft (I-68). Consequently, Gemini VII, piloted by Frank Borman and James Lovell, was launched first on 4 December 1965 to become the rendezvous target for Gemini VI. When Gemini VI was launched on 15 December, piloted by Walter Schirra and Thomas Stafford, the two spacecraft rendezvoused and flew in formation for 5 hours. Their first test of rendezvous had been successful and proved the concept of human involvement in space rendezvous. Gemini VII remained aloft for 14 days to study the effects of long-duration flight. The 330 hours in space had no long-term harmful effects on the crew, but the flight turned into something of an endurance test for the two pilots, confined in their hot, cramped quarters. At the conclusion of the lengthy time cooped up together, Lovell joked to reporters that he and Borman were happy to announce their engagement. It was astronaut humor that said quite a lot about the masculine culture of the fliers (I-69).[147]

145. Buzz Aldrin, "Apollo and Beyond," in Stephen J. Garber, ed., *Looking Backward, Looking Forward: Forty Years of Human Spaceflight Symposium* (Washington, DC: NASA SP-2002-4107, 2002), pp. 91–99; Roger D. Launius, *Frontiers of Space Exploration*, 2nd ed. (Westport, CT: Greenwood Press, 2004), pp. 83–84.

146. V. E. Jones, and J. E. Mangelsdorf, "Preliminary Rendezvous Experiments Definition and Design," Lockheed Missiles & Space Company, Sunnyvale, CA, p. 4-1; D. Chiarrappa, "Analysis and Design of Space Vehicle Flight Control Systems, Volume VIII—Rendezvous and Docking," NASA CR-827, prepared by General Dynamics Corp., San Diego, CA, for the NASA-Marshall Space Flight Center, Alabama, July 1967, p. 31; *NASA Pocket Statistics, 1997 Edition* (Washington, DC: NASA Headquarters Facilities and Logistics Management, 1998), pp. B-100–B-102.

147. "Summary of Telephone Conversations RE Gemini 7/6," 25–27 October 1965; "Gemini Program Mission Report, Gemini VI-A," January 1966. Folder 18674, NASA Historical Reference Collection, NASA History Division, NASA Headquarters, Washington, DC; J. H. Latimer, "Observation of Gemini 6-GEMINI 7 Rendezvous," SAO Special Report #202 (Cambridge, MA: Smithsonian Astrophysical Observatory, 1966); J. R. Burton, "Rendezvous With the Gemini Spacecraft," *Proceedings*

It was perhaps the flight of Gemini VIII in the spring of 1966 that demonstrated more clearly than any other mission the capability of the program to accomplish rendezvous and docking in orbit. Gemini VIII had two major objectives, but was able to complete only one of them. The first objective involved completing the first ever on-orbit rendezvous and docking. Second, the crew was to accomplish an extended EVA. After launch on 16 March 1966, the crew of Neil Armstrong and David Scott approached their Agena target vehicle without difficulty. The crew then docked with it as had been planned. While undertaking maneuvers when attached to the Agena, the crew of Gemini VIII noticed that for some unexplained reason the spacecraft was in a roll. Armstrong used the Gemini's orbital maneuvering system to stop the roll, but the moment he stopped using the thrusters, it started again. They then turned off the Agena and this seemed to stop the problem for a few minutes. Then suddenly it started again. Scott then realized that the problem was with the Gemini capsule rather than the Agena. After transferring control of the Agena back to the ground, they undocked and with a long burst of translation thrusters moved away from the Agena. At that point, Gemini VIII began to roll about one revolution per second. They decided to turn off the orbital maneuvering system and try to regain control of the spacecraft with its reentry control system. If they failed to do so the accelerating rotation would eventually cause the crew to black out and for the mission to the lost, perhaps with loss of life. Even so, the use of the reentry control system would require Armstrong and Scott to return to Earth as soon as possible so as not endanger the mission any further. After steadying the spacecraft they tested each thruster in turn and found that Number 8 had stuck on. This had caused the roll. The mission then returned to Earth one orbit later so that it could land in a place that could be reached by the Navy.

There was no question that astronauts Armstrong and Scott had salvaged the mission, even if they did have to return to Earth earlier than expected. A review of the incident found no conclusive reason for the thruster sticking as it did. But it was obvious that the crew's presence allowed the diagnosis of the anomaly. Reviewers believed it was probably caused by an electrical short that caused a static electricity discharge. Even if the switch to the thruster was off, power could still flow to it. To prevent reoccurrence of this problem, NASA changed the system so that each thruster could be isolated (I-70).[148]

---

of the Symposium on Space Rendezvous, Rescue, and Recovery, Volume 16, Part 2, American Astronautical Society, Edwards Air Force Base, CA, 10–12 September 1963, pp. 173–176.

148. "Gemini VIII Technical Debriefing," 21 March 1966. Folder 18674, NASA Historical Reference Collection, NASA History Division, NASA Headquarters, Washington, DC; S.R. Mohler, A.E. Nicogossian, P.D. McCormack, and S.R. Mohler Jr., "Tumbling and Spaceflight: The Gemini VIII Experience," *Aviation and Space Environmental Medicine* 61 (January 1990): pp. 62–66; Bo J. Naasz, "Classical Element Feedback Control for Spacecraft Orbital Maneuvers," M.S. thesis, Virginia Polytechnic Institute and State University, Blacksburg, VA, 2002, pp. 1–2; M. E. Polites, *An Assessment of the Technology of Automated Rendezvous and Capture in Space* (Huntsville, AL: NASA/ TP—1998–208528, 1998), pp. 3–9; H. J. Ballard, "Agena Target Vehicle for Gemini," *Proceedings of the Symposium on Space Rendezvous, Rescue, and Recovery,* Volume 16, Part 2, American Astronautical Society, Edwards Air Force Base, CA, 10–12 September 1963, pp. 177–187.

Conclusion

By the end of the Gemini program in the fall of 1966, orbital rendezvous and docking had become routine: astronauts could perform spacewalks; it seemed clear that humans could live, work, and stay healthy in space for several weeks at a time. Above all, the program had added nearly 1,000 hours of valuable spaceflight experience in the years between Mercury and Apollo, which by 1966 was nearing flight readiness. In every instance, NASA had enhanced the role of the astronauts as critical fliers of spacecraft, a role that would become even more significant in the accomplishment of the Moon landings between 1969 and 1972. Additionally, as a technological learning program, Gemini had been a success with 52 different experiments performed on the 10 missions. The bank of data acquired from Gemini helped to bridge the gap between Mercury and what would be required to complete Apollo within the time constraints directed by the President (I-72, I-73).[149]

---

149 Robert C. Seamans, Jr., Deputy Administrator, NASA, Memorandum for Associate Administrators, Assistant Associate Administrators, and Field Center Directors, NASA, "Gemini Program; Record of Accomplishments, Attached," 17 January 1967, with attached: "Project Gemini Summary"; George E. Mueller, Associate Administrator for Manned Space Flight, NASA, "Gemini Summary Conference," 1–2 February 1967. Folder 18674, NASA Historical Reference Collection, NASA History Division, NASA Headquarters, Washington, DC; *Gemini Summary Conference* (Washington, DC: NASA SP-138, 1967); Linda Ezell, *NASA Historical Data Book, Vol. II*, pp. 149–170.

Document I-1

Document Title: H. J. E. Reid, Director, Langley Memorial Aeronautical Laboratory to National Advisory Committee for Aeronautics, "Research on Space Flight and Associated Problems," 5 August 1952.

Source: National Archives and Record Administration, Fort Worth, Texas.

As World War II was in its final stages, the National Advisory Committee for Aeronautics (NACA) inaugurated sophisticated studies of high-speed upper atmosphere flight that had significant ramifications for the development of human spaceflight. The Langley Memorial Aeronautical Laboratory (LMAL), now the Langley Research Center, in Hampton, Virginia, led this effort. In early 1945, NACA asked Congress for a supplemental appropriation to fund the activation of the Pilotless Aircraft Research Division (PARD), and a short time later NACA opened the Auxiliary Flight Research Station (AFRS) to launch rockets on Wallops Island, Virginia. On 4 July 1945, PARD launched its first test vehicle, a small two-stage, solid-fuel rocket to check out the installation's instrumentation. The group soon began serious work to learn about the aerodynamics of spaceflight. By 1952 Langley's involvement in rocketry and spaceflight research had transformed the Laboratory into one of the world's leading facilities involved in this entirely new field of flight research. This memorandum sought to capitalize on the work of the PARD and to advance the state of the technology by establishing a formal panel to plan for future research.

Langley Field, Va.
August 5, 1952

From Langley
To: NACA
Ref: NACA Letter, July 10, 1952 MBApep Enc.
Subject: Research on space flight and associated problems

1. The Langley laboratory has carefully considered the subject proposed in letter of reference regarding research on space flight and associated problems. It was recommended that the laboratory assign a three-man group to study and prepare a report covering the various phases of a proposed program that would carry out the intent of the resolution of letter of reference. In order to effect NACA coordination of the program proposed by this study, it is further recommended that a review board with representatives of the 3 laboratories and the NACA High-Speed Flight Research Station at Edwards be appointed.

2. If this plan is approved, the Langley laboratory would appoint to the study group Messrs. C. E. Brown, C. H. Zimmerman, and W. J. O'Sullivan. The recommended members of the review board from Langley and Edwards are Messrs. Hartley A. Soulé and Walter C. Williams, respectively.

3. Results of some preliminary work relative to this subject are already available as a result of consideration given to this matter at Langley and Edwards.

H. J. E. Reid
Director

Document I-2

Document Title: H. Julian Allen and A. J. Eggers, Jr., NACA, "Research Memorandum: A Study of the Motion and Aerodynamic Heating of Missiles Entering the Earth's Atmosphere at High Supersonic Speeds," 25 August 1953.

Source: Folder 18674, NASA Historical Reference Collection, NASA History Division, NASA Headquarters, Washington, DC.

In the later 1940s aerodynamicists began research on the best means of reentry to Earth from space, where the high speeds caused atmospheric heating in excess of 1,800°F. These investigations found that a blunt-nose body experienced much less heating than a pointed body, which would burn up before reaching Earth's surface. The blunt reentry body, discovered in 1951 by H. Julian Allen, an engineer with NACA's Ames Research Center, created a stronger shock wave at the nose of the vehicle and dumped a good deal of the reentry heat into the airflow, making less heat available to heat the reentry vehicle itself. Allen's work led to the design of wide-body bases for spacecraft, giving the capsules their characteristic "teardrop" shape, and to the use of the ablative heat shields that protected the Mercury, Gemini, and Apollo astronauts as their space capsules reentered Earth's atmosphere. This document represents one of Allen's earliest contributions to understanding the reentry problem. Coupled with his later contributions, as well as with the research of others including his early collaborator Alfred Eggers, Allen's research made possible human spaceflight in the 1960s.

NACA

RESEARCH MEMORANDUM

A STUDY OF THE MOTION AND AERODYNAMIC HEATING OF MISSILES
ENTERING THE EARTH'S ATMOSPHERE AT HIGH SUPERSONIC SPEEDS

By H. Julian Allen and A. J. Eggers, Jr.

Ames Aeronautical Laboratory

Moffett Field, Calif.

August 25, 1953

SUMMARY

A simplified analysis is made of the velocity and deceleration history of missiles entering the earth's atmosphere at high supersonic speeds. It is found that, in general, the gravity force is negligible compared to the aerodynamic drag force and, hence, that the trajectory is essentially a straight line. A constant drag coefficient and an exponential variation of density with altitude are assumed and generalized curves for the variation of missile speed and deceleration with altitude are obtained. A curious finding is that the maximum deceleration is independent of physical characteristics of a missile (e.g., mass, size, and drag coefficient) and is determined only by entry speed and flight-path angle, provided this deceleration occurs before impact. This provision is satisfied by missiles presently of more usual interest.

The results of the motion analysis are employed to determine means available to the designer for minimizing aerodynamic heating. Emphasis is placed upon the convective-heating problem including not only the total heat transfer but also the maximum average and local rates of heat transfer but also the maximum average and local rates of heat transfer per unit area. It is found that if a missile is so heavy as to be retarded only slightly by aerodynamic drag, irrespective of the magnitude of the drag force, then convective heating is minimized by minimizing the total shear force acting on the body. This condition is achieved by employing shapes with a low pressure drag. On the other hand, if a missile is so light as to be decelerated to relatively low speeds, even if acted upon by low drag forces, then convective heating is minimized by employing shapes with a high pressure drag, thereby maximizing the amount of heat delivered to the atmosphere and minimizing the amount delivered to the body in the deceleration process. Blunt shapes appear superior to slender shapes from the standpoint of having lower maximum convective heat-transfer rates in the region of the nose. The maximum average heat-transfer rate per unit area can be reduced by [2] employing either slender or blunt shapes rather than shapes of intermediate slenderness. Generally, the blunt shape with high pressure drag would appear to offer considerable promise of minimizing the heat transfer to missiles of the sizes, weights, and speeds presently of interest.

Document I-3

Document Title: Adelbert O. Tischler, Head, Rocket Combustion Section, NACA, Memorandum for Associate Director, NACA, "Minimum Man-In-Space Proposals Presented at WADC, 29 January 1958 to 1 February 1958," 10 April 1958.

Source: Folder 18674, NASA Historical Reference Collection, NASA History Division, NASA Headquarters, Washington DC.

Document I-4

Document Title: Paul E. Purser, Aeronautical Research Engineer, NACA, Memorandum for Mr. Gilruth, "Langley Manned-Satellite Program," 11 April 1958.

Source: Folder 18674, NASA Historical Reference Collection, NASA History Division, NASA Headquarters, Washington, DC.

The spring of 1958 brought to the fore a range of possibilities for advocates of an aggressive spaceflight effort in the U.S. The Soviet successes with Sputniks I and II in the fall of 1957, coupled with the spectacular failure of a televised Vanguard launch on 6 December 1957, ensured that national leaders were under the gun to take positive action. Accordingly, this situation led directly to several efforts aimed at "catching up" to the Soviet Union's space achievements. These included: a) a full-scale review of both the civil and military programs of the U.S. (scientific satellite efforts and ballistic missile development); b) establishment of a Presidential science advisor in the White House who had responsibility for overseeing the activities of the Federal government in science and technology; c) creation of the Advanced Research Projects Agency in the Department of Defense, and the consolidation of several space activities under centralized management by that agency; d) the proposed establishment of a new space agency, NASA, based on NACA to manage civil space operations; and e) passage of the National Defense Education Act to provide federal funding for education in the scientific and technical disciplines. As this was taking place, NACA leaders studied the possibility of launching a human into space. These documents represent the deliberations taking place during this time that explored how initial human spaceflight might be accomplished and suggest the wide variety of concepts being examined. In Document I-3, WADC is the abbreviation for the Wright Air Development Center, Wright-Patterson Field, Dayton, Ohio.

Document I-3

[CONFIDENTIAL] [DECLASSIFIED]
NACA - Lewis

Cleveland, Ohio
April 10, 1958

MEMORANDUM For Associate Director

Subject: Minimum man-in-space proposals presented at WADC, January 29, 1958 to February 1, 1958

    1. The purpose of a series of meetings at WADC under the Chairmanship of Mr. E. Barton Bell was to hear proposals from various contractors on the quickest way to put man in space.

    2. Proposals fell into two rough categories: (a) a blunt-nose cone or near-spherical zero-lift high-drag vehicle of a ton to a ton-and-a-half weight, and (b) a hypersonic glider of the ROBO or Dyna-Soar type. The first category of vehicles used existing ICBM vehicles as boosters; the second used more complex and arbitrary multiplex arrangements of existing large-thrust rocket engines. A number of contractors looked at the zero-lift high-drag minimum weight vehicle as the obvious expedient for beating the Russians and the Army into space. Others, notably Bell, Northrup, and Republic Aviation, set this idea aside as a stunt and consequently these contractors stressed the more elaborate recoverable hypersonic glider vehicle as the practical approach to the problems of flight in space. In the following paragraphs the no-lift minimum weight vehicles are reviewed first without regard for order of presentation and the hypersonic glider vehicles follow. An effort is made to summarize the pertinent gross features of each proposed vehicle at the head of each review and some of the details are discussed thereafter.

3. The proposal configurations were patterned directly after concepts developed by Allen, Eggers, et al of NACA-Ames and the background of data obtained through NACA research was impressively apparent throughout the proposals. Therefore, before going into the individual proposals it seems worthwhile to review briefly the suggestions made by NACA people. This review is covered in the next three paragraphs.

4. Mr. John Becker of NACA-Langley discussed two separate minimum man-in-space proposals. The first of these was a no-lift configuration as diagrammed.

[2]

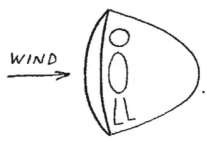

$W_g < 3000$ pounds
$W/S = 35$ pounds per square foot

This discussion considered only vehicles which fly within the atmosphere at perigee so that a small impulse applied anywhere along the flight path will initiate reentry. With zero-lift the drag deceleration will reach a maximum of about 8.5 g's with greater than 5 g's for 20 seconds. Small controls (flaps like air brakes) at the edge of the dish can be used to change angle of attack and thereby produce lift; this will reduce maximum g's to 4.5. The heating rate will approach a maximum of 100-150 Btu per square foot per second. This heat can best be absorbed in a Berylium heat sink between 1/2 and 1-inch thick. The vehicle ultimately requires descent by parachute - this rules out landing in predesignated pin-point areas.

Boost with the ballistic-nose-cone-type vehicle can be accomplished with Atlas. During boost the heat shield is behind the pilot (passenger). Upon injection into orbit the vehicle must be reversed. This requires attitude controls that work in space. Reentry also requires a well-controlled gimbaled retrorocket of nominal impulse and weight (<100 pounds).

Langley is presently making a structural analysis and investigating aerodynamic behavior in hypersonic tunnel.

5. The second proposal was a lift flat-bottomed wing device to enter the atmosphere at an angle of 25° ($C_L = 0.6$).

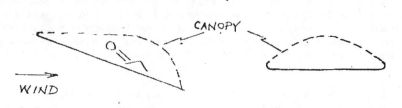

[3]

Blunt leading edge radius was given as 3 inches with W/S = 20 pounds per square foot. The heat transfer rate on the bottom was estimated at less than 12 Btu per square foot per second and the airframe can radiate all heat. Temperature of the skin would reach a maximum of 2000° F at the leading edge. This device would have less than 1 g deceleration at all times during reentry (except possibly at the earth's surface). Range can be controlled somewhat by varying angle of attack (L/D = 3.5 to 1.0).

The weight of this aerodynamic-ski would be less than 5000 pounds, possibly considerably less.

One of the problems of both the NACA devices will be control of velocity and angle at injection into orbit. Present ICBM quality guidance is not good enough. A retrorocket to slow ski down by 200 feet per second at 300,000 feet perigee altitude will lower perigee by 50,000 feet. This will initiate reentry.

6. Mr. Clarence Syvertson of Ames outlined certain configurations that are being studied there. These assume an initial circular orbit at 500,000 feet. If at this altitude the velocity is reduced by 2 percent the altitude will be lowered by 220,000 feet.

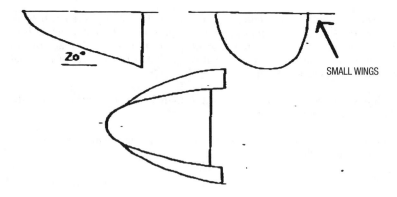

$W_g < 5000$ pounds

The C_L for the above configuration is 0.6. The maximum deceleration is always less than 1 g. Hypersonic wind tunnel studies of this configuration have shown a serious aerodynamic center shift at transonic speeds. This A.C. shift [4] amounts to about 30 percent of the body length. Two suggested solutions are to use thicker wings or no wings at all. Drag-brake-like controls surfaces (about 5 in number) around the periphery of the configuration are being studied.

Essentially this device is one half of an ICBM reentry body. The controllable lift provides control over range. A former idea for parachute landing is being abandoned.

AUTONEUTRONICS

Initial presentation by Mr. Krause who discussed an elaborate scheme of experiments to be accomplished with rocket-propelled vehicles along with an extended series of engine-vehicle assemblies to carry out the program. Even the manned satellite concept was regarded as the end experiment of a series starting with protozoa and bacteria and building up through invertebrates, vertebrates, and finally, man. Only the manned space vehicle is outlined below. The space program discussed is reviewed in the discussion section.

Vehicle: One man zero-lift nose-cone.

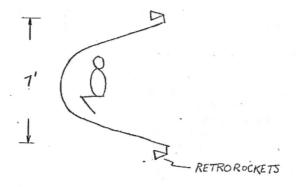

RETROROCKETS

[5]

Weight: Payload 855 pounds:

Man	175 pounds
Oxygen and purifier	85 pounds
Water	20 pounds
Food	10 pounds
Clothing	110 pounds
Temperature controls	130 pounds
Attitude controls	85 pounds
Communication	85 pounds
Navigation	45 pounds
Experiments	65 pounds
Telemetering	—
Total payload	**855 pounds**

Beryllium ablation shield   850 pounds
Parachute   100 pounds
Recovery (?) system   50 pounds
Retrorocket   250 pounds
Structure   300 pounds
Instruments   70 pounds
Capsule   70 pounds
    Total weight   **2545 pounds**

Boosters: Atlas and Rustler ($F_2 - NH_3$)

Schedule: Manned earth satellite late in 1963 (with 3300 pound payload and Atlas-Rustler booster.)

Costs $80 x 10^6.

[6]

Discussion

Concerning the high lift/drag glider configurations as opposed to the high drag zero-lift ballistic reentry concepts Mr. Krause presented the following arguments:

The high L/D configuration entail:

(a) Unsolved aerodynamic and structural problems
(b) Development of new test methods
(c) Major test facilities
(d) Material development
(e) Long development schedule
(f) Structural fabrication problems

The high drag configurations entail:

(a) Nominal test facilities
(b) Development of heat sink materials
(c) Short development schedule
(d) Simple fabrication
(e) Aerodynamic data available

A high altitude research program involves:

(a) Procuring scientific data
(b) Studying environmental effects
(c) Testing components
(d) Developing recovery techniques
(e) Biomedical experiments
(g) Manned flight

The high altitude research and test work can be accomplished with: Thor, Polaris, Atlas, Titan, or special vehicles.

Satellite program mast include:

(a) Study of physical environment
(b) Precision experiments
(c) Reconnaissance

[7]

Satellite flights can be accomplished with the Thor-Hustler, Thor-Rustler, Atlas-Hustler, and Fitan-Rustler. Both the Fitan and the Rustler will use fluorine as oxidant

Lunar flight program might include

(a) Navigation in precision orbit
(b) Lunar impact
(c) Instrument landing

For lunar flight, vehicles might be the Thor-Rustler-Vanguard 3rd stage, Atlas-Rustler, and Fitan-Rustler.

The vehicle development chart for these programs proceeds: (a) Thor-Hustler, (b) Thor-Rustler, (3) Atlas-Rustler, and (4) Fitan-Rustler.

For the earth satellite program estimated payload weights for several configurations were calculated:

Configuration	Payload, pounds	Initial W_0, pounds
Th-6V	201	
Th-6V-V	335	
Th-H	377	112,499
Th-R	858	113,169
A-H	3272	
A-R	3848	
F-R	5459	

The schedule was for four earth-satellite flights in twelve months, first recovery 18-20 months, biomedical experiments 22-25 months. Rustler (F_2, N_2H_4) was predicted to be ready in the twenty-first month (extremely optimistic). This would raise payload over 500 pounds within two years.

The purpose of the series of biomedical experiments is to establish survival limits and determine nervous system behavior. Protozoa-bacteria experiments were to be done with Thor-Hustler, Rhesus monkey experiments with Thor-Rustler, chimpanzee experiments with Atlas-Rustler, and manned flight also with the Atlas-Rustler. The Thor-Hustler program is expected to cost $50 x 10^6, boosters and range free, the Thor-Rustler program an additional $30 x 10^6.

[8]

<u>MARTIN</u>

Introductions by Mr. Bunker, President, and Mr. Merrill, Vice President at the Denver facility. Outline of proposal by Mr. George Trimble, Vice President of engineering. Details by Mr. Demeret, head of the astronautics section(?), Denver.

Pertinent Features

Vehicle: One-man zero-lift body illustrated below. Later vehicles planned to have lift and controls. Later vehicle weights up to 40,000 – 60,000 pounds.

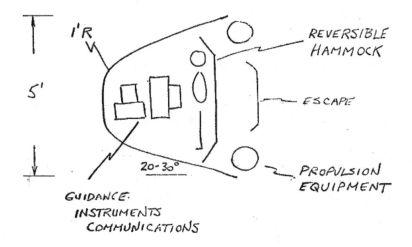

Weight: 3500 Pounds total; ablation shield (phenolic resin), 650 pounds; instruments, 150 pounds.

Boosters: Titan.

Time and altitude: 24-hour trip; 150 - 200 miles at perigee.

[9]

Reentry methods: Retrorocket (-500 feet per second) applied at apogee; ballistic reentry, $W/C_DA = 100 – 150$, maximum g's = 7.5 at 3° reentry angle; maximum temperature = not stated; maximum heat transfer = 5 – 6 Btu per square inch per second; ablation cooling.

Recovery and landing: Parachute within +/- 50 miles.

Tracking: ICBM inertial. Minitrack to provide altitude and position data.

Safety feature: Retrorocket used to separate vehicle from booster if booster fails.

Schedule: Manned flight by mid-1960.

Cost: Not stated.

Discussion

Present ICBM guidance okay for attitude reference for ten days. Integrating accelerometer will gage retrorocket within +/- 10 feet per second. Propose H_2O_2 retrorocket with hot-gas vernier control system for attitude stabilization. Present Titan system fails to lift 3500 pound payload into orbit by 500 feet per second. (Can carry about 2900 pounds into orbit.) Twenty percent additional payload capacity is foreseeable growth by 1960.

No specific design proposed for controlled flight. A second proposal to carry 16,000 pound third stage with a -5000 feet per second retrorocket (weighing 10,000 pounds) was suggested - but not detailed - to avoid heating problem.

Maximum heat transfer can be reduced by same lift.

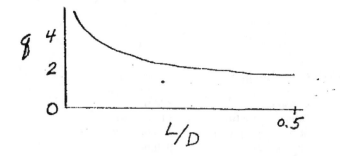

[10]

Martin suggested use of control flaps on reentry vehicle to produce angle of attack and lift.

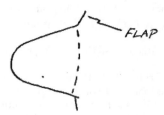

The Martin proposal impressed me as the most thoroughly worked out proposal.

AVCO

Presentation principally by Dr. Arthur Kantrowitz, Director of Research. AVCO has twice previously proposed the following. (The last time in November, 1957.)

Pertinent Features

Vehicle: Eight-foot spherical vessel, zero lift. Parachute equipped to provide drag for reentry.

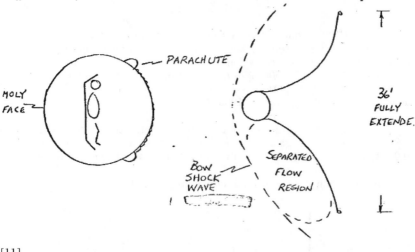

[11]

Weight: 1500 pounds

Capsule	220 pounds
Internal structure	114 pounds
Survival equipment	126 pounds
Parachute	330 pounds
Other	85 pounds
Escape rocket	240 pounds
Man and clothing	150 pounds
	1265 pounds
Contingency	235 pounds
Total Weight	**1500 pounds**

Boosters: Titan (or Atlas) as is. Maximum g's = 11 (second stage).

Time and altitude: Orbit at 110 - 120 miles (circular). Time arbitrarily variable. With parachute furled as many as 40 orbits are possible. Propose a three-day flight.

Reentry method: Thirty-six foot diameter stainless steel parachute creates drag to cause descent in one half orbit. Control of parachute used both to adjust orbit and to control landing point. Grazing angle (1/4°) reentry; maximum g's @ 250,000 feet; maximum temperature, 1800° F; maximum temperature of parachute, 1140°F; maximum heat transfer (?) @ 270,000 feet.

Recovery and landing: Parachute also used to land capsule; landing area about size of Kansas. Landing velocity, 35 feet per second. Fall requires one half hour.

Tracking: Existing minitrack systems. Man is strictly passenger – no control.

Controls: Air-jets to kill angular momentum (attitude)- Liquid paddles to maintain attitude. Parachute extension controllable by expansion of bellows ring at rim; changes drag by 50 x. Used to correct orbit. ICBM-quality guidance good for 0.8°, want 0.25°.

[12]   [illustration deleted here]

Safety features: Solid rocket provided for safety escape if boosters fail; otherwise fires along with second stage. Safety rocket separates from capsule after firing.

Schedule:

	58	59	60	61
Research	————			
Model tests	———			
Capsule tests	——			
Parachute tests	————————————————		
Escape system		————————	
Balloon flights		————————	
Satellite launch				————
Animals				——
Man				-

Cost: $40 x 10^6 plus $12 x 10^6 for flight vehicles excluding launchers.

Discussion

Dr. Kantrowitz looks at this proposal as the quickest way to manned satellite. This he regards as no stunt. A permanent orbit at 110 - 120 miles can be established with a solar propulsion device capable of overcoming the 5 grams drag force. The vehicle described may also offer a means for escape from a disabled large-scale vehicle.

[13]

The g factors and capsule atmosphere (60 percent oxygen, remainder N_2, He at .5 atmosphere pressure) are based on WADC Aeromedical group data. Certain upper atmosphere data on which the calculations are based are in doubt but better data will certainly be available before manned flight schedule.

The stainless steel parachute lies in a region of secondary flow (this is the reason for the shuttlecock shape) and will be subjected to peak temperatures of

1140° F, in contrast to the 1800° F peak at the sphere nose. The capsule will be above 200° F skin temperature for about 20 minutes. With the planned insulated wall construction the amount of heat transferred into the sphere will amount to about 1500 Btu (heat of fusion of a 10 pound block of ice) and is nearly negligible. Construction of the pressure vessel capsule is 0.020-inch stainless steel, insulation thermoflex with 0.001-inch stainless steel radiation shield; the nose is coated with 0.010-inch molybdenum skin.

The parachute has inflatable stainless steel bellows around the rim (about 3-inch diameter) and is made of 400 mesh stainless fabric covered with shim-stock "shingles"; total fabric weight about 60 pounds. Unfurling this chute increases the drag of the vehicle about 50 times. The chute can be opened and collapsed several times provided the bellows expansion doesn't exceed two times its length.

There was some discussion of magneto-hydrodynamic propulsion and means of projecting the shock location away from the vehicle using magnetic fields. This was not part of the proposal.

This presentation impressed me as having the most thorough correlation with available scientific and aerodynamic data.

LOCKHEED

Proposal presented by Dr. Perkins, Development Planning.

Pertinent Features

Weight: Capsule	600 pounds, includes	150 pound man
	Ablating material	600 pounds
	Retrorocket	100 pounds
	Parachute	200 pounds
Total weight	**1500 pounds**	

Boosters: Atlas and Bell Hustler. Hustler fires for 80 seconds between 620 -700 seconds flight time to inject into orbit. Maximum g's 6.6 (first stage). Specific impulse = 280 pounds per second per pound required of Hustler.

Time and Altitude. Six hours; perigee 150 miles; apogee 300 miles.

Reentry method: Retrorocket (-225 feet per second. One mile altitude at perigee roughly equals 1.5 feet per second, Retrorocket weighs less than 5 percent of vehicle.)

$W/C_O A = 0.01$

Maximum g's, 7.5 at 1° reentry angle

Maximum temperature, not stated

Maximum heat transfer, not stated

Vehicle subsonic at 60,000 feet; 600 feet per second at 40,000 feet.
Cooling by ablation material.

[15]

Recovery and landing: Cloth parachute to recover entire capsule. Landing area 20 miles by 400 miles along path. Errors of 150 miles due to lack of knowledge of air density between 125,000 and 250,000 feet. Lockheed proposes water recovery.

Tracking: 108 mc minitrack plus two way voice. Flywheel attitude control. Either gimbal or jet-vanes on retrorocket.

Safety features: Overrides on man's limited control functions. Hustler is fired to escape if Atlas booster fails. If Hustler fails man has had it.
Schedule: Preliminary experimental flights (Thor with 300 pound payload) in late 1958. Experimental centrifuge and vacuum chamber testing in mid-1958. Manned flight in late 1959.

Cost: $10 x10^6 <$cost$ < \$100$ x10^6

Discussion

X-17 tests (28 out of 38 successful) have confirmed hot shot and shock tube high speed data. Lockheed Missile Systems Division (MSD) has three-inch shock tube which uses l-inch model.

Beryllium is better than copper for ablation cooling material. Vehicle has no cosmic or meteor protection.

Lockheed feels lift reentry is best in long run.

CONVAIR

Introduction by Mr. William H. Patterson of Convair Astronautics. Extended discussion by Mr. Krafft A. Ehricke. Much of Ehricke's discussion was concerned with an extensive program of populating the solar system with reconnaissance, radio relay, etc., systems. Pulling that part of the presentation which applies to the minimum man-in-space concept from my notes is difficult. The program reviewed by Mr. Ehricke is contained in the Convair Astronautics report entitled: "A Satellite and Space Development Program", December, 1957 (Copy No. 185 is charged to the Lewis Laboratory Library.).

Some notes on the minimum man-in-space device are outlined in the following.

[16]

Pertinent Features

Vehicle: 60-inch sphere.

Weight: Man plus 1000 pounds, 500 pound heat-sink.

Boosters: Atlas Series D plus third stage. It is conceivable that the third stage need not be used – will require 5 specific impulse units improvement in present rocket performance.

Time and altitude: Time not noted. Altitude – 300 miles.

Reentry method: Retrorocket. Reentry angle, 2° maximum, $C_oA/W = .05$ to .08. Maximum g's = 8, greater than 6 g for 60 seconds.

Schedule: First flight (unmanned), February, 1959. Man in satellite orbit, April, 1961. Vehicle production schedule, one per month. Atlas D vehicle will not be fired before May, 1959.

McDONNELL

Introduction by Dr. Wokansky, coordinator of research. Presentation by Mr. Michael Weeks, Chief, preliminary design. McDonnell Aircraft presentation was an obvious bid for the "payload package" without the propulsion problem. Details were given in the McDonnell report to WADC.

Pertinent Features

Vehicle: One-man capsule. Data below are for a nose-cone device. Winged vehicle was considered possible in 30 months.

Weight:

Crew	270 pounds
Seat	50 pounds
Oxygen (3 hours)	70 pounds
Pressurization	134 pounds

| Canopy | 135 pounds |
| Structure | 345 pounds |

[17]

(Weight – continued.)

Stabilization	202 pounds
Retrorocket	90 pounds
Electronic	400 pounds
	1693 pounds

Landing chute	135 pounds
Heat problem	450 pounds
Incidentals	122 pounds

2400 = 25 percent

Boosters: Atlas and Polaris. Peak g's (Polaris) about 7.

Time and Attitude: One orbit (1.5 hours); apogee not noted; reentry initiated at 300,000⁺ feet.

Reentry method: Retrorocket (-143 feet per second). Ballistic reentry $WC_D/M = .5$ to 1; maximum g's 10 (1°) to 13 (4°), above 7 g's for 50 seconds, above 5 g's for 100 seconds; maximum temperature, 1200° F one foot behind leading edge (?); heat sink (not ablative) cooling.

Recovery and landing: Ejection seat with parachute.

Safety features: Polaris fired if Atlas fails. Ejection seat throws pilot clear.

Tracking: Ground control system for control of vehicle.

Schedule: Manned capsule flight 20-25 months.

Costs: Not stated.

GOODYEAR

After introduction by Mr. Baldwin, Mr. Darrell Romick described Phase 1 of a long-range (six to eight years) program to provide an operational satellite platform. The initial program was to conduct reentry tests (30 pound payload) [18] with Jupiter launcher; with primate (150 - 300 pound payload) with a Thor or Jupiter; manned spherical ball (1500 - 2000 pound payload) with Atlas or Titan. Only the features of the minimum man proposal are given here.

Pertinent Features

Vehicle: One-man 7-foot spherical vessel.

Weight: 1700 pounds; structure, 1050 pounds.

Boosters: Atlas or Titan first stage. Five third stage Vanguards or eight Irises (made by Goodyear Aircraft Corporation) for second stage. Maximum g's 5 to 7.

Time and Altitude:

Perigee = 110-115 miles; and Apogee = 600 - 800 miles.

Reentry method: External retrorocket (H_2O_2 -polyethylene, -800 feet per second, with tankage jettisoned before reentry.) $W/C_D A = 0.028$; maximum g's 7.5, greater than 4 for five minutes; maximum temperature, $3200°$ F, temperature greater than 2000° F for five minutes; maximum heat transfer not stated. Ablative cooling using Goodyear Aircraft Corporation-developed material.

Recovery and landing: Parachute for entire capsule deployed at 15,000 feet, with velocity 300 feet per second. Water landing in Western Pacific Ocean. Velocity at impact – 28 feet per second.

Control features: Extendable skirt for attitude control. Peripheral jet nozzles for attitude stabilization with retrorocket.

Tracking: Minitrack and defense radar.

Schedule: First firing - 23 months; manned flight - 26 months.

Cost: $100 x 10^6.

[19]

<u>NORTHRUP</u>

Northrup's presentation stressed the previously presented Dyna-soar proposal along with the military advantages and capabilities of the reconnaissance-bomber glider vehicle. Their man-in-space proposal is essentially an adaptation of this vehicle.

Pertinent Features

Vehicle: A glider as outlined below:

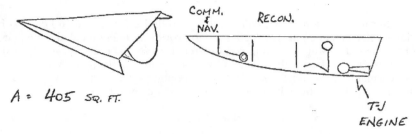

Weight: 10,500 pounds including 400 pounds fuel and a J-83 turbojet engine for final maneuver and landing.

Boosters: Three stages not detailed. The boost requirements vary in carrying out the separate steps in the development program. For the satellite flight the three stages of boost would each require a mass ratio of 3.6 with oxygen - JP-4. Initial gross weight, about 600,000 pounds. Maximum g's - 3.

Range and Altitude- 22,000 miles. Total time 70 minutes, attitude and speed variable with time, maximum altitude 200,000 feet.

[20]

Reentry: By increasing drag as glider descends. Maximum temperatures 3450° F at leading edge; 2150° F on lower surface, 1000° F on upper surface. 100 pounds liquid coolant and radiation cooling.

Recovery and landing: Conventional glide landing.

Tracking: Self-contained navigational system claimed to have 2 miles C.E.P.

Schedule: First glider vehicle 1960, boost flight tests at 4000 - 5000 mile range 1961, near-orbital manned flight - 1964.

Cost: 22 test vehicles.

Year	Megadollars	Year	Megadollars
1958	0.5	1962	92
1959	26.5	1963	59
1960	57	1964	29
1961	116		
Total	**380**		

Discussion

The test sequence for the various stages of vehicle development would be: (1) air launch from a carrier aircraft, (2) unmanned version for boost development testing, and (3) manned tests. Several technologies are needed, particularly the development of reliable boost rockets. Results of the x-15 high-speed flights and development of stellar-sight guidance equipment are also required.

Development of this vehicle brings with it the possibility of bombing or reconnaissance missions. These Northrup has detailed in the "Dyna-soar" proposals. Much of the allotted discussion time was spent in reviewing these weapon system capabilities.

[21]

Bell

Introduction by Dr. Dornberger. Presentation by Mr. Casey Forest. Dr. Dornber[g]er discussed briefly a 3000 pound eight-foot sphere with retrorocket and maximum stagnation temperature of 3500° F; maximum deceleration - 8 g's. Although it could be accomplished one and one half years sooner, he dismissed the sphere as having no growth potential. Bell's proposal therefore hinged strongly on their ROBO concept. The glide vehicle is outlined in the following.

Pertinent Features

Vehicle: Glider.

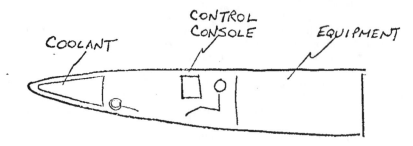

Weight: 18,800 pounds.

Boosters. Several assembled configurations were discussed. The one favored was a three-stage clustered configuration with three Titans, one Titan, One F_2-NH_3 rocket, as diagrammed below:

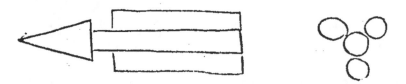

Initial gross weight would be 747,000 Pounds; ϕ 1 = 521,300 pounds, ϕ 2 = 175,300 pounds, ϕ 3 = 31,300 pounds, airplane - 18,800 pounds, research equipment 1500 pounds. Total propellants would be 659,500 pounds. Maximum acceleration - 4 g's.

[22]

Time and altitude: 127 minutes duration. Boost for seven minutes. Altitude at end of boost - 260,000 feet. Follow maximum L/D (Breguet) path.

Recovery and landing: Conventional man-controlled glider landing.

Navigation: Self-contained.

Schedule: Five years. Flight in 1963.

Cost: $889 x 10^6 including everything. Maximum cost in 1961 - $240 x 10^6.

Discussion

A short term (10 hour) satellite glider was also mentioned.

Bell has had a contract for development of a F_2-NH_3 rocket; they are willing to incorporate a $F_2N_2H_4$ rocket in the manned glider as the third stage of boost. In their ROBO studies and F_2NH_3 development work Bell claims to have $2.5 x 10^6 effort, with the USAF also spending $2.5 x 10^6.

A system development team has been in existence for more than a year. This team is organized as follows:

Airplane	Bell Aircraft
Electronics	Bendix Aviation
Navigation	Minneapolis-Honeywell
Boost	——
Ground and Supplies	——
Special Radar	Goodyear Aircraft

REPUBLIC

Introduction by Dr. Alexander Kartveli, Vice-President of Research and Development. The Republic proposal was based on the hypersonic glider concept of Dr. Antonio Ferri.

[23]

Pertinent Features

Vehicle: A blunt leading-edge sled with solid propellant rockets housed in the edges, Surface = 120 square feet. W/S = 25 pounds per square foot. C_L = .7.

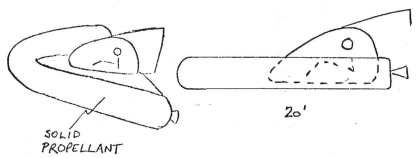

SOLID PROPELLANT

20'

Weight:

Propulsion   1000 pounds

Electrical   375 pounds
H_2O_2 control   175 pounds

Weight Total 3895 pounds (includes other factors)

Boosters: Atlas or Titan with modified propellants.

Time and altitude: 24 hours. Perigee 550,000 feet, reentry altitude – 300,000 feet.

Reentry method: Two retrorockets (-65 feet per second) applied at apogee. This reduces perigee to 300,000 feet. At this altitude produce negative lift (0.0005 g) by flying airplane inverted to prevent increasing altitude. Reaches M = 2 at 80,000 feet. Angle of attack on reentry about 40°. L/D = 1. Maximum temperature 2500° F.

Recovery and landing: Ejection seat for pilot and survival kit. Vehicle destroyed.

[24]

Safety features: H_2-O_2 attitude controls. Three aerodynamic control surfaces.

Guidance and tracking: Inertial guidance with short term rate gyros, accurate within 20 - 25 miles in position. Also UHF communication and beacon transponder.

Schedule: 18 months. Lead item claimed to be inertial guidance system.

Cost: Not estimated.

Discussion

The performance estimates of the booster units are summarized in the following:

	Stage 1	Stage 2	Stage 3
Wo	198,700	74,000	15,000
W_{Bo}	78,600	16,890	4,000
F	304,000	72,200	16,000
IS	230	250	265

CONCLUSION

This was the first round of proposals. Mr. Bell and WADC people expected to review these critically (they solicited comments from the NACA representatives before breakup of the group). A second round of the favored proposals was in the latter part of February and was planned to yield a proposal for WADC to submit to the Pentagon.

[signed]

Adelbert O. Tischler
Head, Rocket Combustion Section

WTO
MG
GM
AOT:jcs

Document I-4

[confidential] [declassified]
NACA- Langley
April 11, 1958

MEMORANDUM For Mr. Gilruth
Subject: Langley manned-satellite program

1.   Langley has been working for several months on the general problems
of manned-satellite vehicles. General studies have led to the choice of a basic
drag-reentry capsule as the most logical first vehicle. Following this choice several more specific studies were undertaken at Langley. The overall program into
which these specific studies fit are:

1) Reduced-scale recoverable satellite

2) Wind-tunnel and flight model studies of capsules and vertical flight
    vehicles

3) Laboratory studies (models, analyses, mock-ups) of structures,
    loads, stability and control, etc.

4) Full-scale vertical flight

5) Full-scale orbiting flight

Studies in item 1 above are summarized on the attached sheet 1 [not included].
Studies in items 2 and 3 are summarized in sheets 2 to 7 [not included]. All of
these studies are pointed directly toward items 4 and 5.

2.   In addition to the above studies the following are underway:

a) The Langley Instrument Research Division is studying attitude-control sys-
    tems for the vertical flight capsule. Some hardware is already on order.

b) The Langley Theoretical Mechanics Division is studying orbits and gen-
    eral space-mechanics problems of satellite flight.

c) The human-factors problems of vertical-flight and satellite reentry capsules
    have been discussed with personnel of the Naval Medical Acceleration

Laboratory, Johnsville, PA. Copies of typical vertical reentry histories of V, g, h, and oscillation have been sent to NMAL for their further study.

[Signed]
Paul E. Purser
Aeronautical Research Engineer

Document I-5

Document Title: Maurice H. Stans, Director, Bureau of the Budget, Memorandum for the President, "Responsibility for 'Space' Programs," 10 May 1958.

Source: Folder 18674, NASA Historical Reference Collection, NASA History Division, NASA Headquarters, Washington DC.

Document I-6

Document Title: Maxime A. Faget, NACA, Memorandum for Dr. Dryden, 5 June 1958.

Source: Folder 18674, NASA Historical Reference Collection, NASA History Division, NASA Headquarters, Washington DC.

Document I-7

Document Title: Clotaire Wood, Headquarters, NACA, Memorandum for Files, "Tableing [sic] of Proposed Memorandum of Understanding Between Air Force and NACA For a Joint Project For a Recoverable Manned Satellite Test Vehicle," 20 May 1958, with attached Memorandum, "Principles for the Conduct by the NACA and the Air Force of a Joint Project for a Recoverable Manned Satellite Vehicle," 29 April 1958.

Source: Folder 18674, NASA Historical Reference Collection, NASA History Division, NASA Headquarters, Washington DC.

Document I-8

Document Title: Hugh L. Dryden, Director, NACA, Memorandum for James R. Killian, Jr., Special Assistant to the President for Science and Technology, "Manned Satellite Program," 18 July 1958.

Source: Folder 18674, NASA Historical Reference Collection, NASA History Division, NASA Headquarters, Washington DC.

Perhaps the most difficult policy question to be resolved in the first half of 1958 revolved around the roles and missions of individual governmental entities in the new space initiative. Virtually every service within the Department of Defense (DOD) sought to control at least the lion's share of the human spaceflight mission. The National Advisory Committee for Aeronautics (NACA), then in the process of being transformed into NASA, thought it should have control of the mission as well. These rivalries led to debate and disagreement, negotiation, and compromise among these various entities. At first, NACA believed it would have to play a supporting role in the spaceflight initiative, yielding human activities to the military, but it became clear as 1958 progressed that the White House wanted NASA to take the lead, with the military supporting its efforts. This set of documents provided a detailed perspective on these deliberations and their results.

Document I-5

EXECUTIVE OFFICE OF THE PRESIDENT
BUREAU OF THE BUDGET
WASHINGTON 25, DC

May 10, 1958

MEMORANDUM FOR THE PRESIDENT

Subject: Responsibility for "space" programs

In your letters of April 2, 1958, you directed the Secretary of Defense and the Chairman of the National Advisory Committee for Aeronautics to review and report to you which of the "space" programs currently underway or planned by Defense should be placed under the direction of the new civilian space agency proposed in your message to Congress. These instructions specifically stated that "the new Agency will be given responsibility for all programs except those peculiar to or primarily associated with military weapons systems or military operations."

It now appears that the two agencies have reached an agreement contemplating that certain space programs having no clear or immediate military applications would remain the responsibility of the Department of Defense. This agreement would be directly contrary to your instructions and to the concept underlying the legislation the administration has submitted to Congress.

The agreement is primarily the result of the determination of the Defense representatives not to relinquish control of programs in areas which they feel might some day have military significance. The NACA representatives apparently have felt obliged to accept an agreement on the best terms acceptable to Defense.

Specifically, Defense does not wish to turn over to the new agency all projects related to placing "man in space" and certain major component projects such as the proposed million pound thrust engine development. The review by your Scientific Advisory Committee did not see any immediate military applications of these projects.

The effect of the proposed agreement would be to divide responsibility for programs primarily of scientific interest between the two agencies. This would be an undesirable and unnecessary division of responsibility and would be highly impractical. There would no be any clear dividing line, and unnecessary overlap and duplication would be likely. The Bureau of the Budget would have an almost hopeless task in trying to keep the two parts of the program in balance, and problems on specific projects would constantly have to come to you for resolution. The net result of the proposed arrangement would be a less effective program at higher total cost.

On the other hand, it will be relatively simple to work out practical working arrangements under which responsibility and control of the programs in question would clearly be assigned to the new agency as contemplated in your instructions, and the military interest would be recognized by the participation of the Department of Defense in the planning and, where appropriate, the conduct of the programs.

In the circumstances, it is recommended that you direct that the two agencies consult with the Bureau of the Budget and Dr. Killian's office to be sure that any agreement reached is in accordance with the intent of your previous instructions. It is especially important that the announcement of the agreement now being proposed be avoided at this stage of the consideration by the Congress of legislation to establish the new space agency.

If you approve this recommendation, there are attached memoranda to the Secretary of Defense and the Chairman of the National Advisory Committee for Aeronautics for your signature.

(Handled orally per President's instructions. (AJG, 5/13/58) [AJG is General Andrew Goodpaster, the President's military assistant.]

[Signed Maurice H. Stans]
Director
5/13/58

5/14/58
I notified the Secretary of Defense (General Randall) and Dr. Dryden.
AJG

Document I-6

[SECRET] [DECLASSIFIED}

Washington DC
June 5, 1958

MEMORANDUM for Dr. Dryden

This memorandum is submitted to review my dealings with ARPA during the past several weeks.

A. Background

1.   I made my first contact with ARPA personnel on May 14, 1958. At that time the NACA was under the impression that it was to work with Dr. Batdorf to prepare a man-in-space program that would be acceptable to both the NACA and ARPA from a technical standpoint. My first visit to the Pentagon revealed that ARPA had a somewhat different impression of what was to take place. I was told that at the request of Mr. Johnson a panel had been formed in ARPA to create a man-in-space program and to advise him how this program could best be managed. ARPA had formed this panel approximately a week earlier from members of the ARPA technical staff under the Chairmanship of Dr. Batdorf. I was told that this was only one of many such working groups that was concurrently attacking various jobs in ARPA and that the membership on these various panels greatly overlapped. Accordingly, my position on the man-in-space panel was a special one resulting from an invitation by Dr. York.

2.   Inasmuch as this situation was not exactly in keeping with my impression of what it should be, I told the panel that while I would sit as a member of their panel I would also consider myself as a liaison representative of the NACA. In this respect, I reminded them that the direct responsibility for the man-in-space program may quite likely be given to the soon to be created civilian space agency. Thus, I would be concerned that the man-in-space program to be formulated would be one that is acceptable to the NASA and that the management responsibility would be one which could be transferred with the least difficulty. I stated further that if there are any final agreements to be reached between ARPA and NACA they would have to be approved by higher authority, presumably Dr. Dryden and Dr. York and quite possibly by knowledgeable people from the White House. Dr. Batdorf concurred with this and stated that the ARPA staff, most of whom work for IDA, serve only in an advisory capacity.

3.   My dealings with the ARPA panel have been quite pleasant and I think fruitful. On the majority of the issues the panel has reached essentially unanimous agreement. On controversial subjects my viewpoints are apparently being given fair consideration. In addition, the panel is quite aware that the NACA has a firm position in the man-in-space business. From this standpoint, I have additional influence when the question of acceptability to NACA arises in certain instances.

4.   While a good number of the ARPA staff have attended the panel discussions and the presentations made to the panels by the services, those who actually serve on the panel are:

1.   Dr. Sam Batdorf, recently from Lockheed
2.   Dr. Arthur Stosick, recently from JPL
3.   Mr. Bob Youngquist, recently from RMI
4.   Mr. Jack Irvine, recently from Convair

    5.   Captain Robert Truax, recently from BMD (117L Project)

    6.   Mr. Dick Cesaro, recently from ARDC and NACA

    5.   The panel has conducted its business by questioning representatives of the Air Force, Navy, Army, and Industry who are familiar with proposed man-in-space programs and by conducting discussions within the panel alone. The Air Force has sent a large number of representatives to two panel meetings to answer questions. These have included people from HQ, ARDC, BMD, and WADC.

B.   Present Situation

    1.   The work of the panel is apparently drawing to a close. We have put together a proposed man-in-space program that is not far different from the Air Force proposal. The essential elements of this program are:

    a.   The system will be based on the use of the Convair Atlas propulsion system. If the expected performance of the Atlas rocket alone is not obtained, then the Atlas 117L system will be used.

    b.   The man-in-space flights will be launched from "Pad-20" at AFMTC.

    c.   Retro-rockets will be used to initiate return from orbit.

    d.   The non-lifting ballistic type of capsule will be used.

    e.   The aerodynamic heating during atmospheric entry will be handled by a heat sink or ablation material.

    f.   Tracking will be carried out primarily with existing or already planned systems. The most important of which will be the G.E. Radio-inertial Guidance System which is highly accurate. [sic] The G.E. system will be in existence at AFMTC, San Salvadore, Australia, and Camp Cook.

    g.   The crew for the orbital flights will be selected from volunteers in the Army, Navy, and Air Force. The crew will be selected in sufficient time to undergo aero-medical training functions.

    2.   The panel is in unanimous agreement that the man-in-space program should begin immediately. The panel feels that in spite of the unsettled status of both ARPA and NACA that this can be accomplished if a national man-in-space program is adopted. ARPA apparently has $10,000,000 to initiate the program. Future funding and management will of course depend on the outcome of present legislation.

    3.   The panel is recommending that the Air Force be given the management of the program with executive control to remain in the hands of NACA and ARPA. This could presumably be accomplished by the creation of an executive committee composed of NACA and ARPA people, plus representatives from the contractors, the Air Force, and perhaps Army and Navy.

    4.   In addition to meeting with the panel I had a short chat with Dr. York on June 4. His views differed from the panel on primarily two issues. He thinks the Atlas alone and the Atlas-117L combination should be considered equally competitive at this time as a propulsion system. The panel considers the 117L

approach as a back up to be dropped as soon as sufficient confidence in the Atlas alone is achieved. Dr. York thinks that contract for the construction of the capsule should be awarded after proposals are received from industry. The panel, although they do not recommend this procedure, believe that it would be much quicker and just as satisfactory to choose a suitable contactor to build a capsule which has been tightly specified.

[Signed]
Maxime A. Faget

Document I-7

Washington, D.C.
May 20, 1958

MEMORANDUM For Files

Subject: Tableing [*sic*] of Proposed Memorandum of Understanding Between Air Force and NACA For a Joint Project For a Recoverable Manned Satellite Test Vehicle

Reference: NACA ltr to DCS/D dtd April 11, 1958

1. On April 11, 1958, Dr. Dryden signed a proposed Memorandum of Understanding for a Joint NACA-Air Force Project for a recoverable manned satellite test vehicle. Minor revisions to the Agreement were discussed and, with Dr. Dryden's approval, agreed on between Colonel Heaton and myself on April 29, 1958.

2. Subsequent to April 29, 1958, it was agreed with Colonel Heaton that the prospective Agreement should be put aside for the time being. The matter may be taken up again when the responsibilities of ARPA and NASA have been clarified.

C. Wood

CloW:dlf

[2]

[April 29, 1958]

MEMORANDUM OF UNDERSTANDING

Subject: Principles for the Conduct by the NACA and the Air Force of a Joint Project for a Recoverable Manned Satellite Vehicle.

A. A project for a recoverable manned satellite test vehicle shall be conducted jointly by the NACA and the Air Force, implementing an ARPA instruction to the Air Force of February 23, 1958. Accomplishment of this project is a matter of national urgency.

B. The objectives of the project shall be:

    a. To achieve manned orbital flight at the earliest practicable date consistent with reasonable safety for the man,

    b. To evaluate factors affecting functions and capabilities of man in an orbiting vehicle,

    c. To determine functions best performed by man in an orbiting weapon system.

C. To insure that these objectives are achieved as early and as economically as possible the NACA and the Air Force will each contribute their specialized scientific, technical, and administrative skills, organization and facilities.

D. Overall technical direction of the project shall be the responsibility of the Director, NACA, acting with the advice and assistance of the Deputy Chief of Staff, Development, USAF.

E. Financing of the design, construction, and operational phases of the project [the following words were in the April 11 version of the memorandum but were deleted on April 29. A handwritten note on the April 29 version of the document says *"this deletion suggested by Silverstein & Gilruth, agreed on by JWC & HCD."* The text continues: as well as of any studies which may be determined necessary to supplement Air Force or NACA studies to permit the accomplishment of the objectives,] shall be the function of the Air Force.

F. Management of the design, construction, and operational phases of the project shall be performed by the Air Force in accordance with the technical direction prescribed in paragraph C. Full use shall be made of the extensive background and capabilities of the Air Force in the Human Factors area.

[Handwritten note at bottom of page – "Col. Heaton advised 2:15 p.m. 4-29-58 that this version agreeable to Director NACA provided that deletion is made as marked in paragraph E, Clo Wood."]

G. Design and construction of the project shall be accomplished through a negotiated contract (with supplemental prime or sub-contracts) obtained after evaluating competitive proposals invited from competent industry sources. The basis for soliciting proposals will be characteristics jointly evolved by the Air Force and NACA based on studies already well under way in the Air Force and the NACA.

H. Flights with the system shall be conducted jointly by the NACA, the Air Force, and the prime contractor, with the program being directed by the NACA and the Air Force. The NACA shall have final responsibility for instrumentation and the planning of the flights.

I. The Director, NACA, acting with the advice and assistance of the Deputy Chief of Staff, Development, USAF, shall be responsible for making periodic progress reports, calling conferences, and disseminating technical information and results of the project by other appropriate means subject to the applicable laws and executive orders for the safeguarding of classified information.

General Thomas D. White
Chief of Staff, USAF
Hugh L. Dryden
Director, NACA

Document I-8

July 18, 1958

MEMORANDUM for Dr. James R. Killian, Jr.
Special Assistant to the President for Science and Technology

SUBJECT: Manned Satellite Program.

1. The current objective for a manned satellite program is the determination of man's basic capability in a space environment as a prelude to the human exploration of space and to possible military applications of manned satellites. Although it is clear that both the National Aeronautics and Space Administration and the Department of Defense should cooperate in the conduct of the program, I feel that the responsibility for and the direction of the program should rest with NASA. Such an assignment would emphasize before the world the policy statement in Sec. 102(a) of the National Aeronautics and Space Act of 1958 that "it is the policy of the United States that activities in space should be devoted to peaceful purposes for the benefit of all mankind."

2. The NASA through the older NACA has the technical background, competence, and continuing within-government technical back-up to assume this responsibility with the cooperation and participation of the Department of Defense. For a number of years, the NACA has had groups doing research on such items as stabilization of ultra-high-speed vehicles, provision of suitable controls, high-temperature structural design, and all the problems of reentry. More recently, the NACA research groups have been working on these problems with direct application to manned satellites. The human-factors problems of this program are not far different from those for the X-15 which the NACA has been studying in cooperation with the Navy and Air Force. Thus, the NACA has enlisted the cooperation of the military services and marshaled the required technical competence. Included in this competence are large, actively-working, staffs in NACA laboratories providing additional technical back-up for the manned-satellite program.

3.The assignment of the direction of the manned satellite program to NASA would be consistent with the President's message to Congress and with the pertinent extracts from the National Aeronautics and Space Act of 1958 given in the appendix to this memorandum.

Hugh L. Dryden
Director
National Advisory Committee for Aeronautics

Attachment [not included]

Document I-9

Document Title: Maxime A. Faget, Benjamine J. Garland, and James J. Buglia, Langley Aeronautical Laboratory, NACA, "Preliminary Studies of Manned Satellites," 11 August 1958.

Source: Folder 18674, NASA Historical Reference Collection, NASA History Division, NASA Headquarters, Washington DC.

Prior to the Mercury, Gemini, and Apollo programs of the 1960s, virtually everyone involved in space advocacy envisioned a future in which humans would venture into space aboard winged, reusable vehicles. That was the vision from Hermann Oberth in the 1920s through Wernher von Braun in the 1950s to the U.S. Air Force's X-20 Dyna-Soar program in the early 1960s. Because of the pressure of the Cold War, NASA chose to abandon that approach to space access in favor of ballistic capsules that could be placed atop launchers developed originally to deliver nuclear warheads to the Soviet Union. This memorandum states the position, one NASA eventually adopted, that advocated using ballistic capsules for human spaceflight. Led by Maxime A. Faget, one of the most innovative thinkers in NACA/NASA, the authors contend that because of the desire to launch humans as soon as possible, moving to a capsule concept represented the only genuine option available for the U.S. A capsule could make use of research on reentry undertaken for ballistic missiles, as well as make possible the ready adoption of ballistic missiles as launchers for spaceflight.

[CONFIDENTIAL] {DECLASSIFIED}

NACA

RESEARCH MEMORANDUM

PRELIMINARY STUDIES OF MANNED SATELLITES

WINGLESS CONFIGURATION: NONLIFTING

By Maxime A. Faget, Benjamine J. Garland, and James J. Buglia

Langley Aeronautical Laboratory
Langley Field, VA.

August 11, 1958

[Note: Only Summary and Introduction are included]

SUMMARY

This paper is concerned with the simple non-lifting satellite vehicle which follows a ballistic path in reentering the atmosphere. An attractive feature of such a vehicle is that the research and production experiences of the ballistic–missile programs are applicable to its design and construction, and since it follows a ballistic path, there is a minimum requirement for autopilot, guidance, or control equipment. After comparing the loads that would be attained with man's allowable loads, and after examining the heating and dynamic problems of several specific shapes, it appears that, insofar as reentry and recovery is concerned, the state of the art is sufficiently advanced so that it is possible to proceed confidently with a manned-satellite project based upon the ballistic reentry type of the vehicle.

INTRODUCTION

This paper is concerned with the simple non-lifting satellite vehicle which follows a ballistic path in reentering the atmosphere. An attractive feature of such a vehicle is that the research and production experiences of the ballistic–missile programs are applicable to its design and construction.

The ballistic reentry vehicle also has certain attractive operational aspects which should be mentioned. Since it follows a ballistic path there is a minimum requirement for autopilot, guidance, or control equipment. This condition not only results in a weight saving but also eliminates the hazard of malfunction. In order to return to the earth from orbit, the ballistic reentry vehicle must properly perform only one maneuver. This maneuver is the initiation of reentry by firing the retrograde rocket. Once this maneuver is completed (and from a safety standpoint alone it need not be done with a great deal of precision), the vehicle will enter the earth's atmosphere. The success of the reentry is then dependant only upon the inherent stability and structural integrity of the vehicle. These are things of a passive nature and should be thoroughly checked out prior to the first man-carrying flight. Against these advantages the disadvantage of large area landing by parachute with no corrective control during the reentry must be considered.

In reference 1, Dean R. Chapman has shown that the minimum severity of the deceleration encountered during a ballistic reentry is related to the fundamental nature of the planet. Thus it can be considered a fortunate circumstance that man can tolerate this deceleration with sufficient engineering margin.

Document I-10

Document Title: Roy W. Johnson, Director, ARPA, DoD, Memorandum for the Administrator, NASA, "Man-in-Space Program," 3 September 1958.

Source: Folder 18674, NASA Historical Reference Collection, NASA History Division, NASA Headquarters, Washington DC.

Document I-11

Document Title: Roy W. Johnson, Director, ARPA, DOD, Memorandum for the Administrator, NASA, "Man-in-Space Program," 19 September 1958, with attached Memorandum of Understanding, "Principles for the Conduct by NASA and ARPA of a Joint Program for a Manned Orbital Vehicle," 19 September 1958.

Source: Folder 18674, NASA Historical Reference Collection, NASA History Division, NASA Headquarters, Washington DC.

One of the issues that NASA, officially established on 1 October 1958, had to work in its first weeks of existence was an agreement on how to manage the human spaceflight program. Laboriously its leadership negotiated with interested organizations in the Department of Defense for transfer of some resources, as well as for support for the conduct of the mission. A constant consideration at the time was the next act of the Soviet Union, which had already several times bested the U.S. in "space firsts." Should more money be allocated to human spaceflight to ensure U.S. primacy in this arena? Should other actions be taken to ensure that the U.S. launched the first human into space? Should the U.S. pursue a capsule approach because of this rivalry with the USSR? The answer to all of those questions was yes, as shown in these documents, but in the end the Soviets still launched Yuri Gagarin first on 12 April 1961.

Document I-10

[SECRET] [DECLASSIFIED]

ADVANCED RESEARCH PROJECTS AGENCY
WASHINGTON 25, D.C.

Sep 3 1958

MEMORANDUM FOR THE ADMINISTRATOR, NATIONAL AERONAUTICS
AND SPACE ADMINISTRATION

SUBJECT: Man-in-Space Program

In accordance with agreements reached at the meeting with the Deputy Secretary of Defense on August 20, 1958 we have taken the following actions:

(1) Designated an ARPA officer to work with NASA to arrange for the transfer, not later than January 1, 1959, of all IGY tracking stations from the DoD to the NASA.

(2) Advised the Assistant Secretary of the Navy for Air of your intent to request the transfer to NASA of those persons at NRL engaged on the VANGUARD program.

(3) Arranged for visits by NASA teams to AOMC and to West Coast installations of the DoD for the purpose of evaluating capabilities to pursue future NASA projects.

(4) Arranged for the stationing of resident representatives of NASA at AOMC, BMD and NOTS to become familiar with the details of those scientific programs (covered by ARPA Orders 1, 2, 3 and 9) being conducted by ARPA, in anticipation of their transfer to NASA on or about October 1, 1958.

(5) Prepared a comprehensive management recommendation to the Secretary of Defense for coordination of all DOD satellite tracking and data reduction facilities designed to complement those facilities; to be operated by NASA.

Preparations have thus been made for an orderly transfer of appropriate programs without interruption.

I am troubled, however, with respect to one of the projects in which there is general agreement that it should be a joint undertaking. [2]

This is the so-called "Man-in-Space" project for which $10 million has been allocated to ARPA and $30 million to NASA. My concern over this project is due (1) to a firm conviction, backed by intelligence briefings, that the Soviets next spectacular effort in space will be to orbit a human, and (2) that the amount of $40 million for FY 1959 is woefully inadequate to compete with the Russian program. As you know our best estimates (based on some 12 - 15 plans) were $100 to $150 million for an optimum FY 1959 program.

I am convinced that the military and psychological impact on the United States and its Allies of a successful Soviet man-in-space "first" program would be far reaching and of great consequence.

Because of this deep conviction, I feel that no time should be lost in launching an aggressive Man-in-Space program and that we should be prepared if the situation warrants, to request supplemental appropriations of the Congress in January to pursue the program with the utmost urgency.

Certain projects planned and financed by ARPA and now underway will contribute to this undertaking. I list them here for ready reference.

(1) The bio-medical project of WS-117 L will attempt the recovery of three primates, thus affording valuable information on space environmental data.

(2) Approximately fourteen ATLAS/117L flights are scheduled during the next eighteen months in connection with the WS-117L program. This would give a capability during this period of achieving 4, 000 pounds in a low orbit and offers the most promising early capability of placing a man in orbit with sufficient safety considerations.

(3) A project to design a high energy upper stage (liquid hydrogen liquid oxygen) rocket for ATLAS/TITAN with an engine thrust of approximately 30,000 pounds has been authorized. This has promise of placing 8,000 to 10,000 pounds in orbit during 1960.

(4) A project to construct a 1 - 1.5 million pound thrust booster utilizing existing hardware in a "cluster" arrangement has been authorized. This should permit placing 25,000 pounds in orbit by 1961. It is our intention that this project be carefully coordinated with the single-chamber super thrust engine being developed by NASA so that much of the booster equipment later could be used on the large engine when it becomes available in 1964-1965. The early capability afforded by the cluster project would make possible a space platform for manned reconnaissance and for a related military space operating base. [3]

With these projects forming a basis for the propulsion requirements, parallel efforts should go forward as a matter of urgency on the recoverable vehicle itself. It is my understanding from talks with members of our staff that the NASA will concentrate on the "capsule" technique. I agree that this offers the earliest promise and urge that the program be pursued vigorously. As an alternate approach it is the intention of the DoD to proceed with a winged vehicle based on the general concept of the DYNA SOAR Weapon System 464L. The winged vehicle approach is believed to most nearly satisfy the military objectives in regard to flexibility of mission and independence from ground guidance and recovery operations during hostilities. Many of the design requirements, especially those relating to human factors, will be similar for the two approaches. Thus we would expect to make maximum use of the NASA capsule data in our alternate approach.

I therefore urge that you join with me at the earliest practicable date to consider every possible step that we might take to achieve a U.S. lead in this important program.

[signed]
Roy W. Johnson
Director

Document I-11

September 1958

[Handwritten: ARPA]

MEMORANDUM FOR THE ADMINISTRATOR, NATIONAL AERONAUTICS
AND SPACE ADMINISTRATION

SUBJECT: Man-in-Space Program

This is to confirm that the agreement reached in our discussions of yesterday that ARPA will join with NASA in a joint Man-in-Space program based on the "capsule" technique.

I consider this program to be of the highest urgency and have directed appropriate members of my organization to work with your staff to outline a detailed program for early implementation. I believe it very desirable that at least an outline of an agreed program be available for presentation to the Space Council on October 20.

As indicated yesterday, ARPA is of the opinion that a follow-on winged maneuverable space vehicle is essential to meet military requirements and it is our intention to initiate a modest program in this direction during FY 1959.

[Signed]
Roy Johnson
Director
[Attachment included]

9/19/58

MEMORANDUM OF UNDERSTANDING

Subject: Principles for the Conduct by NASA and ARPA of a Joint Program for a Manned Orbital Vehicle

1. It is agreed that a program for a manned orbital vehicle will be conducted jointly by NASA and ARPA. It is agreed that accomplishment of this program is a matter of national urgency.
2. The objective of this program is to demonstrate the capability of manned orbital flight at the earliest practicable date consistent with reasonable safety for the man. The program will include constructing and testing in flight a manned-orbital vehicle.
3. It is agreed that this program will be supported by NASA and ARPA until it is terminated by the achievement of manned orbital-flights.
4. Technical direction and management of the program will be the responsibility of the Administrator of the NASA, acting with the advice and assistance of the Director of ARPA.
5. It is agreed that the concurrence of the Administrator of NASA will be required for any parts of this program which are carried out by contract.
6. It is the intent in this program to make full use of the background and capabilities existing in NASA and in the military services. [2]
7. It is agreed that a working committee consisting of members of the staff of NASA and ARPA will be established to advise the Administrator of the NASA and the Director of ARPA on technical and management aspects of this program, and that the chairman of this committee will be a member of the NASA staff.

Document I-12

Document Title: Minutes of Meetings, Panel for Manned Space Flight, 24 and 30 September, 1 October 1958.

Source: Folder 18674, NASA Historical Reference Collection, NASA History Division, NASA Headquarters, Washington, DC.

In August 1958, before NASA was officially established, NACA Director Hugh L. Dryden and Robert R. Gilruth, Assistant Director of Langley Aeronautical Research Laboratory, had both informed Congress of their intent to seek $30 million for the development of a piloted satellite vehicle. One month later NASA Administrator T. Keith Glennan and Roy Johnson, Director of the Advanced Research Projects Agency (ARPA), managed to come to a general agreement concerning a joint NASA-ARPA program for developing a vehicle based upon a ballistic capsule concept that had been proposed by engineers at Langley. (See Documents I-10 and I-11.) The panel for Manned Spaceflight, also referred to as the Joint Manned Satellite Panel, was established by executive agreement between NASA and ARPA on 18 September. It held its first meeting on 24 September. In that and subsequent meetings in the following days, the panel established the basic goals and strategies for the initial U.S. piloted spaceflight program.

National Aeronautics and Space Administration
1520 H Street Northwest
Washington 25, D.C.

MINUTES OF MEETINGS

PANEL FOR MANNED SPACE FLIGHT
September 24, 30, October 1, 1958

A meeting of the Panel for Manned Space Flight was held on September 24 and 30 and October 1, 1958, at NASA Headquarters, Washington, D.C.

The following panel members were present:

Robert Gilruth (Chairman), NASA
Dr. S. B. Batdorf, ARPA
D. A.J. Eggers, NASA (Sept. 24)
Max Faget, NASA
George Low, NASA
Warren North, NASA
Walter Williams, NASA (Sept. 24-30)
Roberson Youngquist, ARPA (Sept. 24)

The objectives of this series of meetings were to set up a basic plan for the manned satellite project, determine a preliminary flight test schedule, and establish a funding program. Attached as Appendix A is the draft of "Objectives and Basic Plan." Appendices B and C include the tentative flight test program and flight test schedule. A cost breakdown of the project is attached as Appendix D. [Appendices B, C, and D not included]

It was decided that no aero-medical experiments will be supported by the Manned Satellite Project except those required for the successful completion of the mission. Dr. Lovelace will aid in establishing the aero-medical and pilot training requirements. Mr. Williams mentioned that aero-medical information obtained from the X-15 project should be applicable to the Manned Satellite Project.

Approval was obtained from General Boushey for use of a C-130 airplane in order to expedite some of the capsule and parachute drop tests. [2]

Mr. Williams will determine the feasibility of using a F-104 launch airplane for a portion of the drogue parachute tests.

Dr. Eggers stressed the fact that the panel should consider a lifting vehicle in planning for future manned space flight projects.

[Signed]
Warren J. North
Secretary

Appendix A

OBJECTIVES AND BASIC PLAN
FOR THE MANNED SATELLITE PROJECT

I. Objective

The objectives of the project are to achieve at the earliest practicable date orbital flight and successful recovery of a manned satellite, and to investigate the capabilities of man in this environment.

II. Mission

To accomplish these objectives, the most reliable available boost system will be used. A nearly circular orbit will be established at an altitude sufficiently high to permit a 24-hour satellite lifetime; however, the number of orbital cycles is arbitrary. Descent from orbit will be initiated by the application of retro-thrust. Parachutes will be deployed after the vehicle has been slowed down by aerodynamic drag, and recovery on land or water will be possible.

III. Configuration

A. Vehicle

The vehicle will be a ballistic capsule with high aerodynamic drag. It should be statically stable over the Mach number range corresponding to flight within the atmosphere. Structurally, the capsule will be designed to withstand any combination of acceleration, heat loads, and aerodynamic forces that might occur during boost and reentry of successful or aborted missions.

B. Life Support System

The capsule will be fitted with a seat or couch which will safely support the pilot during acceleration. Provision will be made for maintaining the pressure, temperature, and composition of the atmosphere in the capsule within allowable limits for human environment. Food and water will be provided.

C. Attitude Control System

The vehicle will incorporate a closed loop control system which consists of an attitude sensor with reaction controls. The reaction controls will maintain the vehicle in a specified orbital attitude, will [2] establish the proper angle for retro-firing, reentry, or an abort maneuver. The pilot will have the option of manual or automatic control during orbital flight. The manual control will permit the pilot to visually observe various portions of the earth and sky.

D. Retrograde System

The retro-rocket system will supply sufficient impulse to permit atmospheric entry in less than ½ revolution after application of retro-thrust. The magnitude and direction of the retro-thrust will be predetermined on the basis of allowable declarations and heating within the atmosphere, and miss distance.

E. Recovery Systems

A parachute will be deployed at an altitude sufficiently high to permit a safe landing on land or water; the capsule will be buoyant and stable in water. Communication and visual aids will be provided to facilitate rescue.

F. Emergency Systems

An escape system will be provided to insure a safe recovery of the occupant after a malfunction at any time during the mission. Parallel or redundant systems will be considered for the performance of critical functions.

IV. Guidance and Tracking

Ground-based and vehicle equipment will be employed to allow the establishment of the desired orbit within satisfactory tolerance, to determine the satellite orbit with the greatest possible accuracy, to initiate the descent maneuver at the proper time, and to predict the impact area.

V. Instrumentation

Medical instrumentation required to evaluate the pilot's reaction to space flight will be incorporated in the capsule. In addition, instrumentation will be provided to measure and monitor the internal and external cabin environment and to make scientific observations. These data will be recorded in flight and/or telemetered to ground recorders.

VI. Communication

Provisions will be made for adequate two-way communications between the pilot and ground stations. [3]

VII. Ground Support

The successful completion of the manned satellite program will require considerable ground support, such as pre-launch support and an elaborate recovery network.

VIII.   Test Program
An extensive test program will be required to implement this project. The test program will include ground testing, development and qualification flight testing, and pilot training.

Document I-13

Document Title: NASA, "Preliminary Specifications for Manned Satellite Capsule," October 1958.

Source: Folder 18674, NASA Historical Reference Collection, NASA History Division, NASA Headquarters, Washington DC.

Most of the design work for what became the Mercury spacecraft took place under the auspices of, and in many cases directly by, Max Faget. This document, written by Faget and his research team, established very detailed specifications for the Mercury spacecraft for the use of industry, necessary for their proposals to build the hardware. These specifications outlined the program and suggested methods of analysis and construction. Faget specifically asked for the construction of a simple, nonlifting vehicle that could follow a ballistic path in reentering the atmosphere without experiencing heating rates or accelerations that would be dangerous to an astronaut. He also called for modest pitch, yaw, and attitude control, as well as a retrorocket pack to bring the capsule down from orbital velocity. Finally, this document established the limits of size, shape, weight, and tolerances of the Mercury spacecraft. This set of specifications became the basis for the capsule's construction by the McDonnell Aircraft Company based in St. Louis, Missouri.

NATIONAL AERONAUTICS AND SPACE ADMINISTRATION
PRELIMINARY SPECIFICATIONS
FOR
MANNED SATELLITE CAPSULE

OCTOBER 1958

TABLE OF CONTENTS

[1]

1.   INTRODUCTION

1.1   This preliminary specification outlines the technical design requirements
         for a manned satellite capsule. This capsule will be used in the initial
         research on manned space flight. The research will be concerned primar-
         ily with man's ability to adapt to and perform in a space environment as
         well as in those environments associated with projection into space and
         with return to the surface of the earth.

1.2   The scope of this specification encompasses the capsule configuration,
         stability and control characteristics, heating and loads environments,
         structural design, onboard equipment and instrumentation. In certain
         areas, specific design approaches are outlined herein. The contractor
         shall follow the outlined approaches except in cases where mutual agree-
         ment is reached between the NASA and the contractor that an alternate
         approach is to be taken. Suggestions by the contractor of improved alter-
         nate approaches are invited.

1.3   The contractor shall undertake and be responsible for the design, fabrica-
         tion or procurement, integration, and installation of all components of
         the capsule system as described herein. Details of the responsibilities for
         the matching of the capsule and the booster vehicle will be clarified at a
         later date.

1.4   The design approach shall emphasize the safety of the mission. Although
         not specified herein in every instance, due consideration shall be given
         to simplicity, redundancy, and the use of backup systems in order to
         improve mission reliability.

[2]

2.   MISSIONS

2.1   General - All missions to be described shall be capable of accomplish-
         ment with and without a human occupant and with appropriate animals
         if desired.

2.2   Primary Mission – The primary mission shall be the launching of a manned
         capsule into a semi permanent orbit and subsequent safe recovery to the
         surface of the earth at a designated time and/or position through use of
         retro thrust and aerodynamic drag. The design mission profile is as indi-
         cated in figure 1 [not included] and from histories of pertinent trajectory
         variables are shown in figure 3 [not included].

2.2.1   The design of the capsule shall be based on the use of a single Atlas D
         missile as the launching booster. The capsule shall replace the missile
         nose cone in a manner which requires a minimum of modifications to
         the booster system.

2.2.2   The launching site shall be Cape Cannaveral, Florida. Launching shall be possible at any azimuth within thirty (30) degrees of due east.

2.2.3   A target value of effective launch weight shall be twenty-four hundred (2,400) pounds. Effective launch weight is defined by the following equation:

$$W_e = W_o + 0.2W_j$$

Where
    W_o is the weight of capsule when projected into orbit
    W_j is the weight of capsule components jettisoned at Atlas staging

2.2.4   The launch booster system shall be capable of projecting the capsule into orbit with the following tolerances:

2.2.4.1 The projection altitude shall be not greater than one hundred and twenty (120) nautical miles.

2.2.4.2 The perigee altitude shall not be less than one hundred and ten (110) nautical miles.

2.2.4.3 The eccentricity shall not be greater than five thousandths (0.005)

[3]

2.2.5   For the initial orbital missions, the number of orbital cycles per mission shall be two (2); however, an arbitrary number of orbital cycles per mission up to eighteen (18) shall be possible.

2.2.6   The following specifications pertain to the recovery of the capsule from orbit:

2.2.6.1 The nominal position of the point at which entry is initiated shall be such that impact occurs in a prescribed area in proximity to the launching station; however, in the event an emergency, it shall be possible to initiate the entry at any point in the orbit.

2.2.6.2 The entry shall be accomplished by application of retro thrust to produce a perigee altitude within the atmosphere. The magnitude and direction of the retro thrust shall be regulated so that angles of entry into the atmosphere at an altitude of sixty (60) miles shall be between one half (1/2) and three (3) degrees.

2.2.6.3 Consideration shall be given to high altitude deployment of a drogue parachute. This drogue parachute would be deployed near a Mach number of one (1) and is intended to provide improved dynamic stability to the capsule.

2.2.6.4 A landing parachute shall be deployed at an altitude sufficiently great to allow time to deploy a second parachute in event of failure of the first and to reduce sinking speed at impact to less than thirty (30) feet per second. Impact shall be considered to take place at an altitude of five thousand (5,000) feet. Commensurate with the above requirements, deployment altitudes shall be low enough to keep drift from winds aloft from seriously affecting the area of impact.

2.2.6.5   The capsule shall be designed for water landing, and shall be buoyant and stable in the water; however, consideration shall be given in the design to emergency landing on land surfaces. Protection from serious injury to the human occupant shall be afforded under conditions of land impact.

2.2.6.6   The capsule and the systems within the capsule necessary for location, recovery, and survival shall be capable of sustained operation for a period of twelve (12) hours after impact with the surface of the earth. This requirement is in addition to the twenty-eight (28) hours requirement associated with the space flight phase of the operation

[4]

2.3   Checkout missions- In order to expeditiously lead up to successful achievement of the primary mission, the requirements of the following checkout missions shall be considered in the capsule design.

2.3.1   Ballistic trajectories of limited velocity and range for entry and recovery simulation- A typical mission profile of this type is illustrated in figure 3 [not included]. The entry and recovery phases of this mission shall be accomplished in the same manner as specified for the primary mission. The peak decelerations achieved during entry shall equal those applicable to the primary mission. As this type of checkout mission may represent the first flight tests of a manned space capsule, a buildup in velocity and range may be required. Rocket motors which are immediately available shall be used for this checkout mission.

2.4   Aborted missions- During various periods of the launch operation, it may become necessary to abort the mission and escape from the vicinity of the rocket booster system. An active escape system shall be an [illegible] of the capsule until five (5) seconds after booster staging. At times greater than booster staging plus five (5) seconds, escape shall be accomplished by shutting down the Atlas sustainer engine and operating the nose cone separations motors which are part of the Atlas system. If desirable, the capsule retro rockets can be used to produce a more rapid separation after staging.

2.4.1   The following requirements apply to the escape system.

2.4.1.1   The occupant shall remain within the capsule, and escape shall be accomplished by the firing of an escape rocket using solid propellants. In event of an abort, provisions shall be made for a thrust cut-off on the booster rocket.

2.4.1.2 The minimum separation distance after one (1) second from escape rocket firing shall be two hundred fifty (250) feet at ground launch.

2.4.1.3 In an escape from the ground launching pad, the maximum altitude achieved shall be greater than twenty five hundred (2,500) feet.

2.4.1.4 Up to booster rocket staging, the capsule shall accelerate to a minimum velocity lateral to the plane of the trajectory of thirty (30) feet per second in one (1) second during an escape maneuver.

[5]

2.4.1.5 During the firing of the escape rocket and until the capsule decelerates to low dynamic pressure, the capsule shall be aerodynamically stable and shall trim in the same attitude as normally exists in flight when mounted on the booster rocket. During the escape when the dynamic pressure approaches zero, the capsule configuration shall be altered (if necessary) in a manner to provide an aerodynamically stable trim condition in the normal reentry attitude.

2.4.1.6 When the escape maneuver takes place outside the atmosphere, the capsule shall be aligned in the reentry attitude by means of the attitude control system to be specified in section 4.2.

2.4.1.7 Special consideration shall be given to selecting a launch trajectory that will minimize deceleration and heating during entry from an aborted mission.

2.4.1.8 Consideration shall be given to providing a system which will detect [illegible] during launch and which will initiate the abort in [illegible] of this system, the independence of the booster guidance system shall be preserved.

[6]

3.   CONFIGURATION

3.1   <u>Configuration requirements.</u> – The configuration selected for the capsule shall fulfill the following requirements:

3.1.1   The external configuration shall have an extremely blunt forebody in the entry attitude.

3.1.2   The countours of the forebody shall be such as to provide the maximum practical wave drag and uniform surface heating consistent with other requirements.

3.1.3   The afterbody shape shall be dictated by requirement for subsonic stability, adequate volume, and low heating as well as requirements for parachute storage and attachment of the escape rocket system.

3.1.4   The overall capsule configuration at the time of entry shall be aerodynam-
        ically stable in one direction only (blunt face leading) and shall exhibit
        no tendency to tumble during entry even in recovery from extreme initial
        angles of attack.

3.1.5   Oscillatory motions of the capsule during any phase of the mission shall
        not be of a character to incapacitate or injure a human occupant. If this
        requirement cannot be met by control of the configuration shape auto-
        matic damping means may be employed.

3.1.6   The shape and internal volume of the capsule shall be amenable to cer-
        tain experiments on manned space flight such as:

3.1.6.1 Limited mobility tests (calisthenics, programmed movements, etc.).

3.1.6.2 Observation tests (external and internal).

3.1.6.3 Manual control tests (open loop and closed loop).

3.1.7   The effect of entry forebody shape on water and land impact loads shall
        be considered in the design.

3.1.8   The configuration shall be stable in the water with blunt face down and
        shall be capable of righting itself from any position.

[7]

3.2   Configuration details – A configuration meeting the requirements of
        these specifications is illustrated in figures 4, 5, and 6. An inboard profile
        of the configuration as it would appear when ready for the launch opera-
        tion is shown in figure 4. Configurations for the different phases of flight
        are shown in figures 5 and 6. [No figures included]

3.2.1   The blunt forebody of the capsule shall incorporate a beryllium heat sink.
        A heat shield of the ablation type may be considered as an alternate form
        of heat protection providing experimental data directly applicable to the
        capsule reentry is obtained which establishes to the satisfaction of the
        NASA that this form of heat shield is applicable. The capsule forebody
        shall be attached to the launch rocket system by a suitable adapter.

3.2.2   The pylon-like framework on the launch configuration (figures 5(a) and
        5(b)) [not included] shall support solid-fuel rocket motors that shall be
        used to accomplish an escape maneuver in the event of a malfunction of
        the launch rocket system. The escape motors shall be mounted on the
        pylon-like structure with enough ballast to give the launch configuration
        static stability in its mounted orientation under all flight conditions to the
        time of staging. On a normal launch, the escape motors and pylon shall
        be jettisoned by small auxiliary motors at five (5) seconds after staging of
        the launch rocket system.

3.2.3   In orbit, the capsule will have the configuration shown in figure 5(e) [not included]. The retrograde maneuver shall be accomplished by firing the spherical rocket motors mounted outside of the heat shield. These motors shall then be jettisoned and the entry phase will be made by the configuration illustrated in figure 5(f) [not included].

3.2.4   The capsule is to enter the atmosphere with the blunt face leading. The aerodynamic heating at this face would be absorbed by the heat shield. The area between this heat shield and the pressure vessel (in addition to containing carry-through structure) would contain equipment which is expendable at the time of deployment of the landing parachute, and the heat shield along with this equipment shall be jettisoned at this time. This operation will produce sizable reductions in the parachute loading and will prevent conduction of heat from the hot shield into the pressure vessel during the descent. The bottom contour of the pressure vessel shall be designed from consideration of water and land impact loads. In addition, an inflatable impact bag shall be used to absorb the shock of landing (figure 5(h)) [not included].

[8]

3.2.5   In the event of a malfunction in the launch rocket system, on the ground or in flight, the escape motors shall propel the capsule out of the danger area in the configuration shown in figure 6(b) [not included]. This configuration shall then coast in the pylon-first attitude until the dynamic pressure approaches zero. At this point, the escape rocket system shall be jettisoned and the capsule, with its new center of gravity, will be rotated by aerodynamic moments (figs. 6(c) and 6(d)) [not included] until the heat shield moves to the windward side. If the escape maneuver takes place outside the atmosphere, the rotation of the capsule to the reentry attitude shall be accomplished by means of an attitude control system of a type to be specified in 4.2. At this point, the capsule configuration is [illegible] illustrated in figure 5(f) [not included] for a normal flight. Parachute deployment and heat shield separation shall then be as programmed for a normal flight (fig. 5(g)) [not included].

[9]

4.   STABILIZATION AND CONTROL

4.1   <u>General</u> - Stabilization and the control of the capsule shall be provided in accordance with the following outline of the various phases of the primary mission.

4.1.1   <u>Launch</u>- The launch trajectory control and guidance shall be considered an integral part of the launching rocket system. This system (or systems) shall make possible the missions described in Section 2 of this specification.

4.1.2   <u>Orbit</u> – After booster burn out and separation, the capsule shall be automatically stabilized in attitude as specified in Section 4.2. An independent manual control system shall be provided as specified in Section 4.3. A passive optical instrument from which attitude information can be obtained shall also be provided as specified in Section 4.3.3.2.

4.1.3   Entry- During the period from retro firing to build up of atmospheric drag, the automatic control system shall provide attitude stabilization according to Section 4.1.2. After drag build up to 0.05 g units, all [illegible] of the automatic control system shall convert to a damper mode. The manual control system shall function throughout the entire phase.

4.2   Automatic Control System

4.2.1   Requirements

4.2.1.1   The stabilized orientation of the vehicle during orbiting and reentry prior to build up of atmospheric drag shall be such that the longitudinal axis (axis of symmetry) is in the orbital plane and normal to the local earth vertical. The blunt face of the capsule shall be leading. The capsule shall be roll stabilized so that the occupants head is up with respect to the local earth vertical.

4.2.1.2   After drag buildup to 0.05 g all channels of the stabilization system shall convert to a damper mode. The contractor shall study the desirability of imposing a low steady roll rate to reduce the impact error resulting from lift components of aerodynamic force.

4.2.1.3   The alignment described in Section 4.2.1.1 shall be attained within three (3) minutes after booster separation is achieved and maintained continuously throughout the orbiting phase and reentry prior to drag buildup except under the conditions described in Sections 4.2.1.5 and 4.3.

4.2.1.4   The accuracy of the stabilization system shall be within plus or minus five (5) degrees about each of the three (3) axis except under the conditions described in Section 4.2.1.5.

[10]

4.2.1.5   Immediately before and during firing of the retrograde rocket the capsule alignment shall be maintained to within plus or minus one (1) degree of the orientation specified in 4.2.1.1. The contractor shall study the desirability of controlling the pitch attitude of the capsule during firing of the retrograde rocket to the value which has minimum [illegible] to attitude error.

4.2.1.6   The specifications given in Sections 4.2.1.3, 1.4, 1.5, and 1.6 may be relaxed if properly justified by the contractor. Consideration shall be given to the limits [illegible] for emergency firing of the retrograde rocket.

4.2.1.7   A study shall be made to determine the propellant utilization during the mission both for automatic and manual control. The expenditure of propellant in limit cycle oscillations shall be minimized by the design of the control system. The use of the deadband and an impulse chain closely matching the velocity perturbation are examples of such design techniques.

4.2.2   Reaction Controls

4.2.2.1   For attitude control of the capsule, consideration shall be given to a dual-mode system consisting of a high-torque mode and a low torque mode of operation.

4.2.2.2   The high torque mode shall employ reaction jets for free-axis control and shall operate during the following periods of high torque demand: (a) Damping of residual motion of the capsule after booster burnout and separation, (b) Stabilization during the firings of the retrograde rockets, (c) Damping during entry, (d) Periods of high torque requirement in the event that the low torque system becomes saturated.

4.2.2.3   The low torque mode shall employ reaction jets or reaction wheels for three-axis control during the orbiting phase and entry phase prior to drag buildup to stabilize the capsule against both external and internal disturbances.

4.2.2.4   The reaction jets shall be so situated that no net velocity change will be given to the capsule as a result of applying control torque.

4.2.2.5   The maximum disturbance torque for the high torque mode of operation may be assumed to be that resulting from firing of the retro rocket specified in Section 4.4. It may be assumed that the pilot will be in the fully restrained condition during the retrograde firing.

[11]

4.2.2.6   High reliability shall be provided in the reaction control designs. Consideration shall be given to the use of redundancy in the automatic system and in addition, the advantage the manual control system on a safeguard against failures shall be determined.

4.2.3   Attitude Sensing

4.2.3.1   Consideration shall be given to roll and pitch attitude sensing accomplished with a horizon scan system, and yaw sensing obtained using rate gyros to determine the direction of orbital precessional rate of the attitude stabilized capsule.

4.2.3.2   As an alternate to the horizon scan system, the contractor shall study the feasibility of utilizing a stable platform with appropriate programming for this purpose. If such a system is proposed, it may be assumed that the pilot, using an optical device described in Section 4.3, can erect the platform to the alignment specified in Section 4.2.1.1, but it shall be a requirement that the safety of the mission shall not be jeopardized in the event the pilot is unable to perform this function.

4.3   Manual Control System

4.3.1   General- The manual control system shall afford the pilot means of con-
         trolling the attitude of the capsule and enable him to achieve a safe re-
         entry in the event of an emergency. The manual control system shall
         meet the following requirements.

4.3.2   Reaction Controls

4.3.2.1 Three-axis control of the capsule shall be achieved from a small
         controller(s) located so it is readily accessible from the pilot's normal
         restrained position.

4.3.2.2 A mechanical linkage shall connect the controller(s) to mechanical valves
         which control the flow of reaction jets. The reaction jets and all compo-
         nents of the manual control shall be independent of the automatic con-
         trol system.

4.3.2.3 The manual control jets shall be capable of overcoming the distur-
         bance torque resulting from firing the retrograde rockets as specified
         in Section 4.4.

4.3.2.4 Adequate safeguard shall be provided to prevent inadvertent operation
         of the manual controls. Positive action shall be required of the pilot to
         activate the manual control and de-activate the automatic control when
         he wishes to change the attitude of the capsule.
[12]

4.3.3   Attitude presentation

4.3.3.1 A display of capsule attitude shall be presented to the pilot to provide a
         reference from which he will initiate manual control action.

4.3.3.2 An optical system which gives an unobstructed view of the earth when
         the capsule is stabilized in orbit (as described in section 4.2.1.1) shall
         be conveniently located in the pilot's field of vision when in his normal
         restrained position.

4.3.3.3 The optical system specified in 4.3.1 shall have features which will allow
         the pilot to derive capsule attitude information within sufficient accuracy
         to enable him to level the capsule within 2 degrees of the orbit attitude
         specified in 4.2.1.1.

4.4   Retrograde rocket system

4.4.1   Description – The entry shall be initiated by the firing of a retrorocket
         system incorporating a cluster of (3) three solid-propellant rockets all of
         which shall be fired simultaneously.

4.4.1.2   The retrorockets shall be mounted external to the heat shield and shall be jettisoned after firing.

4.4.2   Requirements

4.4.2.1   The magnitude of the retro-impulse shall produce a velocity decrement of five hundred (500) feet per second.

4.4.2.2   A study shall be made to determine environmental protection for the retrorockets and adequate protection shall be incorporated in the design.

4.4.3   Method of firing

4.4.3.1   The retrograde rockets shall be fired upon signal from a timer device carried on board. The timer shall be set at launch and reset periodically by command link from ground control.

4.4.3.2   Under emergency conditions, the pilot shall be able to fire the retrograde rockets. Safeguards shall be provided to prevent inadvertent firing. The pilot shall be able to fire the individual rockets simultaneously or individually through use of redundant circuits.

[13]

5.   STRUCTURAL DESIGN

5.1   Design loading and heating requirements

5.1.1   General scope- the requirements of this specification apply to the following:

5.1.1.1   The strength and rigidity of the structure of the capsule and related components which include surfaces and supports provided for reacting aerodynamic, hydrodynamic, and inertial forces.

5.1.1.2   The strength of any control systems and their supporting structure that are provided for use during the launch, orbit, entry, or aborted mission phase including such items as retrorockets, escape rockets, attitude control rockets, and parachutes.

5.1.1.3   The strength of fittings attached to the capsule for the purpose of transmitting forces to the structure.

5.1.2   General Loads requirements

5.1.2.1   Ultimate factor of safety – In lieu of an ultimate factor of safety, design may be based on a specified probability of destructive failure based on the design mission and specified deviations from the design mission.

5.1.2.2   Ultimate strength – Failure shall not occur under design ultimate loads. Excessive leakage of the pressure capsule under ultimate load is considered as a failure.

5.1.2.3   Temperature – The effects of the temperature on loading conditions and allowable stresses shall be considered where thermal effects are significant.

5.1.3   Loading types - The following types of loads are to be considered for all loading conditions:

5.1.3.1   Aerodynamic Loads

5.1.3.1.1   Maneuver (static – dynamic)

5.1.3.1.2   Gust

5.1.3.1.3   Wind shear

5.1.3.1.4   Buffeting

5.1.3.1.5   Flutter

5.1.3.2   Inertial loads

[14]

5.1.3.3   Impact loads (water and land)

5.1.3.4   Loads or stresses induced by vibration including noise effects

5.1.3.5   Loads or stresses induced by heating

5.1.4   Loading conditions - The following trajectory phases must be examined for loading conditions.

5.1.4.1   Ground handling – The effect of all ground handling conditions must be considered such as the strength of fittings attached to the capsule for purpose of transmitting handling loads to the capsule.

5.1.4.2   In-flight conditions

5.1.3.2.1 General – Air loads and inertial loads for all phases of the mission shall be associated with the design trajectories with deviations from the design trajectories to be specified by the proved statistical reliability of the propulsion and control systems.

In addition, certain specified conditions of malfunction of the propulsion and control systems shall be considered specified. The structural weight penalties associated with these malfunctions shall be assessed.

Consideration should also be given to the penalties in mission profile caused by structural weight increases due to malfunction of the propulsion or control system. The mission profile parameters for which a mission will be aborted rather than considered for design shall be designated by the limitations given in 2.2 of this specification for the primary mission.

5.1.4.2.2 Launch phase - Loading conditions shall be considered as indicated in 5.1 of this specification for all phases of launch trajectories including capsule separation.

5.1.4.2.3 Aborted mission - The possibility of an aborted mission during all phases of the launching operation and trajectory shall be considered; however, aborted trajectories which would result in axial accelerations greater than twenty-five (25) g need not be considered. (Safety features will, if necessary, include means for anticipating unsafe launch trajectories so that an abort maneuver can be accomplished to keep the g level below twenty-five (25).)

5.1.4.2.4 Orbital phase – The following effect should be considered: Possibility of meteorite damage – The probability of penetration of the pressure capsule by meteorites such that the pressure loss would prove fatal shall be less than 0.001 for a twenty-eight (28) period.

[15]

5.1.4.2.5 Entry – The loading conditions for entry are specified by the design trajectory with deviations as indicated in 5.1 of this specification. Consideration should also be given to the reactions of the retrorocket.

5.1.4.2.6 Parachute deployment – The loads on the capsule, parachute, and related equipment shall be considered for entry and aborted mission conditions as given in 2.2, 2.4, and [illegible] of this specification.

5.1.4.2   Landing – Consideration shall be given to impact loads for water and land impact conditions.

(a) Water - Consideration shall be given to water impact loads in rough water as well as calm water. The capsule design must be such that the buoyancy and water stability is not affected by impact.

(b) Land - Consideration shall be given to emergency impact on land surfaces. The capsule design must be such that the human occupant will survive without injuries severe enough to prevent his own escape from the capsule.

5.1.5   Loads calculations - The loads on the structure and distribution of air and water loads used in design shall be those determined by the use of acceptable analytical methods and with the use of experimental data which are demonstrated to be applicable. The applicable temperature,

Mach number, and Reynolds number effects must be included for the existing flow regime.

5.2   Assumed methods of construction for preliminary design – For the purpose of a feasibility study, a type of construction has been assumed which is compatible with the environment of anticipated vehicle trajectories. The principle components are a pressure capsule, external heat and micrometeorite shielding, and [illegible] layers of heat and noise insulation. With this arrangement, integrity of the pressure capsule structure and control of the internal environment can be maintained during widely varying external environmental conditions. A summary of major design requirements for each of these components and brief descriptions of possible structural solutions are given in the following sub-sections.

5.2.1   Pressure capsule –A construction is required capable of sustaining internal pressures up to fifteen (15) psi with negligible air leakage after being subjected to the vibratory and sound pressure loadings associated with launch. It must also withstand collapsing pressures up to two (2) psi to withstand a blast wave from booster failure, and be vented to preclude the possibility [16] of greater collapsing pressures during a normal mission. The capsule must be designed to withstand rigid body accelerations of twenty-five (25) g axially and four (4) g laterally corresponding to maximum which might be encountered during launch and entry. The trapped atmospheric pressure may be utilized to enhance structural stability and strength during the launch phase, but structural integrity during all entry phases shall not depend upon internal pressure for stabilization. The resulting design shall not be vulnerable to explosive decompression if punctured. The capsule must be leak resistant after a water impact loading of approximately fifteen (15) g's.

The capsule may be divided into three main sections for descriptive purposes; a bottom which supports the internal equipment and which will be subject to a water or earth impact, a mid-section designed to accommodate an entrance hatch, viewing ports, and a top dome designed to accommodate parachute attachments, and mounts for the escape rocket system. Each of these sections may experience somewhat different temperature time histories, with a possible temperature difference between sections of three hundred (300) degrees Fahrenheit. The maximum temperature of each part shall be held to six hundred (600) degrees Fahrenheit through use of heat shielding. Stresses in the capsule induced by differences in thermal expansion between the capsule and its external heat shielding shall be reduced to tolerable values through suitable flexibility in shield mounting.

These design requirements may be met by a shell of titanium honeycomb sandwich. A vessel of this material provides maximum strength, stiffness, and heat resistance with the least weight. A more conventional construction capable of meeting the requirements is a welded semi-monicoque shell of either titanium or stainless steel. The material shall be chosen for maximum ductility and weldability.

5.2.2   Heat and micro-meteorite shielding – An analysis of the convective heating during atmospheric entry revealed the need for heat protection for both the blunt face and afterbody of the vehicle. In addition, the expected frequency of strikes by micro-meteorites of various sizes indicated that a shield thickness equivalent to 0.010 inch of steel is desirable for protection of the underlying pressure capsule against impacts.

Stagnation heating associated with the probable range of entry angles ½ to 3 degrees, indicates duration of heating as long as 500 seconds and maximum heating rates in the range of 50 to 100 Btu/ft². A total heat input of about 8000 Btu/ft² is associated with the entry angle of ½ degrees with lesser inputs for greater angles. A beryllium heat sink appears feasible for front face heat protection. Recent tests have indicated that this type of heating input may be compatible with the behavior of some of the available ablation materials. Hence, a back up approach for protection is an ablating shield.

[17]

The front shield must be supported on the capsule bottom and/or sidewalls in a manner which permits ready disengagement at parachute deployment to expose the landing bag system. The method of support must not cause excessive stresses in the shield during capsule pressurization. For the heat sink type of shield, thermal expansion capability relative to the capsule must be provided.

Estimates of afterbody heating have led to predictions of radiation equilibrium temperatures on the side shields of one thousand and four hundred (1,400) to one thousand six hundred (1,600) degrees Fahrenheit. The total heat input is in the order of one thousand (1,000) Btu/ft². The simplest and lightest weight form of heat protection for these areas appears to be obtained with radiation shields. These shields must be flutter free and yet be free to expand thermally with respect to the capsule structure. Although they are vented, a conservative design criterion is that they be able to carry the local pressure loading. This criterion insures adequate local stiffness and increased resistance to noise fatigue. They must withstand sound pressure fluctuations caused by boundary-layer noise and booster engine noise.

[Illegible] have been made on various shield configurations and it appears that a shield constructed on a 0.010-inch thick longitudinally corrugated nickel base alloy may be satisfactory. Such a shield provides a low probability of being punctured by micrometeorites in a twenty-eight (28) hour orbital period, and with a proper corrugation depth and support spacing can meet the other design requirements.

5.2.3   <u>Heat and Acoustical Insulation</u> – The shielding arrangement previously described implies the use of insulation between the shields and the capsule structure. This insulation must be able to withstand a transient temperature pulse of fifteen hundred (1,500) degrees Fahrenheit, and not deteriorate due to vibration. Transient heating calculations show that 3/8 lb/ft² of commercially available insulations should provide the required heat protection to the capsule structure during the entry maneuver.

Heat soaked up by the structure must also be prevented form heating the capsule contents. The insulation required on the inner wall must also be effective in damping sound pressure waves. It is estimated that 1/8 lb/ft² of dual-purpose insulation should reduce the total heat transmitted to the capsule contents to twenty-five (25) Btu/ft² of wall area during entry. The combination of two metal walls and two insulation layers should be capable of providing a 30 db reduction in noise at frequencies above six hundred (600) cps.

[18]

6.   ONBOARD EQUIPMENT

6.1   Capsule environment controls

6.1.1   General

6.1.1.1   Equipment shall be provided for control of the pressure, temperature, and humidity within the capsule and within a suitable pressure suit to be worn by the occupant.

6.1.1.2   Equipment shall be provided for the supply of breathing gas for the control of the oxygen partial pressure and carbon dioxide concentration in the breathing gas.

6.1.1.3   Equipment shall be provided for the control of the oxygen partial pressure and the carbon dioxide concentration of the capsule atmosphere.

6.1.1.4   The foregoing equipment shall be as simple and passive in operation as practical.

6.1.1.5   The absorptivity and emissivity of the capsule to radiation in the infrared shall be such that the shell is basically cold and that only small heat addition is required to maintain the internal temperature limits of the capsule; however, a study of the effects of the entry temperature pulse shall be made to establish if any cooling requirements exist.

6.1.1.6   The possibility of buildup of toxic contaminations and objectionable odors in the capsule shall be evaluated and if required, provisions shall be incorporated for their removal.

6.1.1.7   Adequate drinking water and food should be provided for twenty-four (24) hour orbital period and a forty-eight (48) hour post-orbital period. The food should be of the low residue type.

6.1.1.8   Provision shall be made for the disposal and/or storage of human excretions.

6.1.1.9   Protection against failure of the capsule environmental control systems shall be achieved by incorporation of appropriate redundancies.

[19]

6.1.2   Quantitative requirements

6.1.2.1   The capsule temperature shall be maintained between fifty (50) and eighty (80) degrees Fahrenheit.

6.1.2.2   The relative humidity in the capsule shall be maintained between the limits of twenty (20) and fifty (50) percent.

6.1.2.3   The capsule pressure shall never be less than local atmospheric pressure.

6.1.2.4   The partial pressure of the oxygen supplied to the occupant of the capsule shall be maintained between one hundred and fifty (150) and three hundred (300) mm Hg in either the normal or in any emergency condition.

6.1.2.5   The carbon dioxide content of the breathing gas shall be limited to less than one (1) percent.

6.1.2.6   The environmental control systems shall be capable of maintaining the foregoing conditions for: (a) the part of the prelaunch period when the environment cannot be maintained by external supply, (b) for a space flight period of twenty-eight (28) hours, (c) for the landing and recovery period of twelve (12) hours. The last condition can be waived if it can be demonstrated that satisfactory ventilation to the external atmosphere can be achieved in rough seas (through use of a snorkel-type apparatus, for example).

6.1.2.7   The character of the vibrations and the acoustic noise within the capsule shall be considered in the design and alleviation of undesirable conditions shall be provided.

6.1.2.8 Where it can be shown that any quantitative requirement herein severely restricts the design, consideration shall be given to a limited adjustment of the requirements.

6.2   Pilot support and restraint

6.2.1 A couch shall be provided which will safely and comfortably support the human occupant.

6.2.2 As a basis for the design, acceleration environments associated with the launch, the aborted launch, the entry parachute deployment, and the landing impact (land and water) shall be considered. In particular, aborted launch conditions in which peak accelerations of the order of twenty (20) g units shall be withstood by the occupant without incurring serious or permanent injury.

[20]

6.2.3 The support system shall be oriented within the capsule so that the peak accelerations can be withstood without repositioning during flight.

6.2.4 The support system shall distribute the loads over as large an area on the subject as practical and as uniformly as practical (eliminate pressure points).

6.2.5 Shock absorption shall be provided in the support system for the reduction of high but short term accelerations existing under such conditions as parachute deployment and landing impact.

6.2.6 Particular attention shall be paid to the elimination of the possibility of large negative accelerations on the occupant. Such conditions are most likely to occur during asymmetric impacts with water and land surfaces.

6.2.7 The occupant shall be firmly restrained in the support system by a suitable harness that shall provide satisfactory support for the conditions of maximum accelerations in a direction to lift the occupant off the couch. Such a condition will occur after burnout of the escape rocket when the escape takes place at the maximum dynamic pressures.

6.3   Landing system

6.3.1 General

6.3.1.1 A landing system shall be employed which shall utilize two (2) independent parachute systems mounted side by side and a system of air bags for landing impact protection.

6.3.1.2 The two independent parachute systems shall be deployed sequentially, but the reserve system shall be deployed only if the primary system fails to deploy satisfactorily.

6.3.1.3   In addition to the main landing parachute, a drogue parachute for the purpose of capsule stabilization shall be deployed at an altitude of approximately seventy thousand (70,000) feet and a Mach number of one (1).

6.3.1.4   The primary landing-parachute shall be deployed at an altitude of approximately ten thousand (10,000) feet. The primary landing parachute shall be deployed by releasing the drogue parachute from the capsule in such a manner as to serve as a pilot chute. The reserve landing parachute shall be deployed by a normal pilot chute.

[21]

6.3.1.5   At deployment of the primary landing parachute, the heat shield and expendable equipment shall be jettisoned and the landing impact bag shall be inflated.

6.3.2   <u>Drogue and pilot parachutes</u>

6.3.2.1   The drogue parachute canopy shall be a FIST ribbon type and shall be capable of opening at Mach numbers up to one and one-half (1.5). This canopy shall have a diameter large enough to provide adequate dynamic stability to the capsule.

6.3.2.2   This canopy shall be built to conform to applicable military specifications.

6.3.2.3   The parachute shall incorporate a metallic coating in a manner to provide a suitable radar reflector.

6.3.2.4   The drogue parachute shall be forcibly deployed by means a of a mortar tube. The deployment bag and packed drogue chute shall be housed in this mortar tube and shall be capable of withstanding the burning powder charge resulting from firing of the mortar. The bridle between the deployment bag of the main chute and the drogue chute shall be forty-five (45) feet in length. The mortar shall have sufficient force to propel the drogue chute and bag a distance equivalent to the bridle length.

6.3.2.5   The pilot chute for the reserve landing parachute shall be of standard pilot chute construction. This parachute shall be deployed in the same manner as specified in 6.3.2.4. To aid deployment, lead shot may be sewn in at the apex. There shall be a forty-five (45) foot bridle between the deployment bag and the pilot chute.

6.3.3   <u>Main landing parachutes</u> – The two main parachutes shall be of equal size and shall be an extended skirt design (similar to Pioneer Parachute Co. design drawing 1.425). Each of these parachutes shall be a proven type having previously been flight tested under conditions representative of the present application. The parachute shall be constructed to withstand the shock loads of opening at twenty-thousand (20,000) feet.

6.3.3.1   The gore colors shall be natural and international orange alternately arranged.

6.3.3.2   The main canopy and risers shall be packaged in a deployment bag. The main parachute deployment bag shall conform to the interior of the parachute canister.

[22]

6.3.3.3   Actuation of deployment of the drogue chute shall be by reliable and proven barometric switches. Each switch on each chute shall be independent of the other although the secondary chute firing sequence should be arranged such that the primary chute is jettisoned prior to actuating the secondary chute. However, if the primary chute fails to jettison, this should not prevent the secondary chute actuation.

6.3.3.4   Provision should be made for manual override of the automatic system should it fail.

6.3.3.5   Provision shall be made for satisfactory operation of the chutes in case of abort.

6.3.3.6   Provision shall be made for release of the parachutes after impact.

6.3.4   Landing impact bag

6.3.4.1   The landing impact bags shall be constructed of an inflatable material and shall be located behind the heat shield in the deflated condition. On separation of the heat shield, these bags shall be inflated.

6.3.4.2   The bags shall be designed so they will deflate on impact under a constant predetermined load.

6.3.4.3   The bags shall be constructed and located in such a manner that they shall be effective under conditions of drift, parachute oscillation, and uneven landing terrain.

6.3.5   Helicopter pickup provisions – Provision shall be made for a helicopter pickup of the capsule after landing. An attachment point shall be located at approximately the parachute attachment point. Auxiliary attachment points shall also be provided just above the capsule water line.

6.4   Cockpit layout

6.4.1   The contractor shall submit proposed layouts of the capsule interior to the contracting agency for approval. In addition to the environment equipment specified in section 6.1, these layouts shall show the location and approximate appearance of all pilot-actuated controls, instruments, and warning devices.

6.4.2   Consideration shall be given to the restrictions imposed on the pilot by the restraining harness specified in Section 6.2.7 and by acceleration forces in the selection of location and method of actuating all pilot-operated controls and in the grouping and placement of instruments and warning devices so as to provide an optimum display of information.

6.4.3   The contractor shall submit a list of all instruments, pilot actuating devices and warning devices to be displayed to the pilot to the contracting agency for approval. This list shall include those instruments specified or described in Section 7.

6.4.4   Consideration shall be given to the location and operation of the optical instrument for display of capsule attitude and navigational information specified in Section 4.3.3. Consideration shall also be given to a means of displaying capsule attitude information to the pilot during the launch and entry period where the optical presentation may be inadequate.

6.5   Communications. –

6.5.1   This specification is intended to include only the vehicle systems. However, these systems must be completely compatible with the ground station complex. It is intended that wherever practicable the systems of telemetry, tracking, and voice communications now existing will be used.

6.5.2   List of communications systems – The following systems of communications will be required aboard the vehicle:

Two-way voice communication

Command receiver from ground to vehicle

Telemetry from vehicle to ground
Radio tracking beacon (108 megacycles)

Rescue beacons (HF and UHF) and other recovery aids.

S- and X- band beacons for GE Guidance System, with retro-rocket firing command system

C-band radar tracking beacon

Flashing lights, for tracking

[24]

6.5.2.1   The two-way voice communications system will utilize frequencies in both the HF and UHF bands. In the event of failure, a HF-UHF transceiver normally intended for use during the recovery phase may be employed at any time.

6.5.2.2 Two command receivers will be operated continuously on VHF to accept coded commands from ground stations. Verification of the reception of the commands will be transmitted via telemetry. The command receivers will be capable of accepting and decoding retrograde rocket firing commands. Also, it will be used to turn on the telemetry system.

6.5.2.3 Initial guidance and orbit insertion will be accomplished through utilization of the GE Guidance system. Additional tracking data will be obtained from FPS-16 radars, from the 108 Megacycle Minitrack complex and other radio tracking devices, and from visual observations.

6.5.2.4 The 108 megacycle-tracking beacon will have an output of not less than 0.10 watts, and will have frequency stability commensurate with Doppler measuring techniques.

6.5.2.5 The C-band radar tracking beacon is to be compatible with the FPS-16 radar equipment, and will have an output peak power of at least 100 watts. The beacon receiver shall have the capability of triggering the beacon at line-of-sight ranges up to 1000 statute miles.

6.5.2.6 Consideration should be given to the installation of high-intensity flashing lights to aid ground observers in sighting the vehicle during dark phases of the orbit.

6.5.3 <u>Antennas</u> – Antennas will be provided for all systems – voice communications, telemetry, tracking, guidance, command, and rescue. Antennas for each system will provide maximum coverage for each phase of the mission. Design will be simplified somewhat by the vehicle stabilization, in that coverage is required only for one hemisphere, during the orbiting phase. Recovery system antenna will protrude from the upper part of the capsule in such a manner to prevent loss of signal from water or salt spray. Multiplexers will be utilized where necessary to limit the number of antennas. Early developmental flights will determine vehicle skin temperatures, enabling more precise antenna design. This will aid in decisions as to types of antennas.

6.5.4 <u>Recovery</u> – The tracking of the vehicle shall be facilitated, during the landing phase, by the ejection of radar chaff at the opening of the drogue chute.

[25]

The vehicle shall contain a suitable small rescue beacon to facilitate air search. It shall transmit suitable signals on 8.364 and 243.0 megacycles and have a range of at least two hundred (200). In the case of the low frequency signal, a thousand (1000) mile range would be desirable. It shall have self-contained batteries suitable for at least twenty-four (24) hours operation.

A high-intensity flashing light system operating from self-contained batteries and automatically starting upon landing shall be provided with provision for twenty-four (24) hours operation. The lights shall be suitably mounted for maximum sighting distance.

A light weight transceiver shall be used for voice communication backup during the recovery phase. It shall have self-contained batteries, have a range of approximately 200 miles, be suitable for twenty-four (24) hours operation and have a suitable antenna on the vehicle.

So-Far bombs which will automatically fire at reasonable time interval after landing shall be used so that signals received at suitable stations will aid in locating the vehicle.

Dye marker shall be deployed upon landing to aid in the visual location of the vehicle during the search phase.

6.6   Navigational Aids

6.6.1   The pilot shall be provided with a means of navigation. To provide a back up to the ground range tracking facilities, in the event of failure of the capsule tracking beacons or other contingency that would exceed the capability of the ground range system. This operation would entail the determination of altitude, velocity position and local earth vertical, and ground track over the earth.

6.6.2   The optical periscope, or equivalent specified in Section 4.3.3 and a chronometer, shall be provided to fulfill the above requirements. Also manual aids in the form of simplified tables or displays shall be provided to facilitate navigational problems based on observations of earth, sun, moon, or stars. The periscope will be used to indicate the misalignment of the longitudinal axis of the capsule with respect to the flight path over the earth. In the case of failure of the stabilization system, it will allow the pilot to manually align the capsule with the flight path prior to firing the retro rockets.

6.7   Power Supply

6.7.1   The main supply shall be of the silver-zinc type. It shall be suitable for providing the capsules various power-requirements for the twenty-eight (28) hour orbital flights plus the twelve (12) hour recovery phase.

6.7.2   Consideration should be given to the use of an emergency silver-zinc battery to operate vital equipment during the reentry phase in case of failure of the main power supply.

[26]

7.   INSTRUMENTATION

7.1   <u>General</u> - In the design of the various instrumentation components, reliability, weight, and power requirements are to be considered of greatest importance.

7.1.1   The data to be measured are separated into the following categories:

   a)   Aero-Medical Measurements

   b)   Internal Environment

   c)   Vehicle Measurements

   d)   Operational Measurements

   e)   Scientific

7.2   <u>List of Instrumentation</u>.- The following detailed list of required measurements includes the data required on the first orbital manned flight and does not reflect the requirements for the unmanned flight tests. This list is to be considered only tentative and will be altered in accordance with the current needs of the project.

7.2.1

Aero Medical	Pilots Ind.	T.M.	On-Board Recording
Electro Cardiogram	x	x	
Respiratory rate and depth	x	x	
Suit, Pressure	x	x	x
Body Temperature	x	x	
Motion Picture of Pilot	x		
Voice Recording	x	x	
Alarm (May Day)	x	x	x
Mental Activity and Phys.			
Coordination		x	x

7.2.2

Capsule Environment	Pilots Ind.	T.M.	On-Board Recording
O_2 Partial Pressure			
(omit if single gas system)	x	x	x
CO_2 Partial Pressure	x	x	x
O_2 Flow Rate	x		x
CO_2 Filter Status	x		x

O2 Reserve	x	x	x
Cabin Pressure	x	x	x
Air Temperature	x		x
Humidity			x
Motion Pictures Inst. Panel			x
Noise Level			x
Vibration			x

7.2.3 | Vehicle Measurements | Pilots Ind. | T.M. | On-Board Recording |

(1 long.)			
Acc. -3 lin	x	x	x
Time	x	x	x
Q	x	x	x
Static Pressure	x		x
Attitude –3 from Stab Sensors	x	x	x
Structural Temperatures	x	x	x
Pilot Control/ Motions 3			CPT/x
Stabilizer Control /Motions 3			CPT/x

7.2.4   Operational Measurements

Power Supply Voltage	x	x	x
Sequence of Events (Chute, Retro-Sep., etc.)			x
[28]			
Failure Signals for System	x	x	x
Reaction Gas Supply Pressure	x	x	x

7.2.5   Scientific Observations and
Photographic Measurements

Cosmic Radiation	x
Meteorite Impacts	x
Earth and Sky Cameras	x

7.3   Recording – Four methods of data recording shall be employed as follows:

On-board data recording,

Telemetering to ground recorders,

On-board tape recording of voice,

Photographic recording of pilot and instrument panel.

7.3.1   General – It is evident, in the detailed instrument listing, that as many
as three different systems are frequently used to record the output of a
single data sensor or pickup. As it is not desirable from the standpoint
of weight and power to use separate pickups for each system, a satisfactory isolation technique must be employed to avoid cross talk and interference between the several systems being fed from a common pickup.
Where this is not feasible, duplicate pickups may be employed.

Provision shall be made for pre-launch check-out of all the instrument
and communication systems. The pilot shall be provided with a suitable
interphone connection with ground personnel to assist in this check-out
procedure.

[29]

7.3.2   On-Board Data Recording – The on-board recorder shall handle the
measurements as indicated in the detailed data list. This recorder shall
operate on a continuous basis during launch, reentry and abort or
emergency maneuvers. During orbit flight and after landing, the data
recorded may be programmed to operate periodically to conserve the
use of recording medium.

With the exception of EKG and respiratory rate and depth, which have
fairly high frequency content, the data may be sampled at rates as low as
once per second.

7.3.3   Telemetering to Ground Recorders – Data will be telemetered to ground
stations to provide necessary real time information concerning pilot, capsule, and life support system. In addition, telemetry will afford back-up in
the event the on-board recorded data are lost for any reason.

These data will be transmitted via radio lines operated in the 225-260
megacycle telemetry band. Reliability will be improved through the use
of two independent telemetry systems.

In addition to the two UHF links, the 108 megacycle beacon will be mod-
ulated with several channels of physiological and capsule environment
data, for continuous transmission to ground stations.

One UHF system will operate continuously, with output power of at least 0.25 watts. A second UHF system with 4 watts output power will operate only on a coded command signal from the ground. Upon interrogation, the system will operate for a period of 6 minutes, at which time it will turn itself off and be in ready status for the next interrogation.

[30]

All telemetered data will be tape recorded at the ground stations. In addition, certain physiological and other data will be displayed in real time for quick observation by engineering and medical personnel.

7.3.4   On-Board Tape Recording of Voice- The on-board recording of voice will be required continuously during launch, reentry, and abort maneuvers. During orbit and after landing, the voice recorder shall be turned on by the pilot to record comments and observations. In addition, all voice messages sent to ground stations by the pilot shall be recorded by this equipment.

7.3.5   Photographic Recording

7.3.5.1   Pilot and Instrument Panel - Two cameras are to be provided for use within the capsule. One for recording the pilot's appearance and motions and the other for recording the indication of the pilot's instruments. The frame rates may be as low as 3 fr/sec during the launch and reentry and 1 frame every 10 seconds during orbit. The lighting for cameras and general illumination shall be a duplicate system.

7.3.5.2   Photographic Recording of Earth and Sky – Cameras shall be used to record pictures of the earth with a 360 degree horizon coverage. As the line of sight at 120 mile altitude in about 2000 miles, the frame rate may be as low as 1 frame every 3 minutes to provide a 50% overlap of picture coverage.

[31]

8.   TESTING

8.1   The capsule, all subsystems, and components shall be designed to withstand the environmental stresses encountered in the missions previously outlined. Suitable simulated environmental ground tests shall be performed by the contractor to establish proof of operational reliability and performance.

8.2   A program of research and development testing of the capsule will be undertaken by the NASA. This program will include full-scale flight tests of simplified capsules. The simplified capsules are not a part of the present specifications.

8.3   The capsules supplied by the contractor will be used in a qualification test program to be conducted by the NASA. This qualification program will have as its final objective the accomplishment of the mission described in 2.1.

Document I-14

Document title: Paul E. Purser, Aeronautical Research Engineer, NASA, to Mr. R. R. Gilruth, NASA, "Procurement of Ballistic Missiles for Use as Boosters in NASA Research Leading to Manned Space Flight," 8 October 1958, with attached, "Letter of Intent to AOMC (ABMA), Draft of Technical Content," 8 October 1958.

Source: Folder 18674, NASA Historical Reference Collection, NASA History Division, NASA Headquarters, Washington DC.

When NASA was established it had virtually no in-house capability to build its own launch vehicles, and so its leaders quickly moved to procure that capability from other organizations in the federal government. This effort took two forms. First, it met with organizations that were developing ballistic missiles with the intention of acquiring some of them for its use. Second, it sought to acquire and enhance capability to develop its own launchers in the future. One was a short-term fix and the other a more long-term solution. This memorandum documents the short-term fix, reporting on a key meeting at the Army's Redstone Arsenal in Huntsville, Alabama, in which NASA and the Army agreed to acquire eight Redstone ballistic missiles for test and operations during Project Mercury for recompense of approximately $7.5 million.

NASA – Langley
October 8, 1958

MEMORANDUM For Mr. R. R. Gilruth

Subject: Procurement of ballistic missiles for use as boosters in NASA research leading to manned space flight

1. A meeting of NASA, ARPA, and Army personnel was held on October 5, 1958 at the Army Ballistic Missile Agency, Huntsville, Alabama. Personnel involved were: Mr. P. D. Purser, NASA; Mr. M. A. Faget, NASA; Mr. W. J. North, NASA; Dr. S. B. Batdorf, ARPA; Major Dunham (?), Army; Brigadier General Barclay, ABMA; Dr. W. von Braun, ABMA; Mr. Mrazek, ABMA; Mr. Carter, ABMA; Colonel Drewry, AOMC; Lieutenant Colonel James, AOMC; and other ABMA personnel. As a result of this meeting, it appears that the services and facilities of the Army Ballistic Missile Agency should be utilized in the NASA research program leading to manned space flight and that the Army is interested in participating in this program.

2. It appears that ABMA has available, in various stages of completion, some 4 to 6 REDSTONE missile boosters which probably can be used as boosters for sub-orbital reentry tests of manned capsules. Other REDSTONEs can be made available on 12 to 14 month lead time basis.

3. It is anticipated that ABMA would furnish the design, construction, and launching of the boosters and the mating of the boosters and capsules.

Certain wind-tunnel tests and some research and engineering studies on the part of ABMA will also be required.

4. It is recommended, in view of the urgency of the subject program, that a letter-of-intent based on the attached draft be issued to the Army Ordinance Missile Command as soon as feasible. The proposed letter carries a financial obligation of $2,400,000 to ABMA in order to allow their studies to begin immediately. It is anticipated that the total obligation to ABMA under this part of the program will be approximately $7,500,000.

Paul E. Purser
Aeronautical Research Engineer

Enc: Draft of letter-of-intent to AOMC
PEP. Jbs

Letter-of-Intent to AOMC (ABMA)

Draft of Technical Content

October 8, 1958

Commanding General
Army Ordinance Missile Command
Huntsville, Alabama

1. The National Aeronautics and Space Administration intends to request that the Army Ordinance Missile Command participate in a program of research leading to manned space flight. As a part of this program, it is intended that the Army Ordinance Missile Command design, construct, and launch approximately eight (8) research and development launching vehicles utilizing the REDSTONE ballistic missile booster and its associated guidance and control equipment. It is anticipated that these vehicles will be required for launching on or about the following dates:

October 1, 1959	April 1, 1960
December 1, 1959	May 1, 1960
February 1, 1960	June 1, 1960
March 1, 1960	July 1, 1960

Or at such earlier times as may appear feasible following further study and discussion between the National Aeronautics and Space Administration and the Army Ordinance Missile Command. The payloads for these vehicles will be developmental and prototype versions of habitable capsules and will be supplied by the National Aeronautics and Space Administration. Details of the payloads and missions will be determined at a later date.

2. You are requested to submit as soon as possible, for review and approval by the National Aeronautics and Space Administration, detailed develop-

ment and funding plans for the design, construction, and launching of these vehicles. These plans shall include time schedule for the work and estimates of the work to be performed at:

    a.   AOMC
    b.   By contract
    c.   By other Government agencies [2]

3.   There is hereby made available a total of $2,400,000 ($300,000 per vehicle) under appropriation symbol for obligation by the Army Ordinance Missile Command only for purposes necessary to accomplish the work specified herein. These funds are immediately available for direct obligation and for use in reimbursing the Army Ordinance Missile Command for costs incurred under this project. These funds are not available for construction of facilities. Upon approval of detailed development and financial plans, as required herein or in accordance with amendments to this request, these funds will be increased as appropriate.

4.   The Administrator of the National Aeronautics and Space Administration or his designated representatives will provide policy and technical guidance for this project. The Army Ordinance Missile Command will exercise the necessary detailed technical direction. This general relationship may be specified in greater detail at a later time if such action is necessary.

5.   The Administrator of the National Aeronautics and Space Administration will be kept informed of progress on the project by proper management, technical, and accounting reports.

6.   The disposition of equipment and materials procured in connection with this project is subject to direction of the National Aeronautics and Space Administration. All reports, manuals, charts, data, and information as may be collected or prepared in connection with the project shall be made available to the National Aeronautics and Space Administration prior to release to other agencies or individuals under procedures to be approved. Such procedures may include, in the future, simultaneous release to the NASA and to other specified agencies.

7.   AOMC shall be responsible for preserving the security of these projects in accordance with the security classification assigned and with the security regulations and procedures of the Department of the Army.

8.   Notwithstanding any other provisions of this request, AOMC shall not be bound to take any action in connection with the performance of this work that would cause the amount for which the Government will be obligated hereunder to exceed the funds made available, and the authorization of the Army Ordinance Missile Command to proceed with the performance of this work shall be limited accordingly. AOMC shall be responsible for assuring that all commitments, obligations, and [3] expenditures of the funds made available are made in accordance with the statutes and regulations governing such matters provided that whenever such regula-

tions require approval of higher authority such approvals will be obtained from or through the Administrator, National Aeronautics and Space Administration, or his designed representative.

Document I-15

Document title: S. B. Batdorf, ARPA, Memorandum for File, "Presentation of MIS Program to Dr. Glennan," 14 October 1958.

Source: Folder 18674, NASA Historical Reference Collection, NASA History Division, NASA Headquarters, Washington DC.

One of the first decisions T. Keith Glennan had to make after taking office as NASA's first Administrator was to approve Project Mercury. This decision came on 5 October 1958. Glennan wrote in his memoirs that, "I am certain that the allocation of such a program to NASA had been agreed between Dryden, Killian, and DOD before NASA was born," suggesting that the briefing to the new Administrator and his decision to support it was more of a fait accompli than anything else. But Glennan's reflection on the decision is telling. "As one looks back, it is clear that we did not know much about what we were doing," he wrote. "Yet the Mercury program was one of the best organized and managed of any I have been associated with." The decision to invest management of Project Mercury to a Space Task Group based at Langley Research Center, taken at the same time, proved equally auspicious. The hard-driving leader, Robert R. Gilruth, provided critical oversight, loyalty to NASA Headquarters, and technical competence that helped ensure success.

This document describes an early briefing to Keith Glennan about planning for a Man in Space (MIS) mission. It was written by one of the individuals who had led early planning for the mission with the Department of Defense's Advanced Research Planning Agency.

INSTITUTE FOR DEFENSE ANALYSES
ADVANCED RESEARCH PROJECTS DIVISION
Washington 25, D.C

October 14, 1958

MEMORANDUM FOR FILE

SUBJECT: Presentation of MIS [Man in Space] Program to Dr. Glennan

At the rather urgent invitation of Mr. Gilruth, I attended the presentation to Dr. Glennan of the MIS Program at NASA Headquarters, 9:00 p.m., October 7. Those present were Dr. Glennan, Dr. Dryden, Dr. Silverstein, and Messrs. Gilruth, Faget, Low, North, Crowley, and Wood.

At the beginning of the discussion, Dr. Silverstein outlined the history of the MIS Program and showed Dr. Glennan a copy of the proposed memorandum of understanding. Dr. Glennan appeared to accept all of it except the section requiring joint approval on all contracts. He felt that it would not only be clearer from a management point of view, but in addition he would vastly prefer to have ARPA contribute its money to NASA to dispose of as it sees fit. I believe he might accept the section as written as a second best solution to the problem but intends to discuss his preferred solution with Mr. Johnson.

It was brought out that the public relations problem is a particularly difficult one in this project. The possibility of firing from some place other than Canaveral was discussed but does not seem to be feasible. It was decided that the public relations aspect needs to be carefully planned right from the start, and they will probably put a man on this fulltime as soon as possible. Dr. Glennan proposes to present the MIS Program at an early meeting of the space council and possible to solicit OCB advice on the matter of handling public relations.

Dr. Glennan attaches a very great time urgency to this project and agrees with the desirability of seeking application of emergency funds of the Secretary of Defense as proposed by Mr. Johnson last week. Dr. Dryden indicated that the MIS Committee should go ahead and plan on the assumption that the money will be available regardless of the source from which it comes. [2]

The last item of business was a rather lengthy dispute as to how the program should be managed within the NASA. It was decided that Dr. Dryden's recommendation would be followed, namely that the work would be done by a task force under Gilruth, reporting to Silverstein. This task force might have most of its members at the Langley Laboratory, but the Langley management would have no hand or voice in the management of the project. Dr. Glennan appeared very pleased with the project plan and admonished the committee to put it into operation as rapidly as possible.

[Signed]
S.B. Batdorf

Copies to:
MIS Panel
Mr. Johnson
Adm. Clark
Dr. York
Mr. Gise
Mr. Godel
Mr. Smith

Document I-16

Document title: Robert R. Gilruth, Project Manager, NASA, Memorandum for Associate Director, NASA, "Space Task Group," 3 November 1958.

Source: Folder 18674, NASA Historical Reference Collection, NASA History Division, NASA Headquarters, Washington DC.

The creation of the Space Task Group based at Langley Research Center proved a critical decision for the management of Project Mercury. NASA handpicked the members of this group from among the best in the agency and placed Robert Gilruth in charge. Gilruth, perhaps more than any other NASA official, served as the godfather of human spaceflight in the U.S. Under his direction NASA successfully completed Projects Mercury, Gemini, and Apollo. His organization recruited, trained, and oversaw the astronauts and the human spaceflight program throughout the heroic age of spaceflight. Yet, his name is lesser known than many others associated with these projects. He was a contemporary on par with Wernher von Braun, and he certainly contributed as much to human spaceflight as any of his colleagues, yet his name is rarely mentioned as a key person. He is a representative of the engineering entrepreneur, a developer and manager of complex technological and organizational systems, accomplishing remarkably difficult tasks through excellent oversight of the technical, fiscal, cultural, and social reins of the effort.

This memorandum identifies the individuals selected by Gilruth as the original members of the Space Task Group. Many of the Group's original members went on to be central to the development of the U.S. human spaceflight program.

NASA - Langley
November 3, 1958

MEMORANDUM For Associate Director

Subject: Space Task Group

1.   The Administrator of NASA has directed me to organize a space task group to implement a manned satellite project. This task group will be located at the Langley Research Center but, in accordance with the instructions of the Administrator, will report directly to NASA Headquarters. In order that this project proceed with the utmost speed, it is proposed to form this space task group around a nucleus of key Langley personnel, many of whom have already worked on this project.

2.   It is request, therefore, that initially the following 36 Langley personnel be transferred to the Space Task Group:

> Anderson, Melvin S. (Structures)
> Bland, William M., Jr. (PARD)
> Bond, Aleck C. (PARD)
> Boyer, William J. (IRD)
> Chilton, Robert G (FRD)
> Donlan, Charles J. (OAD)
> Faget, Maxime A. (PARD)
> Field, Edison M. (PARD)
> Gilruth, Robert R. (OAD)
> Hammack, Jerome B. (FRD)

Hatley, Shirley (Steno.)
Heberlig, Jack C. (PARD)
Hicks, Claiborne R., Jr. (PARD)
Kehlet, Alan B. (PARD)
Kolenkiewicz, Ronald (PARD)
Kraft, Christopher C., Jr. (FRD)
Lauten, William T., Jr. (DLD)
Lee, John B. (PARD)
Livesay, Norma L. (Files)
Lowe, Nancy (Steno.)
MacDougall, George F., Jr. (Stability)
Magin, Betsy F. (PARD)
Mathews, Charles F. (PARD)
Mayer, John P. (FRD)
Muhly, William C. (Planning)
Purser, Paul E. (PARD)
Patterson, Herbert G. (PARD)
Ricker, Harry H., Jr. (IRD)
[2]
Robert, Frank C. (PARD)
Rollins, Joseph (Files)
Sartor, Ronelda F. (Fiscal)
Stearn, Jacquelyn B. (Steno.)
Taylor, Paul D. (FSRD)
Watkins, Julia R. (PARD)
Watkins, Shirley (Files)
Zimmerman, Charles H. (Stability)

[Signed]
Robert R. Gilruth
Project Manager

PEP.jbs

[Handwritten at bottom of document] To Personnel Officer for admin. This request is OK with the exception of Boyer. On Buckley's recommendation substitute Kyle for Boyer.
    FL Thompson
    Acting Director 11-4-58]

Document I-17

Document title: Abe Silverstein, Director of Space Flight Development, NASA, Memorandum for Administrator, NASA, "Code Name 'Project Mercury' for Manned Satellite Project," 26 November 1958.

Source: NASA Collection, University of Clear Lake Library, Clear Lake, Texas.

Document I-18

Document title: George M. Low, NASA, Memorandum for Dr. Silverstein, NASA, "Change of Manned Satellite Project name from "Project Mercury" to "Project Astronaut," 12 December 1958.

Source: National Archives and Record Administration, Forth Worth, Texas.

In the fall of 1958, NASA was preparing to implement its initial human spaceflight effort. The space agency decided to name the effort "Project Mercury," after the messenger of the gods in ancient Roman mythology. The symbolic associations of this name appealed to Abe Silverstein, NASA's Director of Space Flight Development. On December 1958, the 55th anniversary of the first flight of the Wright brothers at Kitty Hawk, T. Keith Glennan announced the name for the first time. A last-minute attempt by the head of the Space Task Group, Robert Gilruth, to change the name to "Project Astronaut" was not successful.

Document I-17

Washington, D.C.
November 26, 1958

MEMORANDUM For Administrator, NASA

Subject: Code name "Project Mercury" for Manned Satellite Project.
1.   Considerable confusion exists in the press and in public discussions regarding the Manned Satellite Project because of the similarity of this program with other Man-in-Space proposals.

2.   At the last meeting of the Manned Satellite Panel it was suggested that the Manned Satellite Project be referred to as Project Mercury.

3.   It is recommended that the code name Project Mercury be adopted.

[Signed]
Abe Silverstein
Director of Space Flight Development

Cc: Robert Gilruth, Langley Task Group
Dr. S.B. Batdorf, ARPA

Document I-18

Washington, D.C.
December 12, 1958

MEMORANDUM For Dr. Silverstein

Subject: Change of Manned Satellite Project name from "Project Mercury" to "Project Astronaut"

1. Bob Gilruth feels that "Project Astronaut" is a far more suitable name for the Manned Satellite Project than "Project Mercury."
2. If you agree, this should be brought to Dr. Glennan's attention immediately. Present plans call for Dr. Glennan to refer to "Project Mercury" in his policy speech on December 17.

George M. Low

Low:lgs

Document I-19

Document title: George M. Low, Program Chief, Manned Space Flight, NASA, Memorandum for Administrator, NASA, "Status Report No. 1, Manned Satellite Project," 9 December 1958.

Source: National Archives and Record Administration, Fort Worth, Texas.

If there was any one person at NASA who obsessed over the details of each of the human space-flight projects of the agency's heroic years it was George M. Low, in 1958 NASA's Manned Spaceflight Program Chief. Low had been born in Vienna, Austria, and came to the U.S. in 1940. After completing his B.S. in aeronautical engineering he joined NACA in 1949 at Lewis Flight Propulsion Laboratory. He also held important positions in Gemini and Apollo before serving as Deputy Administrator of NASA in 1969 to 1976 and then as acting administrator from 1970 to 1971. Low prepared notes at least weekly on all of the initiatives for which he was responsible. They were always both comprehensive and candid. He heavily focused on the technical issues and, until he came to NASA Headquarters in late 1969, rarely commented on policy, but his regular memoranda on these various programs represent an historical treasure trove. This status report is an example of Low's approach to documentation. He ensured that his superiors understood the key issues at play, but he also had a concern for history by leaving these detailed commentaries, to which he often appended key documents.

Washington, D.C.
December 9, 1958

MEMORANDUM For Administrator

Subject: Status Report No. 1,
Manned Satellite Project

1. This is the first of a series of weekly or biweekly status reports on the Manned Satellite Project. In general, these reports will consist of short statements concerning only the progress made during the reporting period. For completeness, however, this first report will contain a summary of the progress made since the formal inception of the project.

2.   Capsule and Subsystems

    a. Preliminary specifications mailed to prospective bidders on October 23, 1958.

    b. Bidders conference held at Langley Field on November 7, 1958. About 38 firms represented.

    c. Nineteen firms indicated by November 14 that they plan to prepare proposals. Final specifications sent to these firms on November 17.

    d. Proposals must be received by December 11.

    e. Technical assessment will be started by members of the Space Task Group on December 12. Mr. Charles Zimmerman heads the technical assessment team. Concurrently, cost and management assessment will be carried out by Mr. A. E. Siepert's office.

    f. Source Selection Board will meet on December 29. Membership: Messrs. Gilruth, Wyatt and Low, and representatives from the Offices of the General Counsel and of the Director of Business Administration; ARPA has been invited to participate in a non-voting capacity.

3.   Booster Procurement

    a. Little Joe. This booster consists of a cluster of four Sergeant rockets; it is capable of imparting a [2] velocity of 6000 ft/sec to a full-scale capsule, and will be launched from Wallops. A contractor is now being selected.

    b. Redstone. The Redstone vehicle is also capable of achieving a velocity of 6000 ft/sec with a full-scale capsule; it will be used for manned short-range ballistic flights. ABMA has submitted a tentative proposal for 8 boosters at a total cost of $13.179 million. The Redstones will be ordered as soon as a firm proposal is received.

    c. Thor or Jupiter. Either vehicle has a capability of boosting a full-scale capsule to about 16000 ft/sec. A tentative decision to purchase Jupiter was made on December 8; this decision will be firm if proposed flight test schedules can be met. Probable cost for 3 vehicles: $5.634 million.

    d. Atlas. This booster will be used both for sub-orbital and orbital flights. Funds have been transferred to AFBMD for one Atlas C and nine Atlas D boosters.

4. Flight Test Operations

      a. Several full-scale dummy capsules have been dropped from a C-130 airplane. Purpose: to check subsonic stability and parachute deployment. Initial results: parachute deployment is satisfactory.
      b. The first Atlas Flight (Atlas C) is scheduled for June or July 1959. Primary purpose: to check ablation heat shield.

5. Pilot Selection: The aero medical group at Langley (Maj. White, USAF, Lt. Voas, Navy, and Capt. Augerson, Army), have set up a tentative procedure for pilot selection and training. Briefly, the plan calls for a preliminary meeting on [3] December 22 with representatives from the services and industry. These representatives will "nominate" a pool of 150 men by January 21. From this pool, 36 candidates will be selected by February 15. A series of physical and other tests will eliminate all but 12 by the middle of March; these 12 men will then go through a nine months training and qualification program. Six men are finally expected to qualify.

George M. Low
Program Chief
Manned Space Flight

Cc: Dr. Dryden
Dr. Silverstein
Mr. Sanders

Low:lgs

Document I-20

Document Title: Invitation to Apply for Position of Research Astronaut-Candidate, NASA Project A, Announcement No. 1, 22 December 1958.

Source: Folder 18674, NASA Historical Reference Collection, History Division, NASA Headquarters, Washington, DC.

In November 1958 aeromedical consultants working for the Space Task Group at Langley worked out preliminary procedures for the selection of astronauts to pilot the Mercury spacecraft. Their proposal involved meetings with industry and the military services which would result in the nomination of 150 men. This would be narrowed down to 36 to undergo extensive physical and psychological testing. Ultimately, 12 would be selected to undergo training and qualification, of which only 6 were expected to fly.

This plan led Charles Donlan, Technical Assistant to the Director of Langley; Warren J. North, a former NACA test pilot and head of the office of Manned Satellite; and Allen O. Gamble, a psychologist detailed from the National Science Foundation, to draft job specifications for applicants for the astronaut program. Although carefully drawn up, this plan was abandoned when President Eisenhower (during the Christmas holiday) decided that only military test pilots should be allowed to apply. This eliminated the option of including civilians in the civilian manned space program, but greatly simplified the selection process.

Even though NASA Administrator T. Keith Glennan had announced on 17 December that the program would be called "Project Mercury," this document still uses the name preferred by the Space Task Group, "Project Astronaut."

NATIONAL AERONAUTICS AND SPACE ADMINISTRATION
Washington 25, D.C.

NASA Project A
Announcement No. 1
December 22, 1958

Invitation to apply for Position of
RESEARCH ASTRONAUT-CANDIDATE
with minimum starting salary range of $8, 330
to $12, 770 (GS-12 to GS-15) depending
upon qualifications
at the NASA Langley Research Center
Langley Field, Virginia

I. DESCRIPTION OF PROJECT ASTRONAUT

The Manned Satellite Project is being managed and directed by NASA. The objectives of the project are to achieve, at the earliest practicable date, orbital flight and successful recovery of a manned satellite; and to investigate the capabilities of man in a space environment. To accomplish these objectives, a re-entry vehicle of the ballistic type has been selected. This vehicle not only represents the simplest and most reliable configuration, but has the additional advantage of being sufficiently light, so that it can be fitted on an essentially unmodified ICBM booster. The satellite will have the capability of remaining in orbit for 24 hours, although early flights are planned for only one or two orbits around the earth.

Although the entire satellite operation will be possible, in the early phases, without the presence of man, the astronaut will play an important role during the flight. He will contribute to the reliability of the system by monitoring the cabin environment, and by making necessary adjustments. He will have continuous displays of his position and attitude and other instrument readings, and will have the capability of operating the reaction controls, and of initiating the descent from orbit. He will contribute to the operation of the communications system. In addition, the astronaut will make research observations that cannot be made by instruments; these include physiological, astronomical, and meteorological observations.

Orbital flight will be accomplished after a logical buildup of capabilities. For example, full-scale capsules will be flown on short and medium range ballistic flights, before orbital flights will be attempted. Maximum effort will be placed on the design and development of a reliable safety system. The manned phases of the flight will also undergo a gradual increase in scope, just as is common practice in the development of a new research aircraft. [2]

II. DUTIES OF RESEARCH ASTRONAUT-CANDIDATES

Research Astronaut-Candidates will follow a carefully planned program of pre-flight training and physical conditioning. They will also participate directly in the research and development phase of Project Astronaut, to help insure scientifically successful flights and the safe return of space vehicles and their occupants. The duties of Research Astronaut-Candidates fall into three major categories:

a. Through training sessions and prescribed reading of technical reports, they will acquire specialized knowledge of the equipment, operations, and scientific tests involved in manned space flight. They will gain knowledge of the concepts and equipment developed by others and, as their knowledge and experience develops, they will contribute their thinking toward insuring maximum success of the planned flights.

b. They will make tests and act as observers-under-test in experimental investigations designed (1) to develop proficiency and confidence under peculiar conditions such as weightlessness and high accelerations; (2) to enable more accurate evaluation of their physical, mental, and emotional fitness to continue the program; and (3) to help elicit the knowledge necessary to evaluate and enable the final development of communication, display, vehicle-control, environmental-control, and other systems involved in space flight.

c. They perform special assignments in one or more of their areas of scientific or technical competence, as an adjunct to the regular programs of the research team, the research center, or NASA. These assignments may include doing research, directing or evaluating test or other programs, or doing other work which makes use of their special competencies.

Appointees who enter this research and training program will be expected to agree to remain with NASA for 3 years, including up to one year as Research Astronaut-Candidates. During the initial months final selection will be made of about half of the group to become Research Astronauts. Candidates who are not at that point designated Research Astronauts will have the option of continuing with NASA in other important capacities which require their special competence and training, without loss of salary and with other opportunities for advancement, and may remain eligible for future flights. [3]

III. QUALIFICATION REQUIREMENTS

A.   Citizenship, Sex, Age

Applicants must be citizens of the United States. They must be males who have reached their 25th birthday but not their 40th birthday on the date of filing application.

Applicants must be in excellent condition and must be less than 5 feet 11 inches in height.

B.   Basic Education

Applicants must have successfully completed a standard 4-year or longer professional curriculum in an accredited college or university leading to a bachelor's degree, with major study in one of the physical, mathematical, biological, medical, or psychological sciences or in an appropriate branch of engineering or hold a higher degree in one of these fields. Proof of education will be required (see paragraph IV-4, below).

C.   Professional Experience or Graduate Study

In addition to a degree in science or engineering or medicine, applicants must have had one of the following patterns of professional work or graduate study or any equivalent combination:

1.   Three years of work in any of the physical, mathematical, biological, or psychological sciences.

2.   Three years of technical or engineering work in a research and development program or organization.

3.   Three years of operation of aircraft or balloons or submarines, as commander, pilot, navigator, communications officer, engineer, or comparable technical position.

4.   Completion of all requirements for the Ph.D. degree in any appropriate field of science or engineering, plus 6 months of professional work.

5.   In the case of medical doctors, 6 months of clinical or research work beyond the license and internship or residency.

Preference will be given to applicants in proportion to the relatedness of their experience or graduate study to the various research and operational problems of astronautics. [4]

NASA desires to select and train a team of Astronaut-Candidates representing a variety of fields including physical and life sciences and technology.

D. Hazardous, Rigorous, and Stressful Experience

Applicants must have had a substantial and significant amount of experience which has clearly demonstrated three required characteristics: (a) willingness to accept hazards comparable to those encountered in modern research airplane flights; (b) capacity to tolerate rigorous and severe environmental conditions; and (c) ability to react adequately under conditions of stress or emergency.

These three characteristics may have been demonstrated in connection with certain professional occupations such as test pilot, crew member of experimental submarine, or arctic or antarctic explorer. Or they may have been demonstrated during wartime combat or military training. Parachute jumping or mountain climbing or deep sea diving (including with SCUBA), whether as occupation or sport, may have provided opportunities for demonstrating these characteristics, depending upon heights or depths attained, frequency and duration, temperature and other environmental conditions, and emergency episodes encountered. Or they may have been demonstrated by experience as an observer-under-test for extremes of environmental conditions such as acceleration, high or low atmospheric pressure, variations in carbon dioxide and oxygen concentration, high or low ambient temperatures, etc. Many other examples could be given. It is possible that the different characteristics may have been demonstrated by separate types of experience.

Pertinent experience which occurred prior to 1950 will not be considered. At least some of the pertinent experience must have occurred within one year preceding date of application.

Applicants must submit factual information describing the work, sport, or episodes which demonstrate possession of these three required characteristics. See paragraph 5 in next section.

IV. MATERIAL TO BE SUBMITTED

These positions are to be filled through a procedure which requires sponsorship of each candidate by a responsible organization. An indication of this sponsorship and a rating of the candidate will be made on a Nomination Form by a member of the sponsoring organization, preferably a superior well acquainted with the candidate. The Nomination Form is attached to this announcement for distribution to solicited organizations, and will be filled out by them and returned by January 12, 1959, if at all possible, to Personnel Office (Project A), NASA, Langley Field, Virginia. [5]

The following materials must be submitted by the applicant himself no later than January 26, 1959 to:

Personnel Office (Project A)
NASA
Langley Field, Virginia

1. Standard Form 57 (Application for Federal Employment). These forms will be furnished to applicants, but copies can be obtained from any U.S. Post Office or Federal agency.

2. Standard Form 86 (Security Investigation Data for Sensitive Position). This form will be furnished to applicants. Those applicants who are invited to report in person for further testing will be asked to bring with them one copy of this form completed in rough draft.

3. Standard Forms 88 (Report of Medical Examination) and 89 (Report of Medical History). These forms will be distributed to applicants. They should be completed by the applicant (paragraph 1 through 14 on S.F. 88 and all appropriate paragraphs of S.F. 89) and taken to the nearest military hospital, base, or procurement office authorized to administer flight physicals. A special letter addressed to such military installations is attached to this announcement, to be detached for use. Applicants should report for these physicals no later than January 21 in order to allow time for receipt of the forms at Langley by January 26. The examining military agencies will forward the S.F.'s 88 and 89 direct to NASA, Langley.

4. College transcript(s). Each applicant must submit a transcript (not necessarily an official copy) of his college or university record including descriptive course titles, grades and credits. These should accompany the application if possible.

5. A description of hazardous, rigorous, and stressful experiences pertinent to section D, above. This description should not exceed 2 or 3 typed pages. It must be factual (dates, events, etc.) and should be corroborated where practicable.

6. A statement concerning the pertinence of the applicant's professional or technical background to the problems of astronautical research and operations. This should not exceed one typed page.

7. A statement as to why the applicant is applying for this position. This statement should not exceed one typed page. [6]

V. SELECTION PROGRAM

On the basis of evaluations of the above-described applications and supporting material, a group of men will be invited to report to the NASA Space Task Force at Langley Field, Virginia, on February 15, 1959. For about three weeks these men will be given a variety of physical and mental tests on a competitive basis to evaluate their fitness for training for the planned space flights. This will involve trips to Washington, D.C., and other locations and will include tests with such equipment as decompression chambers and centrifuges and also aircraft flights. At the end of this competitive testing program all the candidates will return to their homes and jobs.

During the ensuing period of 2 to 3 weeks, laboratory and other test results will be evaluated and a small group of men will be finally selected to become Research Astronaut-Candidates. These men will be notified to report for duty at NASA, Langley Field, on or about April 1, 1959. Travel and moving expenses for them (and their families, if married) will be provided.

VI. APPOINTMENTS AND PAY

These appointments are to civilian positions in the National Aeronautics and Space Administration. They are excepted appointments due to the unusual nature of the duties and the selection process, but carry the benefits and protections of the U.S. Civil Service System including a high level of insurance and retirement.

Original appointments of Research Astronaut-Candidates will be to pay levels commensurate with their backgrounds of education and experience, within the pay range of $8,330 to $12,770 per year (GS-12 to GS-15).

As these men become proficient in the field, they will become eligible for Research Astronaut positions with salaries commensurate with those of the most highly skilled NASA Research Pilots and Aeronautical and Space Scientists.

Document I-21

Document title: Dr. William S. Augerson, Human Factors Branch, NASA, Memorandum for Chief, Operations Division, NASA, "Scientific Training for Pilots of Project Mercury," 27 March 1959.

Source: Folder 18674, NASA Historical Reference Collection, NASA History Division, NASA Headquarters, Washington, DC.

As the first astronauts were selected for Project Mercury from among the test pilot cadre in existence in the various military services, questions arose about the other types of skills NASA desired of those that would fly in space. Since one of the key components of Mercury was the expansion of scientific knowledge, the scientific community wanted the most qualified people possible to engage in this endeavor. In addition to the pilot training all astronauts received before coming to NASA, consensus quickly mounted to further train the Mercury astronauts to undertake scientific research. The additional scientific training required for Project Mercury was not rigorous, as demonstrated by this memorandum, but enhanced the capability of crewmembers to perform experiments on-orbit.

NASA - Space Task Group, Langley
March 27, 1959

MEMORANDUM For Chief, Operations Division

Subject: Scientific training for pilots of Project Mercury

1. It is recommended that pilots of Project Mercury be given graduate level training in areas relating to astronautics and geophysics. It is further recommended that they receive this training early in the course of the project.

2. Justification

    a. Background information in the area of astronautics is an important requisite to understanding of the environment into which these men will be traveling. It will aid them in understanding the vehicle design and the operational procedures. While some of this information can be provided by Space Task Group engineers, they will not have time to provide more than a minimum of information in this area.

    b. It has been stated officially that Project Mercury will investigate human performance in space environment. Since one of the important scientific and peaceful activities of man in space is scientific observation, simple scientific observations should be made by the astronaut. To make these observations, training will be required. The following areas are possible activities:

        i. Simple astronomical observations; that is, coronal studies.

        ii.Simple meteorological observations; that is, synoptic weather reports from visual observations and photographs.

        iii. Simple biophysical studies.

        iv. Radiation physics studies

    c. These pilots will become important "scientific ambassadors" after completing this mission, and should have a general knowledge in areas related to astronautics. This may be of additional importance in a period when other nations may ridicule our space effort as an unscientific stunt. Even in this country, there are persons who believe this project should be more than an aerodynamic flight study. [2]

    d. By giving training early, there will be less interference with the project and will provide time for individual growth along lines of personal interest.

    e. By equipping pilots with training in these areas, we may provide an extra benefit from this project in terms of useful information obtained. It is believed that their grasp of the whole project may be improved.

3.   Procedure

    a.   To condense the maximum information in the minimum time, it is recommended that a university, such as Harvard, MIT, or Penn., be asked to construct a special (if no appropriate course exists) two-three-week course in July or August in a survey of astronautics and geophysics.

    b.   It is recommended that this be further supplemented by an occasional seminar with local or visiting experts in these areas to help keep the astronauts up-to-date on current research; for example reports on data from cloud-cover satellites.

    c.   Some of the pilots may wish to work with some of the groups doing supporting research; for example, radiation studies administered by the Washington office.

    d.   It is recommended that attempts be made, while performing simulated missions, to find out what observations the men can make. It is believed that by using the synthesizing ability of the individual, good meteorological studies can be made using apparatus already in the vehicle.

4.   It is understood that there is reason for contrary opinions to the above. However, it is believed that the efficiency of the vehicle system will be such that time for scientific observations will be available (especially on 28-hour missions) and that the expense of this operation makes it desirable to obtain all the data we can from it. The training necessary to perform these tasks can be given fairly easily considering the experience and intelligence of these pilots. Even if no observations are permitted, it is believed that training in the area of astronautics and geophysics will aid in the operational accomplishment of Project Mercury.

Dr. William S. Augerson
Human Factors Branch

Document I-22

Document title: George M. Low, Program Chief, Manned Space Flight, NASA, Memorandum for Administrator, NASA, "Pilot Selection for Project Mercury," 23 April 1959.

Source: Folder 18674, NASA Historical Reference Collection, NASA History Division, NASA Headquarters, Washington, DC.

Astronaut selection became a topic of interest to many in the general public early on. Who was chosen and why? What criteria were used? Who might have been excluded from consideration, either intentionally or not? This memorandum documents an issue that arose almost simultaneously with the unveiling of the Mercury Seven astronauts in mid-April

1959. Why were there no Army or civilian pilots selected? NASA was exceptionally conscious of the interservice rivalries extant in the DOD and sought to ensure that Army personnel received consideration, even going so far as to undertake a special screening of some candidates, but in the end found none that met NASA's selection criteria. In accordance with President Eisenhower's December 1958 decision to limit the pool of candidates to military test pilots, civilians were not systematically considered in this first round of astronaut selection, although a few applications were screened.

Washington, D.C.
April 23, 1959

MEMORANDUM for Administrator

Subject: Pilot Selection for Project Mercury

1.   The criteria used for the pilot selection were established at a meeting held at NASA Headquarters on January 5, 1959. This meeting was attended by Dr. Lovelace, General Flickinger, Mr. Gilruth, and others. Capt. Augerson was present and represented the Army. At the time, Capt. Augerson appeared to be in full agreement with the selection criteria, although it was even then apparent that these criteria might exclude Army participation.

2.   At the time of our first briefing of the astronauts on February 9, 1959, Gen. Flickinger informed Dr. Silverstein that no Army men had met all of our selection criteria. He suggested that we should approach the Army for names of men who came close to qualifying. Dr. Silverstein agreed and asked Gen. Flickinger to contact the Army. Gen. Flickinger, in turn, asked Capt. Augerson to supply NASA with names of candidates that he thought would qualify.

3.   Several days later, Capt. Augerson appeared with the files of six Army men. He turned these over to Mr. Donlan and the group participating in the selection proceedings. After it was ascertained that none of these men met our selection criteria, and after another discussion with Dr. Silverstein, it was decided not to consider these Army men as candidates for Project Mercury. Capt. Augerson was informed of this decision.

4.   On the subject of possible civilian participation, approximately ten letters were received by me. Several letters were obviously from cranks, while others were sincere. None of the civilians met our selection criteria. All letters received were answered. Other letters may have been received in other parts of the organization.

5.   The heads of the flight activities at all. NASA Centers and Stations were contacted by either Mr. Gilruth or by myself. They, in turn, sought volunteers for Project Mercury among their pilots. None of the NASA pilots volunteered although several expressed interest in joining the Project at a later date.

[Signed]
George M. Low
Program Chief
Manned Space Flight

cc: Dr. Dryden
Dr. Silverstein
Mr. North

GML:lgs

Document I-23

Document title: **George M. Low, NASA, Memorandum for House Committee on Science and Astronautics, "Urgency of Project Mercury," 27 April 1959.**

Source: **Folder 18674, NASA Historical Reference Collection, NASA History Division, NASA Headquarters, Washington, DC.**

From virtually the beginning of the Mercury program, its leaders at the Space Task Group believed that it should receive the nation's highest priority. This status ensured ready cooperation from other federal entities and streamlined procurement and other regulations. Only programs and projects deemed critical to national defense received this designation. In 1958 numerous spaceflight efforts such as the Minuteman and Polaris ICBM development efforts, the Vanguard program, and satellite reconnaissance were already on what was officially named the DOD Master Urgency List. Admittance to the DX, the part of the DOD Master Urgency List associated with the highest industrial procurement priority, required the approval of the National Security Council, but it had already delegated authority to the Secretary of Defense to approve DX status on space projects. Space Task Group leaders, therefore, had to convince Secretary Neil H. McElroy of the significance of Project Mercury. This did not prove an easy task. While senior officials agreed that Mercury was important, key officials at the White House, Congress, and NASA Headquarters regarded both the development of a one-million-pound-thrust rocket, which eventually became the Saturn I, and space science efforts as equally important. However, a priority list is only useful if some items have less priority than others. Why should Project Mercury receive this special designation?

When NASA Deputy Administrator Hugh L. Dryden initiated the request for DX status to the DOD on 14 November 1958, he specifically requested that both the "manned satellite and the one-million-pound-thrust engine" be added, but because of disagreements, especially within the National Aeronautics and Space Council (NASC) created by the same act that has chartered NASA, consideration of this proposal was deferred until a united position could be crafted. It took several months of discussion during the winter of 1958 to 1959 before consensus could be achieved, and only on 27 April 1959, did Eisenhower approve DX status for Mercury. This memorandum prepared by George Low explains to the Congressional committee overseeing NASA the agency's policy with respect to balancing urgency and astronaut safety.

April 27, 1959
In reply refer
To: DAL

MEMORANDUM: For House Committee on Science and Astronautics

Subject: Urgency of Project Mercury

The primary goal of Project Mercury is to achieve orbital flight, and successful recovery, of a manned satellite at the earliest practicable date, and to study man's capabilities in a space environment. This project is NASA's most urgent program, and is being pursued at a rate that will give this nation a highly reliable space vehicle and completely prepared astronaut at the earliest moment.

We have also a desire to be first, because we realize that much in the way of national prestige comes from space flight achievements. But, we cannot place the prestige of the nation above the safety of the astronaut. With this overriding consideration for the safe return of the pilot, we must recognize that another country may accomplish a manned space mission before we do.

But neither the value nor the success of Project Mercury can be gauged by whether it is the first, second or third manned space flight. Mercury is a stepping-stone in the manned exploration of space. From the Mercury program will develop this nation's plans for more advanced manned satellites, space laboratories and stations, missions to the moon, and interplanetary explorations. The most vigorous pursuit of Project Mercury is required to insure that this nation will enjoy a role of leadership in future manned explorations of space.

Document I-24

Document title: George M. Low, Program Chief, Manned Space Flight, NASA Memorandum for Mr. R. R. Gilruth, Director, Project Mercury, NASA, "Animal Payloads for Little Joe," 19 June 1959, with attached Memorandum from T. K. G [T. Keith Glennan] to George M. Low, 15 June 1959.

Source: National Archives and Record Administration, Fort Worth, Texas.

In preparation for the human flights of Project Mercury, NASA decided to undertake several tests of the spacecraft using the Little Joe booster to launch the capsule on a sub-orbital trajectory. The Little Joe booster was produced specifically for Mercury test usage, and consisted of four Pollux or Castor motors grouped with four smaller Recruit motors. Out of a total of eight Little Joe flights, two carried American-born rhesus monkeys (Macaca mulatta). This memorandum discusses the use of these monkeys, obtained from the School of Aviation Medicine at Brooks Air Force Base, San Antonio, Texas. The Little Joe 2 (LJ-2) mission carried an American-born rhesus monkey (Macaca mulatta) named "Sam," an acronym for School of Aviation Medicine, to the edge of space. The mission launched on 4 December 1959, from Wallops Island, Virginia, and flew 51 miles toward space. Sam was housed in a cylindrical capsule within the Mercury spacecraft. Approximately one minute into the flight, traveling at a speed of 3,685 mph, the Mercury capsule aborted from the Little Joe launch vehicle. It was safely recovered in the Atlantic Ocean after a flight of only 11 minutes, 6 seconds.
A second rhesus flight took place on 21 January 1960, flying only 8 minutes, 35 seconds to an altitude of 9 miles. Its passenger, "Miss Sam," also returned safely after taking part in a Max Q abort and escape test.

NASA Headquarters

June 19, 1959

MEMORANDUM: For Mr. R. R. Gilruth, Director
          Project Mercury

Subject: Animal Payloads for Little Joe

1. I am enclosing a copy of a memorandum from the Administrator requesting that only American-born rhesus monkeys will be used in Mercury flights.

2. I understand that we have been assured by the School of Aviation Medicine that all rhesus monkeys supplied by them for the Little Joe flights meet the above requirements. However, I suggest that SAM be informed that "birth certificates" of these monkeys will be required at the time of each flight.

George M. Low
Program Chief
Manned Space flight

[handwritten at bottom: "Hindoos might object"]

Attachment:
Memo to George Low
Dtd 15 June 1959
GM: mdp
Cc: Dr. Smith
    Dr. Worf
    Dr. Henry – Langley STG
    Mr. Sanders
    Mr. C. Wood without attachment
    Mr. W. Hjornevik without attachment

NATIONAL AERONAUTICS AND SPACE ADMINISTRATION
1520 H STREET NORTHWEST
WASHINGTON 25, DC

15 June 1959

MEMORANDUM TO:

George Low
Office of Space Flight Development

Following the public announcements of the use of the American-born rhesus monkey in the recent Jupiter test, the Secretary of HEW raised questions with the Defense Department and with NASA as to the intention of these agencies with respect to the use of Indian-born rhesus monkeys in the future. A copy of the response of the Department of Defense prepared by Admiral John Clark is attached for your information [not included]. For NASA, I informed [Health, Education, and Welfare] Secretary Flemming that we proposed to use relatively few biological specimens and where we felt a rhesus monkey was indicated as the proper animal, we would use American-bred animals. Please take this as your instruction to abide by this statement on my part.

[Signed]
T.K.G

Cc: Dr. Silverstein
    Dr. Randt

Attachment:
Thermofax copy of Memo dtd 6/11/59
From Adm. Clark, ARPA, to Secy.,
HEW [not included]

Document I-25

Document title: NASA, "Information Guide for Animal Launches in Project Mercury," 23 July 1959.

Source: Folder 18674, NASA Historical Reference Collection, NASA History Division, NASA Headquarters, Washington, DC.

In all, there were four launches of Mercury spacecraft with primates aboard to test the life support systems of the vehicle. The first of these was the Little Joe 2 flight of 4 December 1959 with Sam, an American-born rhesus monkey, aboard. Sam was recovered, several hours later, with no ill effects from his journey. He was later returned to his home at the School of Aviation Medicine at Brooks Air Force Base, San Antonio, Texas, where he died in November 1982. Miss Sam, another rhesus monkey and Sam's mate, was launched on 21 January 1960, on the Little Joe 1B mission. She was also recovered and returned to the School of Aviation Medicine. On 31 January 1961, Ham, whose name was an acronym for Holloman AeroMed, became the first chimpanzee in space, aboard the Mercury Redstone 2 (MR-2) mission on a sub-orbital flight. Ham was brought from the French Camaroons, West Africa, where he was born in July 1957, to Holloman Air Force Base in New Mexico in 1959. Upon the completion the successful flight and a thorough medical examination, Ham was placed on display at the Washington Zoo in 1963 where he lived until 25 September 1980, when he moved to the North Carolina Zoological Park in Asheboro until his death on 17 January 1983. Enos became the first chimp to orbit the Earth on 29 November 1961, aboard Mercury Atlas 5 (MA-5) launched on 29 November 1961, from Cape Canaveral, Florida. This two-orbit, 88 minute, 26 second flight, proved the capability of the Mercury spacecraft. Enos died at Holloman Air Force Base of a non-space related case of dysentery

11 months after his flight. Because of the interest in and the sensitivity about these primate flights, NASA took considerable pains to explain how the animals were treated and what role they played in the program, as shown in this information guide. These guidelines were approved by Deputy Administrator Hugh Dryden, Associate Administrator Richard Horner, and Director of Space Flight Development Abe Silverstein on 23 July 1959.

[7-23-59]

NASA INTERNAL USE

INFORMATION GUIDE FOR ANIMAL LAUNCHINGS

IN PROJECT MERCURY

1. Background

Animals will be used in the Project Mercury developmental program to gain information on the biological response to space flight. Problems facing manned orbital flight essentially are engineering in nature, and the animal program will be relatively simple in scope. Knowledge from animal flights will contribute information to the program in the areas of life support systems; instruments to measure physiological reactions in the space environment; prove out design concepts when they are near known limits in such areas as high-g loads; test equipment and instrumentation under dynamic load conditions, and to develop countdown procedures and train personnel in these procedures prior to manned flight.

NASA has selected three animals for developmental work in the Mercury program: rhesus monkey (Macaca Mulatta), chimpanzee and mouse. Primates were chosen because they have the same organ placement and suspension as man. Both rhesus and chimp have relatively long medical research backgrounds, and the type of rhesus born and bred in American vivarium has a 20-year research background as a breed. The chimp is larger and more similar to man in body systems, and will be used for advanced developmental flights with the McDonnell Aircraft Corporation capsule. A "mouse drum" will be used to study the effects of weightless flight and may be the first biological package to be sent into orbit in the Mercury program.
(Page 2 contains a summary of flights and test objectives.)

Management of the animal program is the responsibility of the NASA Space Task Group. Responsibility for supply, training, installation and post-flight evaluation has been assigned to:

USAF Aeromedical Field Laboratory, Hol[l]oman AFB – Chimps
USAF School of Aviation Medicine, Randolph AFB – Rhesus
USAF Aeromedical Laboratory, WADC, Wright-Patterson AFB- Mice

The U.S. Army and U.S. Navy will provide advice and assistance through out the program.

[2]

STUDY OF LITTLE JOE FLIGHTS AND TEST OBJECTIVES

(All at Wallops Station)

Flight #	Animal	Capsule	Mission*
1.	None	NASA	Escape system-booster quals.
2.	Rhesus	NASA	High-angle re-entry
3.	Rhesus	NASA	High-angle re-entry
4.	Rhesus	NASA	Low-angle re-entry
5.	Chimp	MAC	Maximum load escape

6. Backup booster

OTHER ANIMAL FLIGHTS IN PROJECT MERCURY

(All at Atlantic Missile Range)

Redstone	Chimp	MAC	Ballistic flight quals.
Atlas	Chimp	MAC	High-g escape
Atlas	Mouse drum	MAC	Weightless flight
Atlas	Chimp	MAC	Orbital flight

Notes:

* Mission- physiological measurements and environmental readings will be taken during all animal flights.

There will be two animals available for each flight, one of which will be used as a backup.

The above is not necessarily the order in which flights will take place.

[3]

2. Information Procedures

The press will be permitted to witness two Little Joe launches – one of the early flights to be determined by the Deputy Administrator and the Director of Space Flight Development, and Flight No. 5 identified on Page 2.

Procedures at both Wallops Station and the Atlantic Missile Range will be in accordance with the joint NASA-DOD East Coast Launching Plan. For each launching open to the press, a press kit will be prepared containing handout materials on: 1-Program objectives; 2-Launching vehicle, and 3-Project Mercury background.

Only qualified NASA personnel will be permitted to make public statements on the program. The Defense Department, through the Assistant Secretary of Defense for Public Affairs, will be asked to cooperate in this desire so that objective information goals will be attained.

A. Pre-Launch Activities (At Wallops Island) – The afternoon before the first open launch at Wallops, the press will be permitted to view the Little Joe booster. This activity will take place before the animal subject (in the case of an animal flight) is inserted in the capsule. Photography will be permitted. The press kit will be handed out simultaneously at Wallops and in Washington on a "Hold for Release Until Launched" embargo.

Meanwhile, the Navy will be asked to provide photographic coverage on the recovery ships, and a billet for one NASA OPI representative.

The logistics briefing at Wallops will cover press release details, safety requirements and general test objectives. The Wallops Station will provide bus transportation on Wallops Island. No members of the launching team will be required until phase (c) below.

B. Launch Activities - The press will meet at the Mainland Dock two hours before launch for transportation to arrive at the viewing site one hour before scheduled launch.

C. Post-Launch Activities (Wallops Island) – A brief post-launch briefing will be held either at the viewing site or at the cafeteria building on Wallops. This briefing will discuss the launching phase, mission profile, and any recovery data available at that time. Representatives of the launching team and Space Task Group will participate.

[4]

(Washington) - NASA Washington will be the source of all post-launch scientific information. A Technical press briefing will be conducted about 24 hours after the launch to summarize all information known at that time. Representatives on the panel will be from NASA Headquarters; Space Task Group; STG Biomedical Group; Launching Team, and Recovery Team.

Pre-launch information activities require training and housing still and motion pictures of the animal subjects. The responsible agency (i.e., USAF) will be asked to provide footage and a selection of photographs in these areas.

NASA will take and provide photographs (still and motion picture) of the animals in biopacks and the biopack insertion into the capsule.

At no time will the animal subjects be available to the press either for photography or viewing. NASA will follow this policy for these reasons:

1- Test results are influenced by excitement, particularly since the animal subjects have led sheltered lives. A minimum of crowd activity is justified from both scientific and clinical standpoints.

2- Elimination of all but necessary scientific persons curtails added chances of the primates contacting diseases.

3- Complex handling procedures for the animals will not be required.

4- The undesirable effects of the "Roman Holiday" atmosphere are eliminated.

While the above procedures indicate fairly full-dress coverage, the press will be permitted on the spot viewing at only two of the five Wallops Station launchings. NASA OPI will assure that there will be no interruption of scientific activities and personnel until after the launch.

For the launches not open to the press, a NASA OPI representative will witness the firings and prepare releases on them.

Press activities at AMR are governed by the joint NASA-DOD agreement, but will be supplemented with the requirement that the animal subjects will at not time be available to the press, for the above reasons.

3. Summary

NASA OPI will conduct information activities associated with animal launches in a factual manner which will satisfy requirements [5] for accurate reporting and non-interference with scientific personnel conducting the program.

Two launches at Wallops Island will be open to news media.

NASA and DOD personnel will be requested not to comment on aspects of the technical program outside their cognizance. Lines of responsibility are clear:

Management and overall responsibility – Space Task Group
Boosters- Langley Research Center (Little Joe)
        - Army Ballistic Missile Agency (Redstone, Jupiter)
        - Air Force Ballistic Missile Division – (Atlas)
Medical data correlation – NASA biomedical group
Capsule recovery - U.S. Navy under DOD assignment (Note: the Assistant Secretary of Defense for Public Affairs has been designated sole liaison to NASA for the Mercury project. He is expected to detail cognizant military agencies to act in his name.)
Information - NASA OPI (Headquarters and field)

Before there is any critical deviation from this plan, the NASA Director of Public Information will discuss details with the Director, Space Flight Development and Director, Space Task Group.

-END-

Document I-26

Document title: A. J. Goodpaster, Brigadier General, USA, Memorandum of Conference with the President, 29 September 1959.

Source: Dwight D. Eisenhower Presidential Library, Abilene, Kansas.

From the beginning of his first term in January 1953 President Dwight D. Eisenhower had a strategy for defeating the Soviet Union. It revolved around long-term economic, military, international, and social and moral perquisites that would enhance the U.S. as the world leader. It represented a commitment to constant pressure on the Soviet Union on a broad front, but refrained from a confrontation that would require nuclear war to resolve. A key ingredient of this strategy involved not responding to every situation vis à vis the Soviet Union as a crisis. Accordingly, he resisted the crisis sentiment that Sputnik and the early space race fostered among many policy-makers in Washington. This memorandum captures the spirit of that resistance by reporting on the President's questioning of NASA's proposed budget. Eisenhower's approach to space activities stressed the development of launch vehicles for use in the ICBM program, satellite technology for reconnaissance and communications, infrastructure required to support these activities such as tracking and launch facilities, and utilitarian science that either directly supported those missions or was a natural byproduct of them. Eisenhower's space program, however, did not include any real commitment to, or belief in, the goal of human spaceflight. In Eisenhower's view, human spaceflight did not have a serious national security component, and therefore was probably not worthy of significant federal expenditures. NASA Administrator T. Keith Glennan was largely in sympathy with the President's objectives, but faced pressures from elsewhere to surpass the Soviet efforts, hence the large increase in NASA's budget for fiscal year 1961. Senator Lyndon B. Johnson (Dem-Texas), for one, vowed to put additional funding into any NASA budget submission so that it could do so. Glennan wrote in his diary that "Congress always wanted to give us more money . . . Only a blundering fool could go up to the Hill and come back with a result detrimental to the agency." This memorandum reflects these realities as NASA began undertaking Project Mercury.

[SECRET] [DECLASSIFIED]

September 30, 1959

MEMORANDUM OF CONFERENCE WITH THE PRESIDENT
September 29, 1959

Others present: Dr. Kistiakowsky
                 General Goodpaster

The President began by saying had heard that Dr. Glennan is putting in for some $800 million in the FY-61 budget for space activities. He though this was much too great an increase over the current year and in fact said that he thought a program at a rather steady rate of about a half-billion dollars a year is as much as would make sense. Dr. Kistiakowsky said that he has had about the same figure in his mind, but pointed out that this amount would not allow enough funds for space "spectaculars" to compete psychologically with the Russians, while being a great deal more than could be justified on the basis of scientific activity in relation to other scientific activities.

The President recalled that he has been stressing that we should compete in one or two carefully selected fields in our space activity, and not scatter our efforts across the board. He observed that other countries did not react to the Russian Sputnik the way the U.S. did (in fact, it was the U.S. hysteria that had most affect on other countries), even the United States did not react very greatly to the Soviet "Lunik" – the shot that hit the moon.

The President said he had understood that, through the NASA taking over ABMA, there was supposed to be a saving of money, but that it appeared this would in fact increase the NASA budget. He thought that Dr. Glennan should be talked to about this, away from his staff, who are pushing a wide range of projects, and advised not to overstress the psychological factor. The President thought we should take the "man in space project" and concentrate on it. He added he did not see much sense to the U.S. having more than one "super-booster" project. There should be only one.

[2]

Dr. Kistiakowsky said he strongly agreed on this. He pointed out that, by putting ABMA into NASA, there would be an over-all saving of money. The President reiterated that there is need for a serious talk with Dr. Glennan. He thought ABMA should be transferred to NASA and that we should pursue one big-booster project. Our concentration should be on real scientific endeavor. In the psychological field, we should concentrate on one project, plus the natural "tangents" thereto. He thought perhaps Dr. Glennan is overrating the need for psychological impact projects. Dr. Kistiakowsky said that the Defense Department states that if Dr. Glennan does not push fast enough in space activities, Defense will do so. The President said we must also talk to Dr. York, and call on him to exercise judgment. He asked that a meeting be set up, to be attended by Dr. Glennan, Dr. Dryden, Dr. Kistiakowsky, Dr. York, and Secretary Gates in about ten days. The President stressed that we must think of the maintenance of a sound economy as well as the desirability of all these projects. He thought perhaps NASA sights are being set too high, including too many speculative projects.

Dr. Kistiakowsky said that if Dr. Glennan goes in with a lower budget, there will be need for the President to support him publicly, because there will be a great deal of criticism about this. Dr. Kistiakowsky himself thinks that such a limitation may be wise, however, particularly when one contrasts the $60 million being given to the Science Foundation for research purposes with the $3/4 billion proposed to go into space activity, but a much lower NASA budget may not allow us to "compete" with the USSR. The President commented that the space activity is in the development and production state, which is more expensive, and Dr. Kistiakowsky recognized that, of course, this is true.

[Paragraphs 6-12 not included]

A.J. Goodpaster
Brigadier General, USA

Document I-27

Document title: Wernher von Braun, Director, Development Operations Division, Army Ballistic Missile Agency, to Robert R. Gilruth, Space Task Group, NASA, 9 October 1959.

Source: National Archives and Record Administration, Forth Worth, Texas.

Wernher von Braun (who was working for the U.S. Army at the time but transferred to NASA in1960), together with his German-led rocket team, wrote this memorandum to Robert Gilruth of the Space Task Group, which points out two critical aspects of early relations in NASA. First, it demonstrates the friendly rivalry that existed between competing entities in the Agency. Wernher von Braun, certainly pleased to be of assistance to a colleague and exceptionally mindful of the high quality of work required in building space technology, also enjoyed pointing out the flaws that he saw in the construction of the Mercury spacecraft by McDonnell Aircraft Corporation that Gilruth's group was managing. Gilruth and von Braun demonstrated this type of relationship throughout the era. Second, the memorandum demonstrates the intense level of "contractor penetration" that von Braun's team was famous for in the management of its spaceflight projects. Industry officials sometimes complained that working for von Braun's engineers required acquiescing in a technical take-over in which government inspectors, many of whom were more qualified to do the work than the industry technicians, constantly peered over the shoulders of the company workers and got involved in every aspect of the project. The comment in this memo on "soldering rods" would certainly be considered today to be a governmental intrusion into something that was the proper province of the company. The longstanding debate over "contractor penetration" lasted throughout Project Mercury, and indeed to the present, as NASA sought to strike a balance between necessary oversight and contractor autonomy. The letter was apparently drafted by Joachim Kuettner, von Braun's associate who was the project engineer for the Mercury-Redstone portion of Project Mercury.

Kuettner/vonBraun/bh/4814

ORDAB-D 252

9 October 1959

Mr. Robert R. Gilruth
NASA- Space Task Group
Langley Field, Virginia

Dear Bob:

I am writing this letter in full knowledge that I am poking my nose into something that's none of my business. But I am convinced that projecting a man two hundred miles down-range simply requires the ultimate in teamwork. This team composed of NASA, McDonnell, and ABMA must operate flawlessly to drive on to a touchdown; for this time, there is human life at stake.

It has come to my attention that one of our ball carriers has his shoelaces untied and doesn't know it. If he trips and falls we may all lose the game and our astronaut his life. So I feel that I must pass along to you what has been brought to my attention, at the risk of making a few people sore.

On a recent trip to McDonnell Aircraft Corporation, ABMA personnel were permitted to tour the facilities used to fabricate electrical cable harnesses. They discovered to their great consternation that in MAC's electrical shops procedures long since discarded by ABMA as being inadequate and dangerous are still in practice.

Samples:
-Soldering irons of excessive wattage are being used to make joints in pygmy connectors. (Reason: The shop is not air-conditioned; large cooling fans prohibit the use of correct, smaller soldering irons.)
-Poor connections are being hidden by potting compound (making inspection impossible).

It has been our experience that conventional methods of soldering for aircraft are simply not acceptable in the missile [2] field where any and all component [failures] usually result in an aborted mission. In MERCURY the life of an astronaut and the success of the entire project could be jeopardized by one bad solder connection.

I don't want to blame anyone in particular at MAC. I don't even know who is responsible for this electrical shop. But I should like to suggest that you have someone from Langley look into this. While we would prefer to leave it up to you to take any further actions that you may deem advisable we are at your disposal if we can be of any further help.

Sincerely yours,

[Signed]

WERNHER VON BRAUN
Director
Development Operations Division

Copies furnished:
AB-DSRM (Record)
AB-D (Info)

Document I-28

Document title: Mercury Astronauts, Memorandum For [Mercury] Project Director, NASA, "Exchange of Visits with Russian Astronauts," 21 October 1959.

Source: National Archives and Record Administration, Fort Worth, Texas.

From the start of their careers at NASA, the seven Mercury astronauts were eager to make contact with their Soviet counterparts. These efforts were discouraged by NASA and White House leadership.

It was not until the 1960s that NASA astronauts and Soviet cosmonauts met each other in various places around the world. These visits were arranged for mostly propaganda purposes on both sides. American intelligence officials also foresaw the opportunity to pierce some of the secrecy surrounding the Soviet program if the two sets of pilots could talk with each other. The first such interchange took place following the flight of Gherman Titov, when he visited with John Glenn at a May 1962 technical meeting in Washington, DC. The two men and their wives toured the Capitol and visited President John F. Kennedy in the White House. The next exchange between astronauts and cosmonauts did not take place until June 1965 when astronauts James A. McDivitt and Edward H. White, along with Vice President Hubert H. Humphrey, met Yuri Gagarin at the Paris Air Show. As years passed cosmonauts and astronauts began to meet more frequently and freely.

NASA – Space Task Group
Langley Field, Virginia
October 21, 1959

MEMORANDUM: For Project Director

Subject: Exchange of visits with Russian Astronauts

1. The Russians have recently announced their man-in-space program and have given some publicity to the pilots selected. In the eyes of the rest of the world, it appears that Project Mercury is placed in a competitive position, whether we like it or not. This, of course, sets up for another barrage of unfavorable propaganda when, and if, the Russians achieve space flight before we do.

2. Certain action at this time might place us in a better position to gain information about their program and also take the propaganda initiative away from the Russians with regard to manned space flight. Suggested action is to propose mutual visits between the Astronauts of the two countries with the purpose of sharing information on training and mutual problem areas.

3. Propaganda-wise, we apparently stand to gain a great deal and could lose little or nothing.

    a. The U.S. would have taken the initiative in sponsoring international cooperation in the manned space field.

    b. Such a proposal would support, to the world, our statements of the peaceful intent of Project Mercury as a scientific exploration with no ulterior motives.

    c. It is in keeping with the current political atmosphere engendered by the Khrushchev visit and the proposed presidential visit to Russia.

4. There appears to be little we could lose, in that practically all of the details of Project Mercury are already public domain and have been covered repeatedly in the press. The Russian program, on the other hand, has been secret, so anything we could learn would be new information.

5. Refusal of the Russians to cooperate in such a proposal would certainly reflect unfavorably in the eyes of other countries. These are countries already concerned about where the American-Russian space race is leading.

[2]

6. Timing of such a proposal is very important. If such a proposal is made, it should be done very soon, before either Russia or the U.S. has accomplished a man-in-space mission.

7. If we wait until we make the first orbital flight, and then propose an exchange, it would appear that we are "rubbing it in" a little and are willing to throw a little information to our poor cousins who could not do it themselves. This would probably do us more harm than good in the attitude with the rest of the world.

8. If, on the other hand, we wait until the Russians have made the first orbital flight before we propose such an exchange, it would appear that we are trying to get information on how they did it because we have not been able tot do the same thing. This would also do us harm in the eyes of other countries.

9. To summarize, we stand to gain information in an exchange of visits, while giving little information that is not already known. Propaganda value of such a proposal and visit should be very favorable for us, if the proposal is made from the U.S. and before either country has made an orbital flight.

10. One way to assess the value of such a proposal is to think of our reaction and the reaction of other countries if the Russians make such a proposal first. It appears that we stand to gain by making the proposal first.

11. It is realized that there are many considerations involved in such a proposal. NASA, State Department, Intelligence, and many other government sources concerned must have vital inputs that will determine whether the proposal is not only feasible, but advisable.

12. The proposal is herewith submitted for consideration.

M. Scott Carpenter
Lieutenant, USN

Leroy G. Cooper
Captain, USAF

John H. Glenn
Lt. Col., USMC

Virgil I. Grissom
Captain, USAF

Walter M. Schirra
Lt. Cmdr., USN

Alan B. Shepard
Lt. Cmdr., USN

Donald K. Slayton
Captain, USAF

Document I-29

Document title: Charles L. Wilson, Captain, USAF, ed., WADC Technical Report 59-505, "Project Mercury Candidate Evaluation Program," December 1959.

Source: Folder 18674, NASA Historical Reference Collection, NASA History Division, NASA Headquarters, Washington, DC.

The selection of the astronauts for Project Mercury involved numerous organizations and types of activities for the various candidates. One of the key organizations was the Aeromedical Laboratory of the Wright Air Development Center in Dayton, Ohio. This U.S. Air Force facility was one of the most prestigious in the world and had been involved in aerospace medicine for many years. Its scientists had conducted tests on NASA's astronaut candidates in the spring of 1959 to ascertain which of them might be most appropriate for spaceflight. This technical report discusses how and why the center became involved in the initial astronaut selection process and the work undertaken in choosing the Mercury Seven.

WADC TECHNICAL REPORT 59-505

PROJECT MERCURY CANDIDATE EVALUATION PROGRAM
Charles L . Wilson, Captain. USAF, MC
Editor

Aerospace Medical Laboratory

December 1959

Project No. 7164
Task No. 71832

WRIGHT AIR DEVELOPMENT CENTER
AIR RESEARCH AND DEVELOPMENT COMMAND
UNITED STATES AIR FORCE
WRIGHT-PATTERSON AIR FORCE BASE, OHIO

INTRODUCTION
C. L. Wilson, Capt., USAF, MC

The National Aeronautics and Space Administration (NASA), a U.S. Government civilian agency, has been assigned the task of exploring the feasibility of space travel. As a result of thorough and exhaustive study, NASA has concluded that certain aspects of space travel are feasible and, furthermore, that some will be practicable in the very near future. One profile of space travel envisions that a human pilot, transported in a life support system (capsule), could be thrust into orbit by a liquid fuel rocket, maintained there for several revolutions around the earth, and successfully and safely recovered from orbit. Project Mercury intends to realize this vision.

Among the many strategic questions to be answered is: "Who will the pilot be?" This report describes how and why the Aerospace Medical Laboratory participated in the selection of the seven Mercury Astronauts.

HISTORY

The Human Factors Division of the Air Research and Development Command (ARDC) has been keenly aware of the need for clarification of the parameters of human endurance, safety, and comfort during periods of unusual stress. In 1952 Brig.Gen. Don Flickinger, USAF, MC, began directing biomedical research toward the development of tests to assist in selecting pilots for special research projects. Under his guidance Capt. T. F. McGuire, USAF, MC, of the Aerospace Medical Laboratory, employed a series of physiological, psychological, and biochemical tests which were incorporated into a stress-test program. Dr. McGuire's experience extended over a 4-year period, during which time he tested several special groups. These included USAF pilots and young volunteers from the University of Dayton. In his final months at the Aerospace Medical Laboratory he stress-tested 12 USN underwater demolition men (frogmen) kindly loaned by the Underwater Demolition Unit 11, Little Creels, Virginia. The results of his research are presented in Stress Tolerance Studies, Part I, and Tolerance to Physical Stress, Part II. Part III is being completed and will contain a supportive bibliography. Dr. McGuire rightfully should receive credit for his work in this field and development of early prototype crew selection profiles. Several new tests have been made available since then and are discussed later.

Captain F. J. Leary, USAF, MC, of the Aerospace Medical Laboratory also gained considerable experience in candidate evaluation. His research brought about modification of the cold pressor test to its present form. Previous testing utilized the immersion of one foot, then both feet. He also studied the reproducibility of physiological response on the same subject when tested on different days. He developed early scoring techniques based on physiological response. Modifications of his techniques were employed in the Mercury Candidate Evaluation Program.

Captain W.S. Augerson, USA, MC, was immensely valuable in the development of the final test profile. He assisted in a review of literature, experienced the actual tests, and offered valuable opinions on areas where improvement was indicated.

Two assistant investigators during the period of 1957 to 1958 were Gardner Edwards, M, D. (then a University of Virginia medical student on a USAF-sponsored scholarship), and Robert McAdam, associate professor of physical education, Northern Illinois University.

[2]

APPROACH TO THE PROBLEM OF CANDIDATE EVALUATION

The ultimate purpose of any crew recommendation development program is to devise and validate tests which can be used with reliability in selecting crew members for future projects. The Project Mercury Candidate Evaluation Program was an important stage in this ARDC development program. Since the actual approach to this research problem departs from the ideal approach, it will help to present both the ideal and actual methods of attack.

Ideal Approach to Problem:

1 . The candidates must be medically acceptable and technically capable before they will he considered as potential candidates.

2. Those who are tested must be the actual project candidates. A large candidate population will increase the reliability of the results.

3. The test profile must simulate all aspects of the stresses anticipated during the actual project. The simulated stresses must be combined in the same relationship and intensity as they would occur during the project.

4. A battery of nonsimulating but relevant tests must be included in the testing program. These tests will be used to identify significant correlations between the response to simulating and nonsimulating tests. The ultimate goal is to replace simulating tests with the more easily administered, nonsimulating tests in future programs

5. In the final recommendation of candidates, the investigators must only interpret subject performance on the simulating tests. Nonsimulating test performance will not affect recommendation of this first group of candidates.

6. All candidates, both recommended and not recommended, must enter the project.

7. At the completion of the project all of the participants must be graded on the effectiveness of their performances.

8. The investigators must then seek significant correlations between subject performances on the various simulating and nonsimulating tests and successful mission performances.

9. Those nonsimulating tests bearing significant correlation with successful mission performances may then be used to select future subjects from an identical population for identical projects. These future crew members will be highly reliable risks in successfully completing their missions. This is the goal of all endeavors at crew selection.

Actual Approach to Problem:

Inherent errors are frequently introduced when making a transition from an ideal to an applied test program due, for example, to time limitations, accelerated schedules, or unforeseen changes. The actual approach to the problem is stated below, preceded by an underlined restatement of the ideal approach:

1. The candidates must be medically acceptable and technically capable before they will be considered as potential candidates. The candidates were medically acceptable and technically capable. They met the following requirements: a. were pilots in the Department of Defense, b. had received engineering degrees, c. had successfully graduated from a military test pilot school, d. had achieved at least 1500 hours of total flying time, and e. each man's height was 5'11" or less One hundred and ten men met the above requirements. Sixty-nine of these men were invited to a [3] NASA briefing where the detailed plans of Project Mercury were revealed. The subjects were then asked if they desired to volunteer as competitive candidates. Fifty-five of them volunteered.

2. Those who are tested must be the actual project candidates. A large candidate population will increase the reliability of the results. Those who were tested actually were the Project Mercury candidates. The 55 men who were accepted were given a series of interviews and psychological tests. On the basis of the data thus obtained, 32 were chosen for the final phase of the selection program. The 32 candidates were sent to the Lovelace Foundation, Albuquerque, New Mexico, for extensive medical histories, physical examinations, and biochemical and physiological tests.* A large random candidate population was not used. If the candidate population had been larger it would have been impossible to process them in time to meet the close time schedules of the project.

3. The test profile must simulate all aspects of the stresses anticipated during the actual project. It was impossible to devise a laboratory situation which exactly duplicated the stresses anticipated during Protect Mercury. A rational alternative approach was to list the anticipated stresses and to use what laboratory tools were available.

Anticipated Stresses:

a. The men who were chosen could expect a 2- to 3-year period of intensive training including a study of space-frame structures, propulsion, inertial guidance, systems reliability, aerodynamics, and physiology. They would actively participate in training exercises such as: physical fitness, capsule parachute landings, ballistic trajectory flights, and underwater escape from capsules. These represent a prolonged period of genuine stresses.

The best practical laboratory tools to test these areas were: (1) review their past accomplishments, (2) extract personal histories, and (3) conduct psychiatric interviews and psychological tests. Additional information could be derived from observation of these candidates during moments of calibrated hazing such as: acceleration, pressure suit testing, immersing feet in ice water, and isolating the subject. The accumulated impressions of these trained observers should guarantee highly reliable maturity in those recommended.

b. Psychological and physical stresses will exist before, during, and after each flight. The psychological stresses will include fears and anxiety about possible accidents or death. Although well disguised in the mature test pilot, they will be present. The psychiatric evaluation should reveal those who are stable and reliable.

The physical stresses of blast-off and orbit will include acceleration, noise, vibration, weightlessness, tumbling if stabilization is not achieved, and possible capsule depressurization. Those insults of re-entry will contain deceleration, noise, vibration, and heat if the cooling system fails. Landing will be accompanied by deceleration. Before recovery there is the possibility that the capsule will sink. There is also the possibility of isolation in a remote and uninhabitable climate and topography.

The physical facilities available at the Aerospace Medical Laboratory are able to duplicate the important physical and psychological stresses mentioned above. These facilities include: human centrifuge, extremely low-pressure (high-altitude) chamber, heat-controlled test rooms, equilibrium-vibration chair, intense noise generator, aircraft (C-131B) specially modified to safely fly Keplerian trajectories (weightlessness), tumbling turntable, psychiatric interviewing rooms, and anechoic chamber.

Simulating Tests:

* The tests performed at the Lovelace Foundation are detailed in the Appendix. [not included]

Those tests simulating stresses anticipated during Project Mercury are: transverse g profiles (acceleration tests) and vibration-equilibrium and intense noise profiles (biological acoustical tests). Weightlessness tests were not performed on the candidates for one main reason: it would have been impossible in scheduling always to meet the minimum flying safety requirements for each flight each day for 6 weeks. Tumbling tests are so unpleasant and the nausea so prolonged as to warrant its exclusion for the profile.

4. <u>The simulated tests must be combined in the same relationship and intensity as they would occur during the project</u>. The physical separation of test facilities rendered it highly impractical to improvise superimposed stress. While a multi-stress facility was desirable, it was not mandatory for study of the candidates. In any interpretation, partial data when expertly gathered is much more desirable than no data at all. This reasoning serves to defend the approach that was finally taken.

5. <u>A battery of nonsimulating but relevant tests must be included in the testing program. These tests must be easy to administer and safe</u>. A battery of easily administered and safe nonsimulating tests was incorporated into the program. They were (physical fitness tests): Harvard step, Flack, cold pressor, and tilt table. A battery of more complex nonsimulating tests was also devised. The investigators believed these might correlate significantly with simulating tests. The complex tests cannot be easily and/or safely administered. These tests are: positive g to blackout (acceleration); extensive anthropometric and photogrammetric measurements, somatotyping (anthropological); urinary catacholamines, plasma corticosteroids, urinary 3-methoxy-4-hydroxymandelic acid (biochemical); speech intelligibility (biological acoustical); 2 hours of heat stress (thermal); treadmill, MC-1 partial pressure suit (physical fitness); all tests administered (psychological); and maximum breathing capacity, bicycle ergometer, electrical stimulation of muscles (Lovelace Foundation).

6. <u>In the final recommendation of candidates the investigators must only interpret subject performance on the simulating tests. Nonsimulating test performance will not affect recommendation of this first group of candidates.</u> Some of these nonsimulating tests were interpreted and did affect the recommendation of candidates. This was intentional. The sum total of data gathered from all of the simulating tests, although valuable, was insufficient to render candidate recommendations with confidence. However, the investigators agreed that, if they were also allowed to interpret some of the nonsimulating tests with which they were intimately familiar, they could then attach great confidence to the final recommendations. It was unanimously agreed that each investigator-group would be allowed to interpret the nonsimulating tests which they chose. The main goal of this particular crew selection development program was to recommend outstanding candidates. An important but secondary goal was to discover the existence of significant correlations. It was unsound practice to omit data or impressions which might possibly affect the success of Project Mercury. Those nonsimulating tests which were interpreted and which did affect the final candidate recommendations were: positive g (acceleration); index of strain (thermal); Harvard step, Flack, cold pressor (only if feet were prematurely withdrawn), treadmill, MC-1 partial pressure suit (if subject terminated test for psychological reasons), tilt table (physical fitness tests); and all tests administered (psychological).

Those nonsimulating tests which were not used in the final candidate recommendations were: all measurements (anthropological); all measurements (biochemical); speech intelligibility (biological acoustical); and cold pressor test development of hypertension and/or tachycardia, MC-1 test development of presyncope or tachycardia >160, Valsalva overshoot., and tilt table (physical fitness tests).

7. <u>All candidates both recommended and not recommended must enter the project</u>. All of the candidates did not enter the project. The final selection took into consideration all of the assets of the candidates. These assets included past training, experience, recommendations from the Lovelace Foundation, and recommendations from the Aerospace Medical Laboratory (AML).

8. <u>At the completion of the project all of the participants must be graded on the effectiveness of their performances</u>. The above condition has not been satisfied as this report nears completion. It will require several years to satisfy this condition.

9. <u>The investigators must then seek significant correlations between subject performances on the various simulating and nonsimulating tests and successful mission performances</u>. Since condition [5] <u>8</u>. is not satisfied, this condition also cannot be satisfied. An alternative approach has been used. It has been assumed that the Mercury Astronauts are the best potential group to fulfill the mission of Project Mercury. It has also been assumed that they will carry out the mission successfully. There is confidence that these assumptions will mature into fact. Based upon these assumptions a significant correlation study has been sought. Ideally, it is premature. Practically, it is valuable, since the program has demonstrated tests that should be pursued in future crew recommendation studies.

Each chapter has been written by the appropriate principal investigator. Throughout this report the candidates will be referred to by alphabet letters assigned to their names. There is no relationship between these alphabetical designations and their names or NASA numbers. It is impossible for the reader to identify a particular subject's name or performance. This system was designed to maintain the privileged communication due each candidate.

<u>REFERENCES</u>

0.1. McGuire, T. F. <u>Stress Tolerance Studies</u>, Part I. Unpublished Data. 1958.
0.2. McGuire, T. F. <u>Tolerance to Physical stress</u>, Part II. Unpublished Data. 1958.

[pp. 6-98 not included]

[99]
CHAPTER X

<u>DISCUSSION AND RECOMMENDATIONS</u>

Thirty-one highly selected adult males were the subjects of a crew recommendation study. Data were gathered from the performance of each subject on each test. One hundred and four performance variables were correlated. The following statements represent preliminary impressions from this Project Mercury Candidate Evaluation Program. It is recognized that the investigators were studying a small, highly selected population. Therefore, it is difficult to render conclusions on statistical significance.

1. Psychological stability is the most important consideration in evaluating a candidate. The intelligence, maturity, and motivation of a candidate are vital areas to be assessed before rendering a recommendation.

2. Excellent physiological performance was a secondary consideration in the final Committee recommendations.

3. The main value of a severely stressful physiological test was the interpretation of the psychological response to that stress test. Whenever a subject terminated a severe test for psychological reasons, he was not recommended by the Committee.

4. It is possible to eliminate subjects by use of stressful tests. It is not presently possible to select subjects with confidence, where selection is based entirely upon their excellent physiological performances.

5. No single, nonsimulating test has been identified which will be of great assistance in recommending crew members. A large battery of tests, such as were performed, lends confidence to the final recommendations.

6. Whenever a candidate is being considered for a special mission, it is desirable that a large number of trained observers each have the opportunity to test him and to render an opinion before the final recommendation

7. This study has demonstrated that there is no statistically significant difference in the physiological or biochemical responses of the Mercury Astronauts when compared with the remainder of the NASA candidates.

8. There is no evidence to support a thesis which maintains that visual inspection, biochemical measurements, or physiological responses of a candidate are of principal value in rendering a reliable recommendation of suitable candidates. These are secondary considerations.

9. While the hormones and their metabolites are valuable research tools, this study has demonstrated that they were not significantly different in the Mercury Astronauts when compared with the remaining NASA candidates.

10. There is every reason to suspect that safe, standardized, moderately stressful and severely stressful tests (such as having the subject walk on the treadmill until he voluntarily terminates) would be of great assistance in future crew recommendation programs, since severe stress also tests the candidate's motivation.

11. It is believed that testing of those who did not volunteer as candidates would be valuable, since the nonvolunteer group might lack the same intensity of motivation which was possessed by the volunteers.

Document I-30

Document title: John Glenn, Mercury Astronaut, NASA, to Lieutenant Commander Jim Stockdale, USN, 17 December 1959.

Source: Folder 18674, NASA Historical Reference Collection, NASA History Division, NASA Headquarters, Washington, DC.

This letter from astronaut John Glenn to then Lieutenant Commander James B. Stockdale, United States Navy, offers a personal perspective on the early Mercury program and the role of the astronaut in it. Stockdale, a classmate of Glenn's at the Navy's Test Pilot School at Patuxent River, Maryland, would later gain fame as one of the earliest heroes of Vietnam when he was shot down in 1965 and was held as a prisoner of war in the Hoa Lo prison for seven years. Debilitated by torture and maltreatment, Stockdale could hardly walk upon his return to the U.S. in 1973. He received the Medal of Honor in 1976. Stockdale eventually retired from the Navy as a Vice Admiral. In 1992, he was a candidate for Vice President on a ticket headed by Ross Perot.

December 17, 1959

Lt. Commander Jim Stockdale, USN
VF-24
C/O FPO
San Francisco, California

Dear Jim:

Quite a bit of water over the dam or under the bridge since I saw you last. Saw Phil Bolger at the TPT Reunion a couple of months ago and, in talking about various and sundry subjects, your name came up <u>naturally</u>. I don't know if you fall in the category of various or sundry, but anyhow, Phil reminded me again of something I had already known before and that was of your interest in the space program. So I thought I would give you a short run down on what we have been doing.

How are things doing on the USS Boat incidentally? I don't have any idea if you are still deployed or not, but I think from what Phil said that this letter will probably find you still at sea.

This past 8 or 9 months has really been a hectic program to say the least and by far the most interesting thing in which I have ever taken part, outside of combat. It is certainly a fascinating field, Jim, and growing so fast that it is hard to keep up with the major developments, much less everything in the field.

Following our selection in April, we were assigned to the Space Task Group, portion of NASA at Langley Field, and that is where we are based when not traveling. The way it has worked out, we have spent so much time on the road that Langley has amounted to a spot to come back to get clean skivvies and shirts and that's about all. We have had additional sessions at Wright Field in which we did heat chamber, pressure chamber, and centrifuge work and spent a couple of weeks this fall doing additional centrifuge work up at NADC, Johnsville, Pennsylvania. This was some program since we were running it in a lay-down position similar to that which we will use in the capsule later on and we got up to as high as 16 g's. That's a batch in any attitude, lay-down or not.

With the angles we were using, we found that even lying down at 16 g's, it took just about every bit of strength and technique you could muster to retain consciousness. We found there was quite a bit more technique involved in taking this kind of g than we had thought. Our tolerances from beginning to end of runs during the period we worked up there went up considerably as we developed our own technique for taking this high g. A few runs a day like that can really get to you. Some other stuff we did up there involved what we call tumble runs or [2] going from a +g in 2 seconds to a –g and the most we did on this was in going from a +9g to a –9g. Obviously, a delta of 18. This was using pretty much a standard old A/N seat belt, shoulder harness type restraint system that we have used in Beechcraft for many years. When we first talked about doing this, I didn't think it would be possible at all, but in doing a careful build-up, we happily discovered that this was not so horrible. At +9g to –9g, we were bouncing around a bit but it was quite tolerable.

I guess one of the most interesting aspects of the program has been in some of the people we have been fortunate enough to meet and be briefed by. One of the best in this series was the time we spent at Huntsville, Alabama, with Dr. Wernher von Braun and crew. We were fortunate enough to spend an evening with him in his home until about 2:30 in the morning going through a scrap book, etc., from Peenemunde days in Germany and, in general, shooting the bull about his thoughts on the past, present, and future of space activities. This was a real experience for a bunch of country boys fresh caught on the program and a very heady experience as you can imagine.

We have had a good run-down at Cape Canaveral and got to see one of their shots. I guess that is one of the most dramatic things I have ever seen. The whole procedure they go through for a night launch at the Cape is just naturally a dramatic picture far better than anything Hollywood could stage. When the Big Bird finally leaves the pad, it doesn't have to be hammed up to be impressive.

Much of our work, of course, has involved engineering work on the capsule and systems. My particular specialty area has been the cockpit layout and instrumentation presentation for the Astronauts. This has been extremely interesting because we are working on an area way out in left field where our ideas are as good as any one else's. So, you try to take the best of your past experiences and launch from there with any new ideas you can contrive. This is the kind of development work, as you well know, that is by far the most enjoyable.

We just finished an interesting activity out at Edwards Air Force Base doing some weightless flying in the F-100. This was in the two-place F-100 so that we could ride in the rear seat and try various things such as eating and drinking and mechanical procedures while going through the approximately 60-second ballistic parabola that you make with the TF-100. That started at about 40,000 feet, 30 degree dive to 25,000, picking up about 1.3 to 1.4 Mach number, pull out and get headed up hill again at 25,000 and about a 50 to 60 degree climb angle, at which point they get a zero-g parabola over the top to about 60 degrees down hill.

[3]

You can imagine quite a bit in a full minute in those conditions and contrary to this being a problem, I think I have finally found the element in which I belong. We had done a little previous work floating around in the cabin of the C-131 they use at Wright Field. That is even more fun yet, because you are not strapped down and can float around in the capsule doing flips, walk on the ceiling, or just come floating full length of the cabin while going through the approximately 15-seconds of weightlessness that they can maintain on their shorter parabola. That was a real ball and we get some more sessions with this machine some time after the first of the year.

Before this next year is out, we should get the manned Redstone ballistic shots started which will put us to orbital altitude of 105 nautical miles, but not up to the orbital speeds so that we arc back down off the Cape about 200 miles from the pad. We figure now that the first actual manned orbital shots should follow in mid to late 1961.

If you get back this way, Jim, be sure and give me a call. There is no information available yet at all on follow-on programs and what or who might get involved in them. I know you are probably still interested in that who part.

I don't know if this letter is too informative, but if it gets any longer we will have to grade it like a TPT flight report - by the pound.

Give my regards to the family and I hope you get off that unprivate yacht before too long

Sincerely,
[Signed John Glenn]

Document I-31

Document title: Robert B. Voas, NASA Space Task Group, "Project Mercury Astronaut Training Program," 30 May 1960.

Source: NASA Collection, University of Houston, Clear Lake Library, Clear Lake, Texas.

During the build up to the first U.S. spaceflight, NASA's public affairs staff allowed controlled media access to the Mercury astronauts; numerous photographs abound showing the Mercury Seven engaged in weightless simulations during parabolic flight, centrifuge tests, altitude chamber research, physical fitness training, survival school, pilot proficiency preparation, or a host of other activities. But even so, few understood how these various activities fit together and led to the creation of the astronaut team that flew on Project Mercury. This document helps to explain this process, providing a general outline of the nature of the various activities of the astronauts as they prepared for their missions, and the reasons for undertaking such rigorous activities.

PROJECT MERCURY ASTRONAUT TRAINING PROGRAM

By Robert B. Voas*

NASA Space Task Group
Langley Field, Va.

For Presentation at the Symposium on
Psychophysiological Aspects of Space Flight

San Antonio, Texas
May 26-27, 1960

*Voas- Training Officer, Lt. U.S. Navy Medical Service Corps, assigned to NASA
Space Task Group.

[1]

SUMMARY

This paper gives a general outline of the NASA Project Mercury Astronaut
training program. Basic considerations which entered into the development of
the program are listed. Six primary training areas are described, together with the
training equipment and facilities which have been employed. Problem areas for
future training programs are discussed.

INTRODUCTION

Any training program must be based on three factors: the nature of the
job for which training is required, the characteristics of the men to be trained, and
the facilities and time available in which to do the training. In Project Mercury
the Astronaut's job involves both flight and nonflight tasks. He is expected to
contribute to systems design and to the development of operational procedures
through his daily contact with the project engineers. It was considered that by
virtue of the selection process, the Astronaut had the required skills to make
these contributions; therefore, no training was attempted for these nonflight
tasks. The Astronaut's in-flight activities can be broken down into six areas: (1)
"programming" or monitoring the sequence of vehicle operations during launch,
orbit, and reentry; (2) systems management - the monitoring and operation of
the onboard systems, such as the environmental control, the electrical systems,
the communications systems, and so forth (3) the vehicle attitude control; (4)
navigation; (5) communications; and (6) research and evaluation. In addition to
these in-flight activities, the Astronaut has a number of ground tasks connected
with the flight operations. He has a role in the countdown and preparation of the
vehicle; in communications from the ground to the vehicle; and in the recovery
program following the flight. It is for these activities associated with the flight
itself that a training program was undertaken. More detailed descriptions of the
Astronaut's tasks are available in papers by Slayton (ref. 1) and Jones (ref. 2).
It should be noted that the Astronaut's job is only one of many associated with
space flight for which training is required. Brewer (ref. 3) has outlined the overall
training requirements for Project Mercury.
[2]

The Astronaut selection program was designed to select individuals who would require a minimum of training in order to fulfill the Mercury job requirements. Particularly desired were individuals who had sufficient experience in aircraft development operations to make immediate contributions to the Project Mercury program. On this basis, the following criteria were adopted as the minimum requirements for qualification as a Project Mercury Astronaut:

(1) Age - less than 40

(2) Height - Less than 5 ft 11 in.

(3) Excellent physical condition

(4) Bachelor's degree (or equivalent)

(5) Graduate of test-pilot school

(6) 1,500 hours flying Time

(7) Qualified jet pilot

Records of 508 Air Force, Navy, Marine, and Army pilots who had graduated from test-pilot school were reviewed and screened on the basis of these requirements. Of these, 110 met the seven basic requirements. Forty-one of these pilots were eliminated through further screening based on recommendations from instructors at the test-pilot schools. The remaining 69 pilots were interviewed and given an opportunity to volunteer for the Project Mercury program. Of these, 37 pilots either declined or were eliminated as a result of the initial job interviews. The remaining 32 who were considered to be qualified in education and experience were given detailed medical examinations and were exposed to the physical stresses expected in the space flight. The nature of these tests has been described in more detail in references 4 and 5. On the basis of the medical examination and the stress tests, the number of candidates was reduced to 18, from which were selected the seven who demonstrated the most outstanding professional background and knowledge in relationship to the job requirements. Through this procedure, a group of experienced test pilots with extensive training in engineering, excellent health, and a high motivation in the Mercury Project were selected for the training program. The availability of such individuals makes it possible to utilize to a great extent self-instruction and to minimize the amount of formal group training required.

At the outset, few, if any, facilities were available to support the training program. Both training devices and training manuals have become available in stages throughout the first 12 to 15 months of the training program. The more elaborate and complete training devices were [3] not placed in operation until over a year after its initiation. As a result, the early part of the training program depended upon review of design drawings in vehicle components and on travel to various Mercury production facilities to attend design briefings. Great dependence was put upon verbal presentations by scientists of the NASA Space Task Group and of the prime contractor. In addition, early in the program, extensive use was made of established Armed Forces aeromedical facilities for familiarizing the Astronauts with the conditions of space flight. Thus, the training methods and the order in which topics were presented were, to a great extent, dictated by the resources available at the time the program was initiated.

Since mature, intelligent trainees were selected and since little if any training equipment was available initially, it might have been argued that the Astronauts should be allowed to work completely on their own without any attempt to a group program. There are, however, a number of desirable factors to be gained by such a program. A planned group program facilitates the scheduling of activities with other organizations. In addition, a structured program permits more efficient use of instructor and student time. It also makes possible progress from one aspect of the operation to the next in an appropriate sequence. Sequence in training activities is important, since learning is simplified if material is presented in a logical order. An organized program also insures completeness in that no major training requirement is overlooked. Finally, since this project represents a first effort of its kind, the use of a group program facilitates the collecting of records and the evaluation both of the Astronaut's progress and of the various training activities.

The program which has resulted from these considerations has allotted about one-half of the time to group activities and the other half to individually planned activities in each Astronaut's area of specialization. A review of the Astronauts' travel records provides an example of the relative division of their time between group training and other duties associated with the development of the Mercury vehicle. During the 6-month period from July 1 to December 31, 1959, the Astronauts were on travel status almost 2 months or 1 out of every 3 days. Half of this travel time (28 days) was spent on four group-training activities: a centrifuge program; a trip to Air Force Flight Test Center, Air Force Ballistic Missile Division, and Convair; a weightless flying program; and trips to fly high-performance aircraft during a period when the local field was closed. The other half of their travel time (27 days) was devoted to individual trips to attend project coordination meetings at McDonnell and the Atlantic Missile Range, or for pressure-suit fittings, couch moldings, and viewing of qualification tests at McDonnell, B. F. Goodrich Co., and their subcontractors' plants. These individual activities, while providing important trailing benefits, are primarily dictated by the Project Mercury development program requirements and are not considered part of the group training program.
[4]

The extent to which the Mercury crew space area is "customized" to the seven Astronauts and the time required to fit the man to the vehicle should be noted. Each man has had to travel to B. F. Goodrich for a pressure-suit fitting and to a subcontractor for helmet fittings; then to the Air Crew Equipment Laboratory for tests to the suit under heat and lowered pressure; then to McDonnell for couch molding. Usually, he has been required to return to the suit manufacturer for a second fitting and to McDonnell for final fittings of the couch and studies of his ability to reach the required instruments and controls in the capsule. While the Mercury vehicle is more limited in size than future spacecraft, the cost of space flight and the limited personnel involved will probably always dictate a certain amount of customizing of the crew space. The time required for this hype of activity should not be underestimated.

TRAINING PROGRAM

The Astronaut training program can be divided into six major topic areas. The primary requirement, of course, is to train the Astronaut to operate the vehicle. In addition, it is desirable that he have a good background knowledge of such scientific areas related to space flight as propulsion, trajectories, astronomy, and astrophysics. He must be exposed to and familiarized with the conditions of space flight such as acceleration, weightlessness, heat, vibration, noise, and disorientation. He must prepare himself physically for those stresses which he will encounter in space flight. Training is also required for his duties at ground stations before and after his own flight and during the flight of other members of the Astronaut team. An aspect of the training which might be overlooked is the maintenance of the flying skill which was an important factor in his original selection for the Mercury program.

Training in vehicle operation. – Seven training procedures or facilitates were used in developing skills in the operation of the Mercury capsule. These included lectures on the Mercury systems and operations; field trips to organizations engaged in the Mercury Projects; training manuals; specialty study programs by the individual Astronaut; mockup inspections; and training devices. To provide the Astronaut with a basic understanding of the Mercury system, its components, and its functions, a lecture program was set. up. A short trip was made to McDonnell at which time a series of lectures on the capsule systems was presented. These systems lectures were then augmented by lectures on operations areas by Space Task Group scientists. This initial series of lectures provided a basis for later self-study, in which use was made of written descriptive material as it became available. Individual lectures have been repeated as the developments within project Mercury have required a series of lectures on capsule systems by both Space Task Group and McDonnell personnel have been scheduled to coincide with the delivery [5] and initial operation of the fixed-base Mercury trainer. In these lectures, the same areas are reviewed in an attempt to bring the Astronauts up-to-date on each of the systems as they begin their primary procedures training program.

In addition to this lecture program, indoctrination trips have been made to the major facilities concerned with the Project Mercury operations. Two days were spent at each of the following facilities: McDonnell, Cape Canaveral, Marshall Space Flight Center, Edwards Flight Test Center, and Space Technology Laboratories and Air Force Ballistic Missile Division. One day was spent at Rocketdyne Division, North American Aviation, and five days were spent at Convair/Astronautics. At each site there was a tour of the general facilities together with a viewing of Mercury capsule or booster hardware and lectures by top-level personnel covering their aspect of the Mercury operation. The Astronauts also had an opportunity to hear of related research vehicles such as the X-15 and Discoverer and received a brief discussion of the technical problems arising in these programs and their significance to Project Mercury.

Obtaining current and comprehensive study materials on a rapidly developing program such as Project Mercury is a major problem. McDonnell has

.

been providing manuals covering Project Mercury systems. The first of these was the Indoctrination Manual and was delivered at the time of an early Astronaut visit in May 1959. No attempt was made to keep this manual current and a first edition of a full systems manual (Familiarization Manual) was issued in September 1959. It quickly became out of date, however, and a new manual, a second edition of the Familiarization Manual was issued in December of the same year. A first copy of the Capsule Operations Manual (Astronauts' Handbook) was delivered in June of 1960. During initial phases of the program, the Astronauts have had to depend primarily on capsule specifications and specification control drawings for written information on capsule systems. Copies of these, however, were not always available and they were too large to compile into a single manual.

Valuable aids to the Astronauts in keeping abreast of the status of the development program are the regularly issued reports of the Capsule Coordination Group Meetings. At these meetings, the status of each of the capsule systems is reported and any changes are discussed. Miscellaneous reports on boosters and on programs conducted by cooperating agencies have also been provided to the Astronauts. Maintaining an up-to-date flow of accurate information on vehicle development status is a critical problem not only for the Mercury training program, but in all probability, for most near-future space flight applications since training must proceed during the vehicle development phase.

Another method employed to aid in the dissemination of information to the Astronauts was to assign each a specialty area. These assignments [6] were as follows: M. Scott Carpenter, navigation and navigational aids; Leroy G. Cooper, Redstone booster; John A. Glenn, crew space layout; Virgil I. Grissom, automatic and manual attitude control system; Walter M. Schirra, life support system; Alan B. Shepard, range, tracking, and recovery operations; and Donald K. Slayton, Atlas booster. In pursuing these specialty areas, each man attends meetings and study groups at which current information on capsule systems is presented. Regular periods are set aside for all the men to meet and report to the group. Another important source of information about the vehicle, particularly in the absence of any elaborate fixed-base trainers, bas been the manufacturer's mockup. Each of the men has had an opportunity to familiarize himself with the mockup during visits to McDonnell.

Following the initial familiarization with the Mercury system, the primary training in vehicle operation is being achieved through special training devices developed for the Mercury program. Early training in attitude control was accomplished on the Langley Electronics Associates Computer (fig. 1) [not included] which was combined with a simulated Mercury attitude display and hand controller. This device was available during the simmer of 1959. Later, another analog computer was cannibalized from an F-100F simulator and combined with actual Mercury hardware to provide more realistic displays and controls. This MB-3 trainer (fig. 2) [not included] also included provision for the Mercury couch and the pressure suit.

In addition to these two fixed-base simulators, three dynamic simulators were used to develop skill in Mercury attitude control. The first, of these, the ALFA (Air Lubricated Free Attitude) Simulator (fig. 3) [not included] permits the practice of orbit and retrofire attitude control problems by using external reference through simulated periscope and window displays. A simulated ground track is projected on a large screen which is viewed through a reducing .lens to provide the periscope display. This simulator also permits training in the use of earth reference for navigation. The Johnsville Centrifuge (fig. 4) [not included] was used as a dynamic trainer for the reentry rate damping task because it adds the acceleration cues to the instruments available in the fixed-base trainers. It also provides some opportunity to practice sequence monitoring and emergency procedures during launch and reentry. Another dynamic simulation device used to provide training in recovery from tumbling was the three-gimbaled MASTIF (Multi-Axis Spin Test Inertia Facility) device at the NASA Lewis Laboratory (fig. 5) [not included]. In this device, tumbling rates up to 30 rpm in all three axes were simulated and the Astronaut was given experience with damping these rates and bringing the vehicle to a stationary position by using the Mercury rate indicators and the Mercury-type hand controller.

Two more elaborate trainers became available in the summer of 1960. These trainers provide practice in sequence monitoring and systems management. The McDonnell Procedures Trainer (fig. 6) [not included] is similar to the fixed-base trainers which have become standard in aviation operations. The [7] computer used on the MB-3 has been integrated with this device to provide simulation of the attitude control problem. External reference through the periscope is simulated by using a cathode ray tube with a circle to represent the earth. Provision has been made for pressurizing the suit and for some simulation of heat and noise effects. The environmental control simulator (fig. 7) [not included] consists of the actual flight environmental control hardware in the capsule mockup. The whole unit can be placed in a decompression chamber in order to simulate the flight pressure levels. This device provides realistic simulation of the environmental-control system functions and failures. Effective use of these two simulators is predicted upon adequate knowledge of the types of vehicle systems malfunctions which can occur. A failure-mode analysis carried out by the manufacturer has provided a basis for determining the types of malfunction which are possible and the requirements for simulating them (ref. 2). A record system on which possible malfunctions are listed on cards, together with methods of their simulation, has been set up. On the back of these cards there is space for noting when and under what conditions this failure has been simulated and what action the Astronaut took to correct it. In this way, it is hoped that the experience in the detection and correction of systems malfunctions can be documented.

Training in space sciences. – In addition to being able to operate the Mercury vehicle, the Astronaut will be required to have a good general knowledge of astronomy, astrophysics, meteorology, geophysics, rocket engines, trajectories, and so forth. This basic scientific knowledge will enable him to act as a more acute observer of the new phenomena with which he will come in contact during the flight. It will also provide a basis for better understanding of the detailed information which he must acquire on the Mercury vehicle itself. In order to provide this broad background in sciences related to astronautics, the Training Section of the Langley Research Center set up a lecture program which included the following topics: Elementary Mechanics and Aerodynamics (10 hours); Principles of Guidance and Control (4 hours); Navigation in Space (6 hours);

Elements of Communication (2 hours); Space Physics (12 hours). In addition, Dr. W. K. Douglas, Flight Surgeon on the Space Task Group staff, gave 8 hours of lectures on physiology.

Following this initial lecture program, training in specific observational techniques is planned. The first activity of this program was training in the recognition of the primary constellations of the zodiac at the Morehead Planetarium in Chapel Hill, North Carolina. A Link trainer body was modified with a window and headrest to simulate the capsule external viewing conditions. Using this device, the Astronauts were able to practice the recognition of constellations which the Planetarium was programmed to simulate orbital flight. Future plans call for further training in star recognition together with methods of observing solar and meteorological events, earth and lunar terrain, and psychological and physiological reactions. These activities will be in support of [8] a primary objective of the Project Mercury program which is to determine man's capability in a space environment. The training program contributes to this objective in three ways:

(1) First, by establishing base lines, both for the Astronaut's performance and his physiological reactions. These base lines can then be compared with psychological and physiological. factors in the space environment.

(2) Second, through the program in basic sciences described above, the Astronaut is given sufficient background with which to appreciate the importance of the observations which he can make in the space environment.

(3) Specific training in observational techniques and the use of scientific equipment arms him with the skills with which to collect data of value to science.

Thus, the training program attempts to lay the ground work for the scientific activities of the Astronauts, as well as to provide the specific skills which are required to fly the Mercury vehicle.

Familiarization with conditions of space flight. – An essential requirement of the training program is to familiarize the Astronaut with the novel conditions which man will encounter in space flight. An important part of the Astronaut training program has been to provide the trainees with an opportunity to experience eight types of conditions associated with Mercury flights: high acceleration, weightlessness, reduced atmospheric pressures, heat, disorientation, tumbling, high concentration of CO_2, and noise and vibration.

The Astronauts experienced acceleration patterns similar to those associated with the launch and reentry of the Mercury first at the Wright Air Development Division (WADD) in Dayton, Ohio, and later at the Aviation Medical Acceleration Laboratory at Johnsville, Pennsylvania. During this training, they were able to develop straining techniques which reduced the problem of blackout and chest pain. It was generally the opinion of the Astronauts that the centrifuge activity was one of the most valuable parts of the training program.

The Astronauts were given an opportunity to experience weightless flying both in a free-floating condition in C-131 and C-135 aircraft and strapped down to the rear cockpit of a F-100 F fighter. While the latter is more similar to the Mercury operation, the Astronauts, being experienced pilots, felt that there was little or no difference between this experience and their normal flying activities. The free-floating state, however, they felt was a novel and enjoyable experience. Since the longer [9] period of weightlessness available in the F-100E aircraft is valuable for collecting medical data, while the C-131 aircraft appears to give the most interesting experiential training, both types of operations appear to be desirable in a training program. The fact that the pilots experienced no unusual sensations during weightlessness when fully restrained was an encouraging finding for the Mercury operation and supports the desirability of selecting flying personnel for this type of operation.

The Astronauts experienced reduced atmospheric pressure while wearing full pressure suits first at WADD and later at Air Crew Equipment Laboratory (ACEL); in addition to reduced pressure, they also experienced thermal conditions similar to those expected during the Mercury reentry while wearing a full pressure suit. At the Naval Medical Research Institute (NMRI), they were given an opportunity to become familiar with the body's thermal response and the effect of moderate heat loads on the body's regulatory mechanisms was demonstrated. At the end of March 1960, the Astronauts experienced disorientation in the U.S. Naval School of Aviation Medicine Slowly Revolving Room. As already mentioned, they have also experienced angular rotation up to 50 rpm in all three axes on a gimbaled device with three degrees of freedom at the NASA Lewis Laboratory.

In order to indicate the effects of the high concentration of CO_2 which might result, from a failure of the environmental. control system, the Astronauts were given a 3 hour indoctrination period in a sealed chamber at NMRI In this chamber, they experienced a slow buildup of CO_2 similar to that which they would encounter in the event of failure of the environmental system. None of the men showed any adverse effects of symptoms from this training; As part of the selection program, the Astronauts experienced high noise and vibration levels at WADD. During the second Johnsville centrifuge program, noise recorded of the Mercury test flight was played back into the gondola. Further opportunities to adapt to the high noise levels associated with the Mercury launch will be provided by a sound system connected to the McDonnell Procedures Trainer at Langley Field.

Physical fitness Program. – To insure that the Astronaut's performance does not deteriorate significantly under the various types of stresses discussed in the previous section, it is important that he be in excellent physical condition. Since most of the trainees entered the Project Mercury program in good physical health, a group physical fitness program, with one exception, has not been instituted. SCUBA training was undertaken because it appeared to have a number of potential benefits for the Project Mercury, in addition to providing physical conditioning. It provides training in breathing control and analysis of breathing habits, and in swimming skill (desirable in view of the water landing planned in the Mercury program). Finally, there is, in the buoyancy of water, a partial simulation of weightlessness, particularly if vision is [10] reduced. Aside from this one organized activity, each individual has been undertaking a voluntary fitness program tailored to his own needs. This program has included, for most of the Astronauts, three basic items. First of all, as of December 1959, they have reduced or completely stopped smoking. This was an individual, voluntary decision and was not a result of pressure by medical personnel, but a result of their own assessment of the effect of smoking on their tolerance to the stresses to be encountered in the flight, particularly acceleration. Some of the members of the team who have a tendency to be overweight have initiated weight-control programs through proper diet. Nearly all members make it a habit to get some form of daily exercise.

Training in ground activities. – Frequently overlooked are the extent. and the importance of the ground activities of the Astronauts. Their knowledge of the vehicle and its operation makes them specially qualified for certain ground operations. The training in ground procedures has fallen into the three main areas; countdown procedures, ground flight monitoring procedures, and recovery and survival. The Astronauts are participating in the development of the countdown procedures and will be training themselves in their own part of the countdown through observation of countdown procedures for the initial unmanned shots, and finally, by participating in the preparation procedures for the actual manned flights.

An important aspect of the Astronaut's activities when not actually flying the vehicle will be to aid in ground communications with the Mercury capsule. Since he is fully familiar with the capsule operation and intimately acquainted with the Astronaut who will be in the capsule, he makes a particularly effective ground communicator. Procedures for ground monitoring and communicating personnel are presently being developed with the aid of the Astronauts. At Langley Field, a ground monitoring station simulator will be tied in with the McDonnell procedures simulator. By using this device, ground station activities can be practiced and coordinated with capsule simulator training. They will also participate in training exercises at the Mercury Control Center at Cape Canaveral. Finally, just prior to manned flights, Astronauts not involved in launch activities will be deployed to remote communications stations, at which time they will have an opportunity for some on-site training.

A final area of ground training is in recovery and survival procedures. Study materials such as maps and terrain descriptions of the areas under the Mercury orbits are being obtained. They will be augmented by survival lectures and by field training in survival at sea and in desert areas. Finally, extensive training on egress from the capsule into the water has been given. This activity was accomplished in two stages, using the Mercury egress trainer (fig. 8) [not included]. Phase I made use of a wave-motion simulation tank at Langley Field for initial training followed by a Phase II program in open water in the Gulf of Mexico.

[11] Maintenance of flight skills. – One of the continuing problems in training for space flight is the limited opportunity for actual flight practice and proficiency training. The total flight time in the Mercury capsule will be no more than 4 to 5 hours over a period of 3 years for each Astronaut. The question arises as to whether all the skills required in operating the Mercury vehicle can be maintained purely through ground simulation. One problem with ground simulation relates to its primary benefit. Flying a ground simulator never results in injury to the occupant or damage to the equipment. The penalty for failure is merely the requirement to repeat the exercise. In actual flight operations, failures are penalized far more severely. A major portion of the Astronaut's tasks involves

high-level decision making. It seems questionable whether skill in making such decisions can be maintained under radically altered motivational conditions. Under the assumption that vigilant decision making is best maintained by experience in flight operations, the Mercury Astronauts have been provided with the opportunity to fly high-performance aircraft. The program in this area is a result of their own interest and initiative and is made possible by the loan and maintenance of two F-102 aircraft by the Air Force.

IMPLICATIONS FOR FUTURE PROGRAMS

In conclusion, the problems with implications for future space flight projects which have been encountered in development of the Mercury program can be reviewed. In developing skills in operation of the vehicle, the difficulty of providing up-to-date information on the systems when the training must progress concurrently with the development program has been discussed. Concurrent training and development should tend to be a feature of future space flight programs, since many of these will be experimental in nature, rather than operational.

All spacecraft have in common the problem of systems which must be kept functional for long periods without recourse to ground support. Even in the event of emergency termination of the mission with immediate return to earth, prolonged delay may occur before safe conditions within the atmosphere have been achieved. Thus, emphasis on "systems management" will increase in future space operations. Recognition of malfunctions has always been a part of the pilot's task; usually, however, little in-flight maintenance is attempted. Since aborts are dangerous and, in any event, involve greater delay before return, the Astronaut must do more detailed diagnoses of malfunctions and more in-flight maintenance. This will require extensive knowledge of the vehicle systems and training in malfunction isolation and correction. In order to provide this training as many as possible of the numerous malfunctions which can occur in even a relatively simple space vehicle must be identified and simulated. Considerable effort has been devoted to this area in the Mercury training and [12] development program and it should become an increasingly important feature of future programs.

The physical conditions (heat, acceleration, and so forth) associated with space flight are simulated to permit the trainees to adapt to these stressors in order that during the actual flight such stimuli may be less disturbing. Present measures of the adaptation process are inadequate to provide criteria for training progress. A second purpose for the familiarization program was to give the trainees an opportunity to learn the specific skills required to minimize the effects of these factors on their performance. However, in many cases, the skills required have not been fully identified or validated. For example, in developing straining techniques for meeting increased acceleration, the efficacy of a straining technique has not been fully demonstrated nor has the technique itself been adequately described. As yet, inadequate data are available on the effects of combining physical stress factors. Therefore, it is difficult to determine the extent to which the increased cost and difficulty of providing multiple stress simulation is warranted. In the present program, it has been possible to simulate both reduced atmospheric pressure and acceleration on the centrifuge. Initial experience seems to indicate that this is desirable but not critical. However, further data on the interacting effects of these stresses are required before any final conclusions can be developed.

A factor in space flight not yet adequately simulated for training purposes is weightlessness. Short periods of weightlessness have been used in the present program, as described previously. True weightlessness can be achieved through too short a period to be fully adequate for training purposes. On the other hand, ground simulation methods using water seem to be too cumbersome and unrealistic to be fully acceptable substitutes. At the present time, this lack of adequate simulation does not seem to be critical since the effects of weightlessness on performance appear to be minor and transitory. Should early space flights uncover more significant problems, greater efforts will be justified in developing weightless simulation methods.

Finally, it seems important to reiterate the requirements for reproducing adequate motivational conditions in the training program. The basic task of the Astronaut is to make critical decisions under adverse conditions. The results of the decisions he makes involve not just minor discomforts or annoyances, but major loss of equipment and even survival. Performance of this task-requires a vigilance and decision-making capability difficult to achieve under the artificial conditions of ground simulation. It appears probable that-training in ground devices should be augmented with flight operations to provide realistic operational conditions.

[13]

REFERENCES

1. Slayton, Donald K.: Paper presented at Annual Meeting of Society of Experimental Test Pilots, Los Angeles, Calif., Oct. 1959.

2. Jones, Edward R.: Man's Integration Into the Mercury Capsule. [Preprint] 982-59, presented at the Am. Rocket. Soc. 14th Annual Meeting (Washington, D.C.), Nov. 1959.

3. Brewer, Gerald W.: Operational Requirements and Training for Project Mercury. Presented to Training Advisory Comm. of Nat. Security Ind. Assoc. (Los Angeles, Calif.), Nov. 17, 1959.

4. Douglas William K.: Selection and Training of Space Crews. Lectures in Aerospace Medicine. School of Aviation Medicine, USAF Aerospace Medical Center (Brooks AFB, Texas), Jan. 1960.

5. Wilson, Charles L.: Project Mercury Candidate Evaluation Program. WADC Tech. Rep. 59-505, U.S. Air Force, Dec. 1959.

Document I-32

Document title: Abe Silverstein, Director of Space Flight Programs, NASA, Memorandum for Administrator, NASA, "Astronaut Selection Procedure for Initial Mercury-Redstone Flights," 14 December 1960.

Source: Folder 18674, NASA Historical Reference Collection, NASA History Division, NASA Headquarters, Washington, DC.

For all of NASA's human spaceflight initiatives, the method of crew selection for individual missions has been one of its most contentious and misunderstood. The process outlined in this memorandum for choosing the first astronauts to fly in Project Mercury is instructive of the process. It called for each of the Mercury Seven to identify three top candidates other than themselves, and then for a board to recommend a selection of the top three, with actual selection made by the Project Mercury director. These three astronauts would then be trained for the mission, without any of them knowing who would actually be the prime. Only a few days before the launch the astronaut to fly the mission would be named and an order for future mission assignments would follow from that decision. Designed to give all a voice in the selection, this approach also sought to avoid a leak to the media of the astronaut selected.

[CONFIDENTIAL] [DECLASSIFIED]

In reply refer to:
DA-1 (WJN:rfr)
Dec 14 1960

MEMORANDUM for Administrator

Subject: Astronaut Selection Procedure for Initial Mercury-Redstone Flights

1. The purpose of this memorandum is to document the procedure for Mercury-Redstone Astronaut Selection as discussed in your office December 12.

2. The first possible manned Redstone mission is MR-3; however, manned occupancy of MR-3 is, of course, contingent upon successful accomplishment of MR-1 and MR-2. Since capsule 7, for MR-3, arrived at the Cape December 8, the pilot should be chosen in the near future so that he can become fully familiar with the capsule systems and operational procedures. Capsule 7 is the only manned configuration with a mechanical-latching side hatch, interim clock timer, small windows, and control system which does not include the rate stabilization mode.

3. Since it is impractical to train all 7 Astronauts on the proper Procedures Trainer configuration, three men will be chosen as possible pilots and all three will begin working with the capsule 7 configuration. It is hoped

that the identity of the three men can be kept secure from the press. The first pilot and his alternate will not be selected until approximately one week before the launch date of the first manned Redstone. The identity of the two-man flight crew for this flight would thus not be available for announcement to either the Astronauts or the press until approximately a week before the flight. The assignments for the first two manned Redstone flights might be as follows:

	Astronauts
First manned Redstone	1,3
Second manned Redstone	2,3

[2]

4. An astronaut Flight Readiness Board consisting of five men will be established with Robert Gilruth serving as chairman. Individuals on this board will evaluate the following pertinent areas of Astronaut performance.

    A. Medical
        a. General health
        b. Reaction to physical stress
        c. Weight control

    B. Technical
        a. Proficiency on capsule attitude control simulators
        b. Knowledge of capsule systems
        c. Knowledge of mission procedures
        d. Ability to contribute to vehicle design and flight procedures
        e. General aircraft flight experience
        f. Engineering and scientific background
        g. Ability to observe and report flight results

    C. Psychological
        a. Maturity
        b. Motivation
        c. Ability to work with others
        d. Ability to represent Project Mercury to public
        e. Performance under stress

Expert witnesses in the various areas will be called before the Board. Based on evaluations by the Board, the actual selection of the three men and of the final two-man flight crew will be made by the Director of Project Mercury.

5. The Astronauts themselves will be asked to submit to the Board chairman their recommendations for the three best-qualified pilots, excluding themselves. This input will be known by the chairman only and will be used as an additional factor in the selection.

6. The Flight Readiness Board will meet either during the week of December 26 or January 2.

                                        [Signed]
                                        Able Silverstein
                                        Director of Space Flight Programs

Enclosure:
        Astronaut instruction sheet

Cc: AD/Dryden
    AA/Seamans
    STG (Gilruth)

[3]

ASTRONAUT INSTRUCTION SHEET

        You are asked to submit a list of three Astronauts, excluding yourself, who
in your judgment are best qualified for the first two manned Redstone flights. We
assume you would rate yourself in the group, therefore please omit your name
from the list.

        #1. _____ .

        #2. _____ .

        #3. _____ .

Comments, if any:

                                Signed _____ .

                        _____
                        Document I-33

**Document title: Jerome B. Wiesner, The White House, Memorandum for
Dr. [McGeorge] Bundy, "Some Aspects of Project Mercury," 9 March 1961.**

**Source: Folder 18674, NASA Historical Reference Collection, NASA History
Division, NASA Headquarters, Washington DC.**

*When President Eisenhower chose to use military test pilots as astronauts in 1958 to1959,
it set in motion several discussions in various sectors about the signal it would send to the
world of the U.S.'s peaceful intentions in space. The inauguration of John F. Kennedy as
president in January 1961 brought to the nation's highest political office a different political
party and a different philosophy on such matters. This memorandum sent by Presidential*

science advisor Jerome Wiesner to National Security Advisor McGeorge Bundy, suggests that JFK revise that earlier decision and allow civilians to become astronauts. This plea became policy in 1962, when NASA chose its second groups of astronauts and for the first time civilians were selected for the position. The memorandum also discusses the then-controversial question of live television coverage on the initial Mercury suborbital launch.

THE WHITE HOUSE

WASHINGTON
March 9, 1961

MEMORANDUM FOR DR. BUNDY

SUBJECT: Some Aspects of Project Mercury

　　We have an ad hoc panel which is making a technical review of the National Aeronautics and Space Administration project to put a man in an earth orbit, Project Mercury. The time is now nearing when man will be first introduced into the system in a sub-orbital launch using a modified Redstone booster. Although the interest of the panel is primarily in the technical details, two phases of the operation which fall mostly outside the technical area have caused them considerable questioning, and I would like to take this opportunity to bring them to your attention.

　　1.　Many persons involved in the project have expressed anxiety over the mounting pressures of the press and TV for on-the-spot coverage of the first manned launch. Our panel is very concerned that every precaution should be taken to prevent this operation from becoming a Hollywood production, because it can jeopardize the success of the entire mission. The people in the blockhouse and in the control center are not professional actors, but are technically trained people involved in a very complex and highly coordinated operation. The effect of TV cameras staring down their throats during this period of extreme tension, whether taped or live, could have a catastrophic effect. Similarly, following a manned launch and recovery, the astronaut must be held in a confined area for a considerable time period so that the doctors can accomplish the debriefing which will produce the basic information on possible effects of space flight on man. The pressures from the press during this time period will probably be staggering, but should be met with firmness. The experience with the RB-47 pilots has proven that this can be accomplished. [2]

　　Our panel does not profess to be expert in the field of public relations, but the overriding need for the safety of the astronaut and the importance to our nation of a successful mission make them feel that the technical operation should have first consideration in this program. The sub-orbital launch will, in fact, be man's first venture into space. It is enough different from the X15 program to require special consideration. It is my personal opinion that in the imagination of many, it will be viewed in the same category as Columbus' discovery of the new world. Thus, it is an extremely important venture and should be exploited properly by the Administration.

2. Some members of the panel (and other individuals who have contacted me privately) believe that the decision by the previous Administration, that the astronauts should be military personnel, was wrong. They point out that NASA was created expressly for the purpose of conducting peaceful space missions, and the orbiting of a military astronaut will be identified by the world in general as a military gesture, and is sure to be seized upon by the U.S.S.R. for propaganda purposes.

My personal feeling is that any change in status (such as asking the astronauts to become civilians) at this late date will be recognized for what it is, an artificial maneuver. Nevertheless, it might be desirable for this Administration to review the past decision and perhaps lay plans by which astronauts selected for later manned space programs could be given the option to become civilians. It would seem to me that the following might be the appropriate group to discuss the situation: The President, The Vice President, Dr. Bundy, Mr. Webb and Dr. Dryden, Mr. Murrow, and Secretary Rusk.

/S/
Jerome B. Wiesner

Document I-34

Document Title: "Report of the Ad Hoc Mercury Panel," 12 April 1961.

Source: Folder 18674, NASA Historical Reference Collection, History Division, NASA Headquarters, Washington, DC.

A special President's Science Advisory Committee (PSAC) panel, under the chairmanship of Donald F. Hornig, was formed in February 1961 and charged with reviewing Project Mercury. By April 1961, the Space Task Group had concluded that Mercury was near the end of its development phase and that the first human flight could be planned. Several test flights of Mercury capsules on both Redstone and Atlas boosters were in the process being readied. On 10 April, foreign correspondents in Moscow reported rumors that the Soviet Union had already placed a man in space. At the same time, the Space Task Group heard rumors that the Hornig panel was recommending up to 50 more chimpanzee flights before launching a man into space. This recommendation ultimately did not find its way into the final report of the Ad Hoc Mercury Panel, although the panel did express concern about the limited information available about the likely impact of spaceflight on a human. But the same day that the panel submitted its report, the Soviet news agency Tass reported that Yuri Gagarin had been launched into orbit and successfully returned to Earth. Apparently, that answered the question of whether a human could fly in space and return to Earth in good health.

OFFICIAL USE ONLY

April 12, 1961

REPORT OF AD HOC MERCURY PANEL

I.   Introduction

Project Mercury has reached a stage where manned suborbital flight is being planned within months and manned orbital flight within a year. Manned flight will involve great personal risk to the pilot and political risk to the country. The object of the panel, therefore, was to become as well acquainted with Project Mercury as possible with a view toward giving advice on the future conduct of the project.

In order to gain some understanding of the project, the panel spent an intensive five days visiting McDonnell Aircraft (where the capsule is being built), Cape Canaveral, and Langley Field. We were brief [sic] by McDonnell, the Space Task Group, representatives of Marshall Space Flight Center (Redstone), BMD (Atlas), and had two general discussions with NASA personnel. Subsequently, a sub-group of the panel with biomedical competence met with representatives of CIA, Air Force, and the Army, and spent two more days at Langley and a day in Washington.

We would like to express our gratitude for the excellent cooperation we received from everyone. Information was made available to us freely, and the discussions were frank and to the point. We were impressed by the magnitude of what has been accomplished and the competence which has been exhibited in organizing and executing the program. We naturally tend to focus on the areas about which we are not happy, particularly those which we feel might imperil the success of the mission, but it must be realized that these are a small number of items out of an enormous enterprise.

II.   Purpose of Mercury Program

The objective of Project Mercury is to advance the state of the art of manned space flight technology. In order to achieve this objective, the following goals have been adopted:

    1.   Place a manned space capsule in orbital flight around the earth;

    2.   Investigate man's performance capabilities and ability to survive in true space environment; and [2]

    3.   Recover the capsule and the man safely.

Mercury is an initial step in manned space flight and rests on an article of faith – that man wants to venture into space and that he will be an essential part of future space missions. In this sense, it may be likened to the Wright Brother's [sic] first flight or Lindbergh's crossing the Atlantic. In a real sense, it is a national exploratory program. Its justification lies not in the immediate ends achieved but in the step it provides toward the future.

III.   The Mercury Program

The Mercury Program is a large one and has several major sub-divisions, each involving great effort and the solution of complex problems. These include:

1.   The development of an aerodynamically stable reentry vehicle, parachutes, reliable telemetry, control systems, life support system, escape mechanism, etc.

2.   The adaptation of a booster to the capsule and escape systems.

3.   The establishment of a world-wide net of tracking stations for voice, telemetry, and command communications with the capsule, and for communication with a control center which is in contact with computers to help digest the information received.

4.   The selection and training of pilots (Astronauts) to fly the capsule and to back up the automatic devices.

5.   The development of suitable test procedures to achieve the necessary reliability of all of the preceding.

6.   The development of recovery methods and the organization of recovery and rescue operations over a large part of the globe.

The primary responsibility for the project rests with the Space Task Group of NASA under the direction of R. R. Gilruth, Director, C. J. Donlan, Associate Director for Research and Development, and W. C. Williams, Associate Director of Operations. It is being assisted by the Department of Defense in testing, tracking, recovery, transportations, medical and other supporting activities. All indications are that the various agencies are well [3] integrated into a working relationship to give the project the necessary support. It is essential, however, with such a complex organization that along with the responsibility, the NASA Space Task Group must have full authority over the entire operation at all times.

The program is behind its original schedule but each of its component parts has come along sufficiently well so that we did not ascertain particular bottlenecks which will eventually dominate the schedule. General comments are:

1.   Complete capsules are just now being delivered from McDonnell. One of them (No. 5) was flown on Redstone (MR-2) with the chimpanzee. Two environmental test chambers, simulating high altitudes, have been put into operation and extended tests on the capsules are just beginning. There are still a number of problems to be worked out. For example, the impact bag which absorbs the landing shock and stabilizes the capsule in the water needs further development. There has been a limited amount of testing of the escape system in flight tests. The complete capsule was tested under maximum heating and acceleration on the MA-2 shot (2/21/61). (MA – Mercury Atlas).

Although failure of the Atlas-to-capsule interface occurred on MA-1 (7/29/60), stiffening the Atlas with a "belly-band" resulted in a completely successful test on MA-2 (2/21/61). Future shots will use an Atlas with a thicker upper skin.

2.   The equipment installation at all of the major tracking stations is nearing completion. Stations ready to go include the control center at Canaveral, the computers at Goddard, the communications center at Goddard, the subsidiary center at Bermuda, the Atlantic ship, and Canary Islands. The mainland-Bermuda complex was tested on MR-2 (MR – Mercury Redstone) and MA-2 and the net as far as the Canary Islands will be tested on MA-3. Installation is continuing at other stations and checkouts are being made with the aid of aircraft-borne instrumentation, but the world-wide net will not be ready for full operation until about 5/1/61.

3.   Pilots have been selected. They have been subjected to a training program including performance at high g and under weightless conditions. They have practiced the manual controls in simulators and have flown simulated missions with surprise emergencies occurring. They have practiced leaving the capsule in the water and under water. [4]

4.   Recovery operations have been rehearsed and organization of the world-wide recovery and rescue system is proceeding.

The program is approaching the state where manned suborbital flight is contemplated in the near future. The question of our readiness will be discussed at length later in the report since there are still serious problems. Manned orbital flight is probably about a year away but a number of problems must still be solved before such a launch.

IV.   Assessment of Risk and Probability of Success

1.   The Reliability Problem

In the assessment of probable system performance, there are two distinct analyses to be made:

(a) Probability of mission success;

(b) Probability of astronaut survival.

Both of these depend on subsystem and component reliabilities, but in very different ways; e.g., a booster failure means mission failure but by no means implies an astronaut fatality or injury.

Thus an estimate of either of these probabilities must begin with the failure probabilities of the individual subsystems. Two remarks of somewhat opposing tendency are important here. The first is that subsystem failure probabilities are not really independent numbers. There are always interactions, or correlations, among subsystems and components such that the failure of subsystem "A" will make the failure of subsystem "B" more probable than before. In large missile systems, it is not feasible to assess these interactions numerically with any degree of confidence without a large number of launchings. However, it is well to remember they are there. The second remark is that except for the booster the system, its several subsystems and even the components themselves are provided with a generous degree of redundancy, both at the hardware level and in the provision of alternative modes of procedure. Thus, there is both an automatic control system and an essentially independent manual control system; there are regular and emergency oxygen supplies, two modes of egress, and the like. Such redundancy is a standard and effective way of creating over-all reliability with the use of individually less than perfect components. It is necessary to keep this consideration in mind when examining reliabilities of individual components. [5]

2.   Subsystem reliabilities

The subsystems are common to both the orbital and ballistic missions, and hence at this point it is not necessary to make a distinction between the Redstone and Atlas missions. When the ultimate probabilities of mission success and astronaut survival are considered, the distinction is, of course, a very important one.

There is a considerable body of test data on components and subsystems to which the panel has had access. Also, both NASA and McDonnell perform continuing reliability analyses of the over-all system on the basis of these data. For present purposes, it seems undesirable to present such detailed analyses, even in summary form. What follows, then, represents a distillation into very general terms of the Panel's considered reaction to these data. Reliabilities are aggregated into three categories:

Class	Reliability Range
Class 1	95 – 100%
Class 2	85 – 95%
Class 3	70 – 85%

The following chart indicates the major subsystems and lists our judgment of the reliability of each. It should be emphasized that the numerical categorization in the chart is the result not of calculations but rather of the subjective judgment of the panel. Following the chart is a brief discussion of certain of its entries. Those to be discussed are marked by and asterisk.

Subsystem or Component	Reliability Category
Capsule structure and re-entry properties	Class 1
Separation mechanism and Posigrade Rocket	Class 1
Tower and Abort Rockets	Class 1
Voice Communications	Class 1
*Abort sensing instrumentation system	Class 1

[6]

Subsystem or Component	Reliability Category
Manual Control System	Class 1
Retro-rocket system	Class 1
Parachute Landing System	Class 1
Ground Environment system	Class 1
Recovery operation	Class 1
Pilot and Pilot training	Class 1
*Landing Bag	Class 2
*Environmental control system	Class 2
*Automatic Stabilization and Control System	Class 2
*Booster (Redstone or Atlas)	Class 3
*Telemetry	Class 3

The starred items require a few brief remarks, which follow:

Abort Sensing and Implementation System (ASIS): In light of the necessity to provide maximum safety to the pilot with 80% (approximately) booster reliability, critical attention must be focused on the abort sensing system. This system provides the warning of impending failure and automatic aborting of the flight to avoid danger to the pilot. Within itself it is completely redundant for reliability. Other systems, i.e., ground control and pilot over-ride provide further redundancy in an "after the fact" manner for separating of the pilot from the booster; however, the abort sensing system gives the earliest warning and therefore the maximum capability of safety to the pilot in the event of an abort. Viewed in this manner, it is imperative that this system have as high a reliability as possible.

The discussions during the ad hoc panel meetings left some questions on the reliability of the ASIS which were not included in the NASA summary. It has been flown open and closed loop on Atlas flights with no apparent failures to operate as designed. The slosh problem in the last (MA-2) Atlas/Mercury [7] flight showed roll oscillations which were near the initiating limits of the ASIS. Admittedly, this condition is not extraordinary to expect at this stage of development, but one of the suggested solutions to the roll rate problem may involve changing the roll rate limits on the ASIS. If the ASIS on the Redstone missile is ready for a manned flight and utilizes the same design philosophy as the Atlas ASIS, then the limit settings for abort should not be amenable to change. These functions should have been settled beyond any doubts by impending disaster criteria and not be changed to adapt to a missile fix. It is recommended that NASA probe again into the reliability of the ASIS to insure to themselves that the system reliability is adequate.

Landing Bag: This buffer against landing shock performs its function well. The steel binding straps, however, have been observed to fatigue under prolonged wave action and the capsule has been punctured when struck by the heat shield. Redesign and test are being vigorously pursued to eliminate these problems. This must be a Class 1 item before the date of the MR-3 manned ballistic mission and it seems likely that satisfactory solutions can be obtained by the present research program and engineering drop tests.

Environmental Control System: Due to the critical role of the environmental control system in the success of the Mercury program, a more detailed engineering review of this system was conducted with members of the NASA Space Task Group at Langley Field, Virginia during the days of 15 and 16 March 1961.

The idea of using a single gas, O_2, atmosphere, in both the suit and capsule to simplify the system appears to be reasonable from an engineering standpoint if it meets the biomedical requirements. The environmental control system is capable of operating completely automatically if required and still provide redundancy in many areas against failure. In the automatic mode the only single point of failure without backup appears to be with the emergency oxygen rate valve. However, with man functioning in the system, this valve can be manually operated.

From the limited drawings available for inspection, good mechanical designs practices appear to have been followed. This conclusion also is confirmed by the results of what testing has been done to date. To the best knowledge of the panel at the present writing, the complete unit incorporating the final design of the ECS has not been subjected to full environmental and vibration testing. In the absence of such testing, it is impossible a priori to categorize this critical subsystem as a class 1 item. Such tests should be performed so that any doubts in this area can be removed.

Automatic Stabilization and Control System: This system is not critical for suborbital missions; it is mandatory for the later orbital missions. Both the automatic and manual systems have in the past had peroxide corrosion [8] problems in the valves. It is probable that new drying methods and procedural improvements have corrected this condition. The automatic system is much more complex than the manual, and a Class 2 categorization of the system is probably fair. However, in the event of a failure of the ASCS, a functioning pilot can bring the capsule in under manual control from an orbital flight.

Booster: There is a good deal of flight data on both the Redstone and Atlas boosters. These indicate reliabilities of the order of 75 to 80%. However, the Redstone used in the Mercury program is a modified version and a vibrational problem was observed in the MR-2 flight. Several fixes were applied including a filter in the control system to eliminate control vane flutter and stiffening to dampen airframe vibrations in the control section of the booster. These fixes were tested in the MR-BD flight of 3/24/61 and appear to have been completely successful.

The relative unreliability of boosters is not per se a cause of alarm; booster failure will invalidate the mission, will of necessity reduce the probability of astronaut survival, but will not necessarily reduce it below an acceptable value.

Telemetry: Some failures or outages of telemetry are to be contemplated, most of which do not endanger the astronaut. Absence on the ground of biomedical data (particularly if a simultaneous outage of all three communication channels to the astronaut were to occur) could result in an unnecessary abort. This again results in a reduction, but not necessarily below acceptable limits, of astronaut survivability.

3.   Redundant or Backup Systems

During the presentations a great deal of stress was placed on the many redundant features of the entire system. The basic need for alternate ways to bring the astronaut to safety in case of system failure centers around the desire to provide much more reliable over-all operation than can be assured from the presently available reliability of either the Atlas or Redstone boosters. In addition, many newly designed subsystems are involved, and there is no way to guarantee an acceptable reliability without an inordinate amount of testing, unless the backup or redundant system philosophy is adopted.

The following chart illustrates the multiple possibilities which are available to bring the astronaut to safety in case of a subsystem failure. [9]

Critical Functions or Events	Redundant Modes of Operation or Actuation
Accelerate to altitude	Normal booster operation – Abort by use of escape rockets
Initiation of abort	Radio command from Control Center or Blockhouse; direct actuation by astronaut; radio command by range safety; abort sensing and implementation system.
Release of escape rocket tower, and Separation from booster bolts.	Electrical exploding bolts; alternate electrically exploding bolts; direct activated exploding
Oxygen environment for astronaut	Capsule pressurization; space suit with separate oxygen system; emergency oxygen supply
Monitoring astronaut's condition	Telemetry of biomedical instruments; UHF voice link; alternate UHF voice link; HF voice link; manual key on telemetry link.
Attitude control of Capsule	Automatic stabilization control system with dual sets of jets; manual control system with separate set of jets; switch-over between the two systems.
Retro-fire	Three rockets when two are needed; two firing mechanisms and two power supplies.
Landing	Normal parachutes descent of capsule; emergency chute for the capsule, and in Redstone flight a personal chute for astronaut egress from capsule top hatch; and emergency side hatch
Recovery	Helicopter pickup in water plus numerous ground and sea pickup arrangements. [10]

In a similar way, the astronaut can be recovered from the capsule in a variety of ways in the event of a prelaunch emergency. Depending on the time, he can use the escape mechanism, the gantry can be moved into place, or rescue may be attempted with the remote controlled "cherry picker" egress tower or the armored rescue vehicle.

It must be emphasized that these alternate modes of operation for the main events in the flight are only illustrative of the many redundancies which are built into all of the systems, subsystems, and components. Of course, the redundant design philosophy has not proved to be easy. When a backup system is introduced, extra care must be exercised to insure that there is not some subtle common link in the two which can fail and thereby inactivate both the main and emergency system or that the emergency system is not inadvertently used at the wrong time.

In summary, the Mercury system is heavily dependent on the use of redundant systems and upon the reliability and decision-making ability of the astronaut to achieve the desired degree of over-all systems reliability. As far as the panel could learn, the Space Task Group has given ample attention to the interrelationships between the redundant systems and of the relationship of the astronaut to the system.

4. Fire Hazard

The atmosphere in both the capsule and suit is pure oxygen. Consequently the possibility of fire, with electrical switches or pyrotechnic devices as a source of ignition, has to receive careful attention. A number of precautions have been taken to minimize the risk of fire. These include: (a) All electrical switches are potted or hermatically [sic] sealed; (b) all squibs and shaped charges used in the vehicle have been installed to vent outward; (c) all combustible materials have been eliminated wherever this has been possible. Where this was not possible, materials have been chosen which are incapable of ignition from the hot surfaces of the capsule. It is important to observe, however, that the astronaut's suit is made of the combustible material; (d) the capsule can be depressurized to extinguish a fire if one should start. Despite these precautions, a certain hazard of fire remains. Particular attention is required to the period before launch when capsule and the astronaut's suit are being purged with pure oxygen at atmospheric pressure.

In particular, it was felt by the panel that an experiment should be performed in which the emergency hatch was blown off by its explosive bolts with an internal capsule atmosphere of oxygen. [11]

5. Possibility of Failure with no Redundant Backup

In a system which is heavily dependent upon redundancies to obtain acceptable reliability, it is necessary to single out the situations or devices for which there is no backup and in which a single failure would inevitably result in failure of the mission. Experience has shown that in all major systems, there are some operations for which it becomes virtually impossible to provide a backup. During the presentations, no mention was made of the non-redundant operations. A subsequent cursory analysis of the systems showed that there are several such. For example, three retro-rockets are provided to decelerate the capsule out of orbit. The entire package is held on by a single explosive bolt. There is a possibility that a stray current in the circuit beyond the main switch could fire the bolt prematurely and jettison the rocket package before it had performed its function and thus make it impossible to get out of orbit. Just such unexplained currents have plagued two of the Little Joe shots. Another possible difficulty centers around the necessity for releasing both the main and the emergency parachutes from the capsule upon impact with the water. There is a possibility that a premature spurious signal may make the release at high altitude and drop the capsule with catastrophic results.

The panel is concerned that steps may not have been taken to specifically tabulate the operations or functions in which a signal failure would lead to catastrophe. When such possibilities have been defined, then special testing, inspection, and check out procedures should be adopted in order to obtain the maximum possible reliability for the associated components.

6.   Status of the Quantitative Reliability Studies

The recommendation of the preceding section that the single failure possibilities be critically analyzed serves to emphasize the value which a reliability analysis has in a program of this magnitude and complexity. From the presentations, it was learned that the McDonnell Aircraft Corporation had performed an extensive failure mode analysis and that a separate reliability study had been initiated at NASA Headquarters and, subsequently, had been transferred to the Space Task Group.

Through a misunderstanding, the questions which were asked by the panel about the McDonnell study were deferred until later in the briefing when the reliability studies would be presented in detail. The later presentation proved to be on the NASA study, and consequently, very little was learned about the failure mode analysis studies of McDonnell. Hence, no evaluation of their impact on the program can be made. [12]

When the NASA reliability studies were presented, it became apparent that as yet they had not played an important role in the design. The mean time to failure of each component which was the basic parameter in the analysis did not reflect the changes which had been made to correct obvious early difficulties. The results which are available from a reliability analysis for the Atlas orbiting mission in which the man is assumed to play no role. The analysis was in the process of being revised and results would not be available until approximately July 1,1961. In view of this situation, the panel was left in doubt as to how comprehensive has been the analysis of the possibilities of single mode failures, of simple correlated multiple failures, and of subtle failures in redundant subsystems which might preclude the use of either subsystem for the Atlas shots.

In view of this uncertainty, the panel wishes to express an opinion that an emphasis be placed on having the results of such a systematic analysis available prior to the first launchings of the manned Atlas vehicles. Further, the panel recommends that the Space Task Group review with the Marshall Space Flight Center the Redstone subsystem reliability data prior to first manned flight.

V.   Medical Aspects of Project Mercury

    1.   General Comments

The major medical effort for the Mercury has followed the traditional aeromedical approach. Once the "mission" was determined, the philosophy emphasized selection of outstanding individuals to be the first astronauts. A training program was established to expose these men as realistically as possible to the anticipated stresses of space flight. Medical personnel provided specifications and requirements for the design and construction of life support systems for the ballistic and orbital flights and participated in the testing and training utilization of prototype systems; undertook ground simulation of anticipated stressing situations; developed a medical monitor system for ground control at Cape Canaveral and a series of stations along the intended orbital path; participated in requirements for and extensive flotilla of recovery ships and aircraft and provided medical contributions to recovery plans and debriefing of the astronauts. These efforts by the small dedicated medical staff of the Space Task Group have been exemplary.

Much less medical effort has been directed to understanding the unique features of stress anticipated during space flight. During the program, when a new physiological stress was identified, tests were designed to simulate the conditions. After the astronaut "took" the test, the assumption was that [13] he could endure it as a part of a combination of all the stresses in actual flight. The panel was disappointed to learn that no attempt was made to evaluate the degree of the physiological stress on the body. Thus, no penetrating medical analyses can be made of even those combined stresses which can be simulated in a ground environment. As a result, it is not known whether the astronauts are likely to border on respiratory and circulatory collapse and shock, suffer a loss of consciousness or cerebral seizures, or be disabled from inadequate respiratory or heat control. These uncertainties are awesome. Data from NASA and DOD aircraft and high altitude balloon flight programs demonstrate a demanding constellation of stresses, and yet measurements are not available which would provide assurances of physiological fitness and survivability characteristics of the pilots. When one must predict response in a more demanding situation apparent health and satisfactory performance are not enough. Essential observations which could provide the basis for extrapolation have not been made before, during, or after these flight programs nor during comparable ground simulation tests. How great a risk is being hazarded in the forthcoming Mercury flights is at present a matter for clinical impression and not for scientific projection.

The considered opinion, reluctantly arrived at by the panel, is that the clinical aspects of the Mercury medical program have been inadequate. We find that this opinion is also shared by several Mercury consultants, by individuals contributing to the simulation training program, and by other qualified observers.

2.   Mercury Ballistic Program

The proposed ballistic flight (MR-3) has been scheduled for early May. NASA personnel state that from a medical standpoint, all essential studies are complete. Medical approval is based on experience which includes the apparent ability of pilots to adapt to the increasing stresses generated by the F and X series of aircraft, the balloon programs, the centrifuge trails in the Mercury profile and the successful flight of Ham in MR-2. The increased severity of the several known stresses of ballistic flight are recognized, but it is argued, the individual parameters involved are not greatly in excess of those already experienced.

The panel's uneasiness arises from incidents experienced for which little or no explanation is available. These include the unexplained but apparent medical deaths of three pilots in the F series of aircraft, experimental findings of temporal lobe epilepsy in monkeys at 5g (no comparable [14] detailed studies exist on man), loss of consciousness and seizures in qualified pilots during jet flights and disorientation experienced by an astronaut for two days following a centrifuge run, and lack of comparable animal data on the centrifuge for the profile flown by Ham. Further, although perhaps of less importance, vomiting by one of the early monkeys in ballistic flight and the presence of blood on Ham's harness have not been explained. Finally, the animal experimental program at Johnsville and elsewhere has been limited in scope. Data on maximum stress limits and physiological and neurological observations which would allow one to draw a series of medical graphs represents animal vs man leading to an estimation of man's position are not available.

3.   Mercury Orbital Program

In contrast to the ballistic program, NASA personnel state that much more needs to be done prior to the first manned orbital flight. The combined stresses may be much greater than any heretofore experienced. Thus the requirement for ground and animal flight tests data is stringent. The panel is not aware of any firm programs which will accomplish the necessary medical studies, although brief references were made of plans to obtain metabolic information, blood pressure measurements, and electroencepholographic [sic] tracings during centrifuge and actual flight tests. We are concerned that these plans were not designed to assess the critical parameters in sufficient detail to permit predictions of astronaut reactions to prolonged orbital flight.

The current program is centered on a small in-house group of physicians. Funds are not available to provide for the extensive university support required or expand current work in DOD laboratories. While it is true that Project Mercury cannot be expected to provide the national effort in space medicine, certain problems must be vigorously and intelligently attacked to provide a minimum of clinical data. Certainly, all relevant information, including that obtained by the Soviet Union should be assembled and subjected to critical analysis.

The panel noted with approval plans to fly a second chimpanzee in orbit prior to man; however, it should be realized that this one additional flight can only provide minimum data, and consideration should be given to whatever animal flights including those in the DOD Discoverer series are necessary to insure the safety of man in space. [15]

VI.   Manned Suborbital Flight

As with most development projects, it is desirable and often mandatory that the final mission capability be attained in a series of development steps. The Redstone manned flights provide such steps prior to orbiting a man. The MR-3 is the first in a series of proposed manned suborbital flights. These provide much of the actual flight training for the pilot and qualification of the equipment under realistic condition but with considerable reduction in the severity of flight conditions and consequent dangers which may be encountered when orbiting conditions are possible but not necessarily intended. In particular, the environmental control system of the capsule itself can be used to demonstrate its functional adequacy under limited ballistic flight conditions with reduced risk to the man compared to that of the later Atlas flights.

The Redstone mission is limited in range and the capsule will necessarily land in water (short of possible aborts on the stand). The pilot can commit errors without affecting his landing region dangerously or inadvertently leaving himself in orbit. In the same sense the basic systems, particularly the automatic stabilization and control system and the ground command and data links can be demonstrated for adequacy without undue severe consequences if there are failures. Therefore, if the Redstone booster reliability is equal to or better than that of the Atlas, its use can provide an invaluable step in the progression to an orbital mission.

In addition, in the earlier phases of the Mercury program, concern over the unknown factors involved in having a man perform specific duties under weightless conditions following high acceleration resulted in an approach commands. Training programs since that time have shown that man's tolerance for conditions in flight is considerably higher than early estimates. As a result, it appears that the man may be the most reliable single item in the capsule. The suborbital manned flight will give a better insight into whether this is the case by combining stresses which cannot be adequately simulated and testing the skill of the pilot under these conditions. For all of these reasons, the suborbital flight is a necessary prerequisite to later manned orbital flights.

Nevertheless, for the reasons mentioned in the preceding section, we are concerned that enough data has been accumulated to predict with certainty the margin of physiological safety for the astronaut.

Before further ballistic flights are undertaken, it must be seriously inquired in each case whether the objectives justify the repeated risk of a man's life. [16]

VII.   Conclusions

1.   The program is a reasonable step in attaining manned space flight. It represents the highest degree of technical advancement available at the time of its inception.

2.   The system is a complicated one and is made so largely by the automatic devices, which are often duplicated, plus the alternate manual control and safety devices.

3.   The system is not completely reliable and cannot be made so in the foreseeable future. It is not more unreliable then could have been predicted at its inception. The thought and organization which have gone into making it as reliable as possible have been careful and thorough and most of the problems have been thought through. There does not appear to be any shortage of funds for reliability and safety measures.

4.   Manned Mercury flights will definitely be a hazardous undertaking, although related to such initial efforts as the flights of the Wright Brothers, Lindbergh flight, and the X-series of research aircraft.

5.   A suborbital flight or flights are needed as a prelude to orbital flight. They will check out the pilot's performances, including his ability to orient the capsule in flight, adding elements which cannot be adequately simulated such as the anxiety and stress of a real flight and the extension of weightlessness to a five minute period, all under conditions where the risk is very much less than in orbital flight since descent in a reasonably accessible recovery area is assured under all conditions.

6.   The presence of a man in the capsule will very greatly increase the probability of a successful completion of the mission over uninhabited or primate flights. One of the possible conclusions of the Mercury program is that the design philosophy of the automatic system to designing automatic mechanisms as a backup to the man.

7.   We urge that NASA appoint a group of consultants to plan and implement a full-scale crash effort on the Johnsville centrifuge and at other appropriate laboratories to obtain essential measurements under as many kinds of combined stresses as possible. The measurements should be sufficient to permit correlations between man and primates with enough certainty to estimate the human margin of reserve during the anticipated stresses of space flight. Substantial data should be on hand prior to [17] committing an astronaut to the first Mercury flight. In view of the limited time available, and commitments of the Space Task Group Medical personnel to MR-3, we urge that additional qualified personnel be recruited to accomplish the studies.

8.   We recommend consideration of including a chimpanzee in the forthcoming MA-3 flight. This is designed for an abort of the McDonnell capsule, complete with life support systems, from an Atlas booster just prior to capsule insertion into orbit.

9.   We urge a considerable expansion of the scientific base of the medical program. Working consultants, additional in-house personnel and sufficient funds to permit implementation of a sound program, based on the resources and capabilities of several university laboratories and utilizing additional contracts with DOD and other government facilities, are essential if we are to insure reasonable programs toward orbital flight.

General Conclusion

The Mercury program has apparently been carried through with great care and there is every evidence that reasonable stops have been taken to obtain high reliability and provide adequate alternatives for the astronaut in the event of an emergency. Nevertheless, one is left with the impression that we are approaching manned orbital flight on the shortest possible time scale so that the number of over-all system tests will necessarily be small. Consequently, although it is generally assumed by the public that manned flight will not be attempted until we are "certain" to be able to return the man safely and that we are more conservative in our attitude toward human life than is the USSR, the fact seems to be that manned flight will inevitably involve a high degree of risk and that the USSR will have carried out a more extensive preliminary program particularly in animal studies than we will before sending a man aloft.

It is difficult to attach a number to the reliability. The checkout procedures on individual components and for the flight itself are meticulous. There appear to be sufficient alternative means by which the pilot can help himself if the already redundant mechanical system fails. However, there is no reliable current statistical failure analysis and although we feel strongly that such analyses should be certainly be brought up to date before the first orbital flight we see no likelihood of obtaining an analysis which we would really trust. One can only say that almost everything possible to assure the pilot's survival seems to have been done. [18]

The area of greatest concern to us has been the medical problem of the pilot's response to the extreme physical and emotional strains which space flights will involve. On this score the pilot training has been thorough and it has been demonstrated that a man can perform under the conditions of acceleration and weightlessness to which he will be subjected. Nevertheless, the background of medical experimentation and test seems very thin. The number of animals that will have undergone flights will be much smaller than in the USSR program. Consequently, we are not as sure as we would like to be that a man will continue to function properly in orbital missions although the dangers seem far less pronounced in a suborbital flight.

Altogether, the probability of a successful suborbital Redstone flight is around 75 percent. The probability that the pilot will survive appears to be around 90 to 95 per cent although the NASA estimates are somewhat higher. This does not appear to be an unreasonable risk, providing the known problems are taken care of before the flight, and those of our members who have been very close to the testing of new aircraft fell that the risks are comparable to those taken by a test pilot with a new high performance airplane.

It is too early to say anything as definite for the risks of orbital flight. Nevertheless, if the planned program of tests is carried through it seems probable that the situation at the time of the first flight will be comparable to that for a Redstone flight now – a high risk understanding but not higher than we are accustomed to taking in other ventures. [19]

MEMBERSHIP
of the
AD HOC MERCURY PANEL

Dr. Donald F. Hornig, <u>Chairman</u>
Dr. Paul Beeson
Mr. W. John Beil
Dr. Milton U. Clauser
Mr. Edward H. Heinemann
Mr. Lawrence A. Hyland
Dr. Donald P. Ling
Dr. Robert B. Livingston
Mr. Harrison A. Storms
Dr. Cornelius Tobias

Mr. Douglas R. Lord, <u>Technical Assistant</u>
Dr. James B. Hartgering, <u>Technical Assistant</u>

<u>Special Consultants to the Panel</u>

Dr. Alfred P. Fishman
Mr. Paul Wickham

Document I-35

Document title: MR-3 Technical Debriefing Team, NASA, "Debriefing," 5 May 1961.

Source: Folder 18674, NASA Historical Reference Collection, NASA History Division, NASA Headquarters, Washington DC.

At the conclusion of the suborbital flight of the first American astronaut into space on 5 May 1961, a NASA team debriefed astronaut Alan Shepard on the aircraft carrier U.S.S. Lake Champlain *soon after his recovery from his capsule's water landing. This transcript of his debriefing captured critical information about the mission, the performance of the spacecraft, the ability of the astronaut to function in space, weightlessness, and the success of the various systems that made the flight possible. It offers immediacy to the first American steps into space. It also set the precedent for debriefings after all future space missions. This interaction was followed by a more formal and extensive debriefing that took place upon return of the astronaut to the Langley Research Center.*

SECTION A [Enclosure 4, page A1-4]

CARRIER DEBRIEFING

IMMEDIATELY AFTER FLIGHT

MAY 5, 1961

1.   The following is a transcription of a tape recording made by Astronaut Shepard aboard the aircraft carrier approximately one to two hours after flight. This tape recording constitutes an essential part of the planned debriefing of Shepard and covers the time period from his entrance into the capsule to his arrival aboard the aircraft carrier. The period of the flight between retrojettison and main chute deployment was not described aboard the carrier. A description of this part of the flight was made on the day after the flight and is included herein.

2.   "This is the first flight debriefing, and before I go into the formal debriefing kit, I would like to say, as a general comment, that I quite frankly did a whole lot better than I thought I was going to be able to do. I was able to maintain control of the capsule fairly well throughout all of the manual maneuvers I made. I was able to follow the sequences fairly well throughout the entire flight, and, as a general comment, I felt that even though I did not accomplish every single detail that we had planned for the flight, I still did much better than I had originally thought I would.

3.   "With that general comment as a start, I'll go into the first question of the debriefing kit which says 'What would you like to say first?' and I've just said it.

4.   Question, No. 2 'Starting from your insertion into the capsule and ending with your arrival aboard the recovery ship, tell us about the entire mission.'

"Starting with foot over the sill back at Pad 5, I make these remarks. The preparations of the capsule and its interior were indeed excellent. Switch positions were completely in keeping with the gantry check lists. The gantry crew had prepared the suit circuit purge properly. Everything was ready to go when I arrived, so as will be noted elsewhere, there was no time lost in the insertion. Insertion was started as before. My new boots were so slippery on the bottom that my right foot slipped off the right elbow of the couch support and on down into the torso section causing some superficial damage to the sponge rubber insert – nothing of any great consequence, however. [Page A4-7] From this point on, insertion proceeded as we had practiced. I was able to get my right leg up over the couch support and part way across prior to actually getting the upper torso in. The left leg went in with very little difficulty. With the new plastic guard I hit no switches that I noticed. I think I had a little trouble getting my left arm in, and I'm not quite sure why. I think it's mainly because I tried to wait too long before putting my left arm in. Outside of that, getting into the capsule and the couch went just about on schedule, and we picked up the count for the hooking up of the face plate seal, for the hooking up of the biomedical connector, communications, and placing of the lip mike. Everything went normally.

5.   "The suit purge went longer than usual because of the requirement of telemetry to change the potentiometers on the EKG cards; so, as a result, I got a fairly good long suit purge and comfortable one. The temperature was certainly comfortable during suit purge. Joe[1] seemed to have no trouble with the straps as he was strapping me in. Everything seemed to go as scheduled. I think we would have saved a little time at this point, since we were already in a very long suit purge, if Joe had tightened the straps up immediately rather than going out and coming back in again. However, at this point, he may have been getting a little bit tired, so it was probably just as well that the sequence went as we planned it according to the SEDR. As a result of this very long purge, I was surprised that the suit circuit oxygen partial pressure was only 95 percent.

6.   "The oxygen partial pressure in the suit circuit apparently is not necessarily a function of the length of the purge. If it is, then there is a leveling off point so that 95 percent seems to be a fairly good endpoint for the present system that we are using. After suit purge, of course, the gross suit-pressure check showed no gross leaks; the suit circuit was determined to be intact, and we proceeded with the final inspection of the capsule interior and the safety pins. I must admit that it was indeed a moving moment to have the individuals with whom I've been working so closely shake my hand and wish me bon voyage at this time.

7.   "The point at which the hatch itself was actually put on seemed to cause no concern, but it seemed to me that my metabolic rate increased slightly here. Of course, I didn't know the quantitative analysis, but it appeared as though my heartbeat quickened just a little bit as the hatch went on. I noticed that this heartbeat or pulse rate came back to normal again shortly thereafter with the [page A7-10] execution of normal sequences. The installation of the hatch, the cabin purge, all proceeded very well, I thought. As a matter of fact, there were very few points in the capsule count that caused me any concern.

8.   "As will be noted by members of the medical team, it became apparent that we were going to hold first for lack of camera coverage as a result of clouds. At this point, I decided that I better relieve my bladder, which I did, and felt much more comfortable. It caused some consternation. My suit inlet temperature changed, and it may possibly have affected the left lower chest sensor. We can check back to see if the moment at which the bladder was relieved actually coincided with a loss or deterioration of good EKG signal from that pair. My general comfort after this point seemed to be good. Freon flow was increased from 30 to 45, and although

----

1. Joe Schmidt, NASA Suit Technician

I suspect body temperature may have increased slightly, I at no time really felt uncomfortable. I, of course, shifted around continuously to try to get proper circulation, particularly in the lower limbs, and found that normal upper torso and arm movements and following sequence items were such that proper circulation was provided. The couch fit was fine. The helmet fit and sponge support was fine for the static condition. I'll describe other deviations later.

9.   "The parachute is definitely in the way of a yaw movement. When you make an attempt to yaw left, the wrist seal bearing on the right wrist bumps into the parachute, not to the point where it makes less yaw possible, but it certainly does interfere with it. It also, of course, interferes with the voice-operated relay sensitivity control and voice-operated relay shutoff switch which I did reach later in the flight using the 'window pole'. So then we had several holds during the count, but my general comfort was maintained, and I found as we did finally proceed down to the last part of the count that my pulse rate did appreciably increase.

10.   "I felt no apprehension at any time, but I did find that if I thought that some people were a little slow in reporting that their panel was in GO condition, I started to get a little bit flustered. I think that I was anxious to go at this point after having been in the capsule for some time.[2] The transition from the freon flow to suit capsule water flow was made smoothly even though we were very late in the count at that time.

11.   [Page A11-14] "The transfer from MOPIS circuitry to RF was made smoothly. I was able to transmit and get an RF check with the control center and with the chase planes as well as with the block house in plenty of time prior to T minus one minute, when, of course, attention did naturally shift to the umbilical and the periscope.

12.   "Backtracking here slightly, I see that I have slipped by gantry removal at –55 which, as far as I was concerned, posed no problem to me. I was well tied in by that time, and at –45 the panel check posed no problem. I had no difficulty at any time with the CTC[3] on any of the check-off items – I think primarily due to his foresightedness in reading the check-off lists when he had the opportunity, rather than following the launch count document to the second. Escape tower arming at –22 was no problem – all you had to do was throw a switch, and, as all know, the escape tower did not fire. The T-15 panel check was satisfactory, the –5 status check was satisfactory, and I would say that the countdown right up to the point of umbilical pull indeed was satisfactory. This ties me back in where I was before, to the periscope.

13.   "I noticed the umbilical go out and I saw the head of the boom start to drop away as the periscope retracted electrically. This fact was reported as well as main bus voltage and current over RF prior to lift-off. I had

---

2. About 4 hours by now.
3. Capsule Test Conductor

the feeling somehow that maybe I would've liked a little more over RF with respect to the booster countdown steps. I remember hearing firing command, but it may very well be that although Deke[4] was giving me other sequences over RF prior to main stage and lift-off, I did not hear them. I may have been just a little bit too excited. I do remember being fairly calm at this point and getting my hand up to start the watch when I received the lift-off from the Control Center on RF. The time-zero relays closed properly, the onboard clock started properly, and I must say the lift-off was a whole lot smoother than I expected. I really expected to have to use full volume control on UHF and HF to be able to receive. I did not have to – I think I was legible to Tel 3[5] because all of my transmissions over UHF were immediately acknowledged without any repeats being requested.

14.   "Again, insofar as the mission itself is concerned, lift-off was very smooth. I noticed no vibrations of any consequence at all during the period of about the first 30-45 seconds (I would say as a guess). [Page A14-17] I got an extra transmission in primarily to insure myself of a good voice link and also to let the people on the ground know I was in good shape. The 30-second scheduled transmission went according to schedule, right on time. I did start that a little bit early, I remember, as I wanted to again let people know that I was in good shape. It seemed to me then that somewhere about 45 seconds to a minute after lift-off, I started noticing an increase in vibrations at the couch. It was a gradual increase; there was not any concern. As a matter of fact I'd really been looking for an increase in sound levels and roughness just after one minute because, of course, going transonic, and because of the max q point, so I wasn't too upset by this. I think maybe if we look back at film (the pilot coverage film) we'll be able to see my helmet bouncing around vibrating. Actually there was [sic] vibrations there to the degree where it distorted some of the reading of the instruments. I made the voice report at one minute on schedule and from there on up to max q noticed the increase in the sound level and increase in vibrations.

15.   "The cabin pressure, as we know, sealed properly at 5.5. It seemed to slow down a little bit at 6. As a matter of fact, I almost reported it as being sealed at 6, but it gradually came down to 5.5. A quick glance at the suit circuit absolute-pressure gauge confirmed this. After this, things really started to smooth out. The booster noises seemed to fade away, and booster vibrations got a lot smoother. As a matter of fact, I mentioned that over RF, so we'll have that on the record. There was a very definite transition in vibration, not a sharp one, but a gradual one, nonetheless noticeable. The report at 1 minute and 30 seconds was made on schedule. We, of course, included the main-bus and isolated-battery voltage at that time. I found that my scan pattern was not as good as it might have been, and I don't remember looking at the electrical panel as much I probably should have, paying more attention, of course, to the oxygen panel and the fuel panel.

---

4. Capsule Communicator in Mercury Control Center. [Astronaut Deke Slayton]
5. Mercury Control Center

16.   "At 2 minutes, normal periodic transmission was made, and, of course, I gave all systems '<u>GO</u>' at that point. I remember feeling particularly happy at that point because the flight was proceeding very smoothly here, the capsule was working very nicely as far as I could tell. I also called out an additional acceleration of, I think, 5-1/2 g here.

17.   "Cut-off as far as I could tell on the clock came exactly on schedule, right around 142 seconds, 2 minutes and 22 seconds on the count. The tower jettisoned. Immediately I noticed the noise in the tower jettisoning. I didn't notice any smoke coming by the porthole as I expected I might in my peripheral vision. I think maybe I was riveted on that good old 'tower jettison' green light which looked so good in the capsule. I threw the 'retrojettison' switch to disarm at this point as I noted over RF, and 'capsule separation' came on [page A17-19] green right on schedule at 2 minutes and 32 seconds. Aux damping at this point, I thought, was satisfactory. I don't remember reporting it specifically because I reported the periscope coming out, and I think at this point I was going to report it, but the turnaround maneuver actually started on ASCS.[6] I remember reporting the turn around maneuver, and at that point, at about 3 minutes, I went though hand control motions[7], as was noted, and I started switching to the manual control system. I switched of course to pitch first, pitched to retroattitude, and back to orbit attitude. The ASCS controlled in yaw and roll as I was doing this. I then switched next to manual yaw, and ASCS roll still continued to function. I switched then finally to manual roll. I was in the full manual system and found that controlling the capsule was just about the same as it been on the trainers.

18.   "I did not pickup any noticeable noise of the jets. I think if I'd had time I might have been able to decrease the volume control of the voice radio circuits and picked it up but at this point I didn't have time to investigate it. I remember thinking that I did not hear the noise of the manual jets firing at this time.

19.   "I controlled fairly close to orbit attitude on manual and then switched to the scope, and the picture in the scope certainly was a remarkable picture. Unfortunately, I had a filter in the scope to cut the sunlight down on the pad, and I did not feel that I had the time to reach it and change it on the pad. It was difficult for me to reach the filter-intensity knob with the suit on without bumping the abort handle with the wrist seal bearing of the left arm, so as a result I remember saying, 'Well, I'll leave the periscope filter in this position and try to remember to change it later on even though it may get me in trouble.' Of course, actually, it did, because I had in the medium gray filter which very effectively obliterated most of the colors. Clarifying that last remark, there is no question about being able to distinguish between cloud masses and land masses. This is very easy to do even with a gray filter, and I was able to distinguish the low

---

6. Automatic Stabilization and Control System
7. A psychomotor test of positioning the hand controller at predetermined positions.

pressure area as described[8] in the southeastern part of the United States. As I think I mentioned over RF, Cape Hatteras was obliterated by cloud cover. The cloud cover of 3 to 4 tenths, low scattered on the east coast of Florida, was most apparent. The west coast of Florida and the Gulf were clear. I could see Lake Okeechobee. As I described, I could see the shoals in the vicinity of Bimini. I could see Andros Island. The Bahama Islands, Grand Bahama Island itself, and Abaco [page A19-21] were confusing because there was cloud cover there, just enough to confuse my view. I think if I had a little bit more time with a periscope here, though, I would have been able to definitely distinguish these islands, but the cloud cover was confusing to me at that point. I noticed also that I apparently had in a slow pitch rate because I noticed that I wasn't controlling the manual pitch too much at this point. I think I was paying too much attention looking out at the awe-inspiring sight in the periscope.

20.   "The countdown to retrosequence helped me. It helped me come back to the next sequence which was to occur. The next sequence of course was retro. The onboard timer started retro essentially on schedule; the retrosequence and retroattitude lights came green, as expected. I went manually to retroattitude, and I wasn't quite as happy with the pitch control here as I was with yaw and roll. Somehow I got a little bit behind with my pitch control, and I got down fairly close to 20 to 25 degrees rather than staying up around the 34 degrees. Of course, as we all know, the index of this particular capsule is at 45 degrees, but I don't think this added to the confusion; however, I think the confusion was my own here. Okay, with respect to retrofiring – there is no question about it, when those retros go, your transition from zero g of weightlessness to essentially 5g is noticeable. You notice the noise of the retros and you notice the torque[9] of the retros. I think I did a fairly good job of controlling the retros outside of the pitch deviation which I mentioned, and I thought that I was able certainly to control them within reasonable tolerance.

21.   "At the end of retros, the plan was to go to fly-by-wire, which I did. I switched to fly-by-wire, pulled manual, and then, at this point, the plan was to go to yaw and then roll fly-by-wire, but I noticed I was a little lower in pitch than I wanted to be at the end of retrofire itself, so I started back on the pitch – then , at this point, it was either a yaw or roll maneuver that I made, I'm not sure which one. I think it's probably yaw because that is the one I was supposed to make first – a fly-by-wire yaw maneuver – and1, about the time the retros were to have jettisoned, I heard the noise and saw a little bit of the debris. I saw one of the retropack retaining straps. I checked and there was no light at that time. Deke[10] called up and said he confirmed retrojettison, and about this time I hit the manual override and the light came on. This, as I recall, is the only item sequence-wise in the capsule that did not perform properly. I did not do the specific roll

---

8. In preflight weather briefing
9. Misalignment Torques.
10. Astronaut Donald K. (Deke) Slayton, Capsule Communicator in the Mercury Control Center.

[page A21-23] 20 degrees and back as we had planned, because it took a little extra time to verify that retropack jettison had occurred.

22.   "I went down to reentry attitude on fly-by-wire, and I think I made the general comment already that as far as I am concerned, the trainers – all the trainers that we have – the procedures trainer as well as the ALFA trainer, are all pretty close to the actual case. I say this now, because on these I have a tendency to be able to control these trainers on the manual system better than I can with the fly-by-wire system. And I think it's just a matter, really of not using fly-by-wire very much. By that I mean that normally we're controlling retros manually and normally controlling reentry manually, and when you switch to fly-by-wire as we had been doing here, the first tendency is to over-control in rate – at least for me – because the microswitch distances for the high and low thrust jets are very small, and we've had trouble on this. With these microswitches, particularly capsule seven, you get high torque right away, whether you want it or not, and so I noticed the same thing on the capsule. The first thing I do is over- control and get a higher rate than I thought I should have gotten.

23.   "On fly-by-wire I went to reentry attitude, and switched to ASCS which stabilized at about 40 degrees, then at this point, the periscope came in on schedule, and I remember reporting 'periscope in.' Then I got involved with looking out the windows for the starts and anything else that I could see. At this time in the flight, of course, this window looks generally at the horizon, at the moon and the stars.[11] There was nothing there at all – I couldn't see anything in the way of stars or planets out in that area, and I did move my head around. I got a little confused because I though I ought to get my head up to see the horizon out that window, but I never did get a horizon out that window at this point, and I think it was because of the attitude. We had figured out it was 15 degrees above the horizon as I recall, and I thought I ought to be able to see the horizon but I never did see it. Well, that, plus the fact that I was looking for the stars that I couldn't see out of that window, actually got me behind in the flight – this was the only point in the flight that I felt that I really wasn't on top of things. What happened here was that .05g came quickly, as I reported, and I started switching to manual control, and I thought I had time to get on to manual control, but I <u>didn't</u>. The g-build-up started sooner than I figured it would. I don't know whether it was just that [page A23-26] I was late because of being on the time, or whether we don't have the same time difference between .05g and g-build-up on our trainer that we actually had in flight – we can check this later. What I'm talking about is the time period between .05g and the g-build-up in reentry. As I can remember on the trainer, I would have time to go ahead and get on manual control and get set up before the g's built up, but I was surprised when the g's started building up as soon as they did. I wasn't ready for it, but I thought we were in good shape

---

11. The stars he was to look for.

because we were still on the ASCS when the .05g relay latched in. As a result, the roll[12] started on schedule. . . ." END OF RECORD.

24.   (There is a portion of Astronaut Shepard's report missing from the tape recording at this part of the flight. During a later debriefing at GBI the next day, Shepard described this portion of the flight essentially as follows:)

25.   The acceleration pulse during reentry was about as expected and as was experienced on the Centrifuge during training, except that in flight the environment was smoother. During the early part of g-build-up, Shepard switched to manual-proportional control on all axes. He allowed the roll put in by the ASCS to continue. He controlled the oscillations somewhat in pitch and yaw during g-build-up only. The oscillations during and after the g-pulse were mild and not uncomfortable. He arrived at 40,000 feet sooner than he expected and at that time switched to ASCS in all axes in order to give full attention to observing drogue chute deployment. The drogue came out at the intended altitude and was clearly visible through the periscope. The capsule motions when on the drogue chute were not uncomfortable. The snorkel opened at 15,000 feet which Shepard thought was late. The main chute came out at the intended altitude.

Astronaut Shepard's recording made on the carrier continues:

26.   "As to the chute, I was delighted to see it. I had pushed all hand controllers in so that I noticed that all the peroxide had dumped on schedule. At this point I shifted to the R/T position of the UHF-DF switch. The UHF-select was still normal, and I think at this point I reached over and flipped off the VOX relay switch which was obviously, I realized after I had done a superfluous maneuver because the transmitters were keyed anyway. I was a little confused here, I guess. I felt that the carrier[13] was coming in and out for some reason, so I went over there and threw that VOX power switch off. [Page A26-29] In any event, after going to the R/T positions, shortly thereafter, I established contact with the Indian Ocean Ship[14] and gave them the report of the parachute being good, the rate-of-decent indicator being at about 35 ft/sec and everything looked real good. The peroxide was dumped, the landing bag was green, and, of course the 'Rescue Aids' switch was off at that point. They relayed back shortly after that, as I recall.

27.   "CARDFILE 23, the relay airplane, came in first of all with a direct shot and then with a relay, so that I was able to get the word to the Cape prior to other sources that I was indeed in good shape up to this point. The opening shock of the parachute was not uncomfortable. My colleagues will recognize it was a reassuring kick in the butt. I think I made the hand

---

12. Programmed reentry roll rate of 10 to 12 degrees per second.
13. The hum of the carrier frequency.
14. This ship was being exercised for the MR-3 mission and had been positioned in the landing area.

controller movements after the main chute. I can't vouch for it. The exact times of these sequences I do not recall at this point but we can cross-check again. Altitude-wise, the drogue and main came out right on the money, as far as indicated altitude was concerned.

28.   "I put the transmission through that I was okay prior to impact. I was able to look out and see the water, with the capsule swinging back and forth. It was not uncomfortable at all. As a matter of fact I felt no uncomfortable physiological sensations, really, at any point during the flight. Excited, yes, but nothing uncomfortable at all. Prior to impact, I had removed my knee straps; I had released my face plate seal bottle and had removed the exhaust hose from the helmet. Back to the impact now – the impact itself was as expected. It was a jolt but not uncomfortable. The capsule went over on its right-hand side, down pretty close to the water, and of course stayed at about 60° off the vertical. I reached down and flipped the 'Rescue Aids' switch at this time to jettison the reserve chute and to eject the HF antenna although I <u>did</u> leave my transmit switch in the UHF position. At this point, I could look out the left window and tell the dye marker package was working properly. The right window was still under water. I began looking around for any indication of water inside but did not find any. I had broken my helmet at the neck ring seal at this point, and I did no transmitting here. I left the Switch on R/T because I didn't want any discharge from the UHF antenna.

29.   "The capsule righted itself slowly to a near vertical position, though I thought myself 'It is taking an awfully long time to get up there,' but it did get up there, and about the time it did get up [page A29-31] there, I started to relax a little bit and started to read off my instruments. I had made a report to CARDFILE 23 after impact over UHF that I was indeed all right, and it was relayed back to the Cape. Then, getting back to the point where the capsule was close to the vertical, I was going to get a read-off of the instruments at this time prior to shutting down the power. I got the main bus voltage and current, and I got a call from the helicopter and thought that communicating with him was much more important. So I did. I communicated with him and established contact with the chopper. I am not sure he heard me at first, but I was able to get through to him that I would be coming out as soon as he lifted the door clear of the water. In the meantime, I experienced very little difficulty in getting the cable from the door around the manual controller handle and tightened up so that when I called the helo and told him I was ready to come out and he verified that he was pulling me up and I told him I was powering down and disconnecting communications. The door was ready to go off. I disconnected the biomedical packs. I undid my lapbelt, disconnected the communications lead, and opened the door and very easily worked my way up into a sitting position on the door sill. Just prior to doing this, I took my helmet off and laid it over in the position in the – as a matter of fact, I put it over the hand controller.

30.   "The helo was right there. I waited before grabbing the 'horse-collar' for a few minutes because I hadn't seen it hit the water. They dropped

it down in the water and pulled it back up again, and I grabbed it and got into it with very little difficulty, and shortly thereafter, was lifted right directly from a sitting position out of the capsule up toward the chopper. The only thing that gave me any problem at all, and it was only a minor one, was that I banged into the HF antenna but, of course, it is so flexible that it didn't give me any trouble. I got into the chopper with no difficulty at all, and I must admit was delighted to get there. Of course, the pickup of the capsule went very nicely. The sea conditions were such that they were able to get it up right away, and the next thing I knew we were making a pass on the flat top. My sensations at this time were very easy to describe and very easy to notice. It was a thrill, and a humble feeling, an exultant feeling that everything had gone so well during the flight.

31.   "I have not used the script[15] here, so I will go over it now to be sure I have covered most of these items. Item 3 – the most outstanding impression of the flight in special sensory areas. I think [page A31-34] it is really very difficult to describe any one thing as being more outstanding than the other. It was all fascinating, and interesting, and challenging, and everything, all wrapped up into one. But I don't really remember noticing the weightless condition until I noticed a washer flying by. 'Well,' I thought, 'you are supposed to be making some comment on being weightless.' So I did think about it a little bit. Of course, as we had known before, in the backseat of the F-100's, it is a real comfortable feeling. Being strapped in like that, there is no tendency to be thrown around at all and no uncomfortable sensations. I guess the most outstanding impression that I had was the fact that I was able to do as well as I did. A very good flight.

32.   "Major surprises? No major surprises. Some minor ones which I have described. I expected to be able to see the stars and planets, which I did not do. I think I could have found them with a little more time to look. The fact that I did not hear the jets firing – although I do remember now hearing the control jets working just after reentry, after I went back to ASCS. I remember hearing some of the high-thrust jets going at this time. In reference to the sky and stars, I have described the stars which I did not see and which I tried to see. I described the landing in the water; I described the check points; I remember mentioning over RF that I was able to see Okeechobee, also Andros and the Bimini Atoll which was (the latter) most apparent because of the difference in color between the shoals and the deep water.

33.   "I did not describe the perimeter[16] too well because of cloud cover around the perimeter. The predicted perimeter cloud cover was most accurate. The clouds were such that the ones that had any vertical formation were pretty far away, and I didn't really notice much difference in critical cloud heights. I think had I been closer to them, I would have been able to notice this a little more. They were pretty far from the center

---

15. The debriefing form.
16. The perimeter of the field of view through the periscope

of the scope where some distortion occurs. We talked about the horizon. Essentially, there was only the one haze layer between the cloud cover and the deep blue.

34.   "Weightlessness gave me no problem at all. The last question: 'Describe any sound, smell, or sensory impressions associated with the flight experienced.' Sounds? Of course, the booster sounds, the pyros firing, the escape tower jettisoning and the retros firing. Of course all these sounds were new, although none of them were really loud enough to be upsetting. They were definitely noticeable. The only unusual smell in the capsule was a gunpowder smell after – it seems to me – after main chute deploy. I think this was after the main antenna can [page A34-35] went off. I don't remember smelling it before, but I did get it after main chute and, of course, I didn't get it until after I opened my face plate. It didn't appear to be disturbing to me, so I didn't close the face plate. No other sensory impressions that I noticed that I can recall at this time that we did not have in training. The g-load, the onset and relief of g were familiar during reentry and powered flight. They were not upsetting. They were not unusual.

35.   "I am sorry that I did forget to work the hand controller under g-load during powered flight as we had discussed, but I thought that I was operating fairly well during powered flight. I think the fact that I forgot this is not too significant. Well, I think that's just about the size of it for now. We will continue this on a more quantitative basis later on. This is Shepard, off."

Document I-36

Document Title: Joachim P. Kuettner, Chief, Mercury-Redstone Project, NASA, to Dr. von Braun, 18 May 1961.

Source: Folder 18674, NASA Historical Reference Collection, NASA History Division, NASA Headquarters, Washington DC.

In 1958 Joachim P. Kuettner joined NASA's Marshall Space Flight Center at Huntsville, Alabama, and became the Center's Director of the Mercury-Redstone Project, overseeing efforts at the center for the first spaceflights of U.S. astronauts. Subsequently, he became Director of the Apollo Systems Office, responsible for the integration of the Apollo spacecraft and the Saturn V rocket for the lunar landing. The euphoria surrounding the flight of Alan Shepard on 5 May 1961 prompted him to prepare this bold memorandum to Center Director Wernher von Braun advocating a circumlunar mission using a spacecraft under development. He had found that with a follow-on space capsule, which became the Gemini spacecraft, it might be feasible to undertake the truly significant "space spectacular" of a circumlunar flight. Such an endeavor would steal the march on the Soviet Union and significantly advance U.S. prestige in the space race. At the time Kuettner made this proposal President John F. Kennedy had not yet made his famous Apollo landing speech. That would come only a week later, and because of it this proposal was overcome by events. It would be revisited in the 1963 to1964 timeframe, however, but never adopted.

M-DIR, Dr. von Braun

May 18, 1961

M-S&M-TSM

MERCURY PROJECT

[Section 2 only]

2. Circumlunar "Shortcut"

  a. You will remember the proposal we discussed a few weeks ago, using C-1 and MERCURY hardware. This "Beat-Russia" proposal which you took along envisioned a trip around the moon within three years. You considered the capsule problem as the time-critical item and suggested possible use of the present MERCURY capsule (beefed-up)

  b. In the meantime, I have done some "incidental" digging and exploring among "savvy" STG people (Cooper, Slayton, Grissom, Glenn, Dr. White, Dr. Voss, Gilruth, Williams, and Faget). I learned that a slightly scaled-up MERCURY capsule is already being developed by McDonnell for prolonged orbital flights. Chamberlin, one of Gilruth's Division Chiefs, carries the ball. (He is the only one I missed.)

  c. There was very little deviation in the general reaction: It can be done with almost existing hardware if the astronaut is given enough room to stretch to full length and to do some regular body exercises in order to avoid muscular dystrophy under prolonged weightlessness. This means some additional room around his body so that he can move his extremities freely. There is no walk-around requirement. I will find out whether the scaled up capsule at McDonnell fulfills these conditions.

  d. Some astronauts, like Cooper, would ride the present MERCURY capsule for a week without hesitation, but the doctors may object. The reaction to the whole idea of an early circumlunar flight of this type varied from friendly to most positive. There was no objection raised by anybody except that Williams doubted that C-1 can do the job, payload wise. (Of course, the plan was to augment C-1 by four solids such as SCOUTS or MINUTEMEN [handwritten: Or by a 3rd of the Minutemen)].

  e. Since this is one of the few real possibilities to accomplish an important "first" without requiring excessive funding (most of the hardware is developed anyway), I would like to know if you are interested in pursuing this idea. We may look into the costs and get some more details on the scaled-up capsule.

Joachim P. Kuettner
Chief
MERCURY-REDSTONE Project

Enc:

    Letters of Commendation

Copies to: M-S&M-TSM (Record copy)
               M-S&M-DIR

Document I-37

Document Title: James E. Webb, Administrator, NASA, to James C. Hagerty, Vice President, American Broadcasting Company, 1 June 1961.

Source: Folder 18674, NASA Historical Reference Collection, NASA History Division, NASA Headquarters, Washington DC.

James C. Hagerty had served as President Dwight D. Eisenhower's press secretary between 1953 and 1961, and in that capacity he had dealt often with the media issues brought to the forefront by Soviet "space spectaculars." Upon his departure from Washington with the end of the Eisenhower administration he keenly understood the excitement of spaceflight and in that context tried, as this letter suggests, to play upon the public's interest in the astronauts to aid his new organization, ABC, by organizing a joint television special with the first two humans in space, Yuri Gagarin and Alan Shepard. NASA Administrator James E. Webb's instinct was probably correct in refusing this offer. Even if the Soviets were willing to allow Gagarin's appearance on ABC , the question at the time of this correspondence was, why would the U.S. want to allow the Soviet Union to upstage the 5 May success of Shepard's flight with the joint television appearance? Instead, Shepard spent several months making public and media appearances to bolster confidence in the American space effort vis à vis its rival, the Soviet Union.

June 1, 1961

Mr. James C. Hagerty
Vice President
American Broadcasting Company
7 West 66th Street
New York 23, New York

Dear Jim:

Since our discussion on May 19th I have given a good deal of thought to the proposal outlined in the letter you delivered to me on that date, that Alan Shepard and Yuri Gagarin appear together in New York City on a nation-wide telecast. I cannot see how Shepard's appearance would serve a useful purpose, and I believe it could be detrimental to the best interests of the United States.

Although, as I told you, your proposal involves national policy questions beyond my own direct responsibility, I feel that it is my duty to state my conviction that the whole plan is unwise.

The Mercury flight of Alan Shepard was performed before the eyes of the whole world. He reported his immediate experience and reactions at the press conference on May 8th in Washington. On June 6th Shepard and other members of the Space Task Group will give a full report on the results of the flight at a scientific and technical conference in Washington which will be widely reported and whose proceedings will be published. Further reporting could add nothing significant.

The free and open way in which we have proceeded to share our manned contrast to Soviet secrecy and their unsupported and conflicting descriptions of the Gagarin flight.

If, [2] as you have proposed, Gagarin would be free to tell his story in whatever manner he so desired, it is fair to assume it would not be in a full and complete factual framework but a rather in the same framework as previous reports.

Why then should we permit the Soviet Union to blunt the impact of the open conduct of our program by the use of a nation-wide telecast as a propaganda forum?

From past experience, the Russians might very well use Gagarin's appearance here in the United States to announce and to exploit, again without full facts, and to a large audience, another Russian manned flight, timed to coincide with his appearance here—perhaps a flight of two or three persons. In such a situation, to compare Shepard's sub orbital flight with that of Gagarin, or with some other Russian achievement, would be inconsistent with the reporting of the flight as only one step in the U.S. ten-year program for space exploration.

With regret that I cannot encourage your proposal, and with best wishes,
I am Sincerely yours,

[Signed]
James E. Webb
Administrator

Blind copy:
Mr. Lucius D. Battle, Director
Executive Secretariat
Department of State
Washington 25, D.C.

A
Webb:hhm
Cc: AD- Dryden
         A- Phillips

Document I-38

Document Title: MR-4 Technical Debriefing Team, Memorandum for Associate Director, NASA, "MR-4 Postflight Debriefing of Virgil I. Grissom," 21 July 1961, with attached, "Debriefing."

Source: NASA Collection, University of Houston, Clear Lake Library, Clear Lake, Texas.

The second flight of the Mercury program, astronaut Gus Grissom's suborbital mission on 21 July 1961, proved somewhat less successful than Alan Shepard's because of the loss of the capsule in the ocean. On Grissom's mission, an explosively actuated side hatch was used to blow open seventy 1/4-inch titanium bolts that secured the hatch to the doorsill. During the water recovery effort a premature explosion of the side hatch allowed the capsule to sink in 15,000 feet of water. Grissom vacated the spacecraft immediately after the hatch blew off and was retrieved after being in the water for about four minutes. Much of the debriefing for the mission, as shown in this memorandum, relates to this important mishap. How this incident took place has been a mystery ever since, with numerous theories abounding. Some thought Grissom panicked and prematurely hit the control to blow the hatch, either accidentally or on purpose to escape the capsule sooner. Others, especially test pilots who knew a steely-nerved Grissom, have publicly doubted that explanation. Some thought that seawater might have gotten into the system and somehow shorted it out. There is no definitive explanation, and recovery of his capsule from the Atlantic Ocean in 1999 did not yield any final answer to what happened during Grissom's flight.

NASA-Manned Spacecraft Center
Langely Field, Virginia
[7-21-61]

MEMORANDUM for Associate Director

Subject: MR-4 postflight debriefing of Virgil I. Grissom

    1.   The enclosures to this memorandum constitute Captain Grissom's complete debriefing of MR-4. The first enclosure is a general outline of the three sessions of the MR-4 debriefing. The second enclosure is an index of enclosures four, five and six which are Grissom's comments

relative to capsule engineering, operational procedures, and pilot performance. In these enclosures each answer by Captain Grissom is preceded by the question proposed except for enclosure four. The debriefing questionnaire used as a guide by the astronaut for this portion of the debriefing is included as enclosure three.

2. The basic concept of the debriefing was to allow the pilot to freely discuss the flight on board the recovery ship before entering into the direct question and answer sessions held at Grand Bahama Island and Cape Canaveral. An index was prepared which, it is hoped, will help direct the various systems' specialists to the information pertaining to their areas of interest.

3. To take full advantage of the information gained form the MR-4 pilot debriefing, it is suggested that a copy of this material be distributed to each branch of the Manned Spacecraft Center. It is requested that all comments on the debriefing be forwarded back to the Training Office.

MR-4 Technical Debriefing Team

[Signed]
Sigurd A. Sjoberg
Flight Operations Coordination

[Signed]
Robert B. Voas
Training Office

[Signed]
Helmut A. Kuehnel
Spacecraft Operations Branch

Enc: Debriefing
RGZ:srl
Copies to: All MSC Branches

[Debriefing: Only Paragraphs 11-13 provided]

11. Recovery – On landing, the capsule went pretty well under the water. Out the window, I could see nothing but water and it was apparent to me that I was laying pretty well over on my left side and little bit head down. I reached the rescue aids switch and I heard the reserve chute jettison and I could see the canister in the water through the periscope. Then, the capsule righted itself rather rapidly and it was apparent to me that I was in real good shape, and I reported this. Then I got ready to egress. I disconnected the helmet from the suit and put the neck dam up. The neck dam maybe had been rolled up too long, because it didn't unroll well. It never did unroll fully. I was a little concerned about this in the water because I was afraid I was shipping a lot of water through it. In fact, the suit was quite wet inside, so I think I was. At this point, I thought I was in good shape. So, I decided

to record all the switch positions just like we had planned. I took the survival knife out of the door and put it into the raft. All switches were left just the way they were at impact, with the exception of the rescue aids and I recorded these by marking them down on the switch chart in the map case and then put it back in the map case. I told Hunt Club they were clear to come in and pick me up whenever they could. Then, I told them as soon as they had me hooked and were ready, I would disconnect my helmet take it off, power down the capsule, blow the hatch, and come out. They said, "Roger," and so, in the meantime, I took the pins off both the top and the bottom of the hatch to make sure the wires wouldn't be in the way, and then took the cover off the detonator.

12.   I was just waiting for their call when all at once, the hatch went. I had the cap off and the safety pin out, but I don't think that I hit the button. The capsule was rocking around a little but there weren't any loose items in the capsule, so I don't see how I could have hit it, but possibly I did. I had my helmet unbuttoned and it wasn't a loud report. There wasn't any doubt in my mind as to what had happened. I looked out and saw nothing but blue sky and water starting to ship into the capsule. My first thought was to get out, and I did. As I got out, I saw the chopper was having trouble hooking onto the capsule. He was frantically fishing for the recovery loop. The recovery compartment was just out of the water at this time and I swam over to help him get his hook through the loop. I made sure I wasn't tangled anyplace in the capsule before swimming toward the capsule. Just as I reached the capsule, he hooked it and started lifting the capsule clear. He hauled the capsule away from me a little bit and didn't drop the horsecollar down. I was floating, shipping water all the time, swallowing some, and I thought one of the other helicopters would come in and get me. I guess I wasn't in the water very long but it seemed like an eternity to me. Then, when they did bring the other copter in, they had a rough time getting the horsecollar to me. They got in within about 20 feet and couldn't seem to get it any closer. When I got the horsecollar, I had a hard time getting it on, but I finally got into it. By this time, I was getting a little tired. Swimming in the suit is difficult, even though it does help keep you somewhat afloat. A few waves were breaking over my head and I was swallowing some water. They pulled me up inside and then told me they had lost the capsule.

13.   Before I end this debriefing, I want to say that I'll ever be grateful to Wally [Astronaut Walter Schirra] for the work he did on the neck dam. If I hadn't had the neck dam up, I think I would have drowned before anyone could have gotten to me. I just can't get over the fact that the neck dam is what saved me today.

Document I-39

Document Title: Robert R. Gilruth, Director, Space Task Group, NASA, to Marshall, NASA, (attention: Dr. Wernher von Braun), "Termination of Mercury Redstone Program," 23 August 1961.

Source: National Archives and Record Administration, Fort Worth, Texas.

A typical approach to flight research involves the slow and systematic advancement of the various parameters of the research project until the team completes the task at hand. As an example, Chuck Yeager in 1947 did not just kick the tires of his airplane and then fly the X-1 beyond the space of sound. He and several other research pilots worked with a team of aerospace engineers for months methodically advancing the X-1's flight regime until they were ready to make a supersonic flight. The Space Task Group, all of whom had enjoyed early experience in flight test, took the same approach with Project Mercury. After several missions without astronauts aboard, they then flew two suborbital missions with Alan Shepard and Gus Grissom on 5 May and 21 July 1961. They were quite prepared, and had planned for, a third suborbital mission but, as this memorandum makes clear, it would have been redundant of what had already been accomplished and was unnecessary to the systematic progression of the Mercury program. Accordingly, Space Task Group Director Robert Gilruth announced to his counterpart at the Marshall Space Flight Center, Wernher von Braun, that NASA Headquarters had approved cancellation of the third suborbital mission, so that NASA could move on to the orbital part of the research program. Because of this decision, Mercury would not need any additional Redstone rockets from the von Braun team, since orbital missions would be launched atop Atlas boosters.

Langley Field, Va.
August 23, 1961

From Space Task Group
To Marshall

Attention: Dr. Wernher von Braun

Subject: Termination of Mercury-Redstone Program

1. Approval has been received from NASA Headquarters to cancel the previously scheduled third manned suborbital flight and to terminate the Mercury-Redstone program. The objectives of this program have been achieved.

2. In the near future, personnel from the Space Task Group will visit Marshall to discuss disposition of the remaining boosters and Ground Support Equipment incurred in those activities.

3. I wish to take this opportunity again to thank you and your staff for the fine team effort displayed in accomplishing the Mercury-Redstone program.

[Signed]
Robert R. Gilruth
Director

Copy to: NASA Hq- Attn: Mr G. M. Low, DM

Document I-40

Document Title: Abe Silverstein, Director of Space Flight Programs, NASA, Memorandum for Administrator, "Use of a Television System in Manned Mercury-Atlas Orbital Flights," 6 September 1961.

Source: Folder 18674, NASA Historical Reference Collection, NASA History Division, NASA Headquarters, Washington DC.

NASA officials recognized the general public's keen interest in the human spaceflight missions that took place during Project Mercury. They also recognized the propaganda value of these flights for the U.S. during the cold war rivalry with the Soviet Union. At the same time, they were engineers who made virtually all of their decisions on the basis of technical data. As this memorandum demonstrates, Edward R. Murrow, the most respected journalist in the U.S. and the new Director of the U.S. Information Agency, had requested a television hook-up from space for orbital Mercury missions. Murrow's request reflected a desire to show the world that the U.S. was second to none technologically, something many non-aligned peoples questioned at the time. Reviewing the necessary technical components of such a broadcast capability, NASA's Director of Space Flight Programs Abe Silverstein concluded that reconfiguring the Mercury capsule's power, communications, and weight structures at that time would be detrimental to the overall objectives of the program. When the Gemini program flew in 1965 to1966 it did incorporate television, and the broadcasts from the Moon during the Apollo program became legendary.

In reply refer to:

DM (RJW:vr)

SEP 6 1961

MEMORANDUM for Administrator

Subject: Use of a Television System in Manned Mercury-Atlas Orbital Flights

References: (a)   Memo frm AA/Romatowski to D/Silverstein
                        Dtd 9/5/61, same subject
                (b)   Ltr frm USIA (Murrow) to A/Webb, dtd 8/29/61

1. In accordance with the request made in reference (a), I have prepared the following comments to be used as a basis for a reply to reference (b).

2. The use of a television system in the Mercury capsule has been studied throughout the history of the project. Initially, the weights involved were prohibitive; now, light-weight television systems are considered feasible. As a result, within the last few months the question of a television system for the Mercury capsule has again been raised.

3. Studies of present television systems indicate that a complete system (camera and transmitter) for the Mercury capsule would weigh less than twenty-five pounds. The corresponding power and antenna requirements,

plus the heat exchanger, however, increase the total capsule weight beyond acceptable limits.

Furthermore, the necessary redesign of the antenna and heat exchanger systems will require considerable testing and development before reliability and confidence is increased to that required for manned flight. In addition, the inclusion of a television system in the capsule communication link will raise R.F. compatibility problems which in the past required months of tests and developments to solve. For example, ground tests indicate that extraneous R.F. signals or even incompatible systems can cause an inadvertent abort or improper ground command during flight. [2]

4.   There is no doubt that at this time a change in the communication system of this magnitude will compromise the Mercury schedule, the reliability of the entire system, and the safety of the pilot. The use of television in our manned flight program must await future flight projects when adequate booster capability will be available to carry the increased payload and when an integrated television-communication system can be designed, developed and suitably tested.

[Signed]
Abe Silverstein
Director of Space Flight Programs

Document I-41

Document Title: Dr. Robert B. Voas, Training Officer, NASA, Memorandum for Astronauts, "Statements for Foreign Countries During Orbital Flights," 7 November 1961.

Source: Folder 18674, NASA Historical Reference Collection, NASA History Division, NASA Headquarters, Washington DC.

The importance of the Mercury program to the larger cold war rivalry with the Soviet Union is demonstrated by this memorandum concerning the possibility of radio transmissions relating news about the missions to various foreign nations. One of the key aspects of the early space race involved persuading non-aligned peoples in the cold war of the superiority of the U.S. and its way of life over that offered by the Soviet Union's communism. Directly speaking to some of these peoples from space might help sway their opinions. At the same time, the desire to appear genuine, unscripted as to remarks, and non-propagandistic motivated this discussion of commentary by the astronauts. In the end, the Mercury Seven performed their roles quite well, making interesting remarks via radio that were heard around the world.

NASA Manned Spacecraft Center
Langley Air Force Base, Virginia
November 7, 1961

MEMORANDUM: For Astronauts

Subject: Statements for foreign countries during orbital flights

1. The undersigned has attempted to get guidance within the NASA
    organization on the policy to be pursued in making statements of
    possible political significance from the Mercury capsule. In pursuing this
    question, he was referred to Mr. Goodwin of NASA Headquarters. Mr.
    Goodwin made the following suggestions. These he apparently discussed
    with Mr. Lloyd and the Administrator and they have their approval.

    a. It is essential that any statements made by the Astronauts appear to
        be spontaneous, personal and unrehearsed. He felt that there was a
        general agreement that statements made by the Russian Cosmonauts
        were not effective and backfired. There was a general feeling that they
        were being used inappropriately for propaganda. He agreed strongly
        with our own feeling that any political statement would look out of
        place. Mr. Goodwin also thought that statements in a foreign language
        could be dangerous, because unless there was a good basis to believe
        they were spontaneous, they would appear to be contrived. Thus, if
        the Astronaut spoke in Hindustani during the flight, the inevitable
        question could be raised in the press conference following the flight,
        "How did the Astronaut come to know Hindustani?" Unless he could
        show that it was a course given in the high school or college which he
        attended, it would be obvious that this statement had been politically
        inspired. The one point at which a foreign language might effectively
        be used would be over the Mexican station. Here, a few words of
        Spanish, such as, "Saludos Amigos," might be quite appropriate and
        since simple Spanish phrases are known by many Americans, it would
        not appear contrived.

2. While Mr. Goodwin did not feel that either a political statement as such,
    or statements in foreign languages, would be useful, he did feel that
    descriptions by the Astronaut in English of the terrain over which he
    was passing and personal statements of how he felt and reacted to the
    situation would be highly desirable and effective if released to foreign
    personnel. The primary requirement here on the Astronaut would be
    to be familiar enough with the political boundaries, to be able to relate
    his observation of the ground to the countries over which he is passing.
    This way, he could report, for example, "I see it is a sunny day in Nigeria,"
    or "I can still see Zanzibar, but it looks like rain is on the way." To these
    observations related directly to the country should be added any personal
    observations such as, "I feel fine; weightlessness doesn't bother me a
    bit; it's just like flying in an aircraft, etc." In all such statements, care
    must be taken not to make them appear to be contrived, maudlin or too
    effusive. Rather, they should be genuine, personal and with immediate
    impact. Mr. Goodwin points out that the ideas and words expressed are
    more important to communication than using the actual language of the
    country. If the experience which the Astronaut is having can be expressed
    in personal, simple, meaningful terms, when translated, this will be far

more effective than a few words in a foreign language which, in the long run, might appear contrived.

3.   Mr. Marvin Robinson of Mr. Lloyd's office is preparing a set of very short statements which might be made over each of the range stations. These statements are designed to strengthen the position of the nation in the use of these facilities. These statements will be forwarded through channels and can be considered by the Astronauts for use during the flight. The best use of such statements might be to have the Astronaut extract the general meaning, but to make the statement in his own words and in his own way at the proper time.

4.   In summary, it appears desirable for the Astronaut

    a. To learn to recognize the political boundaries of the countries over which he passes in terms of the geographical features which will be visible to him from orbit, and

    b. To take any time available to him during the flight to describe his view of the earth and his personal feelings in simple, direct, terms.

Dr. Robert B. Voas
Training Officer

RBV.ncl

Document I-42

Document Title: Telegram, NASA—Manned Spacecraft Center, Port Canaveral, Florida, to James A. Webb and others, NASA, Washington, DC, "MA-6 Postlaunch Memorandum," 21 February 1962.

Source: Source: Folder 18674, NASA Historical Reference Collection, NASA History Division, NASA Headquarters, Washington, DC.

The Mercury Atlas 6 flight carrying John Glenn was the first astronaut-carrying orbital flight of the Mercury spacecraft and thus a milestone for the American space program. Despite the dramatic achievement of Glenn's flight, the engineers conducting the program were primarily interested in evaluating the performance of the vehicle and using that information for upcoming flights. This is an initial telegram from the launch site to NASA Administrator James Webb immediately after the launch reported on the performance of the Atlas launch vehicle and Mercury spacecraft.

[Handwritten note: "MA-6 Postlaunch Memo"] [DECLASSIFIED]

FROM: NASA – MANNED SPACECRAFT CENTER
          PORT CANAVERAL, FLA

TO: NASA HEADQUARTERS
          WASHINGTON DC
          ATTN: MR JAMES A WEBB, A

          NASA HEADQUARTERS
          WASHINGTON DC
          ATTN: MR D BRAINERD HOLMES

          DIRECTOR, OFFICE OF MANNED SPACE FLIGHT
          NASA – MANNED SPACECRAFT CENTER
          LANGLEY AFB, VA
          ATTN: MR ROBERT R GILRUTH, DIRECTOR

          INFO
          HQ SPACE SYSTEMS DIV
          LOS ANGELES CAL
          ATTN: LT COL R H BRUNDIN, SSVM

          ZEN
          NASA LOD
          GCMSFC
          TITUSVILLE FLA

          PMFO 24 CONFIDENTIAL.

          SUBJECT: MA-6 POSTLAUNCH MEMORANDUM.

          1.0   GENERAL-
          THE MA-6 VEHICLE, SCHEDULED FOR LAUNCH AT 07:30 EST,
          FEB. 20, 1962, WAS LAUNCHED AT 09:48 EST. THE THREE-ORBIT

                              [signed]
                              Walter C Williams
                              Associate Director

[2]

MISSION WITH ASTRONAUT JOHN GLENN ABOARD THE CAPSULE

WAS ACCOMPLISHED AS PLANNED, AND ALL TEST OBJECTIVES WERE

ACCOMPLISHED. MALFUNCTIONS WERE INDICATED FROM CAPSULE

INSTUMENTATION IN THE INVERTER COLD-PLATES, IN THE AUTOMATIC

CONTROL SYSTEM, AND IN THE LANDING BAG DEPLOYMENT SYSTEM. CONFIRMATION OF THE EXISTENCE AND NATURE OF THESE INDICATED MALFUNCTIONS WILL REQUIRE A THOROUGH EVALUATION OF DATA. THE LANDING BAG PROBLEM RESULTED IN A DECISION TO REENTER WITHOUT JETTISONING THE RETROPACK. THE LANDING OCCURRED WITHIN VISUAL RANGE OF THE DESTROYER NOA STATIONED APPROXIMATELY 45 NAUTICAL MILES UP RANGE OF THE CENTER OF THE THIRD ORBIT LANDING AREA. LANDING OCCURRED AT APPROXIMATELY 1943Z AND THE SHIP WAS ALONGSIDE FOR RETRIEVAL AT APPROXIMATELY 1959Z. MERCURY NETWORK OPERATION WAS HIGHLY SATISFACTORY FOR THE MISSION. THE ONLY MAJOR PROBLEM OCCURRED AT APPROXIMATELY T-12 MINUTES AS A RESULT OF POWER SOURCE FAILURE AT THE BERMUDA COMPUTER.

2.0   MAJOR TEST OBJECTIVES.-

(A) EVALUATE THE PERFORMANCE OF A MAN-SPACECRAFT SYSTEM IN THREE-ORBIT MISSION.

(B) EVALUATE THE EFFECTS OF SPACE FLIGHT ON THE ASTRONAUT.

(C) OBTAIN THE ASTRONAUT'S OPINIONS ON THE OPERATIONAL SUITABILITY OF THE SPACECRAFT AND SUPPORTING SYSTEMS FOR MANNED SPACE FLIGHT.

3.0   LAUNCH OPERATIONS

THE 390 MINUTE COMBINED COUNT BEGAN AT 23:30 EST, [3] FEB. 19, 1962. A TOTAL HOLD TIME OF 227 MINUTES WAS USED DURING THE COUNT. THE INDIVIDUAL HOLDS WERE AS FOLLOWS:

04:00 EST – T-120	90 MINUTES PLANNED HOLD.
05:30 EST – T-120	55 MINUTES TO CHANGE GE RATE BEACON.
07:25 EST – T-60	40 MINUTES TO REMOVE AND REPLACE A CAPSULE HATCH BOLT.
08:20 EST – T-45	15 MINUTES TO TOP OFF THE FUEL TANK AND MOVE THE SERVICE TOWER.
08:58 EST – T-22	25 MINUTES TO COMLETE LOX TANKING. A MAIN LOX PUMP FOILED AND TANKING WAS ACCOMPLISHED WITH THE TOPPING PUMP.
09:39 EST – T-6:30	2 MINUTES TO VERIFY MERCURY COMPUTER IN BERMUDA.

4.0   WEATHER. –

WEATHER IN THE LAUNCH AREA WAS INITIALLY UNSATISFACTORY FOR REQUIRED CAMERA COVERAGE BECAUSE OF LOW OVERCAST. BY APPROXIMATELY 09:00 EST, A CLEARING TREND WAS EVIDENT AND BY LAUNCH TIME CONDITIONS WERE ENTIRELY STATISFACTORY AS FOLLOWS:

CLOUDS – 2/10 ALTO CUMULUS

WIND – 18 KNOTS FROM 360 DEGREES WITH GUSTS TO 25 KNOTS

VISIBILITY – 10 MILES

TEMPERATURE – 70 DEG. F.

WEATHER AND SEA CONDITIONS IN ALL ATLANTIC RECOVERY AREAS WERE REPORTED SATISFACTORY PRIOR TO LAUNCH. THE CONDITIONS REPORTED BY THE RANDOLF IN THE THIRD ORBIT LAND AREA JUST PRIOR TO CAPSULE LANDING AREA AS FOLLOWS: [4]

CLOUDS – 2/10

WINDS – 14 KNOTS FROM 119 DEGREES

WAVE HEIGHT – 2 FT

5.0   TRAJECTORY AND EVENTS.-

(A) FLIGHT-PATH CONDITIONS AT SECO	PLANNED	ACTUAL
INERTIAL VELOCITY, FT/SEC	25,715	25,709
INERTIAL FLIGHT-PATH ANGLE, DEGREES	0	-.05
ALTITUDE, NAUTICAL MILES (PERIGEE)	87	86.7 N.M.
ALTITUDE, NAUTICAL MILES (APOGEE)	144	141.0 N.M.

(B) EVENTS	PLANNED TIME	ACTUAL TIME
BOOSTER ENGINE CUTOFF	00:02:11.3	00:02:11
ESCAPE TOWER JETTISON	00:02:34.1	00:02:33
(B) EVENTS	PLANNED TIME	ACTUAL TIME
SUSTAINER ENGINE CUTOFF (SECO)	00:05:03.8	00:05:04
CAPSULE SEPARATION	00:05:04.8	00:05:05
CAPSULE SEPARATION	00:05:04.8	00:05:05
CAPSULE TURNAROUND COMPLETED	00:05:35.0	APPEARED NORMAL
START OF RETROFIRE	04:32:58	04:33:08
RETROPACK JETTISON	04:33:58	NOT AVAILABLE
START OF REENTRY (.05G)	04:43:53	NOT AVAILABLE
DROGUE CHUTE DEPLOYED (21,000 FT)	04:50:00	04:49:17

| MAIN CHUTE DEPLOYED (10,000 FT) | 04:50:36 | NOT AVAILABLE |
| LANDING | 04:55:22 | NOT AVAILABLE |

[5]

6.0 BOOSTER PERFORMANCE. –

VERNIER, SUSTAINER, BOOSTER IGNITION AND TRANSITION TO MAISTAGE WERE NORMAL. LIFT-OFF WAS CLEAN AND ALL EVENTS OCCURRED AS PLANNED. THERE WAS NO ABNORMAL DAMAGE TO THE STAND.

7.0 CAPSULE SYSTEMS PERFORMANCE.-

PERFORMANCE OF THE CAPSULE SYSTEMS WAS SATISFACTORY WITH THE FOLLOWING EXCEPTIONS:

(A) BOTH INVERTERS (150 AND 250 UA) REACHED TEMPERATURES ABOVE 200 DEG. F., PROBABLY AS A RESULT OF INVERTER COLD-PLATE MALFUNCTION, OR FREEZING OF WATER IN THE LINES SUPPLYING THE COLD PLATES.

(B) THE AUTOMATIC CONTROL SYSTEM AFTER 1 ORBIT WAS NOT ABLE TO MAINTAIN THE FINE YAW CONTROL (ORBIT MODE) BUT THE WIDE TOLERANCE PORTION OF THE SYSTEM FUNCTIONED SATISFACTORILY (ORIENTATION MODE). THE FLY-BY-WIRE SYSTEM HAD A MINOR MALFUNCTION LATE IN THE FLIGHT CAUSING LOSS OF CONTROL OF ONE OF THE ONE-POUND THRUSTERS. THE HORIZON SCANNER SYSTEM APPEARED TO HAVE PROBLEMS PROVIDING THE PROPER GYRO REFERENCE ON THE DARK SIDE OF THE EARTH DURING THE SECOND AND THIRD ORBIT.

(C) AN INTERMITTENT INDICATION THAT THE LANDING BAG WAS DEPLOYED WAS EVIDENT FROM CAPSULE INSTUMENTATION, AND THE STATUS OF THE LANDING BAG COULD NOT BE CONFIRMED. THIS PROBLEM RESULTED IN A DECISION TO REENTER WITHOUT JETTISONING THE RETROROCKET PACKAGE. EXTENSIVE EVALUATION OF CAPSULE [6] SYSTEMS DATA WILL BE REQUIRED TO CONFIRM THE EXISTENCE AND ESTABLISH THE NATURE OF THE INDICATED MALFUNCTIONS.

8.0   ASTRONAUT PERFORMANCE

ASTRONAUT JOHN GLENN PERFORMED WELL AND REPORTED FEELING WELL THOUGHOUT THE MISSION. NORMAL VALUES OF HEART RATE, RESPIRATION RATE AND BLOOD PRESSURE WERE REPORTED BY THE [7] MONITORING STATIONS FOR MOST OF THE FLIGHT. ALL PHYSIOLOGICAL SENSORS OPERATED PROPERLY AND DATA WERE OF GOOD QUALITY.

9.0   THE MONITORING OF THE FLIGHT FROM TELEMETRY AND AIR/ GROUND

VOICE INFORMATION WAS EXCELLENT, PROVIDING THE MERCURY CONTROL CENTER WITH ALL THE INFORMATION REQUIRED TO GIVE THE PILOT TECHNICAL ADVICE AND ASSISTANCE. DURING ALL PHASES OF THE FLIGHT THE FLOW OF DATA TO AND FROM THE NETWORK SITES AND THE CONTROL CENTER WAS RAPID AND ADEQUATE, SO THAT BOTH GROUND PERSONNEL AND FLIGHT CREW WERE CONTINUALLY IN AGREEMENT AS TO STATUS OF TRAJECTORY, CAPSULE, SYSTEMS, AND PILOT. OF PARTICULAR SIGNIFICANCE IN PROVIDING REAL TIME INFORMATION TO THE CONTROL CENTER WAS THE RELAYING OF AIR/

GROUND VOICE FROM ALL SITES WHICH HAVE POINT-TO-POINT VOICE. THE PROBLEMS ENCOUNTERED WERE WELL COORDINATED WITH THE PILOT, AND THIS DEFINITELY AIDED IN THE ULTIMATE SUCCESSFUL COMPLETION OF THE MISSION. [8]

10.0 NETWORK PERFORMANCE.-

NETWORK PERFORMANCE WAS HIGHLY SATISFACTORY FOR THIS MISSION. THE COMPUTER CONTROLLED NETWORK TESTS WERE CONDUCTED DURING THE COUNTDOWN AND CONFIRMED NETWORK READINESS WITH THE EXCEPTION OF RADAR DIFFICULTIES DETECTED AT CAPE CANAVERAL AND BERMUDA. THESE DIFFICULTIES WERE CORRECTED PRIOR TO LAUNCH. A TWO-MINUTE HOLD IN THE COUNTDOWN WAS REQUIRED AT T-2 MINUTES AS A RESULT OF POWER SOURCE FAILURE AT THE BERMUDA COMPUTER AT APPROXIMATELY T-12 MINUTES. DURING LAUNCH AND RADAR HANDOVER WITH BERMUDA, AN UNIDENTIFIED C-BAND RADAR ATTEMPTED CAPSULE TRACK. THIS CAUSED INTERFERENCE WITH THE BERMUDA ACQUISITION PHASE BUT ACQUISITION WAS ACHIEVED ANYWAY. TELETYPE AND VOICE COMMUNICATIONS WITH NETWORK SITES, AND RELAY OF AIR-TO-GROUND COMMUNICATIONS TO MERCURY CONTROL CENTER WERE EXCELLENT. EXCELLENT PERFORMANCE WAS ALSO OBTAINED FROM TRACKING SUBSYSTEMS INCLUDING RADAR, ACQUISITION, TELEMETRY, COMMAND CONTROL, AND AIR/GROUND VOICE.

11.0   RECOVERY.-

RECOVERY FORCES WERE POSITIONED TO PROVIDE A RECOVERY CAPABILITY IN THE END OF ORBIT LANDING AREAS, THE ATLANTIC

ABORT LANDING AREAS, AND IN CONTINGENCY RECOVERY AREAS ALONG THE ORBIT GROUND TRACK. RECOVERY READINESS WAS SATISFACTORY IN ALL RESPECTS AT LAUNCH. [9]

THE LANDING OCCURRED WITHIN VISUAL RANGE OF THE DESTROYER NOA, STATIONED APPROXIMATELY 45 NAUTICAL MILES UPRANGE OF THE CENTER OF THE THIRD ORBIT LANDING AREA. THE NOA SIGHTED THE PARACHUTE DURING CAPSULE DESCENT AT A RANGE OF ABOUT 5 MILES AND ESTABLISHED COMMUNICATIONS WITH THE ASTRONAUT. THE SHIP REPORTED THAT LANDING OCCURRED AT APPROXIMATELY 1943Z AND THE SHIP WAS ALONGSIDE FOR RETRIEVAL AT APPROXIMATELY 1959Z. THE ASTRONAUT REMAINED IN THE CAPSULE DURING PICKUP AND THE CAPSULE WAS ABOARD AT ABOUT 2004Z. THE ASTRONAUT THEN LEFT THE CAPSULE THROUGH THE SIDE HATCH AFTER FIRST ATTEMPTING A TOP EGRESS WITHOUT SUCCESS. FOLLOWING INITIAL DEBRIEFING ON THE NOA THE ASTRONAUT WAS TRANSFERRED BY HELICOPTER TO THE AIRCRAFT CARRIER RANDOLF FOR FURTHER TRANSFER TO GRAND TURK. HE ARRIVED AT GRAND TURK AT ABOUT 0145Z. WALTER C. WILLIAMS, ASSOCIATE DIRECTOR.

SCP-4

RGA.jhr

CCK

COPY TO: NASA Hq – Attn.: Mr. G. M. Low, MS

Goddard SFC – Attn: Dr. H. J. Goett

J. C. Jackson

Flight Operations Div
Data Coordination
Mercury Project Office (6)
Preflight Operations Div
Mercury Atlas Office
E. H. Buller

Document I-43

Document Title: R. B. Voas, NASA, Memorandum for Those Concerned, "MA-6 Pilot's Debriefing," 22 February 1962, with attached, John Glenn, NASA, "Brief Summary of MA-6 Orbital Flight," 20 February 1962.

Source: NASA Collection, University of Houston, Clear Lake Library, Clear Lake, Texas.

Mercury Atlas-6's flight carrying John Glenn on a three orbit mission around Earth proved enormously successful for the U.S. despite several technical problems. In this debriefing, Glenn describes what took place while in Earth orbit. He describes the problem with his low-rate attitude thrusters and his manual correction of the problem, as well as his capsule's reentry with its retrorocket pack attached in case the heatshield had come loose during the mission. This debriefing, analysis of the capsule, and review of the telemetry and other data from this mission led to more rigorous testing of the capsules and procedures used on the three following Mercury orbital flights.

PRELIMINARY

[2-22-62]

MEMORANDUM for Those Concerned

Subject: MA-6 Pilot's Debriefing

The enclosure to this memorandum is an edited transcript of the pilot's debriefing aboard the destroyer Noa and at Grand Turk on February 20, 21, and 22. This transcript is released in a PRELIMINARY form in order to aid in the writing of the postlaunch report. A more finished, edited, and index text of the postflight debriefing similar to the documents on the pilot's debriefings for the MR-3 and 4 flights will be issued at a later date. Request for clarification of any of this material should be sent to the Training Office.

The format of the enclosure is as follows:

    1.   Astronaut's brief narrative account of the flight.

    2.   Specific questions keyed to a chronological review of the flight.

        a.   Prelaunch

      b. Launch and powered flight

      c. Zero G phase

      d. Reentry

      e. Landing

      f. Recovery

3. Miscellaneous questions covering the pilot's evaluation of capsule systems.

4. Description by John Glenn of the special astronomical, meteorological and terrestrial observations.

5. Discussion of the predominant sensations during launch and powered flight.

6. Miscellaneous discussion of flight activities by the astronaut. (This section was taken from recordings of several hours of discussion with personnel at Grand Turk. Time has not permitted organizing this material under appropriate headings.

[Signed]
R.B. Voas

[2]

Brief Summary of MA-6 Orbital Flight*

By John H. Glenn, Jr

[*Based on recorded debriefing onboard the destroyer Noa shortly after the MA-6 mission on February 10, 1962.]

There are many things that are so impressive, it's almost impossible to try and describe the sensations that I had during the flight. I think the thing that stands out more particularly than anything else right at the moment is the fireball during the reentry. I left the shutters open specifically so I could watch it. It got a brilliant orange color; it was never too blinding. The retropack was still aboard and shortly after reentry began, it started to break up in big chunks. One of the straps came off and came around across the window. There were large flaming pieces of the retropack – I assume that's what they were – that broke off and came tumbling around the sides of the capsule. I could see them going on back behind me then making little smoke trails. I could also see a long trail of what probably was ablation material ending in a small bright spot similar to that in the pictures out of the window taken during the MA-5 flight. I saw the same spot back there and I could see it move back and forth as the capsule oscillated slightly. Yes, I think the reentry was probably the most impressive part of the flight.

Starting back with highlights of the flight: Insertion was normal this morning except for the delays that were occasioned by hatch-bolt trouble and by the microphone fitting breaking off in my helmet. The weather cleared up nicely and after only moderate delays, we got off.

Lift-off was just about as I had expected. There was some vibration. Coming up off the pad, the roll programming was very noticeable as the spacecraft swung around to the proper azimuth. There also was no doubt about when the pitch programming started. There was some vibration at lift-off from the pad. It smoothed out just moderately; never did get to very smooth flight until we were through the high q area. At this time – I would guess a minute and fifteen to twenty seconds – it was very noticeable. After this, it really smoothed out and by a minute and a half, or about the time cabin pressure sealed off, it was smooth as could be.

The staging was normal, though I had expected a more sharp cutoff. It felt as though the g ramped down for maybe half a second. For some reason, it was not as abrupt as I had anticipated it might be. The accelerometers read one and a quarter g's when I received a confirmation on staging from the Capsule Communicator. I had been waiting for this message at that point because I was set to go to tower jettison as we had planned, in case the booster had not staged. At this time, I also saw a wisp of smoke and I thought perhaps the tower had jettisoned early. The tower really had not jettisoned at that time and did jettison on schedule at 2+34. As the booster and capsule pitched over and the tower jettisoned, I had a first glimpse of the horizon; it was a beautiful sight, looking eastward across the Atlantic.

[2]

Toward the last part of the insertion, the vibration began building up again. This I hadn't quite expected; it wasn't too rough but it was noticeable. Cutoff was very good; the capsule acted just as it was supposed to. The ASCS damped and turned the spacecraft around. As we were completing the turnaround, I glanced out of the window and the booster was right there in front of me. It looked as though it wasn't more than a hundred yards away. The small end of the booster was pointing toward the northeast and I saw it a number of times from then on for about the next seven or eight minutes as it slowly went below my altitude and moved farther way. That was very impressive.

I think I was really surprised at the ease with which the controls check went. It was almost just like making the controls check on the Procedures Trainer that we've done so many times. The control check went off like clockwork; there was no problem at all. Everything damped when it should damp and control was very easy. Zero-g was noticeable at SECO. I had a very slight sensation of tumbling forward head-over-heels. It was very slight; not as pronounced an effect as we experience on the centrifuge. During turnaround, I had no sensation of angular acceleration. I acclimated to weightlessness in just a matter of seconds; it was very surprising. I was reaching for switches and doing things and having no problem.

I didn't at any time notice any tendency to overshoot a switch. It seemed it's just natural to acclimate to this new condition. It was very comfortable. Under the weightless condition, the head seemed to be a little farther out of the couch which made it a little easier to see the window, though I could not get up quite as near to the window as I thought I might.

The rest of the first orbit went pretty much as planned, with reports to the stations coming up on schedule. I was a little behind at a couple of points but most of the things were going right according to schedule, including remaining on the automatic control system for optimum radar and communications tracking. Sunset from this altitude is tremendous. I had never seen anything like this and it was truly a beautiful, beautiful sight. The speed at which the sun goes down is very remarkable, of course. The brilliant orange and blue layers spread out probably 45-60 degrees each side of the sun tapering very slowly toward the horizon. I could not pick up any appreciable Zodiacal light. I looked for it closely; I think perhaps I was not enough night adapted to see it. Sunrise, I picked up in the periscope. At every sunrise, I saw little specks, brilliant specks, floating around outside the capsule. I have no idea what they were. On the third orbit, I turned around at sunrise so that I could face into the sun and see if they were still heading in the same direction and they were. But I noticed them every sunrise and tried to get pictures of them.

[3]

Just as I came over Mexico at the end of the first orbit, I had my first indication of the ASCS problem that was to stick with me for the rest of the flight. It started out with the yaw rate going off at about one and one-half degrees per second to the right. The capsule would not stay in orbit mode, but would go out of limits. When it reached about 20 degrees instead of the 30 degrees I expected, it would kick back into orientation mode and swing back with the rate going over into the left yaw to correct back into its normal orbit attitude. Sometimes, it would cross-couple into pitch and roll and we'd go through a general disruption of orbit mode until it settled down into orbit attitude. Then yaw would again start a slow drift to the right and the ASCS would kick out again into orientation mode. I took over manually at that point and from then on, through the rest of the flight, this was my main concern. I tried to pick up the flight plan again at a few points and I accomplished a few more things on it, but I'm afraid most of the flight time beyond that point was taken up with checking the various modes of the ASCS. I did have full control in fly-by-wire and later on during the flight, the yaw problem switched from left to right. It acted exactly the same, except it would drift off to the left instead of the right. It appeared also that any time I was on manual control and would be drifting away from the regular orbit attitude for any appreciable period of time that the attitude indications would then off when I came back to orbit attitude. I called out some of these and I remember that at one time, roll was off 30 degrees, yaw was off 35 degrees, and pitch was off 76 degrees. These were considerable errors and I have no explanation for them at this time. I could control fly-by-wire and manual very adequately. It was not difficult at all. Fly-by-wire was by far the most accurate means of control, even though I didn't have accurate control in yaw at all times.

Retrorockets were fired right on schedule just off California and it was surprising coming out of the Zero-g field that retrorockets firing felt as though I were accelerating in the other direction back toward Hawaii. However, after retrofire was completed when I could glance out the window again, it was easy to tell, of course, which way I was going, even though my sensations during retrofire on automatic control. Apparently, the solid-on period for slaving just prior to retrofire brought the gyros back up to orbit attitude, because they corrected very nicely during that period. The spacecraft was just about in orbit attitude as I could see it from the window and through the periscope just prior to retrofire. So, I feel that we were right in attitude. I left it on ASCS and backed up manually and worked right along with the ASCS during retrofire. I think the retroattitude held almost exactly on and I would guess that we were never more than 3 degrees off in any axis at any time during retrofire.

[4]

Following retrofire, a decision was made to have me reenter with the retropackage still on because of the uncertainty as to whether the landing bag had been extended. I don't know all the reasons yet for that particular decision, but I assume that it had been pretty well thought out and it obviously was. I punched up .05g manually at a little after the time it was given to me. I was actually in a small g-field at the time I pushed up .05g and it went green and I began to get noise, or what sounded like small things brushing against the capsule. I began to get this very shortly after .05g and this noise kept increasing. Well before we got into the real heavy fireball area, one strap swung around and hung down over the window. There was some smoke. I don't know whether the bolt fired at the center of the pack or what happened. The capsule kept on its course. I didn't get too far off the reentry attitude. I went to manual control for reentry after the retros fired and had no trouble controlling reentry attitude through the high-g area. Communications blackout started a little bit before the fireball. The fireball was very intense. I left the shutters open the whole time and observed it and it got to be a very bright orange color. There were large, flaming pieces of what I assume was the retropackage breaking off and going back behind the capsule. This was of some concern, because I wasn't sure of what it was. I had visions of them possibly being chunks of heat shield breaking off, but it turned out it was not that.

The oscillations that built up after peak-g were more than I could control with the manual system. I was damping okay and it just plain overpowered me and I could not do anymore about it. I switched to Aux. Damp as soon as I could raise my arm up after the g-pulse to help damp and this did help some. However, even on Aux. Damp, the capsule was swinging back and forth very rapidly and the oscillations were divergent as we descended to about 35,000 feet. At this point, I elected to try to put the drogue out manually, even though it was high, because I was afraid we were going to get over to such an attitude that the capsule might actually be going small end down during part of the flight if the oscillations kept going the way they were. And just as I was reaching up to pull out the drogue on manual, it came out by itself. The drogue did straighten the capsule out in good shape. I believe the altitude was somewhere between 30,000 and 35,000 at that point.

I came on down; the snorkels, I believe, came out at about 16,000 or 17,000. The periscope came out. There was so much smoke and dirt on the windshield that it was somewhat difficult to see. Every time I came around to the sun – for I had established my roll rate on manual – it was virtually impossible to see anything out through the window.

The capsule was very stable when the antenna section jettisoned. I could see the whole recovery system just lined up in one big line as it came out. It unreeled and blossomed normally; all the panels and visors looked good. I was going through my landing check off list when the Capsule Communicator called to remind me to deploy the landing bag. I flipped the switch to auto immediately and the green light came on and I felt the bag release. I was able to see the water coming towards me in the periscope. I was able to estimate very closely when I would hit the water. The impact bag was a heavier shock than I had expected, but it did not bother me.

Communications with the recovery ship Noa were very good. The Noa had me in sight before impact and estimated 20 minutes to recovery which turned out to be about right. When the destroyer came alongside, they hooked on with the Shepard's hook and cut the HF antenna. During the capsule pickup, I received one good solid bump on the side of the ship as it rolled. Once on deck I took the left hand panel loose and started to disconnect the suit hose in order to hook up the hose extension prior to egressing through the upper hatch. By this time I was really hot- pouring sweat. The capsule was very hot after reentry and I really noticed the increase in humidity after the snorkels opened. I decided that the best thing at that point was to come out the side rather than through the top. I am sure I could have come out the top if I had had to, but I did not see any reason to keep working to come out the top. So I called the ship and asked them to clear the area outside the hatch. When I received word that the area was clear, I removed the capsule pin and hit the plunger with the back of my hand. It sprung back and cut my knuckles slightly though the glove. The noise of the hatch report was good and loud but not uncomfortable.

In summary, my condition is excellent. I am in good shape; no problems at all. The ASCS problems were the biggest I encountered on the flight. Weightlessness was no problem. I think the fact that I could take over and show that a pilot can control the capsule manually, using different control modes, satisfied me most. The greatest dissatisfaction I think I feel was the fact that I did not get to accomplish all the other things that I wanted to do. The ASCS problem overrode everything else.

Document I-44

Document Title: Robert C. Seamans, Jr., Associate Administrator, NASA, Memorandum for Robert R. Gilruth, Director, Manned Space Flight, NASA, "Astronaut Activities," 31 May 1962.

Source: Folder 18674, NASA Historical Reference Collection, NASA History Division, NASA Headquarters, Washington DC.

In 1962 the Space Task Group moved from the Langley Research Center in Hampton, Virginia, to found the Manned Spacecraft Center (MSC) in Houston, Texas. This change placed the human spaceflight program of NASA on a more formal and permanent footing. Also in 1962, NASA selected its second class of astronauts who would be involved in the Gemini and Apollo programs. With these changes came the institutionalization of a structure for managing the astronauts, the creation of policies regarding what they could and could not do as a part of their outside activities, and a formalization of crew assignments and other duties. This memorandum discusses the management structure for the astronauts. In September 1962, MSC Director Robert Gilruth selected Deke Slayton, one of the Mercury Seven, to coordinate astronaut activities. The effort became even more structured in November 1963 when Slayton assumed the position of Director of Flight Crew Operations. In that capacity, he became responsible for directing the activities of the astronaut office, the aircraft operations office, the flight crew integration division, the crew training and simulation division, and the crew procedures division. Working directly with Gilruth, Slayton closely managed the astronauts and oversaw their activities.

NATIONAL AERONAUTICS AND SPACE ADMINISTRATION

WASHINGTON 25, DC

May 31, 1962

MEMORANDUM

To: Robert R. Gilruth, Director, MSC

From: Associate Administrator

Subject: Astronaut Activities

With our recent announcement concerning additional astronaut selection, it seems timely to restate my understanding of your responsibilities for astronaut activities and to suggest some guidelines for your consideration.

1. Current and future NASA astronauts are employees of the Manned Spacecraft Center and, therefore, are under your direction. In executing this responsibility, it is a sound procedure to have a key member of your operations group as astronaut supervisor. This individual should be held responsible for day-to-day direction of astronaut activities in the same fashion as any other NASA line supervisor accounts for the activities of personnel reporting to him. This responsibility includes supervision of non-project activities covered in the paragraphs following.

2. In connection with astronaut personal appearances, I know you understand that Mr. Webb, Dr. Dryden, and I are under constant demands to make these individuals available. As in the past, we will continue to restrict such appearances to occasions that have a minimum effect on the program assignments of the astronauts and which, in addition, advance the overall

objectives of the National program for space exploration. As you know, Mr. Webb has assigned the responsibility for planning and approving astronaut appearances to Dr. Cox. I believe it is essential that a close tie is maintained between your office and that of Dr. Cox's on these matters. I believe the most satisfactory organizational arrangement to implement such activities is to have one of Dr. Cox's staff work continuously with an individual in Houston in order to best schedule such appearances.

3.   As is the case with other employees of your Center, you are responsible for controlling the extra-program activities of the astronauts, particularly in such areas as newspaper and journal articles and press appearances. Unlike those of other personnel, however, the [2] press relations of the astronauts present a special circumstance because of the status they have assumed as public figures. As in the case of public appearances, Headquarters is under constant pressure for articles, messages, endorsement of causes, etc., by the astronauts. It is necessary, therefore, that activities of this sort by the astronauts also be closely coordinated with Dr. Cox, and that major activities be specifically approved by him. This approval, obviously, would not apply to day-today press contacts which are related directly to their mission, but rather to significant interviews, articles, or statements which might relate to or reflect on national policy.

Consequently, I wish you would discuss this matter further with Dr. Cox in order that we may agree upon an individual for this assignment and upon his position in your organization. With this position designated, I believe we will have an effective relationship between the astronaut supervisor, and Dr. Cox's office in Headquarters.

We have learned a great deal in the last year about the technical and non-technical problems which face us in manned space flight projects. We are in full agreement that, as Director of the Manned Spacecraft Center, it is your responsibility to direct NASA astronauts in order to maximize their individual and combined contribution to our programs. As in the past, Mr. Webb, Dr. Dryden, Dr. Cox, and I will be happy to discuss any particular question with you and provide whatever guidance you feel is needed.

[Signed]
Robert C. Seamans, Jr.
Associate Administrator

Document I-45

Document Title: W. J. North, Senior Editor, E. M. Fields, Dr. S. C. White, and V. I. Grissom, National Aeronautics and Space Administration, Manned Spacecraft Center, "MA-7/18 Voice Communications and Pilot's Debriefing," 8 June 1962.

Source: NASA Collection, University of Houston, Clear Lake Library, Clear Lake, Texas.

On 24 May 1962 Scott Carpenter flew Mercury Atlas-7 on a three-orbit flight that paralleled the John Glenn mission of the previous February. During Carpenter's second orbit he took manual control of the spacecraft and made changes to the capsule's orientation by movements of his head and arms. He also over-used his attitude control jets and ran short of fuel. This and a mis-timed reentry burn resulted in his spacecraft overshooting the planned landing point by 250 miles. This caused major delays in the water recovery of Carpenter and his Aurora 7 capsule and a nationwide concern for the astronaut's safety. Many people criticized Carpenter's performance on this flight. Chris Kraft, senior flight controller and later director of the Manned Spacecraft Center, blamed Carpenter for the poor reentry and worked to ensure that he never flew in space again. Others were more charitable, concluding that monitoring fuel consumption should be done by Mission Control. This debriefing presents Carpenter's assessment of what had taken place. It was only the beginning of several reviews of less than stellar in-flight performance that embarrassed the astronaut and eventually led to his departure from NASA in 1965.

[CONFIDENTIAL] [DECLASSIFIED]

[Only Section 3 of report provided]

[3-1]

3.0 SHIPBOARD DEBRIEFING

3.1 Introduction

The following is an essentially unedited transcript of the self-debriefing of Astronaut Carpenter which he conducted shortly after arriving onboard the recovery aircraft carrier, Intrepid. This shipboard debriefing consists of the pilot's general impressions of the flight from lift-off to the beginning of the retrosequence. From that point through normal egress of the pilot from the spacecraft, the pilot describes his activities in considerable detail.

3.2 Shipboard Debriefing

I would like to give a good debriefing at this point while the events of the flight are still fresh in my mind. I will be able to cover only the high-points. I can not really do the flight justice until I review the voice tape to refresh my memory.

As a whole, I was surprised that the sensations at lift-off, and throughout the launch phase, were as slight as they were. In retrospect, it was a very, very short period. As a matter of fact, the whole flight was very short. It was the shortest five hours of my life.

My general impression of the flight right now is that I am happy to be back. I feel that I brought back some new information. I hope that the pictures turn out because they are photographs of truly beautiful sights. I think that the MIT film was properly exposed. I hope it brings back some worthwhile information. I realize now that a number of the MIT pictures were taken while the spacecraft was in a 90° roll attitude and the filter in the camera was not oriented properly. So there are a few pictures that may be of no value.

I feel badly about having squandered my fuel and I feel badly about the error in impact. I know that there was an error in pitch and I think there was an error in yaw in the gyro attitude presentation from somewhere in the second orbit on. Because the control fuel supply was low, I did not want to evaluate the ASCS problem until just prior to retrofire when I thought it would probably clear up. I thought for some time that the problem in pitch might have been just a scanner error. Now, as I look back at it, it seems to me that that was wishful thinking, [3-2] because I aligned the gyros correctly and the spacecraft was holding orbit attitude when I first selected ASCS. Later, however, when I would recheck attitude the spacecraft would be pitched way down, about 20°. So ASCS was holding orbit attitude in yaw and roll but pitch attitude was not right. It did not agree with the window and it did not agree with the periscope. I say 20° down when I think of the periscope, but when I think about what I saw in the window when the ASCS was holding retroattitude and indicating 34°, I would say that it might be something like 30° down. I noticed the same problem on the second orbit, or maybe it was the very beginning of the third orbit. I also noticed this prior to retrofire.

I think that one reason that I got behind at retrofire was because, just at dawn on the third orbit, I discovered the source of the fireflies. I felt that I had time to get that taken care of and prepare for retrofire properly, but time slipped away. It really raced during this period, as it did through the whole flight. I really needed that time over Hawaii. The Hawaii Cap Com was trying very hard to get me to do the preretrograde checklist. I had previously been busy with the fireflies. Then was busy trying to get aligned in attitude so that I could evaluate ASCS. I got behind. I had to stow things haphazardly. I think everything was stowed, but not in the planned places. Food crumbling gave me a bad problem because I couldn't use that bag for the camera. As it was, I had to carry the camera with me and almost dumped it in the water.

At retrofire I still had the problem in pitch attitude. I did not have any confidence in ASCS just prior to retrofire. So I told the California Cap Com that the ASCS was bad and that I was committing to a fly-by-wire retrofire. By this time, I had gone through part of the preretro checklist. It called for the manual fuel handle to be out as a backup for the ASCS. I selected the fly-by-wire control system and did not go off of the manual system so that attitude control during retrofire was accomplished on both the fly-by-wire and manual control systems.

I feel that attitude control during retrofire was good. My reference was divided between the periscope, the window, and the attitude indicators. At retroattitude as, indicated by reference to the window and the periscope, the pitch attitude indicator read -10 degrees. I tried to hold this attitude on the instruments throughout retrofire but I cross-checked attitude in the window and the periscopes. I have commented many times that you can not divide your [3-3] attention between one attitude reference system and another, and do a good job in retrofire on the trainer. But that was the way I controlled attitude during retrofits on this flight. I did not notice any gross errors in attitude that persisted throughout retrofire. There was some wandering, but I feel that it was balanced out pretty well.

The initiation of retrofire was just a little bit late, although retrosequence came on time. I got the countdown from the California Cap Com. I waited one more second, which was 99:59:59 and did not get retrofire. I punched the manual retrofire button and one or two seconds after that I felt the first retrorocket fire.

I expected a big boot from the retrorocket. But the deceleration was just a very gentle nudge. The sound of the rockets firing was just audible. Retrorocket Two fired on time, Retrorocket Three fired roughly on time. Each rocket gave me a sensation, not of being pushed back toward Hawaii as reported by John Glenn, but of being slowed down in three increments. So that by the time the retroacceleration was over, I felt that there was just enough deceleration to bring the spacecraft to a stop. I felt that, if I looked down, I would see that the obvious motion that I had seen through the window and the periscope before retrofire had stopped. But, of course, it had not.

I put three 'arm' switches on at this time. Retropack jettison occurred on time and the periscope came in on time. At this time I noticed my appalling fuel state, and realized that I had controlled retrofire on manual and fly-by-wire. I went to rate command at this time, and tried manual <u>and</u> rate command, and got no response. The fuel gauge was reading about 6 percent, but it was empty. This left me with 15 percent on the automatic system to last out the ten minutes to .05g and to control reentry.

If the California Cap Com had not mentioned the retroattitude bypass switch, I think I would have forgotten it, and retrofire would have been delayed considerably longer. He also mentioned an Aux Damp reentry which I think I would have chosen in any case, but it was a good suggestion to have. He was worth his weight in gold for just those two items.

The period prior to the .05g was a harried one, because I did not know whether the fuel was going to hold out. The periscope [3-4] was retracted. The attitude indicators were useless. The only attitude reference I had was the window. I did not have much fuel to squander at this point holding attitude. I did use it, gingerly, trying to keep the horizon in the window so that I would have a correct attitude reference. I stayed on fly-by-wire until .05g. At .05g I think I still had about 15 percent reading on the autofuel gage.

I began to get the hissing outside the spacecraft that John Glenn mentioned. I feel that the spacecraft would have reentered properly without any attitude control. It was aligned within 3 or 4 degrees in pitch and yaw at the start of the reentry period. My feeling is that the gradual increase of aerodynamic damping during the reentry is sufficient to align the spacecraft properly.

Very shortly after .05g I began to pick up the oscillations on the pitch and yaw rate needles. At this time I think roll rate was zero, or possibly one or two degrees. The spacecraft oscillated back and forth about zero, just the way the trainer would do at a -.1 (-.1 damping coefficient set into the trailer computer) reentry. From this I decided that the spacecraft was in a good reentry attitude and I selected Aux Damp.

I watched the rate indicator and the window during this period because I was beginning to see the reentry glow. I was beginning to see a few flaming pieces falling off the spacecraft, although the window did not light up as John Glenn reported. It was just a noticeable increase in illumination. I did not see a fiery glow prior to peak g as John Glenn did.

I noticed one thing during the heat pulse that I had not expected. I was looking for the orange glow. I also saw a long rectangular strap of some kind going off in the distance. It was at this time that I noticed a light green glow that seemed to be coming from the cylindrical section of the spacecraft. It made me feel that the trim angle was not right, and that some of the surface of the recovery compartment might be ablating. I think it must have been the berylium [sic] vaporizing. The fact that the rates were oscillating evenly strengthened my conviction that the reentry was at a good trim angle. The green glow was really brighter than the orange glow around the window. [3-5]

I heard Cape Cap Com up to the blackout. He told me that black-out was expected momentarily. I listened at first for his command transmission, but it did not get through. So I just talked the rest of the way down.

Acceleration peaked at about 6.7g. At this time, oscillations in rate were nearly imperceptible. Aux Damp was doing very, very well. The period of peak g was much longer than I had expected. I noticed that I had to breathe a little more forcefully in order to say normal sentences.

The accelerometer read 2.5 to 3g when the spacecraft passed through a hundred thousand feet. At around 80 or 70 thousand feet, we may have run out of automatic fuel. I do not remember looking at the fuel gage but the rates began to oscillate pretty badly, although the rate needles were still on scale.

I put in a roll rate earlier and after we got down around 70 or 80 thousand feet, I took the roll rate out. So I did have fuel at that point. I took the roll rate out at a point where the oscillations carried the sun back and forth across the window. My best indication of the amplitude of the oscillation was to watch the sun cross the window, and try to determine the angle through which the spacecraft was oscillating. I remember calling off about 40 or 50 degrees. This was around 60,000 feet. At about 50,000 feet, the amplitude of the oscillations increased. I could feel the deceleration as we would go to one side in yaw or pitch. I would feel the spacecraft sort of stop, and then the rate would build up in the other direction. I felt that I had a pretty good indication of the variation in attitude from this change in acceleration. I switched the drogue fuse switch on at about 45 thousand feet. At about 40 thousand feet, I began to feel that the spacecraft oscillations were going past 90 degrees. I would feel a deceleration as the spacecraft would go past the vertical. I knew from the amplitudes that I had previously extrapolated, that the spacecraft attitude had reached at least 90 degrees. Then the spacecraft would apparently slip past 90 degrees. I am convinced that the attitudes were diverging, and that there were times when the spacecraft was 30 or 40 degrees small end down. This I remember occurring two or three times. Each time it was worse. I reported that the oscillations were getting too bad and said, "I'm going to have to chance the drogue now." I did deploy the drogue parachute manually at around 25,000 feet. [3-6]

Although I did not make a concerted effort to deploy the drogue parachute when the spacecraft was properly aligned in attitude, I think that it did come out when the spacecraft was in normal attitude, because there was no marked snap on deployment. There was a sudden shock, but I do not think that it dragged the spacecraft around from bad yaw or pitch angle. The spacecraft moved maybe 10 or 20 degrees. I could see the drogue pulsing and vibrating. It was visible against a cloudy sky. I saw no blue sky at this time. All was gray. The drogue was pulsing and shaking much more than I had expected. I watched the parachute for a while along with some other material that came out at this time.

After the drogue parachute was deployed, I operated the snorkel manually. The rate handle did come up but I reached over and pushed it up, too. I did not notice any more cooling at this time. I also did not notice the suit fan cutting down so I assume it continued to run.

I got the main fuse switch at 15,000 feet and waited for the main parachute to deploy. It did not, and I manually operated the main parachute deploy switch at about 9,500 feet. It was just a little below 10,000 feet. It came out and streamed. It was reefed for a little while. Boy! There is a lot of stress on that parachute! You can see how it is being tried. The parachute unreefed and it was beautiful. I could see no damage whatsoever.

Rate of descent was right on 30 feet per second. Incidentally, prior to retrofire the rate of descent indicator was reading about six or seven feet per second. I was convinced that the main parachute was good and selected the auto position on landing bag switch and the bag went out immediately. I went through the post reentry, post-10K, and post landing checklists and got everything pretty well taken care of.

The impact was much less severe than I had expected. It was more noticeable by the noise than by the g-load. There was also a loud knock at impact. I thought "We have a recontact problem of some kind." I was somewhat dismayed to see water splashed on the face of the tape recorder box immediately after impact. My fears that there might be a leak in the spacecraft were somewhat confirmed by the fact that the spacecraft never did right itself on the water. It continued to stay in a 60 degree attitude on the water. [3-7] The direction of list was about halfway between pitchdown and yaw left. That is the attitude it maintained on the water.

I got everything disconnected and waited for the spacecraft to right itself. We do not have a window in the egress trainer, but the level of the water on the window seemed to be higher than I had expected. The list did not change.

I knew that I was way off track. I had heard the Cape Cap Com transmitting blind that there would be an hour before recovery. I decided to get out at that time and went about the business of egressing from the spacecraft.

Egress is a tough job. The space is tight and egress is hard. But everything worked properly. The small pressure bulkhead stuck a little bit. Pip pins and initiators came out very well. I easily pushed out the canister with my bare head. I had the raft and the camera with me. I disconnected the hose after I had the canister nearly out.

I forgot to seal the suit and I did not put the neck dam up. I was aware at this time that the neck dam was not up. It should have been put up right after impact, but I had forgotten it. I think one of the reasons I did not was that it was so hot. However, it wasn't nearly as hot as I expected it to be. I think after impact I read 105 on the cabin temperature gage. I was much hotter in orbit than I was after impact. I did not notice the humidity. I felt fine.

I climbed out. I had the raft attached to me. I placed the camera up on top of the recovery compartment so that I could get it in the raft with me if the capsule sank. I did not want to take it with me while I inflated the raft.

I slid out of the spacecraft while holding on to the neck. I pulled the raft out after me and inflated it, while still holding on, to the spacecraft. The sea state was very good. Later on the swells may have increased to eight or nine feet. But at impact the swells were only five or six feet. I got in the raft upside down. It was attached to the spacecraft.

The rest of the debriefing I can do later. This is the only part I really need to talk about now. The rest will come back in much clearer detail when I get the voice tapes.

Document I-46

Document title: Richard L. Callaghan, NASA, Memorandum for Mr. James E. Webb, "Meeting with President Kennedy on Astronaut Affairs," 30 August 1962.

Source: Folder 18674, NASA Historical Reference Collection, NASA History Division, NASA Headquarters, Washington, DC.

From the beginning of NASA's human spaceflight effort the activities of the astronauts outside of their official duties had been a source of concern and contention. The public, of course, relished as much information as could be obtained about the Mercury Seven and NASA had facilitated the sale of their personal stories to Life *magazine as a means of both satisfying that thirst and as a form of insurance for the astronauts should any lose their lives in spaceflight. This decision faced numerous criticisms, however, and NASA had to explain and find more equitable approaches to the issue in later years. Moreover, companies sought endorsements and some entrepreneurs offered the astronauts gifts such as homes at no expense so they could use the fact that the astronauts lived in their housing developments as selling points for other buyers. It proved a prickly issue for NASA, much of it the result of the celebrity status of the Mercury Seven. As this memorandum demonstrates, concern for these issues rose all the way to the Oval Office and prompted comments by President John F. Kennedy. NASA worked to refine its policies in this regard, but never found a fully satisfactory solution that balanced the rights and privileges of the astronauts with government regulations on private activities.*

NATIONAL AERONAUTICS AND SPACE ADMINISTRATION

WASHINGTON 25, DC

August 30, 1962

MEMORANDUM For Mr. James E. Webb

Subject: Meeting with President Kennedy on Astronaut Affairs

On August 23, I called Pierre Salinger about 5:30 p.m. to advise him of the discussion Mr. Lingle, Mr. Johnson and I had with Alfred Friendly. It occurred to me that such a call might serve to remind him of our interest in having the reaction of the White House that Mr. Lingle and I sought in our meeting with Salinger nearly two weeks ago. Salinger commented that since meeting with Lingle and me, he had had a long talk with the President and others in the White House about a revised policy relating to the affairs of the astronauts. He stated that "the President tends to agree with you (NASA) and Bundy (McGeorge) agrees with me." He expressed no particular interest in the reaction of Mr. Friendly but indicated that he would try and set up a meeting in a few days to get together with us again. Within a half hour, he called back and said "bring Mr. Webb and Bill Lloyd down tomorrow morning at 10:30 and we'll meet with the President and settle this once and for all." I told him that you were out of town but that I would attempt to bring Mr. Lingle and Mr. Lloyd. This was satisfactory to him.

I checked with Dr. Seamans to determine whether he wished to go to the White House but he felt that Mr. Lingle could handle the problem satisfactorily. We attempted to contact you Friday morning but you were somewhat ahead of your itinerary and were apparently enroute [*sic*] from Norton to Medford. Lingle, Johnson, Lloyd and I discussed the proposed meeting with Dr. Seamans prior to going to the White House.

The White House meeting lasted some 30 minutes. The President at the outset stated generally that he felt the astronauts should be permitted to continue to receive some money for writings of a personal nature inasmuch as they did seem to be burdened with expenses they would not incur were they not in the public eye. He felt there should be stricter control of their investments. He cited the proffer of the homes in Houston as an example of the type of situation that should be avoided in the future. Salinger was rather restrained in presenting his own views and seemed satisfied to take his cue from the President. Mr. Lingle prefaced his remarks by expressing the hope that no firm decision would be made at this particular meeting as to the [2] specifics of the policy inasmuch as no member of the NASA group at the meeting was prepared to delineate in a positive way your views. It was made clear to the assembled group that you wished to have the policy reflect White House desires and that you intended to make it clear to the astronauts that any policy decisions would embrace White House attitudes.

The President showed no disposition to criticize NASA's existing policy. Such observations as were made by Salinger, Bundy, and Ted Sorensen ran to a need for tightening up in the implementation of our policy rather than to a need for changing the policy in any drastic measure. It was the consensus that the refinement of NASA's policy should be achieved through discussions with the Department of Defense and that NASA policy should serve as a model to which a Department of Defense policy would conform. The development of such a policy by the Department of Defense seemed to be left within Salinger's hands.

Without detailing the discussion further, the following portrays my impression of the conclusions reached at the meeting with the President.

1. The President leaves to your discretion the preparation of such refinements in NASA's proposed policy revisions as are necessary to:

    a. Permit the continued sale by the astronauts of their personal stories, whether through a LIFE-type contract or otherwise.

    b. Extend the prohibition against commercial endorsements.

    c. Provide reasonable supervision of the astronauts' investments (although this need not be a specifically stated part of the policy, the astronauts are to understand that such supervision is inherent in the policy).

    d. Serve generally as a model of administration policy.

2. Within the framework of its policy NASA should attempt to:

    a. Make available to all news media at debriefings and press conferences a more comprehensive presentation of the official aspects of space missions in which the astronauts participate.

    b. Afford to the press additional access to NASA personnel (including the astronauts), NASA installations, and NASA facilities to the extent that such access does not impede the agency's programs or activities.

    c. Edit more stringently the material made available by the astronauts for publication.

    d. Restrict extravagant claims by publishers who attempt to overemphasize the exclusive nature of material received from the astronauts for publication.

[Signed]
Richard L. Callaghan

Document I-47

Document title: Dr. Walter C. Williams, Deputy Director, NASA Manned Spacecraft Center, NASA, "Project Review," 3 October 1963.

Source: NASA Collection, University of Houston, Clear Lake Library, Clear Lake, Texas.

The Mercury program officially ended with the flight of Faith 7, Gordon Cooper's orbital mission on 15 and 16 May 1963. Within days of that flight those working on the Mercury began assessing their efforts and developing lessons-learned for the future. This review culminated in a large meeting in Houston on 3 to 4 October 1963, where the leading figures of the program discussed the Mercury project and its accomplishments. This document presents the perspective of Walt Williams, Robert Gilruth's assistant for space operations at the Manned Spacecraft Center.

MERCURY PROJECT SUMMARY CONFERENCE

MUSIC HALL, HOUSTON, TEXAS

October 3 and 4, 1963

PROJECT REVIEW

Address by Dr. Walter C. Williams, Deputy Director, NASA Manned Spacecraft Center

[Note: This review also included a slide presentation. The slides are not provided.]

I think that, perhaps, in reviewing a program such as this, the first step to take is to look at where we started and, principally, what were the objectives and what were our guidelines, and I think you'll find that this group that started five years ago, under Dr. Gilruth, stayed quite closely to these.

[Slide 1]

Let's look at the objectives first (first slide, please). I'm not sure this is exactly the same slide that was used five years ago, but I'm certain that the words are. Objectives were to place a manned spacecraft in orbital flight around the earth, to investigate man's performance capabilities and his ability to function in space, and, obviously, recover the man and spacecraft safely. And we hope, as we move along in these next two days, to show how these objectives were reached.

[Slide 2]

Some of the guidelines in establishing this project are shown on the next slide. We knew, or the team knew, that to do this program at any reasonable length of time, wherever possible, existing technology and off-the-shelf equipment would have to be used, wherever practical, and I think, although it was expected to find much equipment on the shelf, I think many of our problems were really finding which shelf this equipment was on, because, in almost every area, because of the design constraints, some new development had to be undertaken to meet the new requirements that [2] a manned spacecraft would place on a system. Obviously, we wanted to use as simple an approach because this, indeed, would give the most reliable approach. The simplicity, again, is a relative term. Because of the question about man's ability to perform, it was required that this spacecraft be capable of fully automatic flight as well as a flight were the man participated as part of the system. Well, when you automate the system and, indeed, then provide redundancy in the automation, you come out with a rather complex system. The existing launch vehicle would be employed; yes, we felt that we should use a launch vehicle that was well along in its development as a weapons system for this job, and we had some interesting experiences along the way in developing and working with the Space Systems Division in converting from a weapons systems to a man-mated booster and, of course as always, we felt this should be a progressive and logical test program and we will discuss that progression.

We were able, or the team at that time was able, to give some detailed requirements for the spacecraft, in a general sense, and these are shown in the next slide.

[Slide 3]

We knew that the state-of-art of the large rockets, the reliable launch escape system was required. We did feel, even though there was a question mark about the pilot's performance, that he should be able to manually control the spacecraft attitude, and I think it's well-known how much this paid off during the life of the program. Obviously, it had to have a reliable retrorocket system, but it was also a question that this spacecraft [3] should be deorbited by retrorockets, that it just wasn't a short life-time orbit; that the spacecraft would truly be in space flight. The zero-lift shape for reentry was chosen as the least difficult and still meet the mission objectives that we had in mind, and obviously, we had to provide a water-landing capability because, even though we would have good—take on the task of providing land capability for the end of successful missions, the vehicle still had to be amphibious in order to cover the abort cases.

Well this – this was about the way the program got started five years ago. Concepts were available; in fact, considerable research-and-development work had been done on these concepts, but there lay ahead the job of translating these concepts into real hardware, into systems that could be used in manned flight. There was the detailed mission planning that was yet to be done. There was the defining and implementing the world network. There were many of these things. Developing the recovery techniques. All of this was still ahead.

[Slide 4]

Scheduling, I think, is about the best way to describe the progress of the flights and of the program since these are, indeed, tangible milestones. And, although there were many schedules, and you could call them success schedules, or the like, this is the actual schedule as the flying occurred. I realize there's a lot of detail here, but I'd like to talk about this overall schedule first. In accomplishing this, in the period of about 45 months of activity, some 25 flights were made which was an activity of a major flight in something less than every 2 months. [Slide 5] [4] To do this, at various states, three launch vehicles were used and two launch sites. The Little Joe was a research-and-development booster used for the development, testing primarily the escape system; these tests were at Wallops Island. The Redstone booster was used for the ballistic flights to help qualify the spacecraft systems and the crew for orbital flight. And, of course, the Atlas was used for orbital flights. It is interesting to note that one of the first major flights was the BJ-1 up there, which was the Big Joe, which qualified the heat-protection system and verified that this concept was proper. Dr. Gilruth talked about the team getting right to work and I can talk a little about this one because I had nothing to do with it. I think this was an amazing job done in something less than a year from project go-ahead. This was a major activity and it involved a ballistic reentry of a full-scale Mercury like spacecraft.

[Slide 6]

And, so, the first year or so we were concerned with these development flights and it was about the end of 1960 really that the heavy activity in qualifying the actual hardware for the manned orbital flights began and I'd like to look at an expanded scale there and it's on this next slide, on the right side, please. This, I think, was the peak of our highest activity in Mercury. We began with our – really, we should start with the Mercuy-Redstone 1 which was our first full-boosted flight of our production spacecraft. We had problems; we fired the escape tower when it was a premature cutoff, but we won't go into this today. But, then, the program moved along rather rapidly on the ballistic program between December and May when Al Shepard made his flight and followed by Grissom's [5] flight that summer. Meanwhile, the Atlas was also moving along; we had a failure back in July, that Dr. Dryden referred to, which cost us about six months in our Atlas program and it was not until the following February, after suitable modifications had been made to both the spacecraft adapter and the launch vehicle, that we were able to resume the Atlas flights. The first of these qualified our production spacecraft for the reentry heating case. That was followed by another Atlas failure, MA-3, which was an electronic failure, but, by then, we had the team really working together; we solved these problems and made our first orbital flight of a Mercury spacecraft in September and within four or five months of that, we had John Glenn's flight following the flight of Enos in orbital flight. This, to me – to anyone planning schedules – The flight program for this time should look at this one, because there were periods here of major activities, at least once a month, and in a research-and-development program, I feel that this is about the limits of human tolerance. Everybody was working terribly hard on this period; it was a rough one.

Now, this is about all of the detail (will you take those slides off)—detail of the program that I can go into at this time (hold that one).

[Slide 7]

I would like to talk now a little, because we will describe all of this in much more detail – I would like to talk about how we managed this program, because I think this was one of the important things we learned. As you know, this program started on – go-ahead was given on October 7, 1958 and a small organization, the Space Task Group, was set up to handle [6] it. The overall management, of course, was the responsibility of NASA Headquarters, but the project management rested in Space Task Group. And of course, it was recognized from the beginning that this had to be a joint effort of many organizations and of many people, because it was an extremely complex program and it would be probably involve more elements of Government and industry than any similar development program that had been undertaken. So, the task was that of establishing an overall plan that would best fit the program and accomplish the objectives at the earliest date, pulling all of these varied groups together, and the scheme that we used to pull people together and pull organizations together is best shown in this next slide, where we might look at this at three levels: At the policy level, which was the overall management of level where general policy decisions were reached and carried out as to how the two organizations would work together; the next level down which was the approval review and direction level; and then, a third level of implementation where we used a system of working teams, with the specialists and design people from each of the various units concerned with any particular problem, and these were action committees and decisions could be reached at their meetings, with formal documentation to follow at a later date, and teams were set up as required wherever there were interfaces to be solved and common problems involving more than one organization. And I might add, and I think this is very important, teams were set up as they were needed; they were dissolved when they were no longer needed. We did not have committees for the sake of committees.

[7] [Slide 8]

And, I think a matter—To put some names and numbers into a chart such as this, I'd like to show the next slide which shows an arrangement we used in the launch vehicles. The manner whereby NASA could get Atlas launch vehicles for the space program was reached in an agreement at the level of NASA Headquarters and the Department of Defense, and this was spelled out in a working agreement. Then, it became the task of NASA Space Task Group and the then Air Force Ballistic Missiles Division to translate this policy into a launch vehicle we could use and then we brought together at the working level members of Space Task Group, members of the Ballistic Missiles Division, as well as their contractors and our contractors, and out of this evolved the details of things, such as the automatic abort system, the structural interface, the launch complex modifications, the launch countdown, that were required. Now, another bit of management arrangement that was established that also worked very well, and this fell primarily in the operational support area and in the network areas, was the fact that NASA, as such, had very little resources to carry out the program of this nature. For example, for recovery, we didn't have a navy. It's this type of resource I am addressing myself to. We did not have a range; so, in order to effectively provide this support from the Department of Defense, and arrangement was made whereby a Department of Defense Representative for Project Mercury Support was appointed and he was the NASA, the single point contact within the

Department of Defense framework for all Department of Defense support. Also, NASA provided such [8] a single point of contact, so that these two could meet; there was a logical place for the requirements to focus, a logical place for them to go, and logical place for them to be implemented. And rather than many parts of NASA trying to work with many parts of the Department of Defense, we had this single point, and I think, and I don't think, I know, in the operating end of it at least, this contributed greatly to the success.

[Slide 9]

I'd like to show how this worked; for example, in the case of our network and that's on the next slide. Again, we had this type of thing; we had our DOD representative, NASA single point of contact and this is for the establishment of the network. At this level, we reached agreement of what parts of the national ranges would be used which would be modified, where new stations had to be implemented and who would operate new stations, how would we work on the existing ranges. At the direction level, we had our Space Task Group and an element of the Langley Research Center which handled the Western Electric contract on the network that provided the detailed implementation, working directly with the Mercury Support Planning Office and the National ranges. And here, again, we had to break out working teams and these involved not only the obvious units shown there, but for example, our spacecraft contractor had to work with this people so that they would be compatible with the range. And, I think that it was arrangements like this that allowed us to move on as we did and I must say that it was also, as Bob Gilruth pointed out, the dedication of a large number of people that allowed these arrangements to work extremely well.

[9] Now, in these types of systems and at this point in time which was the development phase, we used this arrangement of working teams. As we moved into the operating phase, however, we had to go to a more functional type organization, with direct lines of command, and here again, having this single point within DOD helped considerably. I'll not show the entire operating organization, but I'd like to show an element of it in this next slide to give some idea of how organizations were intermingled in this line of command. This is essentially the blockhouse organization and our total operating complex. [Slide 10] The operations director was a NASA man; however, reporting to him was the launch director from the Air Force's Space Systems Division. In turn, there was a launch vehicle test conductor who was a General Dynamics/Astronautics and in turn, had his associate contractors, reporting to him and, meanwhile, the spacecraft test conductor was a NASA man who, in turn, had his contractors reporting to him. And, I think any part of this organization you would find similar intermingling – intermingling of the Services as well, intermingling of the Civil servants, military personnel, and contractor personnel, but I think that the important thing is that it did work; there were direct-line responsibilities and I think we learned a lot out of that.

Now, I think it's interesting to talk a little about the resources we used. (May I have the next slide, please?) [Slide 11] Manpower reached a total, and this, of course, has to be estimates, even though we've got rather small numbers shown, of about 2,000,000 people. The direct NASA effort, [10] Space Task

Group, never reached a peak of over 650 people on the program and I would say this was reached probably at the time of the Glenn flight [Slide 12]. Supporting NASA work was another 700 people. Obviously a large element was the industrial support of prime, subcontractors, and vendors and we had, of course, many people from the Department of Defense, the largest portion of this 18,000 being the Navy's recovery forces. And I think it's interesting in looking at a map, which spots only the major contractors and the Government agencies and universities involved, without going into the subs and vendors, and as you can see, fairly well covered the country, even at this level.

I think, perhaps, to, we should talk a little about the program cost and I'd like to have the next slide [Slide 13]. These figures, I might point out, are different from those that are in the chronology that is part of the handout for this conference. The chronology figures were not complete and left out some of the essential elements. These figures and this total of $384,000,000 is the best that we can come up with for now; it's our estimate of determinations of contracts and it's not a fully audited figure, but it's the best we have at this time and I think this represents a reasonably correct figure. I think the only thing of real interest here is that the two largest items of this was the development of the spacecraft itself and its operation and the implementation of the world network. These items, like this network, are things that normally aren't thought of as the cost of a program – one will concentrate mostly on the [11] flight hardware, but, as can be seen, this, indeed, was a large part of the total cost of Mercury. However, I may add that, although we're charging all of it to Mercury here, it is an investment in our National capability; it will be used in Gemini, it will be used in Apollo. The operations figure is primarily the cost of the recovery forces.

Now, this, in a nutshell, is about what Mercury consisted of. We will try to fill in detail in the next two days. I think we ought to, before I close though, just summarize what it appears to me we learned in Mercury. One, of course – we did, indeed, accomplish our objectives and we found that man does have a place in space, man can function as part of the spacecraft system or the total flight system and can be effective. I think we learned some very—Very obviously, we learned a lot about spacecraft technology and how a spacecraft should be built, what its systems should be, how they should perform, where the critical redundancies are that are required. I think we learned something about man-rating boosters, how to take a weapons system development and turn it into a manned transportation system. I think, in this area, we found primarily, in a nutshell, that this was a matter of providing a malfunction detection system or an abort system, and, also, we found very careful attention to detail as far as quality control was concerned. I think that some of the less obvious things we learned – we learned how to plan these missions and this take a lot of detail work, because it's not only planning how it goes, but how it doesn't go, and the abort cases and the emergency cases always took a lot more effort than the planned missions. These are things that must be done [12]. We learned what is important in training crews for missions of this type. When the crew-training program was laid down, the program had to cover the entire gamut because we weren't quite sure exactly what these people needed to carry out the missions. I think we have a much better focus on this now. We learned how to control these flights in real time. This was a new concept on a worldwide basis. I think we learned, and when I say we, I'm talking of this as a National asset, not

NASA alone, we learned how to operate the world network in real time and keep it up. And I think we learned a lot in how to manage development programs of this kind and to manage operations of this kind.

I thank you very much.

*Oral presentation transcribed by occ; typed by rhd.

Document I-48

Document title: Christopher C. Kraft, Jr., "A Review of Knowledge Acquired from the First Manned Satellite Program," No date, but 1963

Source: Folder 18674, NASA Historical Reference Collection, NASA History Division, NASA Headquarters, Washington DC.

Christopher C. Kraft, Jr. joined the NACA in 1944 and became a member of the Space Task Group upon its inauguration in 1958 and a close associate of STG Director Robert Gilruth. He served as Senior Flight Director for all of the Mercury missions. Because of his unique perspective on Mercury, his review of the program is especially valuable as an historical document.

NASA FACT SHEET #206

NATIONAL AERONAUTICS AND SPACE ADMINISTRATION
MANNED SPACECRAFT CENTER
HOUSTON 1, TEXAS

A REVIEW OF KNOWLEDGE ACQUIRED FROM THE FIRST MANNED SAT-
ELLITE PROGRAM

By

Christopher C. Kraft, Jr.
NASA Manned Spacecraft Center

SYNOPSIS

With the completion of the Mercury program, science has gained considerable new knowledge about space. In more than 52 hours of manned flight, the information brought back has changed many ideas about space flight. Design problems occupied the first and major portion of the Mercury program. The heat shield, the shape of the Mercury spacecraft, the spacecraft systems, and the recovery devices were developed. Flight operations procedures were organized and developed and a training program both ground and flight crew was followed. Scientific experiments were planned with Man in the loop. These included photography, extra spacecraft experiments, and observation or self-performing types of experiments.

But the real knowledge of Mercury lies in the change of the basic philosophy of the program. At the beginning, the capabilities of Man were not known, so the systems had to be designed to function automatically. But with the addition of Man to the loop, this philosophy changed 180 degrees since primary success of the mission depended on Man backing up automatic equipment that could fail.

[2]

INTRODUCTION

As the first manned space flight project of the United States, Project Mercury in its various aspects have [*sic*] been discussed in great detail by almost all members of the project. The purpose of my discussion today will not be to repeat the technical details of Project Mercury, but to outline and discuss some of the significant contributions the program has made to the area of space technology.

It is important to note that 52 hours of manned orbital flight, and less than five hours of unmanned orbital flight by the Mercury spacecraft have produced a large book of new knowledge. The hours spent on the ground development and training, the preparations for flights, and the ballistic flights cannot be calculated, but it contributed heavily to the knowledge we ultimately gained in space flight.

The three basic aims of Project Mercury were accomplished less than five years ago from the start of the program. The first U.S. manned space flight program was designed to (1) put man into Earth orbit (2) observe his reactions to the space environment and (3) bring him back to Earth safely at a point where he could be readily recovered. All of these objectives have been accomplished, and some have produced more information than we expected to receive from conducting the experiment.

The whole Mercury project may be considered an experiment, in a certain sense. We were testing the ability of a man and machine to perform in a controlled but not completely known environment.

The control, of course, came from the launch vehicle used and the spacecraft systems included in the vehicle. Although we knew the general conditions of space at Atlas insertion altitudes, we did not know how the specific environment would affect the spacecraft and the man. Such conditions as vacuum, weightlessness, heat, cold, and radiation were question marks on the number scale. There were also many extraneous unknowns which would not affect the immediate mission but would have to be considered in future flights. Such things as visibility of objects, the airglow layer, observation of ground lights and landmarks, and atmospheric drug effects were important for future reference.

The program had to start with a series of design experiments. We had little criteria for the space vehicle. If we could find that a certain type of heat shield could make a successful reentry and a certain shape of spacecraft, we would have the basis for further design of systems.

A series of flight tests and wind tunnel tests were conducted to get the answers to some of the basic questions. First, would the ablation [3] principle work in our application? Could we conduct heat away from the spacecraft body by melting the fiberglass and resin material? How thick would the shield have to be for our particular conditions? What temperatures would be encountered and for what time period would they exist? Early wind tunnel test proved in theory that the saucer shaped shield would protect the rest of the spacecraft from heat damage. The flight test on the heat shield must prove the theory. In February 1961, we made a ballistic flight in which the spacecraft reentered at a sharper angle than programmed and the heat shield was subjected to great than normal heating. The test proved the heat shield material to be more than adequate.

The Mercury spacecraft did not start with the familiar bell shape. It went through a series of design changes and wind tunnel tests before the optimum shape was chosen. The blunt shape had proven best for the nose cone reentry. Its only drawback was the lack of stability. We next tried the cone-shaped spacecraft, but wind tunnel testing proved that heating on the afterbody would be too severe, although the craft was very stable in reentry. After two more trial shapes, the blunt bottom cylinder on cone shape came into being. It was a complete cycle from the early concepts of manned space-craft, but it was only the first of a series of changes in our way of thinking of the flight program and its elements.

A second part of design philosophy thinking came in connection with the use of aircraft equipment in a spacecraft. We had stated at the start of the program that Mercury would use as much as possible the existing technology and off-the-shelf items in the design of the manned spacecraft. But in many cases off-the-shelf equipment would just not do the job. Systems in space are exposed to conditions that do not exist for aircraft within the envelope of the atmosphere. Near absolute vacuum, weightlessness and extremes of temperatures makes equipment react differently than it does in aircraft. We had to test equipment in advance in the environment in which it was going to be used. It produced an altered concept in constructing and testing a spacecraft. Although aircraft philosophy could be adapted, in many cases, aircraft parts could not perform in a spacecraft.

The third part of the design philosophy, and perhaps the most important one in regard to future systems is the automatic systems contained in the Mercury spacecraft. When the project started, we had no definitive information on how Man would react in the spacecraft system. To insure that we returned the spacecraft to Earth as planned, the critical functions would have to be automatic. The control system would keep the spacecraft stabilized at precisely thirty-four degrees above the horizontal. The retrorockets would be fired by an automatic sequence under a grogramed [sic] or ground command. The drogue and main parachutes would deploy when a barostat inside the spacecraft indicated that the correct altitudes had been reached. The Mercury vehicle was a highly automatic system and the man essentially was riding along as a passenger, an observer. At all costs, we had to make sure that the systems worked.

[4] But we have been able to take advantage of Man's capability in space. It started from the first manned orbital flights. When some of the thrusters became inoperative on John Glenn's flight, he was able to assume manual control

of the spacecraft in order to fly the full three orbits planned in the mission. When a signal on the ground indicated the heat shield had deployed, Glenn bypassed certain parts of the retrosequence manually and retained the retropack after it had fired. In this way, he insured that the heat shield would stay in place during reentry and the spacecraft would not be destroyed by excessive heating. When oscillations built up during reentry, Glenn utilized his manual capability to provide damping using both the manual and fly-by-wire thrusters. The pilot's role in manned space flight was assuming a more important aspect.

Carpenter's flight again emphasized the ability of the pilot to control the spacecraft through the critical reentry period. Excess fuel was used in both of these orbital flights. Schirra's task was to determine if Man in the machine could conserve fuel for a long flight by turning off all systems in drifting flight. It was a task that could not be accomplished by a piece of automatic equipment in the confined area of the Mercury spacecraft. Schirra also was able to exercise another type of pilot control. It was the fine control necessary to adjust pressure suit air temperature to produce a workable environment. When we flew the mechanical man in MA-4, we did not have the capability of making fine suit temperature adjustments or to realize the problems we might encounter in the suit design. Man could analyze and correct suit temperature, thus pointing out necessary design parameters to follow in future programs.

The MA-4 and MA-5 flights were probably the most difficult of the orbital missions. They had to be flown using only one automatic control system. We had no man along with the ability to override or correct malfunctions in the systems. One of the flights ended prematurely due to malfunctions that we could not correct from the ground. In both cases, a man could have assumed manual control and continued the flight for the full number of orbits. It is no hypothesis or theory; it has been borne out by facts. With this design criteria in mind, the Cooper flight was a fitting climax to the Mercury program. Not only did it yield new information for other spacecraft program, but it demonstrated that Man had a unique capability to rescue a mission that would not have been successfully completed with the automatic equipment provided.

Man serves many purposes in the orbiting spacecraft. Not only is he an observer, he provides and redundancy not obtainable by other means, he can conduct scientific experiments, and he can discover phenomenon not seen by automatic equipment.

But most important is the redundancy, the ability of another system to [5] take over the mission if the primary system fails. Duplicate systems are designed to prevent bottlenecks in the operation of the systems. The single point failure caused the false heat shield signal in Glenn's flight. After the mission was successfully completed, we conducted an intense design review to see if there were any more of these single points in the spacecraft that needed redundancy of design for safe operation. We found many areas where the failure of one component could trigger a whole series of unfavorable reactions. This type of problem had been brought about by the design philosophy originally conceived because of the lack of knowledge of Man's capability in a space environment.

The Mercury program taught us not to stack the components on top of each other. It forces limited access, and the failure of one component during checkout makes it necessary to pull out other functioning systems to replace the malfunctioning part. For instance, in the MA-6 flight the short life carbon dioxide absorber in the environmental control system had to be replaced since checkout took longer than had been planned. This replacement required eight major equipment removals and four revalidations of unrelated subsystems for a total delay of 12 hours. All of these problems of course resulted from weight and space constraints brought about by payload limitations.

For the Gemini and Apollo spacecraft, the equipment will be modular and replaceable, allowing the substitution of alternate parts without tearing out whole subsystems.

We depend quite a bit on the automatic systems for retrosequence but man has proven that he can and does play an important role in the reentry process. The only manned flight in which the automatic system for reentry was used completely was at the end of Walter Schirra's six orbits. In all other flights, the astronaut took over and performed at least one part of the reentry manually because of some malfunction which had occurred during the flight.

As we move into the Gemini and Apollo programs, a maneuvering capability has been built into the spacecraft to allow changes in flight path both while in orbit and during reentry into the atmosphere.

The translation engines provided will allow modifications to the orbit for rendezvous with other vehicles in orbit. Also, by use of an offset center of gravity, the spacecrafts will have and L/D capability not provided in the Mercury vehicle. This will allow the onboard computers to select a particular landing point at any time during the flight and after retrofire or atmospheric reentry the vehicle can be maneuvered within a given footprint to reach this desired landing area. The astronauts will provide the necessary back-up to these complex systems and can at any time assume manual control of the system so that a proper and safe landing can be assured.

[6] Our experience with the Mercury network changed our thinking about the operation of this worldwide tracking system for manned flights. In the initial design of the network, we did not have voice communication to all the remote sites.

But we soon found that in order to establish our real time requirement for evaluating unusual situations, we needed the voice link. When we started the program, the determination of the orbital ephemeris was a process that could take several orbits to establish. We could not tolerate such a condition in a manned flight so we set up a worldwide network which would maintain contact with the astronaut approximately 40 minutes out of every hour. But continuous voice contact with the astronaut has proven unnecessary and in many cases undesirable. While we retain the capability to contact an astronaut quickly, we have tried to reduce the frequency of communications with the spacecraft.

In designing and modifying a spacecraft, it is also possible to learn something more than tangible changes or hardware design. We learned about the reliability requirement and the very important need to check details carefully. It is a requirement that cannot be designed into a system on the drawing board. It actually consists in developing a conscientious contractor team that will take care to follow procedures and deliver a reliable product. Then it takes a careful recheck by the government team to insure that reliability has actually been built into the product. The smallest mistake in a man rated system can bring totally unexpected results. The unexpected is the rule in the unknown, and if Man is going to live in the region beyond our atmosphere, he is going to live under rules or not at all. We have been aware of these new rules from the start of the satellite program, but they have not been brought to our attention so vividly as they have in the manned flight program.

If an unmanned satellite malfunctions we cannot get it back for examination. We can only speculate on the causes and try to redesign it to eliminate the source of the supposed trouble. It is necessarily a slow process of elimination. Here again, if a manned craft malfunctions, it can be returned to the ground by the proper action of the pilot. We knew what had failed in Gordon Cooper's flight, but we did not know why the system had failed until we got the spacecraft back for investigations and tests. Knowing why something occurred will give us the tools to improve spacecraft of the future.

AEROMEDICAL EXPERIMENTS

While we can redesign the equipment to accomplish the mission, we cannot redesign the man who must perform in space. Aeromedical experiments for new knowledge about space must simply answer one question. Can Man adapt to an [7] environment which violates most of the laws under which his body normally operates? The answer to the question at the end of the Mercury program seems to be an unqualified yes, at least for the period of one to two days.

The crushing acceleration of launch was the first concern. We knew he would be pressed into his couch by a force equal to many times the weight of his own body. It was not definitely known whether he would be able to perform any piloting functions under these high "g" forces. The centrifuge program was started and the astronauts tested under this stress proved that Man was not as fragile or helpless as we might have supposed. In addition to being able to withstand heavy acceleration, a method was developed of straining against the force and performing necessary pilot control maneuvers.

Weightlessness was a real aeromedical unknown and it was something that the astronauts could not really encounter on the ground. The ability to eat and drink without gravity was one serious question we had to answer. In the weightless condition, once the food is placed in the mouth, normal digestive processes take over without being affected by the lack of gravity.

The next problem was the effect of weightlessness on the cardiovascular system, that is the heart and blood vessel system throughout the body. All types of reactions were possible in theory. In actual flight, a small and temporary amount of pooling of blood in the veins of the legs has occurred, but it is not serious nor does it appear to affect the performance of the pilot. For all pilots weightlessness has been a pleasant experience. All the senses such as sight and hearing perform normally during space flight. There has been no hallucination, no blackout or any other medical phenomena which might have an effect on Man in space. We even experimented with drifting flight and whether the astronaut would become disoriented when he could not distinguish up from down or have the horizon of Earth for a reference. But each time the answer seemed to be that a man could adapt as long as his basic needs for breathing oxygen and pressure were supplied.

Perhaps the greatest contributions to the program have come in the area of development of aeromedical equipment. Blood pressure measuring systems were developed that would automatically take readings and transmit them by telemetry to the ground. The biosensors were designed to pick up other information such as pulse rate and respiration rate. There were numerous small changes that were made to these systems to increase the accuracy of the data that we got back from the man in space. The in-flight studies of the test pilot's reaction are probably the most complete medical records we have tried to keep on an individual. Their value has been to demonstrate that man functions normally in the space environment.

Related to the aeromedical studies in the environmental equipment that provides life support for the astronaut. We started with the basic Navy pressure suit for aircraft flying and modified it for performance in the spacecraft. We found it was desirable to eliminate as many pressure points as possible and have tailored the suits on an individual basis for each [8] astronaut. There are two areas in life support which presented new problems to be overcome. First, there was the problem of circulation of air. In the absence of gravity, the normal rules of air circulation are cancelled, and the carbon dioxide breathed out by the astronaut would suffocate him. The air in the cabin would also have to be forced through the air conditioning system to keep the cabin area from overheating.

Secondly, there is the problem of the air supply itself and its possible effect on the spacecraft pilot. For conserving weight, a single gas system was desirable. But it was not known if breathing pure oxygen over long periods of time could have harmful effects. The Mercury flights and other research in a pure oxygen environment have proven that no injury to the body's system has been produced by using a one gas system.

SCIENTIFIC EXPERIMENTS

Man's role as a scientific observer and experimenter in space was another unknown in the program. Much of it was based on the ability of man to exist in space. It had to first be determined that he would be able to function normally and then the scientific benefits of the program could be explored. Man as an observer has proven his capabilities from the first orbital flight. The brightness, coloring, and height of the airglow layer was [sic] established. It was something a camera could not record nor could an unmanned satellite perform this mission. Man in space has the ability to observe the unknown and to try to define it by experiment. The particles discovered at sunrise by John Glenn were determined by

Scott Capenter, to be coming from the spacecraft, and this analysis was confirmed by Schirra and Cooper.

We can send unmanned instrumented vehicles into space which can learn much about the space environment and the makeup of the planets. However, the use of Man to aid in making the scientific observations will be invaluable. The old problem of what and how to instrument for the unknown can benefit greatly from Man's capability to pick and chose the time and types of experiments to be performed. We have learned much from the Mercury program through this quality of choice and we will continue to learn if man continues to be an important part of the system.

If we have learned more about space itself, we have also learned about Man's capabilities in space. Many experiments have been conducted which have yielded valuable information for future programs. Aside from aeromedical experiments, Man has been able to distinguish color in space, to spot object at varying distances from the spacecraft, to observe high intensity lights on the ground, and to track objects near him. These observations provide valuable information in determining the feasibility of the rendezvous and navigation in Gemini and Apollo.

[9] Pictures taken with infrared filters have aided the Weather Bureau in determining the type of cameras to use in their weather satellites. Special pictures have also been taken for scientific studies such as geological formations, zodiacal light, and refraction of light through the atmosphere.

CONCLUSION

The manned space flight program has changed quite a few concepts about space, added greatly to our knowledge of the universe around us, and demonstrated that Man has a proper role in exploring it. There are many unknowns that lie ahead, but we are reassured because we are confident in overcoming them by using Man's capabilities to the fullest.

When we started the manned space program five years ago, there was a great deal of doubt about Man's usefulness in space. We have now come to a point which is exactly one hundred eighty degrees around the circle from that opinion. We now depend on Man in the loop to back up the automatic systems rather than using automatic systems alone to insure that the mission is accomplished.

We do not want to ignore the automatic aspects of space flight altogether. There must be a careful blending of Man and machine in future spacecraft which provides the formula for further success. By experience, we have arrived at what we think is a proper mixture of that formula. Man is the deciding element; but we cannot ignore the usefulness of the automatic systems. As long as Man is able to alter the decision of the machine, we will have a spacecraft that can perform under any known condition, and that can probe into the unknown for new knowledge.

--END--

Document I-49

Document Title: Manned Spacecraft Center, NASA, "Project Development Plan for Rendezvous Development Utilizing the Mark II Two Man Spacecraft," 8 December 1961.

Source: Folder 18674, NASA Historical Reference Collection, History Division, NASA Headquarters, Washington, DC.

The development plan for the Mercury Mark II spacecraft underwent a number of modifications throughout 1961. The plan was extensively revised up until 27 October 1961. A key question was the selection of a booster to launch the spacecraft; NASA's preference was a modified Titan II ICBM. The Air Force wanted to develop a Titan III, but NASA was wary of this plan, fearing that the development would take too long. The Air Force countered that NASA's requirements for modifications to the Titan II would lead to what was almost a new booster. These issues were solved by November and it was decided by 5 December that NASA would get the Titan II boosters it desired. On 6 December, Robert Seamans approved the project development plan and identified the development of rendezvous techniques as the project's primary objective. Brainerd Holmes asked for $75.8 million from current Fiscal Year 1962 funds to start the project and Seamans approved that request on 7 December. The final plan was approved the next day.

PROJECT DEVELOPMENT PLAN

FOR

RENDEZVOUS DEVELOPMENT

UTILIZING THE

MARK II TWO MAN SPACECRAFT

Manned Spacecraft Center
Langley Air Force Base, Virginia

December 8, 1961

CLASSIFIED DOCUMENT – TITLE UNCLASSIFIED [DECLASSIFIED]

This material contains information affecting the national defense of the United States within the meaning of the espionage laws, Title 18,U.S.C., Secs. 793 and 794, the transmission or revelation of which in any manner to unauthorized person is prohibited by law.

NATIONAL AERONAUTICS AND SPACE ADMINISTRATION

TABLE OF CONTENTS

[2]

TABLE OF CONTENTS

[3]

PART I – PROJECT SUMMARY

This project development plan presents a program of manned space flight during the 1963 – 1965 time period. The program provides a versatile system which may be used for extending the time of flight in space and for development of rendezvous techniques, but may be adapted to the requirements of a multitude of other space missions at a later date. A two man version of the Mercury spacecraft would be used in conjunction with a modified Titan II booster. The Atlas-Agena B combination would be used to place the Agena B into orbit as the target vehicle in the rendezvous experiments. This use of existing or modified versions of existing hardware minimizes the necessity for new hardware development.

The proposed plan is based on extensive usage of Mercury technology and components for the spacecraft. Therefore, it is proposed to negotiate a sole-source cost-plus-fixed-fee contract with McDonnell Aircraft Corporation for the Mark II Mercury spacecraft.

The launch vehicle procurement will involve a continuation of present arrangements with the Air Force and General Dynamics-Astronautics for the Atlas launch vehicles, and the establishment of similar arrangements with the Martin Company for the Modified Titan II launch vehicles, and with the Lockheed Aircraft Corporation for the Agena stages.

A Project Office will be established to plan, direct and supervise the program. The manpower requirements for this office are expected to reach 179 by the end of Fiscal Year 1962.

The estimated cost of the proposed program will total about 530 million dollars.

PART II – JUSTIFICATION

Upon completion of Project Mercury the next step in the overall plan of manned space exploration is to gain experience in long duration and rendezvous missions. It is believed that the program presented here would produce such information and that it would compliment other programs now underway while not interfering with their prosecution.

PART III – HISTORY AND RELATED WORK

The plans for Project Mercury originally recognized the value to be obtained from 18-orbit missions. However, such missions were later deleted from the Mercury schedule due to systems and network limitations. Early in 1961 it was believed that Project Mercury had progressed to the point where 18-orbit missions might be considered once again. At this time, McDonnell was asked to study how such missions could be accomplished with only a minimum of modifications to the spacecraft being required. This study showed that the 18-orbit mission represents the maximum growth potential of the present Mercury capsule with reasonable modifications. Therefore, McDonnell was asked to study means of providing a more extensively modified spacecraft with an extended mission capability, including multiman occupancy and improved systems accessibility. The Martin Company was asked to provide information as to how the Titan II might be adapted to serve as the launch vehicle for these extended missions. Both the McDonnell and Martin studies have progressed to the point that capabilities for performing the missions have been shown. On the basis of these favorable reports the program plan presented here has been developed.

[5]

PART IV – TECHNICAL PLAN
(Description and Approach)

1.0 INTRODUCTION

Project Mercury is an initial step in a long range program of manned exploration of space. The initial objectives of Project Mercury have already been accomplished; therefore, it now becomes appropriate to consider the steps that should be taken to insure immediate continuation of manned space flights following the successful conclusion of this project. Therefore, a follow-on project, after Project Mercury, is proposed which will provide a continuing source of development information. In the execution of the proposed project, maximum use will be made of vehicle and equipment development which has already been accomplished for other programs.

2.0 MISSION OBJECTIVES

The present Mercury spacecraft cannot be readily adapted to other than simple orbital missions of up to about one day duration, with a corresponding limitation on the objectives of the mission. The proposed project will allow the accomplishment of a much wider range of objectives.

   2.1 Long Duration Flights Experience will be gained in extending the duration of flights beyond the 18 orbit capability of the present Mercury spacecraft. It is recognized that for the longer missions a multiman crew is essential so that the work load may be shared, both in time and volume. There are many areas which require investigations so that the multiman crew may be provided with a suitable environment during the prolonged missions. This project will contribute to the development of the flight and ground operational techniques and equipment required for space flights of extended periods. These flights will also determine the physiological and psychological reactions and the performance capabilities of the new crew while being subjected to extended periods in a space environment.

   2.2 Rendezvous The rendezvous and docking maneuver in space may be compared to aerial refueling in that it makes possible the resupply of a vehicle in space and thus extends its mission capabilities. This maneuver makes it possible to put a much larger "effective" payload in space with a given booster. Since most space projects are "booster limited" at present, the development of techniques for getting the most out of available boosters should undoubtedly be treated as of highest priority. As the frequency of manned orbital flights increases, there will be instances when orbital rescue, personnel transfer, and spacecraft repair will be highly desirable. To accomplish these missions develop[6]ment of orbital rendezvous techniques is mandatory. Among the problem areas which are involved in effecting a successful rendezvous and docking maneuver are the following:

2.2.1   Launch Window The second vehicle involved in the rendezvous must be launched very close to a prescribed time if the operation is to be economical in terms of waiting time and propulsion requirements. This requires a major simplification of the countdown procedure and high reliability of equipment.

2.2.2   Navigation Means must be developed for maneuvers in space, using information supplied by the navigation system.

2.2.3   Guidance and Control Guidance and control techniques must be developed for maneuvers in space, using information supplied by the navigation system.

2.2.4   Docking Rendezvous is not effective until the docking maneuver is accomplished. The space environment makes this operation quite a bit different from the same type of operation within the earth's atmosphere and hence considerable work in developing suitable techniques is to be expected.

2.3   Controlled Land Landings Experience has shown that the magnitude of the effort required to deploy adequate naval forces for the recovery of the Mercury spacecraft at sea is such that any means for avoiding, or at least minimizing, this effort would be highly desirable. The sea has proved to be a more inhospitable environment for recovery than was originally envisioned. If space flights are to be accomplished on anything like a routine basis, spacecraft must be designed to alight on land at specified locations. This requires that the landing dispersion be reduced to a very low figure, and a satisfactory method of touchdown developed.

2.3.1   Dispersion Control To effect control of the landing area, it is fundamental that an impact prediction be made available to the pilot and a means provided for controlling the spacecraft so the desired impact point can be reached.

2.3.2   Landing Impact The attenuation of the impact loads which might result from a land landing of the Mercury spacecraft has presented a very considerable problem. Although it is estimated that in many cases the landing accelerations would be within tolerable limits, the random nature of the landing process has made it impossible to consider a sufficient variety of conditions that could be encountered so as to have adequate assurance [7] of success. In order to guarantee safety in landing, the impact must be made at a relatively low velocity and in a selected area.

2.4 <u>Training</u> Although much can be accomplished by ground simulation training, there does not seem to be any real substitute for actual experience in space. Thus, a by-product of this project would be to provide a means of increasing the number of astronauts who have had actual experience in space. A two-manned spacecraft will be an excellent vehicle for this purpose.

2.5 <u>Project Philosophy</u> In general, the philosophy used in the conception of this project is to make maximum use of available hardware, basically developed for other programs, modified to meet the needs of this project. In this way, requirements for hardware development and qualification are minimized and timely implementation of the project is assured.

Another fundamental concept is that in the design of the spacecraft, all systems will be modularized and made independent of each other as much as possible. In this way, an evolutionary process of product improvement and mission adaptation may be implemented with a minimum of time and effort. Thus, it will be possible to use equipment of varying degrees of sophistication as it becomes available and as the mission requirements are tightened. It is important that a minimum of lead time can be obtained by making use of the latest hardware developments. This concept will make possible the attainment of mission and permits reasonable compromises to be made in the face of difficulties rather than excessive delays that otherwise might be required to meet the full objectives.

This project will provide a versatile spacecraft/booster combination which will be capable of performing a variety of missions. It will be a fitting vehicle for conducting further experiments rather than be the object of experiments. For instance, the rendezvous techniques developed for the spacecraft might allow its use as a vehicle for resupply or inspection of orbiting laboratories or space stations, orbital rescue, personnel transfer, and spacecraft repair.

[Parts V-VIII not included]

PART IX-PROJECT RESULTS

The results to be realized from successful accomplishment of the MK II program include the following:

1. <u>Operational Techniques</u> Rendezvous and docking techniques will become operational, making possible the assembly of vehicles in orbit for extended exploration of space. Techniques for reduction of landing dispersion, through the use of reentry lift and the paraglider landing concept, will be developed and optimized. The relative roles of onboard and ground-based intelligence and optimum man-machine relationships will be established.

2. <u>Long Duration Flight Performance</u> Man's reactions and ability to perform during long duration space flight will be determined. Hardware for sustaining man's physical well-being during such extended missions will be developed.

3. <u>Training</u> A group of pilots will be trained in the techniques required for rendezvous, reentry and controlled land landings. Ground operational forces will acquire experience in the launch, tracking and recovery procedures necessary for long duration and rendezvous missions.

[8]

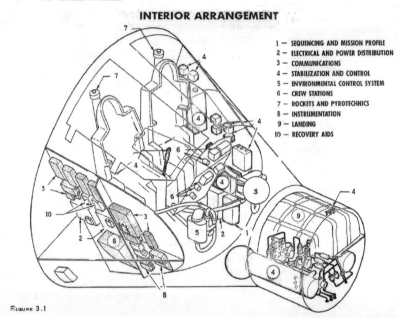

TWO MAN MK II SPACECRAFT

INTERIOR ARRANGEMENT

1 — SEQUENCING AND MISSION PROFILE
2 — ELECTRICAL AND POWER DISTRIBUTION
3 — COMMUNICATIONS
4 — STABILIZATION AND CONTROL
5 — ENVIRONMENTAL CONTROL SYSTEM
6 — CREW STATIONS
7 — ROCKETS AND PYROTECHNICS
8 — INSTRUMENTATION
9 — LANDING
10 — RECOVERY AIDS

Figure 3.1

[9]

ESTIMATED WEIGHT STATEMENT
(18 ORBIT)
TWO MAN MK II SPACECRAFT

	3 ORBIT MERCURY SPACECRAFT	18 ORBIT TWO MAN MK II SPACECRAFT
GROSS WEIGHT AT LAUNCH	4139	6407
EFFECTIVE LAUNCH WEIGHT	3283	5246
WEIGHT IN ORBIT	2886	4755
RETROGRADE WEIGHT	2871	4731
RE—ENTRY WEIGHT	2605	3730
IMPACT WEIGHT	2393	3458
ABORT WEIGHT (ESCAPE ROCKET BURNOUT)	3406	4696

FIGURE 3.2

[10]

EVENTS REQUIRED TO COMPLETE A RENDEZVOUS
MISSION WHEN SPACECRAFT LAUNCH OCCURS WITHIN THE
LIMITS OF THE LAUNCH WINDOW

△ V APPLIED AT NODE TO ROTATE
THE AGENA PLANE INTO THE
SPACECRAFT PLANE

SPACECRAFT
ORBIT

AGENA ORBIT

1. ADJUSTMENT OF PLANE

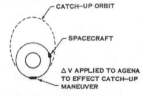

CATCH—UP ORBIT

SPACECRAFT

△ V APPLIED TO AGENA
TO EFFECT CATCH—UP
MANEUVER

2. CATCH—UP MANEUVER

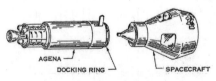

AGENA

DOCKING RING

SPACECRAFT

3. FINAL MANEUVERING & DOCKING

FIGURE 3.9

[11]

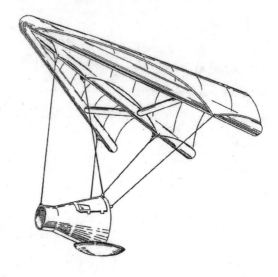

Figure 3.13

[12]

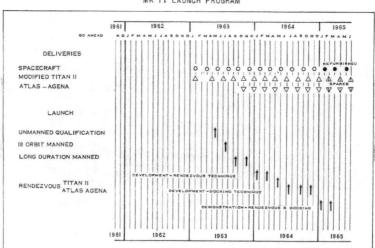

[13]

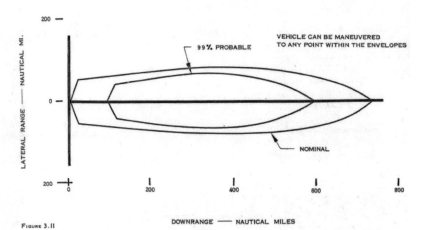

FIGURE 3.11

[14]

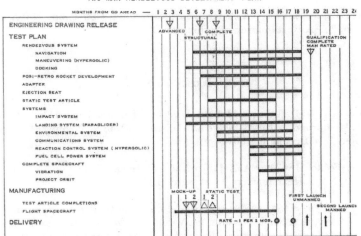

FIGURE 5.2

COST SCHEDULE
ORBITAL FLIGHT DEVELOPMENT

	FY 1962	FY 1963	PROGRAM RUNOUT	TOTAL
SPACECRAFT	42,600	77,500	120,400	240,500
LAUNCH VEHICLES				
TITAN II — MODIFIED	27,000	47,000	39,000	113,000
ATLAS — AGENA	5,200	20,000	62,800	88,000
OPERATIONAL SUPPORT	1,000	14,250	43,700	58,950
SUPPORTING DEVELOPMENT	—0—	5,000	24,000	29,000
TOTAL	75,800	163,750	289,900	529,450

ALL FIGURES SHOWN IN THOUSANDS

Document I-50

Document Title: Al Nagy, NASA, to George Low, NASA, 11 December 1961.

Source: Folder 18674, NASA Historical Reference Collection, NASA History Division, NASA Headquarters, Washington DC.

Document I-51

Document Title: D. Brainerd Holmes, Director of Manned Space Flight Programs, NASA, Memorandum for Associate Administrator, NASA, "Naming Mercury-Mark II Project," 16 December 1961.

Source: Folder 18674, NASA Historical Reference Collection, NASA History Division, NASA Headquarters, Washington DC.

In December 1961 NASA officials began considering what to call the new program planned to follow Mercury. Robert C. Seamans, Jr., NASA's Associate Administrator, wanted to run a competition to name the proposed Mercury Mark II, and offered a token reward of a bottle of good Scotch whiskey to the person suggesting the name finally accepted. In addition to others who recommended the name, Alex P. Nagy, an engineer in NASA's Office of Manned Space Flight, proposed "Gemini," a reference to classical mythology and quite appropriate for the two-astronaut spacecraft. NASA Headquarters officials selected Gemini from a host of other names submitted, including "Diana," "Valiant," and "Orpheus," from the Office of Manned Spaceflight. On 3 January 1962, NASA announced the Mercury Mark II project had been renamed "Gemini."

Document I-50

AP:lgs
December 11, 1961

George:

For the orbital flight development effort, I propose the name "PROJECT GEMINI."

This name, "the Twins" seems to carry out the thought nicely, of a two-man crew, a rendezvous mission, and its relation to Mercury. Even the astronomical symbol (II) fits the former Mark II designation.

[Signed: *Al*]
Al Nagy

Document I-51

In reply refer to:
MS

December 16, 1961

The Office of Manned Space Flight recommends the following names for the project currently referred to as Mercury-Mark II:

Diana [handwritten: *Huntress*]
Valiant
Gemini
Orpheus

These are not listed in any order of preference

[handwritten: *George M. Low*
for]   D. Brainerd Holmes
Director of
Manned Space Flight Programs

Document I-52

Document Title: Flight Crew Operations Division, NASA, "Gemini Familiarization Package," 3 August 1962.

Source: NASA Collection, University of Houston, Clear Lake Library, Clear Lake, Texas.

The Project Gemini Familiarization Manual was a document published by the McDonnell Aircraft Company as a training aid for Gemini astronauts. The first section dealt with a mission description, while a second section related to Major Structural Assemblies. The remaining sections described the Cabin Interior Arrangement, the Sequence System, the Electrical Power System, the Environmental Control System, the Cooling System, the Guidance and Control System, the Communication System, and the Instrumentation System. This "Gemini Familiarization Package" served as a brief summary of the more extensive manual.

[CONFIDENTIAL] [DECLASSIFIED]

GEMINI FAMILIARIZATION PACKAGE

Prepared by the Flight Crew Operations Division
Crew Engineering
August 3, 1962

(This material contains information affecting the National Defense of the United States, within the meaning of the Espionage Laws, Title 18 US. C., Sections 798 and 794, the transmission or revelation of which in any manner to an unauthorized person is prohibited by law)

[1]

1.0   INTRODUCTION

The purpose of this familiarization package is to provide documentation describing the operation, system designs, and crew station arrangement of the two man Gemini spacecraft. These notes are complementary to the contractor furnished pilot's manual which deals primarily with the details of each display and control inside the spacecraft cockpit.

To best appreciate the significance of displays, controls, and manual operational procedures, one should have a thorough knowledge of the mission profile and system functions which are described in detail in the body of this document. First, however, the Gemini program objectives will be listed for reference and a summary description given of the guidelines used to divide crew tasks.

1.1 Program Objectives
        (a) Accomplish 14 day earth orbital flights.
        (b) Demonstrate rendezvous and docking in earth orbit.

(c) Provide for controlled land landing as primary recovery mode.

(d) Develop simplified countdown techniques to aid rendezvous missions (lessens criticality of launch window).

(e) Determine man's capabilities in space during extended missions.

1.1 Crew Tasks

The crew is used as a required integral part of Gemini. The Manned Spacecraft Center philosophy calling for increased crew usage and onboard command and control wherever logical is implemented in this program.

The Pilot-Commander has primary control of spacecraft operation during all phases of flight.

The Co-Pilot/Systems Engineer provides control backup to the pilot and manages operation of spacecraft and Agena systems.

[2]

1.2 Comparison of Mercury and Gemini

While there is similarity between Mercury and Gemini, there are several significant differences in operations and systems design. In summary, the major differences are as follows:

b.   Manual Abort

All aborts will be initiated onboard by the pilot-commander who has launch vehicle system displays on the left hand console, and at least one backup indication of each malfunction situation; (visual, physical, audio, or redundant display).

c.   Maneuvering Capability

Translation capability is provided in Gemini before docking by the OAMS (Orbit Attitude and Maneuver System) and after docking by Agena. Both these systems use similar hypogolic propellants.

d.   Cryogenics

Super-critically stored hydrogen and oxygen are used in the environmental control system and for the fuel cells.

e. Range Control

Modest lift capability is provided during reentry by offsetting the spacecraft center of gravity.

Lift is controlled by rolling the spacecraft about the reentry vector. Greater reentry range and an increased heat load result from this feature which allows point return.

f. Paraglider

An inflatable paraglider and conventional landing gear provide for subsonic flight control and horizontal landing.

[3]

g. Extra-vehicular Operations

The Gemini hatch is designed to permit the crew to leave the spacecraft while in orbit. Specific experiments and extra-vehicular suit provisions have not been defined.

Document I-53

Document Title: Charles W. Mathews, Manager, Gemini Program, "Program Plan for Gemini Extravehicular Operation," 31 January 1964.

Source: Folder 18674, NASA Historical Reference Collection, History Division, NASA Headquarters, Washington, DC.

As the Mark II spacecraft was being designed and redesigned, one of the changes involved the addition of a large mechanical hatch that, in addition to facilitating entry and exit to the spacecraft and allowing the use of ejection seats, would also permit an astronaut to leave the spacecraft in orbit. But the idea was only discussed sporadically for the next few years, since it was not necessary for the Apollo program and it was planned that any extra-vehicular activity (EVA) experiments would be done late in the program. In January 1964, this preliminary plan for EVA operations was developed, but it was not enthusiastically received within NASA. At a press conference in July 1964, Gemini Deputy Manager Kenneth Kleinknecht had suggested that a limited EVA was possible during Gemini IV, but this remark had gone unnoticed. James McDivitt and Edward White, the primary crew for Gemini IV (called GT-4 in this document), and their backups Frank Borman and James Lovell, Jr., lobbied hard for the inclusion of the EVA mission in the Gemini IV flight and ultimately swayed opinions at NASA. An EVA on the Gemini IV mission was approved on 25 May 1965. The fact that the Soviet Union had carried out the first-ever EVA on 18 March 1965 was clearly a factor in that approval, but the intent to do EVAs during Project Gemini had been part of the program plan from the start.

PROGRAM PLAN

FOR

GEMINI EXTRAVEHICULAR OPERATION

January 31, 1964

Approved: _____[signed]_____
Charles W. Mathews
Manager, Gemini Program

[2]

I.   PURPOSE

This program plan has been prepared by the Gemini Program Office to document the Objectives of Gemini extravehicular operation and to outline the program for achieving these objectives. It is intended for use as the basis for overall program control and coordination to ensure proper implementation of program requirements. The plan will be kept current by the Gemini Program Office and revisions will be issued as additional information is developed.

II.   OBJECTIVES OF GEMINI EXTRAVEHICULAR OPERATION

A.   General.   The general objectives to the accomplished are as follows:
1.   Evaluate man's capability to perform useful tasks in a space environment.
2.   Employ extravehicular operation to augment the basic capability of the Gemini spacecraft.
3.   Provide the capability to evaluate advanced extravehicular equipment in support of manned spaceflight and other national space programs.

B.   Phase One. The objectives to be accomplished on the initial extravehicular missions are:
1.   Demonstrate feasibility of extravehicular operation.
2.   Establish confidence in Gemini systems for extravehicular operation.

[2]

3.   Conduct preliminary evaluation of man's ability to perform in free space.

C.   Phase Two. After completion of Phase One, the following objectives are to be accomplished:
1.   Conduct detailed evaluation of man's ability to perform in free space.

    2. Retrieve experimental data packages and equipment from the adapter section and from the Agena.

    3. Conduct preliminary evaluation of advanced extravehicular equipment, including long term life support systems and maneuvering devices.

D.   <u>Phase Three.</u> After completion of Phase Two, the following objectives are to be accomplished:

    1. Evaluate equipment and man's capabilities to operate independent of the spacecraft.

    2. Perform such advanced extravehicular experiments as are approved in the future.

III.   <u>IMPLEMENTATION</u>

A.   <u>Mission Planning.</u>

    1. Mission planning is to be based on a step-by-step progression from the simplest to the more ambitious extravehicular tasks. For planning purposes the following mission scheduling shall be used:
      a. Phase One: GT-4 through GT-6
      b. Phase Two: GT-7 through GT-9
      c. Phase Three: GT-10 and up

[4]

    2. Detailed flight activities planning is being done by the Flight Crew Support Division. Activities for a given mission will be determined on the basis of overall mission requirements and capabilities.

B.   <u>Task Assignments.</u>

    1. Crew Systems Division
      a. Equipment development and procurement
      b. Establishment of ground test program

    2. Flight Crew Support Division
      a. Flight activities planning
      b. Astronaut training

    3. Center Medical Operations Office
    Monitor progress of program to insure fulfillment of medical requirements.

    4. Flight Operations Directorate
    Monitor progress of program to insure fulfillment of flight operations requirements.

5. Gemini Program Office
Overall program direction

IV. <u>EXTRAVEHICULAR EQUIPMENT</u>

A. <u>Portable Life Support System (PLSS).</u>
1. <u>Phase One</u>. The Crew Systems Division (CSD) is developing a PLSS based on the Mercury 7500 psi oxygen bottle. This PLSS is being designed to provide open loop oxygen flow at 5 cfm for a total of 45 minutes. After allowing suitable [5] reserves and time for egress and ingress, this system will be limited to a maximum of 10 minutes outside the spacecraft.

2. <u>Phase Two.</u> In order to accomplish the Phase Two objectives, a PLSS which will provide 30 minutes useful time outside the spacecraft is required. Further study is needed to determine the type of system which will met this requirement. Development of the Phase Two PLSS is to be carried out by CSD.

2[*sic*] <u>Phase Three.</u> It is anticipated that the Phase Two PLSS will be used for egress and ingress during Phase Three operations. More advanced equipment to be used for longer duration periods outside would be stowed in the equipment adapter. This advanced equipment will be defined at a later date.

B. <u>Pressure Suit.</u>
1. A modified version of the Gemini Pressure suit will be used for extravehicular operation. The single wall pressure vessel concept will be retained. The following modifications will be incorporated:
a. An overvisor for glare, ultraviolet, and thermal protection.
b. Gloves modified to incorporate thermal protection.
c. Redundant pressure sealing closure.
2. Development of the Gemini extravehicular suit is to be carried out by CSD.

[5]

C. <u>Thermal Protection.</u>

1. <u>Phase One.</u> The only thermal protection required for Phase One operations consists of local protection against the extreme temperatures of the spacecraft exterior. The gloves, boots, and knees are the primary areas affected.
2. <u>Phase Two and Three.</u> Present studies indicate that a thermal overgarment will be required for extravehicular missions of 30 minutes or more outside the spacecraft. Development of the thermal garment is to be carried out by CSD.

D. <u>Meteoroid Protection.</u>
1. Meteoroid protection will be required to provide a probability of .999 of no puncture of the pressure suit. On the basis of the present

MSC standard meteoroid environment, the following weight of soft goods padding will be required in a protective garment:
a. Phase One (10 minutes) - 2 lb.
b. Phase Two (30 minutes) – 3.5 lb.
c. Phase Three (1 hour) – 4.75 lb.

2.   Development of a meteoroid protective garment is to be carried out by CSD.

E.   Tether.
1.   A tether incorporating a safety line and communications leads is being developed by CSD. Initial planning has been based on no biomedical instrumentation during the extravehicular operation. More recently CSD and Medical Operations have [6] specified minimum desired parameters to be monitored. Provisions for monitoring these parameters will be incorporated in the tether, if possible. The length of the tether is to be sufficient to allow ingress to the equipment adapter section.

F. Maneuvering Unit.
1.   The Air Force has proposed an extravehicular unit (MMU) for use on later Gemini missions under Gemini/DOD Experiment 14C. If this experiment is approved, it is anticipated that the MMU would be used in the latter part of Phase Two and in Phase Three. The MMU would contain propulsion, control, communications, and life support systems. It would be furnished by the Air Force under an independent contract.

V.   SPACECRAFT MODIFICATIONS

A.   Spacecraft modifications will be incorporated to enable the astronaut to move about the exterior of the spacecraft and into the equipment adapter. These modifications are as follows:

1.   Exterior handholds spaced approximately two feet apart from the cockpit to the adapter section interior. The handle configuration will be based on configuration studies by CSD as well as aerodynamic considerations.

2.   Protective cover for the rough edge at the aft end of the adapter section. The astronaut must be able to proceed past this rough edge without the hazard of damage to the pressure suit or the tether.

Document I-54

Document Title: Edward Z. Gray, Director, Advanced Manned Missions Program, Office of Manned Space Flight, NASA, to Director, Gemini Program, NASA, "Gemini Lunar Mission Studies," 30 April 1964.

Source: Folder 18674, NASA Historical Reference Collection, NASA History Division, NASA Headquarters, Washington DC.

Document I-55

Document Title: Eldon W. Hall, Director, Gemini Systems Engineering, NASA, to Deputy Director, Gemini Program, NASA, "Circumlunar Missions," 29 June 1965.

Source: Folder 18674, NASA Historical Reference Collection, NASA History Division, NASA Headquarters, Washington DC.

Document I-56

Document Title: James E. Webb, Administrator, NASA, to Olin E. Teague, Chairman, Subcommittee on NASA Oversight, Committee on Science and Astronautics, House of Representatives, 10 September 1965.

Source: Folder 18674, NASA Historical Reference Collection, NASA History Division, NASA Headquarters, Washington DC.

In the spring of 1964 it appeared to many senior officials at NASA that the Apollo program was stalling and might not be able to make its deadline of a lunar landing by the end of the decade. The last Mercury flight had taken place in May 1963, and Gemini was not scheduled to fly for several months. The Saturn rocket project was having difficulties, and the Apollo spacecraft development effort was lagging behind schedule. Accordingly, Wernher von Braun suggested to a reporter for Missiles and Rockets *that in a contingency he thought Gemini might be reconfigurable for a flight around the Moon. This story, appearing on 18 May 1964, quoted von Braun as saying that Gemini could undertake a circumlunar flight "to salvage this country's prestige if the manned lunar goal proves impossible." Von Braun had voiced something that had been bubbling within NASA for some time, and thereafter pressure mounted to formalize and make public efforts to evaluate the possibility of a Gemini circumlunar flight. Throughout the summer of 1964, as these documents show, NASA undertook internal studies. They were only internal, for on 8 June, NASA Deputy Associate Administrator Robert C. Seamans told NASA Associate Administrator for Manned Spaceflight George Mueller that "any circumlunar mission studies relating to the use of Gemini will be confined to in-house study efforts." In reality, NASA leaders had bet the future of their Agency on the success of Apollo. They intended to make Apollo succeed and any serious effort to reconfigure Gemini as a "quick and dirty" lunar program would detract from that objective. The studies were at best halfhearted. In his 10 September 1965 memorandum to Representative Olin Teague (D-Texas), NASA Administrator James E. Webb said it well,: "Our main objective now is to see that our basic current responsibilities are met effectively . . . the Apollo system now being developed can meet our requirements for knowledge and capability better than the adoption of other courses of action."*

Document I-54

MT-1:JRS:saj

April 30, 1964

MG/Director, Gemini Program

MT/Director, Advanced Manned Missions Program

Gemini Lunar Mission Studies

As you are aware, we have been asked by Dr. Mueller to study the feasibility of using Gemini in a lunar mission and to develop suitable contingency plans to be available by mid-1966, should such a mission be feasible and should it be required. Mr. Taylor's office (MT-1), with the assistance of John Hammersmith from your office, has completed a preliminary review of the feasibility of using Gemini in a lunar mission, based on the work that has been done by your office, MSFC, MSC, and McDonnell Aircraft Corporation. This review has concluded that, although all of the studies are relatively shallow, there are several combinations of hardware which could be used to provide a Gemini lunar mission capability. Enclosure 1 [not included] contains the review results.

I believe that a study should be initiated to more thoroughly investigate the Gemini circumlunar mode, utilizing the Saturn IB with a Centaur as the injection stage, in either a direct ascent or an earth orbit rendezvous trajectory. These modes are summarized in Columns 1 and 3 of the Enclosure.

In addition, I think we should study the Gemini Lunar Orbit mode, as represented in Column 7 of the Enclosure. The purpose of such a study would be to more accurately determine the capability of each configuration, the key technical problems, relative costs, development schedules and key decisions points to provide a basis for possible contingency- type decisions in the 1965-66 time period.

As indicated during our telephone conversation on April 22, I believe these studies should be conducted by McDonnell Aircraft Corporation through existing contracts. These studies should be monitored by MSC, either under your or my jurisdiction. If required, I can make funds available for this study, which I believe will require approximately five (5) man-years of effort. We will be available to work with you in this study to whatever extent you desire.

Edward Z. Gray
Director, Advanced Manned
Missions Program,
Office of Manned Space Flight

Document I-55

[CONFIDENTIAL] [DECLASSIFIED]

UNITED STATES GOVERNMENT MEMORANDUM

DATE: June 29, 1965

TO: MG/Deputy Director, Gemini Program

FROM: MGS/Director, Gemini Systems Engineering

SUBEJCT: Circumlunar Missions

1. On Thursday, June 24, I attended a meeting at MSC in which representatives of Martin-Denver and MAC (including Messrs. McDonnell, Burke, and Yardley) presented a proposal for a circumlunar flight using the Gemini spacecraft and the Titan IIIC booster. In attendance at the meeting was Dr. Gilruth, Messrs. Low, Mathews, Kleinknecht, Evans, and Guild of MSC and myself.

2. In this proposal the Gemini spacecraft modified for circumlunar flight is launched into earth orbit with a GLV. The Titan IIIC launches a stripped down transtage that provides the propulsion for injection to circumlunar velocities after rendezvous with the spacecraft. The general arrangement and flight hardware are summarized in enclosure (1) (Figure 2.1-1 of Attachment C). [not included]

3. The principal changes to the Titan IIIC involve using a double transtage. The first provides propulsion during launch into earth orbit and contains the attitude control and an equipment module for use during rendezvous with the spacecraft. A Gemini Target Docking Adapter is mounted on top of the second transtage.

4. A significant number of changes are proposed for the spacecraft. Weight saving items are summarized in enclosure (2) and enclosure (3) (page 1-8 and Table 1.2-1 of Attachment C). The most significant changes to the spacecraft are summarized as follows:

   a. Addition of a Unified S-Band System.

   b. Additional OAMS tankage and TCA's substituted for the retrograde rockets.

   c. Additional heat protection using coated columbium and ablation shingles.

   d. Shortening of the R&R section by 20 inches.

   e. Use of three fuel cell sections.

   f. "Blow-down" RCS and independent pressurization of fuel and oxidizer.

[2]

5.   Three flights are recommended:

    a.   Heat Protection Qualification (Titan IIC – one transtage on ballistic trajectory);

    b.   Spacecraft Qualification (manned, GLV in earth orbit);

    c.   Manned Circumlunar Orbit.

6.   The Martin schedule, enclosure (4) (last page of Attachment A) [not included], indicates completion by April 1967. The MAC schedule (not available) is even earlier using two refurbished spacecraft and a go-ahead by July 1.

7.   No money estimates were presented by Martin or MAC; however, some preliminary estimates by GPO indicated $350M.

8.   I think the proposal is feasible, but not within the time and effort indicated. The equipment and mission are too marginal to absorb changes and additions that will be required without extensive redesign and testing.

9.   I personally would prefer to see us advance our earth orbital capability. With the same or fewer modifications to the spacecraft advocated in this proposal and additional Agena payloads, we could attain a significant lead in the design and operation of earth-orbital space stations. Gemini is ideally suited to the preliminary determination of problems and to the initial development of techniques and procedures leading to advanced manned earth-orbital missions. The time and money spent in additions or extensions of this type to an earth-orbital Gemini would be more than repaid in time and money saved in later, more expensive, and complicated programs.

[Signed]
Eldon W. Hall

Enclosures: 4 as stated [Not included]

Attachments:

A)   "Configuration, Weight Summary, Performance, Transtage #2 Performance, EOR Operations, Mission Profile, and Related Schedules," by Martin-Denver (Unclassified)

B)   "Rendezvous Concept for Circumlunar Flyby in 1967," by Martin-Denver, P-65-91, June 1965 (Proprietary)

C)   "Gemini Large Earth Orbit (U)," by McDonnell, B743, Vol. I – Technical, June 19, 1965 (Confidential)

Document I-56

NATIONAL AERONAUTICS AND SPACE ADMINISTRATION
WASHINGTON DC 20546

September 10, 1965

OFFICE OF THE ADMINISTRATOR

Honorable Olin E. Teague
Chairman, Subcommittee on NASA Oversight
Committee on Science and Astronautics
House of Representatives
Washington, DC 20515

Dear Mr. Chairman:

With reference to your request for my views on the possibility of a circumlunar flight, using a Gemini system, prior to the Apollo lunar landing, you will note that the enclosed statement, which was submitted to the Senate Committee on Aeronautical and Space Sciences on August 23, indicates that in the process of accomplishing the lunar exploration mission with Apollo, our program will give us experienced crews, operating know-how, and the ground and space equipment to undertake quite a number of other scientific and technological developments. The point is also made that our on-going and approved missions will require, for the next several years, the peak performance of the scientific, engineering, industrial and facilities complex that we have been expanding since 1961.

As indicated to the Senate Committee, we are not ready to recommend major new projects on the order of Gemini or Apollo. Our main objective now is to see that our basic current responsibilities are met effectively. I also feel that the Apollo system now being developed can meet our requirements for knowledge and capability better than the adoption of other courses of action.

The insertion in our program of a circumlunar flight, using the Gemini system, would require major resources. We are now proceeding with many complex, developmental tests, and operational efforts with too thin a margin or resources. Therefore, if additional funds were available, I believe it would be in the national interest to use these in the Apollo program.

As you will remember, I testified in 1961 that the USSR would most likely have the capability and therefore accomplish ahead of us each major milestone in space up to the lunar landing and exploration with manned vehicles. We have clearly stated over the past few years that they will do a lunar fly-by with men before we can accomplish this with the Apollo system. However, there is certainly no assurance that we could do this in advance of them with a modified Gemini system. Further, our main reliance for operating [2] at lunar distances and developing a thorough-going capability that can achieve preeminence in space, and hold it, is the large Saturn V/Apollo system. The fact that this has been under contract for several years; that full duration test runs have been made on each stage of the Saturn V booster; that we now have an eight-day Gemini flight behind us and will shortly have information from a 14-day flight; and the fact that the Apollo ground test equipment has largely been fabricated and the flight line equipment will shortly be constructed and delivered means that we have a growing competence that we and the world can see is considerably beyond anything the Russians have shown us, including Proton One. Therefore, I do not believe a decision not to make the substantial investment that would be required by a modified Gemini lunar fly-by will change the posture which our program has had for a number of years.

Sincerely yours,
[Signed]
James E. Webb
Administrator

Document I-57

Document Title: William C. Schneider, Deputy Director, Gemini Program, NASA, for Deputy Director, Apollo Program, "Gemini Support of Apollo," 25 June 1964 (signed for Schneider by LeRoy Day).

Source: Folder 18674, NASA Historical Reference Collection, NASA History Division, NASA Headquarters, Washington DC.

Document I-58

Document Title: Eldon Hall, Director, Gemini Systems Engineering, NASA, Memorandum for Deputy Director, Gemini Program, NASA, "List of Missions," 17 July 1964 (signed for Hall by John Hammersmith).

Source: NASA Historical Reference Collection, NASA History Division, NASA Headquarters, Washington D.C.

From the very beginning of the Gemini program, it had four major objectives that would support the Apollo effort to reach the Moon by the end of the decade. These included: (1) long duration spaceflight of up to two weeks in duration to demonstrate the human capability to survive such an extended stay in space; (2) rendezvous and docking with another orbiting vehicle; (3) engaging in extra-vehicular activity (EVA) or spacewalks; and (4) developing methods for entering the atmosphere and landing at pre-selected points on land. All of these were skills viewed as necessary for later Apollo missions, and all except the last were accomplished. These two memoranda outline the evolution of efforts on the Gemini missions aimed at satisfying these requirements. The first, signed for Gemini Deputy Director William Schneider by LeRoy E. Day, longtime engineer at NASA, shows a steady progress of achievements in support of the Apollo program, each more complex than the last. The second, signed for Gemini Director of Systems Engineering by Eldon W. Hall, another longtime NASA engineer, offers a shopping list of Gemini "desires" that never came to fruition, such

as propellant transfer in orbit and on-orbit assembly and repair. These initiatives were to
be part of additional Gemini missions that were never approved. Both memos reflect what
S[ch]neider and Hall were thinking about a year in advance of the first Gemini mission and
about the many possibilities for the program.

Document I-57

[CONFIDENTIAL] [DECLASSIFIED]

June 25, 1964

UNITED STATES GOVERNMENT MEMORANDUM

TO: MA/Deputy Director, Apollo Program

FROM: MG/Deputy Director, Gemini Program

SUBJECT: GEMINI SUPPORT OF APOLLO

As you know, one of the primary missions of Gemini is to provide support to Apollo, by developing orbital rendezvous techniques and obtaining data on the effects of long duration weightless flight. We have developed a set of missions which support these objectives. The missions and schedules are outlined below.

Enclosure 1 [not included] shows the launch schedule of Gemini and its relationship to Apollo launch schedules. Enclosure 2 [not included], Gemini Flight Mission Assignments, contains a summary of the Gemini missions.

Flights 4, 5, and 7 will provide experience in long duration orbital flight. A typical mission profile for these long duration flights is shown in Enclosure 3 [not included]. Many measurements and experiments will be performed to assess the effects of orbital weightless flight on man and machine for periods up to 14 days – more than adequate for the Apollo lunar expedition. Among the medical experiments, for example; M-1, Cardiovascular Reflex, will determine the feasibility of using inflatable cuffs to prevent cardiovascular deterioration – evidence of which was noted in Project Mercury flights MA-8 and MA-9. Among the engineering experiments, MSC-1, Electrostatic Charge will determine the buildup of electrostatic charge on spacecraft due to the firing of the rocket engines –a potential hazard due to the possibility of electrical discharge between rendezvous vehicles. These experiments and other are described in the Manned Space Flight document, Description of Gemini Experiments, Flights GT-3 through GT-7, April 13, 1964. In addition to these experiments, we also plan to conduct extravehicular activity to evaluate man's performance outside the spacecraft.

With Flight No. 6, we will establish the feasibility of rendezvous and provide experience for the visual manual docking mode, which is common to both Gemini and Apollo. This flight is outlined in Enclosure 4. The flight plan shown is one of the several proposed to date for this flight; however, the docking procedures shown in the addendum to the enclosure are typical.

[2]

Whereas radar computer guidance will be the primary onboard mode for the terminal rendezvous phase of Flight No. 6; the radar optical and optical guidance modes will be primary for Flights 8 and 9 respectively. The Gemini radar optical and optical guidance modes are very similar to the LEM Manual Alternate guidance modes outlined in Grumman Aircraft Engineering Corporation Report No. LED-540-3, Back-up Guidance Requirements, July 9, 1963. The basic feature of the terminal homing phase in these rendezvous maneuvers is that the LEM and Gemini essentially fly a collision course to their respective rendezvous target. This characteristic is achieved by keeping the inertial rate of the Line of Sight (LOS) to the target below a given threshold value. Following the LOS rate reduction, range rate with respect to the target is measured and thrust applied along the LOS direction until range rate is reduced to a pre-determined value appropriate to the range at which thrust was initiated. This procedure is repeated several times from the initial range of 20 NM down to the docking phase. A mission profile for Flight No. 8, employing radar optical guidance, is shown in Enclosure 5[not included]. The mission profile for Flight No. 9 will be basically the same; however, the optical sight will be used in place of the radar.

When viewed against the malfunctions encountered with the Automatic Stabilization Control System in Project Mercury, it is difficult to over-emphasize the vital importance of simulating and testing the manual alternate modes provided to accomplish critical maneuvers such as rendezvous. The success of Project Mercury was due in large part to its manual modes. Since the Gemini manual modes require the greatest degree of astronaut participation, they will also provide the greatest degree of astronaut training.

In NASA Project Apollo Working Paper No. 1083, Study of Earth Orbit Simulation of Lunar Orbit Rendezvous, July 24, 1963, it is concluded that it would be desirable to perform an earth orbit simulation of lunar orbit rendezvous since this will provide a realistic assessment of the guidance techniques and demonstrate the ability to perform the critical lunar orbit rendezvous maneuver. Enclosures 6 and 7[not included], taken from Working Paper No. 1083, show the close comparisons of earth orbit and lunar orbit rendezvous trajectories and closing times.

By Flights 10 and 11, or earlier, we plan to flight test the feasibility of the LEM lunar orbit direct rendezvous mode in earth orbit if possible. In this mode, the catch up or parking orbits are essentially by-passed and terminal rendezvous is initiated near first apogee as shown in Enclosure 8. In order to insure its successful completion, the astronauts should be ready to take over manual control of the spacecraft at any time should the automatic system falter. This will require a high degree of training and proficiency on the part of the astronauts. While it is true that Gemini does not employ the same guidance hardware as Apollo; Gemini may be in a unique position, based on present plans, to flight test direct rendezvous in earth orbit. In addition, in terms of schedules, Gemini is in a relatively good position to influence Apollo [3] rendezvous techniques with flight test results. Gemini's first rendezvous flight takes place approximately two years prior to the first manned Apollo flight and its first direct rendezvous flight takes place approximately two years prior to the first lunar rendezvous flight.

For Flight No. 12, we plan to simulate LEM abort maneuvers; either abort from an equiperiod transfer orbit as shown in Enclosure 7 or abort from a Hohmann transfer orbit as shown in Enclosure 9[not included].

In conclusion, we believe that Gemini missions as presently planned will make a very significant contribution to Project Apollo. However, in order to insure the most effective Gemini Program, we would appreciate your comments on our mission plans as outlines herein especially with regard to the Apollo support areas of Flights 8 through 12.

L.E Day

[handwritten: *for*] William Schneider
Deputy Director, Gemini Program

Enclosure: 9 as stated [not included]

Copy to:
MSC-DD/Low
MSC-GPO/Mathews
    M/Mueller
    MGO/Edwards
    MGS/Hall
    MGS/HUFF
    MGT/DAY

Document I-58

[FOR INTERNAL USE ONLY]

July 17, 1964

UNITED STATES GOVERNMENT MEMORANDUM

TO: MG/Deputy Director, Gemini Program

FROM: MGS/Director, Gemini Systems Engineering

SUBJECT: List of missions

The following is a quick list of missions (or important experiments), which would be accomplished with a follow-on Gemini program. Certain items may require up-rated GLV launch capability or up-rating of spacecraft performance.

Also enclosed is an equally quick vehicle layout of some of these suggestions. Improved quality will follow.

1. Land landing demonstration.

2. Propellant transfer.

3. Extended duration research (medical, physical, environmental).

4. Apollo rendezvous simulations.

5. Apollo DSIF check out.

6. Rendezvous with empty Apollo Command Module.

7. Rendezvous with LEM.

8. Apollo chaser.

9. Minimum space station.

10. Extended duration at low g's (G-can).

11. MOL-rendezvous – joint Air Force mission.

12. Space assembly and repair.

13. Satellite rendezvous – OAO – photographic adaptor.

14. Satellite recovery (like OSO).

15. Satellite chaser (no velocity match).

16. Space escape, personnel reentry (dummy tests).

17. Spacecraft assembly and checkout for orbital launch of unmanned mission.

18. Gemini deep space guidance and navigation.

19. Gemini circumlunar.

20. Gemini lunar orbit.

21. 3-seat rescue craft.

22. Control of upper stage reentry to reduce hazards.

23. 1-man Gemini and telescope.

[Signed: John L. Hammersmith]
[for] Eldon W. Hall

Document I-59

Document Title: E. C. Welsh, National Aeronautics and Space Council, Executive Office of the President, Memorandum for the President, "Space Rescue," 21 May 1965.

Source: Folder 18674, NASA Historical Reference Collection, NASA History Division, NASA Headquarters, Washington DC.

Document I-60

Document Title: Bill Moyers, Special Assistant to the President, The White House, Memorandum for James Webb, Administrator, NASA, and Robert McNamara, Secretary of Defense, 29 May 1965, with attached: Joseph A. Califano, Jr., Special Assistant to the Secretary and Deputy Secretary of Defense, Memorandum for Mr. Valenti/Mr. Busby, Special Assistants to the President, 29 May 1965, with attached: Cyrus Vance, Office of the Secretary of Defense, Memorandum for Mr. Bill Moyers, The White House, "Comments on Need for Space Rescue," 29 May 1965.

Source: Folder 18674, NASA Historical Reference Collection, NASA History Division, NASA Headquarters, Washington DC.

Document I-61

Document Title: James E. Webb, Administrator, NASA, Memorandum to the President, "Space Rescue," 2 June 1965.

Source: Folder 18674, NASA Historical Reference Collection, NASA History Division, NASA Headquarters, Washington DC.

The safety of the astronauts in orbit has long been a critical concern. In 1941, science fiction author Harry Walton wrote about a rescue vehicle—calling it a "lifeship"—in his novel Moon of Exile. In 1946, science fiction scion Arthur C. Clarke published a version of a space rescue mission in his first short story, titled "Rescue Party," in which aliens on a survey of the solar system try to evacuate humanity from Earth in the face of the Sun exploding. Such dramatic space rescue stories sparked serious concern among advocates as the Space Age dawned. In the 1950s Wernher Von Braun advocated the building of a space station in Earth orbit, and with it individual protective return capsules for its crew. In his scenario a parachute with steel-wire mesh reinforcements and solid rocket booster brings the crewmember to Earth, and a radar beacon would signal the landing location. But when NASA began its human spaceflight programs in earnest in 1958, none of them had the capacity for a rescue of a stranded astronaut in Earth orbit. Concern that this was the case led to the following exchange of correspondence on the subject. In the end, NASA decided to build as much reliability as possible into the system and accept the risk, which its officials believed was minimal. The first true space rescue capability developed by NASA for its astronauts was for the Skylab program, 1973 to 1974. If a crew had to return to Earth from the orbital workshop, an Apollo capsule was available to return the crew.

Document I-59

EXECUTIVE OFFICE OF THE PRESIDENT
NATIONAL AERONAUTICS AND SPACE COUNCIL
WASHINGTON

EXECUTIVE SECRETARY
MAY 21, 1965

MEMORANDUM FOR THE PRESIDENT

Subject: Space Rescue.

The space rescue issue involves the development of a capability to send up a spacecraft to save the life or lives of astronauts whose equipment has failed while in space.

I have discussed the question of developing such a capability with Jim Webb and he feels that it is too early to attempt to develop a practicable competence for such a purpose. In any event, it is something which should be studied, and the President should know that it is being studied and should be prepared to respond as to why we do not have such a capability should a tragedy in space occur.

An unsolicited space rescue proposal has been prepared by the Martin Company. Mr. Earl Cocke, former National Commander of the American Legion and now a consultant to the Martin Company, is representing that Company in attempting to sell their space rescue proposal. He has indicated that he plans to outline his proposal to the President and has left a brief summary and a detailed presentation with the President's office. Such documents have, in turn, been transmitted to me.

In brief, the Martin Company proposes a National Orbital Rescue Service to begin promptly and in a multi-phased manner. This would call for the building of a space rescue capability over the next ten years at an estimated cost of about $50 million per year. That figure would include a provisional system which could be gotten ready in a relatively short period and also a permanent system.

I hold no particular brief for the Martin proposal but, in view of Mr. Cocke's assertions, I thought it advisable that the President know about it. If a study is desired, it would be appreciated to be so advised.

[Signed]
E.C.Welsh

Document I-60

THE WHITE HOUSE
WASHINGTON

May 25, 1965

TO: Honorable Robert McNamara
          Secretary of Defense

          Honorable James Webb
          Administrator, NASA

FROM: Bill Moyers [Signed]

The President asked if he could have your recommendations on the attached memorandum.

Attachment

[SECRET] [DECLASSIFIED]

OFFICE OF SECRETARY OF DEFENSE
WASHINGTON DC 20301

May 29, 1965

MEMORANDUM FOR Mr. Valenti/Mr. Busby
                Special Assistants to the President

Bill Moyers asked me to get the Secretary's comments to the President by the end of this week so that they would be available to the President before the Gemini shot. I am, therefore, sending this out to you by pouch.

                   [Signed]
                   Joseph A. Califano, Jr.
                   The Special Assistant to the
                           Secretary and Deputy
                           Secretary of Defense

     Attach.

[SECRET]

THE SECRETARY OF DEFENSE
WASHINGTON

May 29, 1965

MEMORANDUM FOR MR. BILL MOYERS, SPECIAL ASSISTANT TO
THE PRESIDENT, THE WHITE HOUSE

SUBJECT: Comments on Need for Space Rescue

With regard to Dr. Welsh's memorandum of 21 May 1965, we are familiar
with several proposals by industry for developing separate space rescue systems.
Our view of this subject is the following:

1. If we go ahead with MOL, we will provide crew safety features beyond
    those possible in the earlier manned spaceflight programs. For
    example, the primary mission being performed in the laboratory
    vehicle will always be backed up by the return capsule as a lifeboat.
    In the unlikely event that the laboratory has a major failure, the
    crew can move to the return capsule, separate from the laboratory,
    and then wait up to six hours in orbit before selecting a preferred
    deorbit and landing sequence. In addition, we will employ the
    same practices that have been employed in Gemini and Apollo
    concerning design redundancies, extensive qualification testing of
    parts, and full attention to astronaut abort modes for every phase of
    the flight.

2. It would appear that any genuine rescue service separate from the
    basic flight hardware would be useful only if it could be sustained
    on hold for quick launch throughout the manned program; could
    be capable of rendezvous and docking under uncertain conditions;
    and could be assured of higher reliability than the orbiting vehicle
    requiring help. These essential techniques are among the most
    important objectives of the Gemini, Apollo, and MOL programs.
    Until they are demonstrated, a separate program for space rescue
    could not proceed with reasonable and genuine objectives.

3. It is possible we may strand an astronaut in orbit some day. It is very
    likely that astronauts will be killed, though stranding them is one
    of the less likely ways. The nation must expect such a loss of life in
    the space program. There have been several deaths already in our
    rocket development. We would be untruthful if we were to present
    any different image to our citizens.

4. As the manned space program evolves to a capability and rate of
    operation which might warrant a separate rescue arrangement, I
    expect the Department of Defense to play a large role in the regular
    operation, and correspondingly to participate in any operations to
    rescue from stranded spacecraft, should a decision be made that
    they are justified. For the time being, we consider space rescue
    similar to commercial aircraft or commercial ocean traffic rescue. In
    these cases every realistic precaution is taken to reduce probabilities
    of catastrophic failure, and to insure that effective rescue forces are

available to retrieve passengers should a major failure occur. The extensive ship and aircraft rescue forces which we deploy globally for each manned flight now typifies this practice.

I would point out that [text redacted in document] rescue can take place only to about 400 feet. As a result, a disabling accident in the rather small part of the ocean where the bottom is between 400 [text redacted in document] feet deep would result in a similar "stranding."

I see no advantage for a specific study of the space rescue question at this time. However, I wish to assure you that the matter of crew safety will remain paramount in our manned military space program. In view of the higher public attention to manned spaceflight, I would note that we will continue to provide this program significantly more crew safety precautions that we have in our similarly dangerous aircraft testing programs.

[Signed Cyrus Vance]

Document I-61

June 2, 1965

MEMORANDUM TO THE PRESIDENT

Subject: Space Rescue

With reference to Dr. Welsh's memorandum of May 21, 1965 on the subject of space rescue, our concern for the safety of United States astronauts means that we take steps to reduce risks by every conceivable means. We maintain intense efforts in the fields of reliability, crew training, equipment check-out, design redundancy, safety margins, and the use of short systems. We have also given careful consideration to the practicability of space rescue within the current or immediately predictable state-of-the-art.

It is obvious that we could not have provided a space rescue system in the Mercury Project, which was devoted to demonstrating the feasibility of manned space flight itself.

In the case of Gemini, the equipments and operational techniques essential to space rescue are being developed as part of the Gemini Program. A considerable number of the Gemini experiments are devoted to rendezvous, docking, manned extravehicular activities, tether dynamics, and the use of tools and repair of equipment in space – techniques which must be mastered before a practical space rescue system can be developed. However, in Gemini, we are building on all of the measures for safety that have come from our extensive experience in test flying and such advanced systems as the X-15 – the measures which have also been instrumental in achieving our perfect record of astronaut safety thus far. The redundancy designed into the retro-system for return from orbit is optimized for crew safety. The orbital parameters of the next Gemini mission are planned so that the orbit will decay to reentry within 24 hours after the planned termination of the flight, should all other provisions for initiating the de-orbiting landing sequence fail.

We are actively continuing our studies of all aspects of space rescue. The Mission Analysts Division of our Office of Advanced Research and Technology has evaluated the Martin Company's proposal for the development of a space rescue capability over a ten-year [2] period. It is our judgment that the knowledge needed to begin the design of such a space rescue system is not yet available, but will come from our present developmental and flight program.

You may be assured, Mr. President, that we shall continue to give first priority to considerations of astronaut safety.

[Signed]
James E. Webb
Administrator

Cc: AD/Dr. Dryden
       AA/Dr. Seamans, M/Dr. Mueller, W/Adm. Boone

Document I-62

Document Title: Julian Scheer, Assistant Administrator for Public Affairs, NASA, Memorandum to Mr. Marvin Watson, The White House, 24 May 1965.

Source: Folder 18674, NASA Historical Reference Collection, NASA History Division, NASA Headquarters, Washington DC.

Document I-63

Document Title: Marvin Watson, The White House, Memorandum for the President, 24 May 1965.

Source: Folder 18674, NASA Historical Reference Collection, NASA History Division, NASA Headquarters, Washington DC.

As the Gemini program evolved in 1965, questions about the propriety of lauding the program as a "space spectacular" emerged. NASA, the White House, the media, and the public had treated the various Mercury flights as singular events worthy of intense reporting. Each Mercury launch was exhaustively covered on all three television networks and the astronauts, NASA operational activities, and recovery received considerable exposure. Each astronaut also enjoyed media hype at the time of their mission. But was such involved reporting appropriate for the Gemini program? Julian Scheer, NASA's Director of Public Affairs, did not think so. He advocated a more routine approach to operations, aimed at making spaceflight appear more normal than unusual. While NASA continued to enjoy significant media attention during Gemini, attention to later missions was somewhat less pronounced than for their Mercury counterparts.

Document I-62

NATIONAL AERONAUTICS AND SPACE ADMINISTRATION
WASHINGTON DC 20546

OFFICE OF THE ADMINISTRATOR

May 24, 1965

MEMORANDUM to Mr. Marvin Watson
          The White House

This is in response to your questions about astronaut activities.

During the Mercury program and on into the first manned Gemini flight, space flight was new to this nation and we found a new group of heroes created by the American people. Each flight was a "first" of some kind, we were behind the Russians and our flight program was smaller and more understandable. Both US and Russian space flyers' names became well known.

As a result, New York City always wanted a ticker tape parade and the White House showed, on behalf of the American people, its appreciation of the work the astronauts had done.

We are now entering a new phase of our program. We expect to have gained 2,000 or more hours of space flight between now and the end of the decade when we expect to reach our goal of placing two men on the moon. Each flight, of course, will have new and different elements, but, generally speaking, these are long duration flights of two men (Gemini) and earth orbital Apollo flights.

The image that is, perhaps, best for this nation is that of a nation with this capability, a nation that goes about its work in an orderly and well-planned manner. We will fly these flights as best we can and put these flyers right back into the flight schedule for a future mission.

[2]

We feel that any build-up of personalities resulting from these flights should be spontaneous, based not on the fact that the astronauts flew, but what they accomplished in flight or difficulties they overcame or obvious skills they demonstrated.

Each flight is not going to be spectacular, each astronaut is not going to deserve a medal or award or special recognition. We are at the point, we feel, where we have to very carefully look at each flight and consider it as part of an ongoing program which will be oft-repeated in the months to come.

Therefore, we prefer to have a mechanism built into our Public Affairs program which will enable us to react quickly to given situations and to allow us the flexibility to choose the course that appears best at the time of the completion of a successful mission.

We would expect that you would be interested in this kind of flexibility, too, and would want to consider these things against a day-by-day backdrop.

We would not, of course, move forward on any plans without the most careful consultation with the White House, especially those which may have political implications.

On the upcoming flight, Gemini 4, we must consider that Astronauts Grissom and Young were received at the White House less than ten weeks from this launch date (June 3) and participated in New York and Chicago parades. Similar events 90 days later, unless the flight departs radically from the flight plan, may be too much saturation and repetition.

Therefore, in summary, it is our recommendation that we plan no events in advance of the Gemini 4 flight but be prepared to move rapidly in case there is interest there. We will, however, discourage other activity, such as ticker-tape parades, and will have under consideration a visit by the astronauts to the University of Michigan campus in late June or early July. Both graduated from the University.

[Signed]
Julian Scheer
Assistant Administrator
For Public Affairs

Document I-63

EXECUTIVE MEMORANDUM

THE WHITE HOUSE
WASHINGTON

MAY 25, 1965
Tuesday, 2:15 PM

Mr. President:

Information in the attached memorandum was agreed to by Director James Webb.

NASA suggests that since there will now be frequent space flights, you should reconsider the policy of White House receptions and ceremonies for the astronauts.

The next flight is scheduled for June 3 and will last four days. There will be some six days debriefing in Houston, Texas, which will mean approximately ten days from blast-off until they would be at the White House.

Since both of these astronauts are graduates of the University of Michigan and that the University has asked that both come to the University, Director Webb suggested that you consider not having the White House or Capitol ceremonies and allow it to be handled in this manner.

Do you want a White House ceremony?

Yes_____   No_____

Director Webb also states that the Vice President has become most interested in this program and he would like some guidance from you as to what part the Vice President should play. Do you want the Vice President to receive a lot of credit?

Yes_____   No_____

If you said 'No' on the White House ceremony, Director Webb suggests that since the astronauts will be in Houston for debriefing, and if you are in Texas, you might want to have them come to the Ranch.

Yes_____   No_____

Marvin

Document I-64

Document Title: Robert C. Seamans, Jr., Associate Administrator, NASA, to The Administrator, "Extra Vehicular Activity for Gemini IV," 24 May 1965.

Source: Folder 18674, NASA Historical Reference Collection, NASA History Division, NASA Headquarters, Washington DC.

Document I-65

Document Title: L. W. Vogel, Executive Officer, Memorandum for the Record, "Top Management Meeting on Gemini 4 Extra-Vehicular Activity," 8 June 1965.

Source: Folder 18674, NASA Historical Reference Collection, NASA History Division, NASA Headquarters, Washington DC.

In November 1964 an initial ground simulation for extra-vehicular activity (EVA) was performed by the Gemini III crew in an altitude chamber. But Gemini III was too short for EVA operations and ground controllers and engineers looked to the Gemini IV mission. Manned Space Center Director Robert Gilruth approved altitude chamber tests for the Gemini IV crew on 12 March 1965. But Alexey Leonov made the first spacewalk a week later, spurring a faster schedule for Gemini EVA tests. However, response at Headquarters was still lukewarm, largely due to concerns about the safety of such a new activity. On 14 May 1965, Gilruth arranged for an EVA demonstration for Associate Administrator Robert Seamans. Seamans agreed that it was safe to move the first EVA from the Gemini VI mission to Gemini IV and discussed the matter with Administrator Webb and Hugh Dryden. Webb generally agreed to the proposal, but Dryden was strongly against it. In response to a request from Webb, Seamans drew up a brief making the case for the Gemini IV EVA and delivered it to Webb on 24 May. Webb gave it to Dryden who returned it to Seamans the next day with the words "is recommended" underlined and the handwritten notation "Approved after discussing w. Dryden, J. E. Webb, 5-25-65."

Document I-64

NATIONAL AERONAUTICS AND SPACE ADMINISTRATION
Washington, D.C. 20546

May 24, 1965

MEMORANDUM

To: The Administrator

From: Associate Administrator

Subject: Extra Vehicular Activity for Gemini IV

The Project Approval Document for Gemini, date December 16, 1964, states the following objectives: Development of earth orbital rendezvous techniques, long duration flights of up to two weeks, extra vehicular activity, controlled re-entry, and astronaut operational space flight experience as a prerequisite for the Apollo program. Consequently, extra vehicular activity has been recognized as a primary objective of the Gemini program.

Against Extra Vehicular Activity during Gemini IV

The primary objective of Gemini IV is to extend astronaut and spacecraft time in orbit to four days. Extra vehicular activity reduces by a small but finite amount the chance of success and consequently should not be included.

For Extra Vehicular Activity during Gemini IV

Risk is involved in all manned space missions and consequently we must achieve maximum significant return from each flight, assuming that additional flight operations do not unduly reduce the chance of achieving our primary goals.

Conclusion

The hardware for extra vehicular activity is flight qualified and the astronauts are trained for this operation. Since extra vehicular activity is a primary goal for the Gemini program, it is recommended that this activity should be included in Gemini IV. The flight plan is being carefully planned toward this end and if a decision is reached to proceed a thorough review will be made of the public information releases in order to provide a full understanding

[2] of the care exercised in preparation for this mission and the safeguards available to the astronauts.

[signed]
Robert C. Seamans, Jr.

Document I-65

MEMORANDUM FOR THE RECORD

SUBJECT: Top Management Meeting on Gemini 4 Extra-Vehicular Activity

On May 24, Mr. Webb, Dr. Dryden and Dr. Seamans met with Dr. Mueller and Dr. Gilruth in connection with extra-vehicular activities on the Gemini 4 flight scheduled to take place on June 3.

Concern was expressed about changing the pattern of the flight. Making changes at the last minute always injected the possibility of something being overlooked and not properly considered. Also, if the Gemini 4 flight had to be cut short for any reason, opening the hatch would be blamed. Extra-vehicular activity in Gemini 4 was too obvious a reaction to the Soviet spectacular in this regard.

On the other hand, it was pointed out that suit development to permit extra-vehicular activity was part of the Gemini 4 program all along. Extra-vehicular activity had been originally planned for Gemini 4. One of the basic objectives of extra-vehicular activity was to be able to evaluate the possible utilization of man in space to carry out experiments, repair and adjust scientific satellites, and anything else that would require man to be outside of the spacecraft. The large antenna program was noted as one experiment which would require extra-vehicular activities by man.

It was then stated that there was no questioning of the propriety of having extra-vehicular activity in the Gemini program, but what was being questioned was it being performed on the second manned flight in the program. Since it was not essential to the basic mission of the Gemini 4 flight, which was to check out the reliability of the spacecraft and its systems for a 4-day period, our space posture might suffer if the 4-day period did not materialize.

The counter argument continued with comment about the great concern for the welfare of the astronauts and the fact that in the Gemini 3 flight we had a complete check on all systems. We have confidence in the spacecraft and the astronauts have trained for extra-vehicular activity and-, if nothing than for morale purposes, they shouldn't do anything less than what they can do and have been trained to do. Extensive tests had been conducted under zero-gravity conditions in a K-135. The astronauts practiced getting in and out of the spacecraft under zero-gravity conditions a sufficient number of times so as to build up about an hour of experience. Also, it was pointed out that if we don't accomplish extra vehicular activity (EVA) in Gemini 4 then we must do it on Gemini 5. It is a logical extension of the Gemini program to do EVA on Gemini 4. If

[2]

EVA is successful on GT-4, we will not do it on Gemini 5. If a decision were made today not to have EVA on GT-4, then we could do it on GT-5. However, it would be more of a compromise of the program to do EVA on GT-5 than on GT-4 because of the many other things programmed for GT-5.

The question was raised as to what risk we would be taking on a possible short Gemini 4 flight because of EVA and not finding out as much as we should find out about weightlessness as a problem. Weightlessness can be a problem, even in G-4, and we presumably will be concentrating on this problem in G-5. To this question it was noted that Dr. Berry said that there were no reservations about weightlessness being a problem over a 4-day period. There is no indication that 4 days of weightlessness will hurt man; therefore, this is not a great problem to be considered in the Gemini 4 flight. However, in connection with the Gemini 5 flight of 7 days there possibly are some reservations, primarily because no one has been in space for that period of time. Some medical experts feel that there will be a risk, others do not. Probably a problem just as pressing as the weightlessness problem, is the problem of confinement for 7 days or longer periods.

The question was raised again as to the element of risk to complete the 4-day Gemini flight because of EVA. The reply was that the added risk was simply having to depressurize the spacecraft, open the hatch, seal the hatch, and repressurize the spacecraft. These procedures, involving various systems and sub-systems, of course add a degree of risk because of a possible failure. But these procedures have been done hundreds of times with no failure. Nevertheless, there is always a risk that something will not work, but this is a small risk.

It was noted that one cannot justify EVA in Gemini 4 just because the Russians did it, and one cannot justify EVA in Gemini 4 just because you want to get film out of the Agena rendezvous vehicle on a later Gemini flight. In rebuttal, it was commented that the main reason for EVA in the Gemini program is to further develop the role of man in space. The sophistication of equipment that we put into space is getting ahead of the sophistication of experiments we can do. Experiment sophistication can be increased through the use of man in space, but the use of man in space must be checked out by EVA. The determination as to whether man in space by extra-vehicular activity can repair things, can calibrate satellites, etc. should be looked upon as a significant step forward and not as a stunt.

A strong comment was made that it is no more hazardous to do EVA in Gemini 4 than in later flights. The training for EVA on Gemini 4 was adequate and the only question that was holding up EVA on Gemini 4 was qualification of space suit equipment. This equipment is now fully qualified.

[3] On the other hand, the thought was raised that most of our thinking to date is that man's primary role in space is within the confines of a spacecraft.. We are trying more to qualify the spacecraft in Gemini 4 than we are EVA. However, it was noted that EVA is also important to the Apollo program.

It was acknowledged that everything that had been said was correct, but it still remained a fact that the consequences of failure on Gemini 4 would be more adverse than the consequences of failure on Gemini 5 or 6. There would be no reservation about EVA on Gemini 4 if it was absolutely necessary to accomplish the basic missions of the flight. It is essential to learn more about spacecraft systems over a 4-day period, and therefore we have an obligation to the Government to be sure that we qualify the spacecraft.

It was explained that EVA was planned for the second orbit of Gemini 4 which does create some risk for completing a 4-day flight as opposed to having EVA on one of the latter orbits. However, there is some concern about the ability of an astronaut to undertake EVA after 4 days of flight. The trade-off in risks involved in not knowing the condition of the astronauts after some time in orbit as to what could go wrong with the mechanical systems involved in EVA argued for EVA on an early orbit.

It was again noted that it was more important to check out the spacecraft for 4 days so that it would be possible to extrapolate the guarantee of spacecraft operation for 7 days.

To a comment that in the eyes of the public Gemini 4 would be a success with EVA, a statement was made that Gemini 4 with EVA might not necessarily be considered a success in the eyes of the decision makers. As a guide to risk taking, it was suggested that if there was a 90% chance to have a Gemini 4 flight for 4 days and that with EVA this chance would be only 89%, then we should risk 1% less chance for a 4-day flight for what can be gained from EVA. However, if a chance for a 4-day flight would be only 80% with EVA, then this additional 10% possibility for not having a 4-day flight would not be an adequate trade-off to be gained by EVA and we should not undertake it on Gemini 4.

It was noted that there was no comparison between the risk between the first Mercury flight and the Gemini 4 flight. It was recalled how the Air Force had admonished against the first Mercury flight, but NASA top management decided to go ahead because this flight was absolutely essential to the program. If we take into consideration the risks still inherent in using the rocket as a means of propulsion, then every time we use this means of propulsion we should find out everything that can be found out on the flight.

[4] It was noted that we should not be too concerned about the public reaction in determining what is the best course of action. The decision as to whether or not there would be EVA on Gemini 4 should be made in the light of what is best for the program and should not be influenced by possible public reaction.

After the foregoing discussion, the concern was still raised that the importance of Gemini 4 was to check out the reliability of the spacecraft for 4 days and project this reliability for 7 days. EVA therefore might jeopardize getting everything we should get from Gemini 4. If Gemini 4 does not go for 4 days, then we are in a very difficult position for 7 days on Gemini 5 and presumably we could not go for 7 days on Gemini 5. The real question is whether or not EVA is important enough in view of the risk, no matter how slight, of jeopardizing a 4-day Gemini 4 flight and jeopardizing a 7-day Gemini 5 flight.

Then it was pointed out that if you look at the entire program, EVA is more logical for Gemini 4. If Gemini 4 lasts 3 days then we should not be concerned about spacecraft reliability for 7 days. The basic problems are really to check-out confinement and weightlessness. Therefore, Gemini 5 is more important than Gemini 4 and if there is any chance of reducing total flight time due to EVA, EVA then logically should be accomplished on Gemini 4 rather than on Gemini 5. Every guarantee was given to top management that if EVA were approved for Gemini 4, very firm and adequate instructions would be given covering the procedure.

Mr. Webb, Dr. Dryden and Dr. Seamans then gave careful consideration to the discussions they had with Dr. Mueller and Dr. Gilruth. In their opinion it was important, whatever the decision, that there be an adequate explanation to the public to avoid any unnecessary misunderstanding and to minimize any adverse reactions. There was a strong feeling to ratify EVA for Gemini 4 in order to get the maximum out of the flight. There was unanimity in that EVA eventually would be carried out, but there was some reservation as to whether or not it was the best judgment to have EVA on Gemini 4 as a risk beyond that which has to be taken. It was concluded that Dr. Seamans would discuss the matter further with Dr. Mueller and Dr. Gilruth, in view of the discussions which took place, and that if he did not care to press for EVA on Gemini 4, such EVA would not be undertaken. However, if the final discussion led Dr. Seamans to press for EVA in Gemini 4, then it would be unanimously approved for the flight.

NOTE: Following the meeting, a memorandum from Dr. Seamans to Mr. Webb, dated May 24, 1965, recommending EVA for the Gemini 4 flight was approved by Mr. Webb and Dr. Dryden.

[signed]
L.W. Vogel
Executive Officer

Document I-66

Document title: James E. Webb, Administrator, NASA, Cabinet Report for the President, "Significance of GT-3, GT-4 Accomplishments," 17 June 1965.

Source: Folder 18674, NASA Historical Reference Collection, NASA History Division, NASA Headquarters, Washington DC.

The first two piloted missions of the Gemini program occurred on 23 March 1965 (GT-3) and 3 to 7 June 1965 (GT-4). Both were quite successful. The first mission was a checkout of the Gemini launch system and orbital spacecraft that demonstrated its flight-worthiness. In this mission the crew proved their mischievousness by smuggling a corned beef sandwich aboard. Both Gus Grissom and John Young enjoyed a few bites, but they were reprimanded for their hijinks by project managers because of the fear that crumbs from the bread might float into the spacecraft's systems and damage electronics. It was a lighthearted episode that pointed out the serious nature of the enterprise. There is, not surprisingly, no mention of this incident in this report of the mission by NASA Administrator James E. Webb. What is truly significant about GT-4, however, received considerable treatment here. Edward White's 36-minute extra-vehicular activity (EVA) of spacewalk on the first day of the mission proved successful and a source of pride for the U.S.

June 17, 1965

CABINET REPORT FOR THE PRESIDENT

FROM :   Administrator, National Aeronautics and Space Administration

SUBJECT: Significance of GT-3, GT-4 Accomplishments

On March 22, the first manned Gemini mission, GT-3, served to flight qualify the crew-spacecraft combination as well as checkout the operational procedures. During the course of the four-orbit mission, the two-man crew maneuvered their craft in orbit preparing the way for the rendezvous missions to follow. GT-3 also initiated the use of the Gemini spacecraft as an orbiting laboratory. Astronauts Grissom and Young also executed the first manned, controlled, lifting reentry.

With the success of GT-3, NASA moved forward the time-table for the Gemini program and decided to conduct extra vehicular activity (EVA) on the next mission. GT-4 was launched on June 3, more than 3 weeks earlier than our target date. GT-4 successfully achieved one of the major objectives of Gemini—to demonstrate that two men can carry out extended space flight while performing an extensive series of scientific and operational experiments during the mission.

During the third revolution, Astronaut White executed the first of a series of EVA that will be continued on later Gemini and Apollo flights. This successful experiment of EVA shows that man can maneuver in space for inspection, repair, crew transfer and rescue. All these can have both peaceful as well as military space applications. Tests of GT-4 rendezvous equipment have given important data which is being applied to the remaining eight Gemini missions.

The use of MCC-Houston for control of the GT-4 mission was a major milestone. This new facility worked perfectly and its use is essential in future Gemini and Apollo rendezvous flights. All 11 experiments and all operational checks were accomplished despite significant changes to the scheduling and time phasing. This ability for the ground crews to work with a well-disciplined space crew indicates a growing capacity to make changes in plans while operations are being conducted and, therefore, realize the most from each flight.

[2]

The excellent condition of the crew throughout the entire mission, including their recovery at sea, indicates the effectiveness of the working environment and life support system of the spacecraft. The crew was quite active and this apparently helped keep them in good condition. Medical monitoring during the flight and post-flight examination revealed no requirement for a period of rehabilitation or "decompression."

This second flight of the Gemini spacecraft indicates its excellent handling characteristics and provides strong assurance that more extended missions can be now undertaken. The computer which failed was not critical to the mission and the minor mechanical difficulties encountered were not serious.

It is significant that the first operational flight of Gemini, GT-4, has provided significant experience in each of the major mission areas of Gemini: long duration flight, rendezvous and docking, extra vehicular activity, and the conduct of experiments. The success of the GT-3 and GT-4 missions has proven the design and confirmed the results of the ground tests, has increased our confidence in the reliability of the overall Gemini systems, and has enabled NASA to advance the Gemini Program such that rendezvous and docking are now scheduled during the Calendar Year 1965.

James E. Webb

Document I-67

Document title: NASA Program Gemini Working Paper No. 5038, "GT-4 Flight Crew Debriefing Transcript," No date, but soon after the June 1965 Gemini IV mission.

Source: NASA Collection, University of Houston, Clear Lake Library, Clear Lake, Texas.

The first multi-day mission of the Gemini program took place during the flight of Gemini IV, 3 to 7 June 1965. Since this was the first of the Gemini program's longer missions, it created a new set of challenges both for the astronauts and those in Mission Control. For example, Mission Control divided into a three-shift operation with flight directors for each shift. Chris Kraft acted as both Mission Director for the entire flight and Flight Director for the first shift, while Gene Kranz took charge of the second shift and John Hodge the third. Gemini IV proved a successful mission for many reason, not the least of which was its 36-minute spacewalk by Ed White on the first day of the flight. This transcript provides a vivid first-hand account of the initial U.S. extra-vehicular activity.

[CONFIDENTIAL] [DECLASSIFIED]

NASA Program Gemini Working Paper No. 5038

GT-4 FLIGHT CREW DEBRIEFING TRANSCRIPT

[No date included; declassified Mar 15, 1973 under Group 4 (declassified after 12 years)]

[only pp. 4-19 through 4-66 provided]

[4-19]
4.2 Extravehicular Activity

White
And this was the time I went after the gun.

McDivitt

Okay. At that time we reverted from station-keeping, which we were both attempting to do, to EVA preparation, which we both had to do. That's when Ed went after the gun, and we started our preparation. We weren't really far behind at this time. All we had to do was get the gun out and get the maneuvering unit. The cameras were already out. You had the Zeiss too, didn't you?

White
Yes. The Zeiss came out with the Hasselblad, from the same package as the movie camera. And the storage certainly was a lot easier. What do you think?

McDivitt
That's right.

White
Particularly getting it out of that center thing. You can just zip them out of there with no problem at all.

McDivitt
So, at about 1:30 we started to assemble the gm. If you look at the checklist, you see that we probably got the gun [4-20] assembled in nothing flat.

White
It's no problem to assemble the gun.

McDivitt
We started our egress preparations essentially on time. As a matter of fact, I think we even got started a little earlier.

White
Then, we weren't worrying about anything else.

McDivitt
Then, we weren't worrying about staying with the booster. We probably started it about 1:35 or 1:40. Over the States we started our egress preparation. We went to our other checklist.

White
You were over Ascension, calling off the checklist.

McDivitt
I started reading the checklist off to Ed and we went through it. He unstowed everything. Why don't you tell them what you did there, Ed? I just read the checklist off to you, and you went ahead and did it.

White
Okay. I had to get back into the right-hand box, and I unstowed the items there. The first time I went back in there, I took the first items out, and I did not unstow the full box, I remember I told you, "It's all coming out, Jim. I'm going to bring them all out on the lanyard." Remember?

McDivitt
Right.

White
We'd take them off piece by piece if we need it. At that time I pulled the whole lanyard out and the cockpit was full of little bags. I was quite happy that they had prevailed upon me to put a lanyard, on all this equipment. I had thought at one time that it would be more desirable not to put a lanyard on. We'd been working a lot in our simulations without the lanyard and it seemed pretty easy. But looking at it now, I highly recommend that everybody keep that stuff on a lanyard.

McDivitt
We would have really had a mess if we'd had all those things floating around. It was bad enough as it was.

White
Yes, eight or ten of those little bags, and I was glad they were all tied on to one string. I could control them in that manner. They were quite simple to unsnap. I thought the snap attachment made it pretty easy to unstow and selectively pick out the items that I wanted. I unstowed the pouches that I needed, and then we got ready to take the long umbilical out. I had a little difficulty. It took me about three tries to get it out. It's fairly big package to come through a small hole. It was a good thing that we had taken the Velcro off of the batch, because there no tendency for anything to hang up as we removed it. On the third try I got it out.

[4-21]

McDivitt

I thought you did an extremely good job getting the bag out. You got it out a lot quicker than I'd ever seen you do it in the Crew Procedures Trainer in Houston or in the simulator at the Cape.

White

You didn't know it. It took me three tries.

McDivitt

Well, maybe it did, but it sure looked like it came out a lot easier. I thought you got it out in a big hurry. I didn't notice that it took you three tries. I saw you start, and then just a short time later, it was out.

White

Well, it did come out pretty easy, and I think the storage was satisfactory, but I'd certainly recommend that nothing be on the outside to keep it from coming out. It's a real tough –

McDivitt

Yes, we need the velcro off of there. We're pretty well sure of that.

White

The rest of the equipment - the "Y" connectors, the bag that   contained the "Y" connectors, and the attachments for the chest pack I handed to you. I think you were keeping track of most of those things until the time I needed them.

McDivitt

Yes, I was.

White

The storage of the ventilation module from the floor came off pretty easily. That's when I started going ahead and putting it all on. You read the checklist off to me. I had gone ahead and done a few things anyhow. As you read them off, I checked them off to be sure that I had done them all. I think we had everything out without much problem at all. I think it took us longer actually to put it all together.

McDivitt

That's right. It did. We started going through the checklist here and putting the things on, and we started getting more and more rushed. We were supposed to start the Egress Preparation Checklist at about 1:44. We probably started it at about 1:35 or so. We started it about 10 minutes early, roughly, maybe 5 to 10 minutes early. We were supposed to be ready to start the depressurization at 2:30 over Carnarvon.

[4-22]

White

I think I could have gone through and hooked everything all up, but I felt that we should go through fairly close to the procedure we had set up on the checklist.

McDivitt
That's right.

White
I think this slowed us down.

McDivitt
Well, we set the procedure up so that when we finished with it, it would be right. I think this helter-skelter thing that we were being forced into was for the birds. So as we got farther along, it became apparent to me that the thing to do would be to stop.

White
Right.

McDivitt
Go ahead with the assembly of the stuff. Why don't you comment on that?

White
I've commented in my Self-Debriefing about the equipment and the assembly of it. I thought there was no difficulty at all in connecting the "Y" connectors, the hoses, and the chest pack. I thought the connection of the chest pack to my harness was a good one. With the velcro I could move it in and out whenever I wanted to so that I could make my connections on the inlet side of the ECS hoses. It went along pretty smoothly, as a matter of fact. I think as we progressed along in it though, we felt that we had everything done. I didn't really feel that we had everything done in a thorough manner. And I think you had that same feeling.

McDivitt
That's right. When we got to Kano or Tananarive - I think it was Tananarive - I called whoever I was talking to and said that we were running late and I thought that we would probably not do the EVA on this particular rev. I knew that we had another rev on which we could do it. It looked to me like we had all the stuff hooked up, but we hadn't really had a chance to check it. I also noticed, Ed, that you were getting awfully hot. You were starting to perspire a lot. I didn't like the way you looked to start this whole thing off. So I told them over Tananarive – I believe it was Tananarive – that we would go ahead and continue on, and I would let them know whether or not we were going to depressurize at the next station. We went on ahead and it looked to me like you were all hooked up and about ready to go except for one thing.

White
We forgot the thermal gloves. I did not have my thermal gloves on.

[4-23]

McDivitt
You did not have the thermal gloves on, which is sort of insignificant, but we hadn't really had a chance to check over the equipment to make sure that it was in the right spot.

White

Well, we talked and you said, "What do you think?" We talked it over and I had the same feeling. I thought it sure would be smart if we had about 20 minutes to just sit here real still before we went out.

McDivitt

I think we were in a situation where it would probably have gone all right. We had completed about 80 percent of what we really should have had done as far as the checking went, and I just didn't feel that we were in the right shape. Ed didn't think we were, and besides, I could see Ed. He couldn't see himself. Ed looked awfully hot, and he looked like he was getting a little pooped out from playing around with that big suit. I thought that the best thing for his sake, and I knew he wouldn't admit it, was to let him rest up for another orbit.

White

I agree that was the best judgment.

McDivitt

So, when we got to Carnarvon - I guess it was Carnarvon I called them and said we were not going to come out on that orbit.

White

It was Carnarvon. It was just before we depressurized.

McDivitt

So, we postponed it until the next orbit. As a matter of fact, after that we just sat there. We didn't do a thing for about 10 minutes. I let Ed cool off a little bit. We were on two-fan operation at the time. We just sat there and we were cooled off. We went around for about 20 minutes then.

White

Okay. Then as we went back around, I asked you to go through the checklist again, and we went through item by item this time.

McDivitt

That's right. I might add that we went right back to the beginning checklist, the Egress Preparation Checklist. We started at the top one, and we did every step on it again. We verified every step to make sure we hadn't left anything out.

White

We actually went in and checked this time. Another thing we hadn't really positively checked was the position of all the locks on all of the hose inlets and outlets. This time we [4-24] actually checked all those locked. All of them were locked in, but it was a good thing to do, I believe.

McDivitt

You want to make sure. We did do our Suit Integrity Check before we started all this stuff.

White

That's right. We started before we actually went to the unstowing of the stuff from the right-hand aft food box. We went to the Suit Integrity Check.

McDivitt

Well, I don't know where it is, but we did it when we were supposed to do it.

McDivitt

We did the Suit Integrity check before we started the Egress Preparation Checklist. That's when we did it, over the States.

White

I think we did that just about the time you decided to give up on the booster. We did the Suit Integrity Check. Both suits checked out all right, It went up to 8.5 and it leaked down to about 8.3 or something like that.

McDivitt

Same thing with mine. It went up to 8.5 and leaked down just a little bit. Not enough to be concerned about.

White

No. Oh, one thing that we did do on that extra orbit that we went around – I disconnected the repress system and we went back on the -

McDivitt

Oh, yes. We never even got on the repress system, did we?

White

Yes, I believe we were, but then we turned it off. We were all ready to depressurize, and then we went back on the spacecraft ECS system, full, and went through and reverified the whole checklist again. The only things that I would say we hadn't done to my satisfaction the first time was to check the inlet and outlet positions of the locks, and I didn't have my thermal gloves on. It turned out I didn't need them.

McDivitt

Also, during this period of time I alined [sic] the platform, which was completely misalined. It was probably alined [sic] within a couple degrees, but as we went around in Orbit Rate it got farther and farther out of tolerance. So, I managed to aline [sic] the platform. Here again, I might comment on the fact that our initial flight plan was so optimistic that it was almost unbelievable. The both of us worked full time on doing nothing except preparing for EVA, and we didn't quite get the job done. I can't believe that we could have possibly flown formation with the booster and taken pictures of it and all the [4-25] other things that we had scheduled, and still prepared for this thing and even come close to completing it.

White

Well, the way we would have had to do it, would have been without a checklist. I would have had to just go ahead and hook everything up. I think we could have done it satisfactorily in this manner, but it wouldn't have been the way we would have wanted it.

McDivitt
Yes, that's right. I don't think that's the way it should be done. It was just too bad that we had a time limit on it, but when we did get rid of the booster, or the booster no longer became a part of the flight plan, then the time limit vanished. We found out that we really needed that extra orbit, or probably could have used another 20 minutes.

White
Yes. We went back. And I remember as we came over Carnarvon, we had about a 15 minute chat back and forth - kind of a rest period. We were all hooked up at that time, and that's the time we went on the repress flow, ready for the depressurization. I think they gave us a GO then for our EVA.

McDivitt
That's right. We depressurized the cabin and got down to 2 psi to check our blood pressure. We tried to put our blood pressure plugs in the blood pressure plug port and found out that we didn't have any blood pressure plugs on either suit. This was quite a surprise - an unpleasant one, I might add. Well, we decided that from our past experience and our knowledge of the suit that, even if we did spring a leak in the blood pressure cuff, the size hole that we had in the suit would not be catastrophic, and we decided to go ahead with the EVA.

White
It was within the capability of the system we were using.

McDivitt
At Carnarvon we not only got the go-ahead to start the depressurization, we also got the go-ahead to open up the hatch, the go-ahead that we weren't supposed to get until Hawaii. So, we went ahead and did that.

White
Yes. I'm kind of curious of the whole time. We were out nearly an orbit, I think. We didn't get it closed back again till we got back around to Carnarvon.

McDivitt
We were in a whole orbit depressurized.

White
Yes, I don't think people quite realize that.

[4-26]

McDivitt
We'll remind them. As we got to the hatch opening thing, we had our first difficulties with the hatch. The gain gear, I guess you want to call it – actually I call it the ratchet–didn't want to engage into the UNLOCK position. We fooled with it a few times and it finally engaged in the UNLOCK position, and Ed was able to go ahead and start.

White
The first indication of trouble was when I unstowed the handle to open the hatch. The handle freely moved up and down with no tension on it at all. I knew right away where the trouble was. It was up in that little spring on the gain pawl. So, I went up and manipulated it back and forth in hopes that I could break the lubrication loose in the spring to get it to work. We must have spent several minutes with the hatch. I thought perhaps it might have been stuck in the manner that the hatch got stuck in the Wet Mock, where it just was stuck. You could ratchet it open, but the hatch itself wouldn't open. It was pretty apparent the trouble was in the gain pawl. I jimmied it back and forth, and then I decided to go ahead and try the technique of actuating it in sequence with the hatch handle. If you actually replaced the operation of the spring with mechanically moving the gain pawl up and down, you can do the same work that the spring does.

McDivitt
Your fingers sort of take the place of the spring and rive this little pawl home.

White
This is the first time we actually tried this in a suit. It requires you to press up with your left arm to get at the gain pawl, and at the same time to hold yourself down. And I think later on this was a source of some of our problems which I brought out now so that we can find out later on. I felt it start to engage and start to ratchet the lugs out. Jim also verified that they were coming open. I backed them off, and I remember Jim saying "Ooop! Not so fast!" and at that time it popped. The hatch actually popped open, jumped open about 3 or 4 inches.

McDivitt
I was expecting the hatch to come open with a bang. Although we had the cabin to vent and it had bled on down to where there was nothing indicated on the Cabin Pressure Gage, we still really had the repress valve on. He was bleeding right into the spacecraft. We never got down to a vacuum and, even though we had a cabin pressure of only a tenth of a psi, we spread it over the entire area of that hatch, and that puts a pretty good size force on it. I had a real tight hold on [4-27] the hatch closing device, and when it popped open I was able to snub it.

White
It didn't really open with much force, did it?

McDivitt
Well, it did. It opened with a fair amount. It popped and I couldn't stop it the first inch or so. Then, of course, as soon as it opened, that much pressure bled off. I just sort of snubbed the thing to keep it from flying all the way open. Now if I hadn't been holding onto it, I don't think it would have gone open more than 2 or 3 feet.

White
This is another point too. There's more force on the hatch actuator than I thought. I didn't just flip the door open with my hand. I had to actually forcibly push it open, similar to the force with which I opened the hatch lying on my back under 1-g. That's about the force that I had to on the hatch to open it.

McDivitt
This extra force that we are talking about is due to the 0-rings they put in the pyros that are used for jettisoning the hatch. This is something that they put in just before the flight – something that we'd gone out to the spacecraft to feel. We knew just about what the force was, but it was pretty high.

White
Okay. At this time I had certain things that I had to accomplish. I had to mount the camera on the back of the adapter and mount the umbilical guard on the edge of the door. I elected, as I had planned, to go ahead and mount the camera first and then the umbilical guard. I mounted the camera and it went on without too much difficulty. The three little lugs on the bottom are a good mounting scheme. I think I would make a little easier engaging device for working out in a hard suit. I had familiarity with it, and it did lock up there all right. The umbilical guard for the umbilical on the side of the door took me a little longer to mount. Back to opening the hatch – I had the thermal gloves on when we were opening the hatch, and because of the fine work I had to do with the little gain and the drive lugs up there. I had to remove the thermal gloves so that I could actually actuate those small levers. I couldn't do them with any precision with my gloved hand. So, I took the thermal gloves off at this time, and I handed them to Jim. When I got back out, I didn't notice any temperature extremes. I felt quite confident that there wouldn't be any heat, since we just came out of the dark side, so I decided to do the actual work in putting this equipment on with my plain pressure suit gloves. [4-28] I had much more feel with them. Let me get back now to the umbilical guard on the door. It went on pretty well. It took me a little longer and it took me four or five tries to get the little pin into the hole that actually snubbed the guard down on the door. I did something then that I hadn't planned to do. The bag floated up and out of the spacecraft and now it was above the point where the hose was going through the umbilical guard. I had planned to keep it down inside. I left it there for two reasons: (1) I figured it was there already and I would have had to take the umbilical cord off again and scooted it back down, and (2) I also felt that Jim might have had a better view if it wasn't sitting right in front of him on the hose coming up from the repress valve. I elected to go ahead and leave the bag there. I then reported to Jim that I had everything all mounted and was ready to go. I had planned to take a short series of pictures. Since we had gotten out early, I had a little extra time at this time, so I went ahead and turned the outside EVA camera on. I took a short sequence of pictures that actually gives the egress up out of the seat, I kind of went back down and came out again so they would get an actual picture of it, and then I turned the camera off again. I mounted the camera and I turned it on while it was on the mount. I took a short sequence when I asked Jim to hand me my left thermal glove, which he did. I put the thermal glove on while the camera was running. I turned back around. I wanted to be sure the camera was off, so I took it off the mount, and I turned the camera off and actually visually took a look to see if the switch was off.

McDivitt
Did you knock it off one time? I thought you said the camera fell off.

White
By golly, I did. So I must have mounted it four times. That's right. I knocked it off one time during this time when I was out there. I got the picture of the egress, and then I asked you to hand me the gun. At this time the camera wasn't running. I had the glove on my left hand, and I went ahead and took the gun and made sure that it was ready to go. I had the camera on at that time and the valve was on. I checked the valve to be sure it was on and I was essentially ready to go. I don't know how long this took, but it took me longer than I thought. We had had early egress and it wasn't too much before I got the GO that I was ready to leave the spacecraft.

[4-29]

McDivitt
I'm not sure whether we got that GO from Hawaii or Guaymas. I sort of suspect that we got that GO from Hawaii, not Guaymas as we had originally planned.

White
Well, it sure seemed short from the time I was mounting all that stuff out there to the time you told me to go.

McDivitt
That's right. I'm sure we were talking to Hawaii, and they said you're clear to proceed with EVA.

White
And that's when I went. I bet we went out at Hawaii.

McDivitt
I think we went out at Hawaii.

White
I delayed from the time you gave just a minute, long enough to actuate the camera on the outside. This was kind of interesting. When I actuated that camera, I had my gun tied to my arm with the tether. It floated freely to my right. I turned back around and turned the switch ON on the camera, and listened and made sure the thing was running. I knew it was running, and put it down. I think you'll see this on the film. I wanted to be sure it was running when I mounted it back there. I actually took it off and turned it on, and I remember it jiggling up and down when I was trying to stick it on there. It ought to be a funny looking film. And it might even show the gun floating beside me as I was mounting it. That's when you said, "Slow down. You're getting awfully hot." I was working pretty hard to get that on. I mounted the camera again, and this is where I tried to actually maneuver right out of the spacecraft. I knew right away as soon as I got up – I felt even before – that the technique of holding on to the bar in the spacecraft and sticking a finger in the RCS thruster wasn't going to work. I mentioned that to Jim before – that I didn't think I would be able to do it.

McDivitt
I think you and I both knew how you were going to do, and everybody else was planning for us how we were going to do it, but without any real experience in it. People who didn't know a lot about it were planning this sequence, and it wasn't the way it should have been.

White
I couldn't have done that. I didn't have three hands. I couldn't hold the gun and put a finger in the RCS nozzle, and hold the handle at the same time. I thought it would be more desirable anyhow to actually depart the spacecraft with no velocity, other than that imparted by the gun. This is exactly what I did. I thought that I was free of the spacecraft, and I fired the gun. I realized that my legs were still [4-30] dragging a little bit on the side of the seat, so I pulled myself out until I could see that my feet were actually out of the spacecraft. I think you called me and said I was out of the spacecraft.

McDivitt
I called and told you that you were clear. That's right.

White
And that's when I started firing the gun and actually propelled myself under the influence of the gun. I don't believe I gave any input into the spacecraft when I left that time, did I?

McDivitt
No, you left as clean as a whistle.

White
Later on, I gave you some pretty big ones.

McDivitt
You were really bouncing around then.

White
Now at the time, I left entirely under the influence of the gun, and it carried me right straight out, a little higher than I wanted to go. I wanted to maneuver over to your side, but I maneuvered out of the spacecraft and forward and perhaps a little higher than I wanted to be. When I got out to what I estimate as probably one-half or two-thirds the way out on the tether, I was out past the nose of the spacecraft. I started a yaw to the left with the gun and that's when I reported that the gun really worked quite well. I believe that I stopped that yaw, and I started translating back toward the spacecraft. It was either on this translation or the one following this that I got into a bit of a combination of pitch, roll, and yaw together. I felt that I could have corrected it, but I knew that it would have taken more fuel than I had wanted to expend with the gun, so I gave a little tug on the tether and came back in. This is the first experience I had with tether dynamics and it brought me right back to where I did not want to be. It brought me right back on the top of the spacecraft, by the adapter section. Jim was calling me and said that I was out of his sight. I told him that I was all right, that I was up above the spacecraft, I looked down and I could see attitude thrusters firing, little white puffs out of

each one. I wasn't very close. They looked just like what Chamberlain's report told us. It looked just like about a foot and a half or maybe 2 feet of plume from the spacecraft and certainly didn't look ominous to me at all. In fact it looked kind of like the spacecraft was really alive and working down there. I knew Jim was doing his job holding attitude for me.

[4-31]

McDivitt
Let me comment on the attitude-holding right now. Initially we started out in blunt-end-forward, banked to the left about 30° or so. This happened to be the attitude we were in. We wanted to be blunt-end-forward for the sun, and they told me it didn't make any difference what attitude that we were in when we opened up the hatch. We had originally planned on opening the hatch toward the ground. I was called by some station that said it didn't make any difference what attitude I was in when I opened the hatch. We opened the hatch. We opened it in that particular attitude, and I held the attitude for the first portion of the time that Ed was out. When you had the gun you managed to stay reasonably well out in front. I held the spacecraft essentially stationary with respect to the local horizontal. After you ran out of fuel in the gun you were on top of the spacecraft all of the time, I felt that unless you really had to have the thing stabilized, to maintain your sense of balance or whatever you want to call it, I wouldn't fire the thrusters.

White
You asked that already when I was out.

McDivitt
Yes. I asked you if you needed it and you said no. So, then I felt it would be better not to fire the thrusters, because you were drifting back up over the cockpit. I could see that you were going up over us. I couldn't see back behind me, but I could see by the motions that you had when you went by me that you were going to continue on. I felt that it would be a lot safer if we just let the spacecraft drift unless it got into very high rates. I fired the jets a couple of times just to knock off the rates. I let it start drifting when you got on the tether so that you wouldn't get back there on top of one of those thrusters when I fired them. From about the time you ran out of fuel until you got back in I didn't do much attitude controlling. I did some. Everytime [*sic*] the rates got up pretty high, I'd knock them off. You were able to maneuver around the spacecraft when the spacecraft itself had rates of say +/-2 degrees/second in a couple of the axes at the same time. Here again before the flight we discussed the axis system. Ed selected the spacecraft as his axis system. It didn't appear that he was having a bit of trouble with it. He was maneuvering with respect to it, regardless of what the earth, sun, moon, and stars were doing. It was pretty obvious to me that was exactly what he was doing.

White
Well, when I came back the first time to the spacecraft with the gun –I had used the tether to bring me back – I did go back up on the adapter area. This is the first time it had happened. I said, "All right. I'm coming back out again."

[4-32]

This is one of the most impressive uses of the gun that I had. I started back out with that gun, and I decided that I would fire a pretty good burst too. I started back out with that gun, and I literally flew with the gun right down along the edge of the spacecraft, right out to the front of the nose, and out past the end of the nose. I then actually stopped myself with the gun. That was easier than I thought. I must have been fairly fortunate, because I must have fired it right through my CG. I stopped out there and, if my memory serves me right, this is where I tried a couple of yaw maneuvers. I tried a couple of yaw and a couple of pitch maneuvers, and then I started firing the gun to come back in. I think this was the time that the gun ran out. And I was able to stop myself with it out there that second time too. The longest firing time that I put on the gun was the one that I used to start over the doors up by the adapter section. I started back out then. I probably fired it for 1 second burst or something like that. I used small burst all the time. You could put a little burst in and the response was tremendous. You could start a slow yaw or a slow pitch. It seemed to be a rather efficient way to operate. I would have liked to have had a 3-foot bottle out there – the bigger the better. It was quite easy to control. I feel that with the gun there would be no difficulty in maneuvering back to the aft end of the spacecraft, and this was exactly what I did later on. Just on the tether. I got all the way back. So, I ran out of air with the gun, and I reported this to Jim. I didn't attempt to take any pictures while I was actually maneuvering with the gun. The technique that I used with the gun was the technique that we developed on the air-bearing platform. I kept my left hand out to the side, and the gun as close to my center of gravity as I could. I think that the training I had on the air-bearing tables was very representative, especially in yaw and pitch. I felt quite confident with the gun in yaw and pitch, but I felt a little less confident in roll. I felt that I would have to use too much of my fuel. I felt that it would be a little more difficult to control and I didn't want to use my fuel to take out my roll combination with the yaw. We divided our plan so that I would have a part of it on the maneuver and a part of it on the tether. I don't know how far along we were when the gun ran out.

McDivitt
Right on schedule when the gun ran out. We planned 4 minutes for the gun portion of it. We were just about on schedule.

White
I bet we used a little more than 4, because I think we came out earlier than we thought.

[4-33]

McDivitt
No, I started the event timer to time it.

White
Well, this is where my control difficulty began. As soon as my gun ran out I wasn't able to control myself the way I could with the gun. With that gun, I could decide to go to a part of a spacecraft and very confidently go. I think right now that I wish that I had given Jim the gun and taken the camera off. Now I was working on taking some pictures and working on the tether dynamics. I immediately realized

what was wrong. I realized that our tether was mounted on a plane oblique to the angle in which I wanted to translate, I remember from our air-bearing work that everytime you got at an angle from the perpendicular where your tether was mounted, it gave you a nice arching trajectory back in the opposite direction. You're actually like a weight on the end of a string. If you push out in one direction and you're at an angle from the perpendicular, when you reach the end of a tether, it neatly sends you in a long arc back in the opposite direction. Each time this arc carried me right back to the top of the adapter, to the top of the spacecraft, in fact, toward the adapter section. One time I was so close to the thrusters back there that I called Jim. I said, "Don't fire any more", because I was right on the thrusters. I was even closer than that foot and a half which I noted to be the length of the thruster plumes, and I didn't want to sit on a firing thruster.

White
We were discussing the EVA and I was saying that I spent approximately 70 percent of my time, it seemed, trying to get out of the area back above the spacecraft in the adapter area.

McDivitt
Yes, you intended to go toward the position that was directly over the cockpit. You always arced past it because you were coming from the front.

White
This was exactly right because that's exactly where my tether was connected. Chris had been very emphatic that he wanted me to stay out of this area, and I had agreed to stay out of there, I tell you, I was doing my level best to keep out, but the tether dynamics just put me back there all the time.

McDivitt
Let me interject something here. When we were talking about the control modes and how we were going to control the spacecraft, we decided on the Pulse Mode rather than the Horizon Scan Mode, or anything like that. The Horizon Scan Mode would leave me free to use both hands to take pictures of you, and that way I wouldn't have had to control the spacecraft. But since it was an automatic mode and it fired whenever it felt [4-34] like firing, it didn't give us any flexibility, and this is why I felt that the best mode to be in was Pulse, in case you did get back there.

White
That's exactly what happened.

McDivitt
I didn't have to worry about the thruster going off in your face. I didn't want the thrusters to fire, and they didn't fire because I didn't touch them. It was a wise choice.

White
I think this was good. When you look at it from a picture-taking viewpoint, it gave a wider spectrum of pictures. You got different views of the earth and the horizon. I'm glad we weren't held to a specific mode.

McDivitt
I think that the picture we did take or the attitude that we started out, which is shown in the newspaper, is just about right.

McDivitt
I guess we banked over to the right, I don't know.

White
That must have been just as I came out.

McDivitt
I don't remember, but it had enough of the ground in the background so that it was certainly worthwhile.

White
On one of my passes back to the adapter area I got so far back that I was about 3 or 4 feet from the adapter separation plane, perpendicular to it. It was rather jagged. There did appear to be some sharp edges, but it really didn't look very imposing to me. I took a picture of it. That's one picture I believe was good and should come out.

McDivitt
The trouble is it was probably set on infinity and you were up about 5 feet.

White
No, I set the camera to about 15 feet or so. It might be a little fuzzy because it was too close.

White
No, I didn't see the far side of the adapter. It didn't go all the way around. I think I could have pushed off and gotten back that far.

McDivitt
No. Better to stay away from it.

White
Well, I felt that if I got going I could have swung all the way around and had my umbilical right on the edge, without anything to hold on to or any gun to control myself. This [4-35] didn't seem like it was at all safe, and I had told Chris that I wouldn't go behind the craft. So I didn't go back there.

McDivitt
That must have been just about the time I told you to come back in.

White
No, I would estimate this was about two-thirds of the way, and about this time I was after pictures. I knew that was a part of the flight plan that I had, in my mind, fulfilled satisfactorily. So I tried to get some pictures, and this is where I really imparted some velocities, trying to get away from the spacecraft into a position so I could take a picture. I went out to the end of my tether cord quite a few times

doing this. I seemed like every time I would be completely 180 degrees to the spacecraft. I'd have beautiful views of the ground but I couldn't see the spacecraft. It was a definite mistake to mount the camera on the gun. That made it very difficult to use the camera. I had to point not only the camera but the gun with the long thrusters mounted out on the little arms. I'd want to take a picture of an object like the spacecraft, and there were too many loose items to get tangled up in and block the camera. I know my tie-down strap was floating loose. I had left that out intentionally so that I could get it later on any time I had to pull my helmet down. Occasionally when I got in close to the spacecraft, the bag and strings associated with the bag were tangling up around the vicinity of the gun and the camera. And it seemed like the umbilical was right in front of the camera all the time. So, I think the pictures will verify that I was flicking my right arm quite a bit in the later part of the flight, trying to clear things out from in front of it to get a picture. Whenever I was in a position to get a picture it seemed like I was facing away from the spacecraft. I took a couple of shots in desperation, and I think I might have gotten a piece of the spacecraft. But I never got the picture that I was after, I wanted to get a picture of Jim sitting in that spacecraft, through the open hatch, with the whole spacecraft. I know that I didn't get that. In fact, as time went on I realized that I wasn't going to get much of a picture. I was trying everything I knew to get out there and get stabilized so that I could turn around and get a good picture. I just couldn't do this. This was at the time when I was looking a little into the tether dynamics, and I actually kicked off from the spacecraft pretty hard. I remember Jim saying, "Hey, you're imparting 2 degrees/second rotational velocity to the spacecraft when you depart." I was pushing the spacecraft [4-36] quite vigorously. I wanted to push off at an angle of about 30 or 40 degrees to the surface of the spacecraft. And any time I pushed off from the surface of the spacecraft, my main velocity was perpendicular to the surface. It shot me straight out perpendicular to where the tether was attached. Again, this wasn't in the position that Jim could take a picture of me, and it wasn't too good a position for myself. I usually ended up facing away from the spacecraft.

McDivitt
Let me interject something here. In desperation I took the Hasselblad camera and stuck it over out through Ed's open hatch, and asked him if he could see the camera and if he could tell me which way to point it. He couldn't see the camera so he never really did tell me which way to point it.

White
No. This was the time that you said, "Hey, get in front of my window." It just so happened that I was right up close to the spacecraft and that's when I came over. Do you remember me coming over and actually looking about a foot from your window, Jim?

McDivitt
Yes.

White
Looking right at you.

McDivitt
Yes, I think that was the time the movie camera wasn't going and I was fooling around with it, trying to make sure that it was running.

White
Oh, that would have been a very interesting picture.

McDivitt
I'm not sure it was going, Ed, because, as you know, we had so much trouble making the left-hand one run. We had that trouble throughout the remainder of the flight. You pushed a switch over and it seemed to run sometimes, but sometimes it wouldn't. I kept worrying about whether or not it was running so I would grab a hold of it to see if I could feel it clicking over. I switched the ON-OFF switch on a couple of times to make sure I could tell the change in the feel of it. I'm afraid this time is one of the times that I didn't have the camera going, because I was trying to make sure that it was going. I'm not positive. I hope I got the picture, but I'm not sure about it.

White
That was the time that I came right in, and I couldn't have been more than a foot from your window, looking in, I could actually see you sitting there.

[4-37]

McDivitt
That's probably when you put a mark on my window.

White
I think the way I did that – I could actually see you in there and I pushed away with my hands a little bit. I think this was the time that either my arm or my shoulder contracted the upper part of your window, and you called me a "dirty dog" because I had messed your window up. You know, as you look back in retrospect, I wish you'd handed me a kleenex and I wish I'd cleaned up the outside of those two windows. I think we could have done it.

McDivitt
Yes. We'd have never gotten to the Kleenex at that time, but I think we might have done something about it.

White
I think I might have, but we might have smeared them so irreparably that it might have –

McDivitt
That's right. When you looked at that window of mine from the inside while the sun was shining, it looked like it was a black paint smear, such as if you'd take a piece of white linoleum and a black rubber soled shoe and made a mark on the linoleum. It had that kind of consistency. It was absolutely opaque. Just as black as it could be.

White
Yes, I could tell. When I hit it I could see from the outside that it turned white.

McDivitt
It turned black from the inside.

White
From the outside it was white.

McDivitt
From the inside it was black. When I got the thing turned around a different way with the sun on it, it was perfectly clear as if you had taken the coating off, and what I was seeing was through a perfectly clear surface. So, I don't know really whether the thing was black, that you placed something on the window that would make it black, or whether you'd taken something off that was very white, very thin.

White
I smeared the film that was on your window. I'm quite confident that is what happened.

McDivitt
I looked at our spacecraft windows after they got it onboard and I could still see that little hunk of window. It looks to me like what you did was remove a layer off the window, rather than put something on it. You took something off the window, rather than put something on it. You took something off it. Except I can't possibly imagine why it was so black and opaque with the sun shining on it at certain angles.

[4-38]

White
I'd like to comment on the ease of operation outside on a tether. If you've ever tried to hang on the outside of a water tower, or about an 8-foot diameter tree, you can visualize the problem I had out there. The decision to leave the hatch open was probably one of the very best that we made. I had nothing outside the spacecraft to stabilize myself on. There just isn't anything to hold onto. I think Jim will remember one time when I tried to hook my fingers in the RCS thrusters. I think Jim could see because –

McDivitt
I could see.

White
I was right out in front of Jim's window. This gave me really nothing particularly to hold onto. It didn't stabilize me at all. I had nothing really to hold onto, and so if you have ever tried to grasp an 8-foot diameter tree and shinny up at, you know the kind of feeling that I had outside there. There just wasn't any-thing for me to hold onto. One thing though that I'll say very emphatically – there wasn't any tendency to recontact the spacecraft in anything but very gentle contacts. I made some quite interesting contacts. I made one that I recall on the bottomside of the right door in which I had kind of rolled around. I actually contacted the bottom

of the spacecraft with my back and the back of my head. I was faced away from the spacecraft, and I just drifted right up against it and just very lightly contacted it. I rebounded off. As long as the pushoffs are slow, there just isn't any tendency to get in an uncontrollable attitude.

McDivitt
It seemed Ed did hit it pretty hard at one time. I thinly that was after he pushed off violently; he went out and it seemed, he came back and bashed it pretty hard. I remember a pretty solid thump. It seemed it was over the right-hand hatch or just right behind – .

White
I know a couple of times I kicked off with my feet, and I think I know the time you are talking about. I came in with my foot. It wasn't so much the contact with myself –.

McDivitt
What did you do? Contact and push off?

White
I contacted and pushed with my foot.

McDivitt
I heard a big thump and I think I called you at this time to take it easy.

White
I believe that was on the front end of the R and R Section on my side where you couldn't see me.

[4-39]

McDivitt
It was a position that I couldn't see.

White
One of the pictures that I saw last night in the movies, I think, was made at that time. I was coming in fairly rapidly and I wanted to get back out, so I kicked off again with my foot fairly hard. It was a very good kick. I felt that I certainly could have controlled myself without the gun out there if I had just some type of very insignificant hand-holds or something that I could have held onto. I believe that I could have gone on back to the adapters with a minimum of several hand-holds to go back there, going from one to the other. I was actually looking for some type of hand-holds out there. I remember that the only one that I saw was the stub antenna on the nose of the spacecraft. I could see the ceramic covering over it, I believe it was ceramic, or some kind of covering over it.

McDivitt
Yes, it's white.

White
I felt that this wasn't quite the thing to grab onto; this was at the time when I wanted to get out at about 10 or 12 feet directly in front of the spacecraft. I certainly had the urge to hang onto the antenna and push myself out. But I didn't and there really wasn't anything to hold onto. You really need something to stabilize yourself. I worked around the open hatch.

McDivitt
Let me ask you a question. How about putting the hand-hold inside the nose cone? A fairing is up there for launch, just the fairing. We could mount a hand-hold right inside.

White
I think we could have really made some money if we had had an attachment for the tether out there right on the nose of the spacecraft.

McDivitt
Strung the tether out there and then attached there?

White
Right. Have a second attach point and put it right out there. It would give you something to hold onto out there.

McDivitt
Yes.

White
There wasn't anything to hold onto on the R and R Section.

McDivitt
I know it.

White
It had smooth corner and the only thing I could have grabbed was the antenna, and I didn't want to grasp that. We thought [4-40] one time of holding on out there and thrusting, but –.

McDivitt
There isn't anything to hold onto. I think you probably could have gotten a hold on the antenna and held onto it without hurting it. I examined it pretty closely before the launch, and it looked pretty sturdy.

White
I thought this was something we needed and I didn't want to fool with it.

McDivitt
As it turned out we finally needed that antenna because that was the antenna that we used the whole flight – that stub antenna in the nose.

White
Yes.

McDivitt
When we opened up the spacecraft the hatch came open with a bang. The air that we had inside was obviously of greater pressure than that outside, and we had a great outflow of things including a piece of foam that we had used to pack our maneuvering gun in its box. It was the first thing that we put in orbit. But then throughout the time that Ed was out, he wanted the door wide open. It was pretty obvious that the flow was from the spacecraft to the outside because partway through his maneuvers his glove floated out and floated away from the spacecraft with a reasonably good relative velocity. The entire time he was out, even after we had the hatch open for 20 to 25 minutes, we were still getting particles floating out through the hatch. It was the flow. The streamlines were very obvious. It was from inside the spacecraft to the outside. I guess the spacecraft was out-gassing at a sufficient rate to cause a reasonably large pressure differential from inside to outside, and it was certainly relieving itself. I noticed this even as we were trying to get the hatch closed. There was still a flow from inside to outside.

White
Okay. I think that pretty well covers most of the things that we actually did while I was out there.

McDivitt
Now, as for getting back in – .

White
Yes, let's go all the way back through and come back in. The time really did go fast! I had watches with me, but I didn't look at them.

McDivitt
I was watching the time. I noticed my watch around 4 minutes, 6 minutes, and 8 minutes. And then you got involved in [4-41] floating around as we were trying to get that last picture.

White
The time really flew!

McDivitt
You kept getting behind me all the time and I became distracted from the time we were on VOX, completely blocking out the ground. Our VOX must have been triggered constantly, because whenever we were on it they couldn't transmit to us.

White
That's where the time got away from me.

McDivitt
That's right, and it was 15 minutes and 40 seconds when I looked at my clock. So, I thought that I had better go to the ground. I said to the ground, "Do you have

any message for us?" because I knew it was time to get back in. And they just said, "Yes. Get back in!"

White
Right. I remember hearing Gus say, "Yes, get him back in."

McDivitt
This is what all the fuss was about. They might have been transmitting to us to get back in but we were on VOX and couldn't hear a thing.

White
I did a few things after this time that I wasn't doing to deliberately stay out. But I was deliberately trying to do one last thing. I was trying to get that last picture. And this was one of a couple of times that I kicked off the space craft really hard, to get out to the end of the tether. And I wasn't successful in getting the position so that I could get a picture. I felt this was the one part of the mission that I hadn't completed. Everything else was successful and I wanted very badly to get that picture from outside. I spent a moment or so doing this. This was also the period of time in which I called down to Jim and said, "I'm actually walking on top of the spacecraft." I took the tether held onto it, and used it as a device to pull me down to the space craft. I walked from about where the angle starts to break between the nose section and the cabin section. I walked from there probably about two-thirds of the way up the cabin, and it was really quite strenuous. Could you see me walking along, Jim?

McDivitt
No, I couldn't see but I could feel the thumping on the outside.

White
That's when I got to laughing so hard. This was when Jim was saying to come in.

[4-42]

McDivitt
Yes, I think this is when I got a little stern and said, "Get in here!"

White
When I was walking on the top and was laughing, Jim probably didn't think I thought he was serious. But it was a very funny sensation. Now as far as delaying, there were certain things that I had to do before I came in. And there wasn't anything in the world that was going to hurry me up in doing them. We had just agreed that we'd do things in a slow manner and this is the way we'd do it.

McDivitt
Let me talk about the time here. It is implied in the papers that Ed didn't really want to come back in, and didn't. I think one of the things is that we didn't hear. We didn't have any transmissions from the ground after he stepped outside until I went off VOX at 15:40. They said, "Come back in!", and I told him to come back in. I think that he probably delayed about a minute or 2 minutes.

White
I think so, trying to get the pictures.

McDivitt
And at that time I got a little irritated and hollered at Ed, too. Then he started back in.

White
But when I came back I had things to do.

McDivitt
Yes. I know it. That's what I'm trying to say to get this thing in its proper perspective.

White
Yes.

McDivitt
We were 3 minutes 40 seconds late getting started back in because we just lost track of the time. I couldn't see Ed any longer. I was trying to keep track of what he was doing without being able to see, and I lost track of time. Then I think he delayed probably a minute or a minute and a half before he started back in.

White
That's right.

McDivitt
So, those are the two delays. We'd agreed on that he'd start back in after 12 minutes, From then on all the time was spent just trying to get back in.

White
I had certain things to do. I had to disassemble the camera that was on the spacecraft. I did this very slowly. I had to disconnect the electrical connection to it and hand the camera back in to Jim. Then I had to go out and disconnect [4-43] the umbilical, and this really went pretty well. The little tether that I had them put on the ring, a pull ring, to disconnect the pin worked pretty well. I disconnected the umbilical and discarded the umbilical cord.

McDivitt
That was the last thing Ed put into orbit.

White
Right. I put that in orbit. Earlier, it was really quite a sensation to see the glove floating off. I asked Jim a few minutes before about the glove, or Jim had asked me, "Hey, do you want this other glove?" About a minute later, I saw it go floating out of the hatch.

McDivitt
All I can say, Ed, was about a half hour later I was sure thankful that we had gotten rid of something. We had so much other junk that we didn't want.

White

I saw the glove come floating out of the right-hand hatch, and it was a perfectly clear picture of the glove as it floated out. It floated out over my right shoulder and out – it looked like it was on a definite trajectory going somewhere. I don't know where it was going. It floated very smartly out of the spacecraft and out into space.

McDivitt

I think this had a lot to do with that out-gassing. There was a definite stream –.

White

Yes. It was following the streamline right out of the spacecraft.

McDivitt

It went out perpendicular to the spacecraft, whichever direction that is.

White

Back to getting back in the spacecraft - I had the one thermal glove on the one hand, my left hand. I always wanted my right hand to be free to operate that gun and the camera. The way the camera was mounted on there, I had to use both hands - one hand to actually stabilize it with the gun and the other hand to reach over. Again, I think dynamics played a little bit of a role there. Everytime I brought my hand in from a position out on my left, it tended to turn me a little bit, which is exactly what we found happened on the air-bearing tables. I think that the camera should have been velcroed to my body somewhere and used independently of the gun.

McDivitt

Yes. I got the same impression. I got the impression that what you really should have done was –.

White

Dropped the gun.

McDivitt

Unhooked the camera out there floating around and just thrown the gun away. I don't think you ever should have tried to bring it back.

White

Well, what I should have done was fold the gun and handed it to you.

McDivitt

That would have taken longer. It would have taken precious seconds out of the very few that we had anyway. I think you should have just unhooked it and thrown the gun away.

White

This was probably the thing that I was most irritated with not completing. I didn't feel the pictures were satisfactory with the camera outside. But I think the reason was that my camera was not in a position so I could use it adequately. But coming back in was the last thing. As a matter of fact, before I dismounted the movie camera and dismounted the umbilical, I folded the gun.

White
I took the lanyard off with the camera on it, and handed Jim the gun and the camera.

McDivitt
And I stuck it down between my legs.

White
That was the first thing I handed in. Then I handed in the 16-mm camera, and then I threw away the umbilical. This was where the fun started. I found it was a lot more difficult coming back in that I had remembered in the zero-g training. It seemed like I was contacting both sides of the hatch at the same time, much firmer than I had in the zero-g airplane.

McDivitt
You mean you were hitting the hatch on one side and the hatch opening on the other side.

White
Coming back in, I was contacting the side of the spacecraft on both sides.

McDivitt
Yes, that's right.

McDivitt
You weren't really hitting the hatch on both sides; you were hitting the hatch opening on both sides.

White
Yes. I was coming down through there. I felt a much firmer attachment wedging in there than I'd remembered from the zero-g training. I think this might be associated with the extra 7/10 or 8/10 pound of pressurization on the suit. I just might have been a little fatter. I did notice that the suit was a little harder. I felt this type of suit during my pre-work, so this wasn't a [4-45] surprise to me at all. But I did feel like I was a little fatter getting in and wedged a little tighter.

McDivitt
I really don't think Ed was any fatter. I think that link the suit holds the suit to whatever volume it's going to go to. And I don't think a couple psi are going –.

White
Well, I felt like I was hitting a little more as I came in.

McDivitt
Yes. I think what happened was he was stiffer, and he wasn't bending his legs and his arms any.

White
You mean with the harder suit I was stiffer?

McDivitt
Harder. And your arms were stiffer and you weren't bending them around as much. It looked a lot more rigid.

White
This might have been.

McDivitt
Not semi-rigid – Ed was rigid.

White
All right. This might have been.

McDivitt
And that looked to me like it might have been the problem.

White
This might have been part of the recontact on the side of the spacecraft that I noticed. But as I came back in, I noticed that I had to work a little harder, and I hoped the tape was running because I think we had a very good commentary. We were both talking very clearly back and forth to each other during this time, and I was telling Jim that I was going to come in slow because it was a little tougher than I had thought. We were talking back and forth about being slow and taking it easy.

McDivitt
I actually helped push Ed down in there. I don't know whether he felt it or not in that suit.

White
No, I couldn't.

McDivitt
I reached over and I steered his legs down in, and I sort of got him settled in the seat a little better than what he was getting himself.

[4-46]

White
Yes. Right. I was kind of free wheeling my feet up there.

McDivitt
Yes. It looked to me like Ed was holding on to the top of the open part of the hatch and just swiveling around that part. It looked like he didn't have enough mobility and strength in his arms to actually twist his body down against the force of the suit into the seat.

White
After awhile, I reached my left arm underneath, the same technique we had used in the zero-g training, and actually I had my hands all over the circuit breakers.

McDivitt
Yes. Ed was a real hazard to the switches.

White
Yes, and I pulled myself down in and that's when I really started coming in – when I got hold of the underneath side of the circuit breaker panel and pulled myself in. That's when my first real progress was made toward actually getting down in.

McDivitt
Because, while I could steer Ed from where I was, really didn't have the strength to pull him in.

McDivitt
It was 90 degrees to the way that he really wanted to be pulling. I could steer. I did do a little bit of pushing, but not a heck of a lot. I wasn't really contributing much to the effort there except –.

White
You were guiding me down into the footwells.

McDivitt
Yes. That was about it.

White
But once I got my hands up underneath the instrument panel, I was back pretty well in familiar grounds – the work that we'd done five dozen times in the zero-g airplane, and I knew the technique pretty well.

McDivitt
Ten thousand times! White does check pretty well.

White
I really did it a lot. Maybe the suit was stiffer, or maybe I was fatter, but I wasn't going in quite as easy as I had before – getting into the initial position to pull myself down into the seat. So it took me a little longer. If you recall, I had to go back out again one time. I got back down and started to wedge myself down, and I got two fat cramps at the bottom of my thighs in both legs, where the muscles started to ball up a little.

[4-47]

McDivitt
Oh? Did you get in your thighs or calves?

White

Both of the muscles in the back of my thighs balled up in a ball, and I thought, "Well, I have to go back out and let them straighten up." So I straightened my legs out.

McDivitt

We had that problem before in the zero-g airplane.

White

This is the time Jim said, "Hey while you're up, why don't you throw the visor out?" I hesitated a minute because I thought, "Well, you son-of-a-buck, you might have problems here. You might have to be spending an orbit or so trying to get in."

McDivitt

No, as a matter of fact, I don't think that is when you did throw it out. I think you threw it out when you came back down and you started to close the hatch. You were having trouble. It wouldn't close, and you said, "I'm going to have to take this visor off so that I can see these things." And I said, "Listen, if we get this thing closed we're not going to open it again. Throw the visor away."

White

That's right. That was when I got the cramps, went back up again, and then I came back down again, and said, "Hey, I can't see them. I'm going to have to take the visor off."

McDivitt

No, it was a little bit later than that. You had already started to try to close it, and you were having difficulty closing it.

White

Okay. Let's get the sequence out. We came down in. I got up to straighten my legs a little but, went back up, then I came back down –.

McDivitt

–with all your equipment on –.

White

I hadn't held the handle yet, had I?

McDivitt

No. You hadn't done a thing with it.

White

So I got back down into position –.

McDivitt

–with all your equipment on and pulled the hatch down.

White

The hatch was down far enough to close at this time.

McDivitt
I thought it was.

[4-48]

White
I did too. I felt it was down far enough. I can tell by looking right straight down at the edge–.

McDivitt
Yes. I can tell by looking up underneath the right-hand side to see where the dogs are.

White
Okay. So I thought the hatch was down far enough to close at that time. I reached up and got the handle, but I don't know what I said to you.

McDivitt
You didn't say anything. I don't know whether you said anything to me or not, but you didn't have to say anything to me. I saw you move that handle, and I saw how easy it was going, and I saw that the dogs weren't moving.

White
I think I said something. I don't remember what I said. But I said something, and you knew right away what had happened.

McDivitt
You didn't say a word. I was watching the dogs and that lever, and I knew what the trouble was.

White
Right. So I guess that's when I said, "I'm going to have to take the visor off because I can't see." And then we went back up and Jim said, "Well, we're not going to open the hatch again. Why don't you throw the visor out." I hesitated for a minute to throw it out because I thought that we might have a problem.

McDivitt
Actually, we had a little more difficulty than we had expected. We fooled around for a minute or 2 or maybe even 3 or 4 with the handle. It was pretty apparent to us that we weren't going to get the hatch closed with normal, straight-forward techniques, and that we were going to have to start going to other things. While we say that we came down and moved the handle once or twice, it was over about a 3 or 4-minute period, at least.

White
The normal method of closing the hatch is for me to come down and wedge myself down, hold onto the little canvas handle up there, and actually apply a downward force on the hatch to help close it. Then with my right hand I use the hatch handle to ratchet the hatch down. This is normally our technique to ratchet the hatch down. This is normally our technique we would always use, and never in

the past has Jim had to help me with the hatch-closing device. This wasn't the case this time. As soon as I had gotten up there to operate the gain lever, I couldn't operate the canvas handle anymore. I couldn't apply any torque or pull there because –.

[4-49]

McDivitt
Not only that, but you were actually pushing yourself up off the seat. And I'm not sure that even the first time that we had the hatch closed far enough. It looked like it was closed far enough. As a matter of fact, later on when we got it down to that position it looked like it was closed fine. It really wasn't closed far enough because you never did get those dogs out until we - .

White
No, the dogs came out, Jim, the first time I got torque on it. Those dogs started out, then it closed.

McDivitt
Did they? Okay.

White
Yes. I think we had it down far enough.

McDivitt
It looked to me like we did, and I couldn't understand why they weren't coming out. I knew that the ratchet wasn't engaged, but I got the impression that it was from watching your hand when you came down one time. You had the ratchet engaged and the little tit pin that sticks in the door that doesn't allow things to come closed wasn't there.

White
No, the ratchet wasn't engaged. There was nothing on the handle at all, It was free, completely free. The situation hadn't changed at all. Another thing I'd like to point out now, too, was the chest pack was in the way of bringing the handle down to a full-crank position. And I wanted definitely to do this because you can interrupt the sequence of the dogs if you don't fully stroke the handle each time.

White
We went back up so that I could actually see and observe the levers. This was the time Jim said to throw the visor out because we probably wouldn't open the hatch again, once we get it closed. And this seemed like very good sound advice to me. The only thing I was a little questionable about was that at this time I had the inkling in my mind that we might spend quite a bit of time getting this hatch closed, and I might want the visor when I was back out again. But I thought the judgment to throw the visor out was best, and I threw it out – opened the door about a foot and a half and threw the visor out. The next time we came back down, I was still having the little bit of problem with the cramps, but not nearly the problem I was having with the gain lever.

McDivitt
One superseded the other.

White
That's right. One problem became of much higher magnitude than the other. So this was the time that we started working.

[4-50]

I knew what I had to do. I knew I had to work the gain lever in sequence with the handle again, just like we had when we opened it. We both had an inkling that this was going to happen when we opened it the first time. But this posed the problem of when I reached up with my left arm to work the gain lever. It takes a great deal of force. This isn't the direction that the suit is designed to reach in. And it takes a great deal of force to lift your arms up in the vicinity of your helmet to operate something there. In so doing it pulled me back up out of the seat. And I think this is the time that Jim noticed that I was up higher than I had ever been before, and he actually felt that my helmet was up against the hatch. I tend to agree that I was up in that position.

McDivitt
Yes. I actually pulled Ed down in the seat by pulling on the –.

White
I think so.

McDivitt
I did it in steps. I'd pull down and Ed would come down. Then I'd pull some more, he'd come down some more.

White
I was actually pushing up with my left hand, and my helmet was wedged right up against the hatch. I had a little bit of area in which they actually see the dogs that I was working with up there.

McDivitt
You could see them though?

White
Yes, I could see them. At least I could see what positions they were in. I could see the little lever operating under the spring – where I was actually operating the spring on the gain lever. This is where I think we got some very good teamwork, because it was necessary that Jim pull down in conjunction with the time that I pulled down on the closing handle and operated the gain lever. I just hope that the tape worked because I can remember I was in there. Jim was talking to me, and then when it came to the point when we really had to make the big pull I felt a little torque on the handle. I knew that we had it at that time if we could only get the hatch down close enough so that the dogs would engage. And I can remember giving the old – I think I was yelling HEAVE! HEAVE! Is that what I was yelling?

McDivitt
I think so.

[4-51]

White

And it was perfect timing, because I could see Jim or I could see the hatch come down each time that I was yelling HEAVE! I think it was probably the most–.

McDivitt
The most interesting moment of the flight.

White
Yes. It was the most interesting moment of the flight, but I think it was probably the most, if you want to say, dramatic. I don't know the right word. But it was probably the most dramatic moment of my life – about those 30 seconds we spent right there. The dogs started latching. I could feel them going in, and then I could feel them come over dead-center. Jim called out that the dogs were in.

McDivitt
I knew that once we got them moving we'd be all right.

White
Yes, once they started coming in. As long as we got those dogs to engage, with the little lever that permitted them to come out and lock, I knew that we had it hacked.

McDivitt
Yes. So did I. Even if we would have had to reenter with the hatch in that position, we'd have been all right. I don't think that the heat leaks were that tremendous.

White
I knew we could continue and dog it on in all the way. It seems like whenever you know you're right on something, you want to be darn sure that they fix it. This was going through my mind then. And I remember that I felt I was right in that the bar and the attachment on that bar and lanyard were not strong enough. I remembered that, and I knew how hard you were pulling on that thing. I think, if nothing else, they ought to be sure. Howe many times did we break that attachment at the bar?

McDivitt
We broke the attachment about three or four times on the zero-g airplane. Every time they kept telling us it wasn't made out of the right kind of stuff, and the stuff we were going to have in the spacecraft would be the right material. Well, it didn't break in the spacecraft, just coincidently, or maybe because we both had doubts about the strength of that particular piece. The same thing crossed through my mind. I was thinking that the success or failure of this hatch closure depends on whether this hatch closing device stays hooked onto that spacecraft and doesn't break off.

White
We would have been flat out of luck!

[4-52]

McDivitt
We would have been in deep trouble! I'm not sure we wouldn't have been able to get the hatch closed, because we had put that canvas strap on there and I might have been able to pull you down that way. But I had about all the pull I had in me on that last –

White
I know you did.

McDivitt
–on that last thing and I had a lot of mechanical advantage over it. When we went to that canvas strap we would have had to go with no mechanical advantage – as a matter of fact, a mechanical disadvantage.

White
This is one thing that didn't fail, but I recommend that it be made stronger.

McDivitt
Stronger anyway!

White
I think so.

McDivitt
For nothing else than a psychological purpose.

White
Right. I'd like to take the spacecraft now and see if I could break it, because I had the feeling that I never had been confident that the attachment nor the bar nor the lanyard were strong enough.

McDivitt
When I say I was really pulling as strong as I could, I really had some pull left in me, but I guess what I should have said is that I was pulling about as hard as I dared pull at the time. I guess I could have pulled another few pounds, but I hated to apply more than was needed on there because of the lack of confidence in the strength of it.

White
Everything I had was in it over there. I was pulling down with my legs as hard as I could and operating. I was pulling on the handle. I remember one time you said, "Hey don't pull on that handle so hard! You're going to break it!"

McDivitt
I was cautioning you to take it easy, which you don't usually have to do.

White
This was when we were yelling HEAVE! I was heaving on the handle as I was pulling it down each time. It felt like to me that the handle was giving. But I didn't give a darn! If it broke, it was going to break. So one of the points we learned out of this was we'd like to see the bar and lanyard strengthened.

[4-53]

White
Let me say one thing about the decision to go ahead and open the latch. If we hadn't done so much work together with this hatch and run through just about every problem that we could possibly have had, I would have decided to leave the hatch closed and skip with EVA when we first started having trouble with it. We had encountered just every conceivable problem that we could possibly have with the hatch. If it failed we'd know exactly what it was.

McDivitt
That's right. I personally had disassembled this cylinder and piston and spring combination up at McDonnell prior to the altitude chamber, so I knew exactly what it was made of. I am sure the problem was that the dry lubrication coagulated, or whatever a dry lube does, and was causing the piston to stick. I knew how we could do this thing. Carl Stone and I had dismantled it and put it back together, cleaned it out, put it back together, relubricated it, put it back together, and it operated fine. I figured out how to make the thing work with it not working properly by using you finger as the spring.

White
That's the exact technique we had used.

McDivitt
If we hadn't had the training together that we had, and had not encountered all these problems before, I know darn well I would have decided not to open the hatch.

White
Maybe we sound overdramatic about the effort we made getting me back in, and I'll honestly say it's one of the biggest efforts I ever made in my life, but I don't think we were all done then.

McDivitt
There were a lot of things we could do.

White
We could have gone around several orbits working on closing the hatch. That wasn't the last time we were going to get a chance to close it. So there were things left if we understood, and other procedures we could have used to go ahead and close it. When we got it closed back in, I was completely soaked wasn't I?

McDivitt
Yes. You were really bushed.

White
Sweat was just pouring down. In fact, I could hardly see. It was in my eyes.

McDivitt
So I told you, "Just sit there and I'll get a repress. Don't even move for 30 minutes." I just left the repress valve where it was. I closed the vent valve, and we had a lot of instructions from the ground to close the water seal and a whole bunch of other things that didn't make any sense to me. I knew that the spacecraft was repressurizing. I watched. There wasn't anything else that we had to do right then, and we were both bushed, especially Ed. He was perspiring so that I could hardly see him inside the face plate. So, I just said, "You sit there and I'll sit here and we'll just coast around. When we get the thing repressurized, we'll start doing something." That was exactly what we did. I did finally extend the HF antenna and try to call somebody on HF and let them know that we were back in safely and that thing was repressurizing. I didn't get any response until we got to Carnarvon, which was about 3 minutes later. I called and told them that we were repressurizing and had the hatch closed.

White
You know, that was some pretty good gage reading that we saw when we got the first ½ psi.

McDivitt
The first ½ psi. Ha! Ha!

White
That was a really big one. Since we've described the whole operation, we'd like to go back now and specifically point out the pieces of equipment that we used and our opinions of them, a few features that came out loud and clear to use in operation, general conclusions on EVA as an operation, and what we have to do to make it an operational procedure. So the first thing I'll do is go down through the equipment. As an overall comment on the equipment, I would say I felt very confident the equipment would do the job. And without question the equipment performed as it was advertised. It performed just exactly as it had been designed. There wasn't one thing on them as far as the VCM, the umbilical, the gloves, the gun, and the visor that didn't perform just exactly as it had been designed. There wasn't one thing on them as far as the VCM, the umbilical, the gloves, the gun, and the visor that didn't perform just exactly as it had been designed. I'll take them all one piece at a time, and discuss them a little. I'll start right with the visor. The visor was a rather controversial piece of equipment from the beginning. And I, for one, doubted a little bit the necessity for quite the protection that we were providing, although I had helped right from the beginning in the design with some of our ideas on the visor. It turned out, though and I commented on this during the time that I was out, that I was very happy to have the visor. I was able to look directly into the sunlight. I did so in installing the camera on the back of the adapter. I felt that the vision out of the visor was about as it would be on a normal sunny day [4-55]. This is because it is so bright up there in space. I felt as if my vision was what I would consider normal. I was looking at the different parts of the spacecraft and down at the ground, and the view that I received at this time was what I would expect on a normal sunny day. I was certainly glad to have the visor

and I left it down throughout EVA. I think on a later flight we might recommend going ahead and lifting the visor and observing any changes we might see in visual acuity when looking down at the ground. The ground vision through the visor really didn't seem to me to be degraded at all. Evidently just the intensity, and not what I was seeing, was cut down.

McDivitt

Let me comment a little bit on that visor. I didn't have a visor, and the bright sunlight that was in the cockpit didn't seem to bother me. I imagine that the visor turned out just like a pair of sunglasses. You go outside on a normal day and wear a pair of sunglasses. If you don't have them, you're squinting. But if you start out without them you tend to get accustomed to it. I think I was accustomed to what light there was coming through the spacecraft, admittedly much less than that outside. Ed was accustomed to the sun visor and it turned out just like two people with and without sunglasses. They both could have adapted. I didn't look into the bright sun straight ahead.

White

Well, the first time I looked into the bright sun, the first thought I had was, "Boy! Am I glad I've got this visor on!"

McDivitt

I know you mentioned it on the radio.

McDonnell

– because I was looking straight into the sun. I had to look into it to attach the camera onto the adapter section. I don't normally wear sunglasses. As you know, Jim, I have never worn sunglasses very much, and I didn't notice it from then on, throughout the time I was out. I had no impulse whatever to lift my visor. My vision was as clear as I could have expected it to be without the visor. There are a few design points in the visor that we could make better and I'll briefly go into them right now. When you are seated in the spacecraft one visor slips up underneath the other and back along the back of your helmet, so that instead of resting on your helmet on the headrest you're resting the visor on the headrest. You certainly don't want to do that. The visor should be restrained in some manner from slipping up along the back of the helmet. Also, my visor was quite difficult for me to raise and lower. Once it was down it fit quite snugly, for which I was happy. But it was difficult for me [4-56] to raise and lower. It was actually a two-handed operation, which is one of the reasons why I didn't raise it outside, although I had no impulse to raise it when I was outside. I think that we might be able to design them to be raised up and down more easily.

McDivitt

Let me make a comment on that visor. I never did see any need for the little lexion visor.

White

That's exactly the point I was going to get to next. I think that one single visor made as close to the helmet liner as possible, providing the maximum amount of headroom and a minimum amount of interference, is what we actually need. I don't believe we need the lexion outer visor. As they pointed out to us, it doesn't really protect, because it bows in and it doesn't really give you the protection that it should be affording. I would recommend one visor, one sun visor only. It'll be simpler to operate.

McDivitt
I think so too.

White
Okay. The Ventilation Control Module, I can say without qualification, worked exactly as it was planned to work. There was not one complaint that I had with it. It provided me with the proper flow. The flow was less than with the normal ECS suit system, but it was adequate to keep me cool and ventilated, except for two times during the flight. Those times were when I attached the camera right before departing the spacecraft and reentering the spacecraft. But I think it performed with out fault.

White
The umbilical was another item that I thought performed its part of the flight quite well. I had no complaints about it. I did tend to get it tangled up with the bag and the strings that were attached to the bag during EVA.

White
I am very thankful that we decided to design the gloves in the manner in which we did, the two-piece glove that was easily donned or doffed under pressurized conditions. As it turned out, I took them on and off twice while pressurized. I was quite happy that we had them designed in this manner. As it turned out, the heat on the side of the spacecraft, or the cold on the side of the spacecraft when we came out of the dark side, were not noticeable to the touch at all. I didn't use a right-hand thermal glove at any time during the flight. I took it off when I was opening the hatch and, as I pointed out earlier, it floated off during the EVA operation. I didn't have opportunity to use it again if I had [4-57] wanted to. Coming back in we had difficulty closing the hatch, and I, at this time, removed my left-hand glove and used the plain pressure suit gloves for this operation. The pressure suit gloves were comfortable. In fact, there were no sensations of either hot or cold through my gloves.

White
The gun, I think, was an outstanding point in the flight, a highlight of the flight. It worked just as we had felt it would work and it was, I felt, simple to operate. The training that I had on the air-bearing platform provided me adequate orientation in the use of the space gun. I think that now that we have a little more time to prepare ourselves for the next time we use this gun, training with it on zero-g flights would be appropriate. I don't believe we will have any trouble using it in the zero-g aircraft.

White
One mistake that we made on our EVA equipment was the mounting of the Contarex camera. This camera should have been attached by velcro to me, so that I could use it independently of the gun. It would have been easier for me to use, and I would

have had a much higher probability of getting satisfactory pictures with it. It was a case of lumping too much together - putting the gun and camera together.

White
The attachment of the VCM to the harness was a good type of attachment. It was easy to disconnect the two velcro attachments and move the chest pack in and out. I had to do this both when I opened the spacecraft hatch, so it would clear the hatch handle, and I had to move it out of the way when I closed the spacecraft and pumped the hatch handle.

White
Now we can get into some conclusions. While I was out, I decided to put a piece of velcro strip on the side of the adapter to see if later on we might use this as a method for attaching items on the outside of the spacecraft, if the velcro was still there and if it was in good shape. I think the velcro could be made into a very useful item for a type of tether.   I think you might even be able to do something along the line of just having some female velcro on the gloves and pieces of the male velcro at points along the adapter. This might provide us at least some attachments so that we could maneuver ourselves back to the adapter section. This would be about the simplest kind of handle that we could use. I do believe that we need some type of handles on the outside of the spacecraft. Jim suggested one on the nose and in the cover on the R and R section up there. I think this is an area that we certainly have a possibility of using. I certainly would have found it useful. I would still be a little hesitant, though, of breaking the antenna. You would want to be sure that this wouldn't be broken during EVA. I think the feeling I had out there, again, was like holding onto an 8-foot tree. There wasn't anything to hold onto. You definitely need some kind of hand-holds. The decision to leave the hatch open was one of the best decisions that we made. It provided me with a center of operations for my work. I was able to stabilize myself by holding onto the hatch. It was also surprising to me how much force it took to open the hatch the first time against the preload and the actuators, due to the seals. One other very good decision was to have me wear the heavy suit and Jim the light suit. I think this was one of the things that made our operation easier. It certainly made my getting back in the spacecraft and Jim's assistance in closing the hatch much easier for him. Also, I was handing him things in and out. He was performing quite a bit of coordination in the operation with pieces of equipment that were going in and out of the spacecraft, and I believe that by being in that light suit he was able to do this much easier than if he had been in a heavy suit.

McDivitt
I might make a comment on that suit, too. When we opened up the hatch we were in a vacuum. I noticed that the temperature of the suit dropped slightly so that the suit was a little bit cooler inside. I was wondering if I was going to get too cold through the suit, but the rest of the time we were out the temperature never changed. I don't remember looking at the suit inlet temperature, but the suit itself stayed reasonably warm. I had sun in the cockpit, and I had the cockpit open without the sun in it for a relatively long period of time, 4 or 5 minutes at a time. This didn't seem to affect my temperature inside the suit.

White
I think you felt the temperature more than I did.

McDivitt
I felt the temperature go down, rather than up.

White
I felt that also while outside. I would say it was a very comfortable figure. I figure that I was probably at 68 degrees temperature out there inside the suit, which was cooler than I had been anytime during the flight. It wasn't a cold feeling, just a very natural comfortable temperature.

McDivitt
Suit inlet temperature was running about 55° during most of the flight. It got down around 52°, so it probably might have even been cooler than your 68°.

[4-59]

White
Well, it was cooler inside the suit when I was outside the spacecraft than at any other time during the flight. It wasn't uncomfortably cool there at all.

White
I think that we can go on with some conclusions. Some conclusions that I had were:

> 1. I didn't notice any extremely hot temperatures on the outside of the spacecraft. I also didn't contact surfaces for any period of time to transfer much in the way of a heat load to any part of my suit including the gloves.

> 2. There's a definite requirement for some type of handholds outside the spacecraft.

> 3. We should think a little more on where we want to operate during EVA and where to attach the tether. The tether was not attached at a point that would provide me the capability to operate in the area that I wanted to.

McDivitt
You couldn't get to the nose. It provided great operation for directly above.

White
Straight above.

McDivitt
I just don't know how you would get the thing out there. You would have to run it along the spacecraft, then attach it somewhere at the front.

White
It would preclude operations in other areas. You would either have to accept where we are going to operate or –.

McDivitt
You could have multiple attachment points around the spacecraft.

White
Of course, now, if you have a gun with a good air source, I wouldn't particularly care where it was attached. I think you could go ahead and, maneuver to any point you want if you have a gun. Again, where you're pushing off of surfaces, you tend to go perpendicular to the surface from which you push off. I found when I pushed as hard as I wanted to I'd still tend to go straight up above that hatch instead of out toward the front. I think this is a fairly obvious conclusion, but it proved out. Every time I pushed off I went straight up instead of at an angle to the surface where I wanted to go.

[4-60]

McDivitt
Something that you should bear in mind is that you were pushing off from the front, which tended to make the front go down as you went out.

White
Yes. Everything was working against getting where I wanted to go. Everything I did tended to put me up.

McDivitt
When you started you went in a straight line forward and tended to push the spacecraft down. I think, initially, where I was holding the attitude, you didn't have that much trouble. Of course, you weren't pushing as hard either, because you had the gun.

White
No, I wasn't.

McDivitt
Later on, when we started free drifting, you were back behind me where I couldn't see.

White
Did you feel me stomping around back on the adapter and hitting the adapter?

McDivitt
Well, I felt you hitting things back behind me, and once you went behind the line that was directly overhead the spacecraft. I couldn't see you through your open hatch.

White
I never really had a good contact with the adapter back there.

McDivitt
Just as well. We wouldn't want to disturb those radiator tubes too much,

White
No. Well, now that we're back, we'll have some conclusions on the adapter area. I made it a point right from the beginning to take a look at the thermal lines, the thermal paint on the adapter. It looked like it was in good shape. It was all there. There was discoloration around the attitude thrusters, particularly from the thrusting. The color of the thrusting is just like the RCS thrusting – nice and clear plume. It looked like from outside, though, that I could see a lot more of the plume than I could when I was sitting inside the spacecraft looking out at the RCS thrusters firing. Again, the camera was not attached in an opportune manner to operate.

McDivitt
Which camera? The camera on the spacecraft?

White
I'm really after that camera on the gun. That one wasn't attached good. The camera on the spacecraft was okay. It was a little difficult to attach because of the attachment [4-61] on the bottom of it. You can't have it at any angle to make it engage. It has to be perfectly flat with the mounting plate on the bottom. A big conclusion that I came to – and I'll see how you feel about this one, Jim – I feel that storage in the back of the adapter section was certainly a very high priority for later missions. I feel that we can adequately store equipment in the adapter area, particularly larger pieces of equipment that we don't have room for in the crew station or pieces we don't have particular use for in the early part of the flight. If we can lick the problems in opening and closing of the hatch, we can store equipment in the back of the adapter section as a routine operation.

McDivitt
That's right. I think the extravehicular activities have proved to other people what we already knew a long time ago that EVA is quite simple. I think the thing we've got to iron out is the hatch opening and closing. This is really our problem. I don't think you or I will ever have any doubt about the extravehicular activity. That was, I thought, going to be pretty straightforward. It looked like to me it was pretty straightforward.

White
I felt that I could operate equipment out there. I could assemble equipment. I could put pins in, pull pins out, and screw things in. I did all these things during the flight. I turned the gun on, and I put in the pin to operate the umbilical guide. I attached the camera. I don't think you could do these operations very effectively with big heavy gloves on. Although my gloves operated satisfactorily, I think that for assembly of items you want to have – you ought to look into the glove area a little more thoroughly and try to get a piece of a glove with some type of a surface that will give us some heat protection and gives us a high sensitivity of feel through it. The big conclusions, the final conclusions, that I'd like to draw are that EVA can be made a normal routine operation if the following modifications are made to the spacecraft:

1.The highest priority is that the spring back there on the gain lug has convicted itself, and I don't believe that that's a good design. There should be some way that either the lubrication is made foolproof or the spring made stronger.

McDivitt

I think that we really want to say here is that the locking mechanism is inadequate as it is, completely inadequate. Until it is fixed, I think we should take it easy.

[4-62]

White

That's right. I think we almost had a bad experience with that gain thing. We knew about it ahead of time. We thought we had it fixed, but it's not fixed. I think it convicted itself, and it's guilty, and it has to be fixed.

2. I recommend that at least the egress kit on the right of the crew compartment be removed to provide more room in the spacecraft. I see no reason for it being in there. I think it would be worth the effort and the additional money to provide the extra room in the spacecraft, So, my second recommendation on EVA is to remove the egress kit, at least from the right-hand side, to provide more head room.

McDivit

Yes, that's good. I might add that it's a good thing that we had that egress kit modified to the minimum height, because without that we would have been in deep trouble.

White

That's right.

White

Yes. You and I had been telling each other that that was the biggest thing we did on our whole 9 months prior to the flight - to get that thing cut down. I think it sure paid for itself on our flight.

3. My third item is to make the bar and lanyard completely foolproof in strength. That was a device that provided us with the added force we needed to close the hatch, just as we sat there and said we might need during the SAR of the spacecraft in St. Louis. I think the attachments of the bar and the cable to the spacecraft should probably be at least doubled in strength, so there just isn't any question in the pilots' minds or the engineers' minds. I guess the engineers were convinced that you didn't have Jim and me convinced that those two attachment points –.

McDivitt

We've seen it break too many times, I think.

White

We've broken the bar, and we've broken that attachment point. I had actually physically twisted the attachment right off the spacecraft up in the zero-g airplane.

I certainly wouldn't have put my full strength into it if I knew my life depended on that attachment. It should be made absolutely foolproof.

McDivitt
Well, that was the point I was trying to make earlier when I said I was pulling as hard as I could. Then I said that I really wasn't pulling as hard as I was capable of.

White
You didn't have confidence in that attachment.

[4-63]

McDivitt
I didn't really think that I should pull on it any harder,

White
No. I think that should be the third recommendation and it should be corrected.

McDivitt
I think we could spare a couple of extra pounds of weight there, just for the pilots' peace of mind.

White
That's right. Take the time it takes to put a new attachment on there. They told us they didn't want to do it because they'd have to rerig it. I think they'd better rerig it and take the time to put a good attachment on there.

> 4. The final thing really doesn't fit in with the first three recommendations, but I would sure like to have the opportunity to use that gun again with about a 10-times supply of oxygen in a great big canister. I think that maybe this is one of the items we could carry in the back of the adapter. We could use a small supply to provide the means to go back there to get a great big canister. Then we'd have a unit that we could actually do some maneuvering with.

McDivitt
That's right. I think that, in essence, we proved the usefulness of a self-stabilized or a man-stabilized maneuvering unit –.

White
Yes.

McDivitt
– rather than one that is gyro-stabilized with automatic stability features. I think that although you didn't burn up a lot of fuel, you certainly proved the feasibility of this type of maneuvering unit.

White
We had an awfully small amount. We just had the 6 feet/second.

White
We proved, in my mind, that I had the capability to go from Point A to Point B with that maneuvering unit.

McDivitt
Let me ask you this question, and be honest about it. Would you detach your tether and go without it? Don't be too optimistic, because other people's lives may depend on it,

White
I think that we probably have not done enough investigation to do that at this time, but I feel we are progressing toward the point. We made the first, say 50 percent, of the step toward being able to detach the tether and go. I don't believe that I would detach the tether and go with that 6-feet/second –.

[4-64]

McDivitt
Oh, no. I didn't mean that. I mean with that type of unit.

White
If I had some more change of V in a unit like that I think that I would be willing to detach myself on the next flight, right now, from the spacecraft and go. That's combined with two things, you see. You have two things working for you. You have the capability to maneuver yourself, and, if you should get out of control, the spacecraft still has the capability to come over and get close enough so that you could get yourself back in control and get in the spacecraft.

White
I think that 40 or 50 feet/second would be a minimum, I had 6 and I'd like to see, probably, a capability of about 10 times that. That may be a little –.

McDivitt
It's difficult. I would think it would be difficult to fix a number on it until you fixed the job.

White
Yes.

McDivitt
If you wanted to go to something that was 10 feet away and come back, you'd probably get by with 20 feet/second.

White
If I wanted to get out of the spacecraft and go along to the back of the adapter and get in the adapter without being attached to the spacecraft, I'd only need two or three times the amount. I'd be happy to go with that.

McDivitt
There are some problems in the capability to aline one's self onto an object. I think chasing the booster around points this out. You say you'd be willing to go away because the spacecraft can come and get you. Admittedly it can, but keep in mind the difficulty we had with the booster. I don't really anticipate us ever getting into the situation like that because you'd never get so far away that you're in different orbits, like we were with the booster –.

White
What I visualize is a 25 to 50 foot operation where you're going out to investigate either another spacecraft or another satellite up there, or making a transfer similar to the type of transfer that we visualize as a backup mode for Apollo. I think with the gun I had, if the LEM and the Command Module were there, I'd be satisfied to depart the Command Module and maneuver over to the LEM situated 10 to 20 feet away from the Command Module. I feel I could do that at the present time. I don't think it would be a very smart thing at the present time to go maneuvering off 200 to 300 feet away from the spacecraft with this type of device. I think this device is designed and has its greatest usefulness in close operation around the spacecraft.

[4-65]

McDivitt
That's right. There is no need to maneuver off about 400 or 500 feet away, because if you want to go that far, use the spacecraft. This gun is for a close-working job.

White
I think it's a valuable tool in this manner.

McDivitt
Okay. That's the same conclusion I came to. We'd be willing to do it at close range.

White
I'd be willing to do it right now. I might not go tell somebody else to go do it, but I'd be willing, with the training that I had with it, to transfer 15 or 20 feet without a tether. But I think we should spend some more time with the gun.

McDivitt
I think so too.

White
I also think it would be of value to go in the zero-g airplane with it.

McDivitt
Yes, I think so too.

White
I think the work that we might do in the zero-g airplane doesn't necessarily have to be done in full regalia, with all the pressure suits in a pressurized condition. I think we can go up there and learn a lot about the gun without pressure suits on, in a plain flying suit type operation - perhaps polish the training off with a little

work in pressurized suits. If you work in the zero-g airplane with a pressurized suit, it's pretty awkward.

White
In pitch and yaw I felt I could maintain effectively zero rates. I don't know how it looked to you, Jim, but it looked like I could establish a rate and take the rate out without too much trouble. The yaw is the lowest moment of them all. Pitch was very easy, just to pitch the thing up and down. I'm still a little suspicious of roll. That's the area that I would like to look into a little more. I think that you could get yourself into a kind of balled up situation with pitch, roll, and yaw all coupled up. It might take a little bit of fuel to get yourself straightened back out again. But just in translating from Point A to Point B, you could care less if you rolled, as long as you kept pitch and yaw straight. And that's why I say I think you can translate and correct pitch and yaw very successfully and effectively forget about roll, just as we do in our reentries or our retros.

[4-66]

White
The question is: Was there any problem with the gun of maintaining a fairly well stabilized attitude and still get my translation input? I did this actually three different times, and this was what I had done when I was coming back to the spacecraft the last time. I had to put in both pitch and yaw and had taken them out and I was coming back. I was going to fire my last thrust toward the spacecraft. I got a little burst. I could feel a little burst and then it petered out. But you can put a translation in. I was also surprised that I was able to stop at the time I tried to stop it out there, about one-half or two-thirds of the way out on the end of the lanyard. It seemed to stop pretty well. It was either the gun or the lanyard dampening me. It didn't dampen me in roll, so I then it was the gun that actually did it.

McDivitt
I think that this previous bunch of words just spoken covers a lot of detail of the first three or four orbits of our flight, and it covers that first phase of mission sequences that I first mentioned. I think the next thing we should do is go through the interim orbits, about 50 or 55, or however many there were, where we set about to save up enough fuel to do something constructive, to check on our orbit to see what it was, to see how we were decaying, what our lifetime expectancy would be, and perform the experiments that we'd initially set out to do on our flight plan. Although it's not going to be of much use to go through it in a chronological order, I suppose that is probably the best way. As I just finished saying, we're not going to get an awful lot out of going through the flight plan sequentially, but we'll do it quickly, and then we'll come back and discuss each experiment or operation, check an entity in itself, and we'll discuss the systems as an entity, too. We'll do this, generally, in elapsed time.

McDivitt
Going back to the EVA for just one moment. I'd like to say that the use of the manual heaters on ECS Oxygen bottle was about two 5-minute periods separated by about 10 minutes. We really didn't need an awful lot of manual heater when we were doing the extravehicular activity.

Document I-68

Document Title: "Summary of Telephone Conversations RE Gemini 7/6," 25–27 October 1965.

Source: Folder 18674, NASA Historical Reference Collection, NASA History Division, NASA Headquarters, Washington DC.

Document I-69

Document Title: "Gemini Program Mission Report, Gemini VI-A," January 1966.

Source: Folder 18674, NASA Historical Reference Collection, NASA History Division, NASA Headquarters, Washington DC.

Document I-70

Document Title: "Gemini VIII Technical Debriefing," 21 March 1966.

Source: NASA Collection, University of Houston, Clear Lake Library, Clear Lake, Texas.

The Gemini program consisted of a methodical series of steps intended to develop procedures and experience necessary for the Apollo program. These documents discuss three important Gemini missions. Gemini VI was the first test of rendezvous in Earth orbit. The mission plan called for the spacecraft to maneuver to intercept an Agena target vehicle. When the Agena spacecraft failed, NASA quickly reconfigured the Gemini VI mission plan to rendezvous with another Gemini mission, Gemini VII. The flight proved to be extremely successful and the results were an important validation of the rendezvous concept for Apollo. The Gemini VIII mission successfully docked with an Agena target vehicle but because of problems with the spacecraft control system, the crew was forced to undock after approximately 30 minutes and spent most of the rest of the shortened mission overcoming a failure of the attitude control system or the flight's early return to Earth.

Document I-68

SUMMARY OF TELEPHONE CONVERSATIONS RE GEMINI 7/6

October 25, 1965

3:10 p.m.

Dr. Mueller called Mr. Webb (from the Cape) to report on the accident, giving details of the Agena explosion. He said it would be ten days before the investigation would be completed. He said it did not affect our actual schedule; will probably move some of the flights forward. Press conference held; no need for further information to the press.

October 26, 1965

3:00 p.m.

Mr. Webb called Dr. Mueller about a Herald Tribune article on the accident. He wanted to know if there were any feeling on the part of Lockheed that "this thing" was not ready to fly. Dr. Mueller assured him there was no reservation on the part of either NASA, AF, or Lockheed.

October 27, 1965

5:25 p.m.

Mr. Webb called Dr. Mueller to inquire as to the possibility of scheduling a rendezvous in December, with a view to announcing that we are looking into the possibility of doing it. Mr. Webb said that a possible announcement might state that we re taking down the booster for the Gemini 6 and erecting the one for Gemini 7 because of the experience we have gained in mating the Gemini 6 spacecraft to the booster, we may be able to re-erect Gemini 6 in time for a rendezvous with Gemini 7 during its 14-day mission. It would further state that in reporting this to the President, he has asked us to endeavor to do this rendezvous in December.

> Mr. Webb said that Dr. Seamans was a little more conservative than he was, but Mr. Webb felt it would give the image that we have the resources to retrieve.

> Dr. Mueller said they would have a much better view of the situation by Monday and could tell whether it was a 20-80 chance or a 80-20 chance of succeeding.

> [2]

> Dr. Seamans asked if it would place an undue burden on Chris Kraft, and Mueller said it would not as long as we tell them they don't _have_ to do it.

6:__ p.m.

> Mr. Webb called Mr. Joseph Laitin at the White House. He explained to Mr. Laitin that we were taking down the booster set up to fly Gemini 6 in order to erect the one for Gemini 7. He said we were going to look very carefully over the next several days at whether or not it might be possible to launch Gemini 7 the latter part of November or early December, and if it gets off with no damage to

the launching pad, to launch Gemini 6 before the 14-day trip is over. He said we, would not know for sure until next week whether or not this was possible.

Mr. Webb said he would like some judgment from Mr. Laitin and Mr. Moyers as to whether or not it would not be a good idea for the White House to put out a press release saying that NASA has informed the President that the Gemini 6 booster is not adequate enough to carry Gemini 7 into a 14-day orbit; that, therefore, the Gemini 6 booster is being taken off the pad and Gemini 7 is put on the pad for a launching as early as possible in December. Second, that we have told him that we have already done the work of mating Gemini 6 and the booster and both to the launching apparatus itself, and that there is a possibility that if Gemini 7 got of without damaging the launching pad, Gemini 6 could be launched and have a rendezvous between Gemini 6 and 7.

Mr. Webb said we couldn't promise that we could do it, and the President mustn't tell us to do it but to endeavor to do it.

Document I-69

GEMINI PROGRAM MISSION REPORT

GEMINI VI-A

(U)

CLASSIFICATION CHANGED TO
U   E.O. 11652.
BY AUTHORITY OF
JUN 1 1972
DATE   5-13-75

```
GROUP 4
DOWNGRADED
AT 3 YEAR INTERVALS;
DECLASSIFIED
AFTER 12 YEARS
```

CLASSIFIED DOCUMENT   This material contains
information affecting the National Defense of the United States within
the meaning of the espionage laws, Title 18, U.S.C., Secs. 793 and
794, the transmission or revelation of which in any manner to an
unauthorized person is prohibited by law.

JANUARY 1966

NATIONAL AERONAUTICS AND SPACE ADMINISTRATION   MANNED SPACECRAFT CENTER

[1-1]

1.0 MISSION SUMMARY

The fifth manned mission and first rendezvous mission of the Gemini Program, designated Gemini VI-A, was launched from Complex 19, Cape Kennedy, Florida, at 8:37 a.m. e.s.t., on December 15,1965. The flight was successfully concluded with the recovery of the spacecraft and the flight crew at 23°22.5' N. latitude 67°52.5' W. longitude by the prime recovery ship (U.S.S. Wasp), approximately 1 hour and 6 minutes after landing. This rendezvous mission was launched from Complex 19 within 11 days after the launch of the Gemini VII space vehicle. The spacecraft was manned by Astronaut Walter M. Schirra, command pilot, and Astronaut Thomas P. Stafford, pilot. The crew completed the flight in excellent physical condition and demonstrated excellent control or the rendezvous and competent management of all aspects of the mission.

The primary objective of the Gemini VI-A mission was to rendezvous with spacecraft 7. The secondary objectives of the Gemini VI-A mission were to perform a closed loop rendezvous at M=4 (fourth darkness of the mission), conduct station keeping with spacecraft 7, evaluate the re-entry guidance capability of the spacecraft, conduct visibility tests of spacecraft 7 as a rendezvous target vehicle, conduct 3 experiments, and conduct systems tests. The primary objective and all secondary objectives of the mission were successfully accomplished except for one of the three experiments for which valid data were not received.

The Gemini launch vehicle performed satisfactorily in all respects. The countdown was nominal, resulting in a launch within one-half second of the scheduled time. First-stage flight was normal with all planned events occurring within allowable limits. The first stage offset yaw steering technique was used for the first time on this flight in an attempt to place spacecraft 6 in the same orbital plane as spacecraft 7. The technique results in a "dog-leg" trajectory, and it was used successfully.

Staging was nominal; however, the crew reported that the flame front caused by staging enveloped the spacecraft in such a manner that it deposited a thin burned residue on the windows which affected the visibility through them. The pilot was able to verify this phenomenon as he had been observing a string of cumulus clouds prior to staging and also observed them after staging. He reported that the clearness and whiteness of these clouds was diminished after staging.

The second stage flight was normal and all but 7 ft/sec of the -660 ft/sec out-of-plane velocity achieved during first stage operation was steered out during second stage flight. The spacecraft was inserted into an orbit having an 87.2 nautical mile perigee and an

[1-2]

140 nautical mile apogee. The apogee was about 7 nautical miles below the planned altitude. The slant range to spacecraft 7 from spacecraft 6 at its insertion into orbit was a nominal 1067 nautical miles.

Nine maneuvers were performed by spacecraft 6 during the following 5 hours 50 minutes to effect the rendezvous with spacecraft 7. These maneuvers were all performed using the spacecraft guidance system for attitude reference. Initial radar lock-on with spacecraft 7 occurred at a range of 248 nautical miles. Continuous lock-on started at a range of 235 nautical miles and no losses of lock occurred until the system was turned off at a range of 50 feet from spacecraft 7. The rendezvous phase of the mission was completed at 5:56:00 ground elapsed time when spacecraft 6 was 120 feet from spacecraft 7 and all relative motion between the two vehicles had been stopped.

Station keeping was performed at distances between 1 foot and 300 feet for about 3 1/2 orbits after which a 9 ft/sec separation maneuver was performed.

The relative motion of spacecraft 6 from the separation maneuver was stopped at a range of about 30 miles.

The spacecraft and its systems performed very satisfactorily throughout the mission, except for the delayed-time telemetry tape recorder which failed at 20 hours 55 minutes ground elapsed time because of a bearing seizure. This recorder malfunction resulted in the loss of all delayed-time telemetry data for the remainder of the mission.

The flight progressed nominally to its full duration. All checklists and stowage were completed in preparation for retrofire and reentry and the reentry control system was activated. Retrofire occurred exactly on time at 25:15:58 ground elapsed time for a landing in the West Atlantic landing area (primary). The reentry and landing were nominal, and the landing point achieved was less than 7 nautical miles from the planned landing point. The crew remained in the spacecraft until the spacecraft had been secured on the deck of the recovery ship.

[pp. 2-1 through 7-17 not included]

[7-18]

7.1.2.3 Rendezvous Phase.-

7.1.2.3.1 Radar acquisition of spacecraft 7: At approximately 3 hours g.e.t., the ground update for acquisition of spacecraft 7 was received as an attitude of or yaw and 5.50 pitch up. The ground controllers also indicated that the initial computer readout of range (248 nautical miles) would occur at 3 hours 15 minutes g.e.t. The [7-19] radar was turned on in the standby position at approximately 3 hours 5 minutes g.e.t. The analog meter indication cycled exactly as predicted, and the range and range rate indications oscillated until the set warmed up. The radar was then placed on "ON".

The first radar-range readout on the MDRU was 248.66 nautical miles, which is the maximum range readable. At this time, the radar lock-on light was flickering. The radar lock-on became steady at 246.22 nautical miles, At that time, a radar test was performed with the rendezvous mode of the computer to verify the interface and sequencing of the computer and the radar. This radar-computer test was not conclusive in that the specified 130° angle of orbit travel to rendezvous (wt) was not inserted and the last wt that was loaded was 180°, which had been used for a prelaunch test. Subsequent to the N_{SR} maneuver and the final switching to the rendezvous mode, the correct value of wt (130°) was loaded. The computer cycled properly, holding the range in the register for 100 seconds, and the IVI's corresponded to the computer readout of total-velocity-change for rendezvous. The initial-velocity-change for target intercept was also noted, and the values were found to be decreasing as range decreased. The event timer was synchronized with the initiation of the N_{SR} maneuver. Four minutes after initiation of the N_{SR} maneuver, the computer was switched to the rendezvous mode and continuously monitored by the pilot. A time synchronization revealed that the event timer was approximately 7 seconds ahead of the computer time sequence (for 100-second intervals). The event timer was resynchronized with the computer-time and counted correctly throughout the remainder of the run. After the NSR maneuver, the range was approximately 169 nautical miles. The pilot did not record anything on the data sheet until the values began to match the nominal values at approximately 136 nautical miles range. After that, the values were recorded and data points were frequently called to the ground. The computer solution for the total-velocity-change for rendezvous was very close to nominal. The target-centered coordinate plot (see fig. 7.1.2-1) [not included] showed that the NSR maneuver had placed spacecraft 6 into the nominal trajectory and that the maximum deviation was approximately 0.25 mile high with no ellipticity. During this time, the elevation and azimuth pointers were oscillating approximately ±1.5° from the electrical null. The period of the oscillation was approximately 4 seconds. As the range decreased to 97 miles, there was a noticeable reduction in the amplitude of the oscillation; however, the period remained constant. It should be noted that both the azimuth and elevation readings crossed the null point simultaneously during these oscillations. At a range of 79 miles, all pointer oscillations ceased and remained steady throughout the remainder of the rendezvous operation [7-20] and down to a range of 20 feet. The radar data were continually being plotted and computations made as spacecraft 6 approached the point of terminal phase initiation.

7.1.2.3.2 Visual acquisition: Visual acquisition of spacecraft 7 occurred at 5 hours 4 minutes g.e.t., 54 miles slant-range from spacecraft 6 to spacecraft 7. The target vehicle appeared as a bright star, 0.50 to the right of the boresight line on the optical sight. The target appeared brighter than the star Sirius, and during postflight comparisons, the flight crew believed it was probably brighter than the planet Venus. The target stayed in sight because of reflected sunlight until 05:15:56 g.e.t., or for approximately 12 minutes. Spacecraft 7 was lost in darkness about 3 minutes prior to the transfer thrust, at a range of approximately 30 miles. The crew, however, could have determined a backup solution during the programmed tracking period prior to transfer, and would have been able to perform the maneuver without visual contact.

7.1.2.3.3 Terminal phase: During the terminal phase, the crew used the data provided by the IGS (closed-loop) to perform all maneuvers. However, the pilot did make all backup computations for each maneuver in order to compare them with the results of the closed-loop solution. The target-centered coordinate plot revealed very quickly that the relative trajectory was near nominal and that the transfer thrust would be very close to the planned value of 32 ft/sec along the line of sight. For the backup procedure, the component normal to the line of sight was determined from the time change of the total pitch angle. The ground solution, transmitted from Guaymas, indicated that the value was 31.5 ft/sec. The initial time transmitted to the flight crew for the initiation of the terminal phase was 05:16:54 g.e.t. A short time later this was refined to 05:18:54 g.e.t. The onboard computer solution gave a thrust time of 05:18:58 g.e.t., 4 seconds later than that computed on the ground.

As the point of terminal phase initiate approached, it became evident that the exact- time to initiate the maneuver would be near the halfway point between two of the computer solutions that are 100 seconds apart. At this point the crew discussed the situation and decided to take the second of these solutions, if it still met the basic criteria. This decision was made to insure that transfer would occur from a position that would place spacecraft 6 forward and below spacecraft 7 at final rendezvous, and that braking would occur slightly later than nominal rather than earlier. This was the crew's approach to being conservative with respect to the lighting conditions during the braking maneuver in that, being slightly later, it would insure that the target would be in daylight during the final approach. A pitch angle to spacecraft 7 of 20.8 deg was selected for terminal phase initiate at a range of 41.06 nautical miles. At this time the START COMP button was pressed, [7-21] and the initial computer solution produced a value of 31 ft/sec forward, 7 ft/sec up (this value later decreased to 4 ft/sec up at the time of thrust), and 1 ft/sec right. The backup solution was computed to be 23 ft/sec forward and 2 ft/sec up, and a notation was made of this anomaly. The crew discussed the problem and decided that if a backup maneuver had been necessary they would have applied the nominal thrust of 32 ft/sec. This decision was reached because of the nominal trajectories that were indicated, up to that point, on the onboard target centered coordinate plot. In case the radar or computer had failed, the thrusts that would have been applied were those necessary to achieve changes in velocity of 2 ft/sec up and 32 ft/sec forward.

After completion of the transfer thrust, the fuel remaining was 62 percent. At this point, the time system was reset to zero based on the beginning of the first computer time cycle that occurred 270 seconds after depressing the START COMP button (nominally, this time coincides with the end of the transfer maneuver). The crew used this phase elapsed time (p.e.t.) as a time reference through final rendezvous. The target was not in sight during the tight-tracking period from 3 to 5 minutes after the transfer maneuver. During the 3-to-5 minute tight-tracking period, the analog range rate was 160 ft/sec at 3 minutes 30 seconds p.e.t. Computations from the onboard computer showed

156 ft/sec. At 4 minutes 30 seconds p.e.t., range rate from the analog meter was 155 ft/sec, and the computer value was 152 ft/sec. These comparisons show the close agreement between the analog meter readout and the computer solution and provided the crew with high confidence in the radar-computer interface.

At 5 hours 23 minutes g.e.t., during the 3-to-5 minute tight-tracking period, spacecraft 7 lights were barely visible and not sufficient for tracking. This time corresponds to a range of approximately 24 miles.

Subsequent to 5 minutes p.e.t., the spacecraft was pitched down to horizontal, using the direct attitude-control mode, to align the platform. It was decided that alignment would be conducted during the planned optional alignment period, from 5 minutes to 10 minutes p.e.t. This decision was based on the fact that 1.5 hours had elapsed since the last alignment. During this alignment period (with the platform in SEF, the control mode in pulse, and the flight director indicator displaying platform and attitude), very little motion was detected in the pointers, indicating that the platform had been in good alignment. In addition, the optical sight and the visible horizon also indicated good alignment before starting the align period. This excellent performance of the platform provided the crew with further confidence in the spacecraft IGS system. At 10 minutes 20 seconds p.e.t., direct control was selected and the spacecraft was pitched back up in order to track spacecraft 7. The radar lock-on light had not extinguished;

therefore, lock-on was continuous during the alignment period. The radar was nulled on [7-22] the target, and the target lights appeared very dim in the sight at this time. The target lighting was evaluated as sufficient for subsequent tracking and angular measurements.

At this time, an estimation was made, using the data entered on the target-centered coordinate plot, that the first midcourse correction would require slight forward and up velocities. The IVI's indicated 7 ft/sec forward, 7 ft/sec up, and 5 ft/sec left at a p.e.t. of 11 minutes 40 seconds. This p.e.t. corresponds to 5:31:31 g.e.t. After the midcourse correction thrust was applied, the IVI read zero in all axes. A second tight tracking of the target was required again between 15 minutes and 17 minutes p.e.t. It was not difficult to observe the docking light on the target spacecraft at this time. The acquisition lights did not show clearly, but they could have been tracked for backup solutions from approximately 12 minutes after the transfer maneuver through final rendezvous.

During the second period of tight tracking, the range rate was noted from the analog meter at 15 minutes 30 seconds p.e.t. and indicated 90 ft/sec. The computer data gave a range rate at this time of 91 ft/sec. At 16 minutes 30 seconds p.e.t., the analog meter indicated a range rate of 85 ft/sec and the onboard computed range rate was also 85 ft/sec. At 17 minutes p.e.t. the range to the target was 7.7 nautical miles. After this data point was obtained, the desired velocity changes in guidance axes were zeroed in the computer, and tight tracking was maintained for a period of 3 minutes to determine the backup solution for a normal-to-the-line-of-sight correction. The command pilot remarked that the spacecraft 7 docking light was as bright as the Agena. At 16 minutes p.e.t. (5:35:51 g.e.t.) the pilot remarked that he could see the docking light even though he had a brightly lighted area in the cockpit.

The docking light on spacecraft 7 was displaced 0.50 to the right of the zero position in the optical sight, while using the radar null as the pointing command. Farther to the left, approximately 100, two bright stars, Castor and Pollux, were in sight. These stars provided excellent pitch, roll, and yaw reference. In addition, there were sufficient stars near and around the target to permit good tracking. It was also noted that the docking light obscured the acquisition lights because of its relatively greater brilliance. However, the spacecraft 6 crew requested that the spacecraft 7 docking light be left on.

The target-centered coordinate plot indicated that small up and forward corrections would be required for the second mid-course correction. The backup solution indicated 6 ft/sec up. No backup velocity correction along the line of sight could be obtained because the computer math flow locked out ranges at this time. At 23 minutes 40 seconds p.e.t., the computer solution gave a correction of 4 ft/sec forward, 3 ft/sec up, [7-23] and 6 ft/sec right. When this maneuver was completed, the IVI was zeroed and the computer switched to the catchup mode. The pilot then cleared MDRU addresses 25,26, and 27 (X, Y, and Z, desired velocity changes in guidance axes) and the IVI displayed all zeros.

From this point, the pilot continually called out the pitch angle to spacecraft 7 as it increased and the range decreased. The command pilot, at this point, acquired a very good star pattern to maintain a celestial line of sight. Very little motion was discerned during this period. The target-centered coordinate plot indicated a flight path that was forward of and nearly parallel to the nominal trajectory. At one point, the pilot stated that it appeared as if the target were going *up;* however, the command pilot decided not to make any changes at that time. At a range of 2 miles it again appeared from the pilot's plot that the target was going *up* a small amount, but there was no apparent motion in relation to the star background. At 5 hours 46 minutes g.e.t., no relative motion was observable. The range rate was approximately 42 ft/sec, and at 05:48:11 g.e.t., the target appeared to start moving down a small amount but this relative motion was stopped. At this point, the START COMP button was pressed. This caused all subsequent changes in velocity to be displayed in cumulative totals. At 05:49:06 g.e.t., both the command pilot and the pilot noted that the reentry control system (RCS) heater light came on at the telelight panel. This was at a range of 1 mile. This indicates that the panel was observable to the crew during this critical period. The total pitch angle, from 1.30 nautical miles into station keeping at 120 feet, was approximately 125°.

7.1.2.3.4 Braking maneuver: During the terminal phase a combination of radar display and optical tracking was utilized by the command pilot with the platform continually in orbital rate. The target held steady on the indicator throughout the terminal phase maneuver. At 05:49:41 g.e.t., the command pilot remarked that the docking light was quite bright, and the pilot noted the same thing.

At 0.74 mile range (05:49:58 g.e.t.), the pilot noted that the target appeared to be moving down. This comment was prompted as a result of seeing sunlight reflected off frost particles leaving spacecraft 6 and confusing them with stars. Spacecraft 6 was approaching the BEF attitude (spacecraft 6 was 30° beyond the local vertical). The ballistic number of these particles was such that they trailed the spacecraft, tending to move upward toward the nose of the spacecraft. As the crew observed the frost particles, they appeared to go up in relation to this apparent star field. There were stars still visible beyond these bright particles and these stars confirmed that the target was not moving in relation to the stars. This illusion for the pilot developed from the lighting conditions in the right crew station. This side of the cockpit was lighted sufficiently to permit the pilot to record data and work with the computer throughout this period. As a result, when [7-24] he made an out-the-window observation, he could not see the stars, and the particles appeared as stars to him. (This could have resulted in additional fuel expenditures if both the command pilot and the pilot had reacted identically.) At 0.48 mile range, the crew started decelerating spacecraft 6 from a closing range rate of approximately 42 ft/sec. During this period, there appeared to be no out-of-plane motion. As the braking continued, the velocity was reduced in a continuous thrust. The command pilot peered behind the black shield on the vernier scale until the pointer for range rate just appeared, having determined in the training simulator that this represented approximately 7 ft/sec. At this point, thrust was terminated and the range was approximately 1200 feet. The target had dropped slightly and a downward thrust was also added. At 800 feet range, 32 minutes after the translation maneuver, the closing velocity was approximately 6 ft/sec and the IVI's were cleared. The cumulative velocity changes at this point read 27 ft/sec aft, 14 ft/sec left, and 7 ft/sec down.

The total distance encompassed during the braking maneuver was 0.24 nautical miles (from 0.48 to 0.24 n. mi. from the target). When the range was 0.20 nautical miles, the pilot called the range to spacecraft 7 in feet to the ground and to the command pilot.

At a range of approximately 700 feet, the sunlight illuminated spacecraft 7 and the target was so bright that no stars were visible. The total impact of the brightness was as if a carbon arc lamp had been turned on immediately in front of spacecraft 6. The range decreased nominally, during which time both the pilot and command pilot continually commented on the brightness of the target. Because of the brightness, the radar display and the flight director attitude indicator (FDAI) were then used for tracking. As spacecraft 6 approached a range of 300 feet from spacecraft 7, the pitch angle decreased to 90° and held that value. Spacecraft 6 then continued to approach from directly below spacecraft 7.

At 240 feet, all rates in translation, except the closing velocity, had been reduced to zero. The closing velocity was being reduced by a series of small thrusts to approximately 2 ft /sec. Finally, at a range of 120 feet, all relative motion between the two spacecraft was stopped at approximately 36 minutes after the translation maneuver.

The final braking maneuver was difficult because of (1) the brightness of the reflected sunlight from the target at a range of approximately 700 feet, and (2) the fact that the crew could no longer use stars as a reference. Also, the target spacecraft was changing pitch attitude in order to track spacecraft 6 and, as a visible object, could [7-25] not be used for attitude reference with relation to motion in a pitch maneuver of spacecraft 6.

A very low, relative translation rate remained near the end of the braking maneuver. Spacecraft 6 had moved from a pitch angle of 90° to a pitch angle of 60° by the time the forward relative velocity was reduced to zero. The crew elected to continue this motion at a 120-foot radius, pitching down to the SEF attitude, and holding this position. At this point, spacecraft 6 was in the SEF position, with spacecraft 7 facing it in BEF, and all relative motion was stopped. The attitude control system was placed in SEF platform control mode, and all maneuvers were then performed with the maneuver controller.

The performance of the guidance and control system and radar system during all phases of rendezvous was excellent and the use of radar for rendezvous was shown to be extremely valuable. Throughout the rendezvous phase, the radar maintained positive lock-on and an accurate indication of range was available through the minimum readable value of 60 feet. The attitude indications were steady throughout the entire maneuver.

7.1.2.4 Station keeping.- From the crew's analysis of the timing, spacecraft 6 arrived in formation with spacecraft 7 about 23 seconds earlier than predicted prior to lift-off. In the SEF attitude, the distance between the spacecraft was closed to approximately 6 to 10 feet in order to observe spacecraft 7 in detail. Still photographs and motion pictures were taken and all exposure values were determined with the spot meter. The results of this photography indicate that a spot meter is a valuable aid in photographing objects in space. Initially, station keeping was accomplished in platform mode with minute thrust motions made with the maneuver controller. Shortly after the start of station keeping, the sun striking the command pilot's window completely obscured his view of spacecraft 7. The pilot gave voice positions of the target, and finally, control was passed to the pilot for approximately 1 minute until the spacecraft moved out of this sun angle. (This effect will continue to be a problem for station keeping.) The crew did not elect to do the in-plane fly-around at this point because they wanted to determine the composition of the strap observed hanging from the adapter of spacecraft 7. Shortly thereafter, the Gemini VII crew informed the Gemini VI-A crew that they also had a strap hanging from their adapter. This subsequently was determined to be part of the shaped charge holders. (See section 5.1.9.) [not included]

During the final portion of the first daylight period, station keeping was conducted in platform mode and finally in pulse mode when it was determined to be an easy task. Spacecraft 6 closed to about 1 foot, nose to nose with spacecraft 7, and it was concluded that [7-26] docking would not present any problems. It was also noted during this period that one spacecraft could influence the horizon scanners of the other spacecraft.

During the first night period, station keeping was maintained at ranges varying from 20 to 60 feet and the spacecraft were nose to nose. During the transition from daylight to night, the blurred horizon caused the scanner to lose track; therefore, orbit rate was selected prior to entering this period to avoid any transients that might occur during the period of scanner loss. Station was maintained by first using the docking light and platform mode, then with the docking light and pulse mode, then without the docking light and using the illuminated windows of spacecraft 7 as a reference. During a subsequent night pass, an out-of-plane position was encountered where the crew could not see the window of spacecraft 7. The hand-held penlights were then utilized to illuminate spacecraft 7 at a range of approximately 30 or 40 feet. The crew determined that they had sufficient lighting for station keeping. The most efficient way to conduct station keeping was to maintain station in horizon scan mode, letting the spacecraft drift in yaw. The recommended position for maintaining station is in the out-of-plane position, rather than trying to maintain station above or below the spacecraft. This provides a visual aid in that the horizon relative to the target permits holding pitch and roll relatively steady in the horizon scan mode.

During the second daylight period, spacecraft 7 was scheduled to perform an experiment and conduct a small amount of station keeping. To provide a fixed target for the D-4/D-7 experiment, spacecraft 6 was moved to a nose-to-nose position, 20 feet from spacecraft 7. The amount of fuel remaining in spacecraft 7 did not permit more than about 2 to 3 minutes of station keeping, and both the command pilot and pilot maneuvered o the nose of spacecraft 6 or this period. Subsequent to the station keeping performed by spacecraft 7, spacecraft 6 again picked up the nose position and the command pilot initiated an in-plane fly-around.

The in-plane fly-around was conducted for 20 minutes starting at 7 hours 42 minutes g.e.t. The pilot conducted an out-of-plane fly-around for 11 minutes starting at 8 hours 10 minutes g.e.t.

The command pilot, during the in-plane fly-around, allowed the range between the two spacecraft to increase to an estimated 300 feet. The relative position of spacecraft 6 at that time was above spacecraft 7, and slightly to the rear. This distance appeared to be excessive for proper station keeping and the range was quickly reduced to less than 100 feet. The radar system was not used during the station-keeping period. These ranges were determined both by visual observation in relation to the 10-foot diameter of the spacecraft as viewed through the [7-27] optical sight during the flight and by measurements after the flight of photographs taken with known optical systems.

It is recommended that station keeping not be conducted in-plane above or below the target. The ideal condition for station keeping is SEF or BEF in platform mode; however, station keeping can easily be conducted out-of-plane at ranges up to 60 or 80 feet without losing the perceptive cues that pilots have learned to recognize in formation flying with aircraft.

The smallest distance between spacecraft 6 and spacecraft 7 during station keeping was approximately 1 foot, and both the command pilot and pilot flew at this distance with great ease. This, of course, greatly enhanced the crew's confidence in the control system for subsequent station-keeping operations. The control-system response can be described as perfect. The torque-to-inertia ratios of the attitude control system using the pulse mode, and thrust-to-inertia ratios of the translation system using minute inputs, were excellent for the station keeping performed during this mission. Docking with a target vehicle could have' been easily executed by applying a small burst of forward thrust from the l-foot range.

Document I-70

[CONFIDENTIAL] [DECLASSIFIED]

GEMINI VIII

TECHNICAL DEBRIEFING

March 21, 1966

NATIONAL AERONAUTICS AND SPACE ADMINISTRATION

MANNED SPACECRAFT CENTER

HOUSTON, TEXAS

PREFACE

This preliminary transcript was made from voice tape recordings of the Gemini VIII flight crew debriefing conducted by Captain Schirra immediately after crew recovery, March 18, 1966.

A subsequent debriefing was conducted at the Crew Quarters, Cape Kennedy, Florida, by Mr. J. Van Bockel on March 19-20, 1966.

Although all material contained in this transcript has been rough edited, the urgent need for the preliminary transcript by missions analysis personnel precluded a thorough editorial review prior to its publication.

Note: The section covering the problem area encountered after docking and referred to as the Gemini VIII Self-debriefing is contained within Section 4.0, Orbital Operation.

[pp. 2-54 not included]

[55]

GEMINI VIII SELF DEBRIEFING

Armstrong:

> Okay. Approximately 7 hours 00 minutes in the flight plan, we were in configuration to perform a Platform Parellelism Check and had just completed the yawing of the Agena-Spacecraft combination to spacecraft BEF position, 0-180-0. We were on the night side. We had docked at approximately 6:34, and that was just a couple of minutes into the night side, or thereabouts. In the Flight Plan – at the position where we were sending command 041 with the computer already set up with Addresses 25, 26, and 27 inserted. At the time, the Flight Plan was on the left-hand side and I was reading the commands to Dave, and, at the same time, was working on restoring the cabin into a better configuration after just recently completing the Post-docking Checklist. Then Dave reported that there was some kind of divergence. How did you remember that, Dave?

Scott:

> Well, we had just finished putting the commands in, and the next thing on the Flight Plan was to start the Agena recorder. I had just sent 041 command to the Agena and written down the time at which the recorder started. I looked up and saw the Spacecraft-Agena [56] combination starting a roll. With no horizon, it wasn't apparent until I happened to glance at the ball and I didn't really feel it at first. I called Neil and he suggested turning the ACS off. I turned it off as fast as I could and also in a short period of time turned off the Horizon Sensor and the Geo Rate to give spacecraft control to the combination.

Armstrong:

> I would agree that I could not feel the angular acceleration either. We had the lights up in the cockpit and could not really see outside, since it was night and we had no horizon reference. My initial notice of the acceleration was an increase in rates and attitudes on the attitude ball.

Scott:

> Yes. That was my same indication. With no horizon at all, and it was hard to tell unless you looked at the ball.

Armstrong:

> Since we expected the SPC-loaded yaw maneuver to come sometime within the next 10 minutes and the spacecraft was essentially inactive with the OAMS Attitude Control Power off, it seemed as though the trouble was probably originating with the Agena Control System. So, I turned on the Attitude Control Power, went to RATE COMMAND (we had previously been in PULSE) and attempted [57] to stabilize the combination. It was my impression that after some period of time, perhaps less than a minute, we essentially had the combination stabilized. But, when we'd let go of the stick, we would again start to accelerate.

Scott:

> And, at some point in there when we had almost stabilized the combination, we sent a command to disable the SPC maneuver, too.

Armstrong:

> That is correct. We were at the ....
>
> I guess I read that command out of the book. 340 I think it was, or something - - S240.

Scott:

> Whatever it was, and I checked it in on the card.

Armstrong:

> Right. SPC Disable. Then, noting that the combination was still accelerating and desiring to stop the Agena Control System, we suggested trying to cycle the ACS on in case we could find its Rate Command operative again and help stabilize the combination. We did not see any improvement and later cycled ACS back off. In the meantime, we had sent Power Relay Reset, which I think is 271.

Scott:

> Right. Okay. I think the next thing we both commented on was being able to see the ACS thruster gas, or some gas coming out of there, out of the Agena. [58]

Armstrong:

> This is correct. Since we were approaching a lit horizon, as we would rotate our line of vision through the horizon we could see the cones of ACS

thrust coming out of the Agena pitch thrusters. And they appeared to be on full time to me, at the times I could see them.

Scott:

Yes, I agree. And it was about a 40-degree spread, about 25 feet long.

Armstrong:

That's right. A wide cone that was illuminated by the sunlit horizon or air glow. Okay, we noted at that time that the gas pressures on the Agena were down to approximately 20 percent.

Scott:

Right.

Armstrong:

And we realized then that indeed the ACS was losing gas at a fast rate, either because of a leak or because of all thrusters firing simultaneously. We also had excessive OAMS propellant usage and I called out when we went through 30 percent OAMS propellant on the Propellant Quantity Indicator. At this time, we felt there was some possibility of a spacecraft control system problem at the same time, so we initiated procedures to check out the OAMS system and tried turning the Bias Power off. That did not stop the [59] accelerations. We turned the Motor Valves off and this did not have any apparent affect either. We turned the Attitude Control Power on and switched Bias Power drivers logic and, we think, switched the roll logic to the pitch thrusters.

None of these actions had any apparent affect, and we were simultaneously, whenever possible, trying to use the thrusters to reduce the rates. We never, however, were able to reduce the rates in any axis completely. It was obvious at this time that the only satisfactory way for diagnosing the control system was undocking the vehicle so that we could disengage possible Agena problems from possible spacecraft problems. To do so, we had to get the rates of the combination down to a value that was suitable for undocking with some assurance that we would not have a recontact problem. We, of course, had to have the OAMS on to reduce these rates and it took us quite a bit of time to get the rates down to a value that we both agreed would be satisfactory to try a release. Upon mutual agreement, Dave undocked with the use of the Undocking Switch and I used the forward-firing thrusters to back away from the Agena as quickly as possible, using about [60] a 5 second burst. We did not have excessive rates at separation. What would your analysis have been there, Dave?

Scott:

Yes, it looked like a clean separation to me with very low relative rates, and we backed straight off a good 4 or 5 feet before we started tumbling there and lost sight of the Agena. I might add that before we backed off I sent L-Band ON and UHF Enable to the Agena.

Armstrong:

> Shortly after backing off, we noticed that we were essentially losing control of the spacecraft in roll and yaw and we suspected that we were over the life-time of these attitude thrusters. The spacecraft was continuing, however, to accelerate, and we were obtaining rates in roll at least that approached 200 to 300 degrees per second, or perhaps more.

Scott:

> Yes, I would agree with that. It looked like even more to me, and it was by far more in roll than in yaw. The roll was the most predominate.

Armstrong:

> We realized that physiological limits were being approached, and that we were going to have to do something immediately, in order to salvage the situation. So, we turned off all the OAMS thruster circuit breakers, closed the Attitude Control Power Switch, [61] closed the Motor Valves, armed the RCS, had no effect using the ACME, and went to DIRECT.

Scott:

> I might add in there that the rates were high enough that both of us had trouble seeing the overhead panel due to the vertigo problems and the centrifugal force as we went around.

Armstrong:

> The RCS DIRECT DIRECT was working satisfactorily and as soon as we determined that we were able to reduce the rates using this mode, we turned the A-Ring OFF and reduced the rates slowly with the B-Ring, putting in a pulse to reduce the rate, then waiting awhile, then putting in another pulse, and so on until the rates were essentially zero in all areas. At this time we carefully reactivated the OAMS, found some popped or inadvertently manually actuated circuit breakers, OAMS control and so forth. Upon reactivating the system we found that the Number 8 thruster was failed on, so we left that circuit breaker off. We had no other yaw thrusters with the exception of Number 8 but the pitch was apparently starting to come back in and we ensured that the roll logic was in pitch. We stayed in PULSE, controlling the spacecraft with pitch and roll pulses then to essentially a BEF attitude. [62]

Scott:

> Do you want to add in there about the hand controller, in not getting anything?

Armstrong:

> Yes. When I earlier referred to the fact that I'd lost control completely it appeared to us as though at that time we had no control out of the hand controller in any axis. I might reiterate that we reactivated the OAMS and found no roll or yaw control with the Number 8 circuit breaker off but pitch was slowly coming back then. It was somewhat ineffective at first, but it was usable after awhile. Sometime later we saw the Agena, approxi-

mately a half to a mile below us for a short period of time in daylight. It did not have excessive pitch and yaw rates at this time, nor did it appear to be tumbling end over end. However we were too far away to determine whether there were any roll rates involved in the Agena.

Scott:

Yes, I agree. It went by pretty fast. We did get to see it wasn't tumbling, but it was hard to tell exactly what attitude or rates it had.

Armstrong:

Sometime later, when preparing for retrofire, we were asked by the ground whether we had identified the proper operation of the Reentry Rate Control System. So, in checking that system out, we found that we had [63] regained some yaw control at this time, and guessed at the time that those thrusters may have been cooling down to the point where we were once again getting thrust out of them. So, we used the OAMS then in all three axes to align the platform for retrofire.

Scott:

You might add that the camera was on there during the undocking at some unknown setting.

Armstrong:

Roger, we did have the camera on during this time period – the 16 millimeter camera—but we, of course, could not take time to check the settings, and we could not identify at this time whether it was set for daylight or darkness, or for what configuration. That film may or may not come out. [64]

Scott:

One thing we might add on the stability of the combination – as far as bending we didn't notice any oscillations on the docking or post-docking between the two vehicles after TDA Rigidized. Also during the rolling and yawing maneuvers, when we had the problems with the Agena and spacecraft, I don't believe we noticed any oscillations or bending between the two vehicles. It seemed to be a pretty firm attachment.

Armstrong:

I am certain that we put fairly sizeable bending loads on the combinations as a result of the inertial loads and also the thruster loads which were long time duration and in all sorts of combinations out of both the OAMS and the Agena ACS. There certainly was no evidence of any relative motion between the Agena and the spacecraft or any noticeable deflections of any sort. After being informed by the ground that they were considering a 6-3 landing area, we realized that we had a reasonably short time to get reconfigured from the stowage point of view to an entry configuration. We immediately started to prepare for that possibility. This involved the restowage of the cameras first. (Both our right and left boxes were not yet opened so they did not pose a problem).

Document I-71

Document title: NASA, "Gemini Contingency Information Plan," 11 May 1966.

Source: Folder 18674, NASA Historical Reference Collection, NASA History Division, NASA Headquarters, Washington DC.

Given the inherent riskiness of spaceflight, NASA officials understood that the potential of loss of spacecraft and crew during flight existed. What should be done if this were to happen? The first order, they found, was to impound all technical and other types of data relating to the mission to help reconstruct how and why a failure had occurred. A second action required obtaining statements from all individuals involved in the mission, which would probably be only preliminary to more detailed debriefings to follow. The third step, and one that was virtually as important as these others, required the management of the flow of information to the public and other parties around the world. This plan, one of several prepared prior to 1966 and similar to but less elaborate than those still in use by NASA for more recent missions, seeks to ensure the appropriate release of details and the management of information to the media and others.

GEMINI CONTINGENCY INFORMATION PLAN

MAY 11, 1966

[i]

Although extremely unlikely, situations may occur which could result in aborting a manned mission.

Attached are suggested plans of action should a contingency occur.

Coordination by the Department of State with other governments, should it be necessary, is covered in a DOS airgram of March 9, 1965, to appropriate posts.

NASA will remain the prime source of public information throughout all contingency situations, with support from both the Department of Defense and the State Department.

[ii]

INDEX

[1]

PAD OR CLOSE-IN ABORT

Should an abort occur, the crew would be located and immediately transported to the Bioastronautics Support Unit (BOSU) at Cape Kennedy.

When the Mission Director or the Flight Surgeon (in Mission Control Center-Houston) is advised of the crew's physical condition, he will immediately notify the Gemini Information Director.

Operating Plan

1. Applicable portions of Plan A will be initiated.

2. The White House, State Department, and Department of Defense Public Information Offices will be kept advised of the situation by means of a conference telephone call initiated by the NASA Headquarters Public Information Director or the Senior NASA Public Information Officer present.

3. Two Public Information Office representatives will escort a news-pool team from the Cape press site to BOSU. Upon arrival, one will establish immediate telephonic communications with the Public Information Director at the Mission Control Center-Cape or Houston, the other with the Gemini Information Director and the Director, Public Information.

4. As soon as feasible and with approval of the Assistant Administrator for Public Affairs or his designee, [2] announcement will be made of the time and site of a news conference which will include applicable personnel as recommended by the Gemini Information Director.

[3]

IN-FLIGHT CONTINGENCY

1. The Gemini Information Director and the NASA Mission Commentator will be kept current on any in-flight contingency by means of monitoring applicable circuits and/or being advised of the situation by the Mission Director and/or the Flight Surgeon.

    a.   The White House, State Department, and Department of Defense will be kept advised of the situation by the NASA Headquarters Public Information Director or the Senior NASA Public Information Officer present.

    b.   The NASA Mission Commentator will issue periodic releasable statements advising the press of the situation.

2.   When a probability of crew fatality or serious injury is indicated the portion of Plan A covering family notification will be activated.

3.   The Prime Recovery Zone Senior Public Information Officer will release information regarding crew condition to the news-pool team following coordination with the Gemini Information Director or his designee.

[4]

CONTINGENCY SITUATION IN PRE-DESIGNATED LANDING ZONES (OTHER THAN AREA OF PRIME RECOVERY VESSEL)

1.   The NASA Mission Commentator will be kept current on crew and spacecraft status by monitoring applicable circuits and/or being advised of the situation by the Mission Director, the Flight Surgeon and/or the Recovery Zone Public Information Officer.

    a.   The White House, State Department, and Department of Defense will be kept advised of the situation through a conference call initiated by the NASA Headquarters Public Information Director or the Senior NASA Public Information Officer present.

    b.   Appropriate NASA officials will be alerted to the situation by the Gemini Information Director.

    c.   The NASA Mission Commentator will issue periodic and timely statements advising the press of the situation.

2.   The Prime Recovery Zone Senior Public Information Officer will release information regarding crew condition to the news-pool team following coordination with the Gemini Information Director or his designee.

[5]

CONTINGENCY SITUATION IN OTHER THAN PRE-DESIGNATED LANDING AREAS

The purpose of this section is to provide guidance and direction for Department of Defense Public Information Officers and other Department of Defense and NASA personnel in the event the astronauts make a contingency landing anywhere except in the pre-designated landing areas.

While NASA will immediately dispatch public information representatives to such an area, it is recognized that, for a limited period of time, Department of Defense or NASA personnel may be the only representatives of this government at the scene.

According to the provisions of the "Overall Plan, Department of Defense for Project Gemini Operations," dated November 7, 1963, Section IX, 4b, "when a contingency recovery operation has been initiated, acknowledgement may be made subject to the condition that NASA has made the initial announcement that reentry and landing operations have been initiated. Any other responses to news media will be based upon instructions forwarded through operational communications channels on the basis of particular circumstances involved. Contingency recovery communications channels are the appropriate operational communications channels for this purpose as long as those circuits are maintained in operational status."

It is recommended that the above quoted provisions be applicable to any landing area except the planned landing areas.

In the absence of NASA Information Officers, Department of Defense personnel on the scene will initiate and maintain communication on a priority basis with the Public Information Officer for the Department of Defense Manager for Manned Space Flight Support Operations and will keep him informed of activities at the contingency landing site, including medical examinations and/ or other debriefing activities. He will serve as a point of contact for the Gemini Information Director and relay public information [6] to the Department of Defense personnel on the scene.

Following NASA's announcement that the astronauts are being taken to a specific site, the Department of Defense Public Information Officer there may respond to news inquiries with approval of the Gemini Information Director.

Under no circumstances may he comment on the physical condition of the astronauts or the conditions which resulted in the termination of the flight, with the exception of certain cleared releases which have been forwarded through communications channels from the Mission Control Center.

In the event the astronauts' arrival at any installation precedes that of NASA Public Information personnel, the Department of Defense Public Information Officer may confirm the pilots have arrived on the base. With the concurrence of the Gemini Information Director, the Department of Defense Public Information Officer may authorize news media to photograph the arrival.

Upon arrival of NASA Public Information representatives, the Department of Defense Public Information Officer will be relieved of public information responsibility in connection with the specific mission. He may, however, be requested to assist in accommodating local news media.

As regards NASA personnel:

1. The Senior Recovery Zone Public Information Officer, after coordination with the Gemini Information Director, will issue periodic and timely statements advising the prime recovery news-pool team of the situation.

2. When a probability of fatality or serious injury is indicated, that portion of Plan A covering family notification will be activated. [7]

3. The Gemini Information Director and the Director, Public Information, will be kept current on such information as crew condition, destination, and ETA of the recovery vessel.

4. Public Information personnel designated by the Assistant Administrator for Public Affairs will proceed to the recovery vessel debarkation point.

[8]

GEMINI CONTINGENCY INFORMATION PLAN

(PLAN A)

1. Notification to pilots families by telephone

   a. D.K. Slayton, Assistant Director for Flight Crew Operations (MSC), will notify command pilot's family over an unlisted phone installed by the MSC Public Affairs Office in the home. Dr. Robert R. Gilruth, Director (MSC) will speak to the command pilot's wife following notification by Mr. Slayton.

   b. Capt. A.B. Shepard, Jr., Chief, Astronaut Office (MSC), will notify the pilot's family over an unlisted phone installed by the MSC Public Affairs Office in the home. Dr. Gilruth will speak to the pilot's wife after Capt. Shepard.

2. Suggested Statements:

   a. The NASA Headquarters Public Information Director will recommend to the White House that appropriate statements (as outlined in Attachment 1 herein) be issued.

   b. The Gemini Information Director will recommend to appropriate NASA officials that applicable statements (as outlined in Attachment 1 herein) be issued.

   c. The Department of Defense Manager for Manned Space Flight Support Operations or his representative(s) will recommend to the Department of Defense (Joint Chiefs of Staff) that applicable statements (as outlined in Attachment 1 herein) be issued.

3. Astronaut and flight-controller voice tapes bearing directly on the accident may be impounded pending an investigation of the accident. [9]

4. As soon as possible, the NASA Mission Commentator will confirm the contingency situation and crew condition to news media representatives and will announce that a news conference will be held as expeditiously as circumstances permit. He will also announce the initiation of a special investigation board.

   a. The Prime Recovery Zone Senior Public Information Officer and all other NASA Public Information Office personnel located at sites other than MSC will release information following coordination with the Gemini Information Director.

   b. The NASA Mission Commentator may include the following items in the announcement of a special investigation:

      i. The Mission Director has called a meeting with the following people for the purpose of establishing a special investigation board.

      ii. When chosen, the board will conduct an investigation which will be of a technical, fact-finding nature. Its intent will be to:

         1. Determine the sequence of events related to the contingency

         2. Seek to isolate initial hardware malfunction to system component part level

         3. Seek to determine the failure mechanism and physical cause of the failure

         4. Reproduce the failure in a laboratory if feasible.

[10]

DRAFT STATEMENTS

IN THE EVENT OF CREW FATALITY

(Attachment 1)

The President would contact the command pilot's and/or pilot's wife by telephone to express personal condolence.

President:

"I have conveyed to (_____) and/or (_____) and members of the (_____) and/or (_____) family (ies) my deepest sympathy.

"This nation—indeed, the world – owes (_____) and/or (_____) a great debt of gratitude. He/They gave his/their life/lives in the performance of one of the highest callings of this nation. He/They has/have also contributed immeasurably to the advancement of science and technology. I have been, and will continue to be, deeply impressed by his/their dedication to the nation's space program – his/their insistence that the advancement of manned space flight was a pursuit of the highest order which must be carried out despite personal risks involved.

"The United States of America will ever revere the spirit, dedication, and conviction of (_____) and/or (_____)."

Vice President:

"The death(s) of (_____) and/or (_____) in furthering a space flight program to which he/they has/have dedicated his/their many talents is a profound and personal loss to me. My heart goes out to Mrs. (_____) and/or Mrs. (_____) and her/their wonderful children.

"I propose that in his/their name(s) there be established a permanent scholarship for promising space science students to enhance the space exploration effort for which he/they gave his/their life/lives."

[11]

NASA Administrator:

"I have extended my sympathy and that of all employees of the National Aeronautics and Space Administration to the (_____) and/or (_____) family/families.

"The nation today feels a great sense of loss. That feeling is even greater among those of us who worked with that/those competitive young man/men who was/were so completely devoted to enlarging man's capability in space flight.

"We in NASA know that his/their greatest desire(s) was/were that this nation press forward with manned space flight exploration, despite the outcome of any one flight. With renewed dedication and purpose, we intend to do just that."

Secretary of Defense:

"We in the Department of Defense feel keenly the loss of this/these outstanding young officer(s). His/Their career(s) was/were extraordinary, bridging the jet age and the space age. His/Their work and dedication will forever serve as an inspiration to men who fly."

| Secretary of the Air Force: | Air Force provided. |

| Secretary of the Navy: | Navy provided. |

NASA/MSC Director:

"We of the NASA Manned Spacecraft Center feel the loss of (_____) and/or (_____) very personally. The other astronauts, program people, and I have known and worked with (_____) and/or (_____) day-in and day-out.

"I have already expressed our feelings to Mrs. (_____) and/or Mrs. (_____) in a phone call that I prayed I would never have to make."

[12]

Assistant Director for Flight Crew Operations:

"All of us on the astronaut team have lost good friends in wartime or in flight test work. It's part of the business, and we know that better than anyone else. (_____) and/or (_____) was/were something very special – (an) excellent pilot(s), (a) tireless worker(s), (a) first-rate engineer(s). He/They was/were (a) remarkable man/men."

IN THE EVENT OF SERIOUS PILOT INJURY

All official statements would note the hazardous nature of the work.

NOTE: All NASA officials called upon to make public statements would assure themselves that their statements reflect the on-going spirit of this nation's manned space flight program.

Document I-72

Document title: Robert C. Seamans, Jr., Deputy Administrator, NASA, Memorandum for Associate Administrators, Assistant Associate Administrators, and Field Center Directors, NASA, "Gemini Program; Record of Accomplishments, Attached," 17 January 1967, with attached: "Project Gemini Summary."

Source: NASA Collection, University of Houston, Clear Lake Library, Clear Lake, Texas.

Document I-73

Document title: "Gemini Summary Conference," NASA SP-138, 1–2 February 1967.

Source: NASA Collection, University of Houston, Clear Lake Library, Clear Lake, Texas.

The lessons learned from the Gemini program proved critical to the long-term success of Apollo and the larger cause of human spaceflight. The program succeeded in accomplishing what had

*been intended for it from the outset, and then some. It demonstrated the capability of Americans
to undertake long duration space missions. It provided the opportunity to develop rendezvous
and docking techniques that served NASA's programs well into the future. It pioneered the
ability to leave the spacecraft and perform work outside in an extra-vehicular activity (EVA).
This knowledge is captured in summary form in these two important documents explaining the
results of the Gemini program for both NASA engineers and the general public.*

Document I-72

NATIONAL AERONAUTICS AND SPACE ADMINISTRATION

WASHINGTON DC 20546

January 17, 1967

OFFICE OF THE ADMINISTRATOR

MEMORANDUM FOR Associate and Assistant Administrators

Field Center Directors

FROM: AD/Deputy Administrator

Subject: Gemini Program; Record of Accomplishments, attached

The Gemini flight program, concluded on November 15, 1966, succeeded
in accomplishing all of its pre-planned objectives some of them several times over.
As can be expected in any complex developmental-flight program, some of the
individual flight missions experience difficulties. The successful demonstration
that these difficulties could be overcome in later missions is a tribute to the
program organization, personnel directly involved, and to NASA.

A summary of achievements of the program as a whole, a mission by
mission recap of flight performance in terms of the Agency's pre-stated primary
and secondary objectives for each mission, and, a table recapping the major flight
systems and mission performance on each mission attempt, is appended on the
attachment to this memo. This document has been reviewed and concurred
in by the Office of Manned Space Flight and Public Affairs as containing valid
information to serve as an official reference on Gemini accomplishments.

[Signed]
Robert C. Seamans, Jr.

Attachment

PROJECT GEMINI SUMMARY

FOR INTERNAL NASA USE AND OFFICAL GUIDANCE

With the splashdown of Gemini 12 with astronauts Lovell and Aldrin aboard on November 15, 1966, the Gemini Project came to a successful conclusion. All Gemini Project objectives, including Extravehicular Activity and combined vehicle maneuvers, which were added after the project began, were fully accomplished many times over.

Rendezvous: Ten separate rendezvous were accomplished, using seven different techniques ranging from visual/manual control to ground/computer controlled rendezvous.

Docking: Nine dockings with four different Agenas were performed.

Docked Vehicle Maneuvers: Both Gemini X and Gemini XI demonstrated extensive maneuvers and a new altitude record was set on Gemini XI when the Agena Target carried astronauts Conrad and Gordon 851 miles above the earth.

Extra-vehicular Activity: EVA was conducted on five separate Gemini Missions and during ten separate periods. Total EVA time during the Gemini Project was 12 hrs, 22 min. of which a record time of 5 hours and 37 minutes of EVA was performed by Aldrin on Gemini XII.

Long Duration Flight: Gemini VII demonstrated man's ability to stay in space continuously for up to 14 days; Gemini V for 8 days, and two other missions for 4 days.

Controlled Reentry: Landing accuracies of a few miles from the aim point were demonstrated on every Gemini manned mission except Gemini V.

Conduct Scientific and Technological Experiments: Every manned Gemini mission (Gemini III through XII) conducted many experiments. In total 43 experiments were conducted successfully.

Prior to each Gemini mission, individual primary mission objectives were selected which, if accomplished, would provide full advancement of the project. Accomplishment of these primary objectives were mandatory for stating the mission to be successful. To retain the flexibility to capitalize on success, secondary objectives were also assigned*– as many as appeared feasible within the capability of the equipment and the time and experience of the astronauts.

[2]

Of the 14 Gemini mission attempts, 10 missions accomplished all of the primary mission objectives specified before the launch. The four unsuccessful missions and the reasons why they could not accomplish all of their primary objectives follows:

UNSUCCESSFUL MISSIONS	REASONS
GEMINI VI	The Agena Target Vehicle exploded. The Gemini 6 spacecraft was successfully rendezvoused with the Gemini 7 spacecraft later during the Gemini VI-A mission.
GEMINI VIII	An Orbit Maneuvering Thruster malfunction which ruled out a stated primary objective: EVA.
GEMINI IX	An Atlas booster failure drove the Agena into the Atlantic, and the Gemini 9 spacecraft was not launched until later during the Gemini IX-A mission.
GEMINI IX-A	The shroud did not come loose from the Augmented Target Docking Adapter, precluding docking – a specified primary objective for the mission.

Gemini Launch Vehicles

The modified Titan launch vehicle used as the Gemini Launch Vehicle was 100 percent successful in the Gemini Project. Out of 12 launches, 12 successful vehicle performances were achieved.

Gemini Target Vehicles

Six Gemini Agena Targets were launched and four were successfully placed in orbit, rendezvoused and docked with. The Augmented Target Docking Adapter, launched as a back-up target for the Gemini 9 spacecraft to rendezvous and dock with, functioned properly; however, the shroud failed to separate, thereby making docking impossible.

* A listing of primary and secondary objectives accomplished by mission is attached [not included].

Document I-73

GEMINI SUMMARY CONFERENCE

February 1-2, 1967
Manned Spacecraft Center
Houston, Texas

1. INTRODUCTION

By George E. Mueller, *Associate Administrator for Manned Space Flight, NASA*

The Gemini Program is over. The papers in this report summarizing the program were prepared by some of the people who contributed to the overall success. In each case, the authors were actual participants and provide a cross section of what may be called the Gemini team. As is true in any undertaking of this magnitude, involving many diverse organizations and literally thousands of people, a vital element of the Gemini success may be traced to teamwork. In the purest definition of the word, wherein individual interests and opinions are subordinate to the unity and efficiency of the group, the Gemini team has truly excelled.

Much has already been written concerning the Gemini achievements, and many of the achievements are presented again in greater depth within this report. By way of introduction, and to set the stage for the following papers, a few words are necessary to assess the achievements in the context of the goals of the national manned space-flight program. Only in this way is it possible to evaluate the significance of the Gemini accomplishments.

The Gemini Program was undertaken for the purpose of advancing the United States manned space-flight capabilities during the period between Mercury and Apollo. Simply stated, the Gemini objectives were to conduct the development and test program necessary to (1) demonstrate the feasibility of long duration space flight for at least that period required to complete a lunar landing mission; (2) perfect the techniques and procedures for achieving rendezvous and docking of two spacecraft in orbit; (3) achieve precisely controlled reentry and landing capability; (4) establish capability in the extravehicular activity; and (5) achieve the less obvious, but no less significant, flight and ground crew proficiency in manned space flight. The very successful flight program of the United States has provided vivid demonstration of the achievements in each of these objective areas.

The long-duration flight objective of Gemini was achieved with the successful completion of Gemini VII in December 1965. The progressive buildup of flight duration from 4 days with Gemini IV, to 8 days with Gemini V and 14 days with Gemini VII, has removed all doubts, and there were many, of the capability of the flight crews and spacecraft to function satisfactorily for a period equal to that needed to reach the lunar surface and return. Further, this aspect of Gemini provides high confidence in flight-crew ability to perform satisfactorily on much longer missions. The long-duration flights have also provided greater insight into, and appreciation of, the vital role played by the astronauts, the value of flexibility in mission planning and execution, and the excellent capability of the manned space-flight control system. As originally conceived, the Gemini Program called for completion of the long-duration flights with Gemini VII, which was accomplished on schedule.

One of the more dramatic achievements has been the successful development of a variety of techniques for the in-orbit rendezvous of two manned spacecraft. The preparation for this most complex facet of Gemini missions was more time consuming than any other. That it was performed with such perfection is a distinct tribute to the Gemini team that made it possible: the spacecraft and launch-vehicle developers and builders, the checkout and launch teams, the flight crews and their training support, and the mission-planning and mission-control people.

[2]

The ability to accomplish a rendezvous in space is fundamental to the success of Apollo, and rendezvous was a primary mission objective on each mission after Gemini VII. Ten rendezvous were completed and seven different rendezvous modes or techniques were employed. Nine different dockings of a spacecraft with a target vehicle were achieved. Eleven different astronauts gained rendezvous experience in this most important objective. Several of the rendezvous were designed to simulate some facet of an Apollo rendezvous requirement. The principal focus of the rendezvous activities was, however, designed to verify theoretical determinations over a wide spectrum. Gemini developed a broad base of knowledge and experience in orbital rendezvous and this base will pay generous dividends in years to come.

A related accomplishment of singular importance to future manned space-flight programs was the experience gained in performing docked maneuvers using the target vehicle propulsion system. This is a striking example of Gemini pioneering activities – the assembly and maneuvering of two orbiting space vehicles.

The first attempt at extravehicular activity during Gemini IV was believed successful, and although difficulties were encountered with extravehicular activity during Gemini IX-A, X, and XI, the objective was achieved with resounding success on Gemini XII. This in itself is indicative of the Gemini Program in that lessons learned during the flight program were vigorously applied to subsequent missions. The extravehicular activity on Gemini XII was, indeed, the result of all that had been learned on the earlier missions.

The first rendezvous and docking mission, although temporarily thwarted by the Gemini VI target-vehicle failure, was accomplished with great success during the Gemini VII/VI-A mission. This mission also demonstrated the operational proficiency achieved by the program. The term "operational proficiency" as applied to Gemini achievements means far more than just the acceleration of production rates and compressing of launch schedules. In addition and perhaps more importantly, operational proficiency means the ability to respond to the unexpected, to prepare and execute alternate and contingency plans, and to maintain flexibility while not slackening the drive toward the objective. Time and again Gemini responded to such a situation in a manner that can only be described as outstanding.

A few comments are in order on what the Gemini accomplishments mean in terms of value to other programs. There is almost no facet of Gemini that does not contribute in some way to the Apollo Program. Aside from the actual proof testing of such items as the manned space-flight control center, the manned space-flight communications net, the development and perfection of recovery techniques, the training of the astronauts, and many others which apply directly, the Gemini Program has provided a high level of confidence in the ability to accomplish the Apollo Program objectives before the end of this decade. The Apollo task is much easier now, due to the outstanding performance and accomplishments of the Gemini team.

Similarly, the Apollo Applications Program has been inspired in large part by the Gemini experiments program, which has sparked the imagination of the scientific community. In addition to the contributions to Apollo hardware development which provide the basis for the Apollo Applications Program, it has been discovered, or rather proved, that man in space can serve many extremely useful and important functions. These functions have been referred to as technological fallout, but it is perhaps more accurate to identify them as accomplishments – that is, accomplishments deliberately sought and achieved by the combined hard labor of many thousands of people. Some of these people have reviewed their work in this report.

The Manned Orbiting Laboratory Program has been undertaken by the Department of Defense for the purpose of applying manned space-flight technology to national defense and is making significant use of the Gemini [3] accomplishments. This may be considered as a partial repayment for the marvelous support that NASA has received and continues to receive from the DOD. The success of the NASA programs is in no small measure due to the direct participation of the DOD in all phases of the manned space-flight program. This support has been, and will continue to be, invaluable.

The combined Government/industry/university team that makes up the manned space-flight program totals about 240,000 people. In addition, thousands more are employed in NASA unmanned space efforts, and in programs of the Department of Defense, the Department of Commerce, the Atomic Energy Commission, and other agencies involved in total national space endeavors. These people, in acquiring new scientific knowledge, developing new techniques, and working on new problems with goals ever enlarged by the magnitude of their task, form the living, growing capability of this Nation for space exploration.

For the last quarter century, this Nation has been experiencing a technological revolution. Cooperative efforts on the part of the Government, the universities, the scientific community, and industry have been the prime movers. This cooperation has provided tremendous capability for technological research and development which is available now and which will continue to grow to meet national requirements of the future. The influence of this technological progress and prowess is, and has been, a deciding factor in keeping the peace.

Preeminence in this field is an important instrument in international relations and vitally influences this country's dealings with other nations involving peace and freedom in the world. Political realities which can neither be wished away nor ignored make the capability to explore space a matter of strategic importance as well as a challenge to the scientific and engineering ingenuity of man. This Nation can no more afford to falter in space than it can in any earthly pursuit on which the security and future of the Nation and the world depend.

The space effort is really a research and development competition, a competition for technological preeminence which demands and creates the quest for excellence.

The Mercury program, which laid the groundwork for Gemini and the rest of this Nation's manned space-flight activity, appears at this point relatively modest. However, Mercury accomplishments at the time were as significant to national objectives as the Gemini accomplishments are today as those that are planned for Apollo in the years ahead.

That these programs have been, and will be, conducted in complete openness with an international, real-time audience makes them all the more effective. In this environment, the degree of perfection achieved is even more meaningful. Each person involved can take richly deserved pride in what has been accomplished. Using past experience as a foundation, the exploration of space must continue to advance. The American public will not permit otherwise, or better yet, history will not permit otherwise.

[pp. 4-328 not included]

[329]

22. GEMINI RESULTS AS RELATED TO THE APOLLO PROGRAM

By Willis B. Mitchell, *Manager, Office of Vehicles and Missions, Gemini Program Office, NASA Manned Spacecraft Center;* Owen E. Maynard, *Chief, Mission Operations Division, Apollo Spacecraft Program Office, NASA Manned Spacecraft Center; and* Donald D. Arabian, *Office of Vehicles and Missions, Gemini Program Office, NASA Manned Spacecraft Center*

Introduction

The Gemini Program was conceived to provide a space system that could furnish answers to many of the problems in operating manned vehicles in space. It was designed to build upon the experience gained from Project Mercury, and to extend and expand this fund of experience in support of the manned lunar landing program and other future manned space-flight programs. The purpose of this paper is to relate some of the results of the Gemini Program to the Apollo Program, and to discuss some of the contributions which have been made.

The objectives of the Gemini Program applicable to Apollo are: (1) long-duration flight, (2) rendezvous and docking, (3) post-docking maneuver capability,

(4) controlled reentry and landing, (5) flight- and ground-crew proficiency, and (6) extravehicular capability. The achievement of these objectives has provided operational experience and confirmed much of the technology which will be utilized in future manned programs. These contributions will be discussed in three major areas: launch and flight operations, flight crew operations and training, and technological development of subsystems and components. While there is obvious interrelation among the three elements, the grouping affords emphasis and order to the discussion.

Launch and Flight Operations

Gemini experience is being applied to Apollo launch and flight operations planning and concepts. Probably the most significant is the development and understanding of the rendezvous and docking process. The Apollo Program depends heavily upon rendezvous for successful completion of the basic lunar mission. The Lunar Module, on returning from the surface of the Moon, must rendezvous and dock with the Command and Service Module. In addition, the first Apollo mission involving a manned Lunar Module will require rendezvous and docking in Earth orbit by a Command and Service Module placed in orbit by a separate launch vehicle. During the Gemini Program, 10 rendezvous and 9 docking operations were completed. The rendezvous operations were completed under a variety of conditions and applicable to the Apollo missions.

The Gemini VI-A and VII missions demonstrated the feasibility of rendezvous. During the Gemini IX-A mission, maneuvers performed during the second re-rendezvous demonstrated the feasibility of a rendezvous from above; this is of great importance if the Lunar Module should be required to abort a lunar-powered descent. During the Gemini X mission, the spacecraft computer was programmed to use star-horizon sightings for predicting the spacecraft orbit. These data, combined with target-vehicle ephemeris data, provided an onboard prediction of the rendezvous maneuvers required. The rendezvous was actually accomplished with ground-computed solution, but the data from the onboard prediction will be useful in developing space-navigation and orbit-determination techniques.

[330]

The passive ground-controlled rendezvous demonstrated on Gemini X and XI is important in developing backup procedures for equipment failures. The Gemini XI first-orbit rendezvous was onboard controlled and provides an additional technique to Apollo planners. The Gemini XII mission resulted in a third-orbit rendezvous patterned after the lunar-orbit rendezvous sequence, and again illustrated that rendezvous can be reliably and repeatedly performed.

All of the Gemini rendezvous operations provided extensive experience in computing and conducting midcourse maneuvers. These maneuvers involved separate and combined corrections of orbit plane, altitude, and phasing similar to the corrections planned for the lunar rendezvous. Experience in maneuvering combined vehicles in space was also accumulated during the operations using the docked spacecraft/target-vehicle configuration when the Primary Propulsion

System of the target vehicle was used to propel the spacecraft to the high-apogee orbital altitudes. During the Gemini X mission, the Primary Propulsion System was used in combination with the Secondary Propulsion System to accomplish the dual-rendezvous operation with the passive Gemini VIII target vehicle. These uses of an auxiliary propulsion system add another important operational technique.

In summary, 10 rendezvous exercises were accomplished during the Gemini Program, including 3 re-rendezvous and 1 dual operation (fig. 22-1) [not included]. Seven different rendezvous modes were utilized. These activities demonstrated the capabilities for computing rendezvous maneuvers in the ground-based computer complex; the use of the onboard radar-computer closed-loop system; the use of manual computations made by the flight crew; and the use of optical techniques and star background during the terminal phase and also in the event of equipment failures. A variety of lighting conditions and background conditions during the terminal-phase maneuvers, and the use of auxiliary lighting devices, have been investigated. The rendezvous operations demonstrated that the [331] computation and execution of maneuvers for changing or adjusting orbits in space can be performed with considerable precision.

The nine docking operations during Gemini demonstrated that the process can be accomplished in a routine manner, and that the ground training simulation was adequate for this operation (fig. 22-2) [not included]. The Gemini flight experience has established the proper lighting conditions for successful docking operations. Based on the data and experience derived from the Gemini rendezvous and docking operations, planning for the lunar orbit rendezvous can proceed with confidence.

Extravehicular Activity

Extravehicular activity was another important objective of the Gemini Program. Although extensive use of extravehicular activity has not been planned for the Apollo Program, the Gemini extravehicular experience should provide valuable information in two areas. First, extravehicular activity will be used as a contingency method of crew transfer from Lunar Module to the Command Module in the event the normal transfer mode cannot be accomplished. Second, operations on the lunar surface will be accomplished in a vacuum environment using auxiliary life-support equipment and consequently will be similar to Gemini extravehicular operations. For these applications, the results from Gemini have been used to determine the type of equipment and the crew training required. The requirements for auxiliary equipment such as handholds, tether points, and handrails have been established.

Controlled Landing

From the beginning of the Gemini Program, one of the objectives was to develop reentry flight-path and landing control. The spacecraft was designed with an offset center of gravity so that it would develop lift during the flight through the atmosphere. The spacecraft control system was used to orient the lift vector to provide maneuvering capability. A similar system concept is utilized by the Apollo spacecraft during reentry through the Earth atmosphere.

After initial development problems on the early Gemini flights, the control system worked very well in both the manual and the automatic control modes. Spacecraft landings were achieved varying from a few hundred yards to a few miles from the target point (fig. 22-3) [not included]. The first use of a blunt lifting body for reentry control serves to verify and to validate the Apollo-design concepts. The success of the Gemini guidance system in controlling reentry will support the Apollo design, even though the systems differ in detail.

Launch Operations

The prelaunch checkout and verification concept which was originated during the Gemini Program is being used for Apollo. The testing and servicing tasks are very similar for both spacecraft, and the Gemini test-flow plan developed at the Kennedy Space Center is being applied. The entire mode of operation involving scheduling, daily operational techniques, operational procedures, procedures manuals, and documentation is similar to that used in the Gemini operation. Much of the launch-site operational support is common to both programs; this includes tracking radars and cameras, communications equipment, telemetry, critical power, and photography. The requirements for this equipment are the same in many cases, and the Gemini experience is directly applicable. The Apollo Program will use the same mission operations organization for the launch sequence that was established during Project Mercury and tested and refined during the Gemini Program.

[332]

Mission Control

The Gemini mission-control operations concepts evolved from Project Mercury. These concepts were applied during the Gemini Program and will be developed further during the Apollo missions, although the complexity of the operations will substantially increase as the time for the lunar mission nears. The worldwide network of tracking stations was established to gather data concerning the status of the Mercury spacecraft and pilots. The Mercury flights, however, involved control of a single vehicle with no maneuvering capability.

The Gemini Program involved multiple vehicles, rendezvous maneuvers, and long-duration flights, and required a more complex ground-control system capable of processing and reacting to vast amounts of real-time data. The new mission-control facility at the Manned Spacecraft Center, Houston, was designed to operate in conjunction with the Manned Space Flight Network for direction and control of Gemini and Apollo missions, as well as of future manned space-flight programs. Much of this network capability was expanded for Gemini and is now being used to support the Apollo missions. Gemini has contributed personnel training in flight control and in maintenance and operation of flight-support systems. As the Gemini flights progressed and increased in complexity, the capabilities of the flight controllers increased, and resulted in a nucleus of qualified control personnel.

[333]

The development of experience teams of mission-planning personnel has proved extremely useful in the preparation for future manned missions. Mission plans and flight-crew procedures have been developed and exercised to perform the precise in-flight maneuvers required for rendezvous of two vehicles in space, and to perform flights up to 14 days in duration. The techniques which were evolved during Gemini have resulted in flight plans that provide the maximum probability of achieving mission objectives with a minimum usage of consumables and optimum crew activity. The development of satisfactory work-rest cycles and the acceptance of simultaneous sleep periods are examples of learning which will be carried forward to the Apollo planning. The mission planning procedures developed for Gemini are applicable to future programs, and the personnel who devised and implemented the procedures are applying their experience to the Apollo flight-planning effort.

Flight-Crew Operations and Training

Crew Capability

The results of the Gemini Program in the area of flight-crew operations have been very rewarding in yielding knowledge concerning the Gemini long-duration missions. The medical experiments conducted during these flights have demonstrated that man can function in space for the planned duration of the lunar landing mission. The primary question concerning the effect of long-duration weightlessness has been favorably answered. Adaptation to the peculiarities of the zero-g environment has been readily accomplished. The results significantly increase the confidence in the operational efficiency of the flight crew for the lunar mission.

The Apollo spacecraft is designed for cooperative operation by two or more pilots. Each module may be operated by one individual for short periods; however, a successful mission requires a cooperative effort by the three-man crew. The multiple-crew concept of spacecraft operation was introduced for the first time in the United States during the Gemini Program and cooperative procedures for multi-pilot operations were developed.

The Gemini Program has established that man can function normally and without ill effect outside the spacecraft during extravehicular operations.

Crew Equipment

Most of the Gemini technology regarding personal crew equipment is applicable to Apollo. The Block I Apollo space suit is basically the same as the Gemini space suit. The Block II Apollo space suit, although different in design, will have familiar Gemini items such as suit-design concepts, locking mechanisms for connectors, and polycarbonate visors and helmets. The Gemini spacesuit support facilities at the Manned Spacecraft Center and at the Kennedy Space Center, plus the ground-support equipment, will be fully utilized during Apollo.

A considerable amount of personal and postlanding survival equipment will be used for Apollo in the same configuration as was used for Gemini. Some items have minor modifications for compatibility, others for improvements based upon knowledge resulting from flight experience. Specific examples include food packaging, water dispenser, medical kits, personal hygiene items, watches, sunglasses, penlights, cameras, and data books.

Many of the concepts of crew equipment originated in Gemini experience with long-duration missions and recovery: food and waste management; cleanliness; housekeeping and general sanitation; and environmental conditions such as temperature, radiation, vibration, and acceleration. Although the Apollo approach may differ in many areas, the Gemini experience has been the guide.

Flight-Crew Training

The aspects of crew training important to future programs include preflight preparation of the crews for the mission and the reservoir of flight experience derived from the Gemini Program. Apollo will inherit the training technology developed for the Gemini flight crews. The technology began with Project Mercury, and was developed and refined during the training of the Gemini multi-man crews. There now exists an organization of highly skilled specialists with a thorough understanding of the training task. Adequate crew preparation can be assured in all areas, from the physical conditioning of the individual crewmembers to the complicated integrated mission simulation.

One highly developed aspect of flight-crew training is the use of simulators and simulation techniques. A significant result of the Gemini rendezvous experience was the verification of the ground simulation employed in flight-crew training. The incorporation of optical displays in the Gemini simulations was an important step in improving the training value of these devices. Using high-fidelity mission simulators to represent the spacecraft and to work with the ground control network and flight controllers was instrumental in training the pilots and ground crew as a functional team that could deal with problems and achieve a large percentage of the mission objectives.

The Gemini Program resulted in an accumulated total of 1940 man-hours of flight time distributed among 16 flight-crew members. This flight experience is readily adaptable to future programs since the Gemini pilots are flight qualified for long-duration flights with rendezvous operations, and are familiar with many of the aspects of working in the close confines of the spacecraft. This experience is of great value to future training programs. The experience in preparing multi-man crews for flight, in monitoring the crew during flight, and in examining and debriefing after flight will facilitate effective and efficient procedures for Apollo.

Technological Development of Systems and Components

Gemini and Apollo share common hardware items in some subsystems; in other subsystems, the similarity exists in concept and general design. The performance of Gemini systems, operating over a range of conditions, has provided flight-test data for the verification of the design of related subsystems. These data are important since many elements of Apollo, especially systems interactions, cannot be completely simulated in ground testing. The Apollo Spacecraft Program

Office at the Manned Spacecraft Center, Houston, has reviewed and analyzed Gemini anomalous conditions to determine corrective measures applicable to Apollo. The Apollo Program Director has established additional procedures at NASA Headquarters to promote rapid dissemination and application of Gemini experience to Apollo equipment design.

The Gemini missions have provided background experience in many systems such as communications, guidance and navigation, fuel cells, and propulsion. In addition, a series of experiments was performed specifically for obtaining general support information applicable to the Apollo Program.

In the communications systems, common items include the recovery and flashing-light beacons; similar components are utilized in the high-frequency recovery antennas. Reentry and post landing batteries and the digital data uplink have the same design concepts. The major Apollo design parameters concerned with power requirements and range capability have been confirmed.

In the area of guidance and navigation, the use of an onboard computer has been demonstrated and the Gemini experience with rendezvous radar techniques has been a factor in the selection of this capability for the Lunar Module. The ability to perform in-plane and out-of-plane maneuvers and to determine new space references for successful reentry and landing has been confirmed by Gemini flights. The control of a blunt lifting body during reentry will also support the Apollo concept.

In the electrical power supply, the use of the Gemini fuel cell has confirmed the applicability [335] of the concept. The ability of the cryogenic reactant storage system to operate over a wide range of off-design conditions in flight has verified the design, which is similar for Apollo. The performance of the Gemini system has provided a better understanding of the system parameters over an operating range considerably in excess of the range previously contemplated. The design of the cryogenic servicing system for Apollo was altered after the initial difficulties experienced by early Gemini flights. Consequently, a fairly sophisticated system now exists which will eliminate the possibility of delays in servicing. The ability to estimate the power requirements for the Apollo spacecraft equipment is enhanced by the Gemini operational data.

In the propulsion area, the ullage control rockets of the Apollo-Saturn S-IVB stage are the same configuration as the thrusters used for the Gemini spacecraft Orbital Attitude and Maneuver System; the thrusters of the Apollo Command Module Reaction Control System are similar. Steps have been taken to eliminate the problems which occurred in the development of the Gemini thrusters, such as the cracking of the silicon-carbide throat inserts, the unsymmetrical erosion of the chamber liners, and the chamber burn-through. The tankage of the Reaction Control System is based upon the Gemini design, and employs the same materials for tanks and bladders. The propellant control valves were also reworked as a result of early problems in the Gemini system.

The Lunar Module ascent engine also benefited from the Gemini technology; the contractor for this engine also manufactured the engines for the

Gemini Agena Target Vehicle. Following the in-flight failure of the target-vehicle engine during the Gemini VI mission, a test program verified the inherent danger in fuel-lead starts in the space environment. Consequently, the Lunar Module ascent engine and the Gemini target-vehicle engine were changed so that the oxidizer would enter the engine before the fuel. The problem had been indicated during ascent-engine testing, but was not isolated until the required definitive data were furnished by Project Sure Fire on the target-vehicle engine.

In addition to medical experiments, several other types of experiments were conducted during Gemini and have supplied information and data for use by the Apollo Program. The experiments included electrostatic charge, proton-electron spectrometer, lunar ultra-violet spectrometer, color-patch photography, landmark contrast measurements, radiation in spacecraft, reentry communications, manual navigation sightings, simple navigation, radiation and zero-g effects on blood, and micrometeorite collection. Although the direct effects of these experiments on Apollo systems are difficult to isolate, the general store of background data and available information has been increased.

Concluding Remarks

The Gemini Program has made significant contributions to future manned space-flight programs. Some of the more important contributions include flight-operations techniques and operational concepts, flight-crew operations and training, and technological development of components and systems. In the Gemini Program, the rendezvous and docking processes so necessary to the lunar mission were investigated; workable procedures were developed, and are available for operational use. The capability of man to function in the weightless environment of space was investigated for periods up to 14 days. Flight crews have been trained, and have demonstrated that they can perform complicated mechanical and mental tasks with precision while adapting to the spacecraft environment and physical constraints during long-duration missions.

Additionally, the development of Gemini hardware and techniques has advanced spacecraft-design practices and has demonstrated advanced systems which, in many cases, will substantiate approaches and concepts for future spacecraft.

[336]

Finally, probably the most significant contributions of Gemini have been the training, personnel and organizations in the disciplines of management, operations, manufacturing, and engineering. The nucleus of experience has been disseminated throughout the many facets of Apollo and will benefit all future manned space-flight programs.

[337]

23. CONCLUDING REMARKS

By George M. Low, *Deputy Director, NASA Manned Spacecraft Center*

With the preceding paper, one of the most successful programs in our short history of space flight has ended. The Gemini achievements have been many, and have included long-duration flight, maneuvers in space, rendezvous, docking, use of large engines in space, extravehicular activity, and controlled reentry. The Gemini achievements have also included a host of medical, technological, and scientific experiments.

The papers have included discussions of many individual difficulties that were experienced in preparation for many of the flight missions and in some of the flights. The successful demonstration that these difficulties were overcome in later missions is a great tribute to the program, to the organization, and to the entire Gemini team.

A period of difficulty exists today in the program that follows Gemini, the Apollo Program. Yet, perhaps one of the most important legacies from Gemini to the Apollo Program and to future programs is the demonstration that great successes can be achieved in spite of serious difficulties along the way.

The Gemini Program is now officially completed.

Chapter Two

Project Apollo: Americans to the Moon

John M. Logsdon

Project Apollo, the remarkable U.S. space effort that sent 12 astronauts to the surface of Earth's Moon between July 1969 and December 1972, has been extensively chronicled and analyzed.[1] This essay will not attempt to add to this extensive body of literature. Its ambition is much more modest: to provide a coherent narrative within which to place the various documents included in this compendium. In this narrative, key decisions along the path to the Moon will be given particular attention.

---

1. Roger Launius, in his essay "Interpreting the Moon Landings: Project Apollo and the Historians," *History and Technology*, Vol. 22, No. 3 (September 2006): 225–55, has provided a comprehensive and thoughtful overview of many of the books written about Apollo. The bibliography accompanying this essay includes almost every book-length study of Apollo and also lists a number of articles and essays interpreting the feat. Among the books Launius singles out for particular attention are: John M. Logsdon, *The Decision to Go to the Moon: Project Apollo and the National Interest* (Cambridge: MIT Press, 1970); Walter A. McDougall, *. . . the Heavens and the Earth: A Political History of the Space Age* (New York: Basic Books, 1985); Vernon Van Dyke, *Pride and Power: the Rationale of the Space Program* (Urbana, IL: University of Illinois Press, 1964); W. Henry Lambright, *Powering Apollo: James E. Webb of NASA* (Baltimore: Johns Hopkins University Press, 1995); Roger E. Bilstein, *Stages to Saturn: A Technological History of the Apollo/Saturn Launch Vehicles*, NASA SP-4206 (Washington, DC: Government Printing Office, 1980); Edgar M. Cortright, *Apollo Expeditions to the Moon*, NASA SP-350 (Washington, DC: Government Printing Office, 1975); Charles A. Murray and Catherine Bly Cox, *Apollo: The Race to the Moon* (New York: Simon and Schuster, 1989); Stephen B. Johnson, *The Secret of Apollo: Systems Management in American and European Space Programs* (Baltimore: Johns Hopkins University Press, 2002); Norman Mailer, *Of a Fire on the Moon* (Boston: Little, Brown, 1970); Michael Collins, *Carrying the Fire: An Astronaut's Journeys* (New York: Farrar, Straus, and Giroux, 1974); Andrew Chaikin, *A Man on the Moon: The Voyages of the Apollo Astronauts* (New York: Viking, 1994); W. David Compton, *Where No Man Has Gone Before: A History of Apollo Lunar Exploration Missions*, NASA SP-4214 (Washington, DC: Government Printing Office, 1989); Don E. Wilhelms, *To A Rocky Moon: A Geologist's History of Lunar Exploration* (Tucson: University of Arizona Press, 1993); Donald A. Beattie, *Taking Science to the Moon: Lunar Experiments and the Apollo Program* (Baltimore: Johns Hopkins University Press, 2001); Howard McCurdy, *Space and the American Imagination* (Washington, DC: Smithsonian Institution Press, 1997); Marina Benjamin, *Rocket Dreams: How the Space Age Shaped Our Vision of a World Beyond* (New York: Free Press, 2003); De Witt Douglas Kilgore, *Astrofuturism: Science, Race, and Visions of Utopia in Space* (Philadelphia: University of Pennsylvania Press, 2003); and Andrew Smith, *Moondust: In Search of the Men Who Fell to Earth* (New York: Fourth Estate, 2005). In addition to these accounts, a number of Apollo astronauts, NASA managers and flight operations personnel, and managers from the aerospace industry have published memoirs about their engagement with Apollo. Of particular interest is Robert C. Seamans, Jr., *Aiming at Targets* (Washington, DC: National Aeronautics and Space Administration Special Publication-4106, 1996), and *Project Apollo: The Tough Decisions*, NASA, Monographs in Aerospace History No. 37, SP-2005-4537, 2005, and Glen E. Swanson, *"Before This Decade is Out . . .: Personal Reflections on the Apollo Program* (Washington, DC: National Aeronautics and Space Administration Special Publication-4223, 1999).

Origins of Apollo

When it began operations on 1 October 1958, NASA had already been tasked by the Eisenhower administration with the initial U.S. human space flight effort, soon to be designated Project Mercury. NASA also inherited a number of robotic missions that had been planned by various elements of the Department of Defense (DOD) and was given an agenda of desired missions by the Space Science Board of the National Academy of Sciences. NASA spent much of 1959 integrating these missions into a Long-Range Plan; to do so, it also recognized the need to identify its long-range goals for human space flight and the steps needed to achieve those goals. To undertake this task, in the spring of 1959 NASA created a Research Steering Committee on Manned Space Flight. This committee was chaired by Harry Goett, then of NASA's Ames Research Center but soon to become the Director of the new Goddard Space Flight Center. The committee held its first meeting on 25 and 26 May 1959. Its members included senior representatives from the NASA Field Centers and the Agency's Washington Headquarters.

At this meeting, Bruce Lundin from the Lewis Research Center argued that "the ultimate objective is manned interplanetary travel and our present goal should be for a manned lunar landing and return." Engineer and spacecraft designer Maxime Faget of the Space Task Group of the Langley Research Center "endorsed selecting lunar exploration as the present goal of the committee although the end objective should be manned interplanetary travel." George M. Low, then in charge of human space flight at NASA headquarters, suggested that the committee adopt the lunar landing mission as NASA's present long-range objective with proper emphasis on intermediate steps "because this approach will be easier to sell." Others at the meeting suggested a more modest objective, human flight around the Moon without a landing attempt, be adopted as NASA's stated goal. (II-1)

There was no agreement at this point, but by the committee's next meeting in late June, after George Low had lobbied the group, the committee decided that indeed a lunar landing should be selected as the long-range goal for human space flight, with an orbiting space station and circumlunar flight as intermediate steps. The NASA Long-Range Plan, published in December 1959, thus identified as objectives for the 1965 to 1967 time period the first launches "in a program leading to manned circumlunar flight and to [a] permanent near-earth space station." The objective of "manned flight to the moon" was identified, but only in the "beyond 1970" period (Volume I, III-2). While Low and some of his associates would have preferred a faster-paced effort, at least NASA, after only 15 months of operation, was on record as intending to head to the Moon, if only they could get the White House and Congress to agree.

In mid-1960, NASA's thinking about the intermediate steps in human space flight had matured to the point that the space agency called together representatives of the emerging space industry to share that thinking. At a "NASA-Industry Program Plans Conference" held in Washington on 28 and 29 July 1960, George Low told the audience "at this point it should be stated that official approval of this program has not been obtained. Rather, this presentation includes what we now believe to be a rational and reasonable approach to a long-range development program leading to the manned exploration of outer space." He added "our present planning calls for the development and construction of an advanced manned spacecraft with sufficient flexibility to be capable of both circumlunar flight and useful Earth-orbital missions. In the long range, this spacecraft should lead toward manned landings on the moon and planets, and toward a permanent manned space station. This advanced manned space flight program has been named 'Project Apollo.'" (II-2)

The name Apollo had been suggested by Low's boss, NASA's Director for Space Flight Programs, Abe Silverstein, in early 1960. Silverstein had also chosen the name for Project Mercury, and he wanted to establish a tradition of naming NASA's projects after Greek gods.[2]

NASA, and particularly George Low, in the second half of 1960 continued to move forward in planning Apollo and the lunar landing mission that was its long-term goal. On 17 October, he informed Silverstein "it has become increasingly apparent that a preliminary program for manned lunar landings should be formulated. This is necessary in order to provide a proper justification for Apollo, and to place Apollo schedules and technical plans on a firmer foundation." To undertake this planning, Low formed a small working group of NASA Headquarters staff. (II-3)

That NASA was planning advanced human spaceflight missions, including one to land people on the Moon, soon came to the attention of President Eisenhower and his advisors as NASA submitted a budget request that included funds for industry studies of the Apollo spacecraft. This request was not approved, and the president asked his science advisor, Harvard chemist George Kistiakowsky, to organize a study of NASA's plans by the President's Science Advisory Committee. To carry out such a study, Kistiakowsky established an "Ad Hoc Committee on Man-in-Space" chaired by Brown University professor Donald Hornig. The Hornig Committee issued its report on 16 December 1960. The report called Project Mercury a "somewhat marginal effort," and noted "among the reasons for attempting the manned exploration of space are emotional compulsions and national aspirations. These are not subjects which can be discussed on technical grounds." The Committee estimated the cost of Project Apollo at $8 billion, and suggested that a program to land humans on the Moon would cost an additional $26 to 38 billion. (Volume I, III-3) When President Eisenhower was briefed on the report, he found these projected costs well beyond what he thought reasonable. When a comparison was made to Queen Isabella's willingness to finance the voyages of Christopher Columbus, Eisenhower replied that "he was not about to hock his jewels" to send men to the Moon.[3]

George Low's working group on a manned lunar landing presented its interim findings to a meeting of NASA's Space Exploration Program Council in

---

2. Charles A. Murray and Catherine Bly Cox, *Apollo: The Race to the Moon* (New York: Simon and Schuster, 1989), pp. 54–55.

3. John M. Logsdon, *The Decision to Go to the Moon: Project Apollo and the National Interest* (Cambridge: MIT Press, 1970), pp. 34–35.

early January 1961; the council decided that Low should continue his planning effort. However, outgoing NASA Administrator T. Keith Glennan reminded Low that such a program would require presidential approval, and that approval had not been forthcoming. Indeed, as President Dwight D. Eisenhower left office on 20 January 1961, the future of NASA's program of human spaceflight was extremely uncertain. There were no funds in the President's final budget proposal to support Project Apollo, and it was known that the incoming President, John F. Kennedy, was receiving advice skeptical of the value of launching humans into space. There certainly was no sense that Kennedy would, within four months, decide to send Americans to the Moon.

The Decision to Go to the Moon[4]

As he entered the White House, President Kennedy was aware that he would be faced with decisions that would shape the future of U.S. space efforts. One of his top advisors during the period between the election and his taking office, Harvard professor Richard Neustadt, told Kennedy in December 1960 that the United States had been in a race for dramatic space achievements, a race that the Soviet Union was winning because of their superior space launch capability. Neustadt asked "if we are behind and are likely to stay behind in the race for 'Sputnik-type firsts,' should we get out of the race and divert the resources now tied up in it to other uses which have tangible military, scientific or welfare value?" Neustadt was skeptical of the value of the Saturn rockets,[5] which he noted were needed "only in order to put a man on the moon" before Russia, but he did support the development of a very large rocket motor (the F-1). He asked Kennedy "in the longer run, what proportion of government resources, for what span of years, should go into developing the technology of space travel?" (Volume I, III-4)

Kennedy also appointed during the transition an "Ad Hoc Committee on Space," which was chaired by the man who would become his science advisor, MIT Professor Jerome Wiesner. This committee recognized that "manned exploration of space will certainly come to pass and we believe that the United States must play a vigorous role in this venture," but that "because of our lag in the development of large boosters, it is very unlikely that we shall be first in placing a man into orbit." However, the committee believed that too much emphasis had been placed on Project Mercury in comparison to its actual scientific and technological payoffs, and recommended that "we should stop advertising MERCURY as our major objective in space activities. . . . We should find effective means to make people appreciate the cultural, public service, and military importance of space activities other than [human] space travel." (Volume I, III-5)

---

4. Most of the account of this decision is taken from Logsdon, *The Decision to Go to the Moon.* This study of the Kennedy decision, published in 1970, remains the accepted version of the events leading to Kennedy's 25 May 1961 announcement that "we should go to the moon."

5. At this point, Saturn was the name of the Wernher von Braun-led program to develop a larger booster than anything the United States was otherwise planning, but still far short in lifting power of what was ultimately developed as the Saturn V for the lunar landing program.

In his Inaugural Address, delivered on a wintry Washington afternoon, President John F. Kennedy suggested to the leaders of the Soviet Union that "together let us explore the stars."[6] In his initial thinking about space policy, Kennedy favored using space activities as a way of increasing the peaceful interactions between the United States and its Cold War adversary. Soon after he came to the White House, Kennedy directed his science advisor to undertake an intensive review to identify areas of potential U.S.-Soviet space cooperation, and that review continued for the first three months of the Kennedy administration, only to be overtaken by the need to respond to the Soviet launch of Yuri Gagarin on 12 April. Soviet-U.S. cooperation in space was a theme that Kennedy was to return to in subsequent years.

A first order of business was to select someone to head NASA. After a number of candidates indicated that they were not interested in the position, on the advice of his Vice President, Lyndon B. Johnson, powerful Oklahoma Senator Robert Kerr, the chairman of the Senate Committee on Aeronautical and Space Sciences, and his science advisor Jerome Wiesner, Kennedy turned to James E. Webb on 31 January. The NASA position was one of the last top-level jobs to be filled by the new administration. Webb was from North Carolina, trained as a lawyer and veteran of both congressional staff and senior executive branch positions during the Truman administration, and had business experience working for one of Kerr's companies in Oklahoma.[7] Webb agreed to take the NASA job, but only after meeting with the President, who told Webb that he wanted "someone who understands policy. This program involves great issues of national and international policy." Webb got assurances from the President that respected scientist and manager Hugh Dryden would be allowed to stay on as NASA's Deputy Administrator. Webb also decided to retain Associate Administrator Robert Seamans, who served as the Agency's general manager. Seamans was a Republican, and Webb wanted to present NASA as not being influenced by partisan politics. Webb was sworn in as NASA Administrator on 14 February.[8]

John Kennedy's closest advisor, Theodore Sorenson, was later to comment that "Webb was not what we would call a Kennedy-type individual. He was inclined to talk at great length, and the President preferred those who were more concise in their remarks. He was inclined to be rather vague, somewhat disorganized in his approach to a problem, and the President preferred those who were more precise." However, according to Sorenson, "I don't know that the President ever regretted his appointment of Webb, or wished that he had named someone else." (II-43)

Once Webb arrived at NASA, a first task was to review the Agency's proposed budget for FY 1962 that had been prepared by the outgoing Eisenhower

---

6. *Public Papers of The Presidents of the United States: John F. Kennedy, 1961* (Washington, DC: Government Printing Office, 1962), p. 2.

7. For a perceptive biography of James E. Webb, see W. Henry Lambright, *Powering Apollo: James E. Webb of NASA* (Baltimore: Johns Hopkins University Press, 1995).

8. W. Henry Lambright, *Powering Apollo: James E. Webb of NASA* (Baltimore: The Johns Hopkins Press, 1995), pp. 82–87.

administration. In doing so, Webb and his associates came to the conclusion that NASA's planning had been too conservative, and that the milestones included in the Agency's 10-year plan should be accelerated. One input into this conclusion was the 7 February final report of Low's working group, which concluded that "the present state of knowledge is such that no invention or breakthrough is believed to be required to insure the overall feasibility of safe manned lunar flight," that "manned landings on the moon . . . could be made in the 1968–1971 time period," and that it would be possible to carry out a lunar landing program for a total cost of $7 billion. (II-4)

Based on this and other analyses, NASA requested a 30 percent increase in its FY 1962 budget over what had been proposed by President Eisenhower. The Bureau of the Budget reacted negatively to such a large increase, and on 22 March 1961 Webb, Dryden, and Seamans met with President Kennedy and his staff to discuss how best to proceed. At that meeting, NASA noted that President Eisenhower had eliminated from the NASA budget all funds related to human flight after Project Mercury, including the Apollo spacecraft and heavier lift boosters and rocket motors. Webb told the President that "the Soviets have demonstrated how effective space exploration can be as a symbol of scientific progress and as an adjunct of foreign policy. . . . We cannot regain the prestige we have lost without improving our present inferior booster capability."

At this point Kennedy had not made up his own mind about the future of human space flight, and so he was unwilling to approve NASA's request to restore funds for the Apollo spacecraft; the sense is that decisions on this issue would come during the preparation of the FY 1963 NASA budget at the end of 1961. Support for the importance of human spaceflight, as the President deliberated on its future, came from the Space Sciences Board of the National Academy of Sciences. The chairman of that board, Lloyd Berkner, was a longtime friend of James Webb, and on 31 March he sent Webb and Kennedy's science advisor Jerome Wiesner a letter reporting that the board had agreed that "from a scientific standpoint, there seems little room for dissent that man's participation in the exploration of the Moon and planets will be essential, if and when it becomes technologically feasible to include him." (II-5)

Kennedy and his advisors did agree that the United States, for a variety of reasons, needed to approve its space lift capabilities, and so he approved an additional $114 million for launch vehicle development. There matters were planned to rest until NASA was successful in its initial flights of Project Mercury, planned for later in 1961, and it came time to formulate the NASA budget for FY 1963.

Events forced the President's hand much earlier than he had anticipated. In the early morning hours of 12 April, word reached the White House that the Soviet Union had successfully orbited its first cosmonaut, Yuri Gagarin, and that he had safely returned to Earth. The Soviet Union was quick to capitalize on the propaganda impact of the Gagarin flight; Nikita Khrushchev boasted, "Let the capitalist countries catch up with our country!" In the United States, both the public and Congress demanded a response to the Soviet achievement.

President Kennedy called a meeting of his advisors for the late afternoon of 14 April to discuss what that response might be. Kennedy also agreed to an interview the same afternoon with Hugh Sidey, a top reporter for *Life* and *Time* magazines and someone on friendly terms with the President (as were many journalists). In preparation for that interview, Sidey prepared a set of questions and transmitted them to Presidential Press Secretary Pierre Salinger. Wiesner then prepared a background memorandum for the President's use in responding to Sidey. (II-6, II-7)

Rather than meet separately with Sidey, the President decided to let him join the meeting with Webb, Dryden and Kennedy's top advisors; Sidey later described the meeting in a book about Kennedy. Dryden told the President that catching up with the Russians might require a crash program on the order of the Manhattan Project that developed the atomic bomb; such an effort might cost as much as $40 billion. After hearing the discussions of what might be done, according to Sidey, Kennedy's response was "when we know more, I can decide if it's worth it or not. If someone can just tell me how to catch up. . . . There's nothing more important."[9]

While Kennedy considered his course of action, other events reinforced his need to get something positive in place. On the morning of 17 April, Central Intelligence Agency-trained Cubans landed at the Bay of Pigs in Cuba in an attempt to foment an uprising that would result in forcing Fidel Castro to give up his leadership position. During the following two days, Kennedy and his advisors decided not to offer U.S. military support to this failing invasion; as a result, the United States looked weak and vacillating to much of the rest of the world.

Kennedy had decided in December to give his Vice President, Lyndon Johnson, lead responsibility for advising him on space as the Chairman of the existing National Aeronautics and Space Council. That council had been set up as part of the 1958 Space Act, with the President as Chair. Thus legislative action was needed to give the chairmanship to the Vice President. The President signed the legislation making this change on 20 April, and on that same day wrote a historic memorandum to the Vice President, asking him "as Chairman of the Space Council to be in charge of making an overall survey of where we stand in space." In particular, Kennedy asked, "Do we have a chance of beating the Soviets by putting a laboratory in space, or by a trip around the Moon, or by a rocket to land on the Moon, or by a rocket to go to the Moon and back with a man? Is there any other space program which promises dramatic results in which we could win?" (II-8)

Vice President Johnson quickly organized the review that the President requested. On 21 April, he received a first input from the Department of Defense, which suggested that "dramatic achievements in space . . . symbolize the technological power and organizing capability of a nation" and "major achievements in space contribute to national prestige." (Volume I, III-7) NASA's response came a day later; the space agency told the President that

> There is a chance for the U.S. to be the first to land a man on the Moon
> and return him to Earth if a determined national effort is made. . . . It

---

9. Hugh Sidey, *Kennedy, President* (New York: Scribner, 1963), pp. 121–123.

is doubtful that the Russians have a very great head start on the U.S. in the effort required for a manned lunar landing. Because of the distinct superiority of U.S. industrial capacity, engineering, and scientific know-how, we believe that with the necessary national effort, the U.S. may be able to overcome the lead that the Russians might have up to now.

NASA added "a possible target date for the earliest attempt for a manned lunar landing is 1967, with an accelerated U.S. effort." NASA told the Vice President that the cost to carry out the overall NASA 10-year plan at a pace that would allow a first attempt at a lunar landing in 1967 would be $33.7 billion through 1970. (II-9)

Lyndon Johnson consulted not only government agencies, but also individuals whom he respected, as he carried out his review. One of those individuals was Wernher von Braun, who told Johnson "we have an excellent chance of beating the Soviets to the *first landing of a crew on the moon* (including return capability, of course) [emphasis in original]." He added, "The reason is that a performance jump by a factor 10 over their present rockets is necessary to accomplish this feat. While today we do not have such a rocket, it is unlikely that the Soviets have it. Therefore, we would not have to enter the race toward this obvious next goal in space exploration against hopeless odds favoring the Soviets." Von Braun suggested "with an all-out crash program I think we could accomplish this objective in 1967/68." (II-10)

By 28 April, Johnson could report to the President that "the U.S. can, if it will, firm up its objectives with a reasonable chance of attaining world leadership in space during this decade." In particular, he added, "manned exploration of the moon, for example, is not only an achievement with great propaganda value, but it is essential as an objective whether or not we are first in its accomplishment—and we may be able to be first." (Volume I, III-8)

Johnson continued his review, consulting with leading members of Congress. (Volume I, III-10) The review took place as NASA was preparing to launch the first suborbital flight in Project Mercury, and there was debate within the White House regarding whether to televise the event live, given the chance of a catastrophic failure. The decision was made to do so, and on 5 May Alan Shepard became the first American to enter space on a 15-minute journey. During the same week, President Kennedy asked Johnson to travel to Southeast Asia to get a sense of the situation there and whether direct U.S. military intervention was required. Johnson wanted to get his final recommendations on space to the President before he left Washington on Monday, 8 May; this meant that those preparing the basis for those recommendations would have to work over the weekend.

By the morning of 8 May, James Webb and Secretary of Defense Robert McNamara signed a report titled "Recommendations for Our National Space Program: Changes, Policies, Goals." They transmitted the report to the Vice President, saying "this document represents our joint thinking. We recommend that, if you concur with its contents and recommendations, it be transmitted to the President for his information and as a basis for early adoption and implementation of the revised and expanded objectives which it contains." Johnson later that day did deliver the report to the President, without modification and with his concurrence; incidentally 8 May was the day on which Alan Shepard came to Washington to celebrate the success of his Mercury mission.

The Webb-McNamara report called for an across-the-board acceleration of the U.S. space effort aimed at seeking leadership in all areas, not only dramatic space achievements. As its centerpiece, the report recommended

> our National Space Plan include the objective of manned lunar exploration before the end of this decade. It is our belief that manned exploration to the vicinity of and on the surface of the moon represents a major area in which international competition for achievement in space will be conducted. The orbiting of machines is not the same as the orbiting or landing of man. It is man, not merely machines, in space that captures the imagination of the world.

A very expensive undertaking such as sending humans to the Moon was justified, according to Webb and McNamara, because "*this nation needs to make a positive decision to pursue space projects aimed at enhancing national prestige* [emphasis in original]. Our attainments are a major element in the international competition between the Soviet system and our own. The nonmilitary, noncommercial, nonscientific but 'civilian' projects such as lunar and planetary exploration are, in this sense, part of the battle along the fluid front of the cold war." (II-11)

After a quick review of the report's recommendations by the White House staff, Kennedy approved them. He announced his decisions at the end of an address to a joint session of Congress on 25 May 1961. He told the assemblage, and the nation, "I believe that this nation should commit itself to achieving the goal, before this decade is out, of landing a man on the moon and returning him safely to earth." (Volume I, III-12)

Congress quickly and without significant opposition approved the $549 million addition to NASA's FY 1962 budget that was needed to get started on the accelerated program; this amount when added to the increase already approved in March represented an 89 percent increase of the previous year's budget. With this initial approval in hand, NASA could begin to implement Project Apollo.

Getting Started

Locating the Facilities

It was clear from the start of planning for Apollo that NASA would need a major new installation to manage the effort and new facilities for launching the Apollo missions. Prior to the Apollo decision, NASA had planned to move the Space Task Group, which was managing Project Mercury from its base at Langley Research Center in Hampton, Virginia, to the Goddard Space Flight Center in Greenbelt, Maryland. The thinking was that all NASA missions, human and robotic, could be managed by a single Field Center. But a project of the scope of Apollo would overwhelm other activities at Goddard, and there was high political interest in creating a new NASA center for Apollo. This meant that Governors, Congressmen and Senators, and business representatives from a number of locations around the United States pressured NASA to consider locating the new Center in their area. In response, NASA set up a series of criteria that the new facility would have to meet, and a site survey team visited 23 potential locations. In particular, the Massachusetts political establishment put pressure on the President to consider a location in his home state, even though the proposed site did not meet all NASA's criteria, especially a climate that would permit year-round outdoor operations. (II-14)

On 19 September 1961, NASA announced that a new Manned Spacecraft Center would be located "in Houston, Texas, on a thousand acres to be made available to the government by Rice University."[10] This decision may well have been preordained. Even before President Kennedy announced his decision to go to the Moon, on 23 May, James Webb had written a memorandum to Lyndon Johnson on his return from his inspection trip to Southeast Asia to bring the Vice President up to date on what had happened in the two weeks he had been away from Washington. Webb noted that he had had several interactions with Representative Albert Thomas of Houston, who chaired the House appropriations subcommittee controlling NASA's budget, and that "Thomas has made it very clear that he and George Brown were extremely interested in having Rice University make a real contribution" to the accelerated space effort. (Brown was head of the Houston-based construction company Brown & Root and a major political ally of Lyndon Johnson. Brown had been one of the outsiders consulted by Johnson in April as the space review was underway). (Volume II, III-7) Given the influence of Thomas over the NASA budget and the political links between Johnson and Brown, it would have been difficult to choose another location for the new Center.

It was also clear to NASA that it would need to build new launch facilities for the large boosters needed for Apollo. At the time of the decision to go to the Moon, NASA was already developing the Saturn 1 rocket, with first-stage thrust of 1.5 million pounds coming from a cluster of eight H-1 rocket engines, but it would not have sufficient power to launch human missions to the Moon. NASA in March had gotten White House permission to develop a more powerful Saturn 2 vehicle that added a second stage powered by engines using liquid hydrogen as their fuel. At the start of planning for lunar landing missions, NASA's thinking focused on a new, very large launch vehicle called Nova, which would cluster eight F-1 rocket engines, each with 1.5 million pounds of takeoff thrust, as a means of carrying a spacecraft directly to the lunar surface. As NASA planning moved forward during 1961 (this process is discussed below), variations of an advanced Saturn vehicle, using three, four, and ultimately five F-1 engines in its first stage were considered. While a Saturn 1 or Saturn 2 (which never got beyond the preliminary design stage) could be launched from an existing launch pad on

---

10. Henry C. Dethloff, *Suddenly, Tomorrow Came . . .: A History of the Johnson Space Center* (Washington, DC: National Aeronautics and Space Administration Special Publication-4307, 1993), p. 40.

the Air Force-controlled Atlantic Missile Range at Cape Canaveral, Florida, that range could not accommodate the larger advanced Saturn or Nova boosters.

As a lunar landing decision appeared more and more likely in late April 1961, NASA Associate Administrator Robert Seamans had directed Kurt Debus, a von Braun associate who was in charge of NASA's launch operations at the Atlantic Missile Range, to begin to search for a site to launch much larger boosters. By August, Debus and his associates had examined eight possible locations, including three outside of the continental U.S., and had concluded that Merritt Island, Florida, adjacent to Cape Canaveral, was the preferred site, with White Sands, New Mexico as second choice.[11] On 1 September NASA announced its intention to purchase 125 square miles of property on Merritt Island; on 24 August 1961, NASA and the Department of Defense had signed an interim agreement on the relationship between what was called the Merritt Island Launch Area and the Atlantic Missile Range; that agreement was replaced by a more permanent agreement in January 1963. Anticipating a high launch rate for Saturn vehicles, NASA in 1962 decided to build what was to be called Launch Complex 39; the complex included a huge vertical assembly building where the launch vehicles would be assembled and checked out before being transported to one of two launch pads, designated 39A and 39B. While NASA's launch operations at Cape Canaveral had previously been managed by a division of the Marshall Space Flight Center in Huntsville, for Apollo NASA decided to create a separate Launch Operations Center reporting to NASA Headquarters and named Debus to head the facility.[12] (Volume IV, I-41 and I-42)

In addition to new launch facilities, NASA also needed a site for the assembly of the large first stage of the Saturn 1 and advanced Saturn vehicles. NASA selected a former ship, airplane, and tank factory located in the outskirts of New Orleans, Louisiana. The land had been granted by French King Louis XV in 1763 to a wealthy but eccentric recluse and junk dealer named Antoine Michoud for use as a plantation, and the plant built on the site almost two hundred years later was named after him. For testing the powerful F-1 engine, after considering 34 sites, NASA chose an isolated location in Hancock County, Mississippi, and christened it the Mississippi Test Facility. Site selection for both facilities was subject to political maneuvering as well as technical criteria.[13]

Finally, NASA also had to decide where to locate the control center to manage the Apollo missions once they were underway. The mission control center for Project Mercury was located at Cape Canaveral in Florida, and there was some thought of placing the Apollo control room there. By mid-1962, however, NASA decided that a new Mission Control Center should be created as part of the Manned Spacecraft Center in Houston. (II-25)

---

11. Murray and Cox, *Apollo*, pp. 88–89.

12. Ibid, pp. 90-99. See also Charles D. Benson and William Barnaby Faherty, *Moonport: A History of Apollo Launch Facilities and Operations* (Washington, DC: National Aeronautics and Space Administration Special Publication-4204, 1978).

13. For more on this selection, see Roger Bilstein, *Stages to Saturn: A Technological History of the Apollo/Saturn Launch Vehicles* (Washington, DC: National Aeronautics and Space Administration Special Publication-4208, 1980).

Building the Spacecraft

Meantime, NASA had been thinking about an advanced spacecraft called Apollo since at least 1960. Thus the organization was quickly able to initiate the procurement of the vehicle, even before it was known exactly how it would be used for the lunar landing mission. By 28 July 1961, NASA had an approved procurement plan in place; 12 firms were identified as potential bidders. (II-13) Ultimately, only five bids for the contract were submitted. The competition for the Apollo spacecraft contract took place over the following four months; on 28 November, NASA announced that North American Aviation had been selected to build the vehicle. This turned out to be a controversial decision, particularly after problems with North America's performance became known and it was discovered that the NASA Source Evaluation Board had identified the Martin Company as its preferred choice, with North American Aviation as a "desirable alternative."[14] (II-20)

Selecting the Launch Vehicle

While the basic elements of the Apollo spacecraft, with a three-person crew and two elements, (one housing the crew and the command center for the vehicle and the other housing propulsion and other systems) had been fixed since 1960, it took NASA until the end of 1961 to select the launch vehicle for the Apollo missions to the Moon. There were two reasons for this. One was that the "national space plan" contained in the 8 May Webb-McNamara memorandum had called for a collaborative NASA-Department of Defense effort to define a family of launch vehicles that could meet both agencies' requirements and advance the development of both liquid fuel and solid-fuel propulsion systems. While NASA, and particularly its rocket development team headed by Wernher von Braun, had experience only with liquid-fueled boosters, the Department of Defense was interested in pushing the development of large solid-fuel rocket motors for various advanced military and intelligence uses. The focus of this planning effort was a "NASA-DOD Large Launch Vehicle Planning Group." The group was directed by Nicholas Golovin of NASA; its deputy director was Lawrence Kavanaugh of DOD. The group started work in July 1961, and by the fall had become bogged down in very detailed studies and deadlocked over the relative roles of liquid-fueled and solid-fueled boosters in the lunar landing program. Its final recommendations attempted to satisfy both NASA and DOD, and ended up pleasing neither agency. (Volume II, II-20)

In parallel with the Large Launch Vehicle study, NASA continued to carry out its own analyses of what kind of launch vehicles would be needed for Project Apollo. These analyses were hindered by a basic issue; NASA at the end of 1961 had not yet selected the approach—called the "mission mode"—which it would

---

14. For a history of the Apollo spacecraft, see Courtney G. Brooks, James M. Grimwood, and Loyd S. Swenson, Jr., *Chariots for Apollo: A History of Manned Lunar Spacecraft* (Washington, DC: National Aeronautics and Space Administration Special Publication-4205, 1979).

use to send crews to the Moon. (The process of making that decision is described in the following section.) This was the second reason for the delay in identifying the launch vehicles for Apollo; it was hard to define what kind of launch vehicle would be needed without knowing what requirements it would have to meet.

Still, as the end of the year approached there was a need to make some basic launch vehicle decisions. The NASA-DOD study had come out with a general set of recommendations that did not provide an adequate basis for NASA's decisions. So on 6 November, Milton Rosen of NASA Headquarters organized a two-week study to recommend to the NASA leadership "a large launch vehicle program" which would "meet the requirements of manned space flight" and "have broad and continuing national utility." (Volume IV, I-31) On 20 November, Rosen reported that "to exploit the possibility of accomplishing the first lunar landing by rendezvous," NASA should develop an "intermediate vehicle" that had five F-1 engines in the first stage, four or five J-2 engines in its second stage, and one J-2 in its third stage. (The J-2 was an engine powered by high energy liquid hydrogen fuel that would have the capability to stop and restart in orbit.) Since a direct flight to the Moon was at this point still NASA's stated preference for the lunar landing missions, Rosen also recommended that "a NOVA vehicle consisting of an eight F-1 first stage" should be developed on a "top priority basis." He added "large solid rockets should not be considered as a requirement for manned lunar landing." (Volume IV, I-32)

The recommendation for a five-engine first stage for the advanced Saturn launch vehicle, soon called the Saturn C-5 and ultimately the Saturn V, was quickly accepted by the NASA leadership. That decision, as will be seen later, soon became a key to NASA's choice of how to get to the Moon.[15]

Choosing Apollo's Managers

From the time that Kennedy announced his decision to go to the Moon, it was clear that the responsibility for developing the Apollo spacecraft and training the astronauts to operate it would be assigned to the Space Task Group. This group was headed by Robert Gilruth, a widely respected veteran of the National Advisory Committee on Aeronautics (NACA), NASA's predecessor. As soon as it was decided that NASA would build a new Field Center for Apollo and that it would be located in Houston, Gilruth and his team began to move their base of operations to Houston and to hire the many additional staff who would be needed to carry out the spacecraft development, astronaut training, and flight operations. It was equally clear that Wernher von Braun and his German rocket team, now working for NASA in the new Marshall Space Flight Center in Huntsville, Alabama, would be the core of the group developing the launch vehicles for Apollo.

---

15. For more information on the origins and development of the Saturn launch vehicles, see Bilstein, *Stages to Saturn* and Ray Williamson, "Access to Space: Steps to Saturn V" in John M. Logsdon *et al.*, eds., *Exploring the Unknown: Selected Documents in the History of the U.S. Civil Program* (Washington, DC: National Aeronautics and Space Administration Special Publication-4407, Vol. IV, 1999).

What NASA needed were highly qualified individuals to lead the overall Apollo program at its Washington Headquarters. After considering several other candidates from both inside and outside of the Agency, Webb, Dryden, and Seamans settled on D. Brainerd Holmes, who had managed the very large ballistic missile early warning project for RCA. Webb used his powers of persuasion to convince Holmes to join NASA. Holmes accepted the position and joined NASA in October 1961 in the new position of Associate Administrator for Manned Space Flight. One of Holmes's first identified needs was to find someone to apply a "systems management" approach to the already sprawling Project Apollo; that person turned out to be a dynamic young engineer named Joseph Shea, who came to NASA at the very end of 1961.[16] Over the following year, Holmes and Shea provided the energy and technical management skills to get Apollo started down a path to a lunar landing "before this decade is out," although neither was with NASA by the time that first landing took place.

On the same day, 20 November, that Rosen recommended development of a five-engine first stage Saturn vehicle, White House science advisor Jerome Wiesner prepared a memorandum for the President's close associate Theodore Sorenson, summarizing the state of progress on Project Apollo. Wiesner noted, "Six months have elapsed since the decision was announced to put man on the moon, yet none of these crucial hardware programs have progressed beyond the study phase. Lead times on these development and construction programs are of critical importance." In particular, "Major decisions have not been announced as to what extent rendezvous will be employed, what Advanced Saturn vehicle will be built (probably C-4), and what will be the characteristics of the so-called Nova which could put man on the moon by direct ascent. The relative emphasis of rendezvous versus direct ascent is a key to the entire program." (II-19) It would take almost a year before a decision on how to go to the Moon—by some form of rendezvous or by a direct flight—was final; that decision, as Wiesner noted, was key to getting to the Moon before 1970.

Finding a Way to the Moon[17]

NASA Chooses The Way

In early May 1961, when it appeared likely that President Kennedy would approve sending Americans to the Moon, NASA Associate Administrator Robert Seamans asked one of his senior staff members, William Fleming, to put together

---

16. Murray and Cox, *Apollo*, pp. 120–123; Robert C. Seamans, Jr., *Aiming at Targets* (Washington, DC: National Aeronautics and Space Administration Special Publication-4106, 1996), pp. 93–94. For a discussion of the application of systems management to Apollo, see Stephen B. Johnson, *The Secret of Apollo: Systems Management in American and European Space Programs* (Baltimore: Johns Hopkins University Press, 2002).

17. For more detailed discussions of the decision on what should be the preferred approach to a lunar landing, see John M. Logsdon, "NASA's Implementation of the Lunar Landing Decision," NASA HHN-81, September 1968; James R. Hansen, *Enchanted Rendezvous: John C. Houbolt and the*

a task force to examine "in detail a feasible and complete approach to the accomplishment of an early manned lunar mission." Seamans asked for a report within four weeks; the report was actually delivered in mid-June.[18] The task force considered only one approach to the lunar mission, the "direct ascent" mode, in which the very large Nova launch vehicle would send a complete spacecraft to the lunar surface. This approach had been the basis of NASA's early planning for a lunar landing. But Seamans also recognized that there were other approaches to the lunar landing that would involve rendezvous between two or more elements of a lunar spacecraft. So on the same day as President Kennedy announced the lunar landing goal, 25 May, Seamans asked Bruce Lundin of the Lewis Research Center to head up another group that would examine various rendezvous approaches as a way of getting to the Moon.

Lundin and his associates conducted a rapid assessment of various rendezvous approaches and reported back to Seamans on 10 June. They noted, "mission staging by rendezvous has been the subject of much investigation at Marshall, Langley, Ames, Lewis, and JPL." The group examined four rendezvous concepts: 1) rendezvous in Earth orbit; 2) rendezvous in lunar orbit after take-off from the lunar surface; 3) rendezvous in both Earth and lunar orbit; 4) rendezvous on the lunar surface. They concluded "of the various orbital operations considered, the use of rendezvous in Earth orbit by two or three Saturn C-3 vehicles (depending on estimated payload requirements) was strongly favored." This approach was either the first or second choice of all members of the group.[19] (II-12)

Based on this conclusion, Seamans formed yet another group, this one to examine rendezvous approaches in more depth than had been possible in the rapid Lundin study. This group was headed by Donald Heaton of NASA Headquarters. Following on Lundin's report, the group considered only Earth orbital rendezvous approaches. In its late August report, the group concluded *"rendezvous offers the earliest possibility for a successful manned lunar landing* [emphasis in original]."

NASA continued to consider both a direct ascent and Earth orbital rendezvous approaches for the next several months. Then, on 15 November, "somewhat as a voice in the wilderness," John Houbolt, a NASA engineer at the Langley Research Center, bypassed several layers of management and wrote an impassioned nine-page letter to Robert Seamans, arguing that NASA was overlooking the best way to get to the Moon before 1970, lunar orbital rendezvous. He claimed that "the lunar rendezvous approach is easier, quicker, less costly, requires less development, less new sites and facilities" and that Seamans should "Give us the go-ahead, and C-3, and we will put men on the Moon in very short order—and we don't need any Houston empire to do it." Houbolt told Seamans "it is conceivable that after reading this you may feel that you are dealing with a crank. Do not be afraid of

Genesis of the Lunar-Orbit Rendezvous Concept. NASA Monograph in Aerospace History, No. 4, 1999; and Murray and Cox, pp. 113–143.

    18. Barton C. Hacker and James M. Grimwood, *On the Shoulders of Titans: A History of Project Gemini* (Washington, DC: National Aeronautics and Space Administration Special Publication-4203, 1977), pp. 36–37.

    19. Ibid, p. 38.

this. The thoughts expressed here may not be stated in as diplomatic a fashion as they might be. . . . The important point is that you hear the ideas directly, not after they have filtered through a score or more of other people" (II-15).

Houbolt attached a report to his letter summarizing the results of work done by him and his associates at the Langley Research Center. (While Houbolt was only one of the originators of the lunar rendezvous concept, he was its primary spokesperson.) The report described the proposed mission plan:

> A manned exploration vehicle is considered on its way to the moon. On approach, this vehicle is decelerated into a low-altitude circular orbit about the moon. From this orbit a lunar lander descends to the moon surface, leaving the return vehicle in orbit. After exploration the lunar lander ascends for rendezvous with the return vehicle. The return vehicle is then boosted into a return trajectory to the earth, leaving the lander behind.

The primary advantage of this approach was "the marked reduction in escape weight required; the reduction is, of course, a direct reflection of the reduced energy requirements brought about by leaving a sizable mass in lunar orbit, in this case, the return capsule and return propulsion system." With less mass to carry to the Moon, Houbolt and his associates argued, a lunar landing mission could be accomplished by a single Saturn C-3 launch vehicle with two F-1 engines in its first stage. (II-16)

Houbolt in May had written an initial letter directly to Seamans, and the first reaction of NASA management was to discipline him for twice contacting Seamans outside of approved channels. But George Low, now working for Brainerd Holmes at NASA Headquarters, told Holmes that despite its tone, "Houbolt's message is a relatively sound one and I am forced to agree with many of the points he makes." Robert Gilruth and his associates in Houston were also beginning to see the merits of designing two separate spaceships, one for the journey to lunar orbit and return to Earth, the other only to land on the Moon. They began to do their own studies of the concept. By the end of January, Brainerd Holmes's deputy Joseph Shea, after being briefed by Houbolt on what was becoming known at the Lunar Orbit Rendezvous (LOR) concept, noted that "Brainerd and I agreed that LOR looks sufficiently attractive to warrant further study. He feels that the study should be run from OMSF, rather than either Center, to provide a measure of objectivity." He added "We are also concerned that MSFC will be especially negative with LOR because they have not studied it."[20] (II-17)

Over the next four months, both the Manned Space Craft Center (MSC) at Houston and the Marshall Space Flight Center (MSFC) at Huntsville carried out detailed studies of alternative rendezvous approaches to getting to the Moon. The idea of developing a huge launch vehicle, Nova, to carry astronauts to the Moon had by now lost favor as a feasible approach, mainly because it seemed

---

20. Murray and Cox, *Apollo*, pp. 120, 124–140.

to be too large a jump to go from the launch vehicles with which NASA had experience to something so gigantic. In addition, the concept of designing a single spacecraft to carry out all phases of the mission, particularly the lunar landing and the return into the Earth's atmosphere, looked increasingly difficult as Maxime Faget and other designers at MSC gave detailed attention to that challenge. During the early months of 1962, Houston became convinced that some version of the LOR approach, which involved two separate spacecraft, one specialized only for landing on the Moon and one for the journey to and from lunar orbit, was indeed the best way to proceed. The combined weight of the two spacecraft would allow the mission to be launched with a single Saturn C-5 (Saturn V) booster, although there was very little margin for weight growth. They shared their analyses and reasoning with their colleagues at MSFC, who were continuing to focus their efforts to various approached to Earth Orbital Rendezvous (EOR).

A climactic meeting was held at MSFC on 7 June. For most of the day, the Marshall staff presented their positive findings on EOR to Joseph Shea from NASA Headquarters. At the end of the day, MSFC Director Wernher von Braun provided concluding remarks. He shocked many of his associates by announcing that he had concluded that his first priority choice was the "Lunar Orbit Rendezvous Mode," because "We believe this program offers the highest confidence factor of successful accomplishment within this decade." Von Braun added "we agree with the Manned Spacecraft Center that the designs of a maneuverable hyperbolic reentry vehicle and of a lunar landing vehicle constitute the two most critical tasks in producing a successful lunar spacecraft. A drastic separation of these two functions into two separate elements is bound to greatly simplify the development of the spacecraft system." He noted "the issue of 'invented here' versus 'not invented here' does not apply to either the Manned Spacecraft Center or the Marshall Space Flight Center" because "both Centers have actually embraced a scheme suggested by a third." Von Braun told Shea "personnel of MSC and MSFC have by now conducted more detailed studies on all aspects of the four modes than any other group. Moreover, it is these two Centers to which the Office of Manned Space Flight would ultimately have to look to 'deliver the goods.' I consider it fortunate indeed for the Manned Lunar Landing Program that both Centers, after much soul searching, have come to identical conclusions." (II-18)

The White House Disagrees

With this rather startling announcement, given that the two Centers with primary responsibilities for Apollo were now in agreement, NASA Headquarters had little choice but to accept LOR as its choice for getting Americans to the Moon, and scheduled an 11 July press conference to announce that decision. However, James Webb on 3 July learned that there were strong objections to LOR on the part of the President's science advisor, Jerome Wiesner, and his associates. Later that day, Webb called Joe Shea, saying "Jerry Wiesner just called me and he's in a highly emotional state; he thinks LOR is the worst mistake in the world."[21] NASA was allowed to go ahead with its 11 July press conference, but could only announce the LOR choice as tentative, with more studies to be conducted.

Wiesner spelled out his reservations about the LOR choice in a 17 July letter to James Webb. (II-27) Wiesner was worried that the spacecraft weight limitations imposed by using the Saturn C-5 launch vehicle provided no margins if additional radiation shielding or zero-gravity countermeasures were discovered to be needed. He suggested that

> the matter of which mission mode is most consistent with the main stream of our national space program, and therefore the one most likely to be useful in overtaking and keeping ahead of Soviet space technology, is also one that I believe requires further consideration. . . . the question of which mode is likely to be most suitable for enhancing our military capabilities in space, if doing so should turn out to be desirable, should be reviewed with care.

Wiesner's views were in substantial part based on the views of the Space Vehicle Panel of the President's Science Advisory Committee (PSAC), which was chaired by Brown University chemist, Donald Hornig. Wiesner forwarded to NASA with his letter the panel's preliminary 11 July report. (II-26) The panel had concluded that a better approach to the Moon mission was to send a two-person crew, rather than the three astronauts that NASA had been planning on since 1960, and to use the EOR or direct ascent mode rather than LOR. The staff person supporting the PSAC panel was none other that Nicholas Golovin, who had been replaced by Joseph Shea as Brainerd Holmes's deputy and soon left NASA, unhappy with how he had been treated. He was then hired by Wiesner as his space specialist. Golovin and Shea were both self-confident individuals, with diametrically different approaches to key aspects of their systems analysis work. It is not possible to judge how much Golovin's antagonism towards NASA figured in the NASA-White House dispute over the choice of mission mode, but it certainly was an element in the controversy that was to linger for several months.

James Webb replied to Wiesner on 20 July, saying that NASA would indeed carry out the studies recommended by the Space Vehicle Panel and responding to some of Wiesner's criticisms. (II-28) In an attempt to smooth over the dispute, Webb concluded his letter by saying "this constructive criticism by eminently qualified men is of tremendous value, and I am looking forward to further discussions with you as the results of our present studies begin to crystallize."

However, this polite tone did not last. There were continuing tensions over the next few months between Wiesner and Golovin on one hand and Webb and his associates, particularly Joe Shea, on the other. On 11 September, the dispute became public. On that day President Kennedy flew to Huntsville to be briefed on the progress being made on Apollo at the Marshall Space Flight Center. As

---

21. Ibid, p. 141.

the president toured the MSFC facilities accompanied by Wiesner, Webb, and von Braun, he had to intervene to stop a heated discussion between those three that had broken out within earshot of the accompanying press contingent over the wisdom of the LOR choice.

By 24 October, James Webb had had enough of the White House interventions into what he considered NASA's authority to make its own technical decisions. In a letter to Wiesner, he attached a summary report of the reviews of the mission mode choice; that report noted that it was NASA's conclusion that "the lunar orbit rendezvous mode is the best choice for achieving a manned lunar landing mission before the end of the decade," and that "comparisons of the 2-man lunar mission capsules with the present LOR approach lead to the conclusion that LOR is the preferred mode on the basis of technical simplicity, scheduling and cost considerations." (II-29)

Webb in his letter implied that Wiesner, if he still disagreed with NASA's conclusions, would have to bring the matter before the president for resolution. He told Wiesner

> my own view is that we should proceed with the lunar orbit plan, should announce our selection of the contractor for the lunar excursion vehicle, and should play the whole thing in a low key. If you agree, I would like to get before you any facts . . . you believe you should have in order to put me in position to advise Mr. O'Donnell [the president's appointment secretary] that neither you nor the Defense Department wishes to interpose a formal objection to the above. In that case, I believe Mr. O'Donnell will not feel it wise to schedule the president's time and that the president will confirm this judgment.

In early November, Webb and Wiesner "met in a tense confrontation" before the president. Webb "cast the issue in terms of who was in charge of getting to the moon." According to one account, Kennedy said "Mr. Webb . . . you're running NASA— you make the decision."[22] On 7 November, Kennedy's National Security Advisor McGeorge Bundy asked Wiesner to write Webb a final time, telling Webb that "the president thinks the time is coming for a final recommendation and relies on Director Webb to review all the arguments and to produce that recommendation." He added "what the president has in mind is that we should make Webb feel the responsibility for a definite decision and the importance of weighing all opinions, without trying to make his decision for him." Wiesner was to ask Webb for a letter to be part of the president's files that recorded NASA's reasons for its recommendation. (II-30)

Bundy's memorandum was a bit after the fact; on the same day, 7 November, NASA called a press conference to announce that the choice of the lunar orbital rendezvous approach was final, and that NASA had selected the Grumman Aircraft Engineering Corporation to build the lunar landing spacecraft. Webb did write

---

22. Lambright, *Powering Apollo*, p. 113.

the requested letter, telling the president that "the decision to adopt the Lunar Orbit Rendezvous mode was based on major systems and engineering studies which involved over a million man-hours of effort on the part of government and contractor personnel." He added "despite the very extensive study efforts, however, we are dealing with a matter that cannot be conclusively proved before the fact, and in the final analysis the decision has been based upon the judgment of our most competent engineers and scientists who evaluated the studies and who are experienced in this field." Webb noted "The decision on the mode to be used for the lunar landing had to be made at this time in order to maintain our schedules, which aim at a landing attempt in late 1967." (II-31)

Eighteen months after President John F. Kennedy had announced his decision to send Americans to the Moon, the plan for meeting that goal was now in place. The choice of the "mission mode" was, as Wiesner had told Theodore Sorenson a year earlier, the "key to the entire program."

The Science of Apollo

While NASA's managers and engineers were deciding how to get to the Moon, there was a parallel activity focused on what scientific activities would take place on the lunar surface, and who would carry out those activities.[23] As a first step in linking scientific considerations to Apollo planning, NASA Headquarters in March 1962 established a working group to recommend what scientific tasks lunar explorers should perform. This group was headed by Charles P. Sonnett of NASA's Lunar and Planetary Programs Office. The group held its first meeting on 27 March; one immediate question was whether it would be desirable, perhaps even necessary, to include trained scientists on Apollo crews. After that meeting, Joseph Shea asked the relevant staff in the Office of Manned Space Flight: "Is there any fundamental reason which would prevent the use of one or more professional scientists as crew members?" and "What serious practical problems would result if such personnel were included in the selection training program?" (II-21)

There were no major objections raised to selecting scientists as Apollo astronauts, and over the next three years NASA worked together with the National Academy of Sciences to first set criteria for scientist-astronauts and then recruit a first group of individuals who met those criteria. For the first time in selecting astronauts, prior proficiency in piloting high-performance jet aircraft was not required to apply, although those selected would be required to undergo flight training. Over 1,000 applications were sent to the National Academy of Sciences; after screening, the Academy recommended 16 candidates to NASA. On 28 June 1965, NASA announced that it had selected six men as its first scientist-astronauts. (Of those six, only one, geologist Harrison H. "Jack" Schmitt, would fly an Apollo mission, although three others flew during the 1973 Skylab mission.)[24]

---

23. For the origins of scientific planning for lunar exploration, see W. David Compton, *Where No Man Has Gone Before: A History of Lunar Exploration Missions* (Washington, DC: National Aeronautics and Space Administration Special Publication-4214, 1989), Chaps. 2–3.
24. Ibid, Chap. 5.

Sonnett's group completed its work in early July 1962. Its recommendations were then reviewed during a "Summer Study" of the National Academy's Space Science Board, which was already underway at the University of Iowa. The study report endorsed most of the recommendations of Sonnett's report, and as modified by the Board's review they then became the basis for NASA's planning regarding the scientific aspects of Apollo missions. (II-41 and Volume V, I-22, II-12, II-13)

Another pressing issue as the Apollo missions were being designed was how to obtain the needed information about the lunar environment, such as the radiation environment astronauts would experience on the journeys to and from the Moon, the physical properties of the lunar soil, and the topography of the Moon. Brainerd Holmes and his associates turned to previously approved robotic lunar science programs, Ranger and Surveyor, which were managed by the Jet Propulsion Laboratory, in hopes that they could provide much of this information. Ranger missions would make hard landings on the Moon, sending back images as the spacecraft approached the lunar surface; Surveyor missions would land softly on the lunar surface and send back detailed images and other information about the area surrounding their landing site.

Tensions between the original scientific objectives of these missions and NASA's need for engineering information were inevitable. (II-22, Volume V, II-11) Later Ranger and Surveyor missions were indeed modified to meet Apollo's needs, creating lasting resentment among some members of the scientific community with respect to the intervention of engineering concerns into the setting of scientific priorities for robotic missions. NASA also decided to add a third robotic lunar program, Lunar Orbiter, to obtain high-resolution imagery of the lunar surface. That program was managed by the Langley Research Center, which was less closely linked to the scientific community than was JPL; the program used a camera modified from its original highly classified intelligence satellite mission to obtain the images needed.

Even with all of this information, there was continuing controversy about the character of the lunar surface. One prominent astronomer, Thomas Gold of Cornell University, suggested that the smooth areas of the Moon were likely to be covered with a layer of fine dust several meters deep, raising the possibility that a lunar lander might sink into the dust or topple over after landing. Even after the first Surveyor spacecraft landed on the Moon without problems on 2 June 1966, Gold suggested that his views might still be valid. (II-46)

An early planning issue for NASA was the selection of the locations on the Moon where Apollo would land. NASA did not want to restrict itself to a single location for even the first lunar landing attempt, and of course was planning more than one Apollo mission to the Moon. Engineering and trajectory considerations entered into play, making the choice of landing sites complex. Because of the weight limitations associated with the Apollo spacecraft and lunar module, only a landing at a location on the near side of the Moon and near the lunar equator was feasible; this meant that Apollo could not visit approximately 80 percent of the overall lunar surface. (II-24) Ultimately NASA identified a number of potential landing sites on the near side of the Moon and close to the lunar equator; then the scientific community identified locations of highest scientific interest.[25]

Because an explicit objective of Apollo was the safe return to Earth of astronauts and their spacecraft and of the samples of the Moon they would collect during their stays on the lunar surface, NASA in its planning could not ignore the remote possibility that there could be living organisms on the Moon which, if brought to Earth, might have negative effects.[26] The scientific community through the Space Science Board had pointed out this issue since the start of planning for missions to the Moon, and the Board's 1962 Summer Study recommended that NASA develop "appropriate quarantine and other procedures . . . when handling returned samples, spacecraft, and astronauts [in order to] make the risk as small as possible."[27] (Volume V, II-14) NASA did little in response to this recommendation, and in 1964 the Space Studies Board once again expressed its concerns. By 1965, NASA realized that it would have to develop elaborate Lunar Receiving Laboratory facilities at Houston for quarantining whatever had returned from the Moon and that measures to initiate that quarantine would have to be put in place for the period between when the astronauts and their spacecraft returned to Earth and they were placed in those facilities. In 1966 NASA also established an Inter-Agency Committee on Back Contamination to develop policies on the issue.[28]

That Committee issued its report on the elaborate measures to be taken to prevent contamination of Earth by alien organisms from the Moon in August 1967. (II-52) NASA also developed policies to minimize biological contamination to the Moon by the Apollo astronauts, their spacecraft, and the scientific experiments to be carried out on the lunar surface. (II-53)

As the Apollo 11 mission, the first attempt at a lunar landing, was imminent in March 1969, concerns were raised both through the National Academy of Sciences and in representations to Congress that NASA was not being diligent enough in its application of the measures related to back contamination. (II-54, II-55, II-56) There was even some possibility that NASA might be forced to delay the Apollo 11 launch until it convinced the external scientific community that the way it was preceding did not pose unacceptable risks. Ultimately, NASA was able to allay Congressional concerns, and the mission was launched on the planned date.

---

25. Ibid, Chap. 6

26. There is a long history of belief about life on the Moon—not sophisticated, complex life but certainly life that might harm humans. Esteemed astronomer Patrick Moore suggested as late as 1955 that there may indeed be vegetation in the crater Aristarchus where changing bands of color might signal the possibility of life hanging on near gaseous eruptions from underground. See Patrick Moore, "Life on the Moon?" *Irish Astronomical Journal*, 3, no. 5 (1955): 136.

27. Quoted in Compton, *Where No Man has Gone*, p. 45.

28. See Ibid, Chapter 4, for a discussion of the approach taken to handling lunar samples.

What Priority for Apollo?

As part of NASA's buildup for the Apollo project, James Webb on 13 March 1962 wrote to President Kennedy, asking him to assign the top government priority—called "DX"— to the lunar landing project. To be assigned such a priority, a program had to have objectives of key political, scientific, psychological or military import. Those programs with this priority had first call on the scarce resources needed to achieve their goals. The President approved this request upon the recommendation of the National Aeronautics and Space Council. (II-23)

One scarce resource not covered by the DX priority was money—specifically, funds within the overall NASA budget to be allocated to ensuring that Apollo would meet its goal of landing Americans on the Moon before 1970. And the man in charge of Apollo, Brainerd Holmes, by mid-1962 had come to believe that the project was receiving enough funds, and that with additional funds not only would a lunar landing by the end of 1967 (NASA's planning target at the time) be possible, but even might be accomplished earlier. As Robert Seamans observed, "by the summer of 1962, Jim [Webb] and I knew we had a problem with Brainerd Holmes." Holmes was a "very exciting person for the media. He had a way of expressing himself that made news."[29] Indeed, the 10 August issue of *Time* magazine featured Holmes on its cover and dubbed him "Apollo czar."

Holmes was seeking an additional $400 million for Apollo for the current FY 1963. There were two ways to get these funds. One way was to transfer them from other NASA programs within the overall NASA budget provided by Congress. The other was to request that amount in a supplemental appropriation from Congress. James Webb refused to approve either choice, angering Holmes. Apparently Holmes discussed the situation directly with President Kennedy, probably during the president's inspection tour of the Apollo buildup on 11 and 12 September, with a stress on an earlier date for the first landing attempt. The President then asked Webb whether there was indeed a possibility of making the lunar landing in 1966 rather than 1967. Webb responded in late October, telling Kennedy "the late 1967 target date is based on a vigorous and driving effort, but does not represent a crash program. A late 1966 target would require a crash, high-risk effort." Webb added that NASA was "prepared to place the manned lunar landing program on an all-out crash basis aimed at the 1966 target date if you should decide this is in the national interest," but substantial and immediate budget increases would be required. (II-32)

President Kennedy had asked his Bureau of the Budget during that summer to take a careful look at the actual situation with respect to the overall U.S. space program, focusing on two questions: "the pace at which the manned lunar landing should proceed" and "the approach that should be taken to other space programs in the 1964 budget." Director of the Budget David Bell sent the results of the review to the President on 13 November. (Volume I, III-13) The review examined four options for Apollo. The first was the current NASA plan, with no

---

29. Seamans, *Aiming at Targets*, p. 103.

supplemental budget request for FY 1963 and a late 1967 target for the first lunar landing. The second was to examine the budget implications of an accelerated program along the lines being advocated by Holmes. The final option examined the impact of slipping the landing date target by a year. Bell told the President "I agree with Mr. Webb that alternative 1, the NASA recommendation, is probably the most appropriate choice at this time."

Holmes remained unhappy. He was the apparent source for a second *Time* story that appeared on 19 November, titled "Space is in Earthly Trouble." The magazine's editors had deleted a Holmes quote from the story before it was published that said "The major stumbling block of getting to the moon is James E. Webb. He won't fight for our program."[30]

Given the now-public controversy, President Kennedy scheduled a 21 November meeting in the Cabinet Room of the White House to discuss NASA's plans for Apollo. Like a number of meetings while Kennedy was President, this meeting was tape-recorded; a transcript of the discussion provides a rare insight into the interactions between Kennedy and Webb. (II-33) During the meeting, Kennedy and Webb had the following exchange:

> **President Kennedy:** Do you think this program [Apollo] is the top-priority of the Agency?
>
> **James Webb:** No, sir, I do not. I think it is one of the top-priority programs, but I think it's very important to recognize here . . . and that you have found what you could do with a rocket as you could find how you could get out beyond the Earth's atmosphere and into space and make measurements. Several scientific disciplines that are very powerful are beginning to converge on this area.
>
> **President Kennedy:** Jim, I think it is the top priority. I think we ought to have that very clear. Some of these other programs can slip six months, or nine months, and nothing strategic is gonna happen, it's gonna . . . But this is important for political reasons, international political reasons. This is, whether we like it or not, in a sense a race. If we get second to the Moon, it's nice, but it's like being second any time. So that if we're second by six months, because we didn't give it the kind of priority, then of course that would be very serious. So I think we have to take the view that this is the top priority with us.

Later in the meeting, the President and the NASA head continued their debate:

> **President Kennedy:** Everything that we do ought to really be tied into getting onto the Moon ahead of the Russians.

---

30. Dwayne Day has provided a discussion of these issues which can be found at *http://history.nasa.gov/JFK-Webbconv/pages/backgnd.html* (accessed 25 August 2006).

James Webb: Why can't it be tied to preeminence in space?

President Kennedy: Because, by God, we keep, we've been telling everybody we're preeminent in space for five years and nobody believes it because they have the booster and the satellite. We know all about the number of satellites we put up, two or three times the number of the Soviet Union . . . we're ahead scientifically.

President Kennedy: I do think we ought to get it, you know, really clear that the policy ought to be that this is *the* top-priority program of the Agency, and one of the two things, except for defense, the top priority of the United States government. I think that that is the position we ought to take. Now, this may not change anything about that schedule, but at least we ought to be clear, otherwise we shouldn't be spending this kind of money because I'm not that interested in space. I think it's good; I think we ought to know about it; we're ready to spend reasonable amounts of money. But we're talking about these fantastic expenditures which wreck our budget and all these other domestic programs and the only justification for it, in my opinion, to do it in this time or fashion, is because we hope to beat them and demonstrate that starting behind, as we did by a couple years, by God, we passed them.

James Webb: I'd like to have more time to talk about that because there is a wide public sentiment coming along in this country for preeminence in space.

President Kennedy: If you're trying to prove preeminence, this is the way to prove your preeminence.

As he prepared to leave the meeting, the president asked Webb to prepare a letter stating his position on why space preeminence, and not just being first to the Moon, should be the country's goal: "I think in the letter you ought to mention how the other programs which the Agency is carrying out tie into the lunar program, and what their connection is, and how essential they are to the target dates we're talking about, and if they are only indirectly related, what their contribution is to the general and specific things possibly we're doing in space."

Webb's letter was sent to the president on 30 November. (Volume I, III-14) In it, Webb said that in his view "the objective of our national space program is to become preeminent in all important aspects of this endeavor and to conduct the program in such a manner that our emerging scientific, technological, and operational competence in space is clearly evident." Webb emphasized that "the manned lunar landing program, although of highest national priority, will not by itself create the preeminent position we seek."

Webb's response apparently did not totally satisfy John F. Kennedy. As he visited the Los Alamos National Laboratory on 8 December, he asked his science advisor Jerome Wiesner to again look into the possibility of accelerating the target date for the lunar landing. Wiesner replied on 10 January 1963, telling the president "that approximately 100 million dollars of the previously discussed 326 million dollar supplementary could have a very important effect on the schedule, but that to do so it would have to be available in the very near future." Such a funding increase, said Wiesner, should be used to make sure that the Saturn V launch vehicle (he still called it the C-5) would be available when it was needed. (II-34)

Overall, however, President Kennedy seems to have accepted the basic argument made by James Webb— that preeminence in space should be the guiding objective of the national space program. In a 17 July 1963 press conference, Kennedy responded to a press report that the Soviet Union was not planning to send its cosmonauts to the Moon, saying "The point of the matter always has been not only of our excitement or interest in being on the moon; but the capacity to dominate space, which would be demonstrated by a moon flight, I believe, is essential to the United States as a leading free world power. That is why I am interested in it and that is why I think we should continue."[31]

New Leadership and New Approaches for Apollo

As 1963 began, there were a number of technical problem areas in the Apollo program, particularly with the F-1 engine that would power the first stage of the Saturn V. (Volume IV, I-35, I-36, I-37) In addition, the strained relationship between NASA's top leaders and Brainerd Holmes also was only becoming worse.[32] On 12 June, Holmes submitted his resignation. This meant that Apollo was losing the leader who in the eyes of the public and media had come to personify the effort.

It took NASA a little over a month to settle on a replacement for Holmes. The individual selected, George Mueller, was Vice President for Research and Development of Space Technology Laboratories; his selection was announced on 23 July and Mueller reported to NASA on 1 September. At Space Technologies Laboratories, Mueller had excelled in applying a systems engineering approach to the management of the complex Minuteman ICBM program, and he brought the same approach to NASA. Unlike Holmes, who courted media attention, Mueller focused his attention on relationships between NASA Headquarters, the NASA Field Centers, NASA's contractors, and Congress. For example, he created a NASA-Industry Apollo Executives Group that brought together key NASA personnel working on Apollo and the leaders of the companies building Apollo hardware. One of the leading accounts of the Apollo program describes Mueller as "brilliant," "intellectually arrogant," and "a complex man." Robert Seamans characterized him as "tireless."[33]

---

31. "News Conference 58," John F. Kennedy Library and Museum, *http://www.jfklibrary.org/ Historical+Resources/Archives/Reference+Desk/Press+Conferences/003POF05Pressconference58_07171963. htm* (accessed 25 August 2006).

32. Seamans, *Aiming at Targets*, p. 105.

33. Murray and Cox, *Apollo*, p. 158, 160; Seamans, *Aiming at Targets*, p. 110.

Soon after he entered NASA, the organization implemented a major reorganization in which the heads of the Field Centers working on Apollo reported to the Office of Manned Space Flight (i.e., Mueller), rather than directly to Seamans, the Agency's Associate Administrator and general manager. By a combination of his force of will and this reorganization, Mueller "was the undisputed boss of manned space flight from the day he walked into the office in 1963 until he left six years later."[34]

In the next several months Mueller made a number of key personnel changes. He assigned George Low and Joseph Shea— both of whom welcomed the assignments— to the Manned Spacecraft Center in Houston, Low to become Deputy Director under Robert Gilruth and Shea to head the Apollo Spacecraft Program Office. On 31 December 1963, Air Force Brigadier General Samuel Phillips took over the Apollo Program Office at NASA Headquarters. The team of Mueller and Phillips was to provide strong leadership as the Apollo program encountered both tragedy and triumph.

Soon after he came to NASA, Mueller asked two veteran NASA engineers not directly involved in Apollo, John Disher and Del Tischler, to conduct a discrete independent assessment of the situation within Apollo. They reported to Mueller on 28 September with the troubling conclusions that the "lunar landing cannot likely be attained within the decade with acceptable risk" and that the "first attempt to land men on moon is likely about late 1971." The two estimated that the "program cost through initial lunar landing attempt will approximate 24 billion dollars." (II-36) Mueller had Disher and Tischler present their conclusions to Robert Seamans, who found the briefing "unsatisfactory." According to some accounts, Seamans asked that the briefing material be destroyed to prevent its conclusions from becoming known inside and outside of NASA.[35]

Clearly, bold steps were needed to get Apollo on a schedule that had a good chance of meeting President Kennedy's goal of a lunar landing before 1970, and Mueller soon took them. First he canceled flights of the Saturn 1 booster so that attention could be shifted to the upgraded Saturn 1B, which would use the same upper stage as the Saturn V. At an 29 October meeting of his Management Council, with the senior leadership from Houston and Huntsville present, Mueller announced a new approach to getting ready for missions to the Moon that soon became known as "all-up testing." Mueller "stressed the importance of a philosophical approach to meeting schedules which minimizes 'dead-end' testing, and maximized 'all-up' systems flight tests. He also said the philosophy should include obtaining *complete* systems at the Cape [emphasis in original]." (II-37) Two days later Mueller sent a teletype message to the Apollo field centers proposing a new, accelerated schedule of Apollo flights; in this message, he reiterated that his "desire that 'all-up' spacecraft and launch vehicle flights be made as early as possible in the program. To this end, SA-201 [the first flight of the Saturn 1B] and 501 [the first flight of the Saturn V] should utilize all live stages and should carry complete spacecraft for their respective missions." (II-38)

---

34. Murray and Cox, *Apollo*, p. 160.
35. Ibid, pp. 153–154.

The staff at Marshall Space Flight Center was "incredulous" when they first heard of Mueller's dictate. It violated the step-by-step approach to rocket testing they had been following since their time in Germany.[36] But they could not provide compelling counterarguments, particularly given the pressure to have the first lunar landing attempt come before the end of 1969. Von Braun wrote Mueller on 8 November, saying that "We believe the philosophy of flying live all stages, modules, and systems, beginning with the first R&D launching, to be a worthy objective. There is no fundamental reason why we cannot fly 'all-up' on the first flight." Von Braun hedged his response a bit, saying "Our practical application of this philosophy should recognize this objective, but with the important reservation that clear, alternative, 'fall back' positions are also formally recognized."[37] Von Braun was later to agree "in retrospect it is clear that without all-up testing the first manned lunar landing could not have taken place as early as 1969."[38] Mueller's "all-up" decision thus joined the selection of lunar orbit rendezvous as keys to Apollo's success. According to one account, "the crisis in Apollo leadership that had begun in 1962 with Holmes's mutiny thus ended in 1963 with an astute new manned space flight director, a stronger overall Apollo management team, and decisive steps to get Apollo back on schedule."[39]

1963—A Year of Uncertainty

Increasing Criticisms

Even as internal steps were being taken to get Apollo on track to meet its "before the decade is out" goal, external to the space agency there were several developments that placed the future course of the program in some doubt.

After President Kennedy's 25 May 1961 speech announcing the lunar landing goal, the public and political reception to the president's initiative was in general very positive. Beginning in 1963, however, criticism of Apollo in the context of overall national priorities, as well as scientific ones became much more widespread.[40] Much of this criticism was in the form of newspaper articles and editorials, but there were also the beginnings of dissent regarding the Apollo goal within the political system. On 10 and 11 June, the Senate Committee on Aeronautical and Space Sciences, under its new Chairman, Senator Clinton Anderson of New Mexico (Robert Kerr had died on 1 January 1963), listened as 10 scientists discussed Apollo. The majority complained about the priority that had been assigned to the lunar landing program, and provided dramatic examples of

36. Howard McCurdy discusses this difference in approach to testing in Chapter 2 of his book *Inside NASA: High Technology and Organizational Change in the U.S. Space Program* (Baltimore: Johns Hopkins University Press, 1992).

37. Wernher von Braun to George E. Mueller, 8 November 1963, Folder #18675, NASA Historical Reference Collection, NASA History Division, NASA Headquarters, Washington, DC.

38. Murray and Cox, *Apollo*, p. 162.

39. Lambright, *Powering Apollo*, p. 118.

40. Compton, *Where No Man Has Gone*, Chap. 3.

how the funds could otherwise be used. Philip Abelson, editor of the prestigious journal *Science*, reported that he had conducted a straw poll of "scientists not connected by self-interest to NASA," which had resulted in a 110 to 3 vote against the program. In his testimony, Abelson suggested that "manned space exploration has limited scientific value and has been accorded an importance which is quite unrealistic," and that the "diversion of talent to the space program is having or will have direct and indirect damaging effects on almost every area of science, technology, and medicine," and might "delay conquest of cancer and mental illness."[41] Liberal Senator William Fulbright (D-AK), chairman of the Committee on Foreign Relations, suggested "this allocation of priorities [to the lunar program] is a recipe for disaster." Former President Dwight D. Eisenhower, writing in the widely-read *Saturday Evening Post*, stated that "this racing to the moon, unavoidably wasting large sums and deepening our debt, is the wrong way to go about it."[42]

In addition to these public declarations, there were private criticisms from senior members of the U.S. science and technology community. As one example, on 11 April 1963 Vannevar Bush, a highly respected man who had headed the U.S. scientific effort during World War II and whose recommendations in his famous report *Science, the Endless Frontier* had helped shape post-war government support of science, wrote to James Webb (with whom he had worked when Harry Truman was president and whom he knew well) saying: "the difficulty is that the program, as it has been built up, is not sound. The sad fact is that the program is more expensive than the country can now afford; its results, while interesting, are secondary to our national welfare." He added "while the scientific results of an Apollo program would be real, I do not think that anyone would attempt to justify an expenditure of 40 or 50 billion dollars to obtain them." With respect to the argument that Apollo would enhance national prestige, Bush thought that "the courageous, and well conceived, way in which the president handled the threat of missiles in Cuba advanced our national prestige far more than a dozen trips to the moon. Having a large number of devoted Americans working unselfishly in undeveloped countries is far more impressive than mere technical excellence. We can advance our prestige by many means, but this way is immature in its concept." Bush told Webb "as a part of lowering taxes and putting our national financial affairs in order, we should have the sense to cut back severely on our rate of expenditure on space. As a corollary they could remove all dates from plans for a trip to the moon; in fact, he could announce that no date will be set, and no decision made to go to the moon, until many preliminary experiments and analyses have rendered the situation far more clear than it is today." (II-35)

---

41. Quoted in Logsdon, *Decision*, pp. 175–176.
42. Fulbright and Eisenhower are quoted in Dodd L. Harvey and Linda Ciccoritti, *U.S.-Soviet Cooperation in Space*, (Miami: Center for Advanced International Studies, University of Miami, 1974), p. 113.

Apollo Under Review

In November 1962 President John F. Kennedy had identified beating Russia to the Moon as the country's highest priority in space. Less than five months later, there was some suggestion that the president might have been having second thoughts about that priority and about the impacts of the accelerated space program on the nation's economy and technical activities, although whether this indeed was the case is not clear from the historical record. On 9 April, the president wrote Vice President Lyndon B. Johnson in his role as Chairman of the National Aeronautics and Space Council, saying that "in light of recent discussions, I feel the need to obtain a clearer understanding of a number of factual and policy issues relating to the National Space Program which seem to rise repeatedly in public and other contexts." Kennedy asked Johnson to carry out a quick review of the program to answer a number of specific questions.[43] (Volume I, III-15)

Johnson's report came on 13 May; NASA and DOD had been closely involved in its preparation. In addition to answering the specific questions posed by the President, the report noted "the space program is not solely a question of prestige, of advancing scientific knowledge, or economic benefit or of military development, although all of these factors are involved. Basically, a much more fundamental issue is at stake . . . the future of society." (Volume I, III-16)

To the Moon Together?

If Kennedy was indeed questioning the wisdom of racing Russia to the Moon, one reason may have been the changed nature of U.S.-Soviet relations after the United States had forced the Soviet Union to withdraw its missiles from Cuba in October 1962. Kennedy seems to have concluded that the time was ripe to revisit a notion that had preceded his decision to enter the space race—that a flight to the Moon should be a cooperative U.S.-Soviet undertaking. According to Kennedy advisor Theodore Sorensen, "it is no secret that Kennedy would have preferred to cooperate with the Soviets" in manned missions to the Moon.[44] In an interview shortly after Kennedy's assassination, Sorenson expanded on this idea:

> I think the President had three objectives in space. One was to ensure its demilitarization. The second was to prevent the field to be occupied by the Russians to the exclusion of the United States. And the third was to make certain that American scientific prestige and American scientific effort were at the top. Those three goals all would have been assured in a space effort which culminated in our beating the Russians to the moon. All three of them would have been endangered had the Russians continued to outpace us in their space effort and beat us to the moon.

---

43. Compton, *Where No Man Has Gone*, Chap. 3.
44. Aleksandr Fursenko and Timothy Naftali, *One Hell of a Gamble: Khrushchev, Castro and Kennedy, 1958–1964* (New York: Norton, 1997), p. 121.

But I believe all three of those goals would also have been assured by a joint Soviet-American venture to the moon.

The difficulty was that in 1961, although the President favored the joint effort, we had comparatively few chips to offer. Obviously the Russians were well ahead of us at that time. . . . But by 1963, our effort had accelerated considerably. There was a very real chance we were even with the Soviets in this effort. In addition, our relations with the Soviets, following the Cuban missile crisis and the test ban treaty, were much improved—so the President felt that, without harming any of those three goals, we now were in a position to ask the Soviets to join us and make it efficient and economical for both countries. (II-43)

President Kennedy met Soviet leader Nikita Khrushchev only once, on 3 and 4 June 1961. This was soon after Kennedy had made his speech announcing the lunar landing goal, but twice during the summit meeting, and at the President's initiative, Kennedy and Khrushchev had discussed the possibility of cooperation in going to the Moon.[45] Khrushchev reacted negatively to Kennedy's proposal, and the matter was dropped for the next two years.

In mid-1963, the president began again to float the idea of a joint U.S.-Soviet mission to the Moon. One problem, however, was that there was no evidence from intelligence sources that the Soviet Union was in fact intending to send cosmonauts to the Moon.[46] In fact, it was reported that a leading British scientist, Bernard Lovell, had been told by his Soviet counterparts that there was no Russian program to send people to the Moon. Asked at a 17 July press conference on whether he favored a joint U.S.-Soviet lunar mission, Kennedy, for the first time in a public forum, said "we have said before to the Soviet Union that we would be very interested in cooperation." However, he added, "the kind of cooperative effort which would be required for the Soviet Union and the United States to go to the moon would require a breaking down of a good many barriers of suspicion and distrust and hostility which exist between the Communist world and ourselves." Kennedy concluded that he would "welcome" such cooperation, but that he "did not see it yet, unfortunately."[47]

By September, Kennedy had decided to publicly test the waters with respect to possible U.S.-Soviet cooperation in going to the Moon. During a 18 September meeting with James Webb, Kennedy told the NASA Administrator for the first time that he intended to make such a proposal in a 20 September speech to the General Assembly of the United Nations. (Volume II, I-41) In his 20 September speech, Kennedy said

---

45. John M. Logsdon, "To the Moon Together? John F. Kennedy and U.S. Soviet Space Cooperation," unpublished paper in author's files.

46. The reality was that in 1963 the Soviet leadership had not yet decided to approve a lunar landing mission. See John M. Logsdon and Alain Dupas, "Was the Race to the Moon Real?" *Scientific American* 270, no. 6 (June 1994): 36.

47. *Public Papers of the Presidents of the United States: John F. Kennedy, 1963*, (Washington, DC: Government Printing Office, 1964), pp. 567–568.

in a field where the United States and the Soviet Union have a special capacity— in the field of space— there is room for new cooperation . . . I include among these possibilities a joint expedition to the moon. Why, therefore, should man's first flight to the moon be a matter of national competition? . . . Surely we should explore whether the scientists and astronauts of our two countries— indeed of all the world— cannot work together in the conquest of space, sending some day in this decade to the moon not the representatives of a single nation, but representatives of all our countries.[48]

Kennedy's proposal was greeted with dismay by many of those who had been Apollo's strongest supporters. For example, Congressman Albert Thomas sent a handwritten note to the president the day after the speech, saying that "the press and many private individuals seized upon your offer to cooperate with the Russians in a moon shot as a weakening of your former position of a forthright and strong effort in lunar landings." Thomas asked the president for "a letter clarifying your position with reference to our immediate effort in this regard."[49]

Kennedy replied to Thomas on 23 September. (II-39) He told Thomas "if cooperation is possible, we mean to cooperate, and we shall do so from a position made strong and solid by our national effort in space. If cooperation is not possible—and as realists we must plan for this contingency too—then the same strong national effort will serve all free men's interest in space, and protect us also against possible hazards to our national security."

There were suggestions in the aftermath of the president's speech that it was a public relations move or a way of justifying a withdrawal of the United States from a fast-paced lunar landing program. Countering these suggestions is the fact that in the weeks following the United Nations speech, the White House Office of Science and Technology examined ways to turn the president's proposal into reality, even as Nikita Khrushchev on 26 October told a group of visiting journalists that the Soviet Union had no plans to send people to the Moon. For example, on 29 October, Science Advisor Jerome Wiesner provided a memorandum for Kennedy proposing "a joint program in which the USSR provides unmanned exploratory and logistic support for the U.S. Apollo manned landing." (II-40) Wiesner suggested that such a plan be quickly offered to the Soviet Union in light of Khrushchev's statement. Wiesner noted "if the proposal is accepted we will have established a practical basis for cooperative program. If it is rejected we will have demonstrated our desire for peaceful cooperation and the sincerity of our original proposal." Following on Wiesner's suggestion, on 12 November, President Kennedy signed a National Security Action Memorandum directing James Webb "to assume personally the initiative and central responsibility within the Government for the development of a program of substantive cooperation

---

48. Ibid, p. 695.
49. Letter from Albert Thomas (signed only "Thomas"), 21 September1963. National Security Files, Box 308, John F. Kennedy Presidential Library.

with the Soviet Union . . . including cooperation in lunar landing programs."
(Volume II, I-42)

Uncertainties Resolved . . . In the Worst Possible Way

On 18 November 1963, the Senate voted to cut $612 million from NASA's budget request, leading *The New York Times* to question "whether the Administration can count on the budgetary support necessary to achieve a lunar landing by the 1969 deadline."[50] That was a good question, given what was happening within the White House budget staff. The Bureau of the Budget staff in November 1963 was completing a comprehensive review of the national space program that had been initiated in October; its draft report asked: "<u>Should consideration be given at this time to backing off from the manned lunar landing goal?</u> [Emphasis in original]." The budget office suggested "the review has pointed to the conclusion that in the absence of clear changes in the present technical or international situations, the only basis for backing off from the MLL [manned lunar landing] objective at this time would be an overriding <u>fiscal</u> decision either (a) that the budgetary totals in 1965 or succeeding years are unacceptable and should be reduced by adjusting the space program, or (b) that within present budgetary totals an adjustment should be made shifting funds from space to other programs."[51] (II-42) When Congress passed the NASA FY 1964 appropriations bill, the space agency was allocated a $5.1 billion budget, an increase of $1.3 billion over FY 1963 but a $0.6 billion reduction from what the president had requested for NASA at the start of the year.

President Kennedy visited Cape Canaveral on 16 November, and saw the progress being made on the facilities being developed for Apollo.[52] On 21 November, Kennedy gave the space program a strong endorsement in a speech in San Antonio, where he had started a three-day political trip. That evening the president attended a testimonial dinner for Albert Thomas in Houston, and then flew to Dallas. The next day, Kennedy fell victim to an assassin's gun.

With John F. Kennedy's death and Lyndon B. Johnson becoming president, any chance of the United States "backing off" of the lunar landing program that Kennedy had initiated vanished; instead, the program became in a sense a memorial to the fallen president. Lyndon Johnson was far less interested in cooperating with the Soviet Union in space than had been Kennedy. In January 1964 NASA submitted the report requested by Kennedy's 12 November national security directive (Volume II, I-43), but there had been no sign from Nikita Khrushchev that he was interested in discussing cooperation, and President Johnson did not press the issue. For the next three years, while James Webb fought to maintain congressional support for a budget adequate to meet Kennedy's

---

50. Quoted in Murray and Cox, *Apollo*, p. 161.
51. It is not clear from available sources whether this review was carried to completion. Document II-42 is a labeled draft, and contains only the thoughts of the Bureau of the Budget staff. Room was left for recommendations by various senior officials, but whether those recommendations were made in the aftermath of President Kennedy's assassination is not known.
52. Seamans, *Aiming at Targets*, p. 113--15.

"before the decade is out" goal, the rest of NASA turned to getting the Apollo hardware ready to fly and the Apollo astronauts trained for lunar exploration.

Moving Ahead, but Losing Momentum

During the three year period from 1964 to 1966, there was significant (though troubled, as will be discussed below) progress in the program aspects of Apollo, but during those same years, "NASA stopped growing, and [James] Webb sought to maintain momentum."[53] The NASA budget peaked at $5.25 billion in FY 1965 and then began a gradual decline. While Lyndon B. Johnson was strongly committed to completing Apollo, he found himself constrained by the budget demands of his Great Society programs and the war in Vietnam, and was unwilling to provide significant financial support for major post-Apollo space initiatives. Congress continued to question whether NASA needed all the resources it was requesting to complete Apollo, and was equally unwilling to support major new programs. By 1966, Webb was frustrated by what he perceived as lack of adequate political support from the White House as he battled to hold off congressional attempts to slash the NASA budget. (Volume I, III-19)

NASA by this time was a very different organization than it had been just three years earlier, as the mobilization of human and financial resources needed to carry out Apollo peaked. The Agency's budget had increased by 89 percent in the year after President Kennedy's May 1961 speech, another 101 percent in the following year, and then another 38 percent as the budget approached its peak. The NASA staff had increased from 17,500 civil servants in 1961 to 34,300 at the end of 1965, and the related contractor force from 57,000 to 376,700.[54] Apollo was truly a national effort.

On 24 December 1965, NASA Deputy Administrator Hugh Dryden succumbed to cancer. With his death, NASA lost a respected official and a key participant in the management of the Agency as it had gone through this rapid expansion. Dryden was not replaced; Robert Seamans took on the position of Deputy Administrator while continuing his role as the Agency's general manager.

To the outside observer, all elements of the Apollo program appeared to be moving forward towards a lunar landing before the end of the decade, with the first flight of the Apollo spacecraft with a crew aboard scheduled for early 1967. During 1965 and 1966, a series of 10 mainly successful Gemini launches demonstrated many of the capabilities, particularly rendezvous and docking, that would be needed for Apollo. Four Saturn 1 launches and the first Saturn 1B launch tested various aspects of the lunar mission; two of these launches carried Apollo command and service modules without a crew aboard. Technical problems

---

53. Lambright, *Powering Apollo*, p. 132.
54. Sylvia Kraemer, "Organizing for Exploration," in John M. Logsdon *et al.*, eds., *Exploring the Unknown: Selected Documents in the History of the U.S. Civil Space Program*, (Washington, DC: National Aeronautics and Space Administration Special Publication-4407, Vol. I, 1995), p. 613.

with the F-1 engine that powered the first stage of the Saturn V appeared to have been resolved, and the mammoth booster was moving towards its first test flight.

The technical reality was rather different. There were major problems in the Apollo spacecraft program and the S-II second stage of the Saturn V launcher, both being developed by North American Aviation,[55] and the lunar module being developed by Grumman was running well behind schedule and was overweight.[56] By the end of 1966, Apollo's Washington managers were stressing publicly that it would be difficult to attempt an initial lunar landing mission until sometime in the second half of 1969. Thus Administrator James Webb was quite surprised to read an interview with Wernher von Braun that appeared in the 12 December 1966 issue of *U.S. News & World Report* headlines "A Man on the Moon in '68?" In the interview, von Braun suggested, with a number of caveats, that "there is a distinct possibility that, if everything really clicks and we don't hit any major snags, it [the first landing attempt] may come off in '68," on the fourth flight of the Saturn V launcher.[57]

In response to the interview, Webb fired off an annoyed memorandum to von Braun. (II-47) He told von Braun "there is certainly a very, very low possibility that complete Saturn V systems will be available for flights out as far as the Moon in 1968. Under these circumstances, it seems to me that you will need to be very careful in dealing with the press." Webb's concern was that NASA needed to "take account of all the difficulties we are likely to encounter in this very complex Saturn V-Apollo system, particularly as we are now so hemmed in, have so little room to make adjustments, and have no financial margins." He was also concerned that statements like von Braun's could "undermine the credibility of those of us who are working so hard to get the money to continue this program and to avoid having the vehicles now approved (15 Saturn V's) deleted from the program on the basis that they are not needed to accomplish the mission."

Soon after Lyndon Johnson became President, he had asked NASA to begin to identify post-Apollo options. NASA responded by January 1965 with a "laundry list" of future possibilities. (Volume I, III-18) But by that time, "Johnson did not want to hear about the possibilities, nor did he particularly want Congress to hear them."[58] Recognizing that a second Apollo-like initiative was not in the offing, NASA focused its post-Apollo planning on an interim effort that became known as the Apollo Applications Program. The program initially was ambitious in scope, but never received significant funding. (II-45) Ultimately only one of the proposed Apollo Applications missions was flown; this was the 1973 Skylab, using an upper stage of a surplus Saturn V launch vehicle as an interim space station. Lacking any additional missions for the Saturn V, in August 1968 Webb found himself forced to make the painful decision to begin the process of shutting

---

55. See Murray and Cox, *Apollo*, Chaps. 12–13, for a discussion of these problems.

56. For an account of the development of the lunar module, see Thomas J. Kerlly, *Moon Lander: How We Developed the Apollo Lunar Module* (Washington: Smithsonian Institution Press, 2001).

57. "A Man on the Moon in '68?" *U.S. News & World Report*, 12 December 1966, p. 63.

58. Lambright, *Powering Apollo*, p. 139.

down the production of the heavy lift booster, a decision that became final in 1972. (II-58)

Webb's biographer Professor W. Henry Lambright concludes that Webb's "strategies to maintain NASA, Apollo, and other programs had succeeded *and* failed in the 1964-1966 time frame." Webb had "kept up overall momentum for Apollo" but "NASA's budget was cut back . . . post-Apollo was delayed, and Webb saw his own power to persuade start to slip."[59]

The Apollo 1 Fire

Despite these concerns, there was a fair degree of optimism as 1967 began, with the first crew-carrying flight of Apollo (an Earth-orbital test mission of the Apollo command and service modules designated Apollo 204) scheduled for launch on 21 February. The crew included veteran astronauts Virgil "Gus" Grissom, Edward White, and rookie Roger Chaffee. The spacecraft they were to fly was a "Block A" model, intended only for orbital flight.

At 1:00 p.m. on 27 January, the crew was strapped into the spacecraft as it sat atop an unfueled Saturn 1B launcher on Pad 34 at Cape Canaveral for a lengthy countdown test. At 6:31, as the test neared its end, Roger Chaffee told the control room that "we've got a fire in the cockpit." Within less than a minute, the three astronauts were dead of asphyxiation as they inhaled toxic gases created by the fire within the still-sealed spacecraft.[60]

James Webb, Robert Seamans, and George Mueller learned of the fire soon afterwards. Webb immediately notified President Johnson; later the three huddled at NASA Headquarters to decide how to proceed. They decided to ask the president to let NASA manage the accident investigation rather than have the White House appoint an external investigation board. While Webb worked to convince Johnson and congressional leaders that NASA was best qualified to conduct the investigation, Seamans and Mueller identified the individuals who would compose the investigation board. Apollo program director Sam Phillips flew to Cape Canaveral (by then called Cape Kennedy) to take charge there. By the next day, the Apollo 204 Review Board had been named; it was to be chaired by Floyd Thompson, Director of NASA's Langley Research Center, and had eight other members from both within and outside of NASA. Seamans charged the board to "review the circumstances surrounding the accident to establish the probable cause or causes of the accident" and to "develop recommendations for corrective or other action based upon its findings and determinations." (II-48)

The Review Board went about its work intensively. By 25 February, its preliminary findings were ready to be made public, and James Webb issued a statement summarizing them. (II-49) In this statement, Webb noted that astronaut Frank Borman, a member of the board, had told him that "he would not have been concerned to enter the capsule at the time Grissom, White, and Chaffee

---

59. Ibid, p. 141.
60. See Murray and Cox, *Apollo*, Chaps. 14–15 for a description of the fire and its aftermath.

did so for the test, and would not at that time have regarded the operation as involving substantial hazard. However, he stated that his work on the board has convinced him that there were hazards present beyond the understanding of either NASA's engineers or astronauts."

The Apollo 204 Review Board submitted its final report to Administrator Webb on 5 April. (II-50) The board found that "the test conditions were extremely hazardous." Once the fire started, "the crew was never capable of effecting emergency egress because of the pressurization before rupture and their loss of consciousness soon after rupture." With respect to the spacecraft, "deficiencies existed in Command Module design, workmanship, and quality control."

On 27 February, the Senate Committee on Aeronautical and Space Sciences, chaired by Clinton Anderson of New Mexico, held the first congressional public hearing since the accident. While he supported the space program, Anderson did not get along with James Webb, who had resisted Anderson's attempts to exert influence over NASA activities, and was not willing to wait until the Review Board issued its final report to begin congressional questioning. At the hearing, junior Minnesota Senator Walter Mondale asked Webb about a "Phillips Report" severely critical of North American Aviation's management of its parts of the Apollo effort. Webb did not know what Mondale was referring to, and stonewalled the Senator's inquiry. George Mueller told the committee that no such report existed. Later that day Webb became furious when he discovered that there was indeed such a document, in the form of a set of notes and a cover letter sent to North American Aviation President Leland Atwood after a late 1965 visit to North American by a NASA review team led by Apollo program director Sam Phillips. (II-44) In his cover letter, Phillips had told Atwood that "I am definitely not satisfied with the progress and outlook of either program [the Apollo Spacecraft and S-II stage of the Saturn V]" and that "even with due consideration of hopeful signs, I could not find a substantial basis for confidence in future performance."

Neither Seamans nor Mueller thought that what Phillips had prepared in 1965 constituted a "report," but Webb saw immediately that semantic quibbling would not extricate NASA from appearing to be withholding information from Congress. After discussing how best to give Congress access to the material, NASA decided to have Sam Phillips present its contents to an open hearing of Andersen's committee. As he probed further, Webb discovered that there had been continuing criticism of North American's performance of which he had been unaware. Webb had a developing sense "that the men he trusted the most—his senior officials at headquarters— had let him down." In Webb's view, George Mueller "had deliberately presented a filtered picture of the situation, and Seamans had failed to press him on it." Determined to change this situation, Webb reasserted control of the Apollo program "with a vengeance."[61] One of his moves was to force North American to remove the senior manager of its Apollo efforts, Harrison Storms, from his position as head of the company's space

---

61. Lambright, *Powering Apollo*, p. 161. Lambright's book includes a thorough discussion of how Webb reacted to the Apollo fire and its aftermath.

division; if North American did not make such a move, threatened Webb, he would shift the Apollo contracts to another company.[62]

Relations between Webb and Seamans became strained in the months following the fire, and Seamans submitted his resignation on 2 October 1967. Mueller stayed on; a change at the top of the manned space flight program would likely have resulted in unacceptable delays in fixing the problems revealed by the fire and getting NASA back on track. In Houston, Joseph Shea took the Apollo fire as a personal responsibility, and his associates began to worry about his physical and mental condition. He was persuaded to return to Washington as a deputy to George Mueller, but without significant Apollo responsibilities. By July 1967, Shea decided to leave NASA. In Shea's place in Houston, George Low took over the Apollo Spacecraft Program Office in addition to his duties as Deputy Center Director.

Not only were relations strained between Webb and his senior people within NASA; there were continuing tensions between him and members of Congress, and particularly Senator Clinton Anderson. Webb had always prided himself on maintaining a relationship of mutual trust and personal credibility with senior Congressmen, and now that relationship seemed at risk. He wrote Anderson in advance of a 9 May hearing, saying that "I am deeply troubled by your statement to me last Saturday that members of the committee are not satisfied with our testimony on NASA's actions in follow-up of the deficiences [sic] found by the management review team headed by General Phillips at North American Aviation in 1965." He added "your statement that members of the committee believe NASA is endeavoring to put a disproportionate part of the blame for the Apollo 204 accident on North American Aviation and avoid its proper acceptance of blame troubles me even more." (II-51)

Eventually the furor over the accident quieted. There were no serious suggestions that the Apollo program be halted or the "before the decade is out" goal be abandoned. Under George Low's close supervision, North American set about remedying the deficiencies in the Apollo spacecraft. Grumman was moving ahead with its work on the lunar module, but continuing to confront both schedule and weight problems. The Saturn V had its first test launch on 9 November 1967; all test objectives were met successfully. As 1968 began, there was increasing confidence that the first lunar landing attempt could come before the end of 1969.

Apollo Around the Moon

By the beginning of 1968, NASA was ready to schedule the first launch of the redesigned Apollo Command and Service Module; the date was finally set for 7 October. That Earth-orbiting mission would be the first in a sequence of

---

62. Murray and Cox, *Apollo*, p. 231. For an account of this situation sympathetic to Storms, see Mike Gray, *Angle of Attack: Harrison Storms and the Race to the Moon* (New York: W.W. Norton, 1992).

missions leading up to a lunar landing. The missions were designated by letters of the alphabet:

> C – test of the Apollo Command and Service module in low Earth orbit
> D – test of the Apollo Command and Service and Lunar Modules in low Earth orbit;
> E – test of the Apollo Command and Service and Lunar Modules in a mission beyond Earth orbit, but not headed to the Moon;
> F – test of all equipment in lunar orbit;
> G – lunar landing mission.

It was not clear as the year began whether following this schedule would provide adequate assurance that the United States would reach the Moon before the Soviet Union. Throughout the 1960s, the Central Intelligence Agency (CIA) had closely monitored the progress of the Soviet space program. In the years immediately following the 1961 Kennedy decision to go to the Moon, there was no indication that the Soviet Union was developing the facilities and equipment that would be required for a competitive lunar landing program. When Soviet scientists in mid-1963 said that there was no Soviet lunar landing program, they were correct. But earlier in 1963, U.S. satellites had detected what appeared to be the beginning of a large construction project at the main Soviet launch site, the Baikonur Cosmodrome in the Soviet republic of Kazakhstan. By 1964, construction of a large assembly building and two launch pads could be seen. It was during that year that the Soviet leadership finally approved a Soviet Moon program, but there were continuing bureaucratic battles inside of the Soviet space community that slowed progress. The program also, it has been learned in retrospect, never received adequate funding. By mid-1965, the Intelligence Community had concluded that the Soviet Union did indeed have a lunar program, but that it was not proceeding on a pace that was competitive with Apollo. In December 1967, a U.S. satellite returned an image of a previously unseen large booster on one of the new launch pads.

Throughout this period, James Webb was regularly briefed on the status of the Soviet space effort. In 1964, and then with more frequency in 1966 and subsequent years, Webb said publicly that the Soviet Union was developing a launch vehicle with lifting capabilities larger than those of the Saturn V. The fact that the Soviet Union seemed to indeed be racing the United States to the Moon helped Webb politically as Apollo came under criticism in 1967 and 1968.[63]

In fact, the reality was that by 1967 the Soviet Union was conducting two lunar programs, one aimed at a lunar landing and a second, using a version of

---

63. This account of what the United States knew at the time about the Soviet space program is based on Dwayne A. Day, "Webb's Giant," *The Space Review*, 19 July 2004, *www.thespacereview.com/ article/188/1*, (accessed September 6, 2006), Dwayne A. Day, "The Secret at Complex J," *Air Force Magazine* 87, no. 7 (July 2004): pp. 72–76 and Dwayne A. Day, "From the Shadows to the Stars: James Webb's Use of Intelligence Data in the Race to the Moon." *Air Power History* 51, no.4 (Winter 2004): 30–39. For a discussion of what was later learned about the Soviet lunar effort, see John M. Logsdon and Alain Dupas, "Was the Race to the Moon Real?"

the proven Proton launch vehicle and a modified Soyuz spacecraft called Zond, aimed at flights around the Moon, without the capability to land. In April 1968, the CIA issued an update of a 1967 assessment of the Soviet program. (II-57) The report said that "we continue to estimate that the Soviet manned lunar landing program is not intended to be competitive with the US Apollo program. We now estimate that the Soviets will attempt a manned lunar landing in the latter half of 1971 or in 1972, and we believe that 1972 is the more likely date." However, added the CIA, "the Soviets will probably attempt a manned circumlunar flight both as a preliminary to a manned lunar landing and as an attempt to lessen the psychological impact of the Apollo program. In NIE 11-1-67 [the 1967 estimate], we estimated that the Soviets would attempt such a mission in the first half of 1968 or the first half of 1969 (or even as early as late 1967 for an anniversary spectacular). The failure of the unmanned circumlunar test in November 1967 leads us now to estimate that a manned attempt is unlikely before the last half of 1968, with 1969 being more likely." Senior Apollo managers could not help but have this intelligence estimate in the back of their minds as they moved toward the beginning of crew-carrying Apollo flights, although there is little direct evidence that it influenced their thinking.

In addition to getting Apollo hardware ready to fly, there was an immense amount of detailed effort required to actually design the lunar landing missions. That responsibility was assigned to a veteran NASA engineer named Howard W. "Bill" Tindall, who in August 1967 was named Chief of Apollo Data Priority Coordination, an opaque title that gave no indication of his sweeping responsibilities. Tindall had an exuberant personality and viewed Apollo "as one long stretch of fun that had by some miracle given to him instead of work." One of the results of Tindall's approach to his duties was a series of what became known as "Tindallgrams." While dealing with the myriad of serious issues involved in getting ready to land on the Moon, these communications adopted a breezy, irreverent tone, and "became a sensation" around the Manned Spacecraft Center. As one example, Tindall told George Low on 8 August that "a rather unbelievable proposal has been bouncing around lately"—to delete the rendezvous radar on the lunar module as a weight saving measure. Tindall continued "because it is seriously ascribed to a high ranking official [George Mueller]," it was being taken seriously. He told Low, "I thought I'd write this note in hopes you could proclaim it to be a false alarm or if not, to make it one."[64] (II-59)

While the redesigned Apollo spacecraft seemed ready for a crewed launch, the same could not be said of the Saturn V or the lunar module. The second test launch of the Saturn V took place on 4 April 1968. In contrast to the almost perfect first test launch the preceding November, there were multiple problems with this flight. Each of the three stages of the vehicle had a separate failure. It took all of the skill and experience of the von Braun rocket team to diagnose the

---

64. For a discussion of Tindall's contributions and style, see Murray and Cox, *Apollo*, pp. 292–297.

causes of the failures. This was essential, because NASA's planning called for the next flight of the Saturn V to carry three astronauts.[65]

That mission, designated "D" in NASA's plans, was intended to carry a complete Apollo spacecraft, including both the command and service modules and the lunar module, for a test flight in low Earth orbit. Presuming success of the "C" mission in October, NASA hoped to launch the next flight before the end of the year.

However, there was a major obstacle to overcome. The lunar module scheduled to be flown on the mission had arrived at the Kennedy Space Center with a number of problems to be solved. As NASA attempted to address them, it appeared increasingly unlikely that the module would be ready to fly in 1968, and indeed that the test flight might not be possible until February or March 1969. If that happened, the likelihood of landing on the Moon by the end of 1969 became remote. Faced with this situation, George Low began to consider an alternative flight sequence: "the possibility of a circumlunar or lunar orbit mission during 1968," using only the command and service modules launched by a Saturn V, "as a contingency mission to take a major step forward in the Apollo Program." By 9 August, as problems with the lunar module persisted, he took this idea to the Director of the Manned Spacecraft Center, Robert Gilruth, who immediately saw its benefits. The same morning, according to Low's notes:

> I met with Gilruth, Kraft and Slayton. [Christopher Kraft was head of flight operations and Donald 'Deke' Slayton was head of the astronaut office.] After considerable discussion, we agreed that this mission should certainly be given serious consideration and that we saw no reason at the present time why it should not be done. We immediately decided that it was important to get both von Braun and Phillips on board in order to obtain their endorsement and enthusiastic support. Gilruth called von Braun, gave him the briefest description of our considerations, and asked whether we could meet with him in Huntsville that afternoon. I called Phillips at KSC and also informed him of our activities and asked whether he and Debus could join us in Huntsville that afternoon. Both von Braun and Phillips indicated their agreement in meeting with us, and we set up a session in Huntsville for 2:30 p.m.

At the afternoon meeting in von Braun's office, "all present exhibited a great deal of interest and enthusiasm for this flight." The meeting ended "with an agreement to get together in Washington on 14 August 1968. At that time the assembled group planned to make a decision as to whether to proceed with these plans or not. If the decision was affirmative, Phillips would immediately leave for Vienna to discuss the plans with Mueller and Webb (at that time, Administrator Webb and manned spaceflight head Mueller would be attending a United

---

65. Bilstein, *Stages to Saturn*, pp. 360–363.

Nations Conference on the Peaceful Uses of Outer Space), since it would be most important to move out as quickly as possible once the plan was adopted." (II-60)

With all of the key managers of the Apollo meeting agreed, it would be difficult for NASA's top officials to overturn Low's plan, but it turned out that they also were not willing to give it their total approval. The senior managers from Houston, Huntsville, Cape Kennedy, and the NASA Headquarters Apollo program office met with the new NASA Deputy Administrator, Thomas Paine, on 14 August as planned. (Paine was a newcomer to space; before he came to NASA he had been an executive of the General Electric Company, most recently the manager of GE's Center for Advanced Studies. He had assumed the number two position at NASA in January 1968, following the resignation of Robert Seamans.) At the 14 August meeting, Paine "congratulated the assembled group for not being prisoners of previous plans and indicated that he personally felt that this was the right thing for Apollo and that, of course, he would have to work with Mueller and Webb before it could be approved." There was a decision not to send Sam Phillips to Vienna because his sudden appearance there might compromise what were still considered secret plans. Instead, interactions with Webb and Mueller were by secure telephone and diplomatic couriers.

Webb was "shocked" when he first heard of what his staff was planning, but quickly both he and Mueller saw the logic of what was being proposed. However, they added a note of caution. While the Apollo managers could begin to plan for a lunar mission, they could not commit NASA to undertaking such a bold step until the October C mission, designated Apollo 7, was a success. Following this constrained approval of the plan, Apollo Program Director Sam Phillips on 19 August issued a directive announcing the revised program plan. (II-61) The new mission would be designated C' (C prime) and Apollo 8. Whether it would go to the Moon, stay in low Earth orbit, or follow some other mission plan would not be decided until the results of the Apollo 7 mission were available, said Phillips.

As Low noted, the implications of this tentative decision were dramatic in terms of when the first attempt at a lunar landing could be scheduled. At the 14 August meeting,

> We also discussed the mission sequence to be followed after the proposed mission and proposed that the best plan would be to fly the D mission next, followed by an F mission, which, in turn, would be followed by the first lunar landing mission. In other words, the proposed mission would take the place of the E mission but would be flown before D. MSC also proposed that for internal planning purposes we should schedule the D mission for March 1, 1969; the F mission for May 15, 1969; and the G mission for July or August, 1969. However, dates two weeks later for D, one month later for F, and one month later for G should be our public commitment dates. (II-60)

Following Phillips's 19 August directive tentatively approving the C' mission, Low on 20 August issued his own directive to those working on the Apollo spacecraft and planning the Apollo missions. The launch date for the first attempt at a lunar landing was set for 8 July 1969. (II-66)

The Apollo 7 mission took place from 11 to 22 October 1968; aboard were astronauts Wally Schirra, Donn Eisele, and Walter Cunningham. All objectives of the flight were met, clearing the path for a decision to send the Apollo 8 mission into lunar orbit.

That decision would not be made by James E. Webb. On 16 September, Webb had gone to the White House for a meeting with President Johnson to discuss a variety of issues, including how best to protect NASA and particularly Apollo during the transition to the next President. (Johnson had announced in March 1968 that he would not seek reelection.) Webb knew that he was very unlikely to continue as NASA Administrator, whether Hubert Humphrey or Richard Nixon was elected. He and Humphrey did not get along, and as a committed Democrat he was even more unlikely to be retained by Nixon. Webb was weary after six and a half years running NASA at a frenetic pace, and had been a target of congressional criticism since the Apollo fire. Webb thought that at some point in the fall he should step aside and let Thomas Paine, a non-political person, demonstrate that he was capable of running NASA at least through the first lunar landing.

To Webb's surprise, the president not only took up Webb's offer to resign, but decided that Webb should announce it immediately, even before he left the White House. Obediently, Webb told the White House press corps that he would leave NASA on 7 October, his sixty-second birthday. Webb was not able to contact Paine or his wife before making the announcement.[66]

Although momentum was great after the success of Apollo 7 to take Apollo 8 to lunar orbit, a final decision to undertake that bold step had not yet been made. In particular, George Mueller was worried about whether the overall program gains from the mission justified the fallout from a failure. A final review of the mission was scheduled for 10 and 11 November. In advance of those meetings, Mueller wrote to Gilruth, saying "There are grave risks to the program as a whole, not just to the Apollo 8 mission, in embarking on a lunar orbit mission with the second manned flight of the CSM. We have to face the possibility that this type of mission could appear to the public, and to our peers in government, to be a precipitous, risky venture where the propaganda value is the only gain." Mueller was concerned that the enthusiasm within NASA for flying the mission might have had the effect of suppressing justified concerns about the risks. He told Gilruth "the risks from a purely technical aspect are probably reasonable and acceptable. If such a mission failed, however, the risks to the program as a whole could be significant." (II-62)

The 10 November meeting included the top executives of the companies involved in Apollo. After hearing a series of presentations by NASA managers, the executives were polled on their views of whether Apollo 8 should be approved as a lunar orbit mission. Although there were a few questions raised, according to

---

66. For more on Webb's resignation, see Lambright, *Powering Apollo*, pp. 200–204 and Murray and Cox, *Apollo*, pp. 322–323.

George Low, "the meeting was adjourned with the conclusion that a firm recommendation to fly the Apollo 8 mission to lunar orbit would be made the next day to the Acting Administrator." (II-63) That recommendation came the next day in the form of a memorandum from Sam Phillips to George Mueller. (II-64) On 11 November, there were a series of internal NASA meetings in which Thomas Paine heard the same briefings as had been given the previous day. In a first, large meeting, George Mueller continued to play the devil's advocate. A second meeting involved Paine, Associate Administrator Homer Newell, who had been with NASA since its beginning and was respected for his judgment, Mueller, and the NASA Center Directors. A third meeting involved only Paine, Newell, and Mueller. At its conclusion, Paine announced that he had approved the plan to make Apollo 8 a mission to go into orbit around Moon. (II-65) The launch date was set for 21 December, which meant that the Apollo spacecraft would go into lunar orbit on Christmas Eve.

As the launch preparations for Apollo 8 went forward, there was continuing concerns that the Soviet Union might still launch a flight around the Moon. Soviet Union had modified its new Soyuz spacecraft so that it could carry cosmonauts in a flight around the Moon (but not into lunar orbit). They designated the modified spacecraft Zond. It would be launched on its circumlunar trajectory by a version of the Proton rocket. The original hope was that the first flight with cosmonauts aboard could occur on the fiftieth anniversary of the Bolshevik Revolution in October 1967, but a failure in April 1967 of the Earth-orbital version of the Soyuz spacecraft, resulting in the death of cosmonaut Vladimir Komarov, delayed testing of the Zond spacecraft into 1968. A September 1968 Zond-5 did go around the Moon and returned its passengers— turtles and insects— to Earth, still alive. But a November Zond-6 mission had several failures; if there had been a crew aboard, they would have died. Even so, the cosmonauts scheduled to make the first crewed Zond mission asked permission from the Soviet Politburo to make an attempt at the next launch window in early December, but that permission was never given. The way was thus clear for Apollo 8 to be the first spacecraft to reach the Moon with humans aboard.[67]

The five first stage engines of the Saturn V booster rumbled into action at 7:51 a.m. on 21 December, lifting Frank Borman, James Lovell, and Bill Anders on their historic journey. Less than three hours later, the engine on the third stage of the launch vehicle fired, injecting the Apollo 8 spacecraft on a trajectory that would take it to the vicinity of the Moon three days later. Once it arrived at the Moon, the engine on its service module fired, placing the Apollo spacecraft into lunar orbit, where it remained for 20 hours.

The public highlight of the mission came on Christmas Eve, as the crew televised the view of the lunar surface from their spacecraft back to millions of people on Earth. Then, to the surprise of almost everyone, including the mission controllers back on Earth, the crew took turns reading the first verses from the

---

67. For more information on the failed Soviet circumlunar program, see Logsdon and Dupas, "Was the Race to the Moon Real?" and Marcus Lindroos, "The Soviet Manned Lunar Program," *http://www.fas.org/spp/eprint/lindroos_moon1.htm* (accessed 10 September 2006).

Bible's Genesis account of the creation of Earth. Frank Borman closed their broadcast by saying "goodnight, good luck, a Merry Christmas, and God bless you all—all of you on the good Earth."[68] (II-72)

In addition to this dramatic broadcast, the Apollo 8 crew brought home with them the iconic photograph of the blue Earth rising above the desolate lunar landscape when they landed on 27 December. In addition to its public impact, the successful mission demonstrated that NASA was ready to operate at the lunar distance. The path to a lunar landing had been pioneered; "For many of the people in the Apollo Program, Apollo 8 was the most magical flight of all."[69]

Goal Met: Americans on the Moon

Two missions stood between Apollo 8 and, if they were successful, the first attempt at a lunar landing. On 3 March 1969, for the first time a Saturn V launched the full Apollo spacecraft— the command and service modules and the lunar module. That combination at just over 292 thousand pounds was the heaviest payload ever put into orbit. The crew—James McDivitt, David Scott, and Rusty Schweickart—remained in Earth orbit. Over the course of the 10 day mission, the lunar module spent 6 hours undocked from the command and service modules at distances up to 113 miles before rendezvous and redocking, thereby demonstrating an essential element of the lunar orbital rendezvous approach. Both the descent and ascent engines of the lunar module were fired in a variety of modes. Schweickart performed a 39 minute extra-vehicular activity to test the Apollo portable life support system that would be used for walking on the lunar surface. The mission was extremely complex, and all of its objectives were met successfully.

Apollo 10 would be a dress rehearsal for the lunar landing mission, carrying out all elements of that mission except for the final descent from 47,000 feet above the lunar surface. It was planned to follow the same time line as a landing mission attempt, with the same Sun angles and the same out-and-back trajectory. Some, most notably George Mueller, thought that the mission should actually attempt the landing; Mueller's view was that to reduce risks the lunar landing should be achieved in the fewest possible flights. But this idea was vetoed, both because the lunar module assigned to the mission was too heavy to actually land and because the crew and the mission managers in Houston argued successfully that they needed the experience of this mission under their belts to reduce the risks associated with the first landing attempt.[70]

---

68. Murray and Cox, *Apollo*, pp. 325–334, provide a vivid account of the Apollo 8 mission. See also Robert Zimmerman, *Genesis: The Story of Apollo 8*, (New York: Four Walls Eight Windows, 1998). This essay will not provide detailed accounts of the Apollo missions. For such accounts from the astronauts' perspective, see Andrew Chaikin, *A Man on the Moon: The Voyages of the Apollo Astronauts*, (New York: Viking, 1994).

69. Murray and Cox, *Apollo*, p. 333.

70. Ibid, pp. 338–339.

Once again, Apollo 10 met all of its test objectives. The lunar module undocked from the command and service modules by about 350 miles, and then successfully redocked, once again demonstrating the feasibility of the approach that NASA had chosen in 1962. The mission was launched on 18 May and returned to a safe landing in the Pacific Ocean on 26 May. The crew of Thomas Stafford, John Young, and Gene Cernan had demonstrated that NASA was ready to try to land on the Moon. Apollo 11 was next.

NASA, and particularly the top astronaut official Deke Slayton, had adopted an approach to flight crew assignment that resulted in the backup crew for a particular mission becoming the prime crew for a mission three flights down that line. That meant that the Apollo 11 flight assignment would go to the crew that had been the backup for Apollo 8—Neil Armstrong, Edwin "Buzz" Aldrin, and Fred Haise. In reality, Haise was a replacement for the backup crew for Michael Collins, who had recently had back surgery. On 6 January 1969, Slayton informed Neil Armstrong that he would command the Apollo 11 mission, with Aldrin as his lunar module pilot. Collins had fully recovered from the surgery that had sidelined him for Apollo 8, and would serve as command module pilot.[71]

NASA of course had been planning the lunar landing mission in detail for some months. (Indeed, the Sea of Tranquility had been identified as a possible site for the first landing in 1962.) Because the basic objective of the initial mission was to land on the Moon's surface, get a few samples of lunar material, and return safely to Earth, that planning had been quite conservative. For example, Sam Phillips had proposed in October 1968 that the first landing mission should conduct only one extra-vehicular walk on the Moon's surface of no more than three hours, and that the astronauts should stay within 300 feet of the lunar module. An open item was whether both astronauts, or only one, should leave the lunar module for a walk on the Moon. After some controversy on whether it was feasible, Phillips also recommended that the mission should have the capability of televising the first steps on the Moon back to Earth. (II-67)

Once the Apollo 11 prime crew had been chosen, there followed almost seven months of intensive training to get them ready for the mission.[72] While they and their colleagues at the Manned Spacecraft Center focused on that training, NASA Headquarters considered how best to attend to the symbolic aspects of the mission. Richard Nixon had been elected president in November 1968, and as he took office he named Thomas Paine to continue to serve as Acting NASA Administrator; only after a number of others had turned the job down did he nominate Paine to be Administrator on 5 March. Paine was confirmed by the Senate on 20 March and was sworn in on 21 March.

During the spring Paine appointed one of his top advisors, Associate Deputy Administrator Willis Shapley, to chair a Symbolic Activities Committee to recommend to him how best to recognize the historic character of the first lunar landing. Shapley was a veteran Washington bureaucrat and both Webb and Paine

---

71. For more on the Apollo 11 crew assignment process, see Chaikin, *A Man on the Moon*, pp. 136–140.

72. For a discussion of this training, see Ibid, pp. 163–183.

looked to him for advice on political, policy, and budgetary issues. By mid-April, the Committee had decided that "the intended overall impression of the symbolic activities and of the manner in which they are presented to the world should be to signalize [*sic*] the first lunar landing as an historic forward step of all mankind that has been accomplished by the United States of America." The primary way to indicate that the lunar landing was an American achievement would be "placing and leaving a U.S. flag on the moon in such a way as to make it clear that the flag symbolized the fact that an effort by American people reached the moon, not that the U.S. is 'taking possession' of the moon. The latter connotation is contrary to our national intent and would be inconsistent with the Treaty on Peaceful Uses of Outer Space." (II-70, II-71)

In January 1969, some in NASA Headquarters had interpreted Richard Nixon's words in his inaugural address "as we explore the reaches of space, let us go to the new worlds together" as indicating that the White House might prefer that a United Nations flag, rather than the American flag, be placed on the Moon. When he heard of this suggestion, George Low told NASA Headquarters "my response cannot be repeated here. I feel very strongly that planting the United States flag on the moon represents a most important aspect of all our efforts." (Volume II, I-12)

Another matter of concern was what might be said as the first human stepped onto the Moon. Julian Scheer, the top public affairs official at NASA Headquarters, heard a rumor that George Low was seeking advice on what might be said. Scheer wrote to Low, saying "we have not solicited comment or suggestions on what the astronauts might say. Not only do I personally feel that we ought not to coach the astronauts, but I feel it would be damaging for the word to get out that we were soliciting comment." Scheer added "that the truest emotion at the historic moment is what the explorer feels within himself, not for the astronauts to be coached before they leave or to carry a prepared text in their hip pocket." Low quickly responded, saying that there had been a misunderstanding; Low had sought advice on what should be carried to the Moon, not what should be said. He added, "I completely agree with you that the words said by the astronauts on the lunar surface (or, for that matter, at any other time) must be their own. I have always felt that way and continue to do so." Low had made the point to Neil Armstrong, the Apollo 11 commander, that "whatever words are said must be his own words." (II-68, II-69)

The first human mission to the Moon was launched at 9:32 a.m. EDT on 16 July, 1969. Four days later, at 4:17 p.m., after a perilous descent, the lunar module came to rest on the lunar surface. A few seconds later, Armstrong radioed back "Houston, Tranquility Base here. The *Eagle* [the name assigned to the mission's lunar module] has landed."[73]

The mission plan that had been prepared for the Apollo 11 crew called for them to go to sleep between the time they landed and the time they exited the lunar module for the first Moon walk. (A decision had been made that both

---

73. For a vivid account of the landing and the rest of the Apollo 11 mission, see Ibid, pp. 184–227.

Armstrong and Aldrin would conduct an extra-vehicular activity, rather than have one astronaut stay in the lunar module as a safety measure.) But with the landing safely behind them and the lunar module in good condition, the keyed-up crew suggested that they begin their Moon walk five hours ahead of schedule, without the intervening sleep period. Permission was quickly granted. Getting ready to leave the lunar module went more slowly than had been planned, but finally, at 10:56 p.m. EDT, Neil Armstrong stepped off of the lunar module, saying, "that's one small step for man, one giant leap for mankind." (Armstrong meant to say "a man," but the "a" may have gotten lost in the excitement of the moment.)

Aldrin followed Armstrong 14 minutes later. The two spent two and a half hours carrying out their assigned tasks, including planting the U.S. flag on the lunar surface. During their Moon walk, President Richard Nixon called from the Oval Office, proclaiming it "the most historic telephone call ever made from the White House."

The ascent stage of the *Eagle* performed as planned, and at 1:54 p.m. EDT on 21 July Armstrong and Aldrin were launched from the lunar surface to rendezvous with Michael Collins, who had been circling the Moon in the command and service module *Columbia*. The two spacecraft docked three hours later, and a little less than twelve hours later fired the service module engine to send them on a trajectory for a landing in the Pacific Ocean at 12:50 p.m. EDT on 24 July. The crew, the command module, and the 44 pounds of precious lunar cargo were immediately placed in quarantine, where they were soon greeted by the President, who had flown to the recovery ship, the aircraft carrier *Hornet*, to greet them. The *Hornet* docked in Honolulu, Hawaii, on the afternoon of 26 July; from there, the crew flew back to Houston. (I-73)

The goal set by John F. Kennedy just over eight years earlier had been met; Americans had flown to the Moon and returned safely to Earth. Apollo 11 was a success, technically and politically. (II-81, II-74)

What Do You Do Next,
Once You Have Been to the Moon?

Continuing Exploration

There were a few within NASA, Robert Gilruth among them, who thought that there should be no additional flights to the Moon, given how risky they were and that the program's fundamental goal had been achieved. But the momentum behind additional missions overrode these hesitations. As Apollo 11 concluded its mission in July 1969, there were nine additional flights to the Moon, through Apollo 20, being planned. Apollo 12 through 15 would use the same basic equipment as had Apollo 11, but would land at different locations and stay for increasingly longer times on the lunar surface. Apollo 16 through

20 would carry a lunar rover, a small vehicle that would allow the astronauts to transverse the lunar surface, and could stay on the Moon for up to 78 hours.[74]

Apollo 12, carrying Charles "Pete" Conrad, Alan Bean, and Richard Gordon, was launched during a thunderstorm on 14 November 1969. Lightning struck the spacecraft during its initial ascent and for a moment it appeared that the mission would have to be aborted. But this threat passed, and the lunar module made a precision landing within walking distance of the Surveyor III spacecraft that had landed on the moon in April 1967.

The next mission, Apollo 13, was launched on 11 April 1970. Its crew included James Lovell, Jack Swigert, and Fred Haise. Swigert was a last minute substitute for T. K. "Ken" Mattingly. There was concern that Mattingly had been exposed to measles and might become ill during the mission. More than two days away from Earth on the mission's outbound journey, an oxygen tank in the service module exploded, placing the crew's life in jeopardy. There would be no lunar landing, and it took heroic efforts by the crew and those on the ground to bring the crew back safely by using the lunar module as a life boat for most of the journey around the Moon and back to Earth.

The intended landing site for Apollo 13 had been the Frau Mauro, a location of high scientific interest. When Apollo 14 was launched, on 31 January 1971, it was targeted to land at the same site. The crew—Alan Shepard, who had made the first U.S. spaceflight almost ten years earlier, Stuart Roosa, and Edgar Mitchell—carried out two extended Moon walks. The mission became notorious when at the conclusion of the second walk Shepard used a piece of lunar equipment with an actual head of a six-iron golf club inserted in it to hit (after two misses) the first lunar golf shot.

The Apollo 14 crew was the last to be required to undergo quarantine after their return to Earth. The possibility of lifting the quarantine had been examined after the first two Apollo missions, but the Interagency Committee on Back Contamination refused to do so on the recommendation of a committee set up by the National Academy of Sciences to review the issue. Because Frau Mauro, the planned landing site for Apollo 13, was a very different type of location on the Moon than the places where Apollo 11 and 12 had landed, and because the astronauts would take a deep core sample, "a majority (of the committee) recommend continuance of the 3-week lunar quarantine period. A minority favor discontinuance of quarantine." (II-76) When there was no evidence of possible back contamination after Apollo 14 returned, the requirement for quarantine was lifted.[75]

Missions Canceled

When Richard Nixon became president in January 1969, he was advised of the need for decisions on the character of the U.S. civilian space program once

---

74. For a description of the various stages of lunar explorations, see Compton, *Where No Man Has Gone Before*, Chaps. 10–14.

75. Ibid, p. 223.

Apollo had reached the Moon. In February, he asked his Vice President, Spiro Agnew, to chair a "Space Task Group" to provide him with recommendations on the future in space. The Task Group worked through the summer in the midst of the enthusiasm surrounding Apollo 11, and on 15 September submitted a bullish set of recommendations that called for the United States to accept "the long-range option or goal of manned planetary exploration with a manned Mars mission before end of this century as the first target." (Volume I, III-25) As steps toward this goal, the group recommended a series of increasingly larger Earth-orbiting space stations launched by the Saturn V and continued exploration of the Moon. It also recommended the development of a lower cost Earth-to-orbit space transportation system, which soon became known as the Space Shuttle.

This type of recommendation was not at all what the Nixon administration had in mind; its top goal was reducing government spending. Between September 1969 and January 1970, the NASA budget went through a series of reductions from what had been proposed to get started on the recommendations of the Space Task Group, and George Low, who had become NASA Deputy Administrator in December 1969, announced on 4 January 1970 that NASA was canceling Apollo 20 and stretching out the remaining seven missions so that they would continue through 1974. Ten days later, faced with continuing budget cuts, Administrator Thomas Paine announced that production of the Saturn V would be suspended indefinitely once the fifteenth vehicle had been completed.[76] While NASA tried for several years to retain the option of restarting the Saturn V production line, by 1972 it decided that it had no choice but to give up this possibility. (Volume IV, I-46) Thus within six months of the first landing on the Moon, the United States had essentially abandoned the heavy-lift capability that had been so central to James Webb's vision of what Apollo could create.

Richard Nixon finally responded to the Space Task Group in a statement issued on 7 March 1970, saying, "space expenditures must take their proper place within a rigorous system of national priorities. What we do in space from here on in must become a normal and regular part of our national life and must therefore be planned in conjunction with all of the other undertakings which are important to us."[77] It was clear that there would be no more Apollo-like space goals set while Nixon was in office.

Nixon's intent to reduce the government budget continued to have an impact on Apollo through the rest of 1970, and even into 1971. Thomas Paine, frustrated by his lack of success in getting White House support for ambitious post-Apollo plans and eager to return to General Electric, announced his resignation on 15 August 1970; George Low became Acting Administrator. Low was almost immediately faced with a decision on whether to cancel two of the remaining six Apollo missions, recognizing that NASA could not both fly these missions, launch the Skylab space station in 1973, and begin its preferred new program, the Space

---

76. Ibid, pp. 195–196.
77. Richard M. Nixon, "Statement About the Future of the United States Space Program," 7 March 1970, in U.S. President, *Public Papers of the Presidents of the United States: Richard Nixon, 1970* (Washington: Government Printing Office, 1971), p. 251.

Shuttle, within the budget being proposed for the next several years by the White House Office of Management and Budget. (II-77) Reluctantly, Low agreed that NASA should cancel Apollo 15, the last limited capability mission, and Apollo 19. The remaining flights after Apollo 14 were then renumbered Apollo 15 through 17. The cancellation was announced on 2 September.

Even after this announcement, there was continued White House pressure to reduce the NASA budget. There was a possibility that NASA would have to choose between canceling one or more of the remaining Apollo missions and flying Skylab. Low wrote Nixon's science advisor Edward Davis, Jr., on 30 October, saying "on balance, the weight of evidence seems to favor Skylab over Apollo if a choice must be made." This was the case, said Low, because "the scientific returns from the single Skylab mission promise to be greater than those from a sixth Apollo lunar landing. We have already capitalized on our Apollo investment but not yet on that of Skylab; we will have more new options better developed stemming from Skylab than from Apollo; and, for this increased return, we risk less in earth orbit than at lunar distances." (II-78)

NASA was not forced to make this draconian choice in 1970, but the possibility of canceling Apollo 16 and Apollo 17 was revived by the White House in 1971 as the NASA budget was being prepared; the space agency was also seeking White House permission to begin development of the Space Shuttle. James Fletcher, a former president of the University of Utah and industrial executive, had become NASA Administrator in May 1971. In November, he wrote the Deputy Director of the Office of Management and Budget, Caspar W. Weinberger, recommending "against the cancellation of Apollo 16 and 17 because these flights are scientifically important, and because much of the overall support for NASA's space program depends on our actions with respect to these flights." In his letter, Fletcher listed a number of adverse consequences that could result. If, however, a decision to cancel the missions were made, said Fletcher, the rationale behind the decision should be that "in these times of pressing domestic needs, the manned space program should be earth-oriented instead of exploration and science-oriented." (II-79)

In reality, it was the professional staff of the Office of Management and Budget, not Weinberger, who was pressing for the cancellations. In an 12 August 1971 memorandum to President Nixon, Weinberger had written that his office was proposing to cut the NASA budget "because it is cuttable, not because it is doing a bad job or an unnecessary one." He added, "I believe that this would be a mistake" because "an announcement that we are canceling Apollo 16 and 17 . . . would have a very bad effect." In Weinberger's view, "it would be confirming in some respects, a belief that I fear is gaining credence at home and abroad: That our best years are behind us, that we are turning inward, reducing our defense commitments, and voluntarily starting to give up our super-power status, and our desire to maintain world superiority." The memorandum was returned by Nixon with a hand-written notation: "I agree with Cap." (Volume I, III-28)

It was a political rationale that initiated the Apollo program, and at least in part it was a political rationale that convinced the White House to continue the program. Apollo 16 and 17 would be launched.

The Scientists Are Not Happy

Tensions between those who saw Apollo as an opportunity to gather valuable scientific data and materials and those who saw it as primarily a challenging engineering enterprise intended to demonstrate U.S. technological and organizational might had been present since the start of the program, and persisted through to its conclusion. For example, Donald Wise, the chief scientist of NASA's Apollo Lunar Exploration Office, left the Agency in the immediate aftermath of Apollo 11, telling Associate Administrator Homer Newell that his office, with the responsibility for getting lunar science moving, "was largely wasting its time running in tight circles within the bureaucracy and the various competing elements of NASA." He felt that this situation would persist until the NASA leadership "determines that science is a major function of manned space flight." (Volume V, I-25) Echoing this concern, George Mueller wrote to Manned Spacecraft Center Director Robert Gilruth in September 1969, reminding Gilruth that after Apollo 11 "increased interest and direct participation of the scientific community in Apollo is taxing our capability to the limit. Despite this, we will certainly detract measurably from the success of Apollo 11, and the missions yet to be flown, unless we meet the challenge. Therefore, we must provide the support required in the science area." With respect to criticisms from the scientific community about the scientific aspects of Apollo, Mueller added "some members of the scientific community are impatient and as you know, are willing to air their views without necessarily relating those views to what is practicable and possible. Public discussion aside, it is our policy to do the maximum science possible in each Apollo mission and to provide adequate science support." (II-75)

There were some members of the scientific community who were excited by the potential scientific returns from Apollo, and were very upset as NASA canceled three Apollo missions in 1970. A letter from 39 scientists protesting these cancellations was sent to Representative George Miller, Chairman of the House Committee on Science and Astronautics, soon after NASA announced the cancellation of the Apollo 15 and Apollo 19 missions. The scientists argued that "the Apollo lunar program is intended to supply not merely information of interest to scientists, but to give us finally a clear understanding of the origin of the earth-moon system and with this, an understanding of the origin and mode of construction of our earth." "Because the structure of the Apollo program is one of increasing capabilities," they stated, "the two canceled missions represent much more than one third of the planned scientific program. With this curtailment, the program may fail in its chief purpose of reaching a new level of understanding." (II-80)

The Final Missions

Pressure from the science community had one tangible result. On 13 August 1971 NASA announced that the crew for the last mission to the moon, Apollo 17, would include as lunar module pilot Harrison H. "Jack" Schmitt, a Ph.D. geologist who had come to NASA as a scientist-astronaut in 1965 and had been deeply involved in planning the science to be done on the lunar missions. Assigning Schmitt to this mission meant that Joe Engle, who had been part of the Apollo 14 backup crew together with Gene Cernan and Ron Evans, would not have an opportunity to fly to the Moon.[78]

Schmitt's selection came on the heels of the scientifically most successful mission to date, Apollo 15, launched on 26 July with a crew of David Scott, Alfred Worden, and James Irwin. This was the first mission to carry the lunar roving vehicle, and Scott and Irwin used the vehicle to traverse almost 17 miles of the lunar surface, a distance much greater than that traveled by the previous three crews. They spent three days on the Moon, and conducted three extra-vehicular activities. Most significantly, they identified and brought back to Earth specimens of the primitive lunar crust, the first material that had solidified from the molten outer layer of the young Moon; one of these samples was dubbed the "Genesis rock." [79]

The penultimate Apollo mission, Apollo 16, was launched on April 16, 1972. The mission was commanded by John Young; other crew members were command module pilot Ken Mattingly, who had been bumped from the Apollo 13 mission, and lunar module pilot Charles Duke. The mission was targeted to land in the lunar highlands, an area of the Moon that had not yet been explored. Apollo 16's objectives were similar to those of the preceding mission, with a focus on characterizing a region thought to be representative of much of the lunar surface.[80]

All of the prior Apollo missions had been launched during daylight hours. After an almost three-hour delay, Apollo 17 lifted off at 12:33 a.m. EST on 7 December 1972. The vivid light from the Saturn V's five F-1 engines illuminated the night sky with an unreal brilliance. After they landed on the Moon on 11 December "for the next 75 hours Cernan and Schmitt conducted the longest, and in many ways the most productive, lunar exploration of the Apollo program."[81] As they prepared to leave the lunar surface for the last time, Cernan unveiled a plaque on the descent stage of the lunar module, which would remain on the Moon's surface. It read "Here man completed his first explorations of the moon." As he took a last look at the lunar landscape, Cernan added "As we leave the moon at Taurus-Littrow, we leave as we came, and, God willing, as we shall return, with peace and hope for mankind." The Lunar Module *America* lifted off of the Moon at 5:55 p.m. EST on December 14.

With its departure, a remarkable era in human history came to a close, at least for the next half-century. For the first time, human beings had left their home planet.

---

78. For more on Schmitt's selection, see Chaikin, *A Man on the Moon*, 448–451.
79. Compton, *Where No Man has Gone Before*, pp. 231–242.
80. Ibid, pp. 244–247.
81. Ibid, p. 250.

Document II-1

Document Title: NASA, " Minutes of Meeting of Research Steering Committee on Manned Space Flight," 25- 26 May 1959.

Document Source: Folder 18675, NASA Historical Reference Collection, History Division, NASA Headquarters, Washington, DC.

Within less than a year after its creation, NASA began looking at follow-on programs to Project Mercury, the initial human spaceflight effort. A Research Steering Committee on Manned Space Flight was created in spring 1959; it consisted of top-level representatives of all of the NASA field centers and NASA Headquarters. Harry J. Goett from Ames, but soon to be head of the newly created Goddard Space Flight Center, was named chair of the committee. The first meeting of the committee took place on 25 and 26 May 1959, in Washington. Those in attendance provided an overview of research and thinking related to human spaceflight at various NASA centers, the Jet Propulsion Laboratory (JPL), and the High Speed Flight Station (HSFS) at Edwards Air Force Base. George Low, then in charge of human spaceflight at NASA Headquarters, argued for making a lunar landing NASA's long-term goal. He was backed up by engineer and designer Maxime Faget of the Space Task Group of the Langley Research Center and Bruce Lundin of the Lewis Research Center. After further discussion at its June meeting, the Committee agreed on the lunar landing objective, and by the end of the year a lunar landing was incorporated into NASA's 10-year plan as the long-range objective of the agency's human spaceflight program.

NATIONAL AERONAUTICS AND SPACE ADMINISTRATION
1520 H STREET NORTHWEST
Washington 25, D.C.

Minutes of Meeting of
RESEARCH STEERING COMMITTEE
ON MANNED SPACE FLIGHT

NASA Headquarters Office
Washington, D.C.

May 25-26, 1959

Present:
      Mr. Harry J. Goett, Chairman
      Mr. M. B. Ames, Jr. (part time)
      Mr. De E. Beeler
      Dr. A. J. Eggers
      Mr. M. A. Faget
      Mr. Laurance K. Loftin, Jr.
      Mr. George M. Low

Mr. Bruce T. Lundin
Mr. Harris M. Schurmeier
Mr. Ralph W. May, Jr., Secretary

Observers:
Mr. John Disher
Mr. Robert Crane
Mr. Warren North
Mr. Milton Rosen (part time)
Mr. Kurt Strass

COMMITTEE PURPOSE

The Directors of the Offices of Aeronautical and Space Research and Space Flight Development had planned to attend the beginning of this first meeting to express their interests in and objectives for the Committee. As circumstances prevented their attendance, the Chairman disclosed his interpretation of their views. He reaffirmed that the Committee was formed by the Office of Aeronautical and Space Research and reports to that office. The office desires that the Committee take a reasonably long term look at man-in-space problems leading eventually to recommendations as to what [2] future mission steps should be and to recommendations concerning broad aspects of Research Center (including JPL and HSFS) research programs to assure that they are providing proper information. It is hoped that the Centers will assist the Committee by making general studies for it as deemed necessary and that there will be a healthy relationship between the Centers and Committee with mutual support. Although the Committee needs to do long range thinking about space flight missions and concepts, it should be set significantly beyond Mercury and Dyna Soar. The Chairman further explained, that although the Office of Space Flight Development had no cognizance over the Committee, the director has expressed hopes that the Committee in an interim sort of way could make some recommendations by September 1959 regarding what type of approach Space Flight Development should take in using Fiscal Year 1961 budget money earmarked for Project Mercury follow-up.

Following these statements there was some discussion of our relationship to other committees, in particular the ARPA Man in Space Committee, the ARPA MRS-V Committee and the NASA Long Range Objectives and Program Planning Committees. The first committee was set up for ARPA-NASA relations on Project Mercury and apparently is being disbanded. The MRS-V Committee has just formed, has NASA representation (George Low), and concerns an ARPA manned recoverable satellite vehicle project that has no firm status as yet. The latter committee composed of Dr. Hagen and Messrs. Ames and Clement is concerned with arriving at a general ten year NASA research and development program for Dr. Glennan's use with the Space Council in connection with the 1961 budget.

Each member then gave this views about how this Committee should operate. There was unanimous feeling that we should not be influenced by other committees or groups. NASA is concerned with the national space program so this committee should do long range objective planning, decide what supporting research and to some extent what vehicle recommendations are appropriate, and then take aggressive steps to assure that the work is implemented with proper orientation and coordination among all NASA Research Centers including JPL and HSFS. Certain space flight objectives have to be decided upon early to work toward. The Committee should not get bogged down with justifying the need for man in space in each of the steps but out-rightly assume that he is needed inasmuch as the ultimate objective of space exploration is manned travel to and from other planets. It is felt that the Committee can help put [3] more objectiveness in NASA space research by stressing overall jobs to be done and concepts to be explored. Past experience as with the X-15 and Mercury has shown that research geared to definite objectives is mutually beneficial to both research planning and project development. On the other hand a point was made that the Committee has to assure that NASA research retains enough diversity to avoid overlooking important new ideas. It is questionable, however, as to whether the NASA will be able to develop space research to the degree of systematic coverage that the NACA was able to do previously for example in the case of the aerodynamics of aircraft wing and body configurations.

National Booster Program

Mr. Rosen reviewed the national booster program as presently conceived including Scout, Delta, Vega, Centaur, Saturn (formerly Juno IV) and Nova. This information is largely available in a brochure on the booster program distributed to all NASA centers and is not repeated here. A tabulated synopsis is appended however. [not included] This information is still fairly current except it now appears that the Saturn payload capabilities may be as much as 50 percent higher. The Nova vehicle depends strongly on hydrogen and its design is still very fluid. Lewis is working reasonably on this project and Mr. Lu[n]din agreed to supply reasonably detailed information on it for distrubution [sic] to committee members.

NASA has invited proposals to develop a system to recover the two rocket engines and associated vehicle tail section that is normally ejected from Atlas. The proposals are to be for eight Atlas' if the overall operation is shown to be at a net saving to the Government. The Committee asked for copies of the proposal invitation. Subsequent to the meeting Mr. Rosen has indicated that the contract document is not in a form suitable for distribution and probably would not be of any interest to the members since no specifications are made, Space Flight Development will be glad to make them available to the Committee.

Mr. Crane reviewed the stringent booster requirements of Dyna Soar to assure that the vehicle will not exceed critical load and temperature limits throughout the flight range. This restricts the vehicle to a rather limited altitude-velocity corridor. To accomplish this any of the [4] boosters in the present booster program would have to be modified to a major degree. A 4-barrel modified Titan first stage booster has been proposed by one of the contractors. The Committee asked to be kept informed of the Dyna Soar booster developments.

Dyna Soar

Mr. Ames briefly reviewed the history of Dyna Soar up to its present status of source selection between two contractors – Boeing and Martin-Bell. He also discussed some of the philosophy of why it is considered as a hypersonic research vehicle for exploring the flight corridor at speeds up to orbital and some of the design features.

Some concern was voiced that the Dyna Soar concept utilizes the radiation cooling principle to the limit of existing technology without leaving much room for growth. Thus some members felt that Dyna Soar did not fit in the NASA space exploration mission. Other members, however, recognized the need for continuing to look at vehicle conrecognized [*sic*] the need for continuing to look at vehicle concepts with orbital flight and conventional landing capabilities; Dyna Soar does fit into this picture and also permits exploration of winged vehicles at speeds up to orbital.

Project Mercury

Mr. Faget discussed in considerable detail the Project Mercury concept, its operational and design features, the test and build up development programs, its status and planned schedule. The material he discussed is largely summarized in a document prepared by the Langley Space Task Group entitled "Project Mercury Discussion" dated May 18, 1959, which was distributed to most Committee members. A movie was also shown dealing with the Mercury capsule fabrication, mockup, escape rocket system and orientation control system. Project Mercury has pointed up the need for general research on large parachutes.

PROGRAM REVIEWS BY COMMITTEE MEMBERS

Mr. Loftin: - Sixty percent of Langley's effort is currently related to space and reentry flight broken down approximately as follows:

Satellites and spacecraft 24% (Including 7% on Project Mercury)

Ballistic Missiles 10%

[5]

Anti-Ballistic Missiles 6%

Boost-glide winged reentry 20%

A substantial amount of Langley's work is on new flight concepts and is across the board involving investigations of overall aerodynamic characteristics, stability and control, heat transfer, structural aspects, systems analysis, pilot integration and so forth. Examples of these are work on (1) winged reentry at 90% angle of attack of a vehicle type having folding wing tips and a landing L/D of 8 to 10, (2) a kite type concept utilizing folding high temperature metal cloth between structural members, (3) inflatable wing concepts so that wings can be folded for take-off, and (4) a half pyramid configuration with possible aerodynamic heating advantages but a large base area resulting in low L/D.

Langley's more general research on spacecraft may be categorized along the general lines of –

(1) Aerodnamics [*sic*] and Gas Dynamics: - Examples – Heat transfer, general aerodynamic characteristics, boundary-layer transition, static and dynamic stability investigations.

(2) Structures: - Examples – Investigation of structural design concepts such as radiation cooling, forced cooling, ablation, heat sinks, sandwich construction, environmental effects and aeroelastic characteristics. (The Committee asked for a detailed elaboration of this at the next meeting.)

(3) Materials: - Examples – Studies of emissivity, ablating materials, refractory materials, oxidation and evaporation in high vacuums.

(4) Dynamic Loads: - Examples – Noise, vibration, flutter, fuel sloshing, gust loads and landing loads.

(5) Trajectories, guidance and control: - Examples – Trajectory calculations, static and dynamic computations of manned reentry, flight simulation studies, reaction control investigations, use of ground flight control center, and investigation of such things as solar auxiliary power, horizon scanners, flywheel inertial devices and the like.

[6]

(6) Space Flight: - Langley has a lunar committee that is considering a conceptual approach to a small lunar orbital vehicle that would be of interest to manned space flight. Mr. Loftin felt that we are reaching the point where reentry research is being overemphasized in comparison to research on actual space flight. In true space flight man and the vehicle are going to be subjected to space environment for extended periods of time and there will undoubtedly be space rendezvous requirements. All these aspects need extensive study and Mr. Loftin felt the best means would be with a true orbiting space laboratory that is manned and that can have crew and equipment changes. Langley is starting to look at this step.

Mr. Eggers: - Mr. Eggers reviewed a write-up he had prepared for the meeting summarizing research being conducted at Ames applicable to problems associated with manned space flight. Since copies were distributed to the membership, no reference is made to its content in the minutes.

In addition, Mr. Eggers discussed some preliminary long range research program thinking at Ames geared toward a space flight objective. It is difficult to evolve a good research program without some flight objectives in mind. Man's capabilities for space flight are not known and firsthand experience is needed through use of a broad based vehicle concept that is flexible in operation. With this in mind Ames set down some ground rules for thinking about a space flight mission to work toward, namely – (1) For the time being the planning should be for a concept that can be achieved within 5 to 10 years, (2) the space flight concept should be realistic in terms of money and manpower and (3) the flight concept should be one in which there would be a strong interplay between laboratory research and vehicle development. This has led to Ames' present thinking that the next step NASA manned space flight vehicle should have the following performance objectives: (1) two man occupancy, (2) escape speed capability, (3) lunar orbit capability, and (4) a minimum flight duration of one week. Ames' preliminary estimate is that the minimum weight of such a vehicle would be about 6000 pounds. Saturn could probably boost it. Ablation shielding should be able to handle the hearing satisfactorily. The vehicle should be capable of diversified space research on many problems including investigation of space and atmosphere maneuvering, pilot competence and capabilities, space science experiments, telescopic observations, lunar observations and so forth where man would be a vital link in the operation or experiment. [7]

Mr. Lundin: - Mr. Lundin also had a prepared writeup distributed to the membership which discussed Lewis' research in the categories of (1) trajectory studies, (2) mission analyses, (3) storage of cyrogenics in space, (4) power production, (5) shielding, (6) electronic propulsion, (7) supersonic parachutes and stabilizing devices, (8) jet blast and noise at launch site, (9) control, navigation and guidance, and (10) manned space capsule orientation control. The material in this write-up is not reiterated in the minutes. Lewis research now leans heavily toward rockets of various types and high energy propellants. Lewis is working mostly on applications of the type being considered for the top stages in present national booster program. There is a significant amount of work going into pumps and hydrogen-oxygen auxiliary power systems also.

In conclusion Mr. Lundin expressed his views that the Committee should not concern itself much with flight vehicle concepts well along the way such as Mercury or Dyna Soar but rather should look to longer range objectives. He felt strongly that although the Committee must consider interim space flight programs, we should not set our sights too low for the present long objective. The ultimate objective is manned interplanetary travel and our present goal should be for a manned lunar landing and return as with Nova. If we limit our present objective to manned reconnaissance, we may seriously impair the country's ultimate space flight objective. He mentioned that the Air Force already has a manned lunar landing mission under study under SR-183.

Mr. Beeler: - Mr. Beeler likewise reviewed some prepared circulated material on HSFS research pertinent to space vehicles in the categories of (1) reaction controls, (2) terminal guidance and landing problems, (3) exit and entry research, (4) crew factors, (5) air launch studies, (6) astronomical platform, and (7) space flight.

He listed some major space flight and research objectives that the NASA could work toward as follows:

Space flight objectives	Research areas
Man in space soonest (on way with Mercury)	Maneuvering entry
Lunar reconnaissance	Orbiting laboratory
Lunar landing and return	Rendezvous

[8]

Mars-Venus reconnaissance

Mars-Venus landing and return

HSFS feels that manned lunar reconnaissance is a good goal to work toward and has in mind a vehicle with the same general performance objectives as mentioned by Mr. Eggers.

Mr. Schurmeier: - Mr. Schurmeier in discussing JPL's work on Vega gave the following performance fingers which are somewhat lower than mentioned by Mr. Rosen earlier:

Attitude-circular orbit	Payload
3000 miles	1300 lbs.
1000 miles	3500 lbs.
300 miles	5000 lbs.
100 miles	5700 lbs.

Of the eight vehicles ordered, the first four are primarily for vehicle development with payload interests secondary.

The general type of payloads and approximate firing schedules are, however, as follows:

Vehicle Number	Launch Date	Payload
1	August 1960	Lunar probe
2	October 1960	Mars probe
3	January 1961	Venus probe
4	March 1961	Earth satellite
5	May 1961	Earth satellite
6	July 1961	Space mission
7	September 1961	Earth satellite
8	November 1961	Space mission

Firing pad availability at Canaveral restricts launchings to one every other month. Because of this and the necessary vehicle development required for manned flight reliability, Vega will probably not be available for manned missions before 1962.

JPL is doing a substantial amount of research on mission studies. A report on this work has recently been published which Mr. Schurmeier agreed to send copies of to the Secretary for distribution to all members. JPL considers its primary mission to be that of deep space exploration. At present it is concentrating on unmanned flight concepts although the ultimate objective is [9] manned interplanetary travel to explore life on other planets.

The last item Mr. Schurmeier covered was JPL's general research related to man in space. In particular he mentioned JPL's work in control, guidance, navigation, tracking, communications, solid propellant rockets, storable liquid propellant [*sic*] rockets, fundamental physics, nuclear propulsion and auxiliary power. JPL every two months put out a document summarizing its fundamental research programs and another summarizing its vehicle development programs. Mr. Schurmeier stressed the need for a coordinated national program in the general areas of guidance and tracking much along the lines of the present national booster program.

Mr. Faget: - Mr. Faget endorsed selecting lunar exploration as the present goal of the Committee although the end objective should be manned interplanetary travel. Space rendezvous will very likely be desirable in such operations and equatorial orbits certainly have attractive features. This places the space vehicles in the radiation belt, however, and aggressive research is needed to learn more about the radiation belt, its effects on living beings, anti-radiation medicines and shielding.

The Langley Space Task Group has done some preliminary thinking about Project Mercury follow-ups. Mr. Strauss described three ideas. The first is an enlarged Mercury type capsule (7.5 ft. diameter and 10.6 ft. long) weighing 3550 pounds to put two men in orbit for three days. The second would involve placing a two-man Mercury capsule ahead of an 8-foot diameter 12-foot long cylinder to put two men in space for about two weeks. The third idea was to mount the two-man Mercury capsule at an angle to the cylinder mentioned above, to have all of this attached by adjustable cables to the Vega motor some distance away and to rotate the whole affair about the base of the cylinder to provide artificial gravity. This system would weigh of the order of 6000 pounds.

Mr. Low: - Mr. Low recommended that the Committee –

(1) Adopt the lunar landing mission as its present long range objective with proper emphasis on intermediate steps because this approach will be easier to sell,

[10]

(2) Look into vehicle staging so that Saturn could be used for manned lunar landings without complete reliance on Nova,

(3) Look into whether parachute or airport landing techniques should be emphasized,

(4) Attach importance to research on auxiliary power plants such as hydrogen-oxygen systems.

NEXT MEETING

Committee objectives – In summarizing the present meeting it was concluded that the following is a sensible order of accomplishment:

1. Man in space soonest – Project Mercury
2. Ballistic probes
3. Environmental satellite
4. Maneuverable manned satellite
5. Manned space flight laboratory
6. Lunar reconnaissance satellite
7. Lunar landing
8. Mars-Venus reconnaissance
9. Mars-Venus landing

The committee at this meeting was not in agreement on whether the present long range objective should be number 6 or 7. The Chairman asked each member to give more thought to this before the next meeting.

Agenda – The following agenda was agreed to for the next meeting:

1. Space Flight Structural Concepts – Loftin and a Langley structures man

2. Parachute development – Low, Loftin

3. Space vehicle landing techniques – Faget, Eggers, Beeler, Loftin

4. Mercury lift capabilities – Faget

5. Reentry corridor – Eggers

[11]

6. Man's control functions – Beeler

7. Auxiliary power requirements – Lundin, Schurmeier

8. Propulsion requirements for lunar landing – Lundin, Schurmeier

9. Lift support – Schurmeier, Faget, Beeler

10. Control, guidance and navigation – Schurmeier

11. ABMA Saturn payload plans – Low

Location and Date – It was agreed to hold to next meeting at Ames on June 25 and 26.

ADJOURNMENT

The meeting adjourned at 1:15 p.m. on May 26.

Document II-2

Document Title: George M. Low, Chief, Manned Space Flight, "Manned Space Flight," NASA-Industry Program Plans Conference, 28-29 July 1960.

Document Source: Folder 18675, NASA Historical Reference Collection, History Division, NASA Headquarters, Washington, DC.

On 28 and 29 July 1960, NASA held a "NASA-Industry Program Plans Conference" in Washington to discuss the Agency's plans for future programs and to solicit industry interest in participating in them. NASA announced at this conference that the spaceflight project to follow Project Mercury would be named Project Apollo. This conference was in many ways the beginning of what eventually became the most massive engineering project undertaken since the Manhattan Project. This document is George Low's presentation to the conference.

NASA-Industry
Program Plans
Conference

July 28-29, 1960

Departmental Auditorium
Constitution Ave., N.W.
Washington, D.C.

[2]

MANNED SPACE FLIGHT

By George M. Low *Chief, Manned Space Flight*

Introduction

The benefits that might accrue from the manned exploration of space are, in a large measure, unknown. It is certainly clear that no amount of instrumentation will tell us as much about the moon, or the planets, as man himself will be able to tell, once he has visited those distant places. Only man can cope with the unexpected; and the unexpected, of course, is the most interesting.

We should, therefore, state only one broad objective for the manned space flight program:

"To provide the capability for manned exploration of space."

With this objective in mind, we have developed a program that is broadly outlined in figure 1. [not included] At this point it should be stated that official approval of this program has not been obtained. Rather, this presentation includes what we now believe to be a rational and reasonable approach to a long-range development program leading to the manned exploration of outer space.

MANNED SPACEFLIGHT PROGRAM

Program Outline

The initial step in this program is Project Mercury--a project designed to put a manned satellite into an orbit about 120 miles above the earth's surface, let it circle the earth three times, and then bring it back safely.

Project Mercury, we believe, is an essential step before we can proceed with other, more difficult manned space missions. It is true that all of our plans for the scientific exploration of space assume that eventually man will participate in that exploration. The trouble is that, although all of us think men can be useful in this new environment, none of us know for sure.

If it should turn out that men cannot perform useful work in space, it is quite possible that the direction of a substantial portion of our efforts will have to be changed. So it is important to find out about man's capabilities in space-and soon! Project Mercury is the simplest way to learn what we need to know, at the earliest possible date.

But the determination of man's capabilities in a space environment is only one of the benefits that will be derived from Project Mercury. Of equal importance is the technical knowledge being gained during the design, construction, and operation of the first vehicle specifically engineered for manned flight in space.

The accomplishment of Project Mercury will mark a tremendous step forward; man's venture into space will immeasurably extend the frontiers of flight. The speed of flight will be increased by a factor of 8 over present achievements, and the altitude by a factor of 5; the environment encountered in space flight will be one that heretofore has not even been approached. This extension of the flight envelope has required major technical advancements in many diverse fields, including aerodynamics, [3] biotechnology, instrumentation, communications, attitude control, environmental control, and parachute development-to mention only a few. By its very advanced nature, therefore, Project Mercury has opened the door for the next step in the manned space flight program.

This next step involves the development of a manned spacecraft designed to allow man to perform useful functions in space. This spacecraft should ultimately be capable of manned circumlunar flight, as a logical intermediate step toward future goals of landing men on the moon and the planets. The design of the spacecraft should also be sufficiently flexible to permit its use as an earth-orbiting laboratory, as a necessary intermediate step toward the establishment of a permanent manned space station.

In this decade, therefore, our present planning calls for the development and construction of an advanced manned spacecraft with sufficient flexibility to be capable of both circumlunar flight and useful earth-orbital missions. In the long range, this spacecraft should lead toward manned landings on the moon and planets, and toward a permanent manned space station.

This advanced manned space flight program has been named "Project Apollo."

Flight Missions

Further details of the desired dual mission capability are illustrated in figure 2. [not included] It should be pointed out, however, that these details are merely representative and may well be changed and redefined as the results of further studies become available.

The design for an ultimate circumlunar flight will require the solution of many, but not all, of the problems associated with a manned landing on the moon; this is particularly true of earth reentry and recovery. The mission will require a considerable amount of trajectory control, thereby imposing rather severe requirements on the navigation and control system. Manned circumlunar flight is the ultimate manned mission consistent with our planned booster capability, that is, with the Saturn vehicle.

Before circumlunar missions are attempted, earth-orbital flights will be required for spacecraft evaluation, for crew training, and for the development of operational techniques. In conjunction with, or in addition to, these qualification flights, the spacecraft can be used in earth orbit as a laboratory for scientific measurements or technological developments in space.

Modular Concept

In order to achieve this multiplicity of missions, it may be desirable to employ the so-called "modular concept" in the design of the advanced manned spacecraft. This concept is illustrated in figure 3. [not included]

In this design concept, various building blocks or "modules" of the vehicle system are employed for different phases of the mission. Basically, the spacecraft is conceived to consist of three modules: a command center module, a propulsion module, and a mission module.

The command center would house the crew during the launch and reentry phases of flight; [4] it would also serve as the flight control center for the remainder of the mission. We anticipate that this module will be identical for both the circumlunar and the earth-orbital missions.

The propulsion module would serve the primary function of providing safe return to earth in case of an aborted mission. In this sense, it might be compared with the escape tower and retrorockets on the Mercury capsule. In addition, for circumlunar flight, this component should have the capability of making midcourse corrections: it might also be used to place the spacecraft into an orbit around the moon and eject it from that orbit. In an earth-orbital mission, the propulsion module should permit a degree of maneuverability in orbit or rendezvous with other vehicles. Once again, it may be desirable to provide identical propulsion units for both orbital and circumlunar flights.

The command center and propulsion units together might be considered, for some applications, as a complete spacecraft, even without the mission modules.

The mission module would differ for the various flight missions. For circumlunar flight, it would be used to provide better living quarters than the command center can afford, and some equipment for scientific observations. (Detailed design studies may well indicate that the command center and circumlunar mission modules should be combined into a single package.)

For earth-orbital flight, the mission module can be considerably heavier than for circumlunar flight. Hence this module can usefully serve as an earth-orbiting laboratory, with adequate capacity for scientific instrumentation and reasonably long lifetimes in orbit.

Of all the modules mentioned, only the command center unit would be designed with reentry and recovery capability.

Command Center Module

Figure 4 [not included] illustrates some of the requirements for the command center module.

This module must be designed to reenter the earth's atmosphere at essentially parabolic velocity, or about 36,000 feet per second. It will have to withstand the severe heating encountered at these velocities, and it must be statically stable over the entire speed range from 36,000 feet per second to the landing speed.

A degree of maneuverability will be required to stay within the limits of a rather narrow flight corridor. The boundaries of this corridor are determined by maximum tolerable loads or heating, on the one hand, and minimum aerodynamic loads to cause reentry in a single pass, on the other hand. The amount of maneuverability can be minimized through the provision of adequate midcourse propulsive corrections.

The maneuverability provided for corridor control should also permit a landing at a fixed point (or within a small area) on earth.

A conventional airplane-type landing is not required. Instead, vertical landings using parachutes or other devices are acceptable. Because of the worldwide aspects of these missions, the vehicle must be capable of surviving both ground and water landings.

An important design consideration is that safe recovery must be possible for both normal and aborted missions. As in the case of the Mercury capsule, it is expected that the most severe requirements will stem from some of the off-design conditions.

There has been a great deal of discussion concerning the role to be played by the pilot in a space mission. Under the assumption that Project Mercury will demonstrate that man can indeed perform useful functions in space, we believe that in all future missions the primary control should be onboard.

[5] This guideline is not to be construed as implying that there would be no automatic guidance or control systems on board. Certainly there are many functions that can better be performed automatically than manually. But the basic decision-making capabilities, and some control functions, are to be assigned to the man.

Propulsion Module

Because of the possibility of a catastrophic failure of any of the Saturn stages, the spacecraft must be equipped with sufficient propulsion to permit safe crew recovery from aborted missions. Such capability must be provided for an abort at any speed up to maximum velocity and should be independent of the launch propulsion system.

Some of the requirements for the propulsion module are summarized as follows:

Primary   Secondary
Safe recovery from aborts   Lunar orbit
Course corrections   Maneuvering in earth orbit
Return from orbit

Preliminary studies have indicated that, for a circumlunar mission, roughly one-third of the permissible spacecraft weight will be required for onboard propulsion.

In a normal mission, this same propulsion may be applied for course corrections, both while approaching the moon and when returning to earth. As mentioned earlier, the propulsion that must be carried for emergency considerations may, in a normal mission, be sufficient to place the spacecraft into a satellite orbit around the moon.

For the earth-orbital mission, the propulsion module would again serve the primary function of providing safe return capability from aborted missions. If it is not needed for this purpose, then the available impulse might be used for maneuvering in orbit and for orbital rendezvous with other satellites.

Mission Modules

We have tentatively specified that the advanced manned spacecraft should be designed for a 3-man crew. Our concept is that, during launch and reentry, this crew would be located in the command center unit but, for the remainder of the flight, at least two of the crew members would be in the mission module (fig. 5).[not included]

The use of a pressure suit in the command center module may be acceptable. But the mission module should definitely be designed to permit "shirtsleeve" operations, that is, operations without the use of pressure suits. We believe that pressure suits, as currently envisioned, would not be acceptable for the duration of a circumlunar flight.

The foregoing requirements apply for both the circumlunar mission and the earth-orbital mission. However, there are other requirements that differ widely between the two types of flight. For example, the circumlunar mission module requires an environmental control system that need only provide for about 1 week's life support; on the other hand, it may be desirable to keep the earth-orbiting laboratory in space for periods ranging from 2 weeks to 2 months.

The circumlunar module would carry only a minimum amount of instrumentation required to complete the mission, whereas a great deal of instrumentation for scientific measurements and observations should be provided in the orbiting laboratory.

Required Developments

The advanced manned spacecraft will require many systems and subsystems that must be developed especially for this vehicle. Some of these systems may be entirely new, while others may be growth versions of Mercury components.

[6]Major developments that will be needed are listed as follows:
Basic reentry vehicle
Environmental control system
Attitude control system Power supplies
Communications system
Onboard propulsion
Guidance and control system
Pilot displays

The general specifications for the basic reentry vehicle were mentioned earlier. As yet, no specific recommendations regarding its configuration can be made.

The advanced manned spacecraft will involve the development of perhaps several environmental control systems. These systems would be incorporated into the command center, the orbiting laboratory, and the circumlunar module. Gaseous-, liquid-, and chemical-oxygen systems all deserve consideration for these applications.

A system for sensing and controlling the craft's attitude will have to be developed.

Suitable power supplies will have to be selected. It is estimated that the power required for the circumlunar mission will be of the order of 400 kilowatt hours, with a peak load of roughly 4 kilowatts.

Voice and telemetry communication systems most certainly will be needed. Television may also be desirable.

The onboard propulsion requirement was discussed in connection with the propulsion module. The demands on this system are many and varied, ranging from high-thrust, short-duration requirements for abort maneuvers to the very low thrust needed for course corrections.

An area that deserves considerable attention is that of guidance, control, and displays. Sufficient information must be supplied to the pilot to permit him to make trajectory corrections, to enter and stay within the appropriate corridor, and to land at a preselected [sic] location.

Radiation Considerations

A problem of major concern for flight beyond low earth orbits is that of radiation in space (fig. 6) [not included]

The following types of radiation are pertinent to circumlunar flight:

(1) Trapped radiation (Van Allen)

(2) Cosmic radiation

(3) Solar flare particles

The trapped radiation in the Van Allen radiation belts is of rather high intensity but of sufficiently low energy to make shielding feasible. Because the time spent in the radiation belts will be small, only a small amount of shielding is required for this type of radiation.

The energies of cosmic radiation are so high that shielding becomes impractical. However, the peak intensity is sufficiently low that no danger is expected in a 5-day mission.

The most serious problem results from the particles generated by some solar flares. The energy of these particles is of a magnitude that may require more shielding than is practical from the standpoint of weight; following a major flare, the intensity may be so high as to cause severe biological damage. However, there are some indications that it might be possible to predict major flares (or at least their absence) several days in advance. If, in the future, it should indeed be possible to predict these flares, then the radiation problem could be circumvented by avoiding flights during a time of anticipated major flare activity.

The radiation problem, more than any other, requires a great deal of study before the manned spacecraft can be employed for circumlunar flight. Many of the answers now lacking will be supplied through our scientific satellite and probe programs. The effects of the various types of radiation on living tissues are yet to be determined.

[7]   **Weightless Flight**

Another as yet unresolved problem area is illustrated in figure 7. [not included] A question naturally arises as to whether man will be able to function in a weightless environment for prolonged periods of time.

The answer to this question must await the completion of the first manned orbital flight in Project Mercury. That flight should shed much light on the desirability of incorporating artificial gravity into future manned spacecraft.

Inevitably, the solution to this problem will have a profound effect on the design of the orbiting laboratory module and perhaps also on the circumlunar module.

Manned Space Flight Program

Program Phasing

Our planning thus far has led to a proposed overall timetable for the advanced manned spacecraft program, as presented in figure 8.[not included]

This program is expected to be under the direction of the Space Task Group of the Goddard Space Flight Center--the same group that is currently managing Project Mercury.

Several months ago, very detailed program guidelines were presented to each of NASA's research and space flight centers. As a result of these presentations, the centers have initiated intensive research and study programs, all designed to generate the background information required for the design of the advanced manned spacecraft. This information will be available to industry, of course.

In the near future, industry will be invited to participate, by contract, in a program of system design studies. According to present plans, a systems contract for the design, engineering, and fabrication of the manned spacecraft and its components will probably be initiated in fiscal year 1962.

However, it should be emphasized that this program has no official standing as yet. Provision for the initiation of NASA's manned space flight program, beyond Project Mercury, is expected to be included in the fiscal year 1962 budgetary request to be sent to the Congress in January 1961. With that statement as a basic premise, our present thinking is outlined to indicate the probable course of future flight events (fig. 9). [not included]

Flight Program

Major Mercury flights probably will continue for several years.[8] Research and development, and prototype flights of the advanced manned spacecraft are listed to start in 1962 and to end in 1965. Early flights in this series would be used to verify final design criteria for the spacecraft shape and its heat protection; it is planned to use Atlas-Agena-B as the launch vehicle for these missions. Following the Atlas-Agena flights, the Saturn vehicle will be used for full-scale development and prototype flights.

Earth-orbital missions, using the final spacecraft, could conceivably begin in 1966, with circumlunar missions following as soon as the state of both technical and aeromedical knowledge permits such flights.

Program Costs

The final chart (table I) lists the funding associated with the manned spaceflight program. In fiscal years 1960 and 1961, the majority of the funds allocated for manned space flight will be devoted to Project Mercury.

TABLE I Manned Space Flight Funding [millions]

Fiscal	Project Mercury	Advanced manned spacecraft	Total
1960	87.06	0.10	87.16
1961	106.75	1.00	107.75

In future years, we anticipate that an increasingly larger proportion of manned space flight funds will be allocated to the more advanced programs in this area.

Concluding Remarks

NASA's manned space flight program, for the present decade, calls for the development and construction of an advance manned spacecraft with sufficient flexibility to be capable of both circumlunar flight and useful earth-orbital missions. In the long range, this spacecraft should lead toward manned landings on the moon and planets, and toward a permanent manned space station.

In order to achieve this multiplicity of missions, the use of the modular concept is proposed. In this concept, various building blocks, or modules, of the vehicle system are employed for different phases of the mission. Basically, the spacecraft is conceived to consist of three modules: a command center module, a propulsion module, and a mission module.

In addition to the basic vehicle modules, this program will require other new developments, such as environmental control systems, attitude stabilization devices, power supplies, communications, guidance-and-control systems, onboard propulsion, and pilot displays.

In the current fiscal year, contractors will be invited to participate in a program of systems studies. It is believed probable that a contract for the design, engineering, and fabrication of the complete spacecraft system may be initiated in fiscal year 1962.

Document II-3

Document Title: George M. Low, Memorandum for Director of Space Flight Programs, "Manned Lunar Landing Program," 17 October 1960.

Document Source: Folder 18675, NASA Historical Reference Collection, History Division, NASA Headquarters, Washington, DC.

Given the increasing attention in 1960 to the precursors to a future lunar landing mission, in October 1960 Manned Space Flight Program Chief, George Low, informed Director of Space Flight Programs, Abe Silverstein, that he was forming a working group to address the technical, schedule, and budgetary issues associated with a lunar landing program. The results of Low's group provided a basis for NASA's response the following year as President Kennedy considered a dramatic acceleration of the lunar landing program.

October 17, 1960

MEMORANDUM for Director of Space Flight Programs

Subject: Manned Lunar Landing Program

    1.   It has become increasingly apparent that a preliminary program for manned lunar landings should be formulated. This is necessary in order to provide a proper justification for Apollo, and to place Apollo schedules and technical plans on a firmer foundation.

    2.   In order to prepare such a program, I have formed a small working group, consisting of Eldon Hall, Oran Nicks, John Disher and myself. This group will endeavor to establish ground rules for manned lunar landing missions; to determine reasonable spacecraft weights; to specify launch vehicle requirements; and to prepare an integrated development plan, including the spacecraft, lunar landing and take-off system, and launch vehicles. This plan should include a time-phasing and funding picture, and should identify areas requiring early studies by field organizations.

    3.   At the completion of this work, we plan to brief you and General Ostrander on the results. No action on your part is required at this time; Hall will inform General Ostrander that he is participating in this study.

[signed]
George M. Low
Program Chief
Manned Space Flight

Document II-4

Document Title: George M. Low, Program Chief, Manned Space Flight, Memorandum for Associate Administrator, "Transmittal of Report Prepared by Manned Lunar Working Group," 7 February 1961, with Attached Report, "A Plan for a Manned Lunar Landing."

Document Source: Johnson Space Flight Center Archives.

George Low had been among the first in NASA to openly advocate a lunar landing goal and was a vocal proponent of that goal. In October 1960 he formed a Manned Lunar Working Group Task Force. The task force transmitted its findings to NASA Associate Administrator Robert Seamans on 7 February; its report was the first fully developed plan for how NASA proposed to send humans to the Moon. Low and his group concluded that "The present state of knowledge is such that no invention or breakthrough is believed to be required to insure the over-all feasibility of safe manned lunar flight." This was an important consideration two months later as President Kennedy considered whether to commit the United States to sending Americans to the Moon. The group also estimated that the plan could be carried out over 10 years for an average cost of $700 million per year, for a total cost of $7 billion.

[Originally marked "For Internal Use Only"]

February 7, 1961

MEMORANDUM for Associate Administrator

Subject: Transmittal of Report Prepared by Manned Lunar Working Group

     1. The attached report, entitled "A Plan for Manned Lunar Landing" was prepared by the Manned Lunar Working Group. It accurately represents, to the best of my knowledge, the views of the entire Group.

     2. Copies of a draft of this report were submitted to the Program Directors, NASA Headquarters, and to the Directors of Marshall Space Flight Center and Space Task Group. In cases where comments were submitted, these comments were incorporated in the report.

     3. The Group stands ready to make a presentation of the material presented in the report at any time you might so desire.

     4. No additional work is planned until further instructions are received.

/Signed/
George M. Low
Program Chief
Manned Space Flight

NATIONAL AERONAUTICS AND SPACE ADMINISTRATION

WASHINGTON 25, D. C.

A PLAN FOR MANNED LUNAR LANDING*

INTRODUCTION

In the past, man's scientific and technical knowledge was limited by the fact that all of his observations were made either from the earth's surface or from within the earth's atmosphere. Now man can send his measuring equipment on satellites beyond the earth's atmosphere and into space beyond the moon on lunar and planetary probes. These initial ventures into space have already greatly increased man's store of knowledge.

In the future, man himself is destined to play a vital and direct role in the exploration of the moon and of the planets. In this regard, it is not easy to conceive that instruments can be devised that can effectively and reliably duplicate man's role as an explorer, a geologist, a surveyor, a photographer, a chemist, a biologist, a physicist, or any of a host of other specialists whose talents would be useful. In all of these areas man's judgment, his ability to observe and to reason, and his decision-making capabilities are required.

---

*Prepared by the Lunar Landing Working Group, January 1961.

[2]

The initial step in our program for the manned exploration of space is Project Mercury. This Project is designed to put a manned satellite into an orbit more than 100 miles above the earth's surface, let it circle the earth three times, and bring it back safety. From Project Mercury we expect to learn much about how man will react to space flight, what his capabilities may be, and what should be provided in future manned spacecraft to allow man to function usefully. Such knowledge is vital before man can participate in other, more difficult, space missions.

Project Mercury is the beginning of a series of programs of ever-increasing scope and complexity. The future can be expected to include the milestones shown in Figure 1.

The next step after Mercury is Project Apollo. The multi-manned Apollo spacecraft will provide for the development and exploitation of manned space flight technology in earth orbit; it also provide the initial step in a long-range program [3] for the manned exploration of the moon and the planets. In this paper we will focus on a major milestone in the program for manned exploration of space - lunar landing and exploration. This milestone might be subdivided into two phases:

    1. Initial manned landing, with return to earth;

    2. Manned exploration.

       This report will be limited to a discussion of the initial manned lunar landing and return mission, with the clear recognition that it is a part of an integrated plan leading toward manned exploration of the moon.

       An important element in the manned space flight program is the establishment of a space station in an earth orbit. Present thinking indicates that such a station can be established in the same time period as manned lunar landings can be made, and also that many of the same technological developments are required for both purposes. Although both missions were broadly considered in planning developments for the lunar program, only the lunar requirements are discussed in this paper.

       An undertaking such as manned lunar landing requires a team effort on an exceedingly broad scale. The various elements of [4] this effort are indicated in Figure 2. [not provided] The basic capability is provided through the parallel development of a spacecraft and a launch vehicle. Both of these developments must proceed in an orderly fashion, leading to hardware of increasing capability. Supporting these developments are many other scientific and technical programs and disciplines, as shown in the figure. The implementation of the manned spacecraft program requires information that will be obtained in the unmanned spacecraft and life science programs. The development of launch vehicle capability requires new engines, techniques to launch from earth orbit, and might include launch vehicle recovery developments. Both the spacecraft and the launch vehicle programs can progress only as new knowledge is obtained through advanced research.

       All of these program elements currently exist in the total NASA program. Work is under way in areas that are pertinent to the development of the capability for manned lunar landing. In this report the interrelationship between the various programs will be studied. Key items will be examined in detail, to determine the proper phasing between the development of new systems, and the availability of the background information required for these developments.

[5]

NASA RESEARCH

       Already there exists a large fund of basic scientific knowledge, as a result of the advanced research of the past several years, which permits confidence that the technology required for manned lunar flight can be successfully developed. It would be misleading to imply that all of the major problems are now clearly foreseen; however, there is an acute awareness of the magnitude of the problems. The present state of knowledge is such that no invention or breakthrough is believed to be required to insure the over-all feasibility of safe manned lunar flight.

       An aggressive research program which will insure a sound technological foundation for lunar vehicle system development is currently under way. This research is being carried out as a major part of the programs of the NASA Research Centers and in the supporting research activities of the NASA Space Flight Centers, both internally and by contract. It includes basic research in the physical and biological sciences; and applied research leading to the development of spacecraft, orbital operations, operations at the lunar surface, propulsion and [6] launch vehicles. This research is supported by a wide variety of experimental facilities in being, and new highly advanced facilities that are becoming available.

Consider, for example, one of the major spacecraft problems, that of aerodynamic heating. A lunar spacecraft will reenter the earth's atmosphere at about one and one-half times the reentry speed of a near-earth satellite and with twice the kinetic energy. Research to date has shown that radiative heat of the spacecraft by the hot incandescent gas envelope may become an appreciable percentage of the total heating. For the case of the reentering satellites, this radiative heat transfer had been unimportant. Analytical work and early experimental results have enabled estimates to be made of the gross radiative heat transfer. Continuing experimental research will be carried out in newer, more advanced facilities that are becoming available. Selected flight experiments to progressively higher speeds are needed for verification of the analytical and experimental results. The earliest of these, providing reentry velocities of 30,000 ft/sec, are scheduled for early 1962. All of this research will help to achieve detailed understanding of the heating problem, to allow accurate prediction of the heat [7] transfer, and to find the best materials and methods for spacecraft construction.

Research in this area, as well as in the other areas listed in Figure 3 [not provided], seeks to provide the basic information which should lead to greater simplification and reliability, and to reduced weight. The scope of the work is such that the basic information required in support of a manned lunar landing project should be in hand within three to five years.

LAUNCH VEHICLE DEVELOPMENT

The magnitude of a step in our space flight program, at any given time, will always depend on the capability of our launch vehicles. This capability, both present and projected, is shown in Figure 4 [not provided], where the payload weight at escape velocity is plotted as a function of time. During the current year, we should achieve the possibility of propelling 750 pounds to escape velocity, using the Atlas-Agena vehicle. By 1963, the Atlas-Centaur should increase this capability to 2,500 pounds; this will be doubled when the Saturn C-1 becomes operational in 1964. However, the C-1 is only an interim vehicle that is severely limited because of the lack of a sufficiently large high-energy [8] engine for the second stage. A later version of the Saturn, called the C-2, will more than triple the C-1 payload capability at escape velocity. Because the second stage of the C-2 must await the development of the J-2 (200,000 pound thrust hydrogen-oxygen) engine, it will not be operational until 1967. The Saturn C-2 will be the first launch vehicle giving us the capability of manned flight to the vicinity of the moon; however, a single C-2 cannot provide sufficient energy to complete a manned lunar landing mission.

The required launch vehicle capability can be achieved in several ways. Two promising means are: one, orbital operations, wherein a number of Saturn C-2 launched payloads are rendezvoused, assembled or refueled in earth orbit, and then launched as a single system from earth toward the moon; and two, the direct approach, using a vehicle much larger than Saturn which would have the capability of propelling a sufficiently large payload toward the moon from the surface of the earth. Both methods appear to be technically feasible, and will be discussed.

Orbital operation techniques must be developed as part of the space program, whether or not the manned lunar landing mission is considered. These techniques will be required for [9] resupply and transfer to space stations and orbiting laboratories, for inspections and repair of other satellites, for rescue operations and for military applications. Successful development of these techniques of rendezvous, refueling and launching from orbit could allow us to develop a capability for the manned lunar mission in less time than by any other means. In view of these facts, NASA is planning a vigorous program for developing orbital operations techniques. This program is outlined in Figure 5.

Under present plans, initial rendezvousing, docking and refueling tests would make use of the Atlas-Agena vehicle. In these tests, conventional storable propellants will be used. In order to demonstrate the feasibility of orbital operations with high-energy hydrogen-oxygen propellants, a refueling exercise is planned wherein an Atlas-Centaur will be used to refuel an upper stage of a Saturn C-1 vehicle. This demonstration is expected to be attempted in 1965 or 1966. Following this demonstration, full-scale refueling and orbital launch operations will be conducted using Saturn C-2 vehicles. These operations will involve the launch of several C-2's to refuel an upper stage initially put into orbit. Following the development of this capability in the 1967-68 time period, this system is [10] expected to be available for operational use in 1968-69 time period.

For the purpose of the manned lunar mission, the Saturn C-2 would be used to place into earth orbit an empty upper vehicle stage that would subsequently be used to propel the spacecraft toward the moon. Four or five additional C-2 payloads would be required to fill this empty stage with propellants. The last launching would propel the manned spacecraft together with the lunar take-off stage into earth orbit. Six or seven successful Saturn launchings, therefore, are required in order to place a space vehicle system into earth orbit that will then be capable of propelling an 8,000 pound spacecraft toward the moon, landing on the moon and returning it toward earth.

Orbital operations techniques will probably be required to perform the more difficult planetary mission even with the availability of much larger launch vehicles. Many of the missions shown in Figure 1 indicate the need for vehicles larger than the Saturn C-2. Large earth space stations that may be assembled in orbit will very likely require the launching of larger sub-assemblies into orbit than can be carried with a single Saturn C-2. Exploration of the moon following the initial landing will [11] also require vehicles larger than the Saturn C-2. Also, if the spacecraft weight increases materially as a result of information gained in the areas of weightlessness and radiation, the required number of earth launchings using Saturn could increase to an extent where the orbital operations techniques with this vehicle would no longer be attractive.

It is proposed, therefore, that a vehicle larger than the Saturn C-2 be phased into the launch vehicle program in an orderly fashion following the Saturn development. Such a launch vehicle, called Nova, would use a cluster of 1,500,000 pound thrust F-1 engines in its booster stage. The exact number of F-1 engines will have to be determined later, when a more complete definition of Nova missions is in hand. Nova might be sufficiently large to permit a manned lunar landing with a single launching directly from earth. Or, although substantially larger than the Saturn C-2, it might still not be large enough to approach the moon directly from earth; in this case it would materially reduce the number of rendezvous operations needed in earth orbit for each lunar mission.

A Nova-class vehicle development program, based on an assumed configuration, is given in Figure 6 [not provided]. The program is phased so that major decisions concerning the vehicle size and [12] configuration need not be made until after sufficient background information is available in the spacecraft development program.

The present program for development of the F-1 engine is shown in this figure. Preliminary flight rating tests are now scheduled toward the end of 1963, and further testing should lead to a qualified engine by the end of 1965. Studies are under way to determine possible configurations of the vehicle and its performance capabilities. Preliminary design of the vehicle can be started in 1962 and would continue through 1963. As will be shown later, the spacecraft weight for the manned lunar mission should be firmly established in this time period.

Construction of static test stands and launch facility will be initiated in 1963. Developmental flight tests of the first stage could begin in 1966. Subsequent tests would add various upper stages until a complete launch vehicle should be ready for operational use in 1970.

Comparison of Launch Vehicles

A comparison of the Saturn C-2 and several Nova-class vehicles, as used for the manned lunar mission, is made in Figure 7[not provided]. The numbers under each launch vehicle indicate the [13] successful launchings required for each lunar flight. Spacecraft weights from 8,000 to 16,000 pounds are assumed; corresponding weights that must be propelled to escape velocity are indicated. Uncertainties in these latter weights are a result of uncertainties in the design of the lunar landing and take-off stages. In all cases, the use of storable propellants has been assumed for the return propulsion.

Use of the Saturn C-2 requires minimum of six to seven vehicles successfully completing each orbital operation. Increased spacecraft weight, failures of the launch vehicle, failures in the orbital operations, propellant losses either during transfer or by evaporation during the operation, and extra propulsion for accomplishing the rendezvous would all increase the required number of Saturns.

At this time, operations with six or seven Saturns appear to be feasible. However, if several of the aforementioned eventualities materialize, and if the number of launchings increases appreciably, the orbital operations technique for manned lunar landings may no longer be practical. A better definition of these problems will come during the orbital operations development program and during the spacecraft development program. [14] If, as a result of these programs, it appears that orbital operations are indeed feasible, the Nova development could be slowed down and delayed. Conversely, if the orbital operations become too complex and cumbersome, this work should be de-emphasized and the Nova development could be speeded up.

Use of the Nova-class vehicle offers the possibility of greatly reducing the required number of launchings from earth. It might be possible to provide mission capability without rendezvous with a four-engine Nova; with an eight-engine Nova, this type of mission capability is virtually assured.

Thus, if future difficulties force the use of an unacceptably large number of Saturns for this mission, the availability of a Nova-class vehicle would permit accomplishment of the planned flights. It should be recognized, however, that the development of Nova will undoubtedly bring about many problems, and will not be easy.

It is possible that other propulsion developments could contribute to manned lunar flight capability. Examples are the use of large solid propellant rockets, or nuclear propulsion. In defining a Nova configuration, consideration will be given to both of these types of propulsion. At the present time it [15] appears that nuclear propulsion will not be sufficiently developed for the initial manned lunar landing; however, nuclear propulsion might be very desirable and economically attractive for later exploration of the moon.

Programs in Support of Launch Vehicle Development

Activities presently under way or planned in support of the launch vehicle development are shown in Figure 8 [not provided]. For comparative purposes, major milestones for both the orbital operations and the Nova development are indicated.

Engine Development: The chemical fuel engines currently under development include the F-1, the J-2, and the LR-119. The F-1 engine produces 1,500,000 pounds of thrust using conventional LOX/RP propellants; the J-2 engine will produce 200,000 pounds of thrust using hydrogen-oxygen propellants; the LR-119 produces a thrust of 17,500 pounds and also uses hydrogen-oxygen propellant. Both the LR-119 and the J-2 engine are scheduled for use in the Saturn C-2 vehicle. All three engines could be used in the Nova launch vehicle. The end of each bar in Figure 8 indicates the time when a qualified engine could be available. Also indicated in the figure is a proposed plan for testing a cluster of F-1 engines; cluster testing could be completed [16] during 1966, if test facilities can be made available in time. Nuclear propulsion is currently under development jointly by NASA and the AEC. Although actively under development, the research character of this program precludes the possibility of determining schedules for manned use of this engine at the present time.

The feasibility of using large solid rocket motors in the first stages of launch vehicles of the Nova-class is also being studied. Test firings of rocket motors in the one-quarter to one-half million pound thrust class are planned for the 1961-62 time period.

These firings will be made with segmented motors that could be assembled to provide much larger capability.

Launch Vehicle Recovery: Means to reduce the high cost of launch vehicles are continually being sought. A promising method for possible major-reductions in hardware costs for future missions, is the recovery of launch vehicles. Launch vehicle recovery would also permit postflight inspection of hardware, offering the possibility of reducing vehicle development time and increasing vehicle reliability. Because of these possible advantages, a research and development program in the area of launch vehicle recovery will be implemented as indicated [17] in Figure 8. In this program, it is first planned to recover the booster stage of the Saturn C-2; later, recovery of stages from orbit will be attempted. If these methods prove to be successful, all of the launch vehicle hardware required for the orbital operations phase of this plan could be reused. Information gained during these operations could be applied later to the recovery of Nova vehicle hardware, thus offering the possibility of greatly reducing the cost of future operations.

Hawkeye Program: This country's first program making use of rendezvous techniques will be the Air Force's Hawkeye program. Much of the technology developed for Hawkeye might be applied to the proposed program of orbital launch vehicle operations. Close coordination with Hawkeye is, therefore, being maintained in order to derive the maximum benefits from this program.

SPACECRAFT DEVELOPMENT

    The spacecraft development for the manned lunar landing mission will be an extension of the Apollo program. Before a spacecraft capable of manned circumlunar flight and lunar landing can be designed, a number of unknowns must be answered.

[18]   The two most serious questions are:
1. What are the effects on man of prolonged exposure to weightlessness?
2. How may man best be protected from radiation in space?

    The entire spacecraft design, its shape and its weight, will depend to a great extent on whether or not man can tolerate prolonged periods of weightlessness. And, if it is determined that he cannot, then the required amount of artificial gravity, or perhaps of other forms of sensory stimulation, will have to be specified.

    The spacecraft design and weight will also be greatly affected by the amount of radiation shielding required to protect a man. In this area, a clear definition of the pertinent types of radiation, and their effects on living beings, is needed.

    These two unknowns, radiation and weightlessness, might cause the largest foreseeable changes in spacecraft design. Other unknowns are also important, but will have lesser effects on the vehicle weight. For example, the lunar surface [19] characteristics must be defined before a landing system can be designed; yet it is not expected that any landing device will cause major weight perturbations.

    As will be shown later, the complete answers to these questions will not be available for several years. It is proposed, therefore, to implement the Apollo spacecraft development in two phases. Apollo "A" will provide the capability of multimanned flight in earth orbit; it will also be a test vehicle, perhaps unmanned, for reentry at parabolic velocities. Apollo "B" will be an advanced version of Apollo "A" and will be phased into the development program at a later date, when definitive design decisions can be made. Apollo "B" will have the capability of manned circumlunar flight, and manned landing on the moon.

    It is not suggested that the entire spacecraft development would be implemented in two phases. The Apollo spacecraft is conceived to employ a number of components, or modules, as listed in Figure 9 [not provided]. With the exception of the "command center," these modules will either be common to both Apollo "A" and Apollo "B" or they will be required for only one of the two types of mission.

    [20]The command center will house the crew during the launch and reentry phases of flight; it will also serve as the flight control center for the remainder of the mission. It will be the only spacecraft unit designed with

reentry and recovery capability. Apollo "A" used in conjunction with the Saturn C-1 launch vehicle, will provide the capability of multimanned flight in earth orbit for extended periods of time. It will perform missions beyond the capability of Mercury, with increased sophistication and flight duration, leading to more definitive results concerning manned space flight; and it will provide for continuity in the manned flight program.

Apollo "B," used in conjunction with Saturn C-2, will be an advanced version of Apollo "A" with the capability of manned flight to the moon. It is conceivable that only minor changes in design, together with some improvements of onboard systems, will be desirable or required to modify the Apollo "A" spacecraft for the Apollo "B" mission. On the other hand, it is also possible that future knowledge will dictate a major change from Apollo "A" to Apollo "B."

Proposed development schedules for both the "A" and "B" command center units are shown in Figure 9. Also shown in this figure are the schedules for the design, fabrication [21] and flight testing of two types of onboard propulsion system. The Launch Escape Propulsion System will be used in case of a launch vehicle malfunction in the earth's atmosphere. The Mission Abort Propulsion System will provide return-to-earth capability for the remainder of the mission; it will also provide for maneuverability and course corrections; and, for a lunar landing mission, it will be used as the take-off stage from the moon. These propulsion systems will be used in conjunction with both the "A" and "B" command center units. Both propulsion systems will have to be thoroughly tested and highly reliable. The use of existing engines, such as the Agena engine, for the Mission Abort Propulsion System, appears to be very desirable.

The two remaining modules are the Orbital Space Laboratory and the Lunar Landing System. The Orbital Space Laboratory, to be used initially with Apollo "A," will be used for spacecraft evaluation, for crew training and for the development of operational techniques; it can also serve as a base for scientific measurements and technological developments. The Lunar Landing System will be used only with the Apollo "B" command center; controlled by this command center, the landing module will provide for a manned landing on the moon's surface.

[22] The schedules (Figure 9) for the design, fabrication and flight testing of each module of the Apollo vehicle were developed so as to be consistent with the availability of the required background knowledge.

Spacecraft - Launch Vehicle Phasing

The proposed schedule of spacecraft flights is compared with launch vehicle availability in Figure 10 [not provided]. The first manned flights on Saturn C-1 with the Apollo "A" spacecraft will come a reasonable period of time after this launch vehicle is operational; orbital laboratory flights on C-1 are not scheduled until after two years of operational use of this vehicle have elapsed. First manned flights on Saturn C-2 will be made with the Apollo "B" spacecraft, shortly after the C-2 vehicle is operational.

The first lunar landing, using the orbital operations approach, could occur at the time this approach is developed. Manned flights using Nova could take place not much later, if it is determined that the mission should be performed with the Nova vehicle.

[23]

Support by Unmanned Spacecraft Program

A significant amount of the information required in the design of the manned lunar spacecraft will be derived from unmanned space flight programs. These programs will yield scientific data needed to develop design criteria; and technological advancements that might apply directly to the manned spacecraft.

Some of the areas of interest are listed in Figure 11[not provided]. At the top of this figure, significant milestones in the Apollo "B" development, and in the lunar landing system development, are given. Under these milestones, pertinent areas where information is needed are shown. These include: Information concerning the cislunar and lunar environment, where the several types of radiation will be probed, fields will be measured and meteorite impact probabilities will be assessed; the measurement of lunar surface properties, including terrain texture and features, surface composition, and physical characteristics; and the determination of lunar body properties, such as shape and mass distribution. Technological developments include power systems, tracking and telecommunications, attitude orientation and stabilization, mid-course and terminal guidance and control, retropropulsion, and impact absorbers.

[24] Of all of the areas mentioned above, the information pertaining to cislunar and lunar environment, and to lunar surface characteristics, is the most important. A clear understanding of trapped, cosmic, and solar flare radiation is required before the spacecraft weight can be fully determined. For example, reliable solar flare prediction methods would be required to support a decision that shielding against this type of radiation is not required. Of, if such prediction methods should turn out to be less reliable than is currently anticipated, further information on the directionality of solar proton beams would be helpful. Questions such as: "Do solar flare particles impinge on the dark side of the moon, or in the shadow of a crater?" must be answered. Detailed knowledge about the lunar surface characteristics is required before the design for the lading gear of the manned vehicle can be finalized, and before the exact method of touchdown on the moon (i.e., vertical or horizontal) can be determined.

A detailed analysis of the information presented in Figure 11 has shown that flights are scheduled in ongoing NASA programs which could obtain all the required information; and that this information is expected to be in hand prior to the time of hardware fabrication for either the Apollo "B" command center [25] unit, or the lunar landing system.

The earth satellite programs, using Scout, Delta, and the Atlas-Agena launch vehicles, will significantly increase our store of knowledge concerning the near-earth and cislunar environment. At least 26 firings of scientific satellites are planned between now and the end of 1964. In the same period of time, the Ranger spacecraft will probe the environment between earth and moon, and planetary probes of the Mariner series will obtain additional scientific information. In this time period, it might be desirable to schedule additional Ranger flights for the purpose of fully defining the environment in the vicinity of the moon, and on the moon's surface.

Both the Ranger and the Surveyor spacecraft will obtain information concerning lunar topography, surface characteristics, and body properties. According to present schedules, and assuming reasonable success, sufficient information will be available to design a lunar landing system for the manned spacecraft at the time when such information is required.

The Prospector series of flights will provide final landing system design confirmation. It will also assist in selecting the landing site for the manned craft, and might even [26] bring equipment to the moon's surface that could be used in the manned mission. Close coordination between the Prospector and Apollo projects will be maintained in order to assure maximum utilization of developments; such coordination should greatly benefit both projects.

Advancements in spacecraft technology will be derived from the earth satellite programs, and from the Ranger, Surveyor, and Prospector developments. Some of these advancements will apply directly to the manned lunar landing program.

Weightlessness and Radiation-Biological Tests

Before the Apollo "B" spacecraft design can be completed, the question previously raised concerning weightlessness must be answered. In Figure 12 [not provided], programs that are now planned in this area are listed; for comparison, significant milestones in the Apollo "B" development are also shown.

To date, manned weightless flights have been made for a [27] maximum time duration of one minute.[1] In this short time period, no gross physiological effects were noted. Ongoing programs will soon provide information of the effects of weightlessness on man for several minutes, and then several hours; and the effects on animals for many hours and then for several days. If, in each succeeding step, it is demonstrated that there are no adverse biological effects of weightlessness, then the design of a spacecraft without provision for artificial gravity can proceed with confidence; conversely, if future experiments show marked psychological or physiological changes as a result of prolonged exposure to weightlessness, then artificial gravity will have to be incorporated into the Apollo "B" spacecraft design.

_____

[1] Animals have been subjected to several days of weightless flight in Russian experiments. Although there are indications that these animals suffered no adverse effects, insufficient data are available, in this country, to draw any firm conclusions.

[28] As indicated in Figure 12, a considerable amount of experimental evidence on this subject will have been obtained before the Apollo "B" design is even started; complete information should be available before fabrication of hardware is begun. These conclusions, however, are based on the assumption that all programs that are currently in the planning stage, including the biomedical orbiting satellite program using Mercury capsules, will actually be implemented.

The biological effects of radiation in space will be determined largely from a correlation of the physical measurements previously discussed (Figure 11) with the results of ground measurements on biological specimen. However, a number of selected experiments in space, involving living subjects, will have to be made before shielding requirements for Apollo "B" can be fully defined. Tests of this type that either have been made, or are firmly planned, are indicated in Figure 12. Additional tests are currently being planned by NASA, in cooperation with the Air Force and the Atomic Energy Commission.

Manned Flight Technology

Much of the information required for the design of a spacecraft for manned lunar landing will be derived directly from Project Mercury, and from DynaSoar developments.

[29]The experience gained in developing systems for manned flight in space, and in preparing both the equipment and the men for such flights, will be of major importance. Operational concepts being worked out and applied in Project Mercury and DynaSoar should apply directly to future manned missions.

For example, the Mercury spacecraft will have all the onboard systems - the attitude stabilization and control system, the communications system, the environmental control system, etc. - that will be required in future manned spacecraft. Although some of the systems required for the Apollo spacecraft will be entirely new, their design should, in general, be related to Mercury experience; it is more than likely that many of the systems will be direct growth versions of Mercury equipment.

Extensions of Project Mercury, beyond the present program, are planned as part of the Apollo development. These flights would provide for extended periods of weightlessness, and perhaps for experiments with artificial gravity. Manned rendezvous tests, using the Mercury spacecraft for control, and a version of the Hawkeye vehicle as the controlled craft, can be carried out. The Mercury capsule cap also be used as a test bed for the development of Apollo guidance and control equipment. All of these flights can occur before manned flights with Apollo "A" are scheduled to take place.

[30]   SCHEDULES AND COSTS

A summary of manned space flight missions, leading toward a manned lunar landing, is presented in Figure 13 [not provided]. Starting late in 1961, the Mercury-Atlas combination will give us the capability of orbiting one man for a short period of time. The Apollo "A" spacecraft, using the Saturn C-l launch vehicle, will allow multimanned, long duration, orbital flight in 1965. Later, in 1967, an advanced version of the Apollo spacecraft (Apollo "B") launched by the Saturn C-2, will provide the capability for manned circumlunar flight, and for lunar orbits.

Manned landings on the moon, using the Apollo "B" spacecraft, could be made in the 1968-1971 time period. If orbital operations using the Saturn C-2 vehicles prove to be practicable for this mission, then it might be accomplished toward the beginning of this range of time. On the other hand, if the spacecraft becomes much more complex than now envisioned, and consequently much heavier, a Nova vehicle will most likely be required before man can be landed on the moon. In the latter event, the program goals may not be accomplished as quickly.

[31]The plan presented in this report consists of a number of relatively independent programs. Decisions to implement these programs can be made

as time progresses; no single decision committing NASA to carry out the entire plan is required at this time. The plan is also sufficiently flexible to permit major changes in objectives in later years, without the requirement that earlier phases of the program be repeated.

Some of the major phases of the Launch Vehicle Program are shown in Figure 14 [not provided]. For each of these phases, the year of initiation is shown, together with the total duration of this phase and total funding required to complete the phase. Thus, for example, a decision to go ahead with the Atlas-Agena docking demonstration would be required in FY 1962, in order to meet the total program objectives; the total funding required for these tests would be $80,000,000 distributed over a period of nearly three years.

In the Nova development, only those phases that are not now funded are included in Figure 14. Thus, it is assumed that the F-1 engine development, and the Nova configuration [32] studies that are presently under way, will be continued. No major new commitment will be required until late in FY 1963, when the development of the first stage would be started.

A similar breakdown for the phasing of various components of the spacecraft is given in Figure 15 [not provided]. In order to meet the previously presented program objectives, the development of the Apollo "A" spacecraft, the Launch Escape Propulsion System, and the Mission Abort Propulsion System, would have to be initiated in FY 1963. The development of the Orbital Laboratory, the Apollo "B" spacecraft, and the Lunar Landing System would follow in later years.

The aforementioned flexibility of programming also becomes evident in this figure. Assume that for some now unknown reason it becomes undesirable to explore the moon in the suggested time period, and that a decision is made that a large space station should be developed first. Such a decision could be made as late as 1965, without previously having committed anymore than the design phases of the manned lunar vehicles.

[33]A summary of the development and funding schedules is presented in Figure 16 [not provided], where the various program phases are given as a function of the fiscal year of program initiation. Most of the funds initially committed in 1962 will be for design phases. Major hardware contracts would not be awarded until 1963, with additional hardware developments starting in 1964 and 1965. The average cost per year, over a ten year period, for the total program is of the order of $700,000,000.

A basic ground rule in developing this plan was that the funding for fiscal year 1962 cannot be increased beyond the level that has been submitted to the Congress. However, increased funding in fiscal year 1962, in selected areas, might give increased assurance of meeting the projected flight dates. In particular, acceleration of the Saturn C-2, through earlier funding of the S-2 stage, would make this vehicle operational as much as a year before it is required for manned flight; the present program does not provide for any time between launch vehicle availability and manned spacecraft flights.

Earlier C-2 availability, together with earlier funding for the orbital docking demonstrations, would allow for additional unmanned orbital operations before manned flights [34] to the moon are made. Earlier spacecraft funding, for Apollo "A, "would lead to earlier flights with this vehicle. In the area of life sciences, increased funding in fiscal year 1962 would lead to the earlier availability of information on the effects of prolonged periods of weightlessness, and the biological effects of radiation.

An examination of the required NASA staffing to carry out this plan was not made as a part of this study. However, it must be recognized that neither Marshall Space Flight Center nor Space Task Group, as presently staffed, could fully support these programs. If the program is to be adopted, immediate consideration must be given to this problem.

CONCLUDING REMARKS

In, preparing this plan for a manned lunar landing capability, it was recognized that many foreseeable problems will require solutions before the plan can be fully implemented. Yet, an examination of ongoing NASA programs, in the areas of advanced research, life sciences, spacecraft development, and engine and launch vehicle development, has shown that solutions [35] to all of these problems should be available in the required period of time.

Throughout the plan, allowances were made for foreseeable problems; but it must be recognized that unforeseeable problems might delay the accomplishment of this mission. Nevertheless, the plan is believed to be sound in that it requires, at each point in time, a minimum committment [*sic*] of funds and resources until the needed background information is in hand. Thus, the plan does not represent a "crash" program, but rather it represents a vigorous development of technology. The program objectives might be met earlier with higher initial funding, and with some calculated risks.

[pp. 36- 51 not provided]

Document II-5

Document Title: Letter from L. V. Berkner, Chairman, Space Science Board, National Academy of Sciences, National Research Council, to James E. Webb, Administrator, NASA, 31 March 1961, with attached: Space Science Board, National Academy of Sciences, "Man's Role in the National Space Program."

Document Source: Folder 18675, NASA Historical Reference Collection, History Division, NASA Headquarters, Washington, DC.

The Space Studies Board (SSB) had been formed by the National Academy of Sciences a few months before the creation of NASA in 1958, with the hope that it could be the primary influence on the scientific goals of the nation's space program. NASA resisted such a role, and used the SSB as a source of non-binding advice on scientific priorities. The SSB was chaired by Lloyd Berkner, who had been considered for the position of NASA Administrator and was a personal friend of James Webb. The SSB met on 10 and 11 February 1961 to discuss its position on human spaceflight and presented a preliminary list of its findings to Webb on 27 February. The full report, which was only three pages long, was not sent to Webb until 31 March. Copies were also sent to Jerome Wiesner, the President's science advisor; Herbert York, Director of Defense Research and Engineering; and Alan Waterman, Director of the National Science Foundation.

Unlike the negative perception of NASA's human spaceflight program held by member's of the President's Science Advisory Committee, which was reflected in the advice of President-elect Kennedy's space transition team (chaired by Jerome Wiesner), the Space Science Board policy statement presented a positive view of the scientific value of humans in space. Using this statement, Webb and others in NASA could point to scientific support of a human spaceflight effort aimed at the exploration of the Moon and planets.

NATIONAL ACADEMY OF SCIENCES
NATIONAL RESEARCH COUNCIL
OF THE UNITED STATES OF AMERICA
SPACE SCIENCE BOARD

March 31, 1961

Mr. James E. Webb, Administrator
National Aeronautics & Space Administration
1520 H Street, N.W.
Washington 25, D.C.

Dear Mr. Webb:

I am enclosing two major policy positions that have been developed by the Space Science Board as recommendations to the Government.

The first of these concerns the enunciation of the major objective of space exploration and thus embraces man's role. The Board believes that the enunciation of such a policy would clarify the objectives of the national space effort by clearly focusing upon its goals.

The second document [not included] considers the support of basic research and argues, quite aside from current flight-package and related research, that a major and broad effort is required for the long-range success of our national space efforts. Our recommendations in this area represent careful discussions over a period of some three years.

Sincerely yours,

L. V. Berkner
Chairman

SPACE SCIENCE BOARD
National Academy of Sciences
2101 Constitution Avenue
Washington 25, D.C.

Man's Role in the National Space Program

At its meeting on February 10 and 11, 1961, the Space Science Board gave particular consideration to the role of man in space in the national space science program. As a result of these deliberations the Board concluded that <u>scientific exploration of the Moon and planets should be clearly stated as the ultimate objective of the U.S. space program for the foreseeable future.</u> This objective should be promptly adopted as the official goal of the United States space program and clearly announced, discussed and supported. In addition, it should be stressed that the United States will continue to press toward a thorough scientific understanding of <u>space,</u> of solving problems of manned space exploration, and of development of applications of space science for man's welfare.

The Board concluded that it is not now possible to decide whether man will be able to accompany early expeditions to the Moon and planets. Many intermediate problems remain to be solved. However, the Board strongly emphasized that planning for scientific exploration of the Moon and planets must at once be developed on the premise that man will be included.

Failure to adopt and develop our national program upon this premise will inevitably prevent man's inclusion, and every effort should be made to establish the feasibility of manned space flight at the earliest opportunity.

From a scientific standpoint, there seems little room for dissent that man's participation in the exploration of the Moon and planets will be essential, if and when it becomes technologically feasible to include him. Man can contribute critical elements of scientific judgment and discrimination in conducting the scientific exploration of these bodies which can never be fully supplied by his instruments, however complex and sophisticated they may become. Thus, carefully planned and executed manned scientific expeditions will inevitably be the more fruitful. Moreover, the very technical problems of control at very great distances, involving substantial time delays in command signal reception, may make perfection of planetary experiments impossible without manned controls on the vehicles.

[2] There is also another aspect of planning this country's program for scientific exploration of the Moon and planets which is not widely appreciated. In the Board's view, the scale of effort and the spacecraft size and complexity required for manned scientific exploration of these bodies is unlikely to be greatly different from that required to carry out the program by instruments alone. In broad terms, the primary scientific goals of this program are immense: a better understanding of the origins of the solar system and the universe, the investigation of the existence of life on other planets and, potentially, an understanding of the, origin of life itself. In terms of conducting this program a great variety of very intricate instruments (including large amounts of auxiliary equipment, such as high-powered transmitters, long-lived power supplies, electronics for remote control of instruments and, at least, partial data processing) will be required. It seems obvious that the ultimate investigations will involve spacecraft whether manned or unmanned, ranging to the order of hundreds of tons so that the scale of the vehicle program in either case will differ little in its magnitude.

Important supporting considerations are essential to realization of these concepts:

(a) Development of new generations of space vehicles, uniquely designed for use in space research and not adaptations of military rockets, must proceed with sufficient priority to ensure that reliable vehicles of adequate thrust are available for lunar and planetary research. This program should also include development of nuclear stages as rapidly as possible.

(b) Broad programs designed to determine man's physiological and psychological ability to adapt to space flight must likewise be pushed as rapidly as possible. However, planning for "manned" scientific exploration of the Moon and the planets should be consummated only as fast as possible consistent with the development of all relevant information. The program should not be undertaken on a crash basis which fails to given reasonable attention to assurance of success or tries to by-pass the orderly study of all relevant problems.

(c) Consideration should be given soon to the training of scientific specialists for spacecraft flights so that they can conduct or accompany manned expeditions to the Moon and planets.

[3] The Board strongly urges official adoption and public announcement of the foregoing policy and concepts by the U.S. government, Furthermore, while the Board has here stressed the importance of this policy as a scientific goal, it is not unaware of the great importance of other factors associated with a United States man in space program. One of these factors is, of course, the sense of national leadership emergent from bold and imaginative U.S. space activity. Second, the members of the Board as individuals regard man's exploration of the Moon and planets as potentially the greatest inspirational venture of this century and one in which the entire world can share; inherent here are great and fundamental philosophical and spiritual values which find a response in man's questing spirit and his intellectual self-realization. Elaboration of these factors is not the purpose of this document. Nevertheless, the members of the Board fully recognize their parallel importance with the scientific goals and believe that they should not be neglected in seeking public appreciation and acceptance of the program.

Document II-6

Document Title: **Memorandum to Pierre Salinger from Hugh Sidey, 14 April 1962.**

Document Source: **John F. Kennedy Presidential Library, Boston, Massachusetts.**

Document II-7

Document Title: "Memorandum to the President from Jerome Wiesner Re: Sidney Memorandum," 14 April 1961.

Document Source: John F. Kennedy Presidential Library, Boston, Massachusetts.

President John F. Kennedy was extremely effective in his relations with print and electronic media, and often became personal friends with reporters covering his presidency. One of these individuals was Hugh Sidey, who covered the White House for Life *magazine. In the aftermath of the launching of Yuri Gagarin on 12 April 1961, Sidey requested an interview with the president and provided Kennedy's press secretary Pierre Salinger with a memorandum as background for the interview. On the date of the memorandum, 14 April, Sidey sat in on a meeting between Kennedy and his top space advisors in the Cabinet room; he described the meeting in his 1963 book,* John F. Kennedy, President. *In preparation for Sidey's discussions with the president and, separately, Kennedy's top advisor Theodore Sorenson, Presidential science advisor Jerome Wiesner prepared a response to the Sidey memorandum.*

Document II-6

Office Memorandum

To <u>Pierre Salinger</u>
From <u>Hugh Sidey</u>
Date <u>April 14, 1962</u>

Questions for the President on Space –
    (FYI my initial surveillance of the space problem reveals some ragged dilemmas on the landscape. There are a lot of good minds in NASA and other dusty offices of the space agency that think we still are fiddling, haven't made the necessary decisions. They claim the President isn't getting the range of advice on this problem he should have. Their arguments are damned cogent. They scoff at the theory of some scientists that the Russians have now gone as far as they can for a few years. They hoot equally at the idea that our space effort is "locked in" and can't be accelerated. They claim, with compelling logic, that if we are to get in this race at all we've got to declare a national space goal, go for broke on a big booster (which means plenty of dough, granted). If we don't do this then we are going to sit here over the next eight years and watch the Soviets march right on ahead. I must confess, as near as I can tell on the surface there has been no great urgency attached to this space decision. If it has been made, we don't know it.
    But knowing the President some, I can't believe he hasn't sensed [2] the urgency. Therefore if I could get a little guidance on the following questions it would help)
    1. Why haven't we declared a crash program on one of the big boosters and pulled in our horns on others? Has the President accepted the theory that

we can't move faster? The extra 78 million for Saturn indeed is some boost but Saturn isN't [*sic*] the long range solution and there is no crash program in sight for the big Nova engine or solid fuels. Is the budget consideration and the political climate the confining factor this year?

2. Might there now be a change in the Project Mercury? We get rumbles that this pre-orbital shot coming up late this month has really been rushed in hopes we might beat the Soviets. But there is more hazard in it than there should be and now that shot should be delayed, maybe dropped entirely while we try to leapfrog ahead.

3. How much of the feeling of no decision is due to the newness of the administration and preoccupation with other things so far? Will there be a new and tough look followed by some hard decisions soon?

Document II-7

THE WHITE HOUSE

WASHINGTON

April 14, 1961

MEMORANDUM FOR

THE PRESIDENT

The following points are pertinent to the Sidey memorandum:

First of all, no one in the Administration believes that the Russians are finished with their space exploits or that there aren't exciting space exploits still to be carried out that they will undoubtedly drive hard to accomplish. Extended duration flights of man in an earth orbit, unmanned and manned landings on the moon, manned and unmanned exploration of the planets, manned space stations and a variety of important applications of space are still ahead. Among these are communications satellites, meteorological satellites and a variety of military applications of satellite-based systems. We, of course, have no knowledge at all about future Soviet intentions, but it would be surprising if they didn't pursue vigorously at least some of these possibilities.

Sidey is concerned that there is no long-term and very ambitious large booster program. The previous Administration made the decision not to drive vigorously for such a booster, although it did fund the F-1 engine, which would be needed for the Nova booster, and in our recent budget review we provided $9 million to accelerate that program. Because there was not a well developed program looking beyond the Mercury man in space, this Administration has undertaken to examine the range of possibilities which in turn will determine the future booster program. It should be noted that we did add $14 million to the Rover program to accelerate its research. We deferred a decision on the Rover rocket program, an expensive program (about $1 billion to a flight test) until the national policy on the long-range space goals could be established. We have had a thorough review of the Rover program, and both NASA and the Science Advisory Committee are looking into the range of possibilities in the big booster field, including the relative merits of solid fuels and large chemical boosters, as well as nuclear rockets. It should be noted that these ambitious space systems could not exist for a number of years, and it seems inappropriate to cancel all of the on-going activity until that time. It has become perfectly clear to the Administration that these decisions had to be faced, and it is our intention to do so. On the other hand, it would have been erroneous to commit very large sums of money without first establishing clear-cut national goals that go beyond the present plans. In the end it will be necessary to decide how large a share of the funds available [2] to the Federal Government should be committed to this field.

In regard to the Mercury sub-orbital flight now scheduled for April 28, the following are the facts: The dates were not advanced to compete with the Soviet flight. It has always been a tight schedule, paced by available funds and technical problems. We have analyzed it thoroughly and don't believe that its chances of success would be greatly enhanced by any reasonable delays in the firing schedule or a small number of additional test firings. Some consideration should be given to the question of whether or not the risks involved in a failure don't out-weigh the advantages of carrying out of the shot successfully. There are valid technical reasons for carrying out the experiment in view of the bio-medical and systems test information that will be obtained. It is probably fair to say that the successful orbiting of man has removed many of the bio-medical questions which it was designed to answer.

Jerome B. Wiesner

Document II-8

Document Title: John F. Kennedy, Memorandum for Vice President, 20 April 1961.

Document Source: Presidential Files, John F. Kennedy Presidential Library, Boston, Massachusetts

---

Document II-9

Document Title: NASA, "Do We Have a Chance of Beating the Soviets?" 22 April 1961.

Document Source: Folder 18675, NASA Historical Reference Collection, History Division, NASA Headquarters, Washington, DC.

Document II-10

Document Title: Letter to the Vice President of the United States from Wernher von Braun, 29 April 1961.

Document Source: Folder 18675, NASA Historical Reference Collection, History Division, NASA Headquarters, Washington, DC.

Document II-11

Document Title: Memorandum to the Vice President from James E. Webb, NASA Administrator, and Robert S. McNamara, Secretary of Defense, 8 May 1961, with attached: "Recommendations for Our National Space Program: Changes, Policies, Goals."

Document Source: Folder 18675, NASA Historical Reference Collection, History Division, NASA Headquarters, Washington, DC.

President Kennedy's memorandum on 20 April led directly to the Apollo program. By posing the question "Is there any . . . space program which promises dramatic results in which we could win?" President Kennedy set in motion a review that concluded that only an effort to send Americans to the Moon met the criteria Kennedy had laid out. This memorandum followed a week of discussion within the White House on how best to respond to the challenge to U.S. interests posed by the 12 April1961, orbital flight of Yuri Gagarin.

Both NASA and the Department of Defense gave rapid responses to the president's questions. (The Department of Defense response can be found in Volume I, Document III-7.) While the Low study of a piloted lunar landing (Document II-4) had projected a cost of $7 billion for such an effort the NASA response gave only a cost estimate for acceleration the overall NASA program of between $22 and $33 billion.

Vice President Lyndon B. Johnson, in his new role as Chair of the National Aeronautics and Space Council, provided a preliminary report to the president on 28 April indicating that the most likely recommendation to come from his review was a focus on human missions to the Moon (Volume I, Document III-8). This conclusion had been strongly influenced by Wernher von Braun, who the vice president had consulted independent of NASA's Washington managers. Von Braun told the vice president in his letter that the United States had "an excellent chance" of beating the Russians to a lunar landing.

During the Space Council review, the vice president also contacted congressional leaders to make sure that they would be willing to support a bold space recommendation, should the president make one. He found that those whom he consulted were strongly in favor of an accelerated effort (Volume I, Document III-10).

The final recommendations of the review came in the form of a memorandum signed by NASA Administrator Webb and Secretary of Defense Robert S. McNamara. This memorandum was the hurried product of a weekend of work following the successful suborbital flight of Alan Shepard, the first U.S. astronaut, on Friday, 5 May 1961. The urgency was caused by the vice president's desire to get recommendations to the president before he left on a rapidly arranged inspection tour to Southeast Asia. NASA, the Department of Defense, and the Bureau of the Budget staffs and senior officials met on Saturday and Sunday at the Pentagon to put together the memorandum, which the vice president approved without change and delivered to the President on Monday, 8 May. On that same day, Shepard came to Washington for a parade down Pennsylvania Avenue and a White House ceremony with President Kennedy. The recommendation that the United States undertake space programs aimed at enhancing national prestige, even if they were not otherwise justified by scientific, commercial, or military benefits, because such prestige was part of the "battle along the fluid front of the cold war," provided the underpinning rationale of Project Apollo. Only excerpts from the document directly related to setting the lunar landing goal are included here; the complete memorandum appears as Document II-11 in Volume I of this series.

Document II-8

April 20, 1961

MEMORANDUM FOR
VICE PRESIDENT

In accordance with our conversation I would like for you as Chairman of the Space Council to be in charge of making an overall survey of where we stand in space.

1. Do we have a chance of beating the Soviets by putting a laboratory in space, or by a trip around the moon, or by a rocket to land on the moon, or by a rocket to go to the moon and back with a man. Is there any other space program which promises dramatic results in which we could win?

2. How much additional would it cost?

3. Are we working 24 hours a day on existing programs. If not, why not? If not, will you make recommendations to me as to how work can be speeded up.

4. In building large boosters should we put our emphasis on nuclear, chemical or liquid fuel, or a combination of these three?

5. Are we making maximum effort? Are we achieving necessary results?

I have asked Jim Webb, Dr. Wiesner, Secretary McNamara and other responsible officials to cooperate with you fully. I would appreciate a report on this at the earliest possible moment.

John F. Kennedy

Document II-9

NATIONAL AURONAUTICS [*sic*] AND SPACE ADMINISTRATION

April 22, 1961

1.   "Do we have a chance of beating the Soviets?"

a.   "By putting a laboratory in space?"
There is no chance of beating the Soviets in putting a multi-manned laboratory in space since flights already accomplished by the Russians have demonstrated that they have this capability. The U.S. program must include the development of a multi-manned orbiting laboratory as soon as possible since it is essential for the accomplishment of the more difficult flights to the moon.

b.   "Or by a trip around the moon?"
With a determined effort of the United States, there is a chance to beat the Russians in accomplishing a manned circumnavigation of the moon. The Russians have not as yet demonstrated either the booster capability or the technology required for returning a man from a flight around the moon. The state of their booster technology and other technology required for such a difficult mission is not accurately known. With an accelerated program, it is not unreasonable for the U.S. to attempt a manned circumlunar flight by 1966.

[2]
c.   "Or by a rocket to land on the moon?"
On September 12, 1959, the Russians crash-landed a small package on the moon. This package did not transmit any information from the surface of the moon. The NASA program currently includes impacting instruments on the moon in such a way that they may survive the impact and transmit scientific information back to earth. The first flight in this program is scheduled for January 1962. Close-up television pictures will be obtained of the surface of the moon, as the spacecraft descends to the moon. In August 1963 the current NASA program also includes a soft landing of instruments on the moon. Several flights in succeeding months are included in this program to insure the possibility of success. The Russians can accomplish this mission now if they choose.

d.   "Or by a rocket to go to the moon and back with a man?"
There is a chance for the U.S. to be the first to land a man on the moon and return him to earth if a determined national effort is made. The development of a large chemical rocket booster, the spacecraft for landing and return, and major developments in advanced technology are required to accomplish this most difficult mission. The Russians initiated their earth orbiting program probably as early as 1954 as evidenced by their flight of a dog in November 1957. In the earth orbiting [3] competition the United States was attempting to accomplish in less than three years what the Russians had worked on for seven years. It is doubtful that the Russians have a very great head start on the U.S. in the effort required for a manned lunar landing. Because of the distinct superiority of U.S. industrial capacity, engineering, and scientific know-how, we believe that with the necessary national effort, the U.S. may be able to overcome the lead that the Russians might have up to now. A possible target date for the earliest attempt for a manned lunar landing is 1967, with an accelerated U.S. effort.

e. "Is there any other space program which promises dramatic results in which we could win?"

(1)   The current NASA program provides the possibility of returning a sample of the material from the moon surface to the earth in 1964. An experiment of this kind would have dramatic value and may or may not be a part of the Russian program. The Russians could carry out but such an experiment in the same time period or earlier if they choose.

(2)   The lead the U.S. has taken in developing communications satellites should be exploited to the fullest. Although not as dramatic as manned flight, the direct benefits to the people throughout the world in the long term are clear. U.S. national prestige will be enhanced by [4] successful completion of this program. The current program will provide for the flight of an active communications satellite in mid-1962. The experiment will enable live television pictures to be transmitted across the Atlantic. The continuing program will lead to the establishment of worldwide operational communications systems.

(3)   The U.S. lead established in our successful meteorological experiments with the TIROS satellites, should be maintained with a vigorous continuing program. The whole world will benefit from improved weather forecasting with the possibility of avoiding the disastrous effects of major weather disturbances such as typhoons, hurricanes and tornadoes.

[5]

2.   "How much additional would it cost?"

An estimate of the cost of the 10-year space exploration program as planned under the Eisenhower Administration was 17.91 billion dollars, as shown in Table A-1, attached. [not provided] In this program it was planned that manned lunar landing and return to earth would occur in the time period after 1970 but before 1975. Re-evaluation of the cost of this program based on providing adequate back-ups in all areas of the work has recently been made and the original cost estimate revised to 22.3 billion dollars for the ten-year period through 1970. [not provided] For an accelerated national program aiming toward achieving manned lunar landing in the 1967 period, it is estimated that the cost over the same ten-year period will be 33.7 billion dollars, as shown in Table E-1. [not provided] The additional 10 billion dollar cost of the program is due largely to paying for the program in the shorter time period. The resulting annual costs are naturally higher.

A list of the major items that would be initiated in 1962 with an accelerated program is shown in Attachment F. The total FY-62 funds, $1,744 millions, shown in Table E-1 is $509 million more than the approved current FY-62 budget.

[pp. 6-8 not provided]

[9]

Attachment F

MAJOR ITEMS IN THE ACCELERATED PROGRAM REQUIRING THE ADDITIONAL FUNDS SHOWN IN TABLE E-1 FOR FISCAL YEAR 1962

1.   Increase number of Mercury capsule flights to accelerate acquisition of knowledge on man's behavior under space flight conditions.
2.   Initiate possible additions to Mercury capsules for longer duration flights with intermediate launch vehicles.
3.   Accelerate unmanned lunar exploration to provide fundamental scientific data for manned flight to moon.
4.   Accelerate developments which will provide us with the essential knowledge and information to design spacecraft which can survive a return from the moon into the earth's atmosphere.
5.   Initiate developments of solid propellant rockets which can be used as a first or second stage of a launch vehicle for manned lunar landing missions (Nova).
6.   Initiate engineering design work and experimental development of a cluster of F-I engines for Nova.*
7.   Initiate design and engineering of a Nova vehicle using a cluster of F-1 liquid rocket engines as a first stage.*
8.   Initiate development of the tankage and engines required for a second stage of Nova.
9.   Accelerate supporting technology essential to the attainment of the goals of the program.
10.   Initiate construction of launch pads and other necessary facilities.
11   Provide additional vehicles and spacecraft for accelerating the TIROS meteorological program.

*The F-1 is the liquid rocket engine now under development which will have 1,500,000 pounds thrust in a single chamber.

[10]

3.   "Are we working 24 hours a day and, if not, why not?"

There is not a 24 hour a day work schedule on existing NASA space programs, except for selected areas in Project Mercury, the Saturn C-1 booster, the Centaur engines, and the final launching phases of most flight missions.

a.   Project Mercury at Cape Canaveral has been since October 1960 on a three-shift, seven-day- a-week basis plus shift overtime for all phases of capsule checkout and launch preparations. The McDonnell St. Louis plant, where the capsules are made, has averaged a 54-hour week on Mercury from the beginning, but also employs two or three shifts as needed in bottleneck areas. It now runs three shifts in the capsule test and checkout areas.

b.   SATURN C-1 project operates at Huntsville around-the-clock throughout any critical test periods for the first-stage booster; the remaining Saturn work is on a one-shift basis plus overtime which results in an average 47 hour week.

c.   CENTAUR hydrogen engine, which also is needed for the Saturn upper stages, is on three shifts in Pratt &Whitney's shops and test stands.

d.   Lastly, the final launch preparations of most flight missions require around-the-clock work at the launch sites [11] at Cape Canaveral, Wallops Station, or the Pacific Missile Range. In addition, NASA computer installations at Goddard and Marshall Centers operate continuous shifts in order to handle launch vehicle test analyses promptly, and determine orbital and trajectory data, and provide tracking and telemetry of space vehicles in flight.

NASA and its contractors are not working 24-hour days on the rest of its projects because:

a.   Certain projects are at an early stage of experimental study or design engineering where exchange of ideas is difficult to accomplish through multi-shifts.

b.   The schedules have been geared to the availability of facilities and financial resources. The funding levels for both contractors and government laboratories have been sufficient only for single-shift operations plus overtime (generally from 5 to 20%) as required to keep up the schedules.

c.   The limitations on manpower and associated funding determine the extent to which the NASA flight development centers may employ extra shifts.

In a number of areas in the national space program, the work could be accelerated if more manpower and more facilities were to be provided and funded in the immediate future. Recommendations to accomplish this are made elsewhere in this memorandum.

[12]
4.   "In building large boosters should we put our emphasis on nuclear, chemical or liquid fuel, or a combination of these three?"

In building the large launch vehicles required for the manned lunar landing mission, the immediate emphasis must be on the development of large solid and liquid rockets. It is believed that, in order to provide the necessary assurance that we will have a large launch vehicle for the lunar mission, we must have a parallel development of both a solid and liquid fueled large launch vehicle. The program on nuclear rockets must be prosecuted vigorously on a research and development basis. It is not believed that the nuclear rocket can play a role in the earliest attempt at manned lunar landing. The nuclear rockets will be needed in the even more difficult mission following manned lunar exploration. Use of the nuclear rocket for missions is not expected until after 1970 although flight test for developing the rocket will occur before then.

[13]

5a. "Are we making a maximum effort?"

No, the space program is not proceeding with a maximum effort. Additional capability exists in this country which could be utilized in this task. However, we believe that the manpower facilities and other resources now assigned are being utilized in an aggressive fashion.

5b. "Are we achieving necessary results?"

Our program is directed towards unmanned scientific investigation of space, manned exploration of space, and application of satellites to communication and meteorological systems. The scientific investigation is achieving basic knowledge important for a better understanding of the universe and also provides data necessary for the achievement of manned space flight and the satellite applications. It is generally agreed that our scientific program is yielding most significant results.

The Mercury program is the first and necessary step in an ongoing program leading to the manned laboratory, circumlunar flight, and manned lunar landing discussed under Item 1. A manned ballistic flight is scheduled in May, unmanned orbital flights and orbital flights with chimpanzee are scheduled for the Spring and Summer providing the background for the manned flight planned in 1961.

Future manned flight depends upon improved launch vehicle capability as well as a new spacecraft for the crew. The Saturn will [14] provide our first capability for large payloads but must be followed by a still larger vehicle for manned lunar landing. The launch vehicle for the first manned lunar landing will utilize either clustered F-1 liquid engines or solid propellant motors as discussed in item 4. We are achieving necessary technical data on the liquid engines but not on the large solid rocket engines. Ultimately, nuclear propulsion will be used to carry heavy payloads long distances into space. With our great capacity for engine research we have the capacity in this country to proceed more rapidly towards our objectives.

The TIROS and Echo satellites have provided important background data for meteorological and communication satellite systems. Additional experimentation is required in both fields before operational systems can be completely defined. We are continuing our meteorological program with TIROS flights and will use a newly-designed satellite called Nimbus when it is available in 1962. The first communication satellite (Echo) was a 100-ft. balloon which reflected ultra-high frequency signals between transmitters and receivers. The Echo type experiment is continuing and in addition we are instituting a program called Relay which carries microwave equipment for power amplification. This process decreases the requirements on the ground equipment but requires electronic equipment in the satellite with extremely high reliability compared to present day standards.

[15]

In summary, we are achieving significant scientific and technical results. We welcome the opportunity of reviewing these results with you to ensure that these results are compatible with our national goals.

Document II-10

April 29, 1961

Dear Mr. Vice President:

This is an attempt to answer some of the questions about our national space program raised by The President in his memorandum to you dated April 20, 1961. I should like to emphasize that the following comments are strictly my own and do not necessarily reflect the official position of the National Aeronautics and Space Administration in which I have the honor to serve.

Question 1. Do we have a chance of beating the Soviets by putting a laboratory in space, or by a trip around the moon, or by a rocket to land on the moon, or by a rocket to go to the moon and back with a man? Is there any other space program which promises dramatic results in which we could win?

Answer: With their recent Venus shot, the Soviets demonstrated that they have a rocket at their disposal which can place 14,000 pounds of payload in orbit. When one considers that our own one-man Mercury space capsule weighs only 3900 pounds, it becomes readily apparent that the Soviet carrier rocket should be capable of

— launching *several* astronauts into orbit simultaneously. (Such an enlarged multi-man capsule could be considered and could serve as a small "laboratory in space.")

— soft-landing a substantial payload on the moon. My estimate of the maximum soft-landed net payload weight the Soviet rocket is capable of is about 1400 pounds (one-tenth of its low orbit payload). This weight capability is *not* sufficient to include a rocket for the *return flight* to earth of a man landed on the moon. But it is entirely adequate for a powerful radio transmitter which would relay lunar data back to earth and which would be *abandoned* on the lunar surface after completion of this mission. A similar mission is planned for our "Ranger" project, which uses an Atlas-Agena B boost rocket. The "semi-hard" landed portion of the Ranger package weighs 293 pounds. Launching is scheduled for January 1962.

The existing Soviet rocket could furthermore hurl a 4000 to 5000 pound capsule *around* the moon with ensuing re-entry into the earth atmosphere. This weight allowance must be considered marginal for a one-man round-the-moon voyage. Specifically, it would not suffice to provide the capsule and its occupant with a "safe abort and return" capability, a feature which under NASA ground rules for pilot safety is considered mandatory for all manned space flight missions. One should not overlook the possibility, however, that the Soviets may substantially facilitate their task by simply waiving this requirement.

A rocket about ten times as powerful as the Soviet Venus launch rocket is required to land a man on the moon and bring him back to earth. Development of such a super rocket can be circumvented by orbital rendezvous and refueling of smaller rockets, but the development of this technique by the Soviets would not be hidden from our eyes and would undoubtedly require several years (possibly as long or even longer than the development of a large direct flight super rocket).

a) we do *not* have a good chance of beating the Soviets to a manned "*laboratory in space.*" The Russians could place it in orbit this year while we could establish a (somewhat heavier) laboratory only after the availability of a reliable Saturn C-1 which is in 1964.

b) we *have* a sporting chance of beating the Soviets to a soft-landing of a radio *transmitter station on the moon*. It is hard to say whether this objective is on their program, but as far as the launch rocket is concerned, they could do it at any time. We plan to do it with the Atlas-Agena B- Ranger #3 in early 1962.

[3] c) we have a sporting chance of sending a 3-man crew *around the moon* ahead of the Soviets (1965/66). However, the Soviets could conduct a round-the-moon voyage earlier if they are ready to waive certain emergency safety features and limit the voyage to one man. My estimate is that they could perform this simplified task in 1962 or 1963.

d) we have an excellent chance of beating the Soviets to the *first landing of a crew on the moon* (including return capability, of course). The reason is that a performance jump by a factor 10 over their present rockets is necessary to accomplish this feat. While today we do not have such a rocket, it is unlikely that the Soviets have it. Therefore, we would not have to enter the race toward this obvious next goal in space exploration against hopeless odds favoring the Soviets. With an all-out crash program I think we could accomplish this objective in 1967/68.

Question 2. How much additional would it cost?

Answer: I think I should not attempt to answer this question before the exact objectives and the time plan for an accelerated United States space program have been determined. However, I can say with some degree of certainty that the necessary funding increase to meet objective d) above would be well over $1 Billion for FY 62, and that the required increases for subsequent fiscal years may run twice as high or more.

Question 3. Are we working 24 hours a day on existing programs? If not, why not? If not, will you make recommendations to me as to how work can be speeded up.

Answer: We are *not* working 24 hours a day on existing programs. At present, work on NASA's Saturn project proceeds on a basic one-shift basis, with overtime and multiple shift operations approved in critical "bottleneck" areas.

During the months of January, February and March 1961, NASA's George C. Marshall Space Flight Center, which has systems management for the entire Saturn vehicle and develops the large first stage as an in-house project, has worked an average of 46 hours a week. This includes all administrative and clerical activities. In the areas critical for the Saturn project (design activities, assembly, inspecting, testing), average working time for the same period was 47.7 hours a week, with individual peaks up to 54 hours per week.

Experience indicates that in Research & Development work longer hours are not conducive to progress because of hazards introduced by fatigue. In the aforementioned critical areas, a second shift would greatly alleviate the tight scheduling situation. However, additional funds and personnel spaces are required to hire a second shift, and neither are available at this time. *In this area, help would be most effective.*

Introduction of a *third* shift *cannot* be recommended for Research & Development work. Industry-wide experience indicates that a two-shift operation with moderate but not excessive overtime produces the best results.

In industrial plants engaged in the Saturn program the situation is approximately the same. Moderately increased funding to permit greater use of premium paid overtime, prudently applied to real "bottleneck" areas, can definitely speed up the program.

Question 4. In building large boosters should we put our emphasis on nuclear, chemical or liquid fuel, or a combination of these three?

Answer: It is the consensus of opinion among most rocket men and reactor experts that the future of the nuclear rocket lies in deep-space operations (upper stages of chemically-boosted rockets or nuclear space vehicles departing from an orbit around the earth) rather than in launchings (under nuclear power) from the ground. In addition, there can be little doubt that the basic technology of nuclear rockets is still in its early infancy. The nuclear rocket should therefore be looked upon as a promising means to extend and expand the scope of our space operations in the years beyond 1967 or 1968. It should not be considered as a serious contender in the big booster problem of 1961.

The foregoing comment refers to the simplest and most straightforward type of nuclear rocket, viz. the "heat transfer" or "blow-down" type, whereby liquid hydrogen is evaporated and superheated in a very hot nuclear reactor and subsequently expanded through a nozzle.

There is also a fundamentally different type of nuclear rocket propulsion system in the works which is usually referred to as "ion rocket" or "ion propulsion." Here, the nuclear energy is first converted into electrical power which is then used to expel "ionized" (i.e., electrically charged) particles into the vacuum of outer space at extremely high speeds. The resulting reaction force is the ion rocket's "thrust." It is in the very nature of nuclear ion propulsion systems that they cannot be used in the atmosphere. While very efficient in propellant economy, they are capable only of very small thrust forces. Therefore they do not qualify as "boosters" at all. The future of nuclear ion propulsion lies in its application for low-thrust, high-economy cruise power for interplanetary voyages.

As to "chemical or liquid fuel" The President's question undoubtedly refers to a comparison between "solid" and "liquid" rocket fuels, both of which involve chemical reactions.

At the present time, our most powerful rocket boosters (Atlas, first stage of Titan, first stage of Saturn) are all liquid fuel rockets and all available evidence indicates that the Soviets are also using liquid fuels for their ICBM's and space launchings. The largest solid fuel rockets in existence today (Nike Zeus booster, first stage Minuteman, first stage Polaris) are substantially smaller and less powerful. There is no question in my mind that, when it comes to building very powerful booster rocket systems, the body of experience available today with liquid fuel systems greatly exceeds that with solid fuel rockets.

There can be no question that larger and more powerful solid fuel rockets can be built and I do not believe that major breakthroughs are required to do so. On the other hand it should not be overlooked that a casing filled with solid propellant and a nozzle attached to it, while entirely capable of producing thrust, is not yet a rocket ship. And although the reliability record of solid fuel rocket *propulsion units*, thanks to their simplicity, is impressive and better than that of liquid propulsion units, this does not apply to *complete rocket systems*, including guidance systems, control elements, stage separation, etc.

Another important point is that booster performance should not be measured in terms of thrust force alone, but in terms of total impulse; i.e., the product of thrust force and operating time. For a number of reasons it is advantageous not to extend the burning time of solid fuel rockets beyond about

60 seconds, whereas most liquid fuel boosters have burning time of 120 seconds and more. Thus, a 3-million pound thrust solid rocket of 60 seconds burning time is actually not more powerful than a 1 1/2-million pound thrust liquid booster of 120 seconds burning time.

I consider it rather unfortunate that several solid fuel rocket manufacturers (with little or no background in developing complete missile systems) have recently initiated a publicity campaign obviously designed to create the impression that a drastic switch from liquid to solid rockets would miraculously cure all of this country's big booster ills. I am convinced that if we recklessly abandon our liquid fuel technology in favor of something we do not yet understand so well, we would be heading for disaster and lose even more precious time.

My recommendation is to substantially increase the level of effort and funding in the field of solid fuel rockets (by 30 or 50 million dollars for FY 62) with the immediate objectives of

- demonstration of the feasibility of very large segmented solid fuel rockets. (Handling and shipping of multi-million pound solid fuel rockets become unmanageable unless the rockets consist of smaller individual segments which can be assembled in building block fashion at the launching site.)

- development of simple inspection methods to make certain that such huge solid fuel rockets are free of dangerous cracks or voids

- determination of the most suitable operational methods to ship, handle, assemble, check and launch very large solid fuel rockets. This would involve a series of paper studies to answer questions such as

a. Are clusters of smaller solid rockets, or huge, single poured-in-launch-site solid fuel rockets, possibly superior to segmented rockets? This question must be analyzed not just from the propulsion angle, but from the operational point of view for the total space transportation system and its attendant ground support equipment.

b. Launch pad safety and range safety criteria (How is the total operation at Cape Canaveral affected by the presence of loaded multi-million pound solid fuel boosters?)

c. Land vs. off-shore vs. sea launchings of large solid fuel rockets.

d. Requirements for manned launchings (How to shut the booster off in case of trouble to permit safe mission abort and crew capsule recovery? If this is difficult, what other safety procedures should be provided?)

Question 5. Are we making maximum effort? Are we achieving necessary results?

Answer: No, I do *not* think we are making maximum effort.

In my opinion, the most effective steps to improve our national stature in the space field, and to speed things up would be to

- identify a few (the fewer the better) goals in our space program as objectives of highest national priority. (For example: Let's land a man on the moon in 1967 or 1968.)

- identify those elements of our present space program that would qualify as immediate contributions to this objective. (For example, soft landings of suitable instrumentation on the moon to determine the environmental conditions man will find there.)

- put all other elements of our national space program on the "back burner."

- add another more powerful liquid fuel booster to our national launch vehicle program. The design parameters of this booster should allow a certain flexibility for desired program reorientation as more experience is gathered.

Example: Develop in addition to what is being done today, a first-stage liquid fuel booster of twice the total impulse of Saturn's first stage, designed to be used in clusters if needed. With this booster we could

a. double Saturn's presently envisioned payload. This additional payload capability would be very helpful for soft instrument landings on the moon, for circumlunar flights and for the final objective of a manned landing on the moon (if a few years from now the route via orbital re-fueling should turn out to be the more promising one.)

b. assemble a much larger unit by strapping three or four boosters together into a cluster. This approach would be taken should, a few years hence, orbital rendezvous and refueling run into difficulties and the "direct route" for the manned lunar landing thus appears more promising.

[9]

In addition, relief in certain administrative areas would be mandatory. In my opinion, the two most serious factors causing delays in our space program are:

1.   Lack of flexibility in the use of approved funds and in adapting the program to the changes caused by rapidly acquired new knowledge and experience. After the Congress and The President have established the funding level at which the aforementioned national high-priority objective is to be supported, all restraints as to how these funds are to be applied should be removed. At the present time such restraints include:

• Funds assigned to "Research and Development" may not be used to build facilities in support of R&D, and vice versa.

• Government installations such as the Marshall Space Flight Center are unable to hire more personnel or establish a second shift because "personnel spaces" are lacking. Such "spaces" must, of course, be supported with adequate salary funds, but an increase in such funds alone does not yet provide the spaces.

2.   Contracting procedures. Contracting procedures must be simplified. This probably requires some special directives from the highest level. To illustrate the present dilemma: If NASA plans to let a contract for a new stage of Saturn, the first step is a wide-open invitation to everybody interested to attend a bidder's briefing. Here, the interested parties are told what the stage looks like, that substantial facilities are required to develop it, and that each bidder must prepare a very detailed proposal (which might cost him as much as $300,000 to $500,000 to prepare) before the contractor can be selected. This first go-round will usually discourage 80 per cent of the original bidders, but takes approximately eight weeks. In the meantime, NASA must prepare detailed specifications.

For the actual preparation of the proposal the contractors must be given several weeks. Usually, six to ten companies will participate in the final bid. In order to be competitive, these bids must be prepared by the best scientists and engineers at the contractor's proposal. Evaluation of all these many proposals takes [10] additional weeks. Before the contract can be signed, eight to ten months usually have elapsed since initiation of the contracting procedure, and several million dollars worth of efforts of the best rocket and missile brains have been spent.

While there is certainly some merit in this long, drawn-out competitive procedure, we must realize that our Soviet competitors are not faced with some of these problems, simply because the issue of possible favoritism does not exist in a country where all industry is government-owned.

My suggestion is not to switch to indiscriminate sole source procurement, but to limit the participation in important and difficult technological developments to those few companies who really have the resources, the experience and the available capacity to execute the job effectively. With a hungry aircraft and automotive industry, it is not surprising that at the present time the contracting NASA agency is subjected to all kinds of pressure aimed at giving additional contractors a chance to prove themselves. But the NASA agency involved usually knows very well the few companies which really possess the capabilities needed.

Summing up, I should like to say that in the space race we are competing with a determined opponent whose peacetime economy is on a wartime footing. Most of our procedures are designed for orderly, peacetime conditions. I do not believe that we can win this race unless we take at least some measures which thus far have been considered acceptable only in times of a national emergency.

Yours respectfully,
/s/
Wernher von Braun

Document II-11

8 May 1961

Dear Mr. Vice President:

Attached to this letter is a report entitled "Recommendations for Our National Space Program: Changes, Policies, Goals", dated 8 May 1961. This document represents our joint thinking. We recommend that, if you concur with its contents and recommendations, it be transmitted to the President for his information and as a basis for early adoption and implementation of the revised and expanded objectives which it contains.

Very respectfully,

James E. Webb
Administrator

National Aeronautics and
Space Administration

Robert S. McNamara
Secretary of Defense

[1]   **Introduction**

It is the purpose of this report (1) to describe changes to our national space efforts requiring additional appropriations for FY 1962; (2) to outline the thinking of the Secretary of Defense and the Administrator of NASA concerning U.S. status, prospects, and policies for space; and (3) to depict the chief goals which in our opinion should become part of Integrated National Space Plan. These matters are covered in Sections I, II, III, respectively.

Three appendices (Tabs A through C) [not included] support these sections. Tab A highlights the Soviet space program. The bulk of this Tab (Attachment A) is separated from this report since it bears a special security classification. Tab B includes a description of major U.S. space projects and elements. Tab C provides financial summaries of the present programs, the proposed add-ons, and future costs of the program.

The first joint report contains the results of extensive studies and reappraisals. It is a first and not our last report and does not, of course, represent a complete or final word about our space undertakings.

[pp. 2- 6 not included]

[7] II. NATIONAL SPACE POLICY
The recommendations made in the preceding Section imply the existence of national space goals and objectives toward which these and other projects are aimed. Major goals are summarized in Section III. Such goals must be formulated in the context of a national policy with respect to undertakings in space. It is the purpose of this Section to highlight our thinking concerning the direction that such national policy needs to take and to present a backdrop against which more specific goals, objectives and detailed policies should, in our opinion, be formulated.

a. Categories of Space Projects
Projects in space may be undertaken for any one of four principal reasons. They may be aimed at gaining scientific knowledge. Some, in the future, will be

of commercial or chiefly civilian value. Several current programs are of potential military value for functions such as reconnaissance and early warning. Finally, some space projects may be undertaken chiefly for reasons of national prestige.

The U.S. is not behind in the first three categories. Scientifically and militarily we are ahead. We consider our potential in the commercial/civilian area to be superior. The Soviets lead in space spectaculars which bestow great prestige. They lead in launch vehicles needed for such missions. These bestow a lead in capabilities which may some day become important from a military point of view. For these reasons it is important that we take steps to insure that the current and future disparity between U.S. Soviet launch capabilities be removed in an orderly but timely way. Many other factors however, are of equal importance.

b. Space Projects for Prestige

All large scale space projects require the mobilization of resources on a national scale. They require the development and successful application of the most advanced technologies. They call for skillful management, centralized control and unflagging pursuit of long range [8] goals. Dramatic achievements in space, therefore, symbolize the technological power and organizing capacity of a nation.

It is for reasons such as these that major achievements in space contribute to national prestige. Major successes, such as orbiting a man as the Soviets have just done, lend national prestige even though the scientific, commercial or military value of the undertaking may by ordinary standards be marginal or economically unjustified.

This nation needs to make a positive decision to pursue space projects aimed at enhancing national prestige. Our attainments are a major element in the international competition between the Soviet system and our own. The non-military, non-commercial, non-scientific but "civilian" projects such as lunar and planetary exploration are, in this sense, part of the battle along the fluid front of the cold war. Such undertakings may affect our military strength only indirectly if at all, but they have an increasing effect upon our national posture.

c. Planning

It is vital to establish specific missions aimed mainly at national prestige. Such planning must be aimed at both the near-term and at the long range future. Near-term objective alone will not suffice. The management mechanisms established to implement long range plans must be capable of sustained centralized direction and control. An immediate task is to specify long-range goals, to describe the missions to be accomplished, to define improved management mechanisms, to select the launch vehicles, the spacecraft, and the essential building blocks needed to meet mission goals. The long-term task is to manage national resources from the national level to make sure our goals are met.

It is absolutely vital that national planning be sufficiently detailed to define the building blocks in an orderly and integrated way. It is absolutely vital that national management be equal to the task of focusing resources, particularly scientific and engineering manpower [9] resources, on the essential building blocks. It is particularly vital that we do not continue to make the error of spreading ourselves too thin and expect to solve our problems through the mere appropriation and expenditure of additional funds.

[remainder of p.9 – p. 12 not included]

[13] III. MAJOR NATIONAL SPACE GOALS

It is the purpose of this section to outline some of the principal goals, both long range and short range, toward which our national space efforts should, in our opinion, be directed. It is not the intent to specify all of the goals or even all of the major goals of importance to a National Space Plan. We wish to stress five principal objectives which in our opinion have not been adequately formulated or accepted in the past and which we believe should be accepted as a basis for specific project undertakings in the years ahead.

a. Manned Lunar Exploration

We recommend that our National Space Plan include the objective of manned lunar exploration before the end of this decade. It is our belief that manned exploration to the vicinity of and on the surface of the moon represents a major area in which international competition for achievement in space will be conducted. The orbiting of machines is not the same as the orbiting or landing of man. It is man, not merely machines, in space that captures the imagination of the world.

The establishment of this major objective has many implications. It will cost a great deal of money. It will require large efforts for a long time. It requires parallel and supporting undertakings which are also costly and complex. Thus, for example, the RANGER and SURVEYOR Projects and the technology associated with them must be undertaken and must succeed to provide the data, the techniques and the experience without which manned lunar exploration cannot be undertaken.

The Soviets have announced lunar landing as a major objective of their program. They may have begun to plan for such an effort years ago. They may have undertaken important first steps which we have not begun.

It may be argued, therefore, that we undertake such an objective with several strikes against us. We cannot avoid announcing not only our general goals but many of our specific plans, and our successes [14] and our failures along the way. Our cards are and will be face up-their's are face down.

Despite these considerations we recommend proceeding toward this objective. We are uncertain of Soviet intentions, plans or status. Their plans, whatever they may be, are not more certain of success than ours. Just as we accelerated our ICBM program we have accelerated and are passing the Soviets in important areas in space technology. If we set our sights on this difficult objective we may surpass them here as well. Accepting the goal gives us a chance. Finally, even if the Soviets get there first, as they may, and as some think they will, it is better for us to get there second than not at all. In any event we will have mastered the technology. If we fail to accept this challenge it may be interpreted as a lack of national vigor and capacity to respond.

[remainder of memorandum not included]

Document II-12

Document Title: Bruce Lundin et al., "A Survey of Various Vehicle Systems for the Manned Lunar Landing Mission," 10 June 1961

Document Source: Folder 18675, NASA Historical Reference Collection, History Division, NASA Headquarters, Washington, DC.

Once the decision to go to the Moon had been made, NASA had to decide how to achieve that goal. At the time of President Kennedy's decision, the leading plan was to use a very large launch vehicle called Nova to carry a spacecraft directly to the lunar surface. An alternative to this "direct ascent" approach was to carry out rendezvous operations at some location during the lunar landing mission. This study was the first of several between June 1961 and June 1962 that evaluated various rendezvous approaches and compared them to an approach using a very large booster, designated Nova. Based on its results, some sort of rendezvous in Earth orbit was given increasingly serious consideration as an alternative to the direct ascent approach for the rest of 1961. This study was also the first to examine rendezvous in lunar orbit, which in 1962 emerged as NASA's preferred approach to accomplishing the lunar landing.

A SURVEY OF VARIOUS VEHICLE SYSTEMS FOR

THE MANNED LUNAR LANDING MISSION

by

Bruce T. Lundin – Lewis, Chairman
Walter J. Downhower – JPL
A.J. Eggers, Jr. – Ames
Lt. Col. George W. S. Johnson – USAF
Laurence K. Loftin, Jr. – Langley
Harry O. Ruppe – Marshall
William J. D. Escher – Hdqs., Secretary
Ralph W. May, Jr. – Hdqs., Secretary

June 10, 1961

[no page number]

A SURVEY OF VARIOUS VEHICLE SYSTEMS FOR

THE MANNED LUNAR LANDING MISSION

I. INTRODUCTION

In response to the request of the Associate Administrator on May 25, 1961, it has been undertaken to assess a wide variety of systems for accomplishing a manned lunar landing in the 1967-1970 time period. This study has, as directed, placed primary emphasis on the launch vehicle portions of the systems, including vehicle sizes, types and staging. In addition a number of variations on the use of rendezvous to add flexibility and improve energy management in the lunar mission have been considered. The results of this study are the subject of the present report, and they are discussed in the following order.

First, the use of rendezvous to achieve a manned lunar landing is discussed in terms of rendezvous locations, vehicle types, and mission requirements, and the more attractive types of rendezvous are rated in the light of these considerations. Then a number of alternate Nova's for accomplishing the manned lunar mission are discussed briefly, and some consideration is given to the attendant question of launch sites, booster recovery, and the role of man in the system. Finally, the various methods for achieving manned lunar landing are compared in terms of time phasing, reliability, and approximate cost.

II. MISSION STAGING BY RENDEZVOUS

II. 1. General

Mission staging by rendezvous has been the subject of much investigation at Marshall, Langley, Ames, Lewis, and JPL. The work has concerned itself with analytical and simulator studies of orbital mechanics, and control and guidance problems as applied to rendezvous. Such critical questions as launch timing, and automatic and piloted guidance of the vehicles to a rendezvous have been carefully analyzed. Orbital refueling as well as attachment of self-contained modules have been considered.

Because the use of rendezvous permits the accomplishment of a given mission in a number of different ways employing different launch vehicles, the various groups working on rendezvous have arrived [2] at a number of different concepts for accomplishing the lunar landing mission. The assumptions made by the different groups with regard to such parameters as return weight, specific impulse, etc., were, however, consistent to the extent that meaningful comparisons can be made between the different concepts. In the discussion to follow, the more attractive rendezvous concepts will be summarized, after which the advantages and disadvantages of each will be indicated and a rating system developed.

II 1. a: Mission Types

The rendezvous concepts which will be considered for the lunar landing are as follows:

1. Rendezvous in earth orbit;

2. Rendezvous in lunar orbit after take-off from the lunar surface;
3. Rendezvous in both earth and lunar orbit;
4. Rendezvous on the lunar surface.

Also possible are:

5. Rendezvous in transit to the moon;
6. Rendezvous in lunar orbit before landing.

Although advantages can be claimed for concepts 5 and 6, they are excluded from consideration because they are clearly inferior to concepts 1 through 4.

II.   1. b: Vehicles Considered

The vehicles considered were restricted to those employing engines presently under development. These vehicles are:

(a) Saturn C-2 which has the capability of placing about 45,000 pounds in earth orbit and 15,000 pounds in an escape trajectory;

(b) Saturn C-3 which has the capability of placing about 110,000 pounds in earth orbit and 35,000 pounds in an escape trajectory. The configuration of the C-3 considered here employs the following staging:

[3]

First	2 F-1
Second	4 J-2
Third	6 LR115

II.   1. c: Mission Requirements

The significant requirements employed in this examination of the manned lunar mission are as follows:

1. Return spacecraft weight -- 12,500 lbs.

2. Velocity increments (60 hr. transfer)

earth orbit to escape	10,600 fps
braking into lunar orbit	3,400 fps
lunar landing	6,860 fps
lunar ascent and return	9,930 fps

3. Stage mass fractions

launch and transfer stages	0.90
lunar landing stage	0.87

4. Impulses

hydrogen-oxygen	420 sec
storable propellants	300 sec

These figures and estimates are considered reasonable and consistent with those used in the concurrent Nova studies. With this information we can now match the previously discussed vehicles and rendezvous concepts.

II. 2. Mission Vehicle Matching

II. 2. a: Earth Rendezvous Only.

On the basis of the preceding paragraph, the following weights at different stages of the mission pertain to the case of rendezvous in earth orbit only (based on H202 performance):

[4]

Weight returned to vicinity of earth	12, 500 pounds
Lunar take-off weight	28, 800 pounds
Weight landed on moon	31, 000 pounds
Weight in escape trajectory	73, 000 pounds
Weight in earth orbit	210, 000 pounds

These weights indicate that five C-2's or two C-3's are required in order to accomplish the mission.

II. 2. b: Lunar Rendezvous.

A concept in which a rendezvous is made in lunar orbit only or together with earth orbit rendezvous possesses basic advantages in terms of energy management and thus launch vehicle requirements. This approach involves placing the complete spacecraft in orbit about the moon at a relatively low altitude. One or two of the three-man crew then descends to the lunar surface in a special capsule which detaches from the spacecraft. Upon leaving the lunar surface, the capsule performs a rendezvous with that portion of the spacecraft which remained in lunar orbit. The lunar capsule is, of course, left behind on the return trip of the spacecraft to earth. A variation on this approach involves two lunar landing capsules, one of which remains with the "mother" ship and can be used for rescue operations on the lunar surface.

The basic advantage of the system is that the propellant required for the lunar landing and take-off is reduced which in turn translates into a reduction in the amount of weight which must be put into an escape trajectory. The escape weight saving achieved is related to the fraction of the spacecraft weight which is retained in lunar orbit. The actual weight saving which can be realistically achieved by this method can only be determined after detailed consideration of the design and integration of the complete spacecraft. Calculations suggest, however, that the amount of weight which must be put into an escape trajectory for a given reentry vehicle weight might be reduced by a factor of two by use of the lunar rendezvous technique. The earth booster requirement might therefore be reduced to one C-3 with lunar rendezvous or two to three C-2 's with earth and lunar rendezvous.

II. 2. c. Lunar Surface Rendezvous

This scheme envisions accomplishment of the initial manned lunar mission with C-2 launched vehicles assembled on the lunar surface. An unmanned transport spacecraft launched by a C-2 can deposit an approximately 5, 000 lb. payload on the moon. (No.1 on fig. 7).[not included]   Previous SURVEYOR or RANGER shots would establish the landing spot and provide [5] homing beacons, TV monitoring equipment, and so forth (Items 4, 5, and 6 on fig. 7). [not included]

A number of methods for refueling on the lunar surface may be envisioned. One possible concept may be recounted as follows: Four 5,000 lb. refueler vehicles (Item 2 on fig. 7) [not included]would be landed approximately equally spaced around a spacecraft carrying a capsule suitable for returning one man to the earth, and within 45 feet of it. Solid propulsion units would be transferred from the refuelers to the centrally located spacecraft by means of specially designed transfer tracks. The assembly operation would be monitored by TV, and the assembled vehicle would be checked out before sending man from earth to the area via a second landing capsule (Item 3). [not included] The space station would be capable of maintaining itself in the lunar environment. The astronaut would walk from the landing capsule to the take-off vehicle and depart for earth. The four solid rockets used for launch from the lunar surface would be identical to the four retro-rockets used for each vehicle in landing on the moon. These retro-rockets would be jettisoned before touchdown and a soft landing controlled with liquid vernier rockets. A great deal of the technology developed for SURVEYOR would be utilized in this concept for manned lunar landing. The return vehicle weight would be approximately 5,500 lbs. at lunar injection, 5,200 lbs. as the earth is approached, and 3,500 to 4,000 lbs. at earth reentry. Careful study of Apollo study contractor results has indicated this to be adequate for a full-sized three-man Apollo capsule having only one man aboard. (Further description is given in Appendix II-2-c.) [not included]

Saturn C-2's would be used throughout for earth-based launch vehicles. A minimum of 6 successful launches would be required for the basic mission. The actual number required to accomplish the mission would be a direct function of the success rate of the firings and assembly operations on the lunar surface; however, any failure before manned capsule landing does not affect the success of the manned lunar landing sequence. Identical transport spacecraft would be used in all C-2 launches; also, only the payloads would differ, i.e., capsules or return propellant.

[6]

II.   2.   d: Mixed Nova-Saturn Operations for the Time Period 1966 – 1969

Basic launch vehicles available in the time period of interest to accomplish the manned lunar landing and return missions may be both the SATURN and NOVA. Two basic modes of operations using either SATURN or NOVA are as follows:

1. The NOVA vehicle places the spacecraft with or without capsule in the waiting orbit first. A SATURN vehicle standing by on a launch pad will be launched with the lunar crew after the orbit of the NOVA payload has been established, and will rendezvous with the remainder of the spacecraft in the waiting orbit. If desirable either the entire capsule will be mechanically connected with the spacecraft or the crew changes ships only. The SATURN at this time will have had around 30 flights, and therefore be considerably more reliable than the NOVA. The very first NOVA which successfully orbits the payload will offer the first chance for a manned lunar landing. This procedure is expected to save one year in the total program schedule, and possibly to reduce overall cost as a smaller number of NOVA vehicles is required.

2. Same procedure as outlined under 1., but the entire lunar return vehicle with a payload of approximately 60, 000 lbs. will be orbited by a SATURN C-3 and will rendezvous (including docking) with the NOVA payload, which is a stage used both for acceleration to escape and for the landing maneuver on the moon. This procedure offers a 20 percent performance margin and can be used in case the capsule reentry weight should grow beyond the maximum design weight for which the original NOVA was designed.

[Sections II.3 – II.6 not included] …

[16]   II. 7. Summary Rating

Various combinations of boosters and rendezvous concepts for performing the manned lunar mission were reviewed. Guidelines which were adopted for the rating process placed primary emphasis on (1) ability to accomplish the lunar landing mission as soon as possible and (2) relative reliability of the concept. Considerably lesser importance was attached to cost and/or growth potential for other future space missions.

The results are tabulated in the following table.

Rendezvous Concepts	Order *of* Preference by each committee member						Total
	A	B	C	D	E	F	
Earth RV with 5-7 C-2's	4	4	4	6	3	3	24
Earth RV with 2-3 C-3's	2	1	1	1	2	1	8
Lunar RV with 1 C-3	5	3	2	3	4	4	21
Earth and Lunar RV with C-2's	6	6	3	5	5	5	30
RV on Lunar Surface	3	5	6	4	1	6	25
Earth RV with NOVA and C-1	1	2	5	2	6	2	18

The concept of a low altitude earth orbit rendezvous utilizing Saturn C-3's is a clear preference by the group.

[Sections III-VI not included]

[26]

Mission staging by rendezvous offers two advantages of particular significance to such large, complex, and long-range missions as a manned lunar landing. Because both future payload requirements and vehicle capability are uncertain at best, the ability to increase payload by adding a vehicle to the operation reduces the critical dependance [*sic*] of future mission capability on decisions relating to launch vehicle design and development. The inherently smaller vehicles associated with this method also permit the development of effective and efficient launch vehicles with engines currently in development.

Of the various orbital operations considered, the use of rendezvous in earth orbit by two or three Saturn C-3 vehicles (depending on estimated payload requirements) was strongly favored. This preference stemmed largely from the small number of orbital operations required and the fact that the C-3 is considered an efficient vehicle of large utility and future growth.

The rendezvous technique itself, in terms of launch operations, guidance and control, and orbital operations, is considered feasible of development within the time period of interest. Some justification for this point of view is found in both current technology and in the fact that many of the technological advancements required for the lunar landing and take-off operations are applicable to the rendezvous with an artificial satellite.

[27]

The principal difficulties involved in the development of a new 4,000,000 pound RP-LOX engine for a NOVA vehicle are associated with size and the development time span required through PFRT is estimated at 6 years. If a NOVA vehicle incorporating a new large engine development is contemplated, the Phoenix concept possesses sufficient attractive features to warrant serious study. The utilization of pressurized storable propellants for a large first-stage engine offer important reductions in complexity; solid rocket engines are, however, believed to offer even greater simplification without significant performance differences. The "standpipe" concept of propellant delivery through acceleration-head effects incorporates sufficient difficulties of engine system development and manned abort capability as to render it unattractive for application to a NOVA vehicle.

Document II-13

Document Title: Ernest W. Brackett, Director, Procurement & Supply, to Robert R. Gilruth, Space Task Group, "Transmittal of Approved Project Apollo Spacecraft Procurement Plan and Class Determination and Findings," 28 July 1961, with attached: Robert C. Seamans, Jr., Associate Administrator, "Project Apollo Spacecraft Procurement Plan," 28 July 1961; Robert C. Seamans, Associate Administrator, to Robert R. Gilruth, Space Task Group, "Appointment of Source Evaluation Board," 25 July 1961; James E. Webb, Administrator, "Establishment of Sub-Committees to the NASA Source Evaluation Board Project Apollo," 25 July 1961.

Document Source: Folder 18675, NASA Historical Reference Collection, History Division, NASA headquarters, Washington, DC.

Once President Kennedy had established the lunar goal, NASA had to establish procurement procedures for the necessary equipment, as well as evaluation boards for approving contractors. The first major element of Project Apollo to be put under contract would be the Apollo spacecraft. Procedures and committees for the spacecraft procurement were established just over two months after President Kennedy's announcement of Project Apollo on 25 May 1961.

Washington 25, D.C.
July 28, 1961

From: NASA Headquarters

To: Space Task Group

ATTENTION: Mr. Robert R. Gilruth

Subject: Transmittal of Approved Project Apollo Spacecraft
                      Procurement Plan and Class Determination and Findings

Reference: (a) Director, Space Task Group letter of June 26, 1961 to
                      NASA Headquarters transmitting proposed Project Apollo
                      Procurement Plan

    1.   Reference (a) forwarded a proposed procurement plan for Project Apollo. As a result of reviews made by various offices of NASA Headquarters a

revised procurement plan for Project Apollo Spacecraft has been signed by the Associate Administrator under date of July 28, 1961, and is attached hereto as Enclosure 1. Also attached as Enclosure 2, is the Class Determination and Findings for this project which was signed by the Administrator under date of July 25, 1961.

2.   A paragraph has been included in the approved procurement plan (first paragraph, page 4) which provides that, "Prior to commencement of negotiations, NASA will develop a contract clause which will assure NASA control over the selection and retention of the Contractor's key personnel assigned to the project." For your information, the Associate Administrator interprets this paragraph to mean that the number of key personnel that NASA will exercise control over should be restricted to a number not exceeding ten, and if at all possible, some lesser number.

3.   The Director of Reliability, NASA Headquarters, recommended that Mr. James T. Koppenhaver of his office be added to the Source Evaluation Board as a non-voting member in accordance with NASA General Management Instruction 2-4-3. This has been done. He also recommends that Dr. William Wolman of the Office of Reliability, NASA Headquarters, be made a member of the technical subcommittee. He also suggests that consideration be given to establishing a subcommittee consisting of two members of NASA Headquarters and three members of Space Task Group to evaluate the reliability and quality assurance aspects of the proposals. The last suggestions submitted, without recommendation, for your consideration.

[signed]

Ernest W. Brackett Director,

Procurement & Supply

[2]

PROJECT APOLLO SPACECRAFT PROCUREMENT PLAN

The procurement plan describes in brief the requirements for Project Apollo, the overall procurement program, and specifies the policies and procedures to be utilized in the selection of a qualified contractor who will be responsible for the development of the Command Module and Service Module for all missions and for performing systems integration and systems engineering.

DESCRIPTION OF PROJECT APOLLO SCOPE OF WORK

1.   Missions. Project Apollo will be developed in three separate but related mission concepts:

      a.   Phase "A". A manned spacecraft to be placed in orbit around the earth at 300 nautical mile altitude for a two-week duration for the purpose of developing space flight technology and conducting scientific experimentation. The spacecraft is to be capable of rendezvous in earth orbit.

b. Phase "B".   A manned spacecraft for circumlunar and orbital flight around the moon at an appropriate height and duration to permit the development of flight operations in deep space and provide an assessment of the system for the lunar landing mission.

c. Phase "C". A manned spacecraft to be soft landed on the surface of the moon and returned to earth.

2. Module concept. It is contemplated that the spacecraft for each of these three phases of Project Apollo will consist of separate modules as follows:

a. A Command Module which will serve as a control center for spacecraft and launch vehicle operation, as crew quarters for the lunar mission, and as the entry and landing vehicle for both nominal and emergency mission phases. To the greatest extent possible, the same command module will be used for all three phases cited in paragraph 1a, b, and c above.

b. A Service Module which will house support systems and components and will contain propulsion systems, as required, for emergency aborts, returning from earth orbit, mid-course corrections, lunar orbit and de-orbit, and lunar take-off. It is contemplated that separate contracts will be issued for some of these propulsion systems.

[3]

c. An Orbiting Laboratory Module for use in earth orbiting missions, as a laboratory technological or scientific experiments and measurements.

d. A Propulsion Module to be added for the lunar landing mission, for the purpose of landing the Command and Service Modules on the moon's surface.

PROCUREMENT PROGRAM

It is intended, under the overall procurement program, to award several contracts for each separate phase or sub-phase of the Apollo Project Spacecraft specifically as follows:

1. A contract to a Principal Contractor for the following elements of the Project:

a. A Command Module and Service Module to serve all flight missions.

b. A propulsion system for the earth-orbiting mission, if needed.

     c.   Responsibility for systems engineering and systems integration for all elements being developed by other contractors associated with Project Apollo Spacecraft systems. It is intended that the same Principal Contractor will be retained for all three phases, however, the Government will retain the option of selecting a new Principal Contractor for Phases "B" and "C" if it is considered desirable to do so.

2.   A contract for the development of an Orbiting Laboratory Module and vehicle adapter for the earth orbiting mission.

3.   A contract for the development of a Propulsion Module to provide a propulsion system for lunar landing.

4.   An associate contract for the development of a guidance and navigation system, to be housed in the Command Module, required for use on all manned missions.

RESPONSIBILITIES OF CONTRACTORS

The responsibilities of each of the associated contractors participating in Project Apollo Spacecraft procurement will be different and will be developed and defined separately as each contract is negotiated and written. Each contract will contain reliability requirements for mission accomplishment and flight safety. General types of responsibility, however, can be summarized as follows:

[4]

A.   Command Module and Service Module Contractor. (Principal Contractor)

The contractor assigned responsibility for the development of the command module and service module will also be responsible for systems integration and systems engineering for all missions. The Principal Contractor will serve in the role of principal integrator of all modules of the spacecraft to assure compatibility and timely and complete execution of all requirements of each mission. The contractor will also serve as the principal point of coordination with the launch vehicle developer to assure effective solution of interface problems between launch vehicle and spacecraft components, and with ground support developers to meet all of their requirements.

B.   Space Laboratory Module and Propulsion Module Contractors.

The contractors assigned responsibility for the development of Phase A Space Laboratory Module or Phases C Propulsion Module will be expected to complete all aspects of their subsystem and to work under the general technical direction of the Principal Contractor to assure the full and timely completion of the integrated spacecraft system and its integration with the launch vehicle and relate ground support facilities.

TYPE OF CONTRACT ADMINISTRATION

It is intended that a Cost-Plus-a-Fixed-Fee type of contract will be used initially in the procurement of the spacecraft.

A copy of the necessary Class Determination and Findings authorizing negotiation of contracts for Apollo Spacecraft, pursuant to 10 USC 2304(a)(11), is attached. (Enclosure No. 1)

In view of the magnitude, complexity, and substantial dollar value of the Apollo Spacecraft, a Principal Contractor and Associate Prime Contractor method of procurement is recommended. The spacecraft Principal Contractor will be contractually assigned responsibility and authority for the design of the system and the integration of the performance specification of all sub-systems and components to assure that they fit into a compatible, efficient system, and to manage the day-to-day development and production. The Space Task Group will retain authority to make major decisions, resolve conflicts between the Spacecraft Principal Contractor and the associated contractors; review and/or approve decisions made by the Principal Contractor; approve the make-or-buy policies. In addition, the Space Task Group will control concentration of Principal Contractor "in house" effort; assure competition in the selection of sub-contractors by requiring the Spacecraft Principal Contractor to take full advantage of the facilities and capabilities of existing sub-system manufacturers by subcontracting or the placement of systems direct by NASA with associated contractors.

[5]

Requests for proposals will require, [sic] that companies will furnish a description of the proposed organization and management plan for the spacecraft project including names of personnel to be assigned to key positions within such organization. Prior to commencement of negotiations, NASA will develop a contract clause which will assure NASA control over the selection and retention of the contractor's key personnel assigned to the project.

NASA will reserve the right to issue a separate contract to a qualified organization to assess systems reliability. If it is subsequently determined that such a contract is to be issued, this will be done at about the same time as the principal contractor is selected. Similar reliability assessment contracts may be placed for associated systems, as required.

CONTRACT NEGOTIATION AND AWARD

It is the intention of the Government to select at this time a contractor qualified to perform all the tasks set forth under paragraph 1a, b, and c of the Procurement Program section of this plan and to award a contract broad enough in scope to provide, with subsequent amendment, for the accomplishment of the total tasks required toward the completion of the Project Apollo Spacecraft Program. The initial contract will specifically cover the engineering study, detail design, development of manufacturing techniques, fabrication of breadboards, test hardware, laboratory models, "test" spacecraft, certain long lead items, and a detailed engineering mockup of the Apollo Spacecraft.

The Administrator may determine that negotiation will be conducted with several companies. If such negotiations are directed, the names of the companies selected for negotiations will not be announced. Following completion of such negotiation, the Source Evaluation Board will again report to the Administrator the results of the negotiations at which time he will determine that company with which to negotiate a contract if satisfactory terms can be arranged. Announcement of such selection will then be made in accordance with NASA regulations. Negotiations will be conducted by the Space Task Group procurement and technical staffs, with supplemental assistance from Headquarters management and technical staffs. The contract will be negotiated to spell out as extensively as possible all facets of contractor organization, management, technical performance and cost control to achieve the maximum assurance of protection of the interests of the Government, consonant with the urgency of the work of the Project.

SELECTION OF BIDDERS LIST

The field of contractors suitably qualified to undertake a program of this magnitude is limited. It is intended to solicit proposals from only 12 [6] companies who have indicated their definite interest in the Apollo Program and who have demonstrative competence and capability to successfully perform the procurement under consideration. Any other firms who may request an opportunity to submit a proposal will be required to furnish substantiating evidence as to their ability to perform before a request for proposal will be furnished. Space Task Group will maintain a complete documentation of the reasons for selection of companies invited to receive request for proposal and the reasons for declining to furnish requests for proposal to any company so requesting.

The selected sources are as follows:

1. Boeing Airplane Company
    Seattle, Washington

2. Chance Vought Aircraft, Inc.
    Dallas, Texas

3. General Dynamics Corporation
    San Diego, California

4. Douglas Aircraft Company, Inc.
    Santa Monica, California

5. General Electric Corporation
    Philadelphia, Pennsylvania

6. Goodyear Aircraft Corporation
    Akron, Ohio

7. Grumman Aircraft Engineering Corporation
    Bethpage, Long Island

8. Lockheed Aircraft Corporation
    Sunnyvale, California

9. Martin Company
    Baltimore, Maryland

10. McDonnell Aircraft Corporation
    St. Louis, Missouri

11. North American Aviation
    Los Angeles, California

12. Republic Aviation Corporation
    Farmingdale, New York

[7]

In addition, a synopsis of this procurement will be publicized for the benefit of subcontractors in accordance with NASA Circular No. 131, dated April 17, 1961, Subject: Publicizing of NASA Proposed Research and Development Procurement, Reference 18-2.203-4.

SCHEDULE OF PROCUREMENT ACTION

July 28	Request for Proposals mailed and bidders invited to conference
Aug. 14-15	Bidders Conference
Oct. 9	Proposals due
Dec. 1	Evaluation of proposals completed
Dec. 28	Selection of contractor
Dec. 29	Letter Contract (if desirable)
Apr. 30	Definitive contract

SOURCE EVALUATION

It is proposed that a NASA Source Evaluation Board be appointed by the Associate Administrator, NASA Headquarters, to be chaired by Mr. W. C. Williams, Associate Director of Space Task Group. The members of this Board will be specifically designated from appropriate NASA personnel. The Chairman of the Board will appoint such business and technical committees as may be necessary to assist the Board in the evaluation. The membership of these committees will be drawn from appropriate Government personnel. Recommendations for Board appointments are attached, (Enclosure No. 2) for approval by the Associate Administrator. Committee appointments anticipated at this time are also attached for information, (Enclosure No. 3).

REQUEST FOR PROPOSALS

The Source Evaluation Board will review the Request for Proposals, prior to the release to the selected prospective contractors, to assure that the RFP is complete in all details as to the technical, management and cost aspects and further, that it will adequately serve the intended purposes. The Source Evaluation Board shall be free to comment on and recommend any changes in the RFP considered essential to meet all Project Apollo objectives.

[8]

PROPOSAL EVALUATION

The evaluation of proposals submitted by industry will be the primary responsibility of the Source Evaluation Board. In these evaluations the Board will be free to seek further information from bidders or to offer bidders the opportunity of making further oral or written clarification of their submissions. In addition, the Board should be authorized to establish additional sub-committee, work groups or consulting relationships with other NASA employees or with other Government consultants as required. Where desirable, members of the Source Evaluation Board will be authorized to visit the facilities of bidders to acquaint themselves at first hand with the personnel and facilities with which project work would be carried out. The final product of the work of the Source Evaluation Board will be a presentation of findings to the Administrator.

BIDDERS CONFERENCE

Twelve days after release of the RFP a principal contractors bidders conference will be held at the Space Task Group, Langley Field, Virginia or such other suitable location in this geographical area as considered appropriate, for a detailed briefing on the proposed procurement. Contractors who are invited will be limited to a maximum of 10 representatives each which limitation shall include any subcontractor representatives. Attendance by contractors will be limited to those companies invited to submit proposals.

CONTRACT COST DETERMINATION

In view of the significant nature of this procurement, specific attention, review, and analysis will be given by STG in determining the reasonableness of costs submitted by a contractor. To meet this objective, contractors will be required to prepare comprehensive and extensive cost breakdowns for each category of proposed contract performance. Each contractor will be required to furnish with his proposal his procedures and techniques, as appropriate, for his accounting system, which shall include but not necessarily be limited to; methods of costing labor, material, burdens, etc., to each contract; cost estimation and reimbursement billing procedures. All contracts will incorporate all clauses required by law and regulation, plus other special conditions that may be necessary to adequately protect the Government's interests.

Approved

[signed by Robert Seamans]
Associate Administrator

Enclosure No. 1

NATIONAL AERONAUTICS AND SPACE ADMINISTRATION

SPACE TASK GROUP

LANGLEY FIELD, VIRGINIA

DETERMINATION AND FINDINGS

AUTHORITY TO NEGOTIATE CLASS OF CONTRACTS

Upon the basis of the following determination and findings which I hereby make as agency head, the proposed class of contracts described below may be negotiated without formal advertising pursuant to the authority of 10 U.S.C 2304 (a) (11).

This procurement will consist of more than one contract for the accomplishment of Project Apollo.

Findings

I hereby find that the primary objective of Project Apollo is to safely place a manned vehicle containing a 3-man crew into an earth and lunar orbital flight making a soft lunar landing and return take-off and to effect a safe recovery of the men and vehicle. A secondary objective is to study the capabilities of men for extended periods of approximately 14 days in the environments associated with earth and lunar launchings, orbital flights, lunar landings and recovery. There is an urgent requirement for this program to be completed at the earliest date compatible with reasonable assurance of success and with a high degree or assurance that the human occupants can safely escape from any forseeable [sic] situation which may develop. First attempt at orbital flight will be preceded by a program involving numerous research, experimental, and developmental contracts.

In addition to studies and services essential to the successful operation of Project Apollo, there is a continued need for research and development, design, engineering, fabrication and assembly of material and equipment.

The proposed class of contracts does not call for quantity production of any article.

It is impossible to describe in precise detail, or by any definite drawings or specifications, the nature or the work to be performed; only the ultimate objectives and the general scope or the work can be outlined. The materials, equipment, and services to be procured for Project Apollo will be in quantities requiring the high reliability and performance critical to the successful performance or the Manned Satellite Vehicle program.

[2]

Determination

Based on the findings above made, I hereby determine that the proposed contracts are for experimental, developmental or research work, or for the making or furnishing of property for experiment, test, development and research.

This class determination shall remain in effect until June 30, 1962.

James E. Webb, Administrator

Date   July 25, 1961

Enclosure No. 2

From NASA Headquarters
To Space Task Group   ATTENTION: Robert R. Gilruth

Subject: Appointment of Source Evaluation Board

1.   The following personnel are designated to serve as a Source Evaluation Board in connection with the procurement of development of spacecraft required for the Apollo Project. This Board, operating at the direction of the chairman, will review proposals from prospective contractors and based on its findings recommend source selection to this office for approval:

>   Walter Williams (Chairman) – Asst. to Director, Space Task Group
>   Robert O. Piland – Head, Apollo Project Office, Space Task Group
>   George M. Low – Chief, Manned Space Flight, NASA Headquarters
>   Wesley L. Hjornevik – Asst. Director for Administration, Space Task Group
>   Brooks C. Preacher – Office of Procurement Review, NASA Headquarters
>   Maxime A. Faget – Chief, Flight Systems Division, Space Task Group
>   James A. Chamberlin – Head, Engineering Division, Space Task Group

Charles W. Mathews – Head, Operations Division, Space Task Group
Dave W. Lang – Procurement & Contracting Officer, Space Task Group
* James T. Koppenhaver – Office of Reliability and Systems Analysis, NASA Headquarters

[signed by Robert Seamans]
Associate Administrator

* Non-voting Member (Gen. Management Instruction 2-4-3)

Enclosure No. 3

[10]

ESTABLISHMENT OF SUB-COMMITTEES
TO THE NASA SOURCE EVALUATION BOARD
PROJECT APOLLO

Technical Sub-Committee

Robert O. Piland (Chairman) – Head, Apollo Projects Office, Space Task Group
John B. Becker – Aero Physics Division, Langley Research Center
Andre J. Meyer – Assistant Chief, Engineering Division, Space Task Group
Caldwell C. Johnson, Jr. – Head, Systems Engineering Branch, Space Task Group
Robert G. Chilton – Head, Flight Dynamics Branch, Space Task Group
S. C. White – Chief, Life Systems Division, Space Task Group
William A. Mrazek – Director, Structures and Mechanics Division, George C. Marshall Space Flight Center

Such other technical representation as may be required from other Government activities may be requested or designated as required.

Business Sub-Committee

Glenn F. Bailey – Contract Specialist, Space Task Group
Phillip H. Whitbeck – Special Ass't. to the Assistant Director for Administration, NASA Headquarters
John D. Young – Director of Management Analysis Division, NASA Headquarters
Douglas E. Hendrickson – Budget and Finance Officer, Space Task Group
George F. MacDougall, Jr. – Head, Contract Engineering Branch, Space Task Group

John M. Curran – Procurement Review Office – NASA Headquarters
Wilbur H. Gray – NASA Technical Representative, McDonnell
Aircraft Corp.

Such other business management of financial representation as may be
required from other government activities may be requested or designated as
requested.

Document II-14

Document Title: "Memorandum for the President by James Webb, 14 September 1961.

Document Source: Folder 18675, NASA Historical Reference Collection, History Division, NASA Headquarters, Washington, DC.

With Presidential announcement of the decision to go to the Moon, it was clear that NASA would need to create a new Field Center dedicated to human spaceflight. The chair of NASA's House Appropriations Committee, Representative Albert Thomas of Houston, Texas, made it very clear that he expected the new Center to be located in or near his district (Volume II, Document III-7). Even so, at the very least NASA had to go through the motions of an open competition for the location of the Center, and a number of localities in different states (and their Governors, Senators, and Congressmen) made their interest in being chosen well known. President Kennedy's home state of Massachusetts put particular pressure on NASA and the White House to consider Hingham Air Force Base near Boston as the location.

September 14, 1961

MEMORANDUM FOR THE PRESIDENT

In view of the situation which has arisen in Massachusetts, I believe you should know personally that Dr. Hugh Dryden and I, last night and this morning, have carefully reviewed all the factors relating to the location of the manned space flight center. It included a careful examination of the material brought back from Hingham yesterday by the site survey team. The team was sent without notification to the Governor or anyone in Massachusetts and made its visit and examination without any publicity so far as I know.

Our decision is that this laboratory should be located at Houston, Texas, in close association with Rice University and the other educational institutions there and in that region.

A press release has been prepared announcing this decision, and we are holding it for issue after White House notification of those which your staff feels should have advance information.

The only personal commitment I have in connection with the release is to personally call the Acting Chairman of the House Committee on Science and Astronautics, Congressman George Miller of California, so that he will know in advance of newspaper release what the decision is. He has been very active and concerned on behalf of California.

Attached hereto is the transcript of the talk I gave at the National Press Club on September 12. [not included] On page 15 you will find underlined the reference I made to your instructions. You may need this at your next press conference.

[2]

There are also marked sections in the transcript which refer to you, Vice President Johnson, and the facilities location question on pages 1, 2, 3, 6, 13, 14, 15 and 16.

Incidentally, since we had too little time at our meeting Monday for me to give you as full a report on our activities as I would like, you might wish to take this transcript along for reading, perhaps on the plane. Particularly, the checked paragraph at the bottom of page 10 is an area of thought which you and I need to explore. If we can develop this idea in terms of regional patterns of developing science and technology and feeding them back into economic growth, it may be one of the tremendous accomplishments of your Administration.

/Signed/
James E. Webb
Administrator

Enclosure:

Transcript of National Press Club
Speech, September 12, 1961. [not included]

****

SITE SELECTION PROCEDURE

The procedure established for the selection of a site for the manned space flight laboratory, one of four major facilities required for the manned lunar landing mission on the accelerated schedule set by the President, is as follows:

I.   The selection of the site is to be made by the Administrator of NASA in consultation with the Deputy Administrator.

II. As the first step in the collection of information to aid the Administrator in his selection, the Associate Administrator on July 7, 1961 instructed the Director of the Office of Space Flight Programs to establish preliminary site criteria and to propose the membership for a site survey team. The team, appointed on August 7, 1961 consisted of

> John F. Parsons, Chairman
> Associate Director
> Ames Research Center
> N. Philip Miller
> Chief, Facilities Engineering Division
> Goddard Space Flight Center
>
> Wesley L. Hjornevik
> Assistant Director for Administration
> Space Task Group
>
> I. Edward Campagna
> Construction Engineer
> Space Task Group

[2]

Because of the sudden illness of Mr. Hjornevik on August 12, 1961, he was replaced by

> Martin A. Byrnes
> Project Management Assistant
> Space Task Group

III. The site survey team met on August 11 with the Director of the Office of Space Flight Programs, the Associate Director of the Space Task Group, and the Assistant Director of Space Flight Programs for Manned Space Flight. During this meeting tentative site requirements were developed.

IV. The site requirements were formulated in detail by the site survey team, and at a meeting with the Administrator; Deputy Administrator; Director of Space Flight Programs; Director, Office of Programs; and the Assistant Director for Facilities, Office of Programs, the following criteria were approved by the Administrator:

Essential Criteria

1. Transportation:

Capability to transport by barge large, cumbersome space vehicles (30 to 40 feet in diameter) to and from water shipping. Preferably the site should have its own or have access to suitable docking facilities. Time required in transport will be considered.

Availability of a first-class all-weather commercial jet service airport and a Department of Defense air base installation in the general area capable of handling high-performance military aircraft.

2. Communications:

Reasonable proximity to main routes of the long-line telephone system.

[3]

3. Local Industrial Support and Labor Supply:

An existing well-established industrial complex including machine and fabrication shops to support a research and development activity of high scientific and technical content, and capable of fabricating pilot models of large spacecraft.

A well-established supply of construction contractors and building trades and craftsmen to permit rapid construction of facilities without premium labor costs.

4. Community Facilities:

Close proximity to a culturally attractive community to permit the recruitment and retention of a staff with a high percentage of professional scientific personnel.

Close proximity to a well-established institution of higher education with emphasis on an institution specializing in the basic sciences and in space related graduate and post graduate education and research.

5. Electric Power:

Strong local utility system capable of developing up to 80,000 KVA of reliable power.

6. Water:

Readily available good-quality water capable of supplying 300,000 gallons per day potable and 300,000 gallons per day industrial.

7. Area:

1,000 usable acres with a suitable adjacent area for further development. Suitable areas in the general location for low hazard and nuisance subsidiary installations requiring some isolation.

8. Climate:

A mild climate permitting year-round, ice-free, water transportation; and permitting out-of-door work for most of the year to facilitate operations, reduce facility costs, and speed construction.

[4]

Desirable Criteria

1. Impact on Area:

Compatibility of proposed laboratory with the regional planning that may exist and ability of community facilities to absorb the increased population, and to provide the related industrial and transport support required.

2. Site Development Costs:

Consideration of costs for site development required for the proposed laboratory.

3. Operating Costs:

Consideration of costs for normal operations including utility rates, construction costs, wage scales, etc.

4. Interim Facilities:

Availability of reasonably adequate facilities for the temporary use of up to 1,500 people in the same general area as the permanent site.

V.   The site survey team at the same meeting was instructed to survey possible sites on the basis of published and other available information, selecting on the basis of the approved criteria those which should be visited by the team, visiting these sites and such others as might be directed by the Administrator, and preparing a report, including a listing of the advantages and disadvantages of the sites considered.

VI. A review by the site survey team of climatological data furnished by the United States Weather Bureau and information provided by the Department of the Army, Corps of Engineers, on water-borne commerce in [5] the United States (references 1 and 2), provided the following preliminary list of prospective areas which would fulfill the essential criteria of water transportation and climate:

> Norfolk, Virginia
> Charleston, South Carolina
> Savannah, Georgia
> Jacksonville, Florida
> Miami, Florida
> Tampa, Florida
> Mobile, Alabama
> New Orleans, Louisiana
> Baton Rouge, Louisiana
> Memphis, Tennessee
> Houston, Texas
> Corpus Christi, Texas
> San Diego, California

Los Angeles, California
Santa Barbara, California
San Francisco, California
Portland, Oregon
Seattle, Washington

This preliminary list of possible areas was then reviewed with regard to the other essential site criteria with the assistance of references 3 and 4 and through consultations with the General Services Administration regarding surplus Government property, and the list was reduced on August 16, 1961, to the following nine areas:

Jacksonville, Florida (Green Cove Springs Naval Station)
Tampa, Florida (MacDill Air Force Base)
Baton Rouge, Louisiana
Shreveport, Louisiana (Barksdale Air Force Base)
Houston, Texas (San Jacinto Ordnance Depot)
Victoria, Texas (FAA Airport)
Corpus Christi, Texas (Naval Air Station)
San Diego, California (Camp Elliott)
San Francisco, California (Benicia Ordnance Depot)

[6]

To properly evaluate each area accurately a physical inspection of the area by members of the site survey team was deemed essential. Accordingly, arrangements were made to visit these nine areas. While in certain areas additional sites were brought to the attention of the team and arrangements were made to visit those sites. Hence, the original nine sites were increased to twenty-three by the addition of the following:

Bogalusa, Louisiana
Houston, Texas (University of Houston Site)
Houston, Texas (Rice University Site)
Houston, Texas (Ellington Air Force Base)
Liberty, Texas
Beaumont, Texas
Harlingen, Texas
Berkeley, California
Richmond, California
Moffett Field, California (Naval Air Station)
St. Louis, Missouri (Daniel Boone Site)
St. Louis, Missouri (Industrial Park Site)
St. Louis, Missouri (Lewis and Clarke Site)
St. Louis, Missouri (Jefferson Barracks Site)

Visits to the above twenty-three sites were initiated on August 21, 1961 and completed September 7, 1961.

It will be noted that the team felt that locations north of the freezing line were unlikely to meet the requirements and hence proposed no visits to sites in this area.

VII. While the team was visiting sites, several presentations were made directly to the Administrator, Deputy Administrator, and other officials, notably from proponents of sites in the Boston, Rhode Island, and Norfolk areas. It was agreed to consider these sites in the final review.

[7]

On August 12th the Administrator and Deputy Administrator reviewed the factors which had entered into the approved criterion on climate, i.e.:

> "A mild climate permitting year-round, ice-free, water transportation; and permitting out-of-door work for most of the year to facilitate operations, reduce facility costs, and speed construction."

The considerations leading to this criterion are as follows:

1. The purpose of specifying a mild climate which will permit year-round, ice-free, water transportation is self-evident. It is necessary so that the spacecraft and/or its components can be transported by water to other sites at any time of the year to avoid delays in the overall program.

2. The requirement for out-of-door work most of the year stems from our experience with aircraft and large missiles. Since the spacecraft will be of comparable size it is expected that all work cannot be efficiently done within buildings. An appreciable amount of fitting, checking, and/or calibration work will be accomplished out-of-doors to facilitate the overall operation. Also the possibility of handling much larger spacecraft, such as a 10-15 man space station, must be considered. The climate factor will become more important as such spacecraft become parts of the program.

3. A mild climate avoids the necessity of special protection to the spacecraft against freezing of moisture in the many complicated components while transferring to and from sites and between site buildings. To provide such protection would be time-consuming and costly.

4. A mild climate will facilitate recovery procedure training of the astronauts and other activities which must be conducted out-of-doors.

5. A mild climate permits a greater likelihood of day-to-day access by air to and from the site from other parts of the country.

6. In summary, the selection of a site in an area meeting the stated climate criterion will minimize both the cost and the time required for this project. A mild climate will permit year-round construction activity, thereby accelerating the advancement of the project.

[8]

Sites north of the freezing line fail to meet these requirements. For example, in the case of the Boston area, the U.S. Department of Commerce Weather Bureau report entitled "Local Climatological Data with Comparative Data, 1960, Boston, Massachusetts," states:

> In the year 1960 it rained 114 days for a total amount of 44.46 inches. The rainfall was distributed uniformly throughout the year. The normal total annual rainfall over the years is 38.86 inches falling on 133 days.

> The daily minimum temperature for the months of December, January, February and March ranges from 21.6° F to 30.0 °F well below freezing while the average maximum temperature for December, January, and February is below 40 °F.

> Normal degree days, a measure of the heating required, is 5791 — a high value.

> Approximately 52 inches of snow and sleet fell in 1960, the average over the years is about 40 inches.

> The average hourly wind speed is 12.5 miles per hour.

In addition to the detailed information outlined above, this same report in describing the Boston climate states:

> The city's latitude places it ... in ... large bodies of air from tropical and polar regions resulting in variety and changeability of the weather elements.

> ... assuring an ordinarily dependable precipitation supply.

> Hot summer afternoons are ..."

> The average date of the last killing frost in spring is April 16.

> The average date of the first killing frost in autumn is October 25.

> Boston has no dry season; ...

> Coastal storms, or 'northeasters', are prolific producers of rain and snow. The main snow season extends from November through March.

[9]

> Although winds of 32 m.p.h. or higher may be expected on at least one day in every month of the year, gales are both common and more severe in winter.

By direction of the Administrator, the site survey team visited the Hingham, Massachusetts, site near Boston on September 13 for an inspection of the terrain and existing buildings.

References Used by Site Survey Team

1. Waterborne Commerce of the United States, Calendar Year 1958, Department of the Army, Corps of Engineers.

2. The Intercoastal Waterway, Corps of Engineers, U.S. Army, 1961.

3. Army Map Service Map of Major Army, Navy and Air Force Installations of the United States. 8205 Edition 21-AMS.

4. Education Directory 1959-1960, Part 3, Higher Education, U.S. Department of Health, Education and Welfare, Office of Education

---

Document II-15

Document Title: John C. Houbolt, NASA, Langley Research Center, Letter to Dr. Robert C. Seamans, Jr., Associate Administrator, NASA, 15 November 1961.

Document Source: Folder 18675, NASA Historical Reference Collection, History Division, NASA Headquarters, Washington, DC.

Document II-16

Document Title: Langley Research Center, NASA, "MANNED LUNAR-LANDING through use of LUNAR-ORBIT RENDEZVOUS," Volume 1, 31 October 1961

Document Source: Folder 18675, NASA Historical Reference Collection, History Division, NASA Headquarters, Washington, DC.

Document II-17

Document Title: Joseph Shea, Memorandum for the Record, 26 January 1962

Document Source: Folder 18675, NASA Historical Reference Collection, History Division, NASA Headquarters, Washington, DC.

Document II-18

Document Title: "Concluding Remarks by Dr. Wernher von Braun About Mode Selection for the Lunar Landing Program Given to Dr. Joseph F. Shea, Deputy Director (Systems) Office of Manned Space Flight," 7 June 1962.

Document Source: Folder 18675, NASA Historical Reference Collection, History Division, NASA headquarters, Washington, DC.

John Houbolt of NASA's Langley Research Center, along with several Langley colleagues, had been examining rendezvous concepts in 1959 and 1960. Although not the sole originator of the lunar orbit rendezvous (LOR) mode, Houbolt became its most persistent supporter and in 1961 went outside of normal bureaucratic channels to advocate the approach as a means of accomplishing the lunar landing mission, rather than the Earth orbit rendezvous (EOR) mode that had gained favor in the months following President Kennedy's speech announcing the lunar landing goal. This letter from Houbolt to NASA Associate Administrator Robert Seamans was a catalyst in shifting the thinking within NASA in favor of the LOR mode. The Saturn C-3 that Houbolt refers to in his letter was a configuration of the Saturn booster with two F-1 engines in its first stage. The Saturn C-2 was a less powerful booster using older rocket engines in its first stage that NASA had decided not to develop by the time Houbolt wrote his letter. NOVA was a very large booster with eight F-1 engines in its first stage that was designed to take astronauts directly to the Moon without need for rendezvous. The Fleming, Lundin, and Heaton Committees were groups set up within NASA earlier in 1961 to examine various approaches to the lunar landing mission. The Golovin Committee was a NASA-DOD group attempting to develop a national launch vehicle program. PERT was a chart-based management approach to complex projects.

The 31 October 1961 Langley Study that was the basis for Houbolt's arguments indicates the other key members of the Langley team that developed the lunar orbit rendezvous scheme.

By the end of 1961, the new Manned Spacecraft Center in Houston was beginning to be interested in the LOR concept, while Wernher von Braun and his team at the Marshall Space Flight Center focused their studies on the EOR approach to accomplishing the lunar landing mission. This led Brainerd Holmes, head of piloted spaceflight at NASA Headquarters, and his assistant Joseph Shea to conclude that further study of the LOR concept should be managed by Headquarters to minimize inter-Center rivalries.

Following Shea's memorandum, the choice of an approach to carrying out the lunar landing mission received intensive attention within NASA in the first five months of 1962. Project Apollo leaders at the new Manned Spacecraft Center in Houston, Texas, gradually came to favor the LOR approach. Wernher von Braun and his associates at the Marshall Space Flight Center in Huntsville, Alabama, had based their launch vehicle planning on the use of the EOR approach. At a climactic meeting at the Marshall Space Flight Center on 6 June 1962, von Braun made the concluding remarks. (This document, dated 7 June, is the text of those remarks.) Much to the surprise of many of his associates at Marshall, von Braun endorsed the LOR mode as the preferred approach. With his endorsement, NASA soon adopted this approach in its planning for the lunar landing mission.

The C-1 and C-5 vehicles referred to in von Braun's statement became known as the Saturn I and Saturn IB and the Saturn V. The C-8 was a configuration with eight first-stage engines that were never built. The S-IVB was the third stage of the Saturn V vehicle and the S-II its second stage. Robert "Bob" Gilruth was the Director of the Manned Spacecraft Center and Chuck Matthews a senior manager there. NAA was North American Aviation, the contractor building the Apollo Command and Service Module and the S-II and S-IVB stage of the Saturn V launcher. Rocketdyne was the company building the F-1 and J-2 rocket engines.

Document II-15

National Aeronautics and
Space Administration
Langley Research Center
Langley Air Force Base, Va.

November 15, 1961

Dr. Robert C. Seamans, Jr.
Associate Administrator
National Aeronautics and
Space Administration
1520 H Street, N.W.
Washington 25, D.C.

Dear Dr. Seamans:

Somewhat as a voice in the wilderness, I would like to pass on a few thoughts on matters that have been of deep concern to me over recent months. This concern may be phrased in terms of two questions: (1) If you were told that we can put man on the moon with safe return with a single C-3, its equivalent or something less, would you judge this statement with the critical skepticism that others have? (2) Is the establishment of a sound booster program really so difficult?

I would like to comment on both these questions, and more, would like to forward as attachments condensed versions of plans which embody ideas and suggestions which I believe are so fundamentally sound and important that we cannot afford to overlook them. You will recall I wrote to you on a previous occasion. I fully realize that contacting you in this manner is somewhat unorthodox; but the issues at stake are crucial enough to us all that an unusual course is warranted.

Since we have had only occasional and limited contact, and because you therefore probably do not know me very well, it is conceivable that after reading this you may feel that you are dealing with a crank. Do not be afraid of this. The thoughts expressed here may mot be stated in as diplomatic a fashion as they might be, or as I would normally try to do, but this is by choice and the moment is not important. The important point is that you hear the ideas directly, not after they have filtered through a score or more of other people, with the attendant risk that they may not even reach you.

[2]

Manned Lunar Landing Through Use of Lunar Orbit Rendezvous

The plan. - The first attachment [Document II-16] outlines in brief the plan by which we may accomplish a manned lunar landing through use of a lunar rendezvous, and shows a number of schemes for doing this by means of a single C-3, its equivalent, or even something less. The basic ideas of the plan were presented before various NASA people well over a year ago, and were since repeated at numerous interlaboratory meetings. A lunar landing program utilizing rendezvous concepts was even suggested back in April. Essentially, it had three basic points: (1) the establishment of an early rendezvous program involving Mercury, (2) the specific inclusion of rendezvous in Apollo developments, and (3) the accomplishment of lunar landing through use of C-2's. It was indicated then that the two C-2's could do the job, C-2 being referred to simply because NASA booster plans did not go beyond the C-2 at that time; it was mentioned, however, that with a C-3 the number of boosters required would be cut in half, specifically only one.

Regrettably, there was little interest shown in the idea - indeed, if any, it was negative.

Also (for the record), the scheme was presented before the Lundin Committee. It received only bare mention in the final report and was not discussed further (see comments below in section entitled "Grandiose Plans ").

It was presented before the Heaton Committee, accepted as a good idea, then dropped, mainly on the irrelevant basis that it did not conform to the ground rules. I even argued against presenting the main plan considered by the Heaton Committee, largely because it would only bring harm to the rendezvous cause, and further argued that if the committee did not want to consider lunar rendezvous, at least they should make a strong recommendation that it looks promising enough that it deserves a separate treatment by itself - but to no avail. In fact, it was mentioned that if I felt sufficiently strong about the matter, I should make a minority report. This is essentially what I am doing.

We have given the plan to the presently meeting Golovin Committee on several occasions.

In a rehearsal of a talk on rendezvous for the recent Apollo Conference, I gave a brief reference to the plan, indicating the benefit derivable therefrom, knowing full well that the reviewing committee would ask me to withdraw any reference to this idea. As expected, this was the only item I was asked to delete.

[3] The plan has been presented to the Space Task Group personnel several times, dating back to more than a year ago. The interest expressed has been completely negative.

Ground rules. - The greatest objection that has been raised about our lunar rendezvous plan is that it does not conform to the "ground rules". This to me is nonsense; the important question is, "Do we want to get to the moon or not?", and, if so, why do we have to restrict our thinking along a certain narrow channel. I feel very fortunate that I do not have to confine my thinking to arbitrarily set up ground rules which only serve to constrain and preclude possible equally good or perhaps better approaches. Too often thinking goes along the following vein: ground rules are set up, and then the question is tacitly asked, "Now, with these ground rules what does it take, or what is necessary to do the job?". A design begins and shortly it is realized that a booster system way beyond present plans is necessary. Then a scare factor is thrown in; the proponents of the plan suddenly become afraid of the growth problem or that perhaps they haven't computed so well, and so they make the system even larger as an "insurance" that no matter what happens the booster will be large enough to meet the contingency. Somehow, the fact is completely ignored that they are now dealing with a ponderous development that goes far beyond the state-of-the-art.

Why is there not more thinking along the following lines: Thus, with this given booster, or this one, is there anything we can do to do the job? In other words, why can 't we also think along the deriving a plan to fit a booster, rather than derive a booster to fit a plan?

Three ground rules in particular are worthy of mention: three men, direct landing, and storable return. These are very restrictive requirements. If two men can do the job, and if the use of only two men allows the job to be done, then why not do it this way? If relaxing the direct requirements allows the job to be done with a C-3, then why not relax it? Further, when a hard objective look is taken at the use of storables, then it is soon realized that perhaps they aren't so desirable or advantageous after all in comparison with some other fuels.

Grandiose plans, one-sided objections, and bias.- For some inexplicable reason, everyone seems to want to avoid simple schemes. The majority always seems to be thinking in terms of grandiose plans, giving all sort of arguments for long-range plans, etc. Why is there not more thinking in the direction of developing the simplest scheme possible? Figuratively, why not go buy a Chevrolet instead of a Cadillac? Surely a Chevrolet gets one from one place to another just as well as a Cadillac, and in many respects with marked advantages.

[4]

I have been appalled at the thinking of individuals and committees on these matters. For example, comments of the following type have been made: "Houbolt has a scheme that has a 50 percent chance of getting a man to the moon, and a 1 percent chance of getting him back." This comment was made by a Headquarters individual at 'high level ['] who never really has taken the time to hear about the scheme, never has had the scheme explained to him fully, or possible even correctly, and yet he feels free to pass judgment on the work. I am bothered by stupidity of this type being displayed by individuals who are in a position to make decisions which affect not only the NASA, but the fate of the nation as well. I have even grown to be concerned about the merits of all the committees that have been considering the problem. Because of bias, the intent of the committee is destroyed even before it starts and, further, the outcome is usually obvious from the beginning. We knew what the Fleming Committee results would be before it started. After one day it was clear what decisions the Lundin Committee would reach. After a couple days it was obvious what the main decision of the Heaton Committee would be. In connection with the Lundin Committee, I would like to cite a specific example. Considered by this committee was one of the most hair-brained ideas I have ever heard, and yet it received one first place vote. In contrast, our lunar rendezvous scheme, which I am positive is a much more workable idea, received only bare mention in a negative vein, as was mentioned earlier. Thus, committees are no better than the bias of the men composing them. We might then ask, why are men who are not competent to judge ideas, allowed to judge them?

Perhaps the substance of this section might be summarized this way. Why is NOVA, with its ponderous ideas, whether in size, manufacturing, erection, site location, etc., simply just accepted, and why is a much less grandiose scheme involving rendezvous ostracized or put on the defensive?

PERT chart folly. - When one examines the various program schedules that have been advanced, he cannot help from being impressed by the optimism shown. The remarkable aspect is that the more remote the year, the bolder the schedule becomes. This is, in large measure, due to the PERT chart craze. It has become the vogue to subject practically everything to a PERT chart analysis, whether it means anything or not. Those who apply or make use of it seem to be overcome by a form of self-hypnosis, more or less accepting the point of view, "Because the PERT chart says so, it is so." Somehow, perhaps unfortunately, the year 1967 was mentioned as the target year for putting a man on the moon. The Fleming report through extensive PERT chart analysis then "proved" this could be done. One cannot help but get the feeling that if the year 1966 had been mentioned, then this would have been the date proven; likewise, if 1968 had been the year mentioned.

[5]

My quarrel is not with the basic theory of PERT chart analysis; I am fully aware of its usefulness, when properly applied. I have been nominally in charge of a facility development and know the merits, utility, and succinctness by which it is helpful in keeping a going job moving, uncovering bottlenecks, and so forth. But when it is used in the nature of a crystal ball, then I begin to object. Thus, when we scrutinize various schedules and programs, we have to be very careful to ask how realistic the plan really is. Often simple common sense tells us much more than all the machines in the world.

I make the above points because, as you will see, we have a very strong point to make about the possibility of coming up with a realistic schedule; the plan we offer is exceptionally clean and simple in vehicle and booster requirements relative to other plans.

Booster is pacing item. - In working out a paper schedule we have adopted the C-3 development schedule used by Fleming and Heaton, not necessarily because we feel the schedule is realistic, but simply to make a comparison on a parallel basis. But whether the date is right, or not, doesn't matter. Here, I just want to point out that for the lunar rendezvous scheme the C-3 booster is the pacing item. Thus, we can phrase our lunar landing date this way. We can put a man on the moon as soon as the C-3 is developed, and the number of C-3's required is very small. (In fact, as I mentioned earlier, I would not be surprised to have the plan criticized on the basis that it is not grandiose enough.)

Abort. - An item which perhaps deserves special mention is abort. People have leveled criticism, again erroneously and with no knowledge of the situation, that the lunar rendezvous scheme offers no abort possibilities. Along with our many technical studies we have also studied the abort problem quite thoroughly. We find that there is no problem in executing an abort maneuver at any point in the mission. In fact, a very striking result comes out, just the reverse of the impression many people try to create. When one compares, for example, the lunar rendezvous scheme with a direct approach, he finds that on every count the lunar rendezvous method offers a degree of safety and reliability far greater than that possible by the direct approach. These items are touched upon to a limited extent in the attached plan.

Booster Program

My comments on a booster program will be relatively short, since the second attachment [not included] more or less speaks for itself. There are, however, a few points worthy of embellishment.

[6]

Booster design. - In the course of participating in meetings dealing with vehicle design, I have sometimes had to sit back completely awed and astonished at what I was seeing take place. I have seen the course of an entire meeting change because of an individual not connected with the meeting walking in, looking over shoulders, shaking his head in a negative sense, and then walking out without uttering a word. I have seen people agree on velocity increments, engine performance, and structural data, and after a booster design was made to these figures, have seen some of the people then derate the vehicle simply because they couldn't believe the numbers. I just cannot cater to proceedings of this type. The situation is very much akin to a civil engineer who knows full well that the material he is using will withstand 60,00 psi. He then applies a factor of safety of 2.5, makes a design, then after looking at the results, arbitrarily doubles the size of every member because he isn't quite sure that the design is strong enough. A case in point is the C-3. In my initial contacts with this vehicle, we were assured that it had a payload capability in the neighborhood of 110,000-120,000 lbs. Then it was derated. The value used by the Heaton Committee was 105,000 lbs. By the time the vehicle had reached the Golovin Committee, I was amazed to find that it had a capability of only 82, 570 lbs. Perhaps the only comment that can be made to this is that if we can't do any better on making elementary computations of this type, then we deserve to be in the pathetic situation we are. I also wonder where we will stand after NOVA is derated similarly.

"Quantizing" bad. - One of the reasons our booster situation is in such a sad state is the lack of appropriate engines, more specifically the lack of an orderly stepping in engine sizes. Booster progress is virtually at a standstill because there are no engines available, just as engines were the major pacing item in the development of aircraft. Aside from the engines on our smaller boosters, and the H-1 being used on the [Saturn] C-1, the only engines we have in development are:

Capability	Ratio
15,000	
	13.3
200,000	
	7.5
1,500,000	

The attempt to make boosters out of this stock of engines, having very large ratios in capability, can only result in boosters of grotesque and unwieldy configurations, and which require many, many in-flight engine starts. What is needed are engines which step up in size at a lower ratio. Consideration of the staging of an "ideal" rocket system indicates that whether accelerating to orbit speed or to escape speed, the ratio of engine sizes needed is in the order of 3. Logically then we ought to have engines that step in capability by a factor of around 2, 3, or 4. An every-day analog that can be mentioned is outboard motors. There is a motor to serve nearly every need, and in the extreme cases the process of doubling up is even used.

[7]

Booster Program. - In light of the preceding paragraph, and taking into account the engines under development, we should add the following two:

80,000 - 100,000	H_2 - O_2
400,000 - 500,000	H_2 - O_2

This would then give a line-up as follows:

15,000	H/O
80,000 - 100,000	H/O
200,000	H/O
400,000 - 500,000	H/O
1,500,000	RP/O

with the 15,000-lb. engine really not needed. This array (plus those mentioned immediately below) would allow the construction of almost all types of boosters conceivable. For example, a single 80,000-100,000 engine would take the place of the six L-115 engines being used on S-IV; not only is the arrangement of six engines on this vehicle bad, but these engines have very poor starting characteristics. The 400,000-500,000 would be used to replace the four J-2's on the S-II. Thus, C-3 would change from a messy 12-engined vehicle requiring 10 in-flight engine starts to a fairly simple 5-engine vehicle with only 3 in-flight engine starts.

In addition, the following engines should be included in a program:

1,000,000 - 1,500,000 lb.	Solid	
	5,000,000	Solid
and/or	5,000,000	Storable

The 1,000,000 - 1,500,000 lb. solid would in itself be a good building block and would probably work in nicely to extend the capabilities of vehicles, such as Titan. The 5,000,000 solid and/or storable would also be good building blocks and specifically would serve as alternate first-stage boosters for C-3, aiming at simplicity and reliability.

[8]

It may be said that there is nothing new here and that all of the above is obvious. Indeed, it seems so obvious that one wonders why such a program was not started 5 years ago. But the fact that it may be obvious doesn't help us; what is necessary is putting the obvious into effect. In this connection, there may be some who ask, "But are the plans optimum and the best?" This question is really not pertinent. There will never be an optimized booster or program. We might have an optimum booster for a given situation, but there is none that is optimum for all situations. To seek one, would just cause deliberation to string out indefinitely with little, if any, progress being made. The DynaSoar case is a good example of this.

A criticism that undoubtedly will be leveled at the above suggestion is that I'm not being realistic in that there is just not enough money around to do all these things. If this is the situation, then the answer is simply that's why we have Webb and his staff. That's why he was chosen to head the organization, this is one of his major functions, to ask the question, do we want to do a job or not?, and, if so, then to find out where the gaps or holes are, and then to go about doing what is necessary to fill the gaps to make sure the job gets done. Further, the load doesn't have to be carried by the NASA alone. The Air Force and NASA can work together and share the load, and I'm sure that if this is done, the necessary money can be found. Even if some project, say, for example, the 5,000,000-lb. storable engine has to be dropped for some reason after it gets started; no harm will be done. This happens every day. On the contrary, some good, some new knowledge, will have been uncovered, even if it turns out to be the discovery of the next obstacle which prevents such a booster from being built.

<u>Nuclear booster and booster size</u>. - Although not mentioned in the previous section, work on nuclear engines should, of course, continue. Any progress made here will integrate very nicely into the booster plans indicated in the attachment.

As regards booster size, the following comment is offered. Excluding for the moment NOVA type vehicles, we should strive for boosters which make use of the engines mentioned in the preceding section and which are the biggest that can be made and yet still be commensurate with existing test-stand sites and with the use of launch sites that are composed of an array of assembly buildings and multiple launch pads. The idea behind launch sites of this type is an excellent one. It keeps real estate demands to a minimum, allows for ease in vehicle assembly and check-out, and greatly eases the launch rate problem. Thus, C-3 or C-4 should be designed accordingly. We would then have a nice work-horse type vehicle having relative ease of handling, and which would permit a lunar landing mission, as indicated earlier in the lunar rendezvous write-up section. From my point of view, I would much rather confine my spending to a single versatile launch site of the type mentioned, save money in real estate acquisition and launch site development necessary for the huge vehicles, and put the money saved into an engine development program.

[9]

Concluding Remarks

It is one thing to gripe, another to offer constructive criticism. Thus, in making a few final remarks, I would like to offer what I feel would be a sound integrated overall program. I think we should:

1. Get a manned rendezvous experiment going with the Mark II Mercury.

2. Firm up the engine program suggested in this letter and attachment, converting the booster to these engines as soon as possible.

3. Establish the concept of using a C-3 and lunar rendezvous to accomplish the manned lunar landing as a firm program.

Naturally, in discussing matters of the type touched upon herein, one cannot make comments without having them smack somewhat against NOVA. I want to assure you, however, I'm not trying to say NOVA should not be built. I'm simply trying to establish that our scheme deserves a parallel front-line position. As a matter of fact, because the lunar rendezvous approach is easier, quicker, less costly, requires less development, less new sites and facilities, it would appear more appropriate to say that this is the way to go, and that we will use NOVA as a follow on. Give us the go-ahead, and C-3, and we will put men on the moon in very short order - and we don't need any Houston empire to do it.

In closing, Dr. Seamans, let me say that should you desire to discuss the points covered in this letter in more detail, I would welcome the opportunity to come up to Headquarters to discuss them with you.

Respectfully yours,

John C. Houbolt

Document II-16

NATIONAL AERONAUTICS AND SPACE ADMINISTRATION
LANGLEY RESEARCH CENTER

VOLUME 1

MANNED LUNAR-LANDING through use of LUNAR-ORBIT RENDEZVOUS

[i]

FOREWORD

In the course of conducting research on the problem of space rendezvous and on various aspects of manned space missions, Langley Research Center has evolved what is believed to be a particularly appealing scheme for performing the manned lunar landing mission. The key to the mission is the use of lunar rendezvous, which greatly reduces the size of the booster needed at the earth.

More definitely the mission may be described essentially as follows: A manned exploration vehicle is considered on its way to the moon. On approach, this vehicle is decelerated into a low-altitude circular orbit about the moon. From this orbit a lunar lander descends to the moon surface, leaving the return vehicle in orbit. After exploration the lunar lander ascends for rendezvous with the return vehicle. The return vehicle is then boosted into a return trajectory to the earth, leaving the lander behind.

The significant advantage brought out by this procedure is the marked reduction in escape weight required; the reduction is, of course, a direct reflection of the reduced energy requirements brought about by leaving a sizable mass in lunar orbit, in this case, the return capsule and return propulsion system.

This report has been prepared by members of the Langley Research Center to indicate the research that has been conducted, and what a complete manned lunar landing mission using this system would entail. For further reference, main contacts are John D. Bird, Arthur W. Vogeley, or John C. Houbolt.

J.C.H.
October 31, 1961

[ii]

OUTLINE AND INDEX

[iii]

PART II

PART III

[1]

SUMMARY AND CONCLUSIONS

Studies made at Langley Research Center of various schemes for performing the manned lunar landing mission indicate that the lunar rendezvous
method is the simplest, most reliable, and quickest means for accomplishing
the task. This technique permits a lunar exploration to be made with a single
C-3 booster. A first landing is indicated in March 1966, with a possibility of an
attempt as early as November 1965. These dates do not require changes in
previously established Apollo, C-1, and C-3 development schedules. Further, the

lunar rendezvous approach contains a number of features which tend to raise the schedule confidence level; the most important of these are:

(a) The Apollo vehicle, the lander, and the rendezvous experiment can all proceed on an independent parallel basis, thus avoiding schedule conflicts; further, the overall development is simplified because each vehicle has only a single function to perform.

(b) The lunar rendezvous approach permits complete system development to be done with C-1, which will be available and well developed, and makes the entire C-3 picture exceptionally clean and simple, thus resulting in a minimum cost program.

In amplification of these general remarks, the following specific conclusions are drawn from the technical studies which are summarized in the body of this report:

A. Mission Approach and Scheduling:

1. The lunar rendezvous method requires only a single C- 3 or C-4 launch vehicle. Earth orbital weights required for various system arrangements are summarized in figure 1. (See also tables VI and VII later in the text.) [not provided]

2. The lunar rendezvous method schedules the first landing in March 1966.

3. The lunar rendezvous method does not require that the Apollo vehicle be compromised because of landing considerations.

4. The lunar rendezvous method allows the landing vehicle configuration to be optimized for landing.

5. The lunar rendezvous method requires only C-1 boosters for complete system development.

[2]

6. The lunar rendezvous method provides for complete lander checkout and crew training in the lunar landing, lunar launch, and rendezvous docking operations on the actual vehicle.

B. Funding:

The lunar rendezvous method results in a program cost which will be less than the cost of other methods for the following reasons:

1. Requires fewer (20 to 40 percent) large boosters than other programs.

2. Requires no Nova vehicles.

3. Requires less C-3 or C-4 vehicles than other programs.

4. Programs most flights on best-developed booster (C-1).

　　　　5. Requires a minimum of booster ground facilities, because large boosters are avoided and because of a low launch rate.

　　　　The lunar rendezvous method can be readily paralleled with some other program at least total program cost.

C. Lunar Rendezvous:

　　　　The lunar rendezvous under direct, visual, pilot control is a simple reliable operation which provides a level of safety and reliability higher than other methods as outlined below.

D. Safety and Reliability:

　　　　1. The lander configuration is optimized.

　　　　2. The single-lander system permits safe return of the primary vehicle in event of a landing accident.

　　　　3. The two-lander system provides a rescue capability.

　　　　4. Crews can be trained in lunar landing, lunar launch, and rendezvous docking operations in the actual vehicle.

　　　　5. Requires fewest number of large booster flights.

　　　　6. Provides for most flights on best-developed booster (C-1).

[3]

E. Abort Capability:

　　　　1. An abort capability meeting the basic Mercury-Apollo requirements can be provided.

　　　　2. This abort capability can be provided with no additional fuel or weight penalties.

F. Lunar Lander Development:

　　　　1. Lunar lander design is optimized for landing.

　　　　2. Being essentially separate from Apollo, development can proceed with a minimum of schedule conflict.

　　　　3. Research, development, and checkout can be performed on ground facilities now under procurement and which will be available in time to meet the program schedule.

G. Development Facilities:

　　　　1. The lunar rendezvous method requires no additional booster ground facilities (see item B-5).

2. The ground facilities required for rendezvous-operations development are now being procured and will be ready.

3. The ground facilities for lander development and checkout are now being procured and will be ready.

[remainder of report not provided]

Document II-17

January 26, 1962

MEMORANDUM FOR THE RECORD

Brainerd and I agreed that LOR looks sufficiently attractive to warrant further study. He feels that the study should be run from OMSF, rather than either Center, to provide a measure of objectivity.

Apparently we have to go for an open competition, which I shall try to get under way as quickly as possible. Because of the implications on the overall program, we shall attempt to conduct the study at a secret level. We are also concerned that MSFC will be especially negative with LOR because they have not studied it. I will attempt to define areas in which they can contribute to our overall studies, in order to expose them to the details of the mode.

I am concerned that MSC's weight estimates are quite optimistic. We shall concentrate, in the LOR study, on the detail, conservative estimation of the LEM weight, and the mechanization of rendezvous.

[Signed J. F. Shea]

J.F. Shea

Document II-18

CONCLUDING REMARKS BY DR. WERNHER VON BRAUN ABOUT MODE SELECTION FOR THE LUNAR LANDING PROGRAM GIVEN TO DR. JOSEPH F. SHEA, DEPUTY DIRECTOR (SYSTEMS) OFFICE OF MANNED SPACE FLIGHT

JUNE 7, 1962

In the previous six hours we presented to you the results of some of the many studies we at Marshall have prepared in connection with the Manned Lunar Landing Project. The purpose of all these studies was to identify potential technical problem areas, and to make sound and realistic scheduling estimates. All studies were aimed at assisting you in your final recommendation with respect to the mode to be chosen for the Manned Lunar Landing Project.

Our general conclusion is that all four modes investigated are technically feasible and could be implemented with enough time and money. We have, however, arrived at a definite list of preferences in the following order:

1.   Lunar Orbit Rendezvous Mode - with the strong recommendation (to make up for the limited growth potential of this mode) to initiate, simultaneously, the development of an unmanned, fully automatic, C-5 logistics vehicle.

2.   Earth Orbit Rendezvous Mode (Tanking Mode).

3.   C-5 Direct Mode with minimum size Command Module and High Energy Return.

4.   Nova or C-8 Mode.

I shall give you the reasons behind this conclusion in just one minute.

But first I would like to reiterate once more that <u>it is absolutely mandatory that we arrive at a definite mode decision within the next few weeks, preferably by the first of July, 1962</u>. We are already losing time in our over-all program as a result of a lacking mode decision.

[2]   A typical example is the S-IVB contract. If the S-IVB stage is to serve not only as the third (escape) stage for the C-5, but also as the second stage for the C-1 B needed in support of rendezvous tests, a flyable S-IVB will be needed at least one year earlier than if there was no C-1 B at all. The impact of this question on facility planning, buildup of contractor level of effort, etc., should be obvious.

Furthermore, if we do not freeze the mode now, we cannot layout a definite program with a schedule on which the budgets for FY - 1964 and following can be based. Finally, if we do not make a clear-cut decision on the mode very soon, our chances of accomplishing the first lunar expedition in this decade will fade away rapidly.

I.   <u>WHY DO WE RECOMMEND LUNAR ORBIT RENDEZVOUS MODE PLUS C-5 ONE-WAY LOGISTICS VEHICLE?</u>

a.   We believe this program offers the highest confidence factor of successful accomplishment within this decade.

b. It offers an adequate performance margin. With storable propellants, both for the Service Module and Lunar Excursion Module, we should have a comfortable padding with respect to propulsion performance and weights. The performance margin could be further increased by initiation of a back-up development aimed at a High Energy Propulsion System for the Service Module and possibly the Lunar Excursion Module. Additional performance gains could be obtained if current proposals by Rocketdyne to increase the thrust and/ or specific impulses of the F-l and J -2 engines were implemented.

c. We agree with the Manned Spacecraft Center that the designs of a maneuverable hyperbolic re-entry vehicle and of a lunar landing vehicle constitute the two most critical tasks in producing a successful lunar spacecraft. A drastic separation of these two functions into two separate elements is bound to greatly simplify the development of the spacecraft system. Developmental cross-feed between results from simulated or actual landing tests, on the one hand, and re-entry tests, on the other, are minimized if no attempt is made to include the Command Module into the lunar landing process. The mechanical separation of the two functions would virtually permit completely parallel developments of the Command Module and the Lunar Excursion Module. While it may be difficult to accurately appraise this advantage in terms of months to be gained, we have no doubt whatsoever that such a procedure will indeed result in very substantial saving of time.

[3]   d. We believe that the combination of the Lunar Orbit Rendezvous Mode and a C-5 one-way Logistics Vehicle offers a great growth potential. After the first successful landing on the moon, demands for follow-on programs will essentially center on increased lunar surface mobility and increased material supplies for shelter, food, oxygen, scientific instrumentation, etc. It appears that the Lunar Excursion Module, when refilled with propellants brought down by the Logistics Vehicle, constitutes an ideal means for lunar surface transportation. First estimates indicate that in the 1/6 G gravitational field of the moon, the Lunar Excursion Module, when used as a lunar taxi, would have a radius of action of at least 40 miles from around the landing point of the Logistics Vehicle. It may well be that on the rocky and treacherous lunar terrain the Lunar Excursion Module will turn out to be a far more attractive type of a taxi than a wheeled or caterpillar vehicle.

e.   We believe the Lunar Orbit Rendezvous Mode using a single C-5 offers a very good chance of ultimately growing into a C-5 direct capability. At this time we recommend against relying on the C-5 Direct Mode because of its need for a much lighter command module as well as a high energy landing and return propulsion system. While it may be unwise to count on the availability of such advanced equipment during this decade (this is why this mode was given a number 3 rating) it appears entirely within reach in the long haul.

f.   If and when at some later time a reliable nuclear third stage for Saturn C-5 emerges from the RIFT program, the performance margin for the C-5 Direct Mode will become quite comfortable.

g.   Conversely, if the Advanced Saturn C-5 were dropped in favor of a Nova or C-8, it would completely upset all present plans for the implementation of the RIFT program. Contracts, both for the engines and the RIFT stage, have already been let and would probably have to be cancelled until a new program could be developed.

h. We conclude from our studies that an automatic pinpoint letdown on the lunar surface going through a circumlunar orbit and using a landing beacon is entirely possible. Whether this method should be limited to the C-5 Logistics Vehicle or be adopted as a secondary mode for the Lunar Excursion Module is a matter that should be carefully discussed with the Manned Spacecraft Center. It may well be that the demand for incorporation of an additional automatic landing capability in the Lunar Excursion Module buys more trouble than gains.

[4]   i. The Lunar Orbit Rendezvous Mode augmented by a C-5 Logistics Vehicle undoubtedly offers the cleanest managerial interfaces between the Manned Spacecraft Center, Marshall Space Flight Center, Launch Operations Center and all our contractors. While the precise effect of this may be hard to appraise, it is a commonly accepted fact that the number and the nature of technical and managerial interfaces are very major factors in conducting a complex program on a tight time schedule. There are already a frightening number of interfaces in existence in our Manned Lunar Landing Program. There are interfaces between the stages of the launch vehicles, between launch vehicles and spacecraft, between complete space vehicles and their ground equipment, between manned and automatic checkout, and in the managerial area between the Centers, the Washington Program Office, and the contractors. The plain result of too many interfaces is a continuous and disastrous erosion of the authority vested in the line organization and the need for more coordination meetings, integration groups, working panels, ad-hoc committees, etc. Every effort should therefore be made to reduce the number of technical and managerial interfaces to a bare minimum.

j. Compared with the C-5 Direct Mode or the Nova/C-8 Mode, the Lunar Orbit Rendezvous Mode offers the advantage that no existing contracts for stages (if we go to Nova) or spacecraft systems (if we go to C-5 Direct) have to be terminated; that the contractor structure in existence can be retained; that the contract negotiations presently going on can be finished under the existing set of ground rules; that the contractor build-up program (already in full swing) can be continued as planned; that facilities already authorized and under construction can be built as planned, etc.

k. We at the Marshall Space Flight Center readily admit that when first exposed to the proposal of the Lunar Orbit Rendezvous Mode we were a bit skeptical - particularly of the aspect of having the astronauts execute a complicated rendezvous maneuver at a distance of 240,000 miles from the earth where any rescue possibility appeared remote. In the meantime, however, we have spent a great deal of time and effort studying the four modes, and we have come to the conclusion that this particular disadvantage is far outweighed by the advantages listed above.

We understand that the Manned Spacecraft Center was also quite skeptical at first when John Houbolt of Langley advanced the proposal of the Lunar Orbit Rendezvous Mode, and that it took them quite a while to substantiate the feasibility of the method and finally endorse it.

Against this background it can, therefore, be concluded that the issue of "invented here" versus "not invented here" does not apply to [5] either the Manned Spacecraft Center or the Marshall Space Flight Center; that both Centers have actually embraced a scheme suggested by a third. Undoubtedly, personnel of MSC and MSFC have by now conducted more detailed studies on all aspects of the four modes than any other group. Moreover, it is these two Centers to which the Office of Manned Space Flight would ultimately have to look to "deliver the goods". I consider it fortunate indeed for the Manned Lunar Landing Program that both Centers, after much soul searching, have come to identical conclusions. This should give the Office of Manned Space Flight some additional assurance that our recommendations should not be too far from the truth.

II.   WHY DO WE NOT RECOMMEND THE EARTH ORBIT RENDEZVOUS MODE?

Let me point out again that we at the Marshall Space Flight Center consider the Earth Orbit Rendezvous Mode entirely feasible. Specifically, we found the Tanking Mode substantially superior to the Connecting Mode. Compared to the Lunar Orbit Rendezvous Mode, it even seems to offer a somewhat greater performance margin. This is true even if only the nominal two C-5's (tanker and manned lunar vehicle) are involved, but the performance margin could be further enlarged almost indefinitely by the use of additional tankers.

We have spent more time and effort here at Marshall on studies of the Earth Orbit Rendezvous Mode (Tanking and Connecting Modes) than on any other mode. This is attested to by six big volumes describing all aspects of this mode. Nor do we think that in the light of our final recommendation - to adopt the Lunar Orbit Rendezvous Mode instead - this effort was in vain. Earth Orbit Rendezvous as a general operational procedure will undoubtedly play a major role in our over-all national space flight program, and the use of it is even mandatory in developing a Lunar Orbit Rendezvous capability.

The reasons why, in spite of these advantages, we moved it down to position number 2 on our totem pole are as follows:

a.   We consider the Earth Orbit Rendezvous Mode more complex and costlier than Lunar Orbit Rendezvous. Moreover, lunar mission success with Earth Orbit Rendezvous requires two consecutive successful launches. If, for example, after a successful tanker launch, the manned lunar vehicle aborts during its ascent, or fails to get off the pad within a certain permissible period of time, the first (tanker) flight must also be written off as useless for the mission.

b.   The interface problems arising between the Manned Spacecraft Center and the Marshall Space Flight Center, both in the technical and management areas, would be more difficult if the Earth Orbit Rendezvous Mode was adopted. For example, if the tanker as an unmanned vehicle was handled by MSFC, and the flight of the manned lunar vehicle was [6] conducted by the Manned Spacecraft Center, a managerial interface arises between target and chaser. On the other hand, if any one of the two Centers would take over the entire mission, it would probably bite off more than it could chew, with the result of even more difficult and unpleasant interface problems.

c. According to repeated statements by Bob Gilruth, the Apollo Command Module in its presently envisioned form is simply unsuited for lunar landing because of the poor visibility conditions and the undesirable supine position of the astronauts during landing.

III. WHY DO WE NOT RECOMMEND THE C-5 DIRECT MODE?

It is our conviction that the C-5 Direct Mode will ultimately become feasible - once we know more about hyperbolic re-entry, and once we have adequate high energy propulsion systems available that can be used conveniently and reliably on the surface of the moon. With the advent of a nuclear third stage for C-5, the margin for this capability will be substantively widened, of course.

a. Our main reason against recommending the C-5 Direct Mode is its marginal weight allowance for the spacecraft and the demand for high energy return propulsion, combined with the time factor, all of which would impose a very substantial additional burden on the Manned Spacecraft Center.

b. The Manned Spacecraft Center has spent a great deal of time and effort in determining realistic spacecraft weights. In the opinion of Bob Gilruth and Chuck Mathews, it would simply not be realistic to expect that a lunar spacecraft light enough to be used with the C-5 Direct Mode could be developed during this decade with an adequate degree of confidence.

c. The demand for a high energy return propulsion system, which is implicit in the C-5 Direct Mode, is considered undesirable by the Manned Spacecraft Center - at the present state-of-the-art at least - because this propulsion system must also double up as an extra-atmospheric abort propulsion system. For this purpose, MSC considers a propulsion system as simple and reliable as possible (storable and hypergolic propellants) as absolutely mandatory. We think the question of inherent reliability of storable versus high energy propulsion systems - and their usability in the lunar surface environment - can be argued, but as long as the requirement for "storables" stands, the C-5 Direct Mode is not feasible performance wise.

[7]   d. NASA has already been saddled with one program (Centaur) where the margin between performance claims for launch vehicle and demands for payload weights were drawn too closely. We do not consider it prudent to repeat this mistake.

IV. WHY DO WE RECOMMEND AGAINST THE NOVA OR C-8 MODE?

It should be clearly understood that our recommendation against the Nova or C-8 Mode at this time refers solely to its use as a launch vehicle for the implementation of the President's commitment to put a man on the moon in this decade. We at Marshall feel very strongly that the Advanced Saturn C-5 is not the end of the line as far as major launch vehicles are concerned! Undoubtedly, as we shall be going about setting up a base on moon and beginning with the manned exploration of the planets, there will be a great need for launch vehicles more powerful than the C-5. But for these purposes such a new vehicle could be conceived and developed on a more relaxed time schedule. It would be a true follow-on launch vehicle. All of our studies aimed at NASA's needs for a true manned interplanetary capability indicate that a launch vehicle substantially more powerful than one powered by eight F-l engines would be required. Our recommendation, therefore, should be formulated as follows: "Let us take Nova or C-8 out of the race of putting an American on the moon in this decade, but let us develop a sound concept for a follow-on 'Supernova' launch vehicle".

Here are our reasons for recommending to take Nova or C-8 out of the present Manned Lunar Landing Program:

a. As previously stated, the Apollo system in its present form is not landable on the moon. The spacecraft system would require substantial changes from the presently conceived configuration. The same argument is, of course, applicable to the Earth Orbit Rendezvous Mode.

b. With the S-II stage of the Advanced Saturn C-5 serving as a second stage of a C-8 (boosted by eight F-l engines) we would have an undesirable, poorly staged, hybrid launch vehicle, with a payload capability far below the maximum obtainable with the same first stage. Performance wise, with its escape capability of only 132,000 lbs. (in lieu of the 150,000 lbs. demanded) it would still be too marginal, without a high energy return propulsion system, to land the present Apollo Command Module on the surface of the moon.

c. Implementation of the Nova or C-8 program in addition to the Advanced Saturn C-5 would lead to two grossly underfunded and undermanaged programs with resulting abject failure of both. Implementation [8] of the Nova or C-8 program in lieu of the Advanced Saturn C-5 would have an absolutely disastrous impact on all our facility plans.

The rafter height of the Michoud plant is 40 feet. The diameter of the S-IC is 33 feet. As a result, most of the assembly operations for the S-IC booster of the C-5 can take place in a horizontal position. Only a relatively narrow high bay tower must be added to the main building for a few operations which must be carried out in a vertical position. A Nova or C-8 booster, however, has a diameter of approximately 50 feet. This means that the roof of a very substantial portion of the Michoud plant would have to be raised by 15 to 20 feet. Another alternative would be to build a very large high bay area where every operation involving cumbersome parts would be done in a vertical position. In either case the very serious question arises whether under these circumstances the Michoud plant was a good selection to begin with.

The foundation situation at Michoud is so poor that extensive pile driving is necessary. This did not bother us when we acquired the plant because the many thousands of piles on which it rests were driven twenty years ago by somebody else. But if we had to enter into a major pile driving operation now, the question would immediately arise as to whether we could not find other building sites where foundations could be prepared cheaper and faster.

Any tampering with the NASA commitment to utilize the Michoud plant, however, would also affect Chrysler's S-1 program, for which tooling and plant preparation are already in full swing at Michoud. Raising the roof and driving thousands of piles in Michoud may turn out to be impossible while Chrysler is assembling S-I's in the same hangar.

In summary, the impact of a switch from C-5 to Nova/C-8 on the very concept of Michoud, would call for a careful and detailed study whose outcome with respect to continued desirability of the use of the Michoud plant appears quite doubtful. We consider it most likely that discontinuance of the C-5 plan in favor of Nova or C-8 would reopen the entire Michoud decision and would throw the entire program into turmoil with ensuing unpredictable delays. The construction of a new plant would take at least 2-1/2 years to beneficial occupancy and over 3 years to start of production.

d.   At the Marshall Space Flight Center, construction of a static test stand for S-IC booster is well under way. In its present form this test stand cannot be used for the first stage of Nova or C-8. Studies indicate that as far as the noise level is concerned, there will probably be no objection to firing up eight F-1 engines at MSFC. However, the Marshall [9] test stand construction program would be greatly delayed, regardless of what approach we would take to accommodate Nova/C-8 stages. Detailed studies seem to indicate that the fastest course of action, if Nova or C-8 were adopted, would be to build

— a brand new eight F-l booster test stand south of the present S-IC test stand, and

— convert the present S-IC test stand into an N-II test stand. (This latter conclusion is arrived at because the firing of an N-II stage at Santa Susanna is not possible for safety reasons, the S-II propellant load being considered the absolute maximum permissible.)

The Mississippi Test Facility is still a "cow pasture that NASA doesn't even own yet," and cannot compete with any test stand availability dates in Huntsville. Developments of basic utilities (roads, water, power, sewage, canals, rail spur, etc.) at MTF will require well over a year, and all scheduling studies indicate that whatever we build at MTF is about 18 months behind comparable facilities built in Huntsville. MTF should, therefore, be considered an acceptance firing and product improvement site for Michoud products rather than a basic development site.

e.   In view of the fact that the S-II stage is not powerful enough for the Apollo direct flight mission profile, a second stage powered by eight or nine J-2's or two M-1's is needed. Such a stage would again be on the order of 40 to 50 feet in diameter. No studies have been made as to whether it could be built in the Downey/Seal Beach complex. It is certain, however, that its static testing in Santa Susanna is impossible. As a result, we would have to take an entirely new look at the NAA contract.

f.   I have already mentioned the disruptive effect a cancellation of the C-5 would have on the RIFT program.

g.   One of the strongest arguments against replacement of the Advanced Saturn C-5 by Nova or C-8 is that such a decision would topple our entire contractor structure. It should be remembered that the temporary uncertainty about the relatively minor question of whether NAA should assemble at Seal Beach or Eglin cost us a delay of almost half a year. I think it should not take much imagination to realize what would happen if we were to tell Boeing, NAA and Douglas that the C-5 was out; that we are going to build a booster with eight F -1 engines, a second stage with eight or nine J –2's or maybe two M-1 engines; and that the entire problem of manufacturing and testing facilities must be re-evaluated.

[10] We already have several thousands of men actually at work on these three stages and many of these have been dislocated from their home plants in implementation of our present C-5 program. Rather than leaving these thousands of men suspended (although supported by NASA dollars) in a state of uncertainty over an extended period of new systems analysis, program implementation studies, budget reshuffles, site selection procedures, etc., it may indeed turn out to be wiser to just terminate the existing contracts and advise the contractors that we will call them back once we have a new program plan laid out for them. We have no doubt that the termination costs incurring to NASA by doing this would easily amount to several hundred million dollars.

I have asked a selected group of key Marshall executives for their appraisal, in terms of delay of the first orbital launch, if the C-5 was to be discontinued and replaced by a Nova or C-8. The estimates of these men (whose duties it would be to implement the new program) varied between 14 and 24 months with an average estimate of an over-all delay of 19 months.

h.   In appraising the total loss to NASA, it should also not be overlooked that we are supporting engine development teams at various contractor plants at the rate of many tens of millions of dollars per year for every stage of C-l and C-5. If the exact definition of the stages were delayed by switching to Nova/C-8, these engine development teams would have to be held on the NASA payroll for just that much longer, in order to assure proper engine / stage integration.

i.   More than twelve months of-past extensive effort at the Marshall Space Flight Center to analyze and define the Advanced Saturn C-5 system in a great deal of engineering detail would have to be written off as a flat loss, if we abandoned the C-5 now. This item alone, aside from the time irretrievably lost, represents an expenditure of over one hundred million dollars.

j.   The unavoidable uncertainty in many areas created by a switch to Nova or C-8 (Can we retain present C-5 contractors? Where are the new fabrication sites? Where are we going to static test? etc.) may easily lead to delays even well in excess of the estimates given above. For in view of the political pressures invariably exerted on NASA in connection with facility siting decisions, it is quite likely that even the NASA Administrator himself will find himself frequently unable to make binding decisions without demanding from OMSF an extensive re-appraisal of a multitude of issues related with siting. There was ample evidence of this during the past year.

k.   For all the reasons quoted above, the Marshall Space Flight Center considers a discontinuation of the Advanced Saturn C-5 in favor of Nova or C-8 as the worst of the four proposed modes for implementation of the manned lunar landing project. We at Marshall would consider a decision in favor of this mode to be tantamount with giving up the race to put a man on the moon in this decade even before we started.

[11]

IN SUMMARY I THEREFORE RECOMMEND THAT:

a.   The Lunar Orbit Rendezvous Mode be adopted.

b.   A development of an unmanned, fully automatic, one-way C-5 Logistics Vehicle be undertaken in support of the lunar expedition.

c.   The C-l program as established today be retained and that, in accordance with progress made in S-IV B development, the C-l be gradually replaced by the C-I B.

d.   A C-l B program be officially established and approved with adequate funding.

e.   The development of high energy propulsion systems be initiated as a back-up for the Service Module and possibly the Lunar Excursion Module.

f.   Supplements to present development contracts to Rocketdyne on the F -1 and J -2 engines be let to increase thrust and/ or specific impulse.

> [signed]
> Wernher von Braun, Director
> George C. Marshall Space Flight Center

Document II-19

Document Title: Jerome Wiesner, "Memorandum for Theodore Sorensen," 20 November 1961.

Document Source: John F. Kennedy Presidential Library, Boston, Massachusetts.

This memorandum from President Kennedy's science advisor to the president's top advisor provides a top-level snapshot of the status of Project Apollo at the end of 1961.

THE WHITE HOUSE
WASHINGTON
November 20, 1961

MEMORANDUM FOR

Mr. Theodore C. Sorensen

Outline of major problems related to the NASA Manned Lunar Program:

1.   Required decisions
       Rendezvous
       Advanced Saturn
       Nova

2.   Initiate hardware programs
       Launch vehicles (beyond Saturn C - 1)
       Engines (in addition to F-1, J-2)
       Spacecraft
       Launch pads
       Static test stands (for new stages)
       Rendezvous development

3.   Activate new field stations
       Houston (Spacecraft)
       Michoud (Boosters)
       Pearl River (Static test)
       AMR expansion (Launching)

4. Secure supporting information
       Space environment
       Long-term weightlessness
       Lunar characteristics

5.   Manpower (FY'63 total in-house: 26, 224 $\angle$ JPL)
       Availability
       Competence
       Salaries

[2]

6. University support
Research grants
Facilities
Education

7. DOD support and related programs
Biomedical program
Titan II and III
Dyna Soar

8. Financial support
Supplemental FY '62: $156 M - FY '63: $4238.2 M
Future predictions

Note: NASA responsibilities not directly related to the manned Lunar Program:

9. Space Applications (i.e., Meteorology and Communications)

10. Aeronautics

11. Nuclear technology (Snap and Rover)

12. Other space science including Planetary and Interplanetary

13. Long-range spacecraft and vehicle technology

Comments on outline:

1. The major decisions have not been announced as to what extent rendezvous will be employed, what Advanced Saturn vehicle will be built (probably C-4), and what will be the characteristics of the so-called Nova which could put man on the moon by direct ascent. The relative emphasis of rendezvous versus direct ascent is a key to the entire program.

2. Six months have elapsed since the decision was announced to put man on the moon, yet none of these crucial hardware programs have progressed beyond the study phase. Lead times on these development and construction programs are of critical importance.

[3]

3. It is hoped that there will be no further field stations beyond these already announced. However, there are major problems related to the activation of these centers.

4. These are the major questions related to the lunar undertaking which can only be obtained by a broad supporting space and life science program.

5.   Many people believe that the space program may severely tax our supply of technical manpower for in-house and contractor needs. It is also important that competent leadership be available, and adequate salary scales are a continuing problem.

6.   NASA must support a broad program of basic research related to the space effort in the universities. The impact on the universities and upon the educational requirements must also be considered.

7.   There are still major problems in the NASA-DOD relationship related to booster development and supporting technology.

8.   The total being requested of the next Congress is about 50% greater than was predicted for FY '63 last May. Extrapolation to future years of the funding trend does not lend itself to any optimism as to a leveling-off in the next year or two.

9 - 13.   The major item in here which should be singled out at this time is the Nuclear Rocket Program (Rover). The total NASA-AEC request for FY '63 is about $200 million. Is this level of funding realistic for a program which will probably not produce an operational vehicle until 1970 or later?

[signed]
Jerome B. Wiesner

Document II-20

Document Title: NASA, "Project Apollo Source Evaluation Board Report: Apollo Spacecraft," NASA RFP 9-150, 24 November 1961

Document Source: Folder 18675, NASA Historical Reference Collection, History Division, NASA Headquarters, Washington, DC.

The first element of the system to carry astronauts to the Moon was the Apollo spacecraft. Even before President Kennedy set a lunar landing by the end of the decade as a national goal, NASA had been planning a three-person spacecraft for Earth orbital and circumlunar missions. After Kennedy's 25 May 1961 speech announcing the decision to send Americans to the Moon, the spacecraft requirements were modified to support a lunar landing mission, even though the approach to be taken in carrying out the mission had not yet been chosen.

The Source Evaluation Board ranked the proposal by The Martin Company first among the five companies that submitted a bid, with North American Aviation a "desirable alternative." NASA Administrator James E. Webb, Deputy Administrator Hugh Dryden, and Associate Administrator Robert Seamans reversed this ranking, and the contract to build the Apollo spacecraft was awarded to North American Aviation on 28 November 1961.

PROJECT APOLLO

SOURCE EVALUATION BOARD REPORT

APOLLO SPACECRAFT

NASA RFP 9-150

CLASSIFICATION CHANGE
To Unclassified on 8/17/71

November 24, 1961

Use of this document is restricted to personnel
participating in source selection procedures.
CLASSIFIED DOCUMENT – TITLE UNCLASSIFIED

This document contains information affecting the national defense of the United States
within the [unreadable] of the espionage laws, Title 18, U.S.C., Secs. 793 and 791, the transmission
or revelation of which in any manner to an unauthorized person is prohibited by law.

NATIONAL AERONAUTICS AND SPACE ADMINISTRATION

Manned Spacecraft Center   Langley Air Force Base, Virginia

REPORT OF THE SOURCE EVALUTION BOARD
APOLLO SPACECRAFT
NASA RFP 9-150

Submitted by:

Chairman:	Maxime A. Faget ,Manned Spacecraft Center
Member:	Robert O. Piland, Manned Spacecraft Center
Member:	George M. Low, NASA Headquarters
Member:	Wesley L. Hjornevik, Manned Spacecraft Center
Member:	Kenneth S. Kleinknecht, Manned Spacecraft Center
Member:	Charles W. Mathews, Manned Spacecraft Center
Member:	James A. Chamberlin, Manned Spacecraft Center
Member:	A. A. Clagett, NASA Headquarters
Member:	Dr. Oswald H. Lange, Marshall Space Flight Center
Member:	Dave W. Lang, Manned Spacecraft Center
*Member:	James T. Koppenhaver, NASA Headquarters
Ex Officio:	Robert R. Gilruth, Manned Spacecraft Center

*Non-voting member (reliability representative)

TABLE OF CONTENTS

[pp. 1-5 not included]

[6] REQUEST FOR PROPOSAL DISSEMINATION

Subsequent to the Headquarter's approval of the Statement of Work and a Procurement Plan, the Request for Proposal was disseminated on July 28, 1961 to the following twelve companies.

> Boeing Airplane Company
> Chance Vought Aircraft
> General Dynamics Corporation
> Douglas Aircraft Company, Inc.
> General Electric Company
> Goodyear Aircraft Corporation
> Grumman Aircraft Engineering Corporation
> Lockheed Aircraft Corporation
> The Martin Company
> McDonnell Aircraft Corporation
> North American Aviation, Inc.
> Republic Aviation Corporation

In addition, a synopsis of the proposed procurement was publicized in accordance with NASA Circular No. 131, which deals with the publicizing of NASA-proposed research and development procurements.

Four additional companies were provided with the Request for Proposal upon request.

> Radio Corporation of America
> Space General Corporation
> Space Technology Laboratories
> Bell Aerospace Systems

There were no complaints received from companies not invited to submit proposals. Potential subcontractors requesting Request for Proposals were referred to the potential prime contractors.

A preproposal conference attended by representatives of the sixteen companies was held on August 14, 1961, at which time NASA personnel discussed the technical and business aspects of the Request for Proposal .Approximately four hundred questions were answered orally and subsequently documented and confirmed by mail.

[7] PROPOSALS SUBMITTED

Five companies submitted proposals on October 9, 1961, ten weeks after the Request for Proposal was mailed. No complaints on the time allowed were received and no time extensions were requested. The five companies submitting proposals were as follows:

> General Dynamics/Astronautics
> General Electric Company/MSVD
> The Martin Company
> McDonnell Aircraft Corporation
> North American Aviation, Inc./S and ID

The General Electric Company and McDonnell Aircraft Corporation proposed to form teams to carry out the contract. General Electric proposed to team with Douglas Aircraft, Grumman Aircraft, and Space Technology Laboratories. McDonnell proposed to team with Chance Vought, Lockheed and Hughes. General Dynamics proposed a single team member, AVCO. Martin and North American both proposed the prime-subcontractor approach. Representatives of the proposers made oral presentations on October 11, 1961 to members of the NASA evaluation team.

[8] EVALUATION PROCEDURE

Organization

The evaluation was conducted by the Source Evaluation Board. The Chairman of the board appointed Technical and Business Subcommittees to assist in the evaluation. These subcommittees in turn were assisted by panels of specialists. The detail procedures and membership of the subcommittees and panels is presented in references 2 and 3. A total of 190 personnel representing all major elements of the NASA and including several representatives of the DOD participated in the evaluation.

Assessment Areas

The Technical Evaluation consisted of two major areas, Technical Qualifications and Technical Approach. The Technical Qualifications portion covered experience, facilities, personnel, and the technical ramifications of the proposed project organization. The Technical Approach portion consisted of eleven areas which covered mission and system design; systems integration; development, reliability, and manufacturing plans; and operational concepts. The Business Management and Cost Evaluation consisted of the areas of organization and management, logistics, subcontract administration, and cost.

Weighting Factors

The weighting factors assigned to the major proposal areas were as follows:

Technical Qualifications   30
Technical Approach   30
Business Management and Cost   <u>40</u>
                                  100

[9] Rating Method

Ratings were made on a 0-10 rating system, defined in the following manner.

10
9   Excellent
8

7
6   Good
5

4
3   Fair
2

1   Poor

0   Unsatisfactory

Evaluation Schedule

October 9 – 21   Detail assessment by panels
October 23 – 28   Subcommittee review
November 1 – 22   Source Evaluation Board review

[10] EVALUATION RESULTS

Ratings

The Source Evaluation Board reviewed the reports of the Technical and Business Subcommittees and discussed the reports with the subcommittees. The reports and ratings were accepted with minor modifications. The board examined the sensitivity of the weighting factors used by the various panels during the evaluation. It was determined that the results are not sensitive to moderate changes in the weighting factors. The board considered in further detail the item of applicable experience which had been rated by the Technical Subcommittee in the area of Technical Qualifications to insure that the factor of quality of experience had been adequately considered. The board's findings confirmed the ratings given by the subcommittee. The board recognized that the Cost Panel had not had access to sufficient information to adequately rate the items of Cost Experience and Cost Estimate. The board further found that the Organization and Management Panel had sufficient information to adequately assess the Cost Experience item and so its rating of this item was used. The board through the use of an Ad Hoc Panel analysed the realism of the cost estimates in relation to the work proposed and subsequently rated this item. The rating for each proposal in the three major areas and a summary weighting obtained by applying the weight factors is presented below.

<u>Ratings by Area</u>

	Technical Approach (30%)	Technical Qualifications (30%)	Business (40%)
The Martin Company	5.58	6.63	8.09
General Dynamics/ Astronautics	5.27	5.35	8.52
North American Aviation, Inc.	5.09	6.66	7.59
General Electric Company	5.16	5.60	7.99
McDonnell Aircraft Corporation	5.53	5.67	7.62

<u>Summary Ratings</u>

The Martin Company	6.9
General Dynamics/Astronautics	6.6
North American Aviation, Inc.	6.6
General Electric Company	6.4
McDonnell Aircraft Corporation	6.4

As can be seen, the North American and General Dynamics/Astronautics proposals received the same rating. In assessing the ratings, the board recognized that all the proposals had received high ratings in the Business area, the lowest rating (7.59) being higher than the highest rating (6.66) received in either the Technical Approach or Technical Qualifications area. Since those ratings established [11] that all the companies could more than adequately handle the business aspects of the program, the board turned its consideration to the Technical Evaluation for further analysis of the ratings.

The ratings of the proposals considering only the Technical Evaluation areas of Approach and Qualifications are as follows:

The Martin Company	6.1
North American Aviation, Inc.	5.9
McDonnell Aircraft Corporation	5.6
General Electric Company	5.4
General Dynamics/Astronautics	5.3

It may be concluded that General Dynamics/Astronautics' tie for second place rating is due entirely to its very high rating in the Business area, since it rated last in the Technical Evaluation. In view of the relatively high ratings of all companies in the Business area, and General Dynamics lowest rating in the Technical Evaluation, the board finds North American Aviation the clearly preferred source of the two proposals which received the tie second place ratings.

Assessment of Proposed Costs

The Request for Proposal did not specify a particular program for the development of the Apollo spacecraft. Part of each contractor's responsibility in developing his proposal was to indicate a technical development plan, a program schedule including hardware and testing, and a cost estimate supporting this proposed plan and program. The cost estimates received, therefore, were not subject to direct comparison. The cost estimates received for Phase A, the earth orbital portion of the project, were as follows:

GD/A	GE	Martin	MAC	NAA
550	899	563	629	351
		(Cost in Millions)		

As mentioned above, these costs are based on the particular program proposed by the different offerors. These proposed programs varied to a considerable degree. For purposes of analysing the cost estimate an "adjusted" cost was determined. This "adjusted" cost modified the submitted estimate to a reference number of spacecrafts, flights, and months. These adjusted costs are given below. They do not necessarily represent the negotiated contract cost and were used here for the purpose of cost analysis.

GD/A	GE	Martin	MAC	NAA
431	830	473	702	352

The cost estimates were examined in detail. Particular attention was given to rating the realism, validity, and overall quality of the cost proposals. In this regard both General Dynamics/Astronautics and Martin were considered to have high quality cost estimates, [12] well supported, detailed, and carefully considered. The low estimate of NAA was noted and carefully reviewed by the board, its Ad Hoc Panel, the Business Subcommittee and its Cost Panel. Although the quality of NAA's past cost history was recognized, the overall quality of this estimate was not as high as General Dynamics/Astronautics or Martin and the detail and summary information appeared questionable in areas of engineering, design, and subcontracts cost. NAA, accordingly, was not rated as highly as GD/A and Martin. The General Electric Company and McDonnell Aircraft estimates were rated below the other estimates because of high costs for programs proposed which reflected their proposed management philosophy for the Apollo spacecraft development.

Conflict of Resource Requirement

The board was concerned with the possible conflict of resource requirements between Apollo and other present or anticipated projects within the companies. Of particular concern, since they involved the preferred and alternate source, was the possible conflict between Apollo and the anticipated Titan III at The Martin Company and between Apollo and the S-II stage at North American. In order to further assess the possible conflict with the Titan III, the following request for information was sent to The Martin Company.

"NASA would like to ascertain the degree of conflict in manpower and resources between a possible Titan III program by Martin and Martin's Apollo proposal. Martin is asked to review Section 2.3 of their Management Proposal and inform NASA what changes would result in their proposed manpower and resources."

The Martin reply was as follows: "Martin reaffirms the subject proposal delivered October 9 to NASA and calls particular attention to the statements made in our letter of submittal and Section 2.3 of our Management Proposal concerning the priority Apollo will have if an award is made to us. Follow-on Titan programs have always been included in our future planning and, hence, were considered before developing the Manpower and Resources Sections of our Management Proposal. Titan III effort would be accomplished in our Denver Division and, therefore, does not constitute a conflict with Apollo. In any event, we never contemplated use of Denver-Titan manpower or resources for execution of the Apollo program. Therefore, we contemplate no change in our proposed manpower or resources as a result of a possible Titan III program."

A similar request for information related to the S-II was sent to NAA as follows: "NASA would like to ascertain the degree of conflict in manpower and resources between the Saturn S-II contract and NAA's Apollo proposal. North American is asked to correct figures 2.3-12, 2.3-14, and 2.3-16 of their Management Proposal to include the S-II load. NAA is also asked to reaffirm or correct the names of key personnel on pages 2.3-24 through 2.3-32."

[13] North American presented a detailed reply which has been filed with the original proposal. The reply contained considerable detailed discussion and data which in essence reaffirmed North American's position of "no conflict" as presented in the proposal.

The board considered both replies in detail and has satisfied itself to the degree possible that the manpower and resources proposed for Apollo are not in conflict with those required for Titan III at Martin or the S-II at North American.

Discussion of Results

The Martin Company is considered the outstanding source for the Apollo prime contractor. Martin not only rated first in Technical Approach, a very close second in Technical Qualifications, and second in Business Management, but also stood up well under the further scrutiny of the board.

The Martin Company appears to be well prepared to undertake the Apollo effort. This was evidenced by a Technical Proposal that was complete, well integrated with balanced emphasis in all areas, and of high overall quality with a minimum amount of superflous material. Martin's proposal was first in five of the eleven major Technical Approach areas including Technical Development Plan, Flight Mechanics, Onboard Systems, Manufacturing, and Ground Operational Support Systems and Operations. Martin, therefore, scored high in planning, design, manufacturing and operations, reflecting the quality across the complete scope of the job.

Martin has experience in large technically complex systems such as Titan and Vanguard. The personnel proposed and company have a general background of manned aircraft experience, as well as varied background of experience including airplanes (B-5l and B-57), missiles (Titan, Matador, Mace, Bullpup, and Pershing), and space vehicles (Vanguard). Their inhouse experience in many of the required technical areas results in a high confidence as to their capability as a systems integrator. The individual key technical personnel Martin proposed to assign to the project were evaluated as excellent both in competence and experience. Martin's proposed management arrangement of a prime contractor with subcontractors appears technically to be the most sound both as far as reaching technical decisions quickly and properly and also for implementing these decisions.   Short lines of communications involved in their proposed arrangement will minimize interface problems and required documentation and thereby result in fewer opportunities for error.

Martin proposed a strong project organization for Apollo. They would create a Project Apollo Division managed by a Vice President who reports directly to the President of the company. The parent company would put under the direct control of this [14] division the necessary resources of manpower and facilities for this job.

Martin's cost proposal compared well with the others. Their cost estimate was considered to be both realistic and reasonable.

North American Aviation, Inc. is considered the desirable alternate source for the Apollo spacecraft development. It rated highest of all proposers in the major area of Technical Qualifications. North American's pertinent experience consisting of the X-15, Navajo, and Hound Dog coupled with an outstanding performance in the development of manned aircraft (F-1OO and F-86) resulted in it being the highest rated in this area. The lead personnel proposed showed a strong background in development projects and were judged to be the best of any proposed. Like Martin, NAA proposed a project managed by a single prime contractor with subsystems obtained by subcontracting, which also had the good features described for the Martin proposal. Their project organization, however, did not enjoy quite as strong a position within the corporate structure as Martin's did. The high Technical Qualifications rating resulting from these features of the proposal was therefore high enough to give North American a rating of second in the total Technical Evaluation although its detailed Technical Approach was assessed as the weakest submitted. This relative weakness might be attributed to the advantage of the McDonnell Aircraft Corporation's Mercury experience, and the other three proposers' experience on the Apollo study contracts. The Source Evaluation Board is convinced that NAA is well qualified to carry out the assignment of Apollo prime contractor and that the shortcomings in its proposal could be rectified through further design effort on their part. North American submitted a low cost estimate which, however, contained a number of discrepancies. North American's cost history was evaluated as the best.

The remaining three companies, General Dynamics/Astronautics, General Electric Company, and McDonnell Aircraft Corporation, were not considered to be as desirable as Martin and NAA as a source for Apollo. These offerors supplemented their skills and resources with those of one or more additional companies in order to create a team prepared to carry out the Apollo effort. While the board found that these teams did indeed show adequate, or more than adequate, resources for the job, it was also apparent that the ramifications of a large team were serious. The communications problems created by geographical locations, the complex coordination required which leads to slow process of actions, the overlapping and similar capabilities of several team members which may lead to disagreement, and the committee approach [15] to project decision all tend to detract from the desirability of the team approach. There was also an apparent relation between high project cost and the two large teams.

It should be pointed out that the same degree of management difficulty is not inherent in the three team offerors. GD/A with only one team associate should be expected to suffer very little in this respect. MAC showed awareness of the problem and attempted to invest adequate control responsibility in a strong Project Manager, who is also properly located at a high level in the MAC corporate structure. GE's proposal is particularly vulnerable to this criticism since it emphasizes councils and committees. GE was also found in many important technical respects to be weaker than its other team members. Consequently, it may prove to be in a poor position to direct its team's effort in conflicting situations.

General Dynamics/ Astronautics rated third in Technical Approach and last in Technical Qualifications. They rated excellently in big systems experience of an advanced technological nature (Atlas), but exhibited no manned aircraft or spacecraft experience in the Astronautics Division and their experience was not broad being limited to Atlas and Centaur. Relative to the other companies proposing, GD/A did not rate highly in facilities. While enjoying excellent conventional laboratories, no evidence of large-scale simulation or environmental equipment was noted. The personnel proposed were relatively shy in total experience and project experience.

GD/A submitted a cost proposal that was considered best. Like Martin, its estimate of cost was considered both realistic and reasonable. Although they clearly made the best business management proposal, the other offerors all rated sufficiently high in this area to lead the board to the conclusion that the technical aspects should be the controlling consideration. With the weakest technical showing, GD/A is not considered a desirable source.

General Electric Company was rated fourth overall resulting from a third place rating in Business Management and fourth place ratings in both Technical Approach and Qualifications. GE'S middle rating in Business resulted from having excellent facilities and a willingness to invest heavily of company funds. The GE/MSVD experience in managing systems of the scope of Apollo was lacking, and their management and program organizations were considered weaker than top proposers. GE made the highest cost proposal. This was considered by the board to be a true reflection of the [16] layering of fees, the duplication of effort, and the extra, complexity associated with the far-flung large team organization they proposed. For this reason, the GE team was not considered a desirable source.

McDonnell Aircraft Corporation rated lowest in overall rating. Although MAC rated a close second in Technical Approach, it rated third in Technical Qualifications. MAC proposed the second highest cost. With a team approach quite similar to that of GE's, MAC was also considered by the board to be a high cost producer. For these reasons, MAC did not appear to be a desirable source despite its high relative rating in the Technical Approach area.

Document II-21

Document Title: Joseph F. Shea, Deputy Director for Systems, Office of Manned Space Flight, to Director of Aerospace Medicine and Director of Spacecraft and Flight Missions, "Selection and Training of Apollo Crew Members," 29 March 1962.

Document Source: Folder 18675, NASA Historical Reference Collection, History Division, NASA Headquarters, Washington, DC.

As NASA began detailed scientific planning for Apollo, one issue was whether to expand the astronaut corps, which to that point had been limited to accomplished test pilots, to include professional scientists as astronauts. This memorandum solicits the views of key NASA human spaceflight offices on this question. NASA decided to recruit scientist astronauts for Apollo, but only one, Dr. Harrison Schmitt, flew, on Apollo 17, the final Apollo mission to the Moon.

To: Director of Aerospace Medicine   March 29, 1962
     Director of Spaceflight & Flight Missions

From: Joseph F. Shea

Re: Selection and Training of Apollo Crew Members

At the request of the Office of Systems, the NASA Space Science Steering Committee has established an <u>ad hoc</u> group to recommend scientific tasks to be performed on the moon by Apollo crew members. This group, under the chairmanship of Dr. Sonnett, will include scientific consultants as well as representatives from appropriate NASA groups.

Dr. Sonnett has asked the OMSF to present a briefing to this group at its first formal meeting, establishing a context and the ground rules within which they are to perform their task. One of the topics to be covered is the possible use of one or more professional scientists as crew members on lunar missions.

To assist in the preparation of this briefing, it would be helpful if you furnished this office with memoranda, no later than April 16, directed to the following two questions:

1. Is there any fundamental reason which would prevent the use of one or more professional scientists as crew members?

2. What serio us practical problems would result if such personnel were included in the selection [and] training program?

It is assumed that the NH [Aerospace Medicine] memorandum will cover these questions from the viewpoint of medical selection, and that MS [Spaceflight & Flight Missions] will consider the problem in terms of background skills and training requirements.

Joseph F. Shea
Deputy Director for Systems
Office of Manned Space Flight

Document II-22

Document Title: Owen E. Maynard, Spacecraft Integration Branch, NASA Manned Spacecraft Center, Memorandum for Associate Director, "Comments on Mr. Frank Casey's visit to J.P.L. to discuss Ranger and follow-on programs which could provide information pertinent to Apollo missions," 1 February1962.

Source: Folder 18675, NASA Historical Reference Collection, History Division, NASA Headquarters, Washington, DC.

Since the beginning, it had been clear to those planning for human missions to the Moon that they would need information from robotic lunar missions. Those missions were under the management of the Jet Propulsion Laboratory (JPL), a Federally-Funded Research and Development Center operated for NASA by the California Institute of Technology. JPL's mission designs were aimed at answering scientific questions, not providing support for human spaceflight missions, and JPL was rapidly developing within NASA a reputation for excessive independence from the rest of NASA. This was proven after a visit to JPL to see if the lunar hard landing Ranger missions and potential follow-on robotic missions (which became the Surveyor program) could contribute to the planning for Project Apollo. It is interesting to note that in this early stage of planning for Apollo, the Sea of Tranquility had already been identified as a potential lunar landing site.

At this point in the evolution of Apollo, the program was divided into three elements: Apollo A, Earth-orbital tests of the Apollo spacecraft; Apollo B, flights around the Moon; and Apollo C, flights to land on the Moon.

NASA – Manned Spacecraft Center
Langley AFB, Virginia
February 1, 1962

MEMORANDUM for Associate Director

Subject: Comments on Mr. Frank Casey's visit to J.P.L. to discuss Ranger and follow-on programs which could provide information pertinent to Apollo missions.

1. During a recent visit to J.P.L. at Pasadena, California, a group of NASA employees from Langley Research Center, Ames Research Center and Manned Spacecraft Center had an opportunity to discuss the Ranger program and its follow-on programs with the J.P.L. staff. The purpose of this meeting was to determine if the present series of Ranger payloads and the follow-on payloads could be of value to the Apollo mission.

2. Since both the time and experiments available for obtaining further engineering data for design of Apollo systems and components is limited when viewed in terms of the unknowns, the following question was posed within the NASA group as a basis [*sic*] criterion for the planning of payloads to obtain further information on environmental data for the Apollo program:

“What are the environmental parameters for which additional data must be obtained before the Apollo missions will be attempted”?

In consideration of three Apollo phases, this criterion leads to the following conclusions:

Apollo Phase A   No further environmental data required

Apollo Phase B   Possibly additional data on radiation and meteoroids in cis-lunar and lunar space

Apollo Phase C   The above comments on radiation and meteoroids is [*sic*] appropriate. In addition, more definite data on both the large and small scale lunar surface features, the existence and nature of lunar surface dust, and the physical properties of the lunar surface which constitute its ability to support a vehicle.

3. It was recognized that the limits, accuracy and coverage of environmental data to better establish the physical nature of the lunar surface in terms of Apollo missions requirements are incomplete, and that further inputs should be reminded of this need and attempts should be made to supply available information to plan instrumentation of Ranger follow-on payloads.

[2]

4. On the basis of current knowledge and thinking relative to the nature of the lunar surface environment, and the need for engineering data for the design of Apollo systems and sub-systems, it appears that the selection of Ranger follow-on payloads should be directed primarily on the ability of these payloads to yield data which would permit a better evaluation of:

      a.   The large scale features of the lunar surface such as the locations, magnitude, and slopes of mountains, craters, and protuberances;

      b.   The existence and distribution of small scale features of the lunar surface such as roughness, slopes, faults, sharpness, and vesicularity which will aid in the evaluation of the extent to which the Apollo vehicle must be able to hover and translate prior to landing;

      c.   The existence of a dust layer on the lunar surface and the properties of this layer which will permit it to be entrained in the jet exhaust and form clouds which may foul systems components and obstruct optical and R.F. transmission from the vehicle to the surface and from the vehicle to space and the earth;

      d.   The ability of the lunar surface to support the Apollo vehicle including the existence and bearing strength of dust layers in excess of six inches in depth and the bearing strength and hardness of sub-surface material.

Secondary consideration should be given to the measurement of meteoroid and radiation parameters.

5. In consideration of the difficulty associated with obtaining environmental information over a substantial portion of the lunar surface to the accuracy required by Apollo C missions, it would be extremely helpful in the selection of Ranger and follow-on experiments if MSC and J.P.L. could agree on the landing site. It is not possible to get Ranger payloads over to the western limb of the moon where the Sea of Tranquility is located. This would allow the maximum Ranger payload weight to be used to advantage.

6. Since the design freeze date for Apollo occurs in 1964 it is imperative that lines of communication be established immediately if Apollo is to have an input from Ranger and follow-on programs in time to be used as design.

7. J.P.L. is presently investigating the problems of conducting experiments to obtain direct design data for Apollo. They will investigate [3] launch vehicle capabilities to implement the investigations and report their findings to NASA Headquarters about February 8, 1962.

Owen E. Maynard
Spacecraft Integrating Branch

Document II-23

Document Title: Letter to the President from James E. Webb, 13 March 1962.

Document Source: John F. Kennedy Presidential Library, Boston, Massachusetts.

In this letter, NASA Administrator Webb asks the president to assign the highest national priority to Project Apollo. Such a priority meant that Apollo would get preferred treatment in the allocation of scarce resources. President Kennedy approved this request.

R 13 1962 [stamped]

The President
The White House
Washington, D. C,

My Dear Mr. President:

Programs that enjoy the highest (DX) national priority attain this stature only on approval of the President. In order to meet the objectives of the Nation's space program as stated by you and endorsed by the Congress, I consider it essential that Project Apollo – to effect a manned lunar landing and return in this decade – receive such a priority.

Accordingly, I hereby recommend that the highest national priority be assigned to Project Apollo and in order to assure that you have the advice of the National Aeronautics and Space Council, I have addressed a memorandum, copy attached, to the Vice President asking the Council to consider this matter.

Respectfully yours,

James E. Webb
Administrator

Attachment
[2]

Memorandum for the Chairman, National Aeronautics and Space Council

Subject: Request for Highest National Priority for the Apollo Program

1.   The programs that now enjoy the highest (DX) national priority are: Atlas, Titan, Minuteman, Polaris, BMES, SAMOS, Nike-Zeus, Discoverer, Mercury, and Saturn. Of these, the first eight are managed by the Department of Defense, and the last two by the National Aeronautics and Space Administration. The prescribed criteria under which the President has made these determinations is that these programs have objectives of key political, scientific, psychological or military import.

2.   The NASA is requesting that the Apollo program be added to this list. Recognizing the need to restrict the number of projects on the list to the absolute minimum, NASA is prepared to drop Project Mercury from the list by the end of Calendar Year 1962, at which time its mission should be essentially complete. NASA will also expect to drop the Saturn vehicle project from the list except insofar as it pertains to the Apollo mission. In adding Apollo, the NASA would be requesting a DX priority for all of these elements of the Apollo program that are essential to its ultimate mission: to effect a manned lunar landing and return in this decade. The essential elements of the Apollo program would include development of the spacecraft and launch vehicles as well as the facilities which are required for their development, testing and use. Elements of certain other name projects would thus be included, such as Saturn and Gemini, but only insofar as they are directly applicable to the manned lunar landing.

3.   Decisions on the assignment of highest national priority are made by the President and in the case of space program projects, he takes into consideration the advice of the National Aeronautics and Space Council. Therefore, I ask that this matter be placed before the Council at an early date.

4.   I shall be pleased to supply any further information you think is essential to the Council's consideration.

James E. Webb
Administrator

Document II-24

Document Title: Ted H. Skopinski, Assistant Head, Trajectory Analysis Section, NASA-Manned Spacecraft Center, to Chief, Systems Integration Division, "Selection of lunar landing site for the early Apollo lunar missions," 21 March 1962.

Document Source: Folder 18675, NASA Historical Reference Collection, History Division, NASA Headquarters, Washington, DC.

There were discussions in 1962 about selecting a single site on the Moon for the initial lunar landing, with the selection being made primarily on the scientific interest of that particular location. Ted Skopinski, assistant head of the Trajectory Analysis section at the Manned Spacecraft Center, questioned this approach as he outlined the mission operations criteria for a lunar site. Skopinski's letter highlighted the fact that landing sites were to be determined not only by scientific interest, but by other factors such as the need for daylight operations and facilitating the return to Earth. He suggested the desirability of adding to JPL's planned lunar missions a lunar orbiter that could obtain high quality photographs of the lunar surface. This suggestion was accepted and led to the Lunar Orbiter program, which was managed by the Langley Research Center and built by the Boeing Company using a camera modified from its original mission as part of a highly classified intelligence satellite called SAMOS.

NASA-Manned Spacecraft Center
Langley Station, Hampton, Va.
March 21, 1962

MEMORANDUM for Chief, Systems Integration Division

Subject: Selection of lunar landing site for the early Apollo lunar missions

References: (a) Memorandum for Chief, Flight Operations Division by John E. Dornbach, dated Jan. 23,1962, re meeting on circumlunar photographic experiment

                     (b) Memorandum for Associate Director by Owen E. Maynard, dated Feb. 1, 1962, re comments on Frank Casey's visit to J.P.L. to discuss Ranger and follow-on programs which could provide information pertinent to Apollo missions

        1.   The need for obtaining both [*sic*] photographic, cartographic, and geologic information about the lunar surface in order to select a landing site for the Apollo lunar mission is defined in references (a) and (b). A recommendation was made in reference (b) that MSC and JPL agree on a landing site because of the difficulty of obtaining desired environmental data over a substantial portion of the lunar surface. At the present time JPL has defined their prime area of interest for unmanned lunar impacts as approximately 8°N to 8°S latitude and from 25° to 45°W longitude.

        2.   The purpose of this memorandum is to see if a single lunar landing site is compatible with the techniques which reduce the need of plane changes near the moon. The JPL direct ascent and impact type of trajectories differ from the Apollo trajectories in that the following mission rules have to be adhered to in the selection of the Apollo trajectories:

              a. return to the continental U.S.A. or Australia

              b. daylight reentry and a.

  c. lunar landing in earth reflected or direct sunlight and b.

  d. mission design for immediate insertion checking by tracking

  e. allowance for solar interference with deep space tracking and c.

[2]

  f. adequate tracking immediately prior and subsequent to reentry and follow-up to landing to ensure minimum recovery time.

The above rules have been investigated by the Operational Analysis Section of the Mission Analysis Branch, FOD, to see how they affect the Apollo launch window. These same mission rules will also influence the selection of possible lunar landing sites.

  3. Taking into account the mission rules stated in paragraph 2, the following disadvantages for the selection of a single lunar landing site are noted:

  a. Mission rule of a single lunar landing site imposes a severe restraint on the launching day and time because of its dependency on the lunar declination.

  b. A single lunar landing site is not compatible with the variable launch azimuth and parking orbit scheme which opens the launch window and thus eliminates the need of major plane changes.

  c. Without an extensive investigation this extra restraint might be too restrictive because not knowing the final design weights the amount of fuel needed to make a necessary plane change may be prohibitive. Any plane changes in the vicinity of the moon must result in suitable earth return trajectories compatible with 2f.

  d. The photographic and geophysical data obtained from the Ranger and Surveyor programs and the Apollo manned lunar missions may drastically alter present day concepts of the lunar surface, the single landing site selected now could therefore be worthless.

  4. The present day thinking is to restrict the landing site to a belt approximately 10 degrees on either side of the lunar equator and on the front side of the moon. If this will be true a few years from now then the following suggestions of obtaining lunar surface data prior to manned lunar landings could be followed.

  a. Obtain USAF lunar charts 1:1,000,000 scale of the landing area belt that are based on today's state of the cartographic art using lunar telescopic photography taken on earth.

  b. Augment the JPL Ranger and Surveyor programs to include several landing sites in the ±10 degree latitude belt on the front side of the moon.

[3]

c. Launch a circumlunar photographic satellite with a recoverable package to obtain high quality photographs of the lunar surface in the area of interest.

d. Expect the information obtained from the manned lunar orbit reconnaissance missions to be the most reliable and after comparing it with the data obtained from all other sources select several landing sites for the first Apollo manned lunar landing missions.

[signed]
Ted H. Skopinski
Asst. Head, Trajectory Analysis Section

Copies to: J.P. Mayer
M.V. Jenkins
P.F. Weyers, Apollo Project Office
R.O. Piland, Apollo Project Office
C.C. Johnson, Apollo Project Office
O.E. Maynard, Spacecraft Int. Branch
J.E. Dornbach, Space Physics Div.

Document II-25

Document Title: Memorandum to Administrator from Robert C. Seamans, Jr., Associate Administrator, "Location of Mission Control Center," July 10, 1962.

Source: NASA Historical Reference Collection, History Division, NASA Headquarters, Washington, D.C.

Although the decision to locate a new Manned Spacecraft Center in Houston, Texas had been made in September 1961, it was not decided at that time whether to move the Mission Control Center to Houston or to keep it close to the launch site in Florida, as was being done for Project Mercury. This memorandum records the reasoning behind the decision to move the Mission Control Center to Houston.

MEMORANDUM for Administrator

Subject: Location of Mission Control Center

1.   One of the facilities for which NASA has required funds in our FY 1963 Budget Request is the Mission Control Center. The Mission Control Center would be use to control Gemini and Apollo operations in a similar manner to the control of Mercury operations by the Mercury Control Center. In the FY1963 Budget Request, the Mission Control Center is listed in the section titled "Various Locations".

Considerable thought has been given to the proper geographical location for this most important facility. It is the purpose of this memorandum to advise you of the recommended location and to request your concurrence.

2.   The factors which were determined to be of prime importance in selecting the site are:

a. The Center should be co-located with the Gemini and Apollo Project Offices (so that project personnel would be available for advice and consultation during its construction and during operations. Project personnel would be needed, in case of an emergency situation on-board the spacecraft, to provide immediate and detailed information about the spacecraft.)

b. The Center should be co-located with the Flight Operations Division (so that the Flight Operations personnel can guide the construction of the Facility and have the Facility readily available for training and operation.)

c. The Center should be co-located with the astronauts (so that the facility is readily available for their training, as well as their advice during an operation.)

d. The Center must have good communications (in order to link the Center with the world-wide facilities and forces involved in an operation.)

[2]

e. The Center must be able to keep completely abreast of the status of preparations for an operation (in order to have the information required to make operational decisions.)

3.   The choice of sites narrowed rapidly to either Cape Canaveral or Houston. Both sites have good communication capabilities. At the Cape it is a little easier to keep abreast of preparation for a launch, although good communications between the Cape and Houston reduces this Cape advantage. However, the <u>overriding</u> factor in recommending a specific site is the existing location of the Project personnel, the Astronauts and the Flight Operations personnel at Houston. Therefore, after a careful consideration of pertinent factors and extensive consultation and coordination, it is recommended that Mission Control Center be located in Houston.

4.   It is planned to control the Gemini rendezvous flights, all manned orbital Apollo flights, and all subsequent Gemini and Apollo flights from this Center. Thus the schedule requires the Center to be operational in April, 1964. To keep this schedule, a vigorous development effort must be initiated and maintained. Procurement actions must be undertaken immediately, in which, to provide proper guidance for bidders, it must be stated that this Center will be located in Houston. Therefore, your early concurrence on the Houston location for the Mission Control Center is requested.

[signed]
Robert C. Seamans, Jr.
Associate Administrator

Concurrence____[signed]_____   Date_July 10, 1962
           James E. Webb
           Administrator

cc: M/Holman
       MS/Low
       MSC/Gilruth
       MP-Lilly (handwritten)

Document II-26

Document Title: Memorandum from Donald Hornig, Chairman, Space Vehicle Panel, President's Scientific Advisory Committee, to Dr. Jerome Wiesner, "Summary of Views of Space Vehicle Panel," 11 July 1962.

Document Source: John F. Kennedy Presidential Library, Boston, Massachusetts.

Document II-27

Document Title: Letter to James Webb from Jerome Wiesner, 17 July 1962.

Document Source: John F. Kennedy Presidential Library, Boston, Massachusetts.

Document II-28

Document Title: Letter from James Webb to Jerome Wiesner, 20 August 1962.

Document Source: John F. Kennedy Presidential Library, Boston, Massachusetts.

Document II-29

Document Title: Letter to Jerome Wiesner from James E. Webb with attached Office of Manned Space Flight, NASA, "Manned Lunar Landing Mode Comparison," 24 October 1962

Document Source: Folder 18675, NASA Historical Reference Collection, History Division, NASA Headquarters, Washington, DC.

Document II-30

Document Title: "Memorandum to Dr. [Jerome] Wiesner from McG.B. [McGeorge Bundy]," 7 November 1962.

Document Source: John F. Kennedy Presidential Library, Boston, Massachusetts.

Document II-31

Document Title: Letter from James E. Webb to the President, no date, but November 1962.

Document Source: John F. Kennedy Presidential Library, Boston, Massachusetts.

After its key centers and the Office of Manned Space Flight had agreed that the lunar orbit rendezvous (LOR) approach was NASA's preferred choice for sending people to the Moon, NASA scheduled a 11 July 1962 press conference to announce that choice. However, President Kennedy's science advisor, Jerome Wiesner, and the Space Vehicle Panel of the president's Science Advisory Committee came to the conclusion that the LOR choice was ill-conceived, and insisted that NASA carry out additional studies before finalizing its selection of the LOR mode. Wiesner and the Space Vehicle Panel were influenced in their belief by a senior staff person in Wiesner's office, Nicholas Golovin, who had left NASA at the end of 1961 on unfriendly terms and almost immediately gone to work for Wiesner.

NASA did carry out several additional studies during the summer of 1962, but their results did not change the Agency's thinking. Wiesner and his associates also did not change their position, and the dispute flared into the open as President Kennedy visited the Marshall Space Flight Center on 11 September 1962. Wiesner and Webb got into a somewhat heated discussion in front of the President as the nearby press watched.

After this public conflict, the argument between NASA and the president's science advisor continued through the rest of September and October. Webb on 24 October transmitted what he hoped would be a final comparison of the various ways of accomplishing the lunar landing mission to the White House Office of Science and Technology. It arrived in the midst of the tense week that has come to be known as the Cuban Missile Crisis. In his letter Webb challenged science advisor Wiesner to either accept NASA's decision or force the president to decide between NASA's views and those of Wiesner's office. Despite lingering misgivings, Wiesner did not accept this challenge, and President Kennedy decided that the choice was ultimately NASA's responsibility. The memorandum from McGeorge Bundy, the president's National Security Advisor, to Jerome Wiesner, indicated how the president wanted to bring the controversy to a close. (PSAC is the President's Scientific Advisory Committee.) Webb responded a few days later with the requested letter. NASA announced its final choice of the LOR mode on 7 November 1962.

Document II-26

July 11, 1962

MEMORANDUM FOR

Dr. Jerome B. Wiesner

SUBJECT: Summary of Views of Space Vehicle Panel

The purpose of this memorandum is to summarize for you somewhat informally the Space Vehicle Panel's views as presented during discussions with NASA management on July 6, 1962. A somewhat more detailed Panel report is being prepared for submission to the PSAC.

The Panel has now spent a total of 10 days in meetings trying to understand the Manned Lunar Landing mission and its problems. In particular, we have recently concentrated on the question of which mode of approach offers the previous promise of getting us to the moon (and back) at a very early date and which also contributes most to the development of the national space capability.

One of the early ground rules for the competition was that it would be wise to have three men on the mission, traveling in a "shirt-sleeve environment." No rigorous justification has been made for this requirement, but if it can be met, it is reasonable. However, it is clearly subject to re-examination.

This requirement leads to an estimated weight for a landing and return spacecraft, which, if it is used on a direct ascent mission, approximately size a NOVA rocket. It was realized, and we concur, that the step to such a rocket was too large a single step to be the basis of a sound national program. In fact, it was judged that a rocket of the size of the C-5, which is five times the size of the C-1 and employs previously untried large main engines (F-1) plus large hydrogen-oxygen engines (J-2) in its upper stage, was about as large a development step beyond the C-1 as was reasonable to undertake next. With this conclusion we are also in general agreement.

[2]   Consequently, a second ground rule has been that the mission should start with a C-5. Since the C-5 is incapable of carrying out the originally contemplated direct mission, this condition implies steps such as:

a.   Assembly of components in space, fueling in space, or other means of effectively enlarging the rocket, e.g., EOR.

b.   Cutting the payload by one means or another, e.g., C-5 direct.

c.   Devising more efficient staging arrangements, e.g., LOR. Actually, although rockets larger than C-5 can be built, the prospect of long-range growth solely through bigger and bigger rockets, using bigger and bigger launch facilities, is not an attractive one. Hence, there has been insistent pressure that techniques for orbital assembly, orbital fueling etc., be developed as an integral part of the route to space stations, eventual planetary exploration, and the development of a military space capability. So the present context, such arguments led to the proposal of the EOR mode which Mr. Webb has supported and justified so eloquently in many speeches.

The modes which were analyzed and presented to us were:

1.   EOR, 3-man crew, 14 days total capability (up to 4 on moon), storable propellant for lunar landing and takeoff.

2.   LOR, 2-plus man crew, 14 day total (1 max. on moon), storables.

3.   C-5 direct, 3-man-crew, 10 day (4 on moon), hydrogen-oxygen for landing and takeoff.

In the analysis (as presented) all three can carry out the mission. However, in LOR only two-plus men are involved in the most difficult phases of the mission as compared to three in the others, and the stay time is significantly shorter. It is presumably for this reason, that MSFC has insisted that the choice of LOR be accompanied by the development of a direct ascent C-9 "logistics" vehicle. It was not evident to the Panel that there was any significant difference in the development difficulties which could be anticipated for EOR and LOR. It appeared possible that the direct mode would involve the fewest new developments.

[3]   The analysis of inherent probabilities of mission success, of disasters, and of disasters per success appeared to be carefully done, but offered no basis for distinction within the probable uncertainty of results obtained. As a matter of fact, if one counts critical operations, such as staging and rendezvous, the order of choice from most reliable to least would be: (1) Direct, (2) EOR, and (3) LOR. In addition, a factor which is hard to weigh quantitatively is the fact that all the most difficult operations in the LOR mode take place far from the earth where two of the men have no (earth) abort capability.

And to cost and schedule, it is clear that EOR requires more C-5 vehicles. Hence, if vehicles are, indeed, the pacing item, the EOR approach is more costly and, according to NASA schedule, at least five months slower. It also requires more extensive launch facilities. These conclusions are modified, of course, if a "logistic" vehicle capable of near simultaneous launching is needed to support LOR.

Lastly, the analysis of payload margins also offered no significant basis for choice. LOR is a very ingenious idea which has a fundamental advantage in that the heat shield and re-entry mechanism need not be carried to the lunar surface. However, it must pay the price of carrying an entire life support system, communications system, and navigation (for rendezvous) system. The most recent detailed studies indicate that there is no resultant payload advantage for LOR, and that there is probably a disadvantage if the landing is made more than a few degrees from the moon's equator or the stay time is increased because of the plane change which is introduced as the moon rotates.

The clearest point which came out is that the comparison on all scores involved a mission in which two men stayed on the moon a very short time (LOR) with missions involving three men for longer times. With this background of experiences gained from studies already made, it should be possible to estimate the perturbation on the existing estimates if a two-man capsule were employed for EOR or direct ascent in a very short time, say two to three weeks. It is most strongly recommended that this be done. If possible, optimum trajectories should be considered for each mode since there appears to be no need for lunar orbit in the EOR or Direct modes. It is our guess that EOR will then show a substantial payload margin, and that it will be possible to employ earth storable propellants for the lunar liftoff stages.

[4]   Our further thinking has converged on the following conclusions:

1. LOR is an extremely ingenious but highly specialized mode which does not appear to occupy a central role in the development of a continuing national space program -- at least as compared to orbital fueling of large vehicles.

2. LOR appears to have the largest number of critical operations which must be carried out far from the earth after a period of extreme crew stress.

3. We, therefore, feel that at the present time we would choose the EOR mode with a two-man capsule. It ought to be possible then to gain a substantial weight margin.

4. The history of all ICBM systems has been one of upgrading, even early in their careers. The "official" escape rating of C-5 has already grown from 68,000 lbs. to 90,000 lbs. Consequently, we would press efforts to upgrade C-5 in parallel with its present development. Several possibilities are clearly open. With reasonable success in upgrading, the same (item 3) two-man capsule might be carried on

a Direct Ascent with C-5, using storable propellants for lunar landing and takeoff. Alternatively, hydrogen-oxygen technology may reach the point where it is sufficiently reliable to use for landing. In either case, the way would be open when we are farther down-stream to substitute the Direct Ascent for EOR (although we would not gamble on it alone at the present time). One would thus have alternatives without setting up a full backup program.

5. The LOR is an isolated development from which experience and hardware cannot be so readily transferred to the direct ascent mode.

6. If, nevertheless, LOR is adopted, we feel strongly that the C-5 "Logistic" support vehicle should be carried through in parallel and that studies be promptly instituted on its use as a potential manned vehicle.

[5]

We also have a few other observations:

1. A unmanned lunar orbiter from which the moon's gravitational field can be accurately determined must have very high priority. Otherwise it is impossible to seriously discuss lunar orbits as low as those proposed for LOR.

2. None of the modes should rely on their own reconnaissance of the lunar surface. Unmanned reconnaissance of the lunar surface should have very high priority.

3. Since the mission will be carried out near in time to the other solar flare maximum in 1970, we were distressed at not finding any hard consideration of the radiation problem and its effect on mode selection. This problem requires urgent attention.

Finally, it has been noted that MSFC, MSC, and the Office of Manned Space Flight have all concurred in the choice of the LOR mode. We are impressed by this fact. We can only note that the Panel was originally widely divided in its opinions, but that after hearing and discussing the evidence presented to us, there is no dissent in the Panel to the views presented here.

Donald J. Hornig, Chairman
Space Vehicle Panel

Document II-27

July 17, 1962

Dear Jim:

As I agreed to do during our recent meeting, I am forwarding Don Hornig's informal summary of the Space Vehicle Panel's views as presented to you and your staff at that time. I would also like to take this opportunity to put down, more or less systematically, the substance of my own ideas: these overlap in some respects there of the Panel, and have been at least in part passed on to you verbally during the last two weeks.

First, I think that the final lunar mode choice must provide sufficient payload margins to have a reasonable chance of coping with realistic shielding requirements to meet solar flare radiation hazards which will be approaching their cyclical peak at about the same time that the manned lunar mission will be attempted unless other means are developed to cope with this serious problem. Also, it is possible that exposure to zero-g conditions for the time intervals in this mission may be found to present serious crew problems. Clearly, a mission mode choice at this time must assume that this may turn out to be so, and should, therefore, not exclude sufficient growth capability to offer some chance of dealing with such a difficulty. Accordingly, I feel that both of these potential problem areas should be as thoroughly explored as present scientific knowledge makes possible. It seems to me that a combination of Jim Van Allen's group and of STL could supply a competent team to survey the flare hazard problem.

The matter of which mission mode is most consistent with the main stream of our national space program, and therefore the one most likely to be useful in overtaking and keeping ahead of Soviet space technology, is also one that I believe requires further consideration. For example, if LOR is chosen and the NOVA slipped by two years, then the U.S. will most likely not have an escape capability significantly above 90,000 pounds until 1971 or 1972 at the earliest. With LOR and C-5 Direct, on the other hand, a capability of 160,000 pounds to escape will be available in 1966 or early 1967. Which of these situations, broadly considered, is best for the [2] U.S. posture in space? Similarly, the question of which mode is likely to be most suitable for enhancing our military capabilities in space, if doing so should turn out to be desirable, should be reviewed with care. The Space Vehicle Panel considered this item only casually and, as far as I know, your mode studies had no inputs at all from the DOD in this area. Accordingly, I see a need for an appropriate team of engineers and scientists to explore this area on a time scale compatible with the LOR proposal period.

Thirdly, neither the Space Vehicle Panel nor your staffs, insofar as the data presented to us is made clear, delved adequately into the likely effects of environmental stresses on the crew during the journey, and therefore with the effects on crew capabilities to cope either with the normal or the conceivable emergency conditions to be encountered during various phases of each mode. I would certainly recommend that these matters be reexamined in greater technical depth before you allow final commitment to a mode choice. If you like, we can have the PSAC Bioastronautics Panel assist your staffs in dealing with this job. I might also add that with added time the quantitative analyses of mission mode success probability, and of crew safety, might well be carried to substantially higher level of detail in equipment and crew functional sequencing.

Finally, as has been emphasized by the Space Vehicle Panel, the NASA studies of mission modes did not present the relative advantages and defects of each as a valid basis for comparison principally because some modes involved the use of three men in critical mission phases while others used only two. Payload margins and crew survival probability for the various alternatives are both likely to change substantially, in the Panel's opinion, if the LOR and Direct modes are carried out doing a crew of only two men. Studies along these lines should probably be conducted as direct extensions of previous work at Ames, STL, and MSFC.

I have reported the results of our discussion to the President and assured him that there is ample time to make the additional studies we have agreed upon before the contracts for the lunar landing vehicle need be awarded.

[3]

In closing, you should know that I have instructed the Space Vehicle Panel, as well as my staff, to remain in close touch with the Manned Lunar Landing Program and to be available to you for any purpose you may desire. Since the Panel's future usefulness as to the PSAC, as well as to your agency, will both largely depend on the currency and completeness of their knowledge of the program, I am sure your organizational elements and contractors will do their utmost to be helpful in this regard as they have in the past.

My best wishes to you in your vast and vital undertaking.

Sincerely,

Jerome B. Wiesner

Attachment

Honorable James E. Webb
Administrator
National Aeronautics and Space
    Administration
Washington 25, D.C.

Document II-28

Dr. Jerome B. Wiesner, Chairman
President's Science Advisory Committee
Executive Office Building
Washington 25, D.C.

Dear Jerry:

I was pleased to receive your letter of July 17, 1962, summarizing your thoughts on the lunar mission mode. We are, as I have already told you, conducting several system investigations related to the suggestions of the Space Vehicle Panel. The specific studies currently underway are:

1. An analysis of the North American Aviation studies on the C-5 direct mode, including consideration of a two-man capsule;

2. Continuation of the Space Technology Laboratory effort on a direct ascent utilizing a smaller three-man capsule and a two-man capsule based on the same design approach;

3. Preliminary design by McDonnell Aircraft of a two-man lunar mission capsule.

The results of these studies will be available before the end of September, and their impact on both the C-5 direct and EOR mission profiles will be evaluated by our Office of Manned Space Flight and compared to our current planning of the LOR mode based on the proposal submissions of the Lunar Excursion Module.

We have a continuing concern about the specific items you mentioned. The solar flare radiation problem has been much discussed, and although some data is available, we are keeping in close touch with those performing studies in this area, including Dr. Van Allen. Indeed, data from Dr. Van Allen's latest work is being factored into our radiation hazard effort at Houston. The potential problems from prolonged exposure to [2] zero-g can represent a major problem for any of the modes. Both the mechanization of the spacecraft and the payload requirements of the booster will be seriously affected if artificial gravity is required. As you know, we consider that the Gemini program will be a basic source of information in this area. However, our present feeling is that weightlessness, per se, will not be a limiting problem, and we are not presently compromising the system design to accommodate the generation of artificial gravity.

The implications of the mode decision on our national space capability has been one of our major concerns. We believe that our program provides the basis for a national capability in three major areas:

1. Booster payload capability, both to earth orbit and escape.

2. General spacecraft technology.

3.   Operational capability in space.

It is our considered opinion that the LOR mode, which requires the development of both the C-5 launch vehicle and the rendezvous technique, provides as comprehensive a base of knowledge and experience for application to other possible space programs, either military or civilian, as either the EOR mode or the C-5 direct mode. The decision to delay Nova vehicle is dictated as much by economic considerations, both fiscal and manpower, as by the technical need. The realities of our budget do not allow for the almost simultaneous development of two major launch vehicles. In addition, the redefinition of the Nova for payload capability considerably in excess of the C-5 will, I am convinced, provide us with a better national capability in the long run.

The question of evaluating the effects of environmental stress in the various mission modes is a difficult one. This area is one in which there has been considerable debate, and we are attempting to place the comparative data on a more sound scientific basis. I doubt, however, that this can be accomplished in time to contribute significantly to our present deliberations. Again, it is the considered opinion of our people that there are no significant differences between the modes in the area of stress on the astronauts.

[3]   I appreciate the interest you and your panels continue to show in our program. I have passed your comments and the Report of the Space Vehicle Panel of July 26, 1962, on to Mr. Holmes and Dr. Shea for their consideration. This constructive criticism by eminently qualified men is of tremendous value, and I am looking forward to further discussions with you as the results of our present studies begin to crystallize.

Sincerely,

[signed]
James E. Webb
Administrator

Document II-29

October 24, 1962

Dr. Jerome B. Wiesner
Director
Office of Science and Technology
Executive Office Building
Washington 25, D.C.

Dear Jerry:

In accordance with our conversation, I enclose herewith two copies of our confidential report entitled "Manned Lunar Landing Mode Comparison." My understanding is that you and such members of your staff as you choose will examine this and that you will let me know your views as to whether we should ask for an appointment with the President.

My own view is that we should proceed with the lunar orbit plan, should announce our selection of the contractor for the lunar excursion vehicle, and should play the whole thing in a low key.

If you agree, I would like to get before you any facts, over and above the report, perhaps in a thorough briefing, which you believe you should have in order to put me in position to advise Mr. O'Donnell that neither you nor the Defense Department wishes to interpose a formal objection to the above. In that case, I believe Mr. O'Donnell will not feel it wise to schedule the President's time and that the President will confirm this judgment.

With much appreciation for your assistance, believe me,

Sincerely yours,

James E. Webb
Administrator

Enclosed: two copies of report

Dated October 24, 1962

MANNED LUNAR LANDING MODE COMPARISON

OFFICE OF SYSTEMS

OFFICE OF MANNED SPACE FLIGHT

NATIONAL AERONAUTICS AND SPACE ADMINISTRATION

WASHINGTON 25, D.C.

OCTOBER 24, 1962

TABLE OF CONTENTS

[i]

INTRODUCTION

On July 11, 1962, the National Aeronautics and Space Administration announced its decision to base its studies, planning and procurement for lunar exploration primarily on the lunar orbit rendezvous mode while continuing studies on the earth orbital and direct flight modes, subject to confirmation at the time industry proposals to build the Lunar Excursion Module were finally evaluated. Certain additional studies were also to be completed by that time.

This report summarizes the result of recent studies of the possible application of a 2-man capsule to the earth orbit rendezvous and direct-flight modes. It is concluded that the lunar orbit rendezvous mode is the best choice for achieving a manned lunar landing mission before the end of the decade.

[no page number]

MANNED LUNAR LANDING MODE COMPARISON

One of the major factors in the selection of a mode for the manned lunar landing program is a comparison of the several modes being considered with a series of technical criteria which establish mission feasibility and identify unique considerations. The prime technical criteria are physical realizability [*sic*], mission safety and mission success probability. These technical criteria must be balanced against time and cost to arrive at the mission objectives. The mode selection study of July 30[1] demonstrated that both the Lunar Orbit Rendezvous (LOR) and Earth Orbit Rendezvous (EOR) modes were feasible with adequate weight margins, and that the 3-man C-5 direct ascent mode was undesirable because of small performance margins and high developmental risks. Subsequent studies have been conducted on 2-man capsules which might be used in either the C-5 direct flight or the EOR mode. Results of these studies (summarized in Appendix A) show that the 2-man C-5 direct flight mode is only feasible with cryogenic propulsion systems in all spacecraft stages, or with smaller performance margins than we deem desirable at this point in a program. The 2-man capsule would either increase the weight margins for EOR or allow simpler propulsion systems to be utilized throughout the spacecraft. These improvements are not sufficient to make EOR the preferred mode.

All of the sub-systems required to implement each mode can be developed within the scope of the manned lunar program. Estimates of the degree of developmental difficulty which might be encountered are qualitative, varying with the past experience of those conducting the analysis.

Comparisons of the 2-man lunar mission capsules with the present LOR approach lead to the conclusion that LOR is the preferred mode on the basis of technical simplicity, scheduling and cost considerations.

Mission Safety and Success Probabilities

The Mode Selection Report of July 30 demonstrated only minor differences in mission safety probabilities between EOR and LOR. Although LOR showed a higher probability of mission success than EOR (0.43 for LOR vs. 0.30 for EOR), the number of disasters per mission success for LOR was found to be slightly higher than the EOR figure (0.23 for LOR vs. 0.21 for EOR).

[2] A subsequent analysis was conducted in greater detail, considering the LOR, EOR and C-5 direct flight modes. These studies (summarized in Appendix B) [not provided] show that the overall mission success probability for EOR is 0.30, for C-5 direct 0.36, and for LOR 0.40. The number of disasters per mission success for EOR is 0.38, for C-5 direct 0.46, and the LOR 0.37. In particular, analysis has shown that LOR has the highest safety probability for operations in the vicinity of the moon. We believe that LOR is at least as safe as EOR while still enjoying a considerably higher overall mission success probability.

It could be stated that the LOR mode appears preferable based upon the calculated mission safety and success probabilities. However, the analyses leading to these results involve the estimation of the inherent reliability levels which will be reached by the individual sub-systems, and the detailed mechanization of the particular mode with respect to redundancy. These reliability predictions are not exact during the period when the detailed mechanization of the modes is still evolving. The relative results of both the mission success and safety probability

1. Manned Lunar Landing Program Mode Comparison Report. OMSF, 7/30/62 (CONFIDENTIAL)

calculations are sufficiently sensitive that the assumptions related to equipment performance can change the order of the results.

This leads to the conclusion that the difference between the modes from a mission safety standpoint as known at this point in time is the same order of magnitude as the uncertainty of the analysis. Reliability calculations, per se, are therefore not an adequate basis for choosing among the modes.

<u>Major Differences Between Modes</u>

The major technical differences between the modes lie in the following areas:

1.   Cryogenic vs. storable stages in space;

2.   Weight margin;

3.   Lunar landing configurations;

4.   Rendezvous.

These differences will be discussed in the following paragraphs.

<u>Cryogenic vs. Storable Stages.</u>   The question of cryogenic vs. storable stages in space has two aspects: the reliability of the engines, and the storability of the stage. Most propulsion experts agree that a hypergolic, pressure-fed engine is simpler and, by implication, inherently more reliable than a pumped, regenerative cryogenic engine. Study of engine design confirms this. However, it is also agreed that engines reach inherent reliability only after an extended development program. The RL-10 hydrogen-oxygen engine has been in development for about four years; the storable engines are just starting their development cycle. [3] Hence, at the time of the first lunar missions the cryogenic engine (if the RL-10 could be used in all space stages) might be closer to its inherent reliability than the storable engine. Judgment is again involved. The above arguments nonwithstanding [*sic*], it is believed that storable engines will have reached a higher reliability than cryogenic engines at the time of the initial manned lunar attempts.

Space storability depends on the detailed thermal design of the stage. In space, the cryogenic fuels must be insulated to prevent excessive boil-off, the storable fuels insulated to prevent freezing. On the lunar surface, both cryogenic and storables are subject to boil-off during the lunar day, the problem being more severe for the cryogenics. During the lunar night, the cryogenics are subject to boil-off, the storables to freezing. Either stage will require careful design to insure compatibility with the environment. The problems appear to be more severe for the cryogenic fuels, especially since the storable fuels require an environment more compatible with the rest of the lunar vehicle.

The above considerations have led to the conclusion that storable propellants should be used for the Apollo applications. Storables are also the conservative choice on a performance basis, since it is possible from a weight standpoint to convert from storables to cryogenics at a later date, but the reverse is not true. Only LOR or 2-man EOR are compatible with the choice of storables in all space stages.

Weight Margin. The establishment of a proper weight margin is a factor in the realizability of the C-5 direct modes. Our experience has shown that weight levels for manned space vehicles have grown approximately 25% over initial "hard" estimates. This growth accommodates initial misestimates of hardware weights, equipment additions to increase mission capability, and design changes required by better definition of the environment. As a result of their studies, both Space Technology Laboratories and McDonnell Aircraft Corporation concluded that a 10% weight margin would be adequate to cover initial weight misestimations. Our experience dictates that an additional 15% be included for both increased mission capability and design changes which might result from increased environmental knowledge. The requirement for this increased weight margin does affect the possibility of using a storable return propulsion system for the 2-man C-5 direct mission. Considering all factors, the use of storable return propulsion would not provide sufficient assurance of success for the 2-man C-5 direct mode.

Lunar Landing Configuration. There are important differences in landing configuration between the Lunar Excursion Module (LEM) and the Command Module (CM). Although the landing can be achieved with either module, the LEM can be "optimized" for the lunar operations more readily than the CM which must also accommodate reentry. The main factors are the internal arrangement of the capsules, and the degree of visibility provided the astronauts during the lunar landing phase. Landing the CM (particularly the 2-man version) would undoubtedly require use of television cameras to augment the pilot's field of view.

[4]   In comparing the modes in the vicinity of the moon, both the C-5 direct and the EOR flight configurations must be staged during the terminal descent phase to reduce engine throttling requirements and landing gear loads. This staging requirement and the less desirable module arrangement are the factors in the direct landing mode which must be weighed against the requirement for rendezvous in the LOR mode. Continued study of alternate configurations has indicated that the simplicity of the LOR landing configuration is most desirable for early mission success.

In LOR, the re-entry and flight capsule can be separated from the lunar landing capsule during the course of the development program. Re-entry and flight requirements will affect the mass and moment of inertia of the re-entry and flight capsule, as well as the internal couch arrangement and the pilot displays. Astronaut position during lunar landing will affect the internal arrangement of the lunar landing capsule, and the visibility requirements can profoundly affect both capsule shape and structural integrity.

The industrial firms bidding on the LEM concluded that this separation of function was highly advantageous. (Their comments are summarized in Appendix C.)

Rendezvous. The major concern with respect to the Lunar Orbit Rendezvous arises from the requirement for rendezvous during the return phase of the mission. The mechanization of rendezvous has been studied in detail, and the planned configuration provides a redundant rendezvous capability within the LEM for all equipment failures except those in the main propulsion system. A similar capability exists in the command module. Hence the rendezvous maneuver is backed up with essentially a fourfold redundant mechanization. The duplicate contact, both radar and optical, which can be established between CM and LEM before launch from the lunar surface and maintained until docking, assures adequate relative velocity and position information between the two craft. Although earth tracking will not participate directly in the lunar operation, earth-based antennas will monitor the maneuvers and will aid in certification of the ephemeris of the CM lunar orbit. Studies of the rendezvous implementation, and simulations conducted at NASA centers and industry facilities, have indicated that the rendezvous maneuver is less difficult than the lunar landing. Specifically, the rendezvous in lunar orbit is no more difficult than rendezvous in earth orbit. Indeed, the configuration of the LEM may actually make the lunar rendezvous easier for the astronauts to execute than an earth orbit rendezvous operation involving two C-5 vehicles.

Summary of Technical Considerations. The summation of these considerations leads to the conclusion that the conservative approach to the manned lunar mission dictates the use of a 25% weight margin for any new capsule design and the use of storable engines in space. This conclusion, in conjunction with analyses of the several modes, rules out all modes save LOR and 2-man EOR. After comparison of landing configurations and rendezvous mechanizations, we conclude that the technical trade-offs distinctly favor the LOR mode.

[5]

Human Factors

A factor in the LOR mode which has been frequently mentioned is the effect of mission duration and stress on crew performance during the rendezvous maneuver. Our study of these factors is summarized in Appendix D [not provided], which concludes that "pilot performance is not a limiting factor for either direct or lunar orbit rendezvous missions" based on a survey of the applicable literature and available test data. Another consideration is that the stress which the astronauts will undergo during both lunar landing and earth re-entry is at least equivalent to that experienced during rendezvous. The time constants for both re-entry and landing maneuvers are set by the mission. The time constant for rendezvous is at the astronaut's discretion–several orbits may be used to accomplish the actual docking in an extreme case. Based on these considerations, we conclude that the human factors implications are not significant for purposes of selecting a preferred mode.

National Space Capability

Appendix E [not provided] discusses the implications of the mode choice on National Space Capability. The conclusion is that the only payload requirements exceeding the C-5 escape capability of 90,000 pounds which have presently been defined are for manned space flights, and then only if the EOR mode is utilized for the lunar mission. The operational techniques and the specific hardware developed in either the LOR or EOR mode are similar, with the exception of the tanker and fueling technology required for EOR. LOR does require crew transfer techniques and the development of structural docking mechanisms. The development of fuel transfer techniques which may ultimately be required for a wide class of fluids in space (from earth storables to hydrogen), can be most efficiently carried out in an exploratory development program rather than as an in-line element of the manned lunar landing program. We conclude that, on balance, there is no significant difference between LOR and EOR from a national capabilities viewpoint.

Conclusions

Based on the results of the studies summarized in the Appendices and the above discussion, we conclude that:

(1) The C-5 direct flight mode requires cryogenic fuels and is marginal, even with a two-man capsule;

(2) Both the EOR and LOR modes are feasible;

(3) The reliability differences between LOR and EOR cannot be demonstrated conclusively by analysis at this time; however, LOR does appear to have higher mission probability of success at less risk to the astronauts;

[6]

(4) The capability to design the LEM specifically for the lunar landing, and the desirability of performing the mission with a single C-5 launch are important advantages of the LOR mode, offsetting the lesser problems associated with lunar rendezvous;

(5) Human factor considerations are not significant in the mode selections; the addition of rendezvous to the requirement for lunar landing and re-entry does not add appreciably to crew stress or fatigue, or to the overall hazards of the mission;

(6) Both EOR and LOR provide the basis for projected national space requirements prior to the development of NOVA-class vehicles. The C-5 vehicle capability meets estimated payload requirements. LOR provides experience in personnel transfer between space vehicles as contrasted with fuel transfer in EOR.

The scheduling studies last June demonstrated that the LOR mode could accomplish the lunar mission at least six to fifteen months earlier than the EOR mode. The fact that we have pursued the LOR approach during the intervening months has widened the schedule difference. The reason for the increased schedule difference can be identified in terms of the number of tests which must be completed before a lunar mission can be attempted, and the difference in firing schedules. Because of the requirement for two launchings per mission, EOR can only perform a mission every three months. LOR, on the other hand, can launch a mission every two months, since it requires only a single C-5 launch. We are convinced that the time difference between the EOR and LOR modes is now at the very least one year, and most probably in excess of 18 months.

The original mode selection study indicated that the LOR mode was 10 to 15% less expensive than the EOR approach. This difference arises primarily from the extra cost of launch vehicles for the EOR mode. This conclusion is still valid.

In addition to both schedule and cost advantages, the LOR mode provides the cleanest management structure within the NASA organization. The interface between the spacecraft and launch vehicle is simpler, and the responsibilities of the Manned Spacecraft Center at Houston and the Marshall Space Flight Center at Huntsville are easily defined and provide minimum interfaces between items under development at the two Centers.

In conclusion, the studies conducted since June of this year, and the additional work done within NASA and industry on the LOR approach, have indicated that the LOR mode offers the best opportunity of meeting the U.S. goal of manned lunar landing within this decade.

[Appendices not included]

Document II-30

November 7, 1962

MEMORANDUM TO: Dr. Wiesner

The President's conclusion on the moon method is that he would like a last letter to Webb stating something of the following:

(1) that the choice of a means is obviously a matter of the highest importance rendering the most careful technical reviews;

(2) that serious reservations had been expressed by PSAC panel (with some discussion of its terms of reference and its competence) and that for that reason the President has been glad to know that the matter is being reexamined in NASA;

(3) that the President thinks the time is coming for a final recommendation and relies on Director Webb to review all the arguments and to produce that recommendation.

You may think of other things that should be in such a letter –but what the President has in mind is that we should make Webb feel the responsibility for a definite decision and the importance of weighing all opinions, without trying to make his decision for him.

McG. B.

Document II-31

The President
The White House
Washington 25, D.C.

Dear Mr. President:

In accordance with Dr. Wiesner's suggestion that your file on the Lunar Orbit Rendezvous selection might well contain a letter summarizing the action taken and the reasons therefor [*sic*], the following is set forth.

Early in November, NASA announced that it was reaffirming an earlier tentative decision of July 1962 which selected Lunar Orbit Rendezvous as the mode this nation would adopt in accomplishing the first manned lunar landing. A detailed report on the numerous studies that led to this decision has been submitted previously to the office of your Scientific Advisor and is attached for your file. [not included]

The decision to adopt the Lunar Orbit Rendezvous mode was based on major systems and engineering studies which involved over a million man-hours of effort on the part of government and contractor personnel. Despite the very extensive study efforts, however, we are dealing with a matter that cannot be conclusively proved before the fact, and in the final analysis the decision has been based upon the judgment of our most competent engineers and scientists who evaluated the studies and who are experienced in this field. Because we are dealing in an area where judgment is an important factor, we have held several meetings with Dr. Wiesner and his staff to ensure that their views and opinions could be given most careful consideration. These meetings were constructive and assisted in sharpening the critical factors which would determine the final decision.

Following are the most important conclusions which led to the decision to adopt Lunar Orbit Rendezvous:

    a.   Using the Advanced Saturn C-5, the largest booster which will be available in this decade, the Lunar Orbit Rendezvous (LOR) and Earth Orbit Rendezvous (EOR) approaches are technically feasible and both can be conducted with three-man crews. The direct flight mode would require cryogenic fuels for the lunar landing (which we consider less reliable), and would be marginal with regard to weight limitations even using a two-man capsule.

[2]

    b.   By adopting LOR, the mission can be accomplished at least one year earlier in comparison with the EOR mode.

c. The cost will be 10% to 15% less than for the EOR approach. If it were feasible, a two-man direct mode could be conducted at approximately the same cost as LOR.

d. Although our studies show a slight advantage for LOR in terms of reliability, there was not sufficient difference in the safety and mission success calculations for each mode to consider that this factor could significantly influence mode selection.

e. Touch down on the Moon is the most difficult maneuver of the entire mission. Since the LOR mode is the only one which includes a vehicle which will be used for the lunar landing without having to consider earth re-entry problems, it will be possible to design the lunar excursion vehicle to maximize the probability of success in the lunar touch down operation.

f. The techniques and the spacecraft which will be developed for EOR and LOR are similar with the exception that refueling technology is required for the earth orbital mode and a crew transfer for the lunar orbital mode. On balance, there appears to be no significant difference between these modes from a national capability viewpoint. The third mode, a two-man direct ascent, would not provide an opportunity for testing space rendezvous and docking techniques.

If future missions are undertaken which would require a longer stay on the lunar surface, it is probable that a lunar logistics system would be required regardless of the mode chosen for the initial landings. We are well along in the study phase of this supporting system and believe it holds promise as a backup mode for the LOR in a later time period. Successful development of this backup potential depends heavily on whether sufficient weight reductions can be made in the spacecraft system to permit a direct ascent flight using the Advanced Saturn C-5.

The decision on the mode to be used for the lunar landing had to be made at this time in order to maintain our schedules, which aim at a landing attempt in late 1967. We are confident that the decision is the correct one, but recognize that in any matter in which judgment plays an important role, we must be prepared to change our concepts in the light of convincing new evidence. For this reason, we are conducting the program in a manner which will permit us to react promptly to any new factors introduced by the new information we are gaining every day.

[3] We intend to drive forward vigorously on every segment of the manned lunar landing program. To do so, we have marshaled a major segment of this country's finest resources for the effort. We have working with us a group of outstanding industrial firms. Additionally, we are being supported by many of our finest universities as well as by the Department of Defense and other government agencies. Within NASA, the three field centers you visited this past September–the Marshall Space Flight Center under Dr. von Braun; the Manned Spacecraft Center under Dr. Gilruth; and the Launch Operations Center under Dr. Debus–devote their full capabilities to this task. We believe that this team, under the leadership of Mr. Holmes, the Director of Manned Space Flight, provides a cohesive network of research and development resources which can achieve the objective you have established.

Respectfully yours,

[signed]
James E. Webb
Administrator

Document II-32

Document Title: Letter from James E. Webb, NASA Administrator, to President John F. Kennedy, 29 October 1962.

Document Source: Folder 18675, NASA Historical Reference Collection, History Division, NASA Headquarters, Washington, DC.

Document II-33

Document Title: Transcript of Presidential Meeting in the Cabinet Room of the White House, 21 November 1962.

Document Source: *http://history.nasa.gov/JFK-Webbconv/* (accessed 29 January 2007).

Document II-34

Document Title: "Memorandum to President from Jerome Wiesner Re: Acceleration of the Manned Lunar Landing Program," 10 January 1963.

Document Source: John F. Kennedy Presidential Library, Boston, Massachusetts.

During his 11 though 12 September visit to the three NASA installations most involved in Project Apollo, there were suggestions made to President Kennedy (apparently by manned spaceflight head Brainerd Holmes) that the first lunar landing, at that point tentatively scheduled for late 1967, might actually be accomplished up to a year earlier if additional funds were provided to the Apollo program. Holmes and NASA Administrator James Webb disagreed on the wisdom of seeking additional funds from Congress, but Webb told the president in a 29 October letter that with additional funding it might indeed be possible to accelerate the Apollo program.

However, Webb did not press aggressively enough for such an increase to satisfy Holmes. Tensions between him and Webb had been festering since at least August 1962, when Holmes was featured on the cover of Time *magazine and labeled "Apollo czar." Another* Time *story appeared on 19 November, this time suggesting that the program was in trouble and badly needed the extra funds. Holmes was the apparent source of the story.*

Following Webb's 29 October letter, the president had asked his Bureau of the Budget to take a careful look at the financial and schedule aspects of Apollo. The results of that review were sent to the president by budget director David Bell on 13 November (Volume I, Document

III-13). The White House called a 21 November meeting in the Cabinet room to try to understand exactly what was going on at NASA. Like some other White House meetings during the Kennedy presidency, this meeting was secretly recorded; the John F. Kennedy Presidential Library released a copy of the recording in 2001, and space historian Dwayne Day later prepared a transcript of the tape.[1]

As he left the 21 November meeting, President Kennedy asked James Webb to prepare a letter summarizing Webb's view on the appropriate position that the White House should take on Apollo and the NASA program overall. Webb did so, and sent Kennedy the letter on 30 November (Volume I, Document III-14).

Kennedy's interest in accelerating the date for the first lunar landing continued even after the 21 November meeting and Webb's response. During an 8 December visit to Los Alamos National Laboratory, he asked his science advisor Jerome Wiesner to look again into the potential for a lunar landing earlier than NASA was planning.

Document II-32

NATIONAL AERONAUTICS AND SPACE ADMINISTRATION

Washington 25, D.C.

October 29, 1962

Office of the Administrator

The President
The White House
Washington 25, D.C.

Dear Mr. President:

In accordance with the request you made during your recent tour of selected NASA installations, a preliminary analysis has been completed to determine the feasibility and the resources implications of accelerating the manned lunar landing program in order to establish a target date for the first landing in late 1966, one year earlier than the present target.

The late 1967 target date is based on a vigorous and driving effort, but does not represent a crash program. A late 1966 target would require a crash, high-risk effort. The nature of a development program such as the manned lunar landing, however, makes the possibility of achieving target dates set this far in advance no better than fifty-fifty. In contrast, the odds that we can accomplish the landing within this decade are excellent. You might, therefore, think of this matter of target dates as one in which we fix a date which is difficult, but not

---

1. The tape of the meeting can be found at John F. Kennedy Library, President's Office files, Presidential recordings collection tape #63.

impossible to attain. We schedule the work against this date and thereby insure a driving effort. However, until later in the development cycle, target dates cannot be viewed as certain forecasts of when the mission will be accomplished.

The depth of this special analysis on a late 1966 target date is in no way comparable with the detailed analysis which formed the basis of the operating plan for a late 1967 target date. A definitive study of time and resources requirements for the many sequential events involved in the accomplishment of this mission by late 1966 would necessitate a much more intensive and detailed review by the NASA headquarters and field centers, our prime contractors and the principle subcontractors. However, the preliminary analysis which follows permits a gross evaluation of the possibilities currently available.

Current Plan - Mission in late 1967

The NASA operating plan of $3.7 billion for FY 1963 and the requested budget level of $6.2 billion for FY 1964 are aimed at the target date of late 1967 for the manned lunar landing. Within these budget levels, the amounts planned for the manned lunar landing are $2.4 billion in FY 1963 and $4.2 billion in FY 1964. These funds include $2.0 billion and $3.4 billion respectively for propulsion systems, launch vehicles, spacecraft, facilities, and flight operations; and $.4 billion and $.8 billion respectively for necessary supporting effort in unmanned scientific investigations, advanced technology, and improvements to the tracking network. These funds do not cover the personnel costs of NASA employees or amounts for the operation of the NASA centers for which the totals are $446 million in FY 1963 and $579 million in FY 1964.

The major program segments are funded at the following rate under this plan:

	(In Billions)	
	1963	1964
Spacecraft and Flight Missions	$ .7	$1.5
Development of Launch Vehicle and Propulsion Systems	.7	1.0
Facilities, Launch Operations, Integration and Checkout, Systems Engineering and Aerospace Medicine	.6	.9
	$2.0	$3.4

Alternative Plan - Mission in mid-1967

In preparing an operating plan for FY 1963 based on Congressional appropriations, it was estimated from detailed studies that the first landing might be possible six months earlier if an additional $427 million were available early in FY 1963. Thus, the late 1967 target date in the current plan is six months later than a date possible with optimum FY 1963 funding. The additional funds in FY 1963 would provide (1) heavier contractor effort on launch vehicles at Chrysler,

Boeing, North American Aviation, and Douglas (2) procurement of hardware associated with Apollo spacecraft which is now deferred until FY 1964, and (3) accomplishment of the Gemini rendezvous mission nine months earlier than the current plan with resulting benefit to Apollo. The revised program would then be as follows:

	(In Billions)	
	1963	1964
Spacecraft and Flight Missions	$ .9	$1.5
Development of Launch Vehicle and Propulsion Systems	.8	1.0
Facilities, Launch Operations, Integration and Checkout, Systems Engineering and Aerospace Medicine	.7	.9
	$2.4	$3.4

Analysis indicates that if a mid-1967 target date were approved and the additional $427 million were made available in a FY 1963 supplemental appropriation in the early days of the 88th Congress, NASA could revise its target date to mid-1967. NASA would also require deficiency authority to cover total agency operations until receipt of the supplemental, since it would be necessary to commence operation at a higher level immediately in order to attain this schedule.

Alternative Plan - Mission in late 1966

In analyzing the actions which would be necessary to establish a target date for manned lunar landing in late 1966, the following major milestone changes would have to be accomplished relative to the current plan:

1. Advance the first manned Apollo command module flight on the Saturn launch vehicle six months to November 1964 from May 1965.

2. Move the first manned Apollo command and service module flight on the Saturn C-1B launch vehicle forward seven months to October 1965 from May 1966.

3. Accelerate the first Advanced Saturn development flight seven months to September 1965 from April 1966.

4. Change the first manned Apollo command and service module flight on Advanced Saturn 12 months to June 1966 from June 1967.

If these new milestones could be achieved, the first manned lunar landing would be late 1966. To achieve these milestone changes, a number of departures would have to be made from the present development plan. (1) The extremely tight schedule would require heavy sub-system effort very early in the development cycle and would leave no room for any significant test or flight failures. (2) Parallel testing of all stages and an increased rate of development on the Advanced Saturn first stage would be necessary. (3) Concurrent development would have to be initiated on alternative components and subsystems to give better assurance that schedules could be met. (4) The current contractor overtime rate and amount of double and triple shifting would be markedly increased and extensive overtime and multiple shifting would be necessary. (5) A crash contractor manpower buildup and heavy NASA effort would be required to reschedule and execute the new plan.

The runout cost from FY 1965 through FY 1967 for the late 1966 target date is estimated to be 10-15% higher than the funds required for a late 1967 date. The funds required in FY 1963 and FY 1964 to meet this schedule are approximately $900 million and $800 million more respectively than the current FY 1963 availability and FY 1964 budget request. The total would be distributed as follows:

	(In Billions)	
	1963	1964
Spacecraft and Flight Missions	$1.1	$1.9
Development of Launch Vehicle and Propulsion Systems	1.0	1.2
Facilities, Launch Operations, Integration and Checkout, Systems Engineering and Aerospace Medicine	.8	1.1
	$2.9	$4.2

Summary

On the basis of our current analysis, we believe that we can maintain the late 1967 target date for the manned lunar landing with $3.7 billion in FY 1963 funds and $6.2 billion in FY 1964. A budget increase of $427 million to $4.1 billion in FY 1963 and $6.2 billion in FY 1964 is required for a mid-1967 target date; and total resources of $4.6 billion in FY 1963 and $7.0 billion in FY 1964 are required for a late 1966 target date.

Let me emphasize again the preliminary nature of our conclusion that a target date of late 1966 could be established for the manned lunar landing with the indicated funding levels. This conclusion is not based on detailed programmatic plans. With this qualification, however, we are prepared to place the manned lunar landing program on an all-out crash basis aimed at the 1966 target date if you should decide this is in the national interest.

Respectfully yours,

[Signed James E. Webb]
James E. Webb
Administrator

Document II-33

Present at the meeting:
President John F. Kennedy
James Webb, NASA Administrator
Dr. Jerome Wiesner, Special Assistant to the President for Science and Technology
Edward Welsh, Executive Secretary, National Aeronautics and Space Council
David E. Bell, Director, Bureau of the Budget
Dr. Hugh Dryden, Deputy Administrator, NASA
Dr. Robert Seamans, Associate Administrator, NASA
Dr. Brainerd Holmes, Associate Administrator for Manned Space Flight, NASA
Elmer Staats, Deputy Director, Bureau of the Budget
Willis H. Shapley, Deputy Division Chief, Military Division, Bureau of the Budget

President Kennedy: What I understand, it is a question of whether we need four hundred million dollars more to maintain our present schedule, is that correct?

James Webb: Well, it's very hard to say what our present schedule is. I think the easiest way to ... to understand what has happened is to say that we started out after you made the decision in May to come forward with a driving program. We used the best information we had and we settled on late '67 or early '68 as the landing date. We wanted to have some leeway within the decade. Now this was a target date-we recognized we might have some slippage. We had some financial estimates at that time, which have proved to be too small, that the...the increased cost estimated by the contractor is partly because each of them has added to the cost that he submitted on his contract proposals to us. And second, we have added requirements to each of these vehicles.

[2] Now the combination of the increased cost now estimated by the contractors, plus our own increased knowledge as result of about a year's work, has led us first to confirm the fact that the late '67 or '68 date is a good date for us to have as our target date.

Second, that to accomplish that now and to run that kind of program that you want run, we have to go through a real strong, vigorous management period to shake down these things. Obviously you can make it an Apollo that would include a tremendous number of things that would cost a lot of money and probably are not necessary. On the other hand, you could make one that was too marginal and that we would not want to entrust [unknown]. We have to find a place in between as we go along with these projects.

[Additional discussion not included]

[3] **President Kennedy:** Now, let me just get back to this, what is your ... uh, your view is we oughta spend this four hundred forty million?

Brainerd Holmes: My view is that if can strictly spend, it would accelerate the Apollo schedule, yes, sir. Let me say I was very ... I oughta add that I'm very sorry about this ... I have no disagreement with Mr. Webb ... he says with the policy, oh, I think my job is to say how fast I think we can go for what dollars.

James Webb: Well, I think it's fair to say one other thing, Mr. President, that after your visit when you were saying how close this was, the speech you made. I think Brainerd and Wernher von Braun and Gilruth all felt, "We've got to find out how fast we can move here. The President wants to move." So they went to the contractors and said, "How fast can you move, boys, if money were not a limit?" Now, this sort of got cranked up into a feeling that this money was going to be made available, that a policy decision had already been [4] made to ask for the supplemental. And I think, to a certain extent, then, the magazines like *Time*, they picked this up in order to make a controversy.

President Kennedy: Well, as I at least hear, it wasn't so much that we wanted to speed it up as it was how much we were gonna slip ... you don't like that word, but that's what we're talking about.

James Webb: Well, no, sir, I don't think so. The reason I don't like the word is that those schedules were never approved by Dryden, Seamans, or me. They were not officially scheduled flights in the Agency. But they were tagged as the schedule in order to ask the contractors how much they could do, for Brainerd to ... to really get moving. When he came into this program.

President Kennedy: Are you saying that these dates were not ever set?

James Webb: They were not officially set by me or Dryden or Seamans....

President Kennedy: Were set....

James Webb: We were waiting to determine what the Budget Director was going to give us on the '64 budget to definitely set our dates. Because this made a big difference.

President Kennedy: You mean, what part of '67 was never set?

James Webb: Well, the '67 date has been set. And we're going to make it.

President Kennedy: What part of '67 was never set, is that correct?

James Webb: We talk all the time of late '67 or early '68.

Hugh Dryden: You never set a month....

James Webb: That's right.

President Kennedy: So now, when we talk about four hundred million, well now, tell me what's happened here. You had a date in your mind which unless you get the four hundred million you feel that's a good chance it'll go back to the end of until about six months. And, ah, Mr. Webb says that there was [n]ever a date in '67.

Brainerd Holmes: What's happened is this, I think. First of all, we didn't have a definitized program; we had to decide what size booster it would be, for instance, at the very end last year. So as soon as we could, we'd definitize all of the elements of the program but then still until one decides the mission which you are going to go you couldn't [5] interweave these schedules, you couldn't decide *really* what kind of a program you're gonna have and what kind of funding you're gonna have. So once we assumed what the mission would be in June going with this LOR, and I am not here talking about one mission versus another, but *a* mission to justify schedules. So I'm gonna put down all the details hundreds of schedules that interweave, we came up with costs associated with those schedules, and these costs and dates came out to be this first schedule which appears to be a not unreasonable schedule done on a crash basis. Further than that, just as Mr. Webb has said, the contractor estimates were low; our estimates of what they required were low; all that information was pouring in. We put the two together to go versus this time with these dollars that we had as estimates, it came out that we were short in Fiscal '63. So we didn't know that before that.

James Webb: So then we started talking to the Budget Director.

[Many people talking all at once.]

Unknown speaker: August and September.

Hugh Dryden [?]: Mr. President, may I say one more thing which I think you should keep in mind. Practically every program at this point that we've ever had has grown by a factor between two and three in cost from the beginning to end. The Mercury was what? About two and a half ... three. I think you have to bear in mind that these program costs are still going to grow. I'm not sure that Jim or Brainerd will agree with me. On any schedule you pick, you're going to have to face increasing cost year after year, in my opinion. And if we find some trouble, which undoubtedly we're going to find, intangibles stretch and go up in cost. And depending on the level you select now, the rate in which the costs are going to accelerate on you in the future years will be determined.

Unknown speaker: Mr. President....

Unknown speaker: Compared to future years....

Hugh Dryden: I think we learned a great deal from Mercury. As far as the so-called increase in Mercury. For the [honest] definition of what Mercury included. We started an estimate of what the McDonnell contract would be to build a capsule. But Mercury involves not only the capsule, it involves a worldwide tracking network; it involves ground support equipment for handling the capsule on the ground, check-out equipment. And we were learning with Mercury we kept adding new elements, new revisions to the cost, so that it did wind up Mercury cost five hundred million dollars all total. Two dollars and a quarter for each person in the United States, seventy-five cents a year for three years, if you want to look at it that way. And there's no question that it cost a large sum. Now in this analysis, the number of man-hours and years is inexpensive; again working out these numbers, it looks fantastic compared with the corresponding figures on Mercury.

[6] **James Webb:** We know a great deal more.

Unknown speaker: I think this is a much sounder basis. I would be surprised if the cost went up by three...

Robert Seamans: I would be surprised if it went up more than sixty percent.

Unknown speaker: But that's still a lot of money!

James Webb: Well, let me make a statement on that I have made to the Budget Director. You remember when I first talked to you about this program, the first statement I made to Congress was that the lunar program would cost between twenty and forty billion dollars. Now I am able to say right now it's going to be under the twenty billion, under the lower limit that we used. The question is how rapidly do you spend the money and...and how efficiently you manage this so as to get the most possible for the money. This can be speeded up at the expense of...of certain things which I outlined in this letter to you. It can be slowed up if, a year from now, we find that we don't have to proceeded at this basis. But this is a good, sound, solid program that would keep all of the governmental agencies and the contractors and the rest moving ahead. But we're prepared to move if you really want to put it on a crash basis.

President Kennedy: Do you put.... Do you put this program.... Do you think this program is the top-priority of the Agency?

James Webb: No, sir, I do not. I think it is *one* of the top-priority programs, but I think it's very important to recognize here...and that you have found what you could do with a rocket as you could find how you could get out beyond the Earth's atmosphere and into space and make measurements. Several scientific disciplines that are the very powerful and being to converge on this area.

President Kennedy: Jim, I think it is the top priority. I think we ought to have that very clear. Some of these other programs can slip six months, or nine months, and nothing strategic is gonna happen, it's gonna... But this is important for political reasons, international political reasons. This is, whether we like it or not, in a sense a race. If we get second to the Moon, it's nice, but it's like being second any time. So that if we're second by six months, because we didn't give it the kind of priority, then of course that would be very serious. So I think we have to take the view that this is the top priority with us.

James Webb: But the environment of space is where you are going to operate Apollo and where you are going to do the landing.

[7] **President Kennedy:** Look, I know all these other things and the satellite and the communications and weather and all, they're all desirable, but they can wait.

James Webb: I'm not putting those.... I am talking now about the scientific program to understand the space environment within which you got to fly Apollo and make a landing on the Moon.

President Kennedy: Wait a minute-is that saying that the lunar program to land the man on the Moon is the top priority of the Agency, is it?

Unknown speaker: And the science that goes with it....

Robert Seamans: Well, yes, if you add that, the science that is necessary....

President Kennedy: The science.... Going to the Moon is the top-priority project. Now, there are a lot of related scientific information and developments that will come from that which are important. But the whole thrust of the Agency, in my opinion, is the lunar program. The rest of it can wait six or nine months.

James Webb: The trouble ... Jerry is holding up his hand.... Let me say one thing, then maybe you want to [unknown] the thing that troubles me here about making such a flat statement as that is, number one, there are real unknowns as to whether man can live under the weightless condition and you'll ever make the lunar landing. This is one kind of political vulnerability I'd like to avoid such a flat commitment to. If you say you failed on your number-one priority, this is something to think about. Now, the second point is that as we can go out and make measurements in space by being physically able to get there, the scientific work feeds the technology and the engineers begin to make better spacecraft. That gives you better instruments and a better chance to go out to learn more. Now right all through our universities some of the brilliant able scientists are recognizing this and beginning to get into this area and you are generating here on a national basis an intellectual effort of the highest order of magnitude that I've seen develop in this country in the years I've been fooling around with national policy. Now, to them, there is a real question. The people that are going to furnish the brainwork, the real brainwork, on which the future space power of this nation for twenty-five or a hundred years are going be to made, have got some doubts about it and....

President Kennedy: Doubts about what, with this program?

James Webb: As to whether the actual landing on the Moon is what you call the highest priority.

President Kennedy: What do they think is the highest priority?

[8] **James Webb:** They think the highest priority is to understand the environment and ... and the areas of the laws of nature that operate out there as they apply backwards into space. You can say it this way. I think Jerry ought to talk on this rather than me, but the scientists in the nuclear field have penetrated right into the most minute areas of the nucleus and the subparticles of the nucleus. Now here, out in the universe, you've got the same general kind of a structure, but you can do it on a massive universal scale.

President Kennedy: I agree that we're interested in this, but we can wait six months on all of it.

James Webb: But you have to use that information to....

President Kennedy: Yeah, but only as that information directly applies to the program. Jim, I think we've gotta have that....

[Unintelligible.]

Jerome Wiesner: [Unintelligible – 'If you got enough time?"] Mr. President, I don't think Jim understands some of the scientific problems that are associated with landing on the Moon and this is what Dave Bell was trying to say and what I'm trying to say. We don't know a damn thing about the surface of the Moon. And we're making the wildest guesses about how we're going to land on the Moon and we could get a terrible disaster from putting something down on the surface of the Moon that's very different than we think it is. And the scientific programs that find us that information have to have the highest priority. But they are associated with the lunar program. The scientific programs that aren't associated with the lunar program can have any priority we please to give 'em.

Unknown speaker: That's consistent with what the President was saying.

Robert Seamans: Yeah. Could I just say that I agree with what you say, Jerry, that we must gather a wide variety of scientific data in order to carry out the lunar mission. For example, we must know what conditions we'll find on the lunar surface. That's the reason that we are proceeding with Centaur in order to get the Surveyor unmanned spacecraft to the Moon in time that it could affect the design of the Apollo.

President Kennedy: The other thing is I would certainly not favor spending six or seven billion dollars to find out about space no matter how on the schedule we're doing. I would spread it out over a five- or ten-year period. But we can spend it on.... Why are we spending seven million dollars on getting fresh water from saltwater, when we're spending seven billion dollars to find out about space? Obviously, you wouldn't put it on that priority except for the defense implications. And the second point is the fact that the Soviet Union has made this a test of the system. So that's why we're doing it. So I think we've got to take the view that this is the key program. The rest of this ... we can find out all about [8] it, but there's a lot of things we can find out about; we need to find out about cancer and everything else.

James Webb: But you see, when you talk about this, it's very hard to draw a line between what....

President Kennedy: Everything that we do ought to really be tied into getting onto the Moon ahead of the Russians.

James Webb: Why can't it be tied to preeminence in space, which are your own....

President Kennedy: Because, by God, we keep, we've been telling everybody we're preeminent in space for five years and nobody believes it because they have the booster and the satellite. We know all about the number of satellites we put up, two or three times the number of the Soviet Union ... we're ahead scientifically. It's like that instrument you got up at Stanford which is costing us a hundred and twenty-five million dollars and everybody tells me that we're the number one in the world. And what is it? I can't think what it is.

Interruption from multiple unknown speakers: The linear accelerator.

President Kennedy: I'm sorry, that's wonderful, but nobody knows anything about it!

James Webb: Let me say it slightly different. The advanced Saturn is eighty-five times as powerful as the Atlas. Now we are building a tremendous giant rocket with an index number of eighty-five if you give me Atlas one. Now, the Russians have had a booster that'll lift fourteen thousand pounds into orbit. They've been very efficient and capable in it. The kinds of things I'm talking about that give you preeminence in space are what permits you to make either that Russian booster or the advanced Saturn better than any other. A range of progress possible it is so much different [unknown].

President Kennedy: The only.... We're not going to settle the four hundred million this morning. I want to take a look closely at what Dave Bell.... But I do think we ought get it, you know, really clear that the policy ought to be that this is *the* top-priority program of the Agency, and one of the two things, except for defense, the top priority of the United States government. I think that that is the position we ought to take. Now, this may not change anything about that schedule, but at least we ought to be clear, otherwise we shouldn't be spending this kind of money because I'm not that interested in space. I think it's good; I think we ought to know about it; we're ready to spend reasonable amounts of money. But we're talking about these fantastic expenditures which wreck our budget and all these other domestic programs and the only justification for it, in my opinion, to do it in this time or fashion, is because we hope to beat them and demonstrate that starting behind, as we did by a couple years, by God, we passed them.

[9] **James Webb:** I'd like to have more time to talk about that because there is a wide public sentiment coming along in this country for preeminence in space.

President Kennedy: If you're trying to prove preeminence, this is the way to prove your preeminence.

James Webb: It's not if you've got an advanced Saturn rocket ... [unintelligible].

President Kennedy: We do have to talk about this. Because I think if this affects in any way our sort of allocation of resources and all the rest, then it is a substantive question and I think we've got to get it clarified. I'd like to have you tell me in a brief ... you write me a letter, your views. I'm not sure that we're far apart. I think all these programs which contribute to the lunar program are ... come within, or contribute significantly or really in a sense ... let's put it this way, are *essential*, put it that way...*are essential* to the success of the lunar program, are justified. Those that are not essential to the lunar program, that help contribute over a broad spectrum to our preeminence in space, are secondary. That's my feeling.

James Webb: All right, then let me say this: if I go out and say that this is the number-one priority and that everything else must give way to it, I'm going to lose an important element of support for your program and for your administration.

President Kennedy [interrupting]: By who? Who? What people? Who?

James Webb: By a large number of people.

President Kennedy: Who? Who?

James Webb: Well, particularly the brainy people in industry and in the universities who are looking at a solid base.

President Kennedy: But they're not going to pay the kind of money to get that position that we are [who are] spending it. I say the only reason you can justify spending this tremendous ... why spend five or six billion dollars a year when all these other programs are starving to death?

James Webb: Because in Berlin you spent six billion a year adding to your military budget because the Russians acted the way they did. And I have some feeling that you might not have been as successful on Cuba if we hadn't flown John Glenn and demonstrated we had a real overall technical capability here.

President Kennedy: We agree. That's why we wanna put this program.... That's the dramatic evidence that we're preeminent in space.

[10] **James Webb:** But we didn't put him on the Moon ... [unintelligible].

Unknown speakers: [Unintelligible] ... we did what we needed to do.

David Bell: I think, Mr. President, that you're not as far apart as this sounds. Because the budget that they have submitted, 464....

President Kennedy: I know we're not far apart, I'm sure, and the budget we may not be apart at all. But I do think at least we're in words somewhat apart. And I'd like to get those words just the same.

James Webb: It's, it's perfectly fine. I think....

President Kennedy: How about you writing me and telling me how you assign these priorities. And perhaps I could write you my own....

James Webb: But I do think it ... it certainly doesn't hurt us to have this *Time* article that shows we are really going ahead with the program. I don't think that hurts the Agency; I don't think it hurts at all. You have tried several times to say that's number one. But I also think that as Administrator, I've got to take a little broader view of all the budgets here including those that are [unintelligible] appropriation in the Congress. I don't think we've got to use precisely the same word.

Robert Seamans: Could I state my view on this? I believe that we proceeded on Mercury, and we're now proceeding on Gemini and Apollo as the number-one program in NASA. It has a DX priority. Nothing else has a DX priority.

James Webb: And recommended four-point-seven billion funds for it for 1962! That's a....

Robert Seamans: At the same time, when you say something has a top priority, in my view it doesn't mean that you completely emasculate everything else if you run into budget problems on the Apollo and the Gemini. Because you could very rapidly completely eliminate you[r] meteorological program, your communications program, and so on. If you took that to too great of an extreme....

James Webb: And the advanced technology on which military power is going to be based.

Hugh Dryden: Mr. President, I think this is the issue. Suppose Apollo has an overrun of five hundred million dollars, to reprogram five hundred million dollars for the rest of the space program would just throw the whole thing all away. And I think this is the worry in Jim's mind about top priority.

[11] **President Kennedy:** Listen, I think in the letter you ought to mention how the other programs which the Agency is carrying out tie into the lunar program, and what their connection is, and how essential they are to the target dates we're talking about, and if they are only indirectly related, what their contribution is to the general and specific things [unknown-possibly "we're doing"] in space. Thank you very much.

[Kennedy gets up to leave the room.]

[Rest of discussion not included].

Document II- 34

January 10, 1963

MEMORANDUM FOR THE PRESIDENT

Subject: Acceleration of the Manned Lunar Landing Program

On the recent trip to Los Alamos I agreed to look further into the possibility of speeding up the manned lunar program. We have done this and are convinced that approximately 100 million dollars of the previously discussed 326 million dollar supplementary could have a very important effect on the schedule, but that to do so it would have to be available in the very near future.

The November 28, 1962 NASA letter to the Director of the Bureau of the Budget specified Fiscal Year 1963 supplemental appropriations which could be utilized to accelerate the Apollo Program. The data contained in this letter, as well an [*sic*] additional information obtained subsequently from the NASA Office of Manned Space Flight, suggest the following estimates of possible schedule changes and associated funding requirements–assuming that the additional funds would become available for obligation beginning January 1, 1963:

	Without Supplemental For FY 1963	With FY 1963 Supplemental Available Jan. 1. 1963
Apollo Spacecraft Available at AMR for the first manned flight	November 1964	September1964
First Manned Flight -C-1	February 1965	December 1964
First C-1B Launch	August 1965	April 1965
First C-5 Launch	March 1966	October 1965
First Lunar Landing Attempt	October 1967	May 1967

[2] Supplemental funds required for the above:

	(In millions of dollars)
Apollo	$125.2
C-1	23.4
C-1B	27. 2
C-5	103.8
Construction of Facilities	47.1
Total–	$326.7

I have reviewed the arguments contained in NASA's November 28 letter, as well as the general technical situation in the over-all Manned Lunar Landing Program. My principal conclusions are as follows:

1. Although some doubts are present that additional funds at this time will expedite the Apollo system proper, there is no doubt that the date of the first lunar landing attempt can be accelerated only if C-5 rocket availability is advanced.

2. The C-5 has been under development for a longer time than any other major system in the Manned Lunar Landing Program and estimates for what additional funds could or could not do for it are, therefore, more likely to be realistic

than for other systems. The estimate that an additional $103.8 million, available beginning January 1, 1963, could advance the date of the first C-5 launch by some five months, appears well founded. This conclusion is reinforced by the fact that the Marshall Space Flight Center has been relieved in recent month of responsibilities for several vehicles and may be expected, therefore, to exercise effective technical and managerial control over the C-5 development and its funding.

3. In view of the many engineering uncertainties with respect to the eventual reliability of systems as complex as the C-5, any advancement in the date of first launch will enable more extensive testing, and therefore earlier elimination of design inadequacies and faster growth in the reliability of the vehicle to be used for the first manned lunar landing attempt.

4. Although it cannot be argued at this time that an advance in the C-5 launch schedule will necessarily result in an earlier date for the first lunar landing attempt, it is quite certain that time lost now on the C-5 cannot be regained later. Accordingly, if future successes in the spacecraft development program should promise earlier availability of the Apollo system, it would be possible to take advantage of this only if earlier availability of the C-5 has been previously assured.

[3] In view of the above, it appears to be important to proceed immediately with the acceleration of the C-5 development and to provide the $103.8 million in FY 1963 supplemental appropriations as soon as possible. As I point out earlier, this step would only be effective if it can be taken very soon. If authorized these funds would be used by the NASA as follows:

	(In millions of dollars)
First Stage (S-1C at Boeing)	$ 25.8
Second Stage (S-II and F-1, J-2 engines at NAA)	$68.0
Guidance, ground support, etc.	10.0
Total—	$103.8

Jerome B. Wiesner

Document II- 35

Document Title: "Letter to James Webb from Vannevar Bush," 11 April 1963.

Document Source: Folder 18675, NASA Historical Reference Collection, History Division, NASA Headquarters, Washington, DC.

Dr. Vannevar Bush was the head of the World War II Office of Scientific Research and Development and in 1945 authored the seminal report "Science: the Endless Frontier," which was the charter for the post-war involvement of the Federal Government in the support of research. He was thus for many years one of the leaders of the U.S. scientific community. Bush and Webb knew each other well, dating back to their work together during the Truman administration. This 1963 letter expressed Bush's misgivings about the commitment to sending Americans to the Moon; during 1963, similar criticism of the lunar landing program emerged from within the scientific community and from those who have preferred that money being spent on space would instead be allocated to other social priorities.

11 April 1963

Mr. James E. Webb
Administrator
National Aeronautics and
Space Administration
3200 Idaho Avenue
Washington 16, D.C.

Dear Jim:

I have pondered the subject of this letter for a long time. Now I think I should write it out for you.

Early in the space program, I testified to a Senate Committee. As was my duty, I gave my considered judgment, critical of the program.

But since then I have made no public statements. This has been due to a number of reasons. First, I have felt that, being nearly alone in criticism, I would be regarded as an old fogy who could not appreciate the efforts of young men. More important, I hesitated to oppose a program ordered by the President after full advice.

You and I understand this well. During the war I took the position strongly that my job was to transmit to the President the best scientific advice available, and to carry out his orders loyally and without question. I know you have this point of view intensely, for I have seen you respond to the President's wishes many times when it involved hardship or risk on your part.

A part of this attitude has been involved in my relations at M.I.T. There I have taken the point of view that, when duly constituted government called for aid on a program, which aid only M.I.T. because of its unique position could supply, there was a duty to respond, and that my personal estimate of the advisability of the program should not interfere with it doing so.

Now the scene is changing. There are an increasing number of critical editorials and articles. It could change abruptly.

[2]   No great program of this sort can proceed without occasional disasters. We have been lucky, and very careful thus far. But, some one of these days, a couple of young attractive men are going to be killed, with the eyes of millions upon them. Worse, they may be caught in space to die, still talking to us, who are helpless to aid them.

It is often said the public is fickle. It is also said that there is, unfortunately, a measure of bull fight complex in the peoples' following of flights. I mean something deeper than either of these. The American public often fails for a long time utterly to grasp a situation, and, when it finally does, its reversal of attitude can be sobering or terrifying. A prime example is the prohibition experiment. A better example is the attitude in 1916. At first unconcerned about a war in Europe, electing a president who would keep us out of war, it suddenly reversed itself and plunged in to halt the Kaiser.

Thus far the public attitude has been one of national pride, enthusiasm over a good show, wonder at the accomplishments of science. It has been uninformed on, or has chosen to ignore, the adverse aspects. It can change its attitude in a month's time. When it does it can be utterly unreasonable, and it can be cruel. I do not know when this will occur; I do not even know that it will occur. But I fear it.

Now do not misunderstand me. If I were sure the program were sound I would applaud your driving it forward in spite of any amount of criticism, or any amount of personal risk. And I know you well enough to be sure that is just what you would do. The difficulty is that the program, as it has been built up, is not sound.

The sad fact is that the program is more expensive than the country can now afford; its results, while interesting, are secondary to our national welfare. Moreover the situation is one on which the President, and the people, cannot possibly have adequate unbiased advice.

Our national budget has been unbalanced for many years. We have a serious problem in the outflow of gold. Our taxes are so high that they impede commercial vigor, [3] and our rate of growth is hence low compared to recovered nations with which we compete. We have by no means halted the wage-price spiral. We have genuine danger of inflation. The strength of the dollar is questioned. This calls for vigorous, courageous measures to avert disaster. I will not comment here on the nature of the measures I would advocate. But I believe it is crystal-clear that this is no time at which to make enormous - and unnecessary - expenditures.

While the scientific results of an Apollo program would be real, I do not think that anyone would attempt to justify an expenditure of 40 or 50 billion dollars to obtain them. The Academy report was weak on this point. The justifications given are of quite a different nature. First, it is said we are in a race and our national prestige is at stake. I believe we can disregard the matter of race. I do not know whether there is a race to the moon or not; I doubt it. But national prestige is a far more subtle thing than this. The courageous, and well conceived, way in which the President handled the threat of missiles in Cuba advanced our national prestige far more than a dozen trips to the moon. Having a large number of devoted Americans working unselfishly in undeveloped countries is far more impressive than mere technical excellence. We can advance our prestige by many means, but this way is immature in its concept.

I hear that the program will be justified by its by-products. We might get a billion dollars worth of benefit that way. I doubt if it would exceed this.

I also hear, and some of my good friends advance this argument in all seriousness, that the program is inspiring the youth of the country, and spurring us on to great accomplishment. It inspires youth all right, and it also misleads them as to want is really worthwhile in scientific effort. In fact, it misleads them as to what science is. It is well to inspire a child, and the use of fairy tales is legitimate as this is done. But when a child becomes a man he should be inspired to judge and choose soundly, to avoid being carried away by mass enthusiasms, to understand the tough world in which he will play his part, technically and economically. It is wrong to inspire him to have an exciting adventure at his neighbors' expense.

I also hear that this is a form of pump priming, that it is a shot in the arm to industry. Anyone who still [4] believes in pump priming should read again about the 1929 debacle, and the sorry following years when we long failed to emerge from the resulting depression.

In other words, I hear excuses and rationalization, not cold analysis.

A most serious point about this whole affair is that the people of this country, and the President with his appalling responsibilities, cannot possibly have adequate sound scientific, engineering, and economic advice regarding it. This is due to the very vast size of the project. Nearly every man who could speak with authority on the subject has a conflict of interest. Now do not misinterpret this to mean that the scientists of the country are all feeding at the trough, and so selfish they would subordinate their judgment as to what is true to what is advantageous to them. There are some of these of course. I even hear rumors of artificial pressures being brought to bear on individuals and companies to ensure conformity, but such rumors always float about when there are great undertakings, and in any very large organization there are always subordinates of little sense, as we have seen exemplified often.

I do not mean this sort of thing at all. The scientist or engineer in a university or a company is in a difficult quandary. He may honestly believe the program as a whole is highly fallacious. But it has been decided upon at the top level of government. It is supported by his colleagues, many of whom have enthusiasm. His organization has been urged to participate. Who is he to stand out against this powerful trend? He consoles himself by Cromwell's admonition, "I beseech you, bethink you that you may be mistaken", and sides along with the crowd.

We pride ourselves that, in this democracy, the minority has opportunity to speak. Yet it takes courage and an unusual sort of detachment, to stand against a nearly unanimous opinion of friends and colleagues, and to risk one's reputation in a futile attempt to halt an avalanche. I know this whole program has never been evaluated objectively by an adequately informed and disinterested group, and I fear it never will be.

The whole problem is in the hands of the President, and he has many problems on his mind today. He leans on [5] you, to steer him straight. As we now go there is danger ahead for the program, and danger to his prestige. I hope he will alter his handling of this whole affair before a balky Congress, or public opinion, forces him to do so.

You and I think alike on the tough problem of the relation to the President of a man on his team; we have discussed it a number of times. Your creed and mine depends on two main principles. First, the President on a problem should have the best advice this country can afford, with differences of opinion where there are any faithfully transmitted, and it is the job of the man who reports to him in an area to see that he gets it. Second, when the President, with full grasp of a subject and thus advised, makes a decision and issues an order, it is the job of his lieutenant on a subject to carry it out loyally and effectively whether or not he agrees with it. This is especially true in time of war, but it is also true of a key subject in time of cold war. The only exception would be a situation in which the lieutenant's disagreement was so complete that he found himself unable to perform well, in which case he should step aside, and, incidentally, say nothing.

I believe the President could alter his attitude and his orders without a reversal of form which would embarrass him.

I know what I think should be done. As a part of lowering taxes and putting our national financial affairs in order, we should have the sense to cut; back severely, on our rate of expenditure on space. As a corollary they could remove all dates from plans for a trip to the moon; in fact, he could announce that no date will be set, and no decision made to go to the moon, until many preliminary experiments and analyses have rendered the situation far more clear than it is today. He could lop off, without regret, marginal programs that cannot be soundly supported, and continue only where results are clearly attainable and worthwhile, in weather and communication satellites for example. He could order experimentation concerned with long space flights confined to those features which are clearly central and determining, avoiding hardware except where it is necessary. Then, after a year or so, the entire program could be reviewed through a professional dis-[6]interested board, made up of scientists, engineers, economists, financial men, and men with keen judgment of public attitudes here and abroad.

By so doing, he could reduce the rate of current expenditure at a time when any such cutback would help him in his tax program. He could avoid commitment to vast expenditures until such time as economic prosperity justified them, and thorough analysis had shown them to be warranted. And I believe he could do this without real damage to an overall logical sound space program.

There were times when you and I both reported to the President, and we worked closely together in so doing, even when we did not totally agree. Today you are still doing so while I have dropped out of the active picture.

But, whatever you do, and however the program may work out, you have my best wishes, my deep personal regard.

Cordially yours,

[Signed]

V. Bush

Document II-36

Document Title: John Disher and Del Tischler, "Apollo Cost and Schedule Evaluation," 28 September 1963.

Document Source: Folder 18675, NASA Historical Reference Collection, History Division, NASA Headquarters, Washington, DC.

Document II-37

Document Title: Clyde B. Bothmer, "Minutes of Management Council Meeting, October 29, 1963, in Washington, D.C." 31 October 1963.

Document Source: Folder 18675, NASA Historical Reference Collection, History Division, NASA Headquarters, Washington, DC.

Document II-38

Document Title: George E. Mueller, Deputy Associate Administrator for Manned Space Flight, NASA, to the Directors of the Manned Spacecraft Center, Launch Operations Center, and Marshall Space Flight Center, "Revised Manned Space Flight Schedule," 31 October 1963.

Document Source: Folder 18675, NASA Historical Reference Collection, History Division, NASA Headquarters, Washington, DC.

When George Mueller joined NASA in September 1963, replacing D. Brainerd Holmes, he was concerned that the existing schedule for Project Apollo would not result in an initial lunar landing before the end of the decade, the goal that had been set by President Kennedy. Mueller asked two veteran NASA engineers, John Disher and Del Tischler, to conduct a two-week assessment of the situation. The two presented their findings to Mueller on 28 September. Their findings, as presented in the excerpts from their briefing included here, were troubling. After he had heard their briefing, Mueller took the two to present it to NASA Associate Administrator Robert Seamans. According to a 22 August 1976 hand-written note by NASA Historian Eugene Emme on the copy of the briefing sent to the NASA History Division, Seamans asked that all copies of the Disher/Tischler briefing be "withdrawn"; some accounts suggest that because the findings were so at variance with the official schedule that Seamans suggested that all copies of the briefing be destroyed. This briefing was a catalyst to Mueller's rethinking of the Apollo schedule that led to the "all-up" testing concept, in which the Saturn 1B and Saturn V launch vehicles would be tested with all of their stages active, rather than the stage-by-stage testing that was then the plan.

The "all-up" approach was first announced by Mueller at a management meeting on 29 October 1963; more details were provided in a teletyped memorandum two days later.

Mueller's approach was strongly resisted by both the Marshall Space Flight Center and the Manned Spacecraft Center, but Mueller, who soon after this memorandum was written became Associate Administrator for Manned Space Flight, was a strong-willed individual whose views eventually prevailed. The "all-up" decision is regarded by many as key to the United States being able to reach the Moon "before this decade is out."

Document II-36

APOLLO SCHEDULE AND COST EVALUATION

OBJECTIVES

- PROVIDE A REALISTIC ESTIMATE WITH MODERATE CONFIDENCE (~50%) OF THE EARLIEST DATE FOR THE FIRST LUNAR LANDING ATTEMPT
- PROVIDE A CORRESPONDING PROGRAM COST ESTIMATE
- ASSESS TIME AND COST INCREASES REQUIRED TO RAISE CONFIDENCE TO A HIGH LEVEL (~90%)
- ESTIMATE ADDITIONAL COSTS OF WORK WHICH COULD INCREASE CONFIDENCE LEVEL OF EARLIEST DATE

APOLLO SCHEDULE AND COST EVALUATION GROUND RULES

- NO BASIC CHANGE IN TECHNICAL CONCEPT OR APPROACH
- PERSONNEL CEILING FIXED AT FY 65 LEVEL
- FY 64 AND FY 65 FUNDING AT GUIDLINE LEVELS
- FY 66 AND SUBSEQUENT R&D FUNDING CEILING OF $3.00 BILLION PER YEAR (INCLUDES ADVANCED PROGRAM)
- CONTINUATION OF DX PRIORITY
- NORMAL PROCUREMENT LEAD TIMES
- TWO SCHEDULED FLIGHTS REQUIRED FOR ACCOMPLISHMENT OF EACH FLIGHT MISSION
- MAXIMUM FREQUENCY FOR MANNED FLIGHTS OF FOUR PER YEAR
- INFLATION FACTORS NOT CONSIDERED
  <u>CONCLUSIONS AND RECOMMENDATIONS</u>

- If funding constraints assumed herein prevail, lunar landing cannot likely be attained within the decade with acceptable risk.
- First attempt to land men on moon is likely about late 1971 under study guideline funding and constraints.
- Program cost through initial lunar landing attempt will approximate 24 billion dollars (R&D Direct only)
- Progress on program inadequate to provide schedule associated with 90% confidence.
- Projection of lunar landing attempt on early manned Saturn V unrealistic in terms of probable technical problems.

- Late manned spacecraft availability, plus resource diversion to Saturn I from IB and V would strongly indicate cancellation of Saturn I manned flights.
- Funding increases of $400M to $700M in FY 65 and $700M to $1100M each in FY 66 and 67 could accelerate the program by one to two years with a decrease in total program cost.

Document II-37

October 31, 1963

MEMORANDUM FOR DISTRUBUTION LIST

Subject: Management Council Meeting, October 29, 1963 in Washington, D.C.

The subject meeting convened at 8:30 a.m. All members were present with the exception of Mr. Elms.

The Program Review portion of the meeting was conducted from 8:30 a.m. - 2:40 p.m. as scheduled, and the action minutes for that portion of the session are attached. [not included]

The following additional items were considered outside the Program Review.

1. Dr. Mueller stressed the importance of a philosophical approach to meeting schedules which minimizes "dead-end" testing, and maximized "all-up" systems flight tests. He also said the philosophy should include obtaining complete systems at the Cape (thus minimize "re-building" at the Cape), and scheduling both delivery and launch dates. (In explaining "dead-end" testing he referred to tests involving components or systems that will not fly operationally without major modification.)

[remainder of minutes not included]

Document II-38

NATIONAL AERONAUTICS AND SPACE ADMINISTRATION
WASHINGTON 25, D.C.

IN REPLY REFER TO:
M-C M 9330.186

OCT 31, 1963 [stamped]

TO: Director, Manned Spacecraft Center
                        Houston 1, Texas
                        Director, Launch Operations Center
                        Cocoa Beach, Florida
                        Director, Marshall Space Flight Center
                        Huntsville, Alabama
FROM: Deputy Associate Administrator for Manned
                        Space Flight

SUBJECT: Revised Manned Space Flight Schedule

Recent schedule and budget reviews have resulted in a deletion of the Saturn I manned flight program and realignment of schedules and flight mission assignments on the Saturn IB and Saturn V programs. It is my desire at this time to plan a flight schedule which has a good probability of being met or exceeded. Accordingly, I am proposing that a flight schedule such as shown in Figure 1 [not included], with slight adjustments as required to prevent "stack-up," be accepted as the official launch schedule. Contractor schedules for spacecraft and launch vehicle deliveries should be as shown in Figure 2.[not included] This would allow actual flights to take place several months earlier than the official schedule. The period after checkout at the Cape and prior to the official launch date should be designated the "Space Vehicle Acceptance" period.

With regard to flight missions for Saturn 1, MSC [the Manned Spacecraft Center] should indicate when they will be in a position to propose a firm mission and spacecraft configuration for SA-10. MSFC [The Marshall Space Flight Center] should indicate the cost of a meteoroid payload for that flight. SA-6 through SA-9 missions should remain as presently defined.

[2] It is my desire that "all-up" spacecraft and launch vehicle flights be made as early as possible in the program. To this end, SA-201 and 501 should utilize all live stages and should carry complete spacecraft for their respective missions. SA-501 and 502 missions should be reentry tests of the spacecraft at lunar return velocity. It is recognized that the Saturn IB flights will have CM/SM [Command Module/Service Module] and CM/SM/LEM [Command Module/ Service Module/Lunar Excursion Module] configurations.

Mission planning should consider that two successful flights would be made prior to a manned flight. Thus, 203 could conceivably be the first manned Apollo flight. However, the official schedule would show the first manned flight as 207, with flights 203-206 designated as "man-rating" flights. A similar philosophy would apply to Saturn V for "man-rating" flights with 507 shown as the first manned flight,

I would like your assessment of the proposed schedule, including any effect on resource requirements in FY 1964, 1965 and run-out by November 11, 1963. My goal is to have an official schedule reflecting the philosophy outlined here by November 25, 1963.

George M. Low [signed for]
George E. Mueller
Deputy Associate Administrator
for Manned Space Flight

Enclosures:
Figure 1
Figure 2

Document II-39

Document Title: Letter to Representative Albert Thomas from President John F. Kennedy, 23 September 1963.

Document Source: John F. Kennedy Presidential Library, Boston, Massachusetts.

Document II-40

Document Title: Memorandum from Jerome B. Wiesner to the President, "The US Proposal for a Joint US-USSR Lunar Program," 29 October 1963.

Document Source: John F. Kennedy Presidential Library, Boston, Massachusetts.

Speaking before the United Nations General Assembly on 20 September 1963, President Kennedy suggested that the United States and the Soviet Union might cooperate in a "joint mission to the moon." Given that Project Apollo originated in 1961 in a desire to beat the Soviet Union to the Moon, and that the president had reiterated in 1962 that this was his primary motivation for funding the undertaking at a high level, this proposal came as a surprise to many. But President Kennedy had been interested in space cooperation with the Soviet Union since he had come to the White House, and according to his top advisor, Theodore Sorenson, he would have preferred to cooperate with the Soviet Union rather than compete with them. The reaction to the 12 April 1961 Soviet launch of Yuri Gagarin, however, suggested to Kennedy that competition was his only option. When he suggested cooperation to Soviet Premier Nikita Khrushchev at a June 1961 summit meeting, Khrushchev rebuffed the idea, and this reinforced Kennedy's belief that competition was the only path open to him. By September 1963, Kennedy tried once again to raise the possibility of cooperation.

Kennedy's proposal angered those in the Congress who had been strongest in support of Apollo as a competitive undertaking. In a letter to Representative Albert Thomas, who chaired the House Appropriations Subcommittee that controlled the NASA budget, Kennedy explained how his proposal was consistent with a strong Apollo effort. There were also a number of suggestions that Kennedy's proposal was primarily a public relations move, or a way of gracefully withdrawing from the Moon race after the U.S. success during the Cuban Missile Crisis. Balanced against such suggestions are a memorandum from Kennedy's science advisor, Jerome Wiesner, suggesting a detailed approach to cooperation, and a 12 November 1963 National Security Action Memorandum signed by Kennedy asking NASA to take the lead in developing an approach to U.S.-Soviet cooperation in missions to the Moon (Volume II, Document I-42).

Document II-39

THE WHITE HOUSE

September 23, 1963

Dear Al:

I am very glad to respond to your letter of September 21 and to state my position on the relation between our great current space effort and my proposal at the United Nations for increased cooperation with the Russians in this field. In my view an energetic continuation of our strong space effort is essential, and the need for this effort is, if anything, increased by our intent to work for increasing cooperation if the Soviet Government proves willing.

As you know, the idea of cooperation in space is not new. My statement of our willingness to cooperate in a moon shot was an extension of a policy developed as long ago as 1958 on a bipartisan basis, with particular leadership from Vice President Johnson, who was then the Senate Majority Leader. The American purpose of cooperation in space was stated by the Congress in the National Aeronautics and Space Act of 1958, and reaffirmed in my Inaugural Address in 1961. Our specific interest in cooperation with the Soviet Union, as the other nation with a major present capability in space, was indicated to me by Chairman Khrushchev in Vienna in the middle of 1961, and reaffirmed in my letter to him of March 7, 1962, which was made public at the time. As I then said, discussion of cooperation would undoubtedly show us "possibilities for substantive scientific and technical cooperation in manned and unmanned space investigations." So my statement in the United Nations is a direct development of policy long held by the United States government.

Our repeated efforts of cooperation with the Soviet Union have so far produced only limited responses and results. We have an agreement to exchange certain information in such limited fields as weather observation and passive communications, and technical discussions of other limited possibilities are going forward. But as I said in July of this year, there are a good many barriers of suspicion and fear to be broken down before we can have major progress in this field. Yet our intent remains: to do our part to bring those barriers down.

At the same time, as no one knows better than you, the United States in the last five years has made a steadily growing national effort in space. On May 25, 1961, I proposed to the Congress and the nation a major expansion of this effort, and I particularly emphasized as a target the achievement of a manned lunar landing in the decade of the 60's. I stated that this would be a task requiring great effort and very large expenditures' the Congress and the nation approved this goal; we have been on our way ever since. In a larger sense this is not merely an effort to put a man on the moon; it is a means and a stimulus for all the advances in technology, in understanding and in experience, which can move us forward toward man's mastery of space.

This great national effort and this steadily stated readiness to cooperate with others are not in conflict. They are mutually supporting elements of a single policy. We do not make our space effort with the narrow purpose of national aggrandizement. We make it so that the United States may have a leading and honorable role in mankind's peaceful conquest of space. It is this great effort which permits us now to offer increased cooperation with no suspicion anywhere that we speak from weakness. And in the same way, our readiness to cooperate with others enlarged the international meaning of our own peaceful American program in space.

In my judgment, therefore, our renewed and extended purpose of cooperation, so far from offering any excuse for slackening or weakness in our space effort, is one reason the more for moving ahead with the great program to which we have been committed as a country for more than two years.

So the position of the United States is clear. If cooperation is possible, we mean to cooperate, and we shall do so from a position made strong and solid by our national effort in space. If cooperation is not possible—and as realists we must plan for this contingency too—then the same strong national effort will serve all free men's interest in space, and protect us also against possible hazards to our national security. So let us press on.

Let me thank you again for this opportunity of expressing my views.

With warm personal regards,

Sincerely,

/s/
John F. Kennedy

The Honorable Albert Thomas
House of Representatives
Washington, D/C.

Document II-40

October 29, 1963

MEMORANDUM FOR

The President

Subject: The US Proposal for a Joint US-USSR Lunar Program

I believe that Premier Khrushchev's statement of October 26 that the USSR does not plan to land a man on the moon gives us a unique opportunity to follow through on your UN proposal for a joint US-USSR program in a way that will not only be in accord with U.S. objectives for peaceful cooperation if accepted by the USSR, but will also decisively dispel the doubts that have existed in the Congress and the press about the sincerity and feasibility of the proposal itself. Specifically, I would propose a joint program in which the USSR provides unmanned exploratory and logistic support for the U.S. Apollo manned landing. I believe such a program would utilize the combined resources of US and USSR in a technically practical manner and might, in view of Premier Khrushchev's statement, be politically attractive to him.

The manned lunar program encompasses much more that the manned landing vehicle itself. The PSAC space panels have consistently emphasized the importance of the unmanned lunar exploration program to develop technical information about the lunar surface. This information appears critical to a successful manned landing. The U.S. unmanned program hinges around the Surveyor program which at best is a marginal one. At the present time its estimated payload has dropped to 65 pounds and its schedule is unreliable. The Soviet Union, however, apparently has a substantial capability at this time for this type of exploratory mission. A joint program which would use this capability would be very valuable to us.

More directly involved with the manned landing itself is a vehicle and spacecraft for placing a large stock of supplies and equipment at [2] the site of the manned landing. NASA and the PSAC space panels all agree that the 24-48 hours staytime provided by Apollo does not permit the astronauts to conduct significant scientific exploration. It is agreed that to make Apollo a useful scientific endeavor an additional 7000 pounds of equipment and supplies must be landed at [t]his site to permit him 5 to 7 days of useful scientific exploration before he returns to earth. This logistic support requires another large vehicle and spacecraft to be available on about the same time schedule as Apollo. The U.S. development program to provide this capability has not yet been initiated. If the Soviet Union could be convinced that the logistical support was indeed an essential and integral part of the manned landing and persuaded to provide this support system, the resulting program would again result in an effective use of combined resources. The Apollo program would remain a purely U.S. technical program without modification of present plans. A Russian could easily be included as a member of the landing team without complicating the engineering effort. In addition, the proposal would have the practical value of minimizing requirements for complicated joint engineering projects and launching operations and would emphasize the exchange of plans, information and possibly people.

If we assume that Premier Khrushchev is telling the truth (and I believe that he is), this proposal will give the USSR the opportunity of sharing the credit for a successful lunar mission without incurring major expenditures much beyond those that they probably plan to undertake as a part of their present space program. By not including joint engineering and launching activities, the proposal minimizes the security impact on the USSR that undoubtedly acts as a restraint on joint activities because of the close association of the Soviet space and military missile programs.

It is true that the above proposal assumes that the USSR would be willing to follow the now well established U.S. operational plan for manned lunar exploration. This did seem reasonable as long as it appeared likely that Russia has a well developed program of her own. Now, however, Premier Khrushchev's statement, whether it is true or not, makes such a proposal by the United States reasonable from [3] every standpoint. The proposal now not only offers a program which truly enhances the manned lunar exploration effort while leaving the Apollo program intact, but also one which ought to be acceptable to the USSR.

It might be extremely advantageous for you to publicly offer this plan to the USSR as a specific proposal for a joint program, formulated in the light of Premier Khrushchev's statement and designed to effectively combine the resources of both countries. The effectiveness of the offer would be enhanced if it were made while Khrushchev's statement is still fresh in the mind of the public. If the proposal is accepted we will have established a practical basis for cooperative program. If it is rejected we will have demonstrated our desire for peaceful cooperation and the sincerity of our original proposal.

If you believe this proposal has merit, I suggest that you request that NASA prepare as soon as possible a specific plan along these lines for your consideration.

Jerome B. Wiesner

Document II-41

Document Title: Memorandum to Robert R. Gilruth, Director, Manned Spacecraft Center from Verne C. Fryklund, Jr., Acting Director, Manned Space Sciences Division, Office of Space Sciences, NASA Headquarters, "Scientific Guidelines for the Project Apollo," 8 October 1963.

Source: Folder 18675, NASA Historical Reference Collection, History Division, NASA Headquarters, Washington, DC.

The National Academy of Sciences Space Science Board held a 1962 summer study on the campus of the University of Iowa to address all issues of space science (Volume V, Document I-22). Two working groups, one on lunar and planetary exploration and the other on the scientific role of humans in space, addressed the scientific aspects of the Apollo missions. The latter group recommended that astronomical observations from the Moon be relegated to later flights. These views were adopted by NASA as the basic scientific guidelines for early Apollo flights to the Moon. The Apollo Logistics Support System was a proposed extension of the basic Apollo capabilities to enable more extensive exploration of the Moon; it was never developed.

[stamped "OCT 8 1963"]

To: Director, Manned Spacecraft Center
                       Attention: Robert R. Gilruth

From: SH/Acting Director, Manned Space Sciences Division
                       Office of Space Sciences

Subject: Scientific Guidelines for the Apollo Project

Reference: Scientific Guidelines for Apollo Logistic Support System

The following general and preliminary guidelines are being used by the Office of Space Sciences and should be used by the Manned Spacecraft Center in the consideration of scientific investigations to be done by means of the Apollo project. As defined herein, Apollo refers only to the approved project with restricted stay time. The guidelines for the Apollo Logistic Support Systems (ALSS) previously Sent to MSFC are enclosed for your information. [not included]

These guidelines, unless modified in writing, should be followed in the preparation of your plans.

The Office of Space Sciences has established that the primary scientific objective of the Apollo project is acquisition of comprehensive data about the Moon. The steps that resulted in this decision are, I am sure, of interest to you. The Office of Space Sciences formed the Ad Hoc Working Group on Apollo Experiments and Training at the request of the Office of Manned Space Flight in March 1962. This working group issued a draft report (the Sonett Report) on July 6, 1962, that was immediately made available to the various subcommittees of the Iowa Summer Study, which was sponsored by the National Academy of Sciences. The reviews of the Subcommittees were extensive and though the general conclusions of the Sonett Report were accepted, the final report of the Iowa Summer Study ("A Review of Space Research" National Academy of Sciences-National Research Council Publication 1079) recommends that the scope of Apollo scientific investigations be more restricted that those proposed in the Sonnet [2] Report. The Officer of Space Sciences has concurred with the recommendations of the Academy and they are incorporated in those guidelines.

As the moon itself is the primary subject of observation, it follows that the structure of the moon's surface, gross body properties and large-scale measurements of physical and chemical characteristics, and observation of whatever phenomena may occur at the actual surface will be prime scientific objectives.

The guidelines that follow are intended to place some specific constraints on studies in keeping with the paragraphs above.

Guidelines:

1.   The principal scientific activity will be observation of the moon.

2.   The use of the moon as a platform for making astronomical and other observations is, in general, not a function of the Apollo project. (See ALSS Guidelines for additional comment on this subject.)

3.   We may assume that Apollo activities will be largely reconnaissance in nature. The intention is to acquire knowledge of as large an area as possible, and by as simple a means as possible, in the limited time available.

4.   The three functional scientific activities listed in order of decreasing importance, will be:

> a. Comprehensive observation of lunar phenomena;
>
> b. Collection of representative samples; and
>
> c. Replacement of monitoring equipment.

5.   Quantitative analytical chemistry will not be done on the moon by the Apollo project.

6.   Qualitative and semi-quantitative analytical chemistry should be planned for, though there is not yet an obvious need for such data to be obtained on the moon by the Apollo project.

[3]   7. Seismometers, scintillometers, and magnetometers, among other instruments intended to determine the physical properties of the moon, will be studied for inclusion in payloads.

8. Sample collecting, for geological and biological purposes, will be an important activity and possible special equipment requirements should be studied.

Verne C. Fryklund, Jr.

Document II-42

Document Title: Bureau of the Budget, "Special Space Review," Draft Report, 29 November 1963.

Source: Lyndon B. Johnson Presidential Library, Austin, Texas.

This draft report summarizes a 1963 "special review" of the U.S. space program that began under President John F. Kennedy and continued after his assassination under President Lyndon Johnson. This report suggests that consideration was being given, at least within the Bureau of the Budget staff, to "backing off from the manned lunar landing goal." How seriously this possibility was taken at this point in time is not clear from the historical record. This report was a draft; there were no recommendations in it, since they would have had to come from senior officials. It is not clear whether a final version of this report, with such recommendations, was ever prepared, or whether any thought of not following through on the goal that had been set by President Kennedy was quickly abandoned after his death.

SPECIAL SPACE REVIEW

DRAFT REPORT

Bureau of the Budget
November 29, 1963

[2]

SPECIAL SPACE REVIEW - 1965 BUDGET
INTRODUCTION

This report summarizes the principal results of the special review of the goals, nature, and pace of the space programs in the light of 1964 and 1965 budget pressures, which has been undertaken by the Bureau of the Budget in conjunction with the 1965 budget review and in response to the decisions at the October 8, 1963, meeting of the Secretary of Defense, the Administrator of the National Aeronautics and Space Administration, and the Special Assistant to the President for National Security Affairs.

The purposes of the review have been to consider the goals of the space programs and the minimum requirements of a national program to achieve them, and to identify the policy questions, alternatives, and other major issues to be dealt with in the 1965 budget decisions.

The draft report has been prepared by Bureau of the Budget staff in consultation with senior representatives of NASA and the Department of Defense, and others, on the basis of information submitted by the agencies and discussions with agency officials in 1965 budget reviews now in process. The views expressed in the draft are necessarily those of the Bureau of the Budget staff. It is expected that the recommendations of the Director of the Bureau of the Budget, and the concurrences or[hand-written] differing views of the Secretary of Defense; the Administrator, NASA; the Director of OST; and the Executive Secretary of the NASC will be inserted as appropriate after discussion.

Section I covering the Manned Lunar Landing Program and Section II covering Military Space Objectives (including the proposed manned earth orbit experiments) are attached. Problems relating to communications satellites, meteorology, geodesy, space sciences, and technological development are being handled separately.

[3]
[budget table omitted]

[4]

I. MANNED LUNAR LANDING PROGRAM

A. STATEMENT OF PRESENT GOAL:

To attempt to achieve a manned lunar landing and return by the end of this decade, on a high priority but not "crash" basis, with prudent regard for the safety of the astronauts, for the principal purposes of (a) demonstrating an important space achievement ahead of the USSR, (b) serving as a focus for technological developments necessary for other space objectives and having potential significance for national defense, and (c) acquiring useful scientific and other data to the extent feasible.

B. QUESTIONS, DISCUSSION, ALTERNATIVES, AND RECOMMENDATIONS:

1. Should consideration be given at this time to backing off from the manned lunar landing goal?
Discussion: The review has pointed to the conclusion that in the absence of clear changes in the present technical or international situations, the only basis for backing off from the MLL objective at this time would be an overriding fiscal decision either (a) that the budgetary totals in 1965 or succeeding years are unacceptable and should be reduced by adjusting the space program, or (b) that within present budgetary totals an adjustment should be made shifting funds from space to other programs.
Alternatives
a. Adhere to the present goal as stated above. The arguments supporting this alternative include:
(1) That the reasons for adopting the manned lunar landing goal are still valid;
[5]
(2) That in the absence of clear and compelling external circumstances a change in present policies and commitments would involve an unacceptable "loss of face" both domestically and internationally; and
(3) That it is doubtful if budgetary reductions in the manned lunar program would in fact reduce criticism of the total magnitude of the budget or increase support for other meritorious programs to which the funds might be applied.
b. Decide now to abandon current work directly related to the manned lunar landing objective but to continue development of the large launch vehicle (Saturn V) so that it will be available for future space programs. It is estimated that cancellation in January 1964 of Apollo and other programs supporting the manned lunar landing only would result in NOA and expenditure savings in FY 1965 of about $1 billion, less amounts required for any new objectives that might be substituted. The arguments supporting this alternative could include:
(1) The overriding need for economy in the 1965 budget;
(2) The doubts that Congress will provide adequate support for the manned lunar landing program in 1965 and succeeding years, regardless of the administration's recommendations; and
(3) The apparent absence of a competitive USSR manned lunar landing program at this time.

c. Decide now to abandon both current work toward the manned lunar landing objective and the development of the Saturn V large launch vehicle. If the programs involved are cancelled or adjusted in January 1964, savings approaching $2.5 billion in 1965 NOA and expenditures could be anticipated. The arguments supporting this alternative could include, [6] in addition to those for alternative "b" above:

(1) That proceeding with development of the Saturn V launch vehicle is not justified in the absence of approved goals requiring its use; and

(2) That an adequate continuing space program can be built around the use of the Saturn IB (and perhaps the Titan III) launch vehicle.

Recommendations

(Recommendation of Director, Bureau of the Budget, and concurrences or differing recommendations of Secretary of Defense; Administrator, NASA; Director, OST; and Executive Secretary, NASC, to be inserted after discussions)

* * * * * * * * *

2. Does the present-program represent the minimum necessary for achieving the MLL goal?

Discussion: The review has pointed to the conclusions:

a. That the elements comprising the present program (with Saturn I manned flights eliminated) are required for achieving the goal (recognizing the somewhat indirect contribution of the Gemini program), except for certain construction and other relatively minor items in which adjustments are under consideration in the regular budget review; and

b. That the current NASA 1964 and 1965 estimates represent the minimum funding level required to continue the program on the schedule now planned, except for the possible adjustments being considered in the regular budget review.

[7]

Alternatives

a. Approve the program and cost estimates as submitted by NASA, subject to separate resolution of the adjustments under consideration in the budget review. (The question of a 1964 supplemental estimate is considered in Item 3 below.)

b. Decide now that the program should be geared to a schedule slipping the first manned lunar landing attempts one or two years later than now planned to the very end of the decade (i.e., end of CY 1969 or 1970, depending on the definition of "decade"). This alternative might permit reductions in the range of $100 to 200 million in the 1965 budget. Other things being equal, the total cost of the MLLP to the achievement of the first manned lunar landing would probably be greater by at least $200 million because of the need to maintain the same engineering and other overhead costs over a longer period. However, this would probably not mean a corresponding increase in total annual budgets over what they would be under the present schedule, since expected successor programs would then consume the funds that would otherwise go for completing the stretched out MLLP.

In support of this alternative it could be argued that it would recognize the need for minimizing outlays in the 1965 budget without necessitating a decision at this time to abandon the goal of attempting to achieve manned lunar landing in this decade.

Opposing it, the point can be made that some degree of slippage in present schedules is recognized to be inevitable, so that eliminating the present margin between the current scheduled first manned lunar landing [8] attempts (late CY 1968) and the end of the decade would be tantamount to and generally recognized as an admission that achievement of the goal has been deferred beyond the end of the decade.

Recommendations

(Recommendation of Director, Bureau of the Budget, and concurrences or differing recommendations of Secretary of Defense; Administrator, NASA; Director, OST; and Executive Secretary, NASC, to be inserted after discussions)

* * * * * * * * *

3. Should a 1964 supplemental estimate be submitted to Congress in January for restoration in part or in full of the $250 million congressional reductions below the total legislative authorizations for NASA in 1964?

Alternatives

a. Decide to submit a 1964 supplemental in the amount required to keep the MLLP on the current schedules. Arguments that can be made for this course include:

(1) That restoration of 1964 funds is necessary to avoid forced slippage in the program; and

(2) That submission of a supplemental estimate would once again place the question of maintaining the pace in the MLLP squarely before Congress.

b. Decide not to submit a 1964 supplemental estimate to Congress, and accept as the will of Congress whatever slippage in the MLLP is caused by insufficient funds in 1964. Arguments for this course include:

(1) It would avoid placing the administration in the untenable [9] budgetary posture of seeking restoration of the NASA reduction so soon after congressional action without making similar requests for other important programs reduced by Congress;

(2) There is no reason to believe that the Congress will look with more favor on a supplemental estimate than it did on the regular 1964 request;

(3) The outcome of a supplemental request is likely to be uncertain for several months, and the uncertainty will create operating difficulties which will tend to offset the advantages even if the supplemental is ultimately approved; and

(4) Congress has taken the responsibility for slippage in the MLL-program because of insufficient funds in 1964.

c. Decide not to submit a 1964 supplemental estimate but to seek to make up in the 1965 budget the amounts required to adhere to the current MLLP schedules insofar as practicable. The arguments for this alternative are:

(1) It avoids the problems of a 1964 supplemental estimate referred to above [Items (1), (2), and (3) under "b" above];

(2) It offers a possibility of minimizing the impact of congressional 1964 reductions on the MLLP through adjustments in the timing of obligations between 1964 and 1965; and

(3) It may be feasible without increasing previously expected 1965 budget totals for NASA because of possible offsetting 1965 reductions that have been identified in the regular budget review, as follows:
[9]

		(NOA in millions)		
NASA estimates:		MLLP	All Other	Total NASA
	1964	4,129.7	1,220.3	5,350.0
	1965	4,197.5	1,377.5	5,575.0
	Total	8,327.2	21,597.8	10,925.0

Adjustments:

	MLLP	All Other	Total NASA
Possible 1965 adjustments in			
Budget review	–88.4*	–313.8*	–402.2*
Congressional 1964 reductions	–190.0*	–60.0*	–250.0
Restoration in 1965 for MLLP	+190.0*	-	+190.0*

Revised totals:

1964	3,939.7*	1,160.3*	5,100.0
1965	4,299.1*	1,063.7*	5,362.8*
Total	8,238.8*	2,224.0*	10,462.8*

*Tentative numbers; subject to change in final budget recommendations.

Recommendations
(Recommendation of Director, Bureau of the Budget, and concurrences or differing recommendations of Secretary of Defense; Administrator, NASA; Director, OST; and Executive Secretary, NASC, to be inserted after discussions)

* * * * * * * * *

4. Should our posture on the manned lunar landing program attribute a greater degree of military significance to the program?
Discussion: The review points to the conclusions that:
a. The facts of the situation justify the position that the launch vehicle, spacecraft, facilities, and general technology being developed by NASA in the MLLP do have important potential future military significance;

[11]   b. That overplaying this point unduly could have the effects of undercutting the general peaceful image of the program, jeopardizing possibilities for international cooperation, or calling into question the need for a large-scale NASA non-military space program; and

     c. That the question of public posture on potential military significance is separable from, but must be considered in relation to the questions of the composition of the NASA and Defense programs and of possible transfers of projects from NASA to Defense or vice versa.

     Alternatives

     a. Decide (1) to place greater stress on the potential military significance of the capabilities being developed in the MLLP in domestic public statements, exercising due restraint to avoid undesirable international effects; (2) to emphasize that NASA programs are being relied on by Defense for general technological capabilities and developments; and (3) to point to Defense use of Gemini (on whatever basis is decided separately below) as a tangible example of how NASA technological advances contribute to potential Defense needs. The principal advantage of this alternative is that if properly handled it would permit greater use of potential military applications in securing and maintaining congressional and public support for the administration's manned lunar landing program without creating demands for an unwarranted expansion in military space programs in addition to or in lieu of the approved NASA programs.

     b. Decide (1) to play down the potential military significance of the capabilities being developed in the MLLP; (2) to emphasize that all clearly established military requirements are being met by Department of [12] Defense programs coordinated with NASA and drawing on NASA's experience; and (3) to point to the DOD use of Gemini (on whatever basis is decided separately below) as indicating that prompt attention is being given to the exploitation of possible military uses of space. The advantage of this alternative is that it would avoid possible international complications and unwarranted demands for a larger military space program that might result from too much stress on the potential military significance of the MLLP,

     Recommendations

     (Recommendation of Director, Bureau of the Budget, and concurrences or differing recommendations of Secretary of Defense; Administrator, NASA; Director, OST; and Executive Secretary, NASC, to be inserted after discussions)

* * * * * * * * * * * * * * * *

     5. What should be the posture with respect to a joint effort with the USSR?

     Discussion: The review points to the conclusion that in the present situation we must necessarily take the posture that we are prepared to enter into any constructive arrangement which will not jeopardize vital national security interests and which will not delay or jeopardize the success of our MLL program. We will necessarily have to wait to see what proposals, if any, the USSR may make, and then expect an extended series of negotiations.

     Recommendation: That the posture be as indicated above.

(remainder of document not provided)

Document II-43

Document Title: "Oral History Interview w/Theodore Sorensen," 26 March 1964.

Source: John F. Kennedy Presidential Library, Boston, Massachusetts.

A few month after President John F. Kennedy's assassination, his top advisor, Theodore Sorenson, was interviewed by Carl Kaysen, another Kennedy associate who had worked for the National Security Council during the Kennedy presidency. Sorenson provides a fascinating insider's view of the space issues facing President Kennedy.

Oral History Interview

with

THEODORE C. SORENSEN

March 26, 1964

By Carl Kaysen

For the John F. Kennedy Library

KAYSEN: Ted, I want to begin by asking you about something on which the President expressed himself very strongly in the campaign and early in his Administration, and that is space. What significance, in your mind, did the President attach to the space race in terms of, one, competition with the Soviet Union and, two, the task which the United States ought to do whether or not the element of competition with the Soviet Union was important in it?

SORENSEN: It seems to me that he thought of space primarily in symbolic terms. By that I mean he had comparatively little interest in the substantive gains to made from this kind of scientific inquiry. He did not care as much about new breakthroughs in space medicine or planetary exploration as he did new breakthroughs in rocket thrust or humans in orbit. Our lagging space effort was symbolic, he thought, of everything of which he complained in the [Dwight D.] Eisenhower Administration: the lack of effort, the lack of initiative, the lack of imagination, vitality, and vision; and the more the Russians gained in space during the last few years in the fifties, the more he thought it showed up the Eisenhower Administration's lag in this area and damaged the prestige of the United States abroad.

[2] KAYSEN: So that your emphasis was on general competitiveness but not specific competitiveness with the Soviet Union in a military sense. The President never thought that the question of who was first in space was a big security issue in any direct sense.

SORENSEN: That's correct.

KAYSEN: Now the first big speech and the first big action on space was taken in a special message on extraordinary needs to the Congress in May. What accounted for this delay? What was the President doing in the period between his inauguration and May? He didn't really say much about space in the State of the Union message. He mentioned the competition with the Soviet Union in his State of the Union message, but he didn't really say much or present any programs. What was going on in this period between the inauguration and the inclusion of space in a message which was devoted to extraordinary, urgent was the word, urgent national need?

SORENSEN: There was actually a considerable step-up in our space effort in the first space supplementary budget which he sent to the Congress. You can check that against the actual records, but my recollection is that it emphasized more funds for the Saturn booster. Then came the first Soviet to orbit the earth – [Yuri] Gargarin, I believe that was – and the President felt, justifiably so, that the Soviets had scored a tremendous propaganda victory, that it affected not only our prestige around the world, but affected our security as well in the sense that it demonstrated a Soviet rocket thrust which convinced many people that the Soviet Union was ahead of the United States militarily. First we had a very brief inquiry – largely because the President was being interviewed by Hugh Sidey of Time magazine and wanted to be prepared to say where we stood, what we were going to do, what we were unable to do, how much it would cost and so on–in which he asked me and [Jerome B.] Wiesner and others to look into our effort in some detail.

I do not remember the exact time sequence, but I believe it was shortly after that he asked the Vice President, as the chairman of the Space Council, to examine and to come up with the answers to four or five questions of a similar nature: What were we doing that was not enough? what could we be doing more? [3] where should we be trying to compete and get ahead? what would we have to do to get ahead? and so on. That inquiry led to a joint study by the Space Administration and the Department of Defense. Inasmuch as that study was going on simultaneously with the studies and reviews we were making of the defense budget, military assistance, and civil defense, and inasmuch as space, like these other items, obviously did have some bearing upon our status in the world, it was decided to combine the results of all those studies with the President's recommendations in the special message to Congress.

KAYSEN: Was the moon goal chosen as the goal for the space program because it was spectacular, because it was the first well-defined thing which the experts thought we could sensibly say we ought to pick as a goal we could be first in, because it was far enough away so that we could have a good chance of being first? What reason did we have for defining this as the goal of the space program and making it the center of the space element of the message?

SORENSEN: The scientists listed for us what they considered to be the next series of steps to be taken in the exploration of space which any major country would take, either the Soviet Union or the United States. They included manned orbit, two men in orbit, laboratory in orbit, a shot around the moon, a landing of instruments on the moon, etc. In that list, then, came the sending of a man, or a team of men, to the moon and bringing them back safely. After that was exploration of the planets and so forth.

Looking at that list, the scientists were convinced–on the basis of what they assumed to be the Russian lead at that time – that with respect to all of the items on the list between where we were then, in early 1961, and the landing of a man on the moon, sometime in the late 1960's or early 1970's, there was no possibility of our catching up with the Russians. There was a possibility, if we put enough effort into it, of being the first to send a team to the moon and bringing it back. And it was decided to focus our space effort on that objective.

KAYSEN: Now, as early as the inaugural message, the President talked about making space an area of cooperation instead of conflict. He repeated this notion [4] in his speech to the U. N. September '61, although with a rather narrow set of specifics on weather and communications satellites and things like that. At various times in the course of '61 and '62, I think the record suggests that there was a division of emphasis between the competitive element with the Soviet Union and the notion of offering to cooperate in space in the President's 1963 speech to the U.N., he made a specific suggestion that we cooperate in going to the moon. Do you think this represented a change in emphasis, do you think it represented a change in the assessment of our relations with the Soviets, or do you think it represented a change in the assessment of the feasibility and desirability of trying to meet the goal set of getting to the moon in 1970 and being the first on the moon?

SORENSEN: I don't believe it represented the latter. It may have had an element of the first two in it. I think the President had three objectives in space. One was to assure its demilitarization. The second was to prevent the field from being occupied by the Russians to the exclusion of the United States. And the third was to make certain that American scientific prestige and American scientific effort were at the top. Those three goals all would have been assured in a space effort which culminated in our beating the Russians to the moon. All three of them would have been endangered had the Russians continued to outpace us in their space effort and beat us to the moon. But I believe all three of those goals would also have been assured by a joint Soviet-American venture to the moon.

The difficulty was that in 1961, although the President favored joint effort, we had comparatively few chips to offer. Obviously the Russians were well ahead of us at that time in space exploration, at least in terms of the bigger, more dramatic efforts of which the moon shot would be the culmination. But by 1963, our effort had accelerated considerably. There was a very real chance that we were even with the Soviets in this effort. In addition, our relations with the Soviets, following the Cuban missile crisis and the test ban treaty, were much improved – so the President felt that, without diminishing our own space that effort, and without harming any of those three goals, we now were in a position to ask the Soviets to join with us and make it more efficient and economical for both countries.

[5]
KAYSEN: In this last element, was the President persuaded, as some people argued, that the Soviets weren't really in the race; that, for example, we were developing the Saturn, our intelligence suggested to us that the Soviets had no development of comparable thrust and character; and that, in a sense, we were

racing with ourselves, and we'd won, because once we'd make the commitment to develop the Saturn and it looked as if this was feasible, although maybe the schedule wasn't clear, that we could do it and the Soviets really didn't have anything that could match that; and that, therefore, the psychological moment had come to sort of make it clear to them that we knew it?

SORENSEN:　I don't know if that was in his mind. I did not know that.

KAYSEN:　Now, this is a speculative question, but do you think once an offer of cooperation that was more than trivial, that went beyond the kind of things we had agreed, about exchange of weather information or other rather minor and technical points about recovery of parts and all that kind of thing, that once any offer of cooperation of that sort was made and accepted and some cooperation actually started to take place, do you think space would have become politically uninteresting?

SORENSEN:　Politically, in domestic politics?

KAYSEN: Yes.

SORENSEN:　It probably would have been less interesting, that's right.

KAYSEN:　I'm assuming, and I take it you're assuming, that in the initial exchanges there'd be static and the right wing of the Republicans would shout and so on, but I'm assuming we'd get past all that and some actually useful cooperation would result?

[6]

SORENSEN:　I think it would be less interesting. Even though the President would stress from time to time that the idea of a race or competition was not our sole motivation, there was no doubt that that's what made it more interesting to the Congress and to the general public.

KAYSEN:　Was there any indication that you are aware of that in '63, that in the process of assembling the budget for '63, at the time of the first review, midyear review – that is, I'm talking about the '65 budget, of course, which took place in '63–just the size of this program and its rate of growth were beginning to worry the President, and that he was more eager to stress the cooperative issue because he was dubious about either the wisdom or the possibility of maintaining the kind of rate of increase in this program that NASA [National Aeronautics and Space Administration] talked about?

SORENSEN:　I think he was understandably reluctant to continue that rate of increase. He wished to find ways to spend less money on the program and to cut out the fat which he was convinced was in the budget. How much that motivated his offer to the Russians, though, I don't know.

KAYSEN:　What would you assign it to? You'd say that the political interest in trying to find positive things we could do together was much more important than any budgetary concern about the space program or any feeling that this was not the most important effort that ought to be maintained.

SORENSEN: Right.

KAYSEN: Let me ask a couple more, rather narrower questions. What led the President to pick [James E.] Jim Webb as administrator of NASA? What kind of a man was the President looking for, and two years later did he think he'd gotten the kind of man he was looking for in this rather difficult area?

SORENSEN: My recollection here is not very good, and I'm sure my participation in that decision was remote. I believe that Webb was highly recommended, not only [7] by the Vice President and, I would assume, by Senator [Robert S.] Kerr and others who knew him well, but also by [David E.] Dave Bell and Elmer Staats who had known him when he'd been in the government previously. I also have a dim recollection that the President had tried to get others to take the job although I do not now remember any names, whom he tried or why they turned it down.

The President never expressed any specific dissatisfaction with Webb as space administrator. I think Webb was not what we would call a Kennedy type individual. He was inclined to talk at great length, and the President preferred those who were more concise in their remarks. He was inclined to be rather vague, somewhat disorganized in his approach to a problem, and the President preferred those who were more precise. From time to time, the President would check with him on progress he was making –whether the President's own commitments would be upheld. The President was willing to see a large chunk of the space program developed within the Department of Defense, undoubtedly because he had more confidence in [Robert S.] McNamara's managerial ability than he did in Webb's. But even taking all of these qualifications, I don't know that the President ever regretted his appointment of Webb, or wished that he had named someone else.

[rest of discussion omitted, not related to space]

Document II-44

Document Title: Letter to J. Leland Atwood, President, North American Aviation, Inc. from Major General Samuel C. Phillips, USAF, Apollo Program Director, with attached "NASA Review Team Report," 19 December 1965.

Source: Folder 18675, NASA Historical Reference Collection, History Division, NASA Headquarters, Washington, DC.

In late 1965, at the request of NASA Associate Administrator for Manned Space Flight George Mueller, Major General Samuel Phillips, Apollo Program Director at NASA Headquarters, initiated a review of the work of North American Aviation, Inc. (referred to in this document as NAA) to determine why the company was behind schedule and over budget on both the Apollo Command and Service Module and the second (S-II) stage of the Saturn V launch vehicle. This highly critical review was transmitted to North American's president Lee Atwood on 19 December. The review took on added significance in the aftermath of the fatal Apollo 204 fire on 27 January 1967 when it was discovered that NASA Administrator James E. Webb was apparently unaware of its existence.

IN REPLY REFER TO: MA   December 19, 1965

Mr. J. L. Atwood
President
North American Aviation, Inc.
1700 E. Imperial Highway
El Segundo, California

Dear Lee:

      I believe that I and the team that worked with me were able to examine the Apollo Spacecraft and S-II stage programs at your Space and Information Systems Division in sufficient detail during our recent visits to formulate a reasonably accurate assessment of the current situation concerning these two programs.

      I am definitely not satisfied with the progress and outlook of either program and am convinced that the right actions now can result in substantial improvement of position in both programs in the relatively near future.

      Enclosed are ten copies of the notes which was [*sic*] compiled on the basis of our visits. They include details not discussed in our briefing and are provided for your consideration and use.

      The conclusions expressed in our briefing and notes are critical. Even with due consideration of hopeful signs, I could not find a substantial basis for confidence in future performance. I believe that a task group drawn from NAA at large could rather quickly verify the substance of our conclusions, and might be useful to you in setting the course for improvements.

[2]   The gravity of the situation compels me to ask that you let me know, by the end of January if possible, the actions you propose to take. If I can assist in any way, please let me know.

Sincerely,

SAMUEL C. PHILLIPS
Major General, USAF
Apollo Program Director

[Attachment p. 1]

NASA Review Team Report

I.   Introduction
      This is the report of the NASA's Management Review of North American Aviation Corporation management of Saturn II Stage (S-II) and Command and Service Module (CSM) programs. The Review was conducted as a

result of the continual failure of NAA to achieve the progress required to support the objective of the Apollo Program.

The scope of the review included an examination of the Corporate organization and its relationship to and influence on the activities of S&ID [Space and Information Systems Division of North American Aviation], the operating Division charged with the execution of the S-II and CSM programs. The review also included examination of NAA offsite program activities at KSC and MTF [Mississippi Test Facility].

The members of the review team were specifically chosen for their experience with S&ID and their intimate knowledge of the S-II and CSM programs. The Review findings, therefore, are a culmination of the judgements [sic] of responsible government personnel directly involved with these programs. The team report represents an assessment of the contractor's performance and existing conditions affecting current and future progress, and recommends actions believed necessary to achieve an early return to the position supporting Apollo program objectives.

The Review was conducted from November 22 through December 6 and was organized into a Basic Team, responsible for over-all [3] assessment of the contractor's activities and the relationships among his organizational elements and functions; and sub-teams who [sic] assessed the contractor's activities in the following areas:

> Program Planning and Control (including Logistics)
> Contracting, Pricing, Subcontracting, Purchasing
> Engineering
> Manufacturing
> Reliability and Quality Assurance.

Review Team membership is shown in Appendix 7. [not provided]

Team findings and recommendations were presented to NAA Corporate and S&ID management on December 19.

II. NAA's Performance to Date-Ability to Meet Commitments

At the start of the CSM and S-II Programs, key milestones were agreed upon, performance requirements established and cost plans developed. These were essentially commitments made by NAA to NASA. As the program progressed NASA has been forced to accept slippages in key milestone accomplishments, degradation in hardware performance, and increasing costs.

A. S-II

1. Schedules
As reflected in Appendix VI [not provided] key performance milestones in testing, as well as end item hardware deliveries, have slipped continuously in spite of deletions of both hardware and test content. The fact that the delivery [4] of

the common bulkhead test article was rescheduled 5 times, for a total slippage of more than a year, the All System firing rescheduled 5 times for a total slippage of more than a year, and S-II-1 and S-II-2 flight stage deliveries rescheduled several times for a total slippage of more than a year, are indicative of NAA's inability to stay within planned schedules. Although the total Apollo program was reoriented during this time, the S-II flight stages have remained behind schedules even after this reorientation.

2. Costs

The S-II cost picture, as indicated in Appendix VI, [not provided] has been essentially a series of costs escalations with a bow wave of peak costs advancing steadily throughout the program life. Each annual projection has shown either the current or succeeding year to be the peak. NAA's estimate of the total 10 stage program has more than tripled. These increases have occurred despite the fact that there have been reductions in hardware.

3. Technical Performance

The S-II stage is still plagued with technical difficulties as illustrated in Appendix VI. [not provided] Welding difficulties, insulation bonding, continued redesign as a result of component failures during qualification are indicative of insufficiently aggressive pursuit of technical resolutions during the earlier phases of the program.

[5]   B.   CSM

1. Schedules

A history of slippages in meeting key CSM milestones is contained in Appendix VT. [not provided] The propulsion spacecraft, the systems integration spacecraft, and the spacecraft for the first development flight have each slipped more than six months. In addition, the first manned and the key environmental ground spacecraft have each slipped more than a year. These slippages have occurred in spite of the fact that schedule requirements have been revised a number of times, and seven articles, originally required for delivery by the end of 1965, have been eliminated. Activation of two major checkout stations was completed more than a year late in one case and more than six months late in the other. The start of major testing in the ground test program has slipped from three to nine months in less than two years.

2. Costs

Analysis of spacecraft forecasted costs as reflected in Appendix VI [not provided] reveals NAA has not been able to forecast costs with any reasonable degree of accuracy. The peak of the program cost has slipped 18 months in two years. In addition, NAA is forecasting that the total cost of the reduced spacecraft program will be greater than the cost of the previous planned program.

[6] 3. Technical Performance

Inadequate procedures and controls in bonding and welding, as well as inadequate master tooling, have delayed fabrication of airframes. In addition, there are still major development problems to be resolved. SPS engine life, RCS performance, stress corrosion, and failure of oxidizer tanks has resulted in degradation of the Block I spacecraft as well as forced postponement of the resolution of the Block II spacecraft configuration.

III.   **NASA Assessment-Probability of NAA Meeting Future Commitments**

A. S-II

Today, after 4 1/2 years and a little more than a year before first flight, there are still significant technical problems and unknowns affecting the stage. Manufacture is at least 5 months behind schedule. NAA's continued inability to meet internal objectives, as evidenced by 5 changes in the manufacturing plan in the last 3 months, clearly indicates that extraordinary effort will be required if the contractor is to hold the current position, let alone better it. The MTF activation program is being seriously affected by the insulation repairs and other work required on All Systems stage. The contractor's most recent schedule reveals further slippage in completion of insulation repair. Further, integration of manual GSE has recently slipped 3 weeks as a result of configuration discrepancies discovered during engineering checkout of the system. Failures in timely [7] and complete engineering support, poor workmanship, and other conditions have also contributed to the current S-II situation. Factors which have caused these problems still exist. The two recent funding requirements exercises, with their widely different results, coupled with NAA's demonstrated history of unreliable forecasting, as shown in Appendix VI, [not provided] leave little basis for confidence in the contractor's ability to accomplish the required work within the funds estimated. The team did not find significant indications of actions underway to build confidence that future progress will be better than past performance.

B. CSM

With the first unmanned flight spacecraft finally delivered to KSC, there are still significant problems remaining for Block I and Block II CSM's. Technical problems with electrical power capacity, service propulsion, structural integrity, weight growth, etc. have yet to be resolved. Test stand activation and undersupport of GSE still retard schedule progress. Delayed and compromised ground and qualification test programs give us serious concern that fully qualified flight vehicles will not be available to support the lunar landing program. NAA's inability to meet spacecraft contract use deliveries has caused rescheduling of the total Apollo program. Appendix VI [not provided] indicates the contractor's schedule trends which cause NASA to have little confidence that the S&ID will meet its future spacecraft commitments. While our management review indicated that some progress is [8] being made to improve the CSM outlook, there is little confidence that NAA will meet its schedule and performance commitments within the funds available for this portion of the Apollo program.

[9]   **IV.   Summary Findings**

Presented below is a summary of the team's views on those program conditions and fundamental management deficiencies that are impeding program progress and that require resolution by NAA to ensure that the CSM and S-II Programs regain the required program position. The detail findings and recommendations of the individual sub-team reviews are Appendix to this report.

A.   NAA performance on both programs is characterized by continued failure to meet committed schedule dates with required technical performance and within costs. There is no evidence of current improvement in NAA's management of these programs of the magnitude required to give confidence that NAA performance will improve at the rate required to meet established Apollo program objectives.

B.   Corporate interest in, and attention to, S&ID performance against the customer's stated requirements on these programs is consider[ed] passive. With the exception of the recent General Office survey of selected functional areas of S&ID, the main area of Corporate level interest appears to be in S&ID's financial outlook and in their cost estimating and proposal efforts. While we consider it appropriate that the responsibility and authority for execution of NASA programs be vested in the operating Division, this does not relieve the Corporation of its responsibility, and accountability to NASA for results. [10] We do not suggest that another level of program management be established in the Corporate staff, but we do recommend that the Corporate Office sincerely concern itself with how well S&ID is performing to customer requirements and ensure that responsible and effective actions are taken to meet commitments.

C.   Organization and Manning

We consider the program organization structure and assignment of competent people within the organization a prerogative of the manager and his team that have been given the program job to do. However, in view of what we consider to be an extremely critical situation at S&ID, one expected result of the NASA review might be the direction of certain reorganizations and reassignments considered appropriate, by NASA, to improve the situation. While we do have some suggestions for NAA consideration on this subject, they are to be accepted as such and not considered directive in nature. We emphasize that we clearly expect NAA/S&ID to take responsible and thoroughly considered actions on the organization and assignment of people required to accomplish the S-II and CSM Programs. We expect full consideration, in this judgement [*sic*] by NAA, of both near and long term benefits of changes that are made.

Frankly stated-we firmly believe that S&ID is overmaned [*sic*] and that the S-II and CSM Programs can be done, and done better, with fewer people. This is not to suggest that an arbitrary [11] percentage reduction should be applied to each element of S&ID, but we do suggest the need for adjustments, based on a reassessment and clear definition of organizational responsibilities and task assignments.

It is our view that the total Engineering, Manufacturing, Quality, and Program Control functions are too diversely spread and in too many layers throughout the S&ID organization to contribute, in an integrated and effective manner, to the hard core requirements of the programs. The present proliferation of the functions invites non-contributing, "make-work" use of manpower and dollars as well as impediments to program progress.

We question the true strength and authority of each Program Manager and his real ability to be fully accountable for results when he directly controls less that 50% of the manpower effort that goes into his program. This suggests the need for an objective reappraisal of the people and functions assigned to Central versus Program organizations. This should be done with full recognition that the Central organization's primary reason for existence is to support the requirements of the Program Managers. Concurrently, the Program Manager should undertake a thorough and objective "audit" of all current and planned tasks, as well as evaluate the people assigned to these tasks, in order to bring the total effort down to that which truly contributes to the program.

[12]   It is our opinion that the assignment of the Florida Facility to the Test and Quality Assurance organization creates an anomaly since the Florida activities clearly relate to direct program responsibilities. We recognize that the existence of both CSM and S-II activities at KSC may require the establishment of a single unit for administrative purposes. However, it is our view that the management of this unit is an executive function, rather than one connected with a functional responsibility. We suggest NAA consider a "mirror image" organizational relationship between S&ID and the Florida operation, with the top man at Florida reporting to the S&ID President and the two program organizations reporting to the S&ID Program Managers.

D.   Program Planning and Control
Effective planning and control from a program standpoint does not exist. Each organization defines its own job, its own schedules, and its own budget, all of which may not be compatible or developed in a manner required to achieve program objectives. The Program Managers do not define, monitor, or control the interfaces between the various organizations supporting their program.

Organization-S&ID's planning and control functions are fragmented; responsibility and authority are not clearly defined.

[13]   Work Task Management-General Orders, task authorizations, product plans, etc., are broad and almost meaningless from a standpoint of defining end products. Detailed definitions of work tasks are available at the "doing level"; however, these "work plans" are not reviewed, approved, or controlled by the Program Managers.

Schedules-Each organization supporting the programs develops its own detailed schedules; they are not effectively integrated within an organization, nor are they necessarily compatible with program master schedule requirements.

Budgeting System-Without control over work scope and schedules, the budget control system cannot be effective. In general, it is an allocation system assigning program resources by organizations.

Management Reports-There is no effective reporting system to management that evaluates performance against plans. Plans are changed to reflect performance. Trends and performance indices reporting is almost nonexistent.

E.   Logistics

The CSM and S-II Site Activations and Logistic organizations are adequately staffed to carry out the Logistics support. The problems in the Logistics area are in arriving at a mutual agreement, between NAA and NASA, clearly defining the tasks required to support the programs. The areas requiring actions are as follows:

[14]   1. Logistics Plan
       2. Maintenance Manuals
       3. Maintenance Analysis
       4. NAA/KSC Relationship
       5. Common and Bulk Item Requisitioning at KSC
       6. Review of Spare Parts, Tooling, and Test Equipment Status

F.   Engineering

The most pronounced deficiencies observed in S&ID Engineering are:

1. Fragmentation of the Engineering function throughout the S&ID organization, with the result that it is difficult to identify and place accountability for program-required Engineering outputs.

2. Inadequate systems engineering job is being done from interpretation of NASA stated technical requirements through design release.

3. Adequate visibility on intermediate progress on planned engineering releases is lacking. Late, incomplete, and incorrect engineering releases have caused significant hardware delivery schedule slippages as well as unnecessary program costs.

[15]   4. The principles and procedures for configuration management, as agreed to between NAA and NASA, are not being adhered to by the engineering organizations.

G. Cost Estimating

The "grass roots" estimating technique used at S&ID is a logical step in the process of arriving at program cost estimates and developing operating budgets. However, there are several aspects of the total process that are of concern to NASA:

1. The first relates to the inadequate directing, planning, scheduling, and controlling of program work tasks throughout S&ID. While the grass roots estimates may, in fact, represent valid estimates (subject to scrubbing of "cushion") of individual tasks by working level people, we believe that the present deficiencies in Planning and Control permit, and may encourage, the inclusion in these estimates of work tasks and level of efforts that are truly not required for the program.

2. The second concern is that the final consolidation of grass roots estimates, developed up through the S&ID organization in parallel through both Central functional and Program organizations, does not receive the required [16] management judgements [*sic*], at successive levels for (a) the real program need for the tasks included in the estimate, or (b) adequate scrubbing and validation of the man-hours and dollars estimates.

3. The third concern, which results from I and 2 above, is that the final estimate does not represent, either in tasks to be done or in resources required, the legitimate program requirements as judged by the Program Manager, but represents total work and dollars required to support a level of effort within S&ID.

Several recommendations are made in the appended reports for correcting deficiencies in the estimating process. The basic issue, however, is that an S&ID Management position must be clearly stated and disciplines established to ensure that the end product of the estimating process be only those resources required to do necessary program tasks. In addition, the Program Management must be in an authoritative position that allows him to accept, reject, and negotiate these resource requirements.

H. Manufacturing Work Force Efficiency

There are several indications of less than effective utilization of the manufacturing labor force. Poor workmanship is evidenced by the continual high rates of rejection and MRB actions which result in rework that would not be necessary if the workmanship [17] had been good. This raises a question as to the effectiveness of the PRIDE program which was designed to motivate personnel toward excellence of performance as a result of personal responsibility for the end product. As brought out elsewhere in this report, the ability of Manufacturing to plan and execute its tasks has been severely limited due to continual changing engineering information and lack of visibility as to the expected availability of the engineering information. Recognizing that overtime shifts are necessary at this time, it is our view that strong and knowledgeable supervision of these overtime shifts is necessary, and that a practical system of measuring work accomplished versus work planned must be implemented and used to gauge and to improve the effectiveness of the labor force. The condition of hardware shipped from the factory, with thousands of hours of work to complete, is unsatisfactory to NASA. S&ID must complete all hardware at the factory and further implement, without delay, an accurate system to certify configuration of delivered hardware, properly related to the DD 250.

I.   Quality
NAA quality is not up to NASA required standards. This is evidence[d] by the large number of "correction" E.O.'s and manufacturing discrepancies. This deficiency is further compounded [17] by the large number of discrepancies that escape NAA inspectors but are detected by NASA inspectors. NAA must take immediate and effective action to improve the quality of workmanship and to tighten their own inspection. Performance goals for demonstrating high quality must be established, and trend data must be maintained and given serious attention by Management to correct this unsatisfactory condition.

J.   Following are additional observations and findings that have resulted from discussions during the Review. Most of them are covered in most detail in the appended sub-team reports. They are considered significant to the objective of improving NAA management of our programs and are therefore highlighted in this section of the report:

1.   S&ID must assume more responsibility and initiative for carrying out these programs, and not expect step-by-step direction from NASA.
2.   S&ID must establish work package management techniques that effectively define, integrate, and control program tasks, schedules, and resource requirements.
3.   S&ID must give concurrent attention to both present and downstream tasks to halt the alarming trend of crisis operation and neglect of future tasks because of concentration on today's problems.
4.   A quick response capability must be developed to work critical "program pacing" problems by a short-cut route, with follow-up to ensure meeting normal system requirements.

[19] 5.   S&ID must maintain a current list of open issues and unresolved problems, with clear responsibility assigned for resolving these and insuring proper attention by Program and Division Management.
6.   Effort needs to be applied to simplify management systems and end products. There must be greater emphasis on making today's procedures work to solve today's problems, and less on future, more sophisticated systems. The implementation and adherence to prescribed systems should be audited.
7.   NAA must define standards of performance for maintaining contracts current then establish internal disciplines to meet these standards. Present undefinitized subcontracts and outstanding change orders on the S-II prime contract must be definitized without delay.

CONCLUSIONS AND RECOMMENDATIONS
The NASA Team views on existing deficiencies in the contractor's management of the S-II and CSM Programs are highlighted in this section of the report and are treated in more detail in the appended sub-team reports. The findings are expressed frankly and result from the team's work in attempting to

relate the end results we see in program conditions to fundamental causes for these conditions.

[20]   In most instances, recommendations for improvement accompany the findings. In some cases, problems are expressed for which the team has no specific recommendations, other than the need for attention and resolution by NAA.

It is not NASA's intent to dictate solutions to the deficiencies noted in this report. The solution to NAA's internal problems is both a prerogative and a responsibility of NAA Management, within the parameters of NASA's requirements as stated in the contracts. NASA does, however, fully expect objective, responsible, and timely action by NAA to correct the conditions described in this report.

It is recommended that the CSM incentive contract conversion proceed as now planned.

Incentivization of the S-II Program should be delayed until NASA is assured that the S-11 Program is under control and a responsible proposal is received from the contractor.

Decision on a follow-on incentive contract for the CSM, beyond the present contract period, will be based on contractor performance.

It is recommended that NAA respond to NASA, by the end of January 1966, on the actions taken and planned to be taken to correct the conditions described in this report. At that time, NAA is also to certify the tasks, schedules, and resource requirements for the S-II and CSM Programs.

[21] It is further recommended that the same NASA Review Team re-visit NAA during March 1966 to review NAA performance in the critical areas described in this report.

Document II-45

Document Title: Memorandum to Assistant Administrator, Office of Planning, from William E. Lilly, Director, MSF Program Control, "Saturn Apollo Applications Summary Description," 3 June 1966.

Source: Folder 18675, NASA Historical Reference Collection, History Division, NASA Headquarters, Washington, DC.

Beginning in 1964, NASA began planning for missions to follow a lunar landing using the systems developed for Apollo. This program started out with a great deal of ambition, as this document suggests, but neither President Lyndon B. Johnson nor President Richard M. Nixon was willing to provide the funding needed to implement NASA's ideas. The program was formally named the Apollo Applications Program in 1968. Ultimately, only one Apollo Applications mission was flown, the interim space station known as Skylab.

Willis Shapley was a policy advisor to NASA Administrator James Webb; PSAC was the acronym for the President's Science Advisory Committee; OSSA, for NASA's Office of Space Science and Applications; and OART, NASA's Office of Advanced Research and Technology.

UNITED STATES GOVERNMENT
MEMORANDUM

Dir., Office of
Program Review

DATE: Jun 3, 1966

TO   :   P/Assistant Administrator
              Office of Programming

From :   MP/Director, MSF Program Control

Subject: Saturn Apollo Applications Summary Description

Attached per Mr. Shapley's request of May 31 is a summary description for PSAC of current Saturn Apollo Applications planning as reflected in the May 1966 NASA submission to the Bureau of the Budget. This paper has been coordinated with and included inputs from OSSA (Mr. Foster) and OART (Mr. Novik).

/Signed/

William E. Lilly

Attachment:

Saturn Apollo Applications
Summary Description (CONFIDENTIAL) [DECLASSIFIED]
3 copies

[2]

June 9, 1966

SATURN APOLLO APPLICATIONS PROGRAM

SUMMARY DESCRIPTION

SCOPE
      This document summarizes the assumptions, objectives, program content, hardware availability and flight schedules for the proposed Saturn Apollo Applications program, as currently planned and as reflected in the May 1966 NASA submission to the Bureau of the Budget. The plans described in this document are under active consideration within NASA but have not been approved at this time and are subject to further review and change.

PROGRAM ASSUMPTIONS

1.   Prior to 1970, the Gemini and Apollo programs, building on results of Mercury and Saturn I, will have provided:

      a.   The capability to explore space out to 250,000 miles from earth and to conduct manned operations and experiments on flights of up to two weeks duration.

      b.   The Saturn IB and Saturn V boosters, which will have injected 20 and 140 tons of payload per launch, respectively, into near-earth orbit. The Saturn V will have sent 48 tons to the vicinity of the moon.

      c.   The Apollo spacecraft, which will have sustained a three-man crew for two weeks in a two-compartment, modular, [3] maneuverable vehicle and will have landed two men on the moon and returned them, with samples of lunar material, to earth.

      d.   A U.S. manned space flight log of over 500 man days in space, during which data and experience will have been acquired from approximately 100 in-flight experiments in response to the needs of the scientific and technological communities. (To date, U.S. astronauts have logged approximately 75 man-days in space.)

2.   The currently approved Apollo mission objectives can be accomplished with the currently funded flight vehicles.

      a.   If the approved Apollo objectives can be achieved with fewer flights, the remaining flight vehicles can be used for alternate missions during 1968-71. Follow-on missions requiring procurement of flight hardware beyond that now funded would continue the manned space flight effort, based on Apollo systems, beyond that time.

      b.   If all of the presently funded hardware is required for the basic Apollo lunar missions, the program content of the alternate missions can be appropriately phased into the follow-on period.

PROGRAM OBJECTIVES

The basic purposes of the Saturn Apollo Applications Program are to continue without hiatus an active and productive [4] post Apollo Program of manned space flight and to exploit for useful purposes and further develop the capabilities of the Saturn Apollo System. The major flight mission objectives of the proposed Saturn Apollo Applications flight program fall into two principal categories of essentially equal importance as follows:

A.   Long Duration Flights

     1.   Man

     2.   Systems

   B.   <u>Space flight experiments</u> in the following areas:

     1.   Life Sciences (both biomedical and bioscience/technology)

     2.   Astronomy and space physics

     3.   Extended Lunar Exploration

     4.   Applications (including meteorology, communications, earth resources)

     5.   Technology (spent stage utilization, advanced EVA, propellant handling in space, orbital assembly and maintenance, etc.)

A careful review of future mission requirements in relation to long range objectives has shown that extended duration manned flight experience as early as possible is required to establish the basic capabilities required for any of the projected next generation of manned space flight goals (earth orbital space station, lunar station or manned planetary [5] exploration). Flights of up to a year's duration would be attained in Apollo Applications through the use of modified Apollo hardware with resupply. Such an adaptation of Apollo hardware might be used as a long duration Manned Orbital Research Facility.

The experiments in the areas listed above would be responsive to specific needs, as defined by the scientific and engineering communities and as reviewed and approved by the Office of Space Science and Applications in the case of scientific or applications experiments; by the Office of Advanced Research and Technology in the case of technology experiments, and by the Office of Manned Space Flight in the case of operations experiments and experiments on biomedical effects on man. All experiments proposed for flight on manned missions will be reviewed and approved by the joint NASA/DOD Manned Space Flight Experiments Board.

Experiment areas two and three above would support the National Astronomical Observatories objectives proposed in the 1965 Woods Hole Summer Study and extended lunar exploration as recommended at the 1965 Falmouth Summer Study.

<u>PROGRAM CONTENT</u>

Attachment 1 summarizes the Saturn Apollo Applications mission objectives and indicates the planned target dates for flights to meet these objectives. The black triangular symbols represent planned missions, each of which requires one or more Saturn Apollo launches. Most of the missions [6] are planned to accomplish more than one objective, as indicated by the vertical alignment of the mission symbols for the same launch date. The following paragraphs summarize current plans for each of the objectives listed on Attachment 1.

The <u>long duration flight objectives </u>are (1) to measure the effects on men and on manned systems of space flights of increasing duration, (2) to acquire operational experience with increasingly longer manned space flights, and (3) to accomplish this (a) through modifications and adaptations of existing systems without a major new launch vehicle or spacecraft development, and (b) in such a way that the equipment used as modified for this program can serve as important elements of the systems that would be required for one or more of the projected next generation of manned space flight goals.

During 1968-69, extended mission duration can be achieved by adding expendable supplies to each flight and by rendezvous resupply using a second spacecraft launched two to four weeks after the first launch. By this means, missions of up to 56 days duration are possible in 1968-69, each consisting of two flights employing the basic Apollo 14-day spacecraft. Beginning in 1970, up-rated Apollo spacecraft subsystems (primarily electrical fuel cells and cryogenic oxygen and hydrogen storage tanks) are planned to provide a 45-day capability for a single flight. (Attachment 2 describes the various extended Apollo spacecraft capabilities planned for the Saturn Apollo Applications missions.) During 1970, a [7] double-rendezvous mission involving three 45-day spacecraft is planned to achieve a total mission duration of approximately 135 days. In 1971, the objective is a one-year mission involving a Saturn V launch of a crew module derived from the Saturn S-IVB stage, with re-supply by up-rated Apollo spacecraft launched on Saturn IB's. The objectives for 1972-73 are missions of greater than one year's duration as precursors of later earth orbital space stations or manned planetary flights. Suitable biomedical instrumentation is planned to monitor the effects on the crews of these long duration flights.

The same series of flights that is planned for these long duration flight objectives will be used for important space flight experiments in a variety of fields. The present planning for the experiments is discussed below.

Life sciences experiments during 1968-69 will concentrate on the biomedical effects of long duration flight on men, as discussed above. A biomedical laboratory is planned for flight in 1970 in conjunction with the 135-day mission. This laboratory will consist of an Apollo spacecraft module equipped with biomedical and behavioral apparatus to test and record human responses to various stresses (e.g., physical exercise, variable gravity, complex task performance, etc.) during long duration space flights. During 1971-72, bioscience and biotechnology laboratories are planned to extend earlier investigations on various sub-human life forms, ranging from simple cells to primates. In these laboratories, [8] greater stresses can be applied to sub-human life specimens than are normally planned for human subjects, and the results can benefit both the bio-scientific community and later manned spaceflight technology.

Orbital astronomy mission objectives are planned around use of the Apollo Telescope Mount (ATM) concept during 1968-72 (see Attachment 3). During the 1968-70 period of maximum solar activity, emphasis will be on solar astronomy using the ATM in low altitude earth orbit. These first ATM missions, in addition to providing valuable scientific data, will provide an experimental basis for developing the techniques of manned astronomical observations in space and assessing their value and possibilities. Stellar astronomy missions are being studied for the 1971-73 period. Based on experience gained from the early ATM flights, an orbital astronomy mission involving a large aperture telescope (60" to

100") is scheduled for late 1973. This may be a test of a large mirror leading to development of the National Astronomical Observatories.

Space physics experiments are planned generally for flight on astronomy missions. During 1968-69, instrumentation flown on the 1966-67 short duration Apollo earth orbital missions will be reflown to acquire more extensive data in such fields as X-ray astronomy, ultra-violet spectroscopy, ion wake physics and investigations of particles and fields. Beginning in 1971, advanced space physics experiments are planned.

[9]   The extended lunar exploration missions planned for Saturn Apollo Applications include both orbital mapping missions and extended lunar surface explorations. The objective is to extend knowledge of the moon beyond that achievable in the earlier Ranger, Surveyor, unmanned Lunar Orbiter and early Apollo missions, and to provide the basis for possible establishment in the mid or late 1970's of semi-permanent or permanent manned stations on the moon. The lunar orbital missions are planned to acquire high quality mapping and survey photography from polar or near-polar lunar orbits for study of geological features over wide areas of the lunar surface exploration missions. The lunar surface missions surface, and to aid in selection of sites for extended duration are planned to provide up to two weeks stay at selected exploration. Equipments planned for these missions include lunar sites for extensive geological, geophysical and biological small, wheeled vehicles to permit traverses within line-of sight of the landed spacecraft; drills for sub-surface sampling and vertical profile measurements; and instrumentation for acquiring geophysical data to be transmitted back to earth by RF link for up to a year after departure of the astronauts. One such lunar surface mission per year is planned, beginning in 1970. For the 1973 mission, an objective is to provide optical and radio telescopes to evaluate the lunar surface environment for astronomical experiments.

[10]   Applications experiments are planned to develop techniques and to measure the effectiveness of man's participation in such fields as orbital meteorology (see Attachment 4), communications, and remote sensing of earth resources. Low altitude orbits at medium and high inclinations have been studied for meteorology and natural resources missions during 1969-70. An initial synchronous orbit mission is planned to test man's ability to operate effectively in that environment and to test operational techniques for linking low altitude manned spacecraft to central ground control stations. The later synchronous orbit missions are planned for continued operational use as well as for possible experiments in astronomy, space physics, meteorology and advanced communications techniques.

Technology experiments planned for Saturn Apollo Applications missions are focused generally toward the development of equipment and techniques which appear fundamental to the accomplishment of the next generation of post-Apollo space flight missions. During 1968-69, emphasis will be placed on the use as an orbital laboratory of the spent S-IVB stage, which injects an Apollo spacecraft into orbit. Advanced EVA experiments are planned, for example, to retrieve micrometeorite panels from a Pegasus spacecraft orbited in 1965 by a Saturn I vehicle. Resupply and crew transfer techniques are planned, both to extend mission duration, rotate crews, and to test orbital rescue operations. An orbital fluids laboratory is [11]scheduled for flight in 1970 to extend knowledge of propellant behavior and transfer techniques under zero gravity conditions. Orbital assembly of complex structures and in flight maintenance of vehicles and experiment apparatus are planned. Most of the technology experiments are integrally combined with experiments planned to meet other objectives.

HARDWARE AVAILABILITY AND FLIGHT SCHEDULES

The reference baseline for Saturn Apollo Applications mission planning is the flight hardware delivery schedule which has been established to meet the requirements of the Apollo lunar landing program. This schedule provides for delivery of 12 Saturn IB's, 15 Saturn V's, 21 Command and Service Modules and 15 Lunar Excursion Modules for launch during 1966 through early 1970.

Attachments 5 and 6 depict the two alternate Saturn Apollo Applications launch schedules for the period 1968-73 which formed the basis for NASA's May 1966 submission to the Bureau of the Budget. Both cases are based on the assumption that the last four Saturn IB's (AS 209-212) and the last six Saturn V's (AS 510-515) with their associated spacecraft from the approved Apollo program might become available for alternate missions as the initial phase of the Saturn Apollo Applications flight program. Both in Case I (Attachment 5) and in Case II (Attachment 6), the launch dates for these alternate missions using approved Apollo vehicles are planned to occur as much as one year later than the launch date scheduled for those vehicles under the basic Apollo program. This stretch-out of launch schedules [12] for the early Saturn Apollo Applications missions allows time for development and integration of suitable experiment apparatus under the limited funding available in FY 1966-67.

Case I differs from Case II primarily in the rate at which follow-on Saturn IB vehicles are delivered for launch to meet the Saturn Apollo Applications mission objectives. It represents the lowest rate of follow-on vehicle deliveries which could permit accomplishing the basic program objectives, and it would require phasing down both production and launch operations activity during 1969-70, followed by a partial build-up of both activities beginning in 1971. Case I (Attachment 5) provides for carrying out the experiments discussed previously on approximately the schedule shown in Attachment 1, although the Saturn IB missions (low altitude earth orbit) after AS 212 would be delayed from 3 to 9 months. Funding estimates associated with Case I make no provision prior to FY 1972 for developing post-Apollo space vehicles or modules for the next generation of manned space flight objectives.

Case II, starting from the same baseline of approved Saturn Apollo flight vehicles, provides for earlier delivery of follow-on vehicles, and the associated funding estimates would permit a start in FY 1969 on the development of next generation subsystems and modules for flights beyond 1971 on the schedule shown on Attachment 6. Because of the earlier delivery of follow-on vehicles, the Saturn IB missions planned for 1969 and beyond can be scheduled from 3 to 9 months earlier in Case II than in Case I. Thus, Case II has been planned to [13] permit an early and extensive utilization of Saturn Apollo capabilities, with an earlier focus on a post-Apollo national space objective involving the development of new space modules (such as a prototype of a space station or a planetary mission module) for initial earth orbital flight in the early 1970's.

Attachment 7 lists the objectives of planned Saturn Apollo Applications missions scheduled on the flight indicated on Attachment 6. These missions are under continual study to identify and trade-off alternative modes of accomplishing the mission objectives. Approximately two years prior to the scheduled launch date for each mission, the objectives and flight assignments for that mission to be firmly established and a period of intensive mission planning must begin throughout the NASA organization and its contractors. The Saturn Apollo Applications missions planned for 1968-69 will enter this two-year mission preparation phase during FY 1967, while the post-1969 missions will be the subject of further definition studies and long lead item development effort. The total process of identification, definition, selection, hardware development, flight qualification and procurement of experiments can take a total of 3 to 4 years and must be initiated long enough in advance to be in phase with the schedule requirements for detailed mission planning and launch. Similarly, adequate lead times must be allowed for procurement of basic space vehicle hardware.

[14]   Attachments: [not provided]

1.   Saturn Apollo Applications Mission Objectives

2.   Extended Capability of Apollo Space Vehicles Planned for Saturn
     Apollo Applications Missions

3.   Apollo Telescope Mount (ATM) Concept

4.   Applications A Experimental System (AAP A)
     (Primarily Meteorology)

5.   Saturn Apollo Applications Launch Schedule, Case I

6.   Saturn Apollo Applications Launch Schedule, Case II

Document II-46

Document Title: Letter from Thomas Gold to Harold Urey, 9 June 1966.

Source: Archives of the Royal Society, London, England (reprinted with permission).

Professor Thomas Gold was a well-known astronomer at Cornell University, noted for his unconventional views. Gold had suggested that the lunar surface was covered with a deep layer of fine dust, and thus might not support the weight of a landing spacecraft with astronauts aboard. In this letter to equally well-known astronomer Harold Urey, Gold reflects of the results of the Surveyor 1 mission, which landed on the moon on 2 June 1966.

June 9, 1966

Dr. Harold C. Urey

School of Science and Engineering

University of California

La Jolla, California

Dear Harold:

Thank you for your nice letter. I completely agree with you that the Surveyor results are quite incompatible with a frozen lava explanation of the surface. Not only is the imprint of the foot clearly visible and I believe about five inches deep despite the fact that the peak impact loading seems to have been no more than about four pounds per square inch, there apparently is even a detailed fine structure visible in the imprint matching precisely the corrugation on the side of the foot. It is therefore a compressible and slightly cohesive material which one can mold. That is of course just what we have always said it would be, and I recall we had given such examples as that one could build igloos with it by cutting it like chunks of snow with a knife.

Many of the objects that are seen on the surface – the so called rocks – very likely are rocks or harder material, but they are evidently not mostly lying on the surface. In several cases one can clearly see that the contact line with the surface is about the largest perimeter that the object possesses, a very unlikely situation for a stone lying on the surface. One of the objects is a neat pyramid and reminds one of the New Yorker cartoon of the two archaeologists in the middle of the desert who had just brushed away the sand from the tiniest little peak of a pyramid sticking out and one says to the other, "Well, there's no telling how much work it's going to be". It is clear the majority of these objects are mainly submerged with just the tops sticking out. The shore line around each such object meets up with it quite neatly, which it would of course not do if it were merely thrown into a softish material and [2] partly submerged. It will then have a trench some places around it just like the foot of the Surveyor does. This is not the case. The only explanation that I can see for that is to suppose that the filler material fills in flat gradually over the course of time and while this is occurring stones of various sizes are thrown by major explosions so that they accumulate at the same time as the fine stuff sediments. I would guess that such chunks are distributed throughout the interior of the sediment and the radar evidence in fact has been in favor of a substantial amount of scattering being derived from many feet below the surface.

I want to emphasize again that I have always said that I believe the fine particles to be very considerably cohesive, especially in the lunar conditions (most fine dust is already quite cohesive on the earth, too) and that I regarded the material as crunchy and certainly not as flowing. I have also never said that I believed it would not hold the Surveyor or the Apollo, but only that this was uncertain, while of course all the lava experts never even contemplated the possibility it could sink in at all. I regard the amount of sinkage that has occurred as quite within the range of slightly cemented powders.

You mention the word "sand" in your letter. From the pictures of course one cannot directly tell the grain size, but one does see little crumbles thrown out from the foot. Dry sand would not go in crumbles and the chunks that you see are certainly too big to be the individual particles. A slightly cohesive powder on the other hand is well known to result in just such crumbs. Also, a sand could not be cohesive enough to maintain the vertical surface, though of course a sand mixed with a lot of much smaller particles could, as could an aggregate of small particles only. I personally think that the majority of the particles will in fact be very small since I know that the packing fraction must be rather low; not much more than one-third of the volume can be occupied by pieces. If there is a great spread in the size of the particles then usually denser packing results. Yet the radar and the thermal evidence are quite clear about the lower packing that is required. For that reason I believe that the small stuff is all in particle sizes of not much more than a few microns and in that case I can understand as I have always stressed that there are ways in which they can be deposited flat. Sand particles could be [3] scattered among all this but it would be very hard to find any mechanism that would tend make them bed down flat in the absence of wind and water.

I have just read again what I have said in the past on this subject to see why everyone keeps assuming that a deep deposit of dust would make everything disappear out of sight. I find that I have always stressed in print, already in 1956, that vacuum welding will be important and that such a material will not flow. However, that I am not quite sure that it will everywhere be strong enough to suppor [sic] the weights and that I would still maintain now. I quite agree with you that there are many signs of subterranian [sic] holes into which overburden has fallen and I would still be very worried about that also. The plain fact of the matter is that while on earth wind and water has tested most areas of ground and human interference is only a trivial addition, this is not so on the moon. Just like on a glacier after fresh snow, there can easily be structures that are weak and untested as yet.

I expect to be coming out to Los Angeles next week, possibly also to take part in the CBS program to be taped on the Saturday. I understand you will be present there, too. I am looking forward to seeing you there and having a little more discussion on the side.

With best regards,

    Yours sincerely,

    T. Gold

TG:vs

Document II-47

Document Title: Memorandum to Dr. Wernher von Braun, Director of NASA Marshall Space Flight Center, from James E. Webb, NASA Administrator, December 17, 1966

Source: Folder #18675, NASA Historical Reference Collection, History Division, NASA Headquarters, Washington, D.C.

After its budget peaked in 1965, the Apollo Program had to fight in the White House and Congressional budget processes for the funds needed to make sure that both President Kennedy's "end of the decade goal" was met and the program could be carried through to its planned conclusion. NASA had estimated in 1962 that 15 Saturn V launchers would be needed to ensure this outcome.

Perhaps the most recognizable personality within NASA during the development phase of Apollo was Wernher von Braun. He was an optimist by nature, and was frequently sought out by the media for comments on Apollo's progress. In a late 1966 interview with the magazine U.S. News and World Report, von Braun suggested that the first lunar landing might come relatively early in the Saturn V sequence, with the implication that Apollo's objectives might be accomplished with the use of less than the 15 vehicles that had been ordered. NASA Administrator James E. Webb, who was working hard in Washington to sustain support for Apollo, was not happy with von Braun's remarks.

December 17, 1966

MEMORANDUM FOR: Dr. Wernher von Braun
                         Director, Marshall Space Flight Center

THROUGH: Dr. Seamans / AD
                         Dr. Mueller / M

Following our recent Program Review and the exchange I had with Joe Shea about the difficulty our senior officials are having sustaining the credibility of their public statements and the Congressional testimony we gave last year when those responsible for program management were giving optimistic statements about the time when 504 might "well go to the Moon," I thought we had pretty well established the policy that we would not make those kinds of statements. I had this in mind particularly because I testified last year that we had no extra vehicles in the program, and I made the strongest representation I know how to make in the Bureau of the Budget this year that we should not cancel any of the 15 Saturn V's with the high risk of this program. Therefore, you can imagine my surprise when I read the <u>U.S. News and World Report</u> statements in your interview. While I recognize that they overplayed your statements, it does seem to me that your answers to the questions made it possible for them to do so and that you could have given answers which would have made the situation clearer.

In any event, I have now examined the Apollo program adjustments established on December 7 by George Mueller and they clearly indicate that there is certainly a very, very low possibility that complete Saturn V systems will be available for flights out as far as the Moon in 1968. Under these circumstances, it seems to me that you will need to be very careful in dealing with the press not to return to the kind of statements you made in the U.S. News and World Report interview. I hope you can find a way to backtrack to a position that is more consistent with the official estimate established by George Mueller in these recent adjustments.

[2]

Even this series of adjustments does not, in my view, take account of all the difficulties we are likely to encounter in this very complex Saturn V-Apollo system, particularly as we are now so hemmed in, have so little room to make adjustments, and have no financial margins. We could lose several hundred million dollars of badly needed funds for 1968 under conditions as they exist. I know you don't want to contribute to this.

While we have been talking about the SII stage as the pacing item, I understand there is even doubt as to whether the complete LEM system test can take place on AS-206. If the complete test has to go over to 209, and this is not flown until late 1967, it would certainly not seem realistic to take any position publicly that did not indicate we are going to have very great difficulty making the lunar landing in this decade – within the last quarter of 1969.

I know you will understand I am writing in this detail because it is of deep concern to me that statements such as your own in the interview mentioned do have an impact on the credibility of the official statements Seamans and I have made and will have to make again in our Congressional testimony. I know you do not wish to undermine the credibility of those of us who are working so hard to get the money to continue this program and to avoid having the vehicles now approved (15 Saturn V's) deleted from the program on the basis that they are not needed to accomplish the mission.

One possible course you could take in future statements is the same which I took last year before Congress when asked if we had closed the gap with the Russians during the year. You will remember this came about when Congressman Davis asked if I expected to find "Russians" on the Moon when we arrived, as predicted sometime ago by Edward Teller. I stated that a year ago I had felt we would be there first but that during the year I had developed more doubts and now felt much less assurance about it. You might recede from the positions you have taken publicly by saying at the next opportunity, and then repeating in the future, the fact that up until recently you thought we would be able to have the Saturn V-Apollo system so perfected and tested that the experience from the Saturn I-B program added in would permit early Saturn V flights to be released toward the Moon with a good chance that one of the early ones would land, but that the difficulties encountered have now caused you to have much more concern and doubt as to whether this will be possible. If you do not have these doubts, Wernher, than I think Mueller, Seamans, and I should get together and find out how your own views could differ so markedly from our own.

[Signed James E. Webb]

James E. Webb
Administrator

Enclosure:
  Excerpt, 1967 House Authorization Hearings

Document II-48

Document Title: Robert C. Seamans, Jr., Deputy Administrator, "Memorandum for the Apollo 204 Review Board," January 28, 1967.

Source: Folder #18675, NASA Historical Reference Collection, History Division, NASA Headquarters, Washington, D.C.

Document II-49

Document Title: NASA, Office of the Administrator, "Statement by James E. Webb," February 25, 1967

Source: Folder #18675, NASA Historical Reference Collection, History Division, NASA Headquarters, Washington, D.C.

Document II-50

Document Title: Apollo 204 Review Board, "Report of Apollo 204 Review Board to the Administrator, National Aeronautics and Space Administration," April 5, 1967.

Source: Folder #18675, NASA Historical Reference Collection, History Division, NASA Headquarters, Washington, D.C.

A 27 January 1967 fire during a launch pad spacecraft test resulted in the deaths of astronauts Virgil "Gus" Grissom, Edward White, and Roger Chaffee. NASA Administrator James Webb was able to convince President Lyndon B. Johnson that NASA could and should conduct an objective review of what came to be known as the Apollo 204 accident. On 28 January, NASA constituted an internal Review Board for the accident investigation after Administrator Webb convinced the White House that NASA could conduct its own review on an impartial basis. The Review Board was chaired by the Director of the Langley Research Center, Floyd Thompson, to investigate the accident and suggest remedial measures. During the investigation, NASA hoped to damp down Congressional criticism by sharing with the

Congress information on the progress of the review. The report of the Apollo 204 Review Board was released on 5 April 1967.

Document II-48

NATIONAL AERONAUTICS AND SPACE ADMINISTRATION

Washington D.C. 20546

Office of the Administrator   January 28, 1967

MEMORANDUM For the Apollo 204 Review Board

1. The Apollo 204 Review Board is hereby established in accordance with NASA Management Instruction 8621.1, dated April 14, 1966, to investigate the Apollo 204 accident which resulted in the deaths of Lt. Col. Virgil I. Grissom, Lt. Col. Edward H. White and Lt. Cmdr. Roger B. Chaffee on Launch Complex 34, on January 27, 1967.

2. The Board will report to the Administrator of the National Aeronautics and Space Administration.

3. The following are hereby appointed to the Board:

Dr. Floyd L. Thompson, Director, Langley Research Center,
          NASA, Chairman

Lt. Col. Frank Borman, Astronaut, Manned Spacecraft
          Center, NASA

Maxime Faget, Director, Engineering & Development,
          Manned Spacecraft Center, NASA

E. Barton Geer, Associate Chief, Flight Vehicles &
          Systems Division,

Langley Research Center, NASA

George Jeffs, Chief Engineer, Apollo, North
          American Aviation, Inc.

Dr. Frank A. Long, PSAC Member, Vice President for Research
          and Advanced Studies, Cornell University

Col. Charles F. Strang, Chief of Missiles & Space Safety Division
          Air Force Inspector General
          Norton Air Force Base, California

George C. White, Jr., Director, Reliability & Quality, Apollo Program
Office, Headquarters, NASA

John Williams, Director, Spacecraft Operations,
Kennedy Space Center, NASA

[2]

4. George Malley, Chief Counsel, Langley Research Center, will serve as counsel to the Board.

5. The Board will:

   a. Review the circumstances surrounding the accident to establish the probable cause or causes of the accident, including review of the findings, corrective action, and recommendations being developed by the Program Offices, Field Centers, and contractors involved.

   b. Direct such further specific investigations as may be necessary.

   c. Report its findings relating to the cause of the accident to the administrator as expeditiously as possible and release such information through the Office of Public Affairs.

   d. Consider the impact of the accident on all Apollo activities involving equipment preparation, testing, and flight operations.

   e. Consider all other factors relating to the accident, including design, procedures, organization, and management.

   f. Develop recommendations for corrective or other action based upon its findings and determinations.

   g. Document its findings, determinations, and recommendations and submit a final report to the Administrator which will not be released without his approval.

6. The Board may call upon any element of NASA for support, assistance, and information.

Document II-49

NATIONAL AERONAUTICS AND SPACE ADMINISTRATION

WASHINGTON. D.C. 20546

OFFICE OF THE ADMINISTRATOR   February 25, 1967

STATEMENT BY

JAMES E. WEBB

NASA is releasing today a third interim report on the work of the Apollo 204 Review Board resulting from two days of meetings with the Board by Deputy Administrator Robert Seamans at Cape Kennedy. These meetings took place on February 23 and 24.

This statement and Dr. Seamans' third interim report have been reviewed with Chairman Clinton Anderson and Senior Minority Committee Member Senator Margaret Chase Smith and with Congressman George Miller. In continuation of the Senate Committee's review of the Apollo 204 accident. Senator Anderson has announced that the Senate Committee will hold an open hearing on the preliminary findings of the Board and actions to be taken by NASA at 3 p.m., Monday, February 27.

In addition to the information set forth by Dr. Seamans in his three interim reports, I have had the benefit of a review by three members of the Board – the Chairman, Dr. Floyd Thompson, Astronaut Frank Borman, and Department of Interior combustion expert Dr. Robert Van Dolah. This included the preliminary views of the Board as to the most likely causes of ignition, the contributing factors in the rapid spread of the fire, the inadequacy of the means of emergency egress for the astronauts, and the need to recognize that all future such tests be classified as involving a higher level of hazard.

The following emerges from the preliminary views of the Board and the Board's preliminary recommendations:

(1)   The risk of fire that could not be controlled or from which escape could not be made was recognized when the procedures for the conduct of the test were established. Our experience with pure oxygen atmospheres included not only the successful Mercury and Gemini flights but a number of instances where a clearly positive source of ignition did not result in a fire. In one such instance an electric bulb was shattered, exposing the incandescent element to the oxygen atmosphere without starting a fire.

(2)   Our successful experience with pure oxygen atmospheres in Mercury and Gemini, our experience with the difficulty of storing and using hand–held equipment under zero–gravity conditions, and our experience with the difficulty of making sure before flight that no undiscovered items had been dropped or found their way into the complex maze of plumbing, wiring, and equipment in the capsule, led us to place in the Apollo 204 capsule such items as Velcro pads to which frequently used items could be easily attached and removed, protective covers on wire bundles, nylon netting to prevent articles dropped in ground testing from being lost under or behind equipment in the capsule, and a pad or cushion on which, in the planned escape exercise, the hatch could be placed without damage to the hatch itself or to the equipment in the spacecraft. While most of these were constructed of low-combustion-potential material, they were not so arranged as to provide barriers to the spread of a fire. Tests conducted in an Apollo-type chamber since the accident have

shown that an oxygen fire in the capsule will spread along the surface of Velcro and along the edges of nylon netting much faster than through the material itself.

[2]   (3) Soldered joints in piping carrying both oxygen and fluids were melted away, with resultant leakage contributing to the spread of the fire.

(4) The bursting of the capsule happened in such a way that the flames, as they rushed toward the rupture and exhausted through it, traveled over and around the astronauts' couches. Under these conditions, and with just a few seconds of time available, the astronauts could not reach the hatch and open it.

(5) This fire indicates that a number of items to the design and performance of the environmental control unit will require the most careful examination and may require redesign.

Astronaut Borman, in commenting on his reactions to the conditions surrounding the Apollo 204 test and the subsequent knowledge he has gained as a result of serving on the Review Board, stated to Dr. Seamans, Dr. Thompson, and to me that he would not have been concerned to enter the capsule at the time Grissom, White and Chaffee did so for the test, and would not at that time have regarded the operation as involving substantial hazard. However, he stated that his work on the Board has convinced him that there were hazards present beyond the understanding of either NASA's engineers or astronauts. He believes the work of the Review Board will provide the knowledge and recommendations necessary to substantially minimize or eliminate them.

Dr. Thompson, Astronaut Borman, and Dr. Van Dolah have returned to Cape Kennedy are proceeding with the work of the Board. This will require several weeks to complete.

Chairman George Miller, of the House Committee on Science and Astronautics, has announced that as soon as the Board's work is complete, the Committee's Oversight Subcommittee, chaired by Congressman's Olin Teague, will conduct a complete investigation of all factors related to the accident and NASA's actions to meet the conditions disclosed. Chairman Teague spent Friday and Saturday at Cape Kennedy with members of the Manned Space Flight Subcommittee, of which he is also Chairman, reviewing progress in the Apollo Program. Dr. Seamans, Dr. George Mueller, and I will report further to him at 10 a.m., Monday, February 27.

Document II-50

REPORT OF

APOLLO 204

REVIEW BOARD

TO

THE ADMINISTRATOR

NATIONAL AERONAUTICS AND SPACE ADMINISTRATION

[Following text in Box on Un-numbered Page]

APOLLO SPACECRAFT

The spacecraft (S/C) consists of a launch escape system (LES) assembly, command module (C/M), service module (S/M) and the spacecraft/lunar module adapter (SLA). The LES assembly provides the means for rapidly separating the C/M from the S/M during pad or suborbital aborts. The C/M forms the spacecraft control center, contains necessary automatic and manual equipment to control and monitor the spacecraft systems, and contains the required equipment for safely and comfort of the crew. The S/M is a cylindrical structure located between the C/M and the SLA. It contains the propulsion systems for attitude and velocity change maneuvers. Most of the consumables used in the mission are stored in the S/M. The SLA is a truncated cone which connects the S/M to the launch vehicle. It also provides the space wherein the lunar module (L/M) is carried on lunar missions.

TEST IN PROGRESS AT TIME OF ACCIDENT

Spacecraft 012 was undergoing a "Plugs Out Integrated Test" at the time of the accident on January 27, 1967. Operational Checkout Procedure, designated OCP FO-K-0021-1, applied to this test. Within this report this procedure is often referred to as OCP-0021.

TESTS AND ANALYSES

Results of tests and analyses not complete at the time of publication of this report will be contained in Appendix G, Addenda and Corrigenda.[not provided]

CONVERSION OF TIME

Throughout this report, time is stated in Greenwich Mean Time (GMT). To convert GMT to Eastern Standard Time (EST), subtract 17 hours. For example, 23:31 GMT converted is 6:31 p.m. EST.

NATIONAL AERONAUTICS AND SPACE ADMINISTRATION

APOLLO 204 REVIEW BOARD

IN REPLY REFER TO   April 5, 1967

The Honorable James E. Webb
Administrator
National Aeronautics and Space Administration
Washington, D. C. 20546

Dear Mr. Webb:

Pursuant to your directive as implemented by the memorandum of February 3,1967, signed by the Deputy Administrator, Dr. Robert C. Seamans, Jr., the Apollo 204 Review Board herewith transmits its final, formal report, each member concurring in each of the findings, determinations, and recommendations.

                                    Sincerely,

                                    /Signed/
                                    Dr. Floyd L. Thompson
                                    Chairman

/Signed/   /Signed/
Frank Borman, Col., USAF   Dr. Robert W. Van Dolah

/Signed/   /Signed/
Dr. Maxime A. Faget   George C. White, Jr.

/Signed/   /Signed/
E. Barton Geer   John J. Williams

/Signed/
Charles F. Strang, Col., USAF

[Parts 1-5 of report not included]

[6-1]

Part VI BOARD FINDINGS, DETERMINATIONS AND RECOMMENDATIONS

In this Review, the Board adhered to the principle that reliability of the Command Module and the entire system involved in its operation is a requirement common to both safety and mission success. Once the Command Module has left the earth's environment the occupants are totally dependent upon it for their safety. It follows that protection from fire as a hazard involves much more than quick egress. The latter has merit only during test periods on earth when the Command Module is being readied for its mission and not during the mission itself. The risk of fire must be faced; however, that risk is only one factor pertaining to the reliability of the Command Module that must received adequate consideration. Design features and operating procedures that are intended to reduce the fire risk must not introduce other serious risks to mission success and safety.

1.   FINDING:
        a.   There was a momentary power failure at 23:30:55 GMT.
        b.   Evidence of several arcs was found in the post fire investigation.
        c.   No single ignition source of the fire was conclusively identified.
        DETERMINATION:
        The most probable initiator was an electrical arc in the sector between
the -Y and +Z spacecraft axes. The exact location best fitting the total
available information is near the floor in the lower forward section of
the left-hand equipment bay where Environmental Control System (ECS)
instrumentation power wiring leads into the area between the Environmental Control Unit (ECU) and the oxygen panel. No evidence was discovered that suggested sabotage.
2.   FINDING:
        a.   The Command Module contained many types and classes of
combustible material in areas contiguous to possible ignition sources.
        b.   The test was conducted with a 16.7 pounds per square inch absolute,
100 percent oxygen atmosphere.
        DETERMINATION:
        The test conditions were extremely hazardous.
        RECOMMENDATION:
        The amount and location of combustible materials in the Command
Module must be severely restricted and controlled.
3.   FINDING:
        a.   The rapid spread of fire caused an increase in pressure and
temperature which resulted in rupture of the Command Module and creation
of a toxic atmosphere. Death of the crew was from asphyxia due to inhalation of
toxic gases due to fire. A contributory cause of death was thermal burns.
        b.   Non-uniform distribution of carboxyhemoglobin was found by autopsy.
        DETERMINATION:
        Autopsy data leads to the medical opinion that unconsciousness occurred
rapidly and that death followed soon thereafter.
4.   FINDING:
        Due to internal pressure, the Command Module inner hatch could not
be opened prior to rupture of the Command Module.
        DETERMINATION:
        The crew was never capable of effecting emergency egress because of the
pressurization before rupture and their loss of consciousness soon after rupture.
        RECOMMENDATION:
        The time required for egress of the crew be reduced and the operations
necessary for egress be simplified.
5.   FINDING
        Those organizations responsible for the planning, conduct and safety of this
test failed to identify it as being hazardous. Contingency preparations to permit escape
or rescue of the crew from an internal Command Module fire were not made.
        a.   No procedures for this type of emergency had been established either
for the crew or for the spacecraft pad work team.
        b.   The emergency equipment located in the White Room and on the
spacecraft work levels was not [6-2] designed for the smoke condition resulting
from a fire of this nature.
        c.   Emergency fire, rescue and medical teams were not in attendance.

d. Both the spacecraft work levels and the umbilical tower access arm contain features such as steps, sliding doors and sharp turns in the egress paths which hinder emergency operations.

DETERMINATION:

Adequate safety precautions were neither established nor observed for this test.

RECOMMENDATIONS:

a. Management continually monitor the safety of all test operations and assure the adequacy of emergency procedures.

b. All emergency equipment (breathing apparatus, protective clothing, deluge systems, access arm, etc.) be reviewed for adequacy

c. Personnel training and practice for emergency procedures be given on a regular basis and reviewed prior to the conduct of a hazardous operation.

d. Service structures and umbilical towers be modified to facilitate emergency operations.

6. FINDING:

Frequent interruptions and failures had been experienced in the overall communication system during the operations preceding the accident.

DETERMINATION:

The overall communication system was unsatisfactory.

RECOMMENDATIONS:

a. The Ground Communication System be improved to assure reliable communications between all test elements as soon as possible and before the next manned flight.

b. A detailed design review be conducted on the entire spacecraft communication system.

7. FINDING:

a. Revisions to the Operational Checkout Procedure for the test were issued at 5:30 pm EST January 26, 1967 (209 pages) and 10:00 am EST January 27, 1967 (4 pages).

b. Differences existed between the Ground Test Procedures and the In-Flight Check Lists.

DETERMINATION:

Neither the revision nor the differences contributed to the accident. The late issuance of the

revision, however, prevented test personnel from becoming adequately familiar with the test procedure prior to its use.

RECOMMENDATIONS:

a. Test Procedures and Pilot's Checklists that represent the actual Command Module configuration be published in final form and reviewed early enough to permit adequate preparation and participation of all test organization.

b. Timely distribution of test procedures and major changes be made a constraint to the beginning of any test.

8. FINDING:

The fire in Command Module 012 was subsequently simulated closely by a test fire in a full-scale mock-up.

DETERMINATION:

Full-scale mock-up fire tests can be used to give a realistic appraisal of fire risks in flight-configured spacecraft.

RECOMMENDATION:

Full-scale mock-ups in flight configuration be tested to determine the risk of fire.

9.   FINDING:

The Command Module Environmental Control System design provides a pure oxygen atmosphere.

DETERMINATION:

This atmosphere presents severe fire hazards if the amount and location of combustibles in the Command Module are not restricted and controlled.

RECOMMENDATIONS:

a.   The fire safety of the reconfigured Command Module be established by full-scale mock-up tests.

b.   Studies of the use of a diluent gas be continued with particular reference to assessing the problems of gas detection and control and the risk of additional operations that would be required in the use of a two gas atmosphere.

[6-3] 10. FINDING:

Deficiencies existed in Command Module design, workmanship and quality control, such as:

a.   Components of the Environmental Control System installed in Command Module 012 had a history of many removals and of technical difficulties including regulator failures, line failures and Environmental Control Unit failures. The design and installation features of the Environmental Control Unit makes removal or repair difficult.

b.   Coolant leakage at solder joints has been a chronic problem.

c.   The coolant is both corrosive and combustible.

d.   Deficiencies in design, manufacture, installation, rework and quality control existed in the electrical wiring.

e. No vibration test was made of a complete flight-configured spacecraft.

f. Spacecraft design and operating procedures currently require the disconnecting of electrical connections while powered.

g. No design features for fire protection were incorporated.

DETERMINATION:

These deficiencies created an unnecessarily hazardous condition and their continuation would imperil any future Apollo operations.

RECOMMENDATIONS:

a.   An in-depth review of all elements, components and assemblies of the Environmental Control System be conducted to assure its functional and structural integrity and to minimize its contribution to fire risk.

b.   Present design of soldered joints in plumbing be modified to increase integrity or the joints

be replaced with a more structurally reliable configuration.

c.   Deleterious effects of coolant leakage and spillage be eliminated.

d.   Review of specifications be conducted, 3-dimensional jigs be used in manufacture of wire bundles and rigid inspection at all stages of wiring design, manufacture and installation be enforced.

e.   Vibration tests be conducted of a flight-configured spacecraft.

f.   The necessity for electrical connections or disconnections with power on within the crew compartment be eliminated.

g. Investigation be made of the most effective means of controlling and extinguishing a spacecraft fire. Auxiliary breathing oxygen and crew protection from smoke and toxic fumes be provided.

11. FINDING:

An examination of operating practices showed the following examples of problem areas:

a. The number of the open items at the time of shipment of the Command Module 012 was not known. There were 113 significant Engineering Orders not accomplished at the time Command Module 012 was delivered to NASA; 623 Engineering Orders were released subsequent to delivery. Of these, 22 were recent releases which were not recorded in configuration records at the time of the accident.

b. Established requirements were not followed with regard to the pre-test constraints list. The list was not completed and signed by designated contractor and NASA personnel prior to the test, even though oral agreement to proceed was reached.

c. Formulation of and changes to pre-launch test requirements for the Apollo spacecraft program were unresponsive to changing conditions.

d. Non-certified equipment items were installed in the Command Module at time of test.

e. Discrepancies existed between NAA and NASA MSC specifications regarding inclusion and positioning of flammable materials.

f. The test specification was released in August 1966 and was not updated to include accumulated changes from release date to date of the test.

DETERMINATION:

Problems of program management and relationships between Centers and with the contractor have led in some cases to insufficient response to changing program requirements.

RECOMMENDATION:

Every effort must be made to insure the maximum clarification and understanding of the responsibilities of all the organizations involved, the objective being a fully coordinated and efficient program.

Document II-51

Document Title: Letter to Senator Clinton P. Anderson from James E. Webb, NASA Administrator, May 8, 1967

Source: Folder #18675, NASA Historical Reference Collection, History Division, NASA Headquarters, Washington, DC

The attempt by NASA Administrator James E. Webb to limit congressional and public criticism of NASA following the Apollo 204 fire by carrying out a thorough internal investigation was not totally successful. On 27 February, in testimony before the Senate Committee on Aeronautical and Space Sciences, the existence of the critical review of North American Aviation carried out by Apollo Program Manager General Sam Phillips was brought to Webb's attention; he had apparently not been previously aware of its existence.

Later testimony before the same committee did not reassure the Committee's chair, Senator Clinton Anderson, that NASA was being totally forthcoming. Webb valued his credibility with Congress very highly, and in this letter suggested an approach to rebuilding confidence between members of the Committee and NASA.

National Aeronautics and Space Administration
Washington, D.C. 20546

Office of the Administrator

May 8, 1967

Honorable Clinton P. Anderson
United States Senate
Washington, D. C. 20510

Dear Mr. Chairman:

I am deeply troubled by your statement to me last Saturday that members of the Committee are not satisfied with our testimony on NASA's actions in follow-up of the deficiences [*sic*] found by the management review team headed by General Phillips at North American Aviation in 1965. Your statement that members of the Committee believe NASA is endeavoring to put a disproportionate part of the blame for the Apollo 204 accident on North American Aviation and avoid its proper acceptance of blame troubles me even more.

On April 13, 1967, General Phillips testified before your Committee and summarized the actions of his team and the responses made by North American Aviation during the following several months. He answered all questions that were asked. The Oversight Subcommittee of the House Committee on Science and Astronautics, because it had not pressed this line of questioning, immediately asked for a summary of the team report, which was furnished to Chairman Teague on April 15, 1967, and publicly released by him.

Over the past six years, NASA has placed contracts with American industry for more than 22 billion dollars of work. To do this kind of advanced aeronautical and space research and build flight hardware, American industry has had to introduce new, very difficult fabrication and test capabilities. It has had to learn to use new management systems. In this process, NASA has provided a technical interface and technical monitoring function as an addition to the normal or standard process of contract monitoring, much of which is performed for us by the Department of Defense contract administration service. In cases where contractors have encountered serious technical or management difftcu1ties, it has been our policy [2] to assist them to develop strengths they did not have and to utilize our knowledge of the factors which brought success to one contractor to help others take advantage of this experience. We and most of our contractors have cooperated fully in approaching problems in whatever manner was best calculated to solve them and get on with the work, rather than to try to fix blame. At the same time, we have had to find new ways to reward efficiency and penalize poor performance. We and our contractors have placed a high premium on self-analysis and self-criticism, as painful as it has had to be in many cases.

The plain fact is that our U.S. industrial system has in the past generally made its profits from large-scale production and the initial learning period on complex space development projects has not had the incentive of anticipated profits from large production orders. However, after six years, the process we have developed is in its final stages and demonstrating efficiency in most companies with large contract obligations to NASA.

In Apollo, we are very near to a flight demonstration of all the equipment that will prove that six large companies could take contracts for major segments and that the resulting vehicle provides for this country the space capability we have needed since the USSR flew Sputnik in 1957 and Gagarin in 1961. In the Saturn V Apollo system, Boeing makes the first stage, North American the second, Douglas the third, International Business Machines the instrumentation unit, Grumman the lunar excursion module, and North American the command and service module. The General Electric Company provides the automated checkout equipment. Even the smallest of these projects runs into tens or hundreds of millions of dollars.

Almost without exception each company has encountered serious difficulties at one time or another. Many NASA management review teams have had to work with prime or sub-contractors to move the work ahead. The end result is going to be success for Apollo, but it is going to be much harder to achieve if every detail of every difficulty is now to be put on the public record as a failure of either the contractor or NASA.

It is a hard fact of life in this kind of research and development that success cannot be achieved without a certain amount of experimentation in design to find the limits that can be safely reached. This means a high initial rate of failure on inspection and test, [3] and consequent redesign. We are still in or near many areas of the unknown.

As I have pondered the meaning of your statements to me on Saturday, I have tried to think of ways through which the Committee could reestablish the confidence in NASA it formerly had and in the system we are using. I have tried to find some way this could be done without violating the basic commitments we have made to individuals and companies to regard information given as confidential and also without having the Committee undertake the enormous task of forming a judgment about at least a sample of the management review criticisms we have recorded with respect to every major unit in the program.

With the pressure of time to get the program moving, now that we have established a basic plan which will bring us to the next manned flight at an early date, which we will be presenting to your Committee tomorrow, May 9, and with the limited investment of time which the Committee is able to make in understanding the complexities which alone permit valid judgments, I can think of nothing better than to request an executive session of the Committee, to which I would bring General Phillips and all the members of the review team which made the study of

North American Aviation in December of 1965. In such major matters, it is our practice to include on a management review team a knowledgeable senior person from outside NASA. In this case, the member was General G. F. Keeling of the Air Force Systems Command. The NASA members, other than General Phillips, are Dr. J. F. Shea, Dr. E. F. M. Rees, and General E. F. O'Connor.

In such an executive session, this group can lay on the table all of the documentation which it used in its analysis of the situation at NAA and the six volumes of responses made by North American. These responses show the actions taken by North American between December 1965 and April 1966. In an executive session, General Phillips and other appropriate officials will also be prepared to present and answer questions on the actions taken by both North American and NASA in the 1966-1967 period following the April reviews. Statements of most of these actions will be referenced to the management review team materials. Examples are enclosed in order that you may see that NASA and NAA have continued to take vigorous action in the period since the management review.

[4] To answer the questions you have raised, there is no way to exclude from the documentation we are prepared to present in executive session such business confidential data on North American as indirect cost rates, burden rates, direct and indirect employees, general and administrative expenses, bidding expense, independent research and development expense, and other similar information. This material falls within the purview of section 1905 of Title 18, United States Code, which means that the Committee must restrict this information to use by Committee members.

At the end of your executive session, it will be my purpose to gather up the materials referred to above and return them to NASA files, unless the Committee takes action to the contrary.

Through the expenditure of about 25 billion dollars over the last six years, NASA has brought the efforts of over 400,000 men and women working in American laboratories and factories into the development of the space capabilities our nation needs. Our success is shown by the fact that we are now laying off from this work force 5,000 workers per month. We have utilized the American industrial system flexibly and in ways that have added vast new strengths that have permeated practically every segment of our national economy. We have created within NASA's developmental centers such as Huntsville, Houston, and Cape Kennedy, an ability to work with contractors to do new and almost impossible tasks. To make public every detail of the difficulties we have encountered out of the context of the total program efforts involved will do grave injustice to many individuals in private life and many outstanding industrial units, and undoubtedly will destroy our ability to continue this system on the cooperative basis essential to its success. However, after you have inspected the attached materials and we have answered your questions in executive session, that decision must become the responsibility of the Committee. I can only give you my judgment as to what is in the best interests of the country.

Because time is so short, I am sending you sufficient copies of this letter to permit distribution to the members of the Committee should that be considered desirable.

Sincerely yours,

[Signed James E. Webb]

James E. Webb
Administrator

Enclosure

Document II-52

Document Title: Interagency Committee on Back Contamination, "Quarantine Schemes for Manned Lunar Missions," no date, but probably August 1967

Source: Folder #18675, NASA Historical Reference Collection, History Division, NASA Headquarters, Washington, D.C.

Document II-53

Document Title: NASA, "Policy Directive RE Outbound Lunar Biological Contamination Control: Policy and Responsibility," September 6, 1967.

Source: Folder #18675, NASA Historical Reference Collection, History Division, NASA Headquarters, Washington, DC

Document II-54

Document Title: Letter to Thomas Paine, Administrator , NASA, from Frederick Seitz, President, National Academy of Sciences, March 24, 1969

Source: Folder #18675, NASA Historical Reference Collection, History Division, NASA Headquarters, Washington, D.C.

Document II-55

Letter from Senator Clinton Anderson, Chairman, Committee on Aeronautical and Space Sciences, U.S. Senate, to Thomas Paine, Administrator, NASA, May 15, 1969

Source: Folder #18675, NASA Historical Reference Collection, History Division, NASA Headquarters, Washington, D.C.

Document II-56

Document Title: Letter to Senator Clinton Anderson, Chairman, Committee on Aeronautical and Space Sciences, from Homer Newell, Acting Administrator, NASA, June 4, 1969

Source: Folder #18675, NASA Historical Reference Collection, History Division, NASA Headquarters, Washington, D.C.

During the early years of the Apollo program NASA recognized that planetary protection, particularly protection against back contamination of Earth by hypothetical lunar organisms, was a critical issue that had to be addressed before lunar landing missions could go forward. In 1966, the Interagency Committee on Back Contamination (ICBC) was established to determine what measures were needed to preserve public health and protect agricultural and other resources against the possibility of contamination by lunar organisms conveyed in returned sample material or other material exposed to the lunar surface (including astronauts), and to preserve the biological and chemical integrity of lunar samples and scientific experiments with minimal compromise to the operating aspects of the program. This report summarizes the conclusions of the ICBC with respect to quarantine requirements for both returning astronauts and lunar samples.

The contamination of the lunar surface during visits by Apollo astronauts was also of concern, and NASA in September 1967 adopted a policy in this regard.

As the first lunar landing attempt grew closer, concerns were raised about both the readiness of the Lunar Receiving Laboratory (LRL) at the Manned Spacecraft Center to receive the Apollo astronauts and the material they would return from the Moon and the adequacy of the measures being taken to ensure that the astronauts and their spacecraft would not carry back alien organisms to Earth. In particular, NASA a few months before the Apollo 11 mission decided that the astronauts would leave the command module shortly after it landed in the ocean, rather than stay aboard the capsule until it was placed in quarantine aboard the recovery aircraft carrier. Concerns that this approach would undercut other quarantine measures were brought to the attention of Congress in May 1969. NASA was able to convince most members of Congress and scientists that the protections it had put in place were adequate, but even up to a few days before the mission was launched on 16 July 1969 there were a few individuals seeking to delay the launch until more stringent protections were put in place.

Document II-52

QUARANTINE SCHEMES

FOR

MANNED LUNAR MISSIONS

BY: INTERAGENCY COMMITTEE
ON BACK CONTAMINATION

TABLE OF CONTENTS

TABLES

QUARANTINE SCHEMES FOR MANNED LUNAR MISSIONS

Introduction

Presented herein are the fundamental quarantine and sample release plans for manned lunar missions as established by the Interagency Committee on Back Contamination. Obviously, the scheme does not contain all possible finite technical details about quarantine test methods and containment provisions, but it provides the necessary framework for action by the Interagency Committee on Back Contamination and substantive methods for satisfying the quarantine requirements of the Regulatory Agencies.*

---

*In this document the U. S. Department of Health, Education and Welfare, the U. S. Department of Agriculture, and the U. S. Department of the Interior are referred to as the Regulatory Agencies.

It is, of course, impossible in any set of quarantine plans to anticipate every eventuality. Therefore, it is necessary that the schemes include a contingency provision that gives the Interagency Committee and the Regulatory Agencies adequate opportunity to provide requirements and suggestions for situations not covered in the formal plans. It is likewise necessary to emphasize that in spite of efforts being made to assure aseptic collection and return of lunar samples, there is no certainty of the complete absence of earth microbial contaminants. And certainly, the potential of earth contaminants in returned lunar samples will be significantly greater after the first Apollo mission.

Astronaut Release Scheme

Table I provides the general scheme for the quarantine and release of the astronauts and medical support personnel in the Crew Reception Area (CRA) of the Lunar Receiving Laboratory (LRL). The scheme covers three possible results and indicates the course of action for each. Implicit in each is an appropriate review by the Interagency Committee and the accomplishment of any formal action and recommendation that might be required.

Proposition I is the most likely with release of the astronauts and medical support personnel from the CRA after approximately 21 days. This action will accrue if there are no alterations in the general health of the quarantined people and no other indications of infectious disease due to lunar exposure.

[2] Should a definite alteration in the health of one or more persons in the CRA occur (Proposition II), release of the people would probably not be delayed if the alteration is diagnosed as non infectious or is of terrestrial origin. If the source of the alteration cannot be readily diagnosed, however, some prolongation of the quarantine may be necessary. In either case, under Proposition II, review of the data and recommendations by the Interagency Committee are required.

Proposition III establishes the requirement that laboratory personnel from the sample laboratory of the LRL be housed in the CRA following a severe rupture of a cabinet system containing lunar material suspected of containing harmful or infectious materials. While precise specification of events for Proposition III are not outlined in Table I, the NASA medical team should consider all available information and make recommendations concerning release of the laboratory people. These recommendations should be reviewed and approved by the Interagency Committee. If it is decided that the laboratory personnel must undergo quarantine, the medical observations would identify Propositions I or II in Table 1. It must be recognized that this situation could result in prolonged quarantine of the astronauts.

Phase I Sample Release Scheme

The scheme outlined in Table II provides a general plan for each of three sets of circumstances resulting from quarantine testing of lunar samples. Examination and review of the quarantine data by the Interagency Committee before release or non release of the sample is provided in each case. In other words, in each case the Interagency Committee would have identified an appropriate time for coordinating their position and making their recommendations to the National Aeronautics and Space Administration.

Proposition I of Table II shows the course of action for what should be the most probable result of sample quarantine testing, the situation in which the protocol is carried out in the LRL with completely negative results: no viable organisms being isolated and no pathogenic effects being noted in the animals and plant systems tested. For this eventuality, Proposition I calls for the Interagency Committee to meet, examine, and review the quarantine data, and if satisfied as to its validity and reliability, recommend to NASA the release of samples from that returned mission. Formal clearance by the Regulatory Agencies is effected as a part of this plan.

Proposition II of Table II prescribes the course of action to be followed in the event that a replicating organism is detected in the lunar sample <u>without</u> any deleterious effects being noted on the life systems or terrestrial niches tested in the LRL. Should this result materialize, the aim of the flow chart under Proposition II is to [3] determine: (1) if the organism isolated is of terrestrial origin, unmodified by any lunar exposure and generally considered as "non pathogenic", or (2) if the organism is not readily classified as being of terrestrial origin and therefore or potential hazard to terrestrial ecology.

In regard to statement (1) above, demonstration that the organism in question is identical with organisms collected from the spacecraft, from spacecraft equipment, or from the astronauts during preflight sampling, or classification of the organism as a harmless terrestrial microbe would be adequate reason for neither extending nor expanding the quarantine. The inability to recover a common, identifiable, and non pathogenic organism a second time from a duplicate lunar sample would further indicate that an earth contaminant rather than an organism indigenous to the lunar sample was involved. In this same regard, lunar sample contamination could result following a break in the primary barrier of the LRL. If the organism isolated cannot be readily classified or otherwise shown to be of terrestrial origin, there then would be the need for initiation of a contingency quarantine plan.

Under Proposition II, Table II, the scheme requires review by the Interagency Committee at the points indicated. Adequate demonstration that the organisms are terrestrial, unchanged, and usually regarded as "non pathogenic" would be considered by the Interagency Committee as sufficient reason for not requiring challenge of additional terrestrial niches before sample release. Failure of the protocol tests to provide this information about organisms isolated from the lunar sample, however, would signal the need for further quarantine testing (indicated as Phase II quarantine) and/or release of sample according to conditions* then specified by the Regulatory Agencies, and/or release of samples after sterilization.

---

*Release to certain specified laboratories for further study; or sterilization before release, but only after consultation with investigators to determine if this is satisfactory to their specific experiment; or release to the LRL so that visiting scientists (Principal Investigators) can work in the LRL under containment conditions to carry out early experiments.

Proposition III of Table II covers the situation where definite deleterious effects are noted on one or more of the life systems tested in the LRL. Should this occur, the effects observed may be due to chemical toxicity rather than to invasion by a replicating organism. This would be indicated if sterilized lunar material (the control) produced the same deleterious effects and if no replicating organisms were found. It is always possible, however, that replicating contaminants will be uncovered along with a toxic chemical. In such cases,

[4] it will be necessary to identify the organisms as of terrestrial origin and to classify them as "harmless" in order to avoid testing additional terrestrial niches or life systems.

Finally, if replicating organisms are indicated as the cause of definite deleterious effects on tested life systems, Phase II quarantine will be indicated with the possibility of a subsequent conditional release and/or only sterilized samples will be released. Under Proposition III appropriate places for review and action by the Interagency Committee are indicated.

Phase II Sample Release Scheme

The probability is very remote of a contingency quarantine of a lunar sample due to the presence of unidentified replicating organisms or because of non-explained deleterious effects on life systems that are not due to chemical toxicity. Nevertheless, it is necessary that the prevention of possible terrestrial back contamination be specific with regard to these remote probabilities in order that the intent of the Interagency Committee on Back Contamination Terms of Reference* be fulfilled and that the legal requirements of the Regulatory Agencies be satisfied. The Phase II quarantine scheme for these eventualities is specified in Table III.

---

*Interagency Agreement between the National Aeronautics and Space Administration, the Department of Agriculture, the Department of Health, Education and Welfare, the Department of the Interior, and the National Academy of Sciences on the protection of the Earth's biosphere from lunar sources of contamination: Attachment A: Interagency Committee on Back Contamination Terms of Reference.

Phase II requires a prolongation of the quarantine for an unspecified time interval. However, even at the outset of Phase II, the Interagency Committee could recommend release of some portions of the lunar samples to non-biological institutions under specific conditions of handling. The conditions would, for the most part, relate to the use of the sample inside biological barriers.

Otherwise, Phase II quarantine involves continued testing of animal and plant species in the LRL. As indicated in Table III, the scheme could also provide for conditional release of cultures isolated in the LRL or specimens to certain biological laboratory institutions in the United States for more detailed study of possible pathogenic effects. These laboratories, however, must meet existing specifications of the Regulatory Agencies for handling potentially virulent pathogens.

[5] (Phase II quarantine could take advantage of visiting scientists in the LRL as bioscience specialists to carry out specific tests for pathogenicity, should such talents be available.)

Contingency Landings

The release schemes outlined above assume that a nominal or near nominal landing of the crew, spacecraft, and related equipment has been achieved. In the event of a contingency landing – off nominal – the details and method of quarantine must be adapted to the exigencies of the situation. Immediate authoritative decisions must be made as they apply to quarantine and back contamination as well as other time critical problems.

For such cases, the quarantine aspects will be represented by a Quarantine Control Officer.* To the extent possible during a disaster, he will obtain direction from the Regulatory members of the Interagency Committee before initiating disaster control procedures. Prior to the first returned lunar mission it will be the responsibility of the Quarantine Control Officer to prepare and have approved by the NASA medical team and the Science and Applications Director (Manned Spacecraft Center), and the Regulatory Agencies a document outlining typical courses of action for several types of contingency landings.

---

*Manned Spacecraft Center Management Instruction 8030.1, dated January 9, 1967: Assignment of Responsibility for the Prevention of Contamination of the Biosphere by Extraterrestrial Life.

Release of Film and Data Tapes

The film and data tapes will be returned to the LRL in the same manner as the lunar samples, admitted to quarantine, and maintained behind a biological barrier. The data tapes will then be played through the biological barrier for outside processing.

The film will be processed inside the quarantine facility and printed through the biological barrier with an optical printer for outside use.

If current studies indicate that ethylene-oxide sterilization of the film is possible when the film is contaminated with bacterial spores and that no degradation of the film occurs, there is the possibility that immediate release of sterilized film will be allowed without printing through the barrier. The statistical reliability of the ethylene-oxide process should be such that the treatment will fail to give sterility no more than 1 in 10,000 times ($P = 1 \times 10^{-4}$).

[6]

Spacecraft Release

The spacecraft will enter the LRL in a sealed configuration and be placed in isolation near the CRA (this area can become a part of the quarantine facility if necessary). It will follow the same time constraints as the sample – 30 days – prior to release if all results are negative. It will, however, be available for additional bio-sampling if deemed necessary by Quarantine Control Officer. At his discretion, it may also be, entered for technical inspection provided that it is placed inside the biological barrier and the personnel and spacecraft become an integral part of the quarantine facility and scheme of release at that time.

Summary

The Interagency Committee has prepared this document in order that all agencies and persons involved in returned lunar samples may have a clear understanding of the procedures the Interagency Committee feels are necessary for the realistic program to protect this planet from possible back contamination. Moreover, the Interagency Committee presents this document as one that will satisfy the requirements of the Regulatory Agencies of Government without undue hardship on NASA. Although the Interagency Committee feels that very few alternates to this plan are possible, it wishes to acknowledge a speedy and unconditional release of the sample; a minimum of expense and delay is highly advantageous to the scientific community.

The schemes proposed may be summarized as follows:

1.   Astronauts and Medical Support Personnel

a.   Release after 21 days if no alternations in general health are observed and in the absence of an infectious disease attributable to lunar exposure.

b.   If significant alterations in general health occur, release is still indicated if alterations are diagnosed as of terrestrial origin or as non communicable.

c.   If alterations are apparent and not diagnosed, some delay in release would be indicated with the final action to be recommended by the NASA medical team.

[7]
Conditions for Lunar Sample Release

a.   It is expected that prompt release of lunar samples after completion of the protocol tests can be recommended by the Interagency Committee to the Administrator of NASA or NASA's designated representative. The nominal results expected would obviously not impose any unusual conditions upon the release.

b.   Interagency Committee conditional release could result if there is sufficient doubt regarding the presence of pathogenic organisms in the lunar samples. In this instance, release of sterilized samples would be possible, or some samples might be released providing they are used only behind a suitable biological barrier. In the case of a conditional release, Phase II quarantine testing will proceed as rapidly as possible in an attempt to clarify the data regarding possible pathogenic effects.

3.   Validity Constraints for Sample Release

It is in the interest of all concerned that the quarantine testing procedures be designed to avoid events that would produce invalid results. To insure that "lunar pathogens" will not be falsely detected, the sample release scheme contains the following constraints.

a.   If replicating organisms are found in the sample and no deleterious effects are noted in any of the terrestrial niches tested in the LRL, release will not be delayed beyond the time needed to identify the organisms as terrestrial contaminants.

b.   If deleterious effects from lunar material are noted with the terrestrial life systems tested in the LRL, release will not be delayed beyond the time needed to show that the effects were due to chemical toxicity and that any replicating organisms isolated from the sample were of terrestrial origin, harmless, and not responsible for the effects.

c.   Should Phase I quarantine procedures indicate the presence of a substance pathogenic to terrestrial life, Phase II procedures will be initiated to verify or more adequately explain the Phase I results.

[8]

INTERAGENCY COMMITTEE ON BACK CONTAMINATION

Membership

Primary	Alternate
David J. Sencer, M. D. (Chairman) National Communicable Disease Center U.S. Public Health Service	
Dr. John Bagby, Jr. (Co-Chairman) National Communicable Disease Center U.S. Public Health Service	
Dr. Wolf Vishniac University of Rochester (National Academy of Sciences representative)	Dr. Allan Brown University of Pennsylvania
Dr. Ernest Saulmon Department of Agriculture	Dr. A. B. Park Department of Agriculture
Mr. Howard H. Eckles Department of the Interior	Dr. John Buckley Department of the Interior
Dr. Harold P. Klein Ames Research Center, NASA	Dr. Adrian Mandel Ames Research Center, NASA
Charles A. Berry, M.D. Manned Spacecraft Center, NASA	Walter W. Kemmerer, M.D. Manned Spacecraft Center, NASA

Dr. Wilmot N. Hess
Manned Spacecraft Center, NASA

Mr. Joseph V. Piland
Manned Spacecraft Center, NASA

Mr. Lawrence B. Hall
Office of Space Science
and Applications, NASA

Captain Arthur H. Neill
Office of Space Science
and Applications, NASA

Dr. James Turnock
Office of Manned Space Flight, NASA

Colonel John E. Pickering
(Executive Secretary)
Office of Manned Space Flight, NASA

Dr. G. Briggs Phillips
U.S. Public Health Service Consultant

[9]

TABLE I. ASTRONAUT QUARANTINE SCHEME FOR MANNED LUNAR MISSIONS

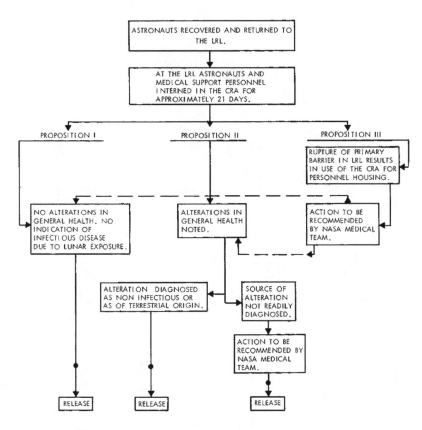

● INDICATES:
    (A) REVIEW OF DATA AND PROPOSED ACTION BY THE INTERAGENCY
        COMMITTEE ON BACK CONTAMINATION, AND
    (B) FORMAL CLEARANCE BY THE REGULATORY AGENCIES, WHEN
        NECESSARY.

[10]

TABLE II. QUARANTINE SCHEME FOR RETURNED LUNAR SAMPLES [PHASE I]

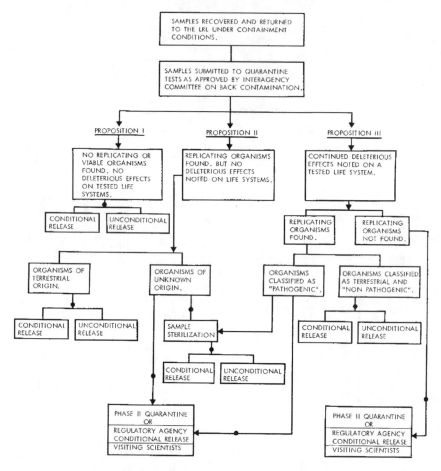

● INDICATES:
   (A) REVIEW OF DATA AND PROPOSED ACTION BY THE INTERAGENCY
       COMMITTEE ON BACK CONTAMINATION, AND
   (B) FORMAL CLEARANCE BY THE REGULATORY AGENCIES, WHEN
       NECESSARY.

[11]

Document II-53

Policy Directive

SUBJECT: OUTBOUND LUNAR BIOLOGICAL CONTAMINATION
             CONTROL: POLICY AND RESPONSIBILITY

1.   PURPOSE

    This Directive establishes the operational responsibilities for manned and automated lunar missions with regard to the amount of biological contamination and its placement on the lunar surface.

2.   SCOPE AND APPLICABILITY

    This Directive applies to all NASA Installations with respect to all outbound missions intended to or which may encounter the Moon.

3.   DEFINITION

    For the purpose of this Directive, the Apollo Landing Zone (ALZ) is defined as that portion of the Moon located between 5° north latitude and 5° south latitude, and between 45° east longitude and 45° west longitude.

4.   SPACE SCIENCE BOARD– NATIONAL ACADEMY OF SCIENCES
      RECOMMENDED POLICY

    During the early phases of lunar exploration, NASA undertook to minimize contamination on Ranger probes in order to avoid depositing terrestrial organisms on the Moon. Eventually, it became apparent that, although the objective was complete sterility, each probe that impacted on the Moon carried a number of microorganisms. In its review of NASA's experience of three years with lunar probe contamination control, the Space Science Board of the National Academy of Sciences made the following pertinent recommendations concerning spacecraft programmed to land on the Moon:

    "(i)   Minimize contamination to the extent technically feasible. By appropriate selection of components (favoring those which are inherently sterile internally) and the use of surface sterilants, it should be possible to achieve a cleanliness level to approximate that which prevails in most hospital surgery rooms.

[2]

    "(ii)   Inventory all organic chemical constituents. This will permit the interpretation of analytical results from future collections of lunar material.

* * *

"(iv) Undertake the development of the sterile drilling system to accompany an early Apollo mission to return an uncontaminated sample of the lunar subsoil. Samples aseptically collected from this subsoil will be of both biological and geochemical interest. Should life exist on the Moon, it might be expected at some depth below the surface where temperatures never exceed 100°C and below the zone of ultraviolet radiation. Every effort should be made to keep this level free of contaminants until it can be sampled by drilling."

5.   POLICY

a. Landings: Unless otherwise authorized by the Deputy Administrator, all manned landings will be confined to the Apollo Landing Zone.

b. Biological Loading

(1) Contamination of the manned landers will be held to the minimum practical level consistent with achieving the major mission objectives as specified in the appropriate mission assignment document as approved.

(2) Contamination on the surface of automated landers and orbiters will be kept below a level such that, if contamination is confined to an area of 2.59 square kilometers (1 square mile) around the lunar impact point, there will not be more than one viable organism per square meter.

c. Biological Inventory: An inventory of probable post-landing biological contamination levels at each Apollo and automated landing site and a total inventory for the Moon will be obtained and maintained for future reference in the event sites are revisited and to aid in the interpretation of data obtained in subsequent experiments.

[remainder of document not provided]

Document II-54

NATIONAL ACADEMY OF SCIENCES

March 24, 1969

Dr. Thomas O. Paine
Administrator
National Aeronautics and
  Space Administration
Washington, D. C. 20546

Dear Tom:

As you know, Professor Wolf Vishniac serves as our representative on the Interagency Committee on Back Contamination, which is concerned with precautions to be taken in connection with lunar materials and vehicles returning from the moon.

In connection with his responsibilities on this committee, and as noted in the enclosed copy of his letter to me, Professor Vishniac has identified apparent weaknesses in the quarantine procedures to be followed at the time of recovery of the lunar vehicle. Although these have been discussed with representatives of the Public Health Service and the Department of Agriculture, Professor Vishniac feels that these questions should be given immediate consideration by the Academy and the National Aeronautics and Space Administration. I am sure that he will be willing to discuss this matter in further detail with you and/or other appropriate representatives of the Administration.

Sincerely yours,

[signed]

Frederick Seitz
President

Enclosure

THE UNIVERSITY OF ROCHESTER
COLLEGE OF ARTS AND SCIENCE
RIVER CAMPUS STATION
ROCHESTER, NEW YORK 14627

Department of Biology   March 5, 1969

Dr. Frederick Seitz
President
National Academy of Sciences
2101 Constitution Avenue
Washington, D.C. 20418

Dear Dr. Seitz:

It is my unpleasant duty to report to you the present unsatisfactory status of the quarantine program which has been the concern of the Interagency Committee on Back Contamination. At the time of this writing there is a six week simulation in progress at the Lunar Receiving Laboratory, previous simulations having shown substantial faults in the functioning of various Lunar Receiving Laboratory components. One simulation had to be called off within two days after several neoprene gloves in the glove boxes gave way.

On February 12 and 13 the representatives of the Regulatory Agencies met at Houston to review the operation of the LRL and the retrieval scheme of the Apollo astronauts. I shall not bore you with a long list of technical faults that were found in the operation of the LRL, let me just mention a few significant samples. It has as yet been impossible to keep colonies of mice alive in the LRL. The mouse colonies, being behind biological barrier, are at reduced pressure with respect to the atmosphere outside the biological barrier. So far every single mouse colony has died, even without being intentionally infected with any pathogenic agent. Routine apparatus does not seem to work properly, so for instance autoclaves tend to fill with water. This was still true on February 27. There seems to be no way of carrying out rapid minor emergency repairs. Although there is a list of spare parts to be kept at the LRL, no parts are actually available. I could continue this unhappy list at great length.

Our major concern which I wish to report is the recurring problem of controlling the spacecraft atmosphere after re-entry. When the Interagency Committee first met it was presented with a procedure whereby the spacecraft, immediately after splash-down, would be ventilated with fresh air for the necessary comfort of the astronauts. Such uncontrolled outventing does not, naturally, impose any biological restraint on whatever particles or microorganisms may be suspended in the spacecraft atmosphere. At that time the Committee directed NASA to investigate the feasibility of installing biological filters in the air vents. The engineering response was that the installation of filters would require larger fans and more power to drive them than could be accommodated, and that filters were therefore not practical.

Meanwhile, calculation had been carried out that the re-cycling of air through the barium hydroxide canisters would remove free floating particles from the atmosphere during the return from the Moon. This calculation has since been shown to be in error by several orders of magnitude. Eventually a compromise solution was reached whereby the divers, in attaching the floatation collar around the craft, would place a biological filter over the vent holes from outside, and also provide a power pack to drive sufficiently powerful fans. On February 13 the members of the regulatory agencies were told that no such filtration was intended and that upon splash-down the capsule would have to be vented without any control. The reaction of the representatives of the various agencies was mixed. In the opinion of Dr. Bagby (PHS) this procedure did not seem to pose a direct threat to human beings and therefore the Public Health Service was not too concerned. Dr. Park (Department of Agriculture) felt that no immediate threat to agricultural crops was presented, and therefore he too was willing to go along with this procedure or at least not make a strong opposition to it.

This left matters up to Mr. Eckles of the Department of the Interior. Dr. Eckles would have to answer to his colleagues for the safety of the marine environment. Dr. Eckles was most unhappy about this procedure, and repeated the frequent complaint voiced by all members of the Interagency Committee, that at every meeting the ground rules previously given to us by MSC have been changed. Dr. Eckles suggested that a meeting might be arranged with a few competent biological oceanographers, in particular he had in mind Dr. Francis Haxo of the Scripps Oceanographic Institution, Dr. Luige Provasoli, of Haskens Laboratories and Dr. Carl Oppenheimer at Florida State University. My feeling is that in such a conference a few experts on atmospheric circulation should also be involved. Mr. Eckles feels that he is not in a strong position to object to the flight of Apollo 11 in its present configuration nor does he see any way in which he could influence a change in spacecraft design or recovery procedure.

My own reaction is based entirely on whether we consider back contamination a matter of concern or not. I believe that this question has been answered in the affirmative since NASA has gone to the expense of constructing a quarantine facility and working out an elaborate system by which astronauts could be transported behind biological barriers from the recovery area to the Lunar Receiving Laboratory. Once we have committed ourselves to this course it would be irresponsible to leave a large breach in the biological barrier in any part, of the recovery procedure. The uncontrolled outventing of the spacecraft is such a breach. I do not believe that either the Department of Agriculture or the Public Health Service should be as unconcerned about the problem as they appear to be. Should pathogenic organisms be brought back, and I will grant readily that the likelihood of this event is small, and should they infect organisms in the ocean, which is Mr. Eckles' concern, then there is the same danger that they may spread to land and become simultaneously a very great concern to the Public Health Service and the Department of Agriculture. If Apollo 11 is allowed to return in the currently contemplated manner, and if the atmosphere on the Spacecraft is to be vented to the outside without any control or restraint, then I see little reason for maintaining the biological isolation garments, the elaborate mobile quarantine units, the transport to Ellington Air Force base, and the entire Lunar Receiving Laboratory quarantine. If we abandon the entire quarantine then we may as well admit that we do so. However, if the quarantine is to be taken seriously then it must be enforced at every link in the chain of events. Another breach of quarantine appears to be the insistence of spacecraft engineers of entering the spacecraft immediately or at least after a very short time, without the three week quarantine that had been contemplated for it.

I am frankly at a loss to suggest what should be done at this mement [*sic*]. Clearly the Apollo Program is moving at a pace which we cannot stop. It is equally clear that this irresistible progress is being used to brush aside the inconvenient restraints which the Interagency Committee has considered to be an essential part of the Quarantine Program. The least I can do is to price [*sic*] you of the facts as they stand at the moment and to call them to the attention of the Space Science Board.

Sincerely yours,

[signed]
Wolf Vishniac

Document II-55

United States Senate
Committee on Aeronautical and Space Sciences
Washington, DC 20510

May 15, 1969

The Honorable Thomas O. Paine, Administrator
National Aeronautics and Space Administration
Washington, DC 20546

Dear Dr. Paine:

Recent news articles say that NASA is considering plans to relax its precaution against the spread of alien organisms that might be brought back from the moon by the Apollo 11 space flight because of recommendations made by an interagency committee. Under the change, it is my understanding that the Astronauts would be permitted to leave the spacecraft while it is still in the water instead of, as previously planned, bringing the spacecraft with the Astronauts inside aboard the carrier and releasing them into a biologically isolated vehicle. It is said that the changes are being proposed so as to air out the aircraft, save the Astronauts some discomfort, and avoid the hazard of bringing the tossing capsule near the hull of the carrier.

I wonder if it is wise to go to a procedure that can be regarded as having less concern for possible contamination of the earth; and hope that you will very carefully consider that if the Agency is to err it ought to err on the side of caution. The program has come a long way and is about to meet its objective. I would not like to see people start to criticize the program on the basis that all necessary and practical caution has not been taken to prevent the spreading of any possible harmful pathogens brought back from the moon.

Sincerely yours,

Clinton P. Anderson
Chairman

Cc: Mr. Robert Allnutt

Document II-56

National Aeronautics and Space Administration
Washington 25, D.C.

Jun 4 1969

Honorable Clinton P. Anderson
Chairman
Committee on Aeronautical and
  Space Sciences
United States Senate
Washington, D.C. 20510

Dear Mr. Chairman:

This is further in response to your letter of May 15 regarding the recovery procedures for the Apollo 11 mission.

The subject of possible back-contamination of the earth's biosphere through Apollo operations has, of course, received our very serious attention for some time. NASA and other agencies of Government have spent considerable effort to insure that everything possible is done to prevent such contamination consistent with safe accomplishment of the mission. To this end, in 1964 an Interagency Committee on Back Contamination (ICBC) was established to provide expert guidance to us on all matters concerning possible back-contamination. This Committee, composed of members from the Departments of Agriculture, Interior, Health, Education, and Welfare (U.S. Public Health Service), National Academy of Sciences and NASA, has the responsibility of insuring that our Apollo mission plans do no violate the integrity of the Earth's biosphere. Hence the preventive procedures we plan to employ must have the ICBC's approval before implementation.

A very difficult problem and decision we and the ICBC have had to resolve is the one you mention, that is, all constraints considered, determining the optimum recovery procedure which would protect the lives of the returned astronauts while at the same time providing the lowest practicable possibility of back-contamination. Our efforts have been directed toward both recovery procedures and methods to prevent uncontained lunar material from entering and leaving the Lunar Module and the Command Module.

The current astronaut-recovery procedure, which has been approved by the ICBC for the Apollo 11 mission, involves egressing them from the spacecraft into a raft and transferring them by helicopter to the recovery ship where they will enter the Mobile Quarantine Facility. The astronauts will don Biological Isolation Garments prior to departing the spacecraft if sea conditions permit; otherwise the garments will be donned in the life raft.

We had considered having the astronauts remain inside the Command Module while it was hoisted onto the recovery ship. Since this represented a departure from the present recovery procedures which have been developed over a period of several years and which are based on the cumulative experience of Mercury, Gemini, and Apollo, a thorough review was made of the difficulties involved in transferring the Command Module to the carrier deck, particularly in a heavy sea. The hazards demonstrated in actual practice led to our decision to transfer the astronauts to the carrier deck by helicopter. The current astronaut recovery procedures received ICBC approval for the Apollo 11 mission only after the ICBC became convinced that (1) there was a real hazard involved in sea retrieval of a manned spacecraft and (2) any increased risk of biosphere contamination was not significant. The former concern has been validated in both tests and previous end-of-mission recoveries. Test data has dictated the installation by a swimmer of a recovery loop or sling onto the spacecraft prior to lifting it from the water because the integral loop on the Command Module will not accommodate all possible recovery loads. Such a procedure is acceptable to us for use only on an unoccupied spacecraft. At the conclusion of the Apollo 9 mission, for example, the spacecraft was dropped back into the water due to a mechanical failure of the crane.

The increase in the contamination potential from extracting the astronauts has been minimized by programming improved housekeeping procedures by the astronauts and by recognition of the scrubbing action of the Lunar Module and Command Module lithium hydroxide (LiOH) canisters on the cabin atmosphere. The astronauts will now bag all items exposed to the lunar surface prior to transfer to the Command Module. They plan to vacuum the cabin at frequent intervals during the return trip from lunar orbit. Of additional significance, however, is recently developed data which indicates that the LiOH canisters will remove essentially all of the particulate effects of minimizing cabin interior contamination and understanding LiOH filtering capabilities have led us to conclude, and the ICBC to concur, that the recovery procedure described does not materially increase the probability of earth contamination.

In these few paragraphs I have not described all the detailed procedural steps we plan to take to reduce the possibility of Earth back-contamination. For instance, the maximum number of items possible which have contacted the lunar surface will either be left on the lunar surface or in the LM. This and many other steps we are taking represent a heavy concentration of effort to tighten our procedures to minimize the possibility of back-contamination of the earth's biosphere.

If we can provide any additional information, please let me know.

Sincerely yours,

Homer E. Newell
Acting Administrator

Document II-57

Document Title: Director of Central Intelligence, "The Soviet Space Program,"
4 April 1968.

Source: Central Intelligence Agency Historical Review Program.

*Throughout the 1960s the Central Intelligence Agency (CIA) used its various capabilities
to track the Soviet space program. This update to a November 1967 National Intelligence
Estimate gives a sense of what the CIA was saying about the Soviet lunar landing program.
According to this estimate, the United States was well in the lead in achieving the first lunar
landing. Of particular note , however, is the estimate that the Soviet Union might attempt
a manned circumlunar flight before the end of 1968. Senior NASA officials were certainly
aware of this possibility as they considered whether to approve sending the Apollo 8 mission
into orbit around the moon in December 1968.*

TOP SECRET [DECLASSIFIED]
CONTROLLED DISSEM
[declassified 1/16/1997]

NIE 11-1-67
4 April 1968
TS 0089284/1

MEMORANDUM TO HOLDERS

NATIONAL INTELLIGENCE ESTIMATE

NUMBER 11-1-67

The Soviet Space Program

Submitted by

[Signed Richard M. Helms]

Director of Central Intelligence

Concurred in by the

UNITED STATES INTELLIGENCE BOARD

As indicated overleaf

4 April 1968

Authenticated

[Signed]

EXECUTIVE SECRETARY USIB   Pages 10

[1]

THE SOVIET SPACE PROGRAM

THE PROBLEM

To examine significant developments in the Soviet space program since the publication of NIE 11-1-67, "The Soviet Space Program," dated 2 March 1967, TOP SECRET, and to assess the impact of those developments on future Soviet space efforts with particular emphasis on the manned lunar landing program.

DISCUSSION

1. In the year since publication of NIE 11-1-67, the Soviets have conducted more space launches than in any comparable period since the program began.[1] Scientific and applied satellites, particularly those having military applications, largely account for the increased activity. The Soviets also intensified efforts to develop what we believe to be a fractional orbit bombardment system (FOBS).[2] The photoreconnaissance program continued at the same high rates of the previous two years.

2. In general, the Soviet space program progressed along the lines of our estimate. It included the following significant developments: new spacecraft and launch vehicle development, rendezvous and docking of two unmanned spacecraft, an unsuccessful manned flight attempt (which ended in the death of Cosmonaut Komarov), the successful probe to Venus, an unmanned circumlunar attempt which failed, and a simulated circumlunar mission. Evidence of the past year indicates that the Soviets are continuing to work toward more advanced missions, including a manned lunar landing, and it provides a better basis for estimating the sequence and timing of major events in the Soviet space program.

3. Considering additional evidence and further analysis, we continue to estimate that the Soviet manned lunar landing program is not intended to be competitive with the US Apollo program. We now estimate that the Soviets will

1 See Annex for a detailed breakdown of launches during the past year.
2 For a discussion of FOBS, see NIE 11-8-67, "Soviet Capabilities for Strategic Attack," dated 26 October 1967, TOP SECRET.

attempt a manned lunar landing in the latter half of 1971 or in 1972, and we believe that [2] 1972 is the more likely date. The earliest possible date, involving a high risk, failure-free program, would be late in 1970. In NIE 11-1-67 we estimated that they would probably make such an attempt in the 1970-1971 period; the second half of 1969 was considered the earliest possible time.

4.   The Soviets will probably attempt a manned circumlunar flight both as a preliminary to a manned lunar landing and as an attempt to lessen the psychological impact of the Apollo program. In NIE 11-1-67, we estimated that the Soviets would attempt such a mission in the first half of 1968 or the first half of 1969 (or even as early as late 1967 for an anniversary spectacular). The failure of the unmanned circumlunar test in November 1967 leads us now to estimate that a manned attempt is unlikely before the last half of 1968, with 1969 being more likely. The Soviets soon will probably attempt another unmanned circumlunar flight.

5.   Within the next few years the Soviets will probably attempt to orbit a space station which could weigh as much as 50,000 pounds, could carry a crew of 6-8 and could remain in orbit for a year or more. With the Proton booster and suitable upper staging they could do so in the last half of 1969, although 1970 seems more likely. Alternatively, the Soviets could construct a small space station by joining several spacecraft somewhat earlier—in the second half of 1968 or 1969—to perform essentially the same functions. We previously estimated that the earliest the Soviets could orbit such a space station was late 1967 with 1968 being more likely.

6.   We continue to believe that the Soviets will establish a large, very long duration space station which would probably weigh several hundred thousand pounds and would be capable of carrying a crew of 20 or more. Our previous estimate, which gave 1970-1971 as the probable date and late 1969 as the earliest possible, was based primarily upon launch vehicle capacity. We now believe that the pacing item will be the highly advanced life support/environmental control technology required, and that such a station will probably not be placed in orbit before the mid-1970's.

[remainder of estimate not provided]

Document II-58

Document Title: Memorandum to Associate Administrator for Manned Space Flight from James E. Webb, Administrator, "Termination of the Contract for Procurement of Long Lead Time Items for Vehicles 516 and 517," 1 August 1968.

Source: Folder #18675, NASA Historical Reference Collection, History Division, NASA Headquarters, Washington, DC.

To ensure that there were enough heavy-lift boosters to complete the Apollo program, NASA had contracted for the elements of 15 Saturn V vehicles. George Mueller, Associate Administrator for Manned Spaceflight, hoped to keep open the various production lines involved in the Saturn V program, anticipating that there would be other uses for the giant vehicle— extended lunar exploration and launching a space station, for example—that would require a heavy-lift capability during the 1970s. The program to carry out such activities was known as the Apollo Applications Program. The first step in ensuring that this could be done was to contract for those components of the vehicle's S-IC first stage that required the longest time to manufacture. In mid-1968, Mueller requested authorization from James Webb to enter into such contracts.

Webb's answer was negative— no uses for Saturn Vs beyond the original 15 had been approved, and the budget outlook for such approval was gloomy. This memorandum, issued even before the initial lunar landing, was thus the first step in a process that led to a 1970 decision to terminate the Saturn V program.

Memorandum to Associate Administrator for Manned Space Flight

SUBJECT: Termination of the Contract for Procurement of Long lead
                  Time Items for Vehicles 516 and 517

REFERENCE: M memorandum to the Administrator, dated June 2, 1968,
                  same subject
                  D memorandum to the Administrator, dated July 31, 1968
                  AD memorandum to M dated July 13, 1967

          After reviewing the referenced documentation and in consideration of the FY 1969 budget situation, your request to expend additional funds for the procurement of long lead time items for the S-IC stages of the 516 and 517 vehicles is disapproved. The decision, in effect, limits at this time the production effort of Saturn through vehicle 515. No further work should be authorized for the development and fabrication of vehicles 516 and 517.

                              James E. Webb
                              Administrator

Document II-59

Document Title: Memorandum to Manager, Apollo Spacecraft Program from Chief, Apollo Data Priority Coordination, "Re: LM rendezvous radar is essential," 1 August 1968.

Source: Folder #18675, NASA Historical Reference Collection, History Division, NASA Headquarters, Washington, DC.

One of the key managers of Project Apollo at the Manned Spacecraft Center was Howard W. "Bill" Tindall. He became famous throughout the program for his "Tindallgrams," messages expressed in direct, often pithy terms. This is an example. The high official referred to was George Mueller, NASA Associate Administrator for Manned Space Flight.

TO   :   PA/Manager, Apollo Spacecraft Program   DATE: AUG 1 1968
                                                          [stamped]

FROM :   PA/Chief, Apollo Data Priority Coordination

SUBJECT: LM rendezvous radar is essential

A rather unbelievable proposal has been bouncing around lately. Because it is seriously ascribed to a high ranking official, MSC and GAEC are both on the verge of initiating activities - feasibility studies, procedures development, etc. - in accord with it. Since effort like that is at a premium, I thought I'd write this note in hopes you could proclaim it to be a false alarm or if not, to make it one. The matter to which I refer is the possibility of deleting the rendezvous radar from the LM.

The first thing that comes to mind, although not perhaps the most important, is that the uproar from the astronaut office will be fantastic - and I'll join in with my small voice too, for the following reason. Without rendezvous radar there is absolutely no observational data going into the LM to support rendezvous maneuvers. This would be a serious situation both during the major rendezvous maneuvers (CSI, CDH, and TPI) and during terminal braking. Please let me discuss these separately.

First of all, let it be clearly understood the MSFN <u>cannot</u> support rendezvous maneuver targeting during lunar operations. That must be an entirely onboard operation due to limitations in MSFN navigation (i.e., orbit determination) using short arcs of data on a maneuvering spacecraft and because much of the rendezvous is conducted out-of-sight - and - voice of the earth. In other words, we couldn't tell them what to do if we knew!

Therefore, without the LM radar the <u>only</u> source of maneuver targeting is the CSM. Using what? A VHF ranging device to be flown for the first time on the lunar mission and a spacecraft computer program (Colossus), which does not have the CSI and CDH targeting programs in it. Thus, the CSM pilot would have to use charts! I'd like to emphasize the fact, though, that the CSM pilot is so busy making sextant observations (which are mandatory - VHF alone is not adequate) and performing mirror image targeting, etc. along with routine spacecraft management that it has been concluded he can not and will not perform onboard chart computations.

[2] And - even if we were to think negative schedule-wise and assume we will get a flight qualified VHF ranging device <u>and</u> CSI/CDH targeting in Colossus, Jr. in time for the lunar mission, I can't believe we'd be willing to fly a rendezvous with no backup or alternate data source for comparison. The ΔV margins are too small and the consequence of failure is unacceptable!

Now, let me speak of terminal phase braking. Range and range rate information are essential for this operation. This can be obtained crudely by visual means and without radar that's the only source. (Lighting conditions must be satisfactory - although poor CSI/CDH targeting will cause TPI time slippage almost certain to mess it up.) The DSKY displays of range and range rate from the computers are based on the state vectors obtained by the rendezvous navigation and they degrade badly at close ranges. That is, their usefulness is highly questionable. (Unless lunar operations are better than "earthal," they are worthless; I'm not sure if lunar is better or not.) So it's the eyeballs then and plenty of RCS.

If I sound like I'm on some higher energy level about this, it's cause I am. I'm sure most will agree that a rendezvous radar failure is the worst that can happen in the PGNCS (and AGS) during rendezvous since without it all data is lost. (For example, the current "D" rendezvous mission rule is that rendezvous radar failure dictates aborting the rendezvous exercise, the CSM goes active for TPI and midcourse corrections, using the sextant, and whoever can see best will give a try at braking.)

Please see if you can stop this if it's real and save both MSC and GAEC a lot of trouble.

[signed]

Howard W. Tindall, Jr.

Document II-60

Document Title: George M. Low, "Special Notes for August 9, 1968, and Subsequent," 19 August 1968.

Source: Papers of George M. Low, Rensselaer Polytechnic Institute, Troy, New York.

Document II-61

Document Title: Sam C. Phillips, Apollo Program Director, "Apollo Mission Preparation Directive," 19 August 1968.

Source: Folder #18675, NASA Historical Reference Collection, History Division, NASA Headquarters, Washington, DC.

Document II-62

Document Title: Letter to Robert Gilruth, Director, NASA Manned Spacecraft Center, from George E. Mueller, NASA Associate Administrator for Manned Space Flight, 4 November 1968.

Source: NASA Manned Spacecraft Center Archives.

Document II-63

Document Title: George M. Low "Special Notes for November 10 and 11, 1968," 14 November 1968.

Source: Papers of George M. Low, Rensselaer Polytechnic Institute, Troy, New York.

Document II-64

Document Title: Memorandum to Associate Administrator for Manned Space Flight [George Mueller] from Apollo Program Director [Sam C. Phillips], "Apollo 8 Mission Selection," 11 November 1968.

Source: Papers of George M. Low, Rensselaer Polytechnic Institute, Troy, New York.

Document II-65

Document Title: Memorandum to Associate Administrator for Manned Space Flight [George Mueller] from Acting Administrator [Thomas Paine], 18 November 1968.

Source: Papers of George M. Low, Rensselaer Polytechnic Institute, Troy, New York.

One of the boldest decisions made during the Apollo program was to send astronauts into lunar orbit on the first Saturn V launch with a crew aboard. The result was the 21–27 December 1968 Apollo 8 mission which carried Frank Borman, James Lovell, and Bill Anders into lunar orbit on Christmas eve and produced the iconic "Earthrise" picture of the blue Earth rising over the desolate lunar surface.

This series of documents illustrates how this decision was made. Apollo program manager George M. Low periodically dictated what he called "special notes" as a chronicle of the Apollo program from his central perspective. These notes, which were supplemented by official documents, form an invaluable record of space policy and management actions from 1967 until Low left NASA in 1976. In his August and November 1968 notes, Low narrates the series of events and decisions that led to the decision to fly Apollo 8 around the Moon. Perhaps most remarkable were the events of 9 August, which began with a brief conversation about the desirability of such a decision between Low and the Director of the Manned Spacecraft Center Robert Gilruth and, by the time the day was over, involved key Apollo decision makers in Houston, Huntsville, and Washington. When NASA Administrator James Webb and Head of Manned Space Flight George Mueller, who were attending a United Nations Conference in Vienna, Austria, heard of the Apollo 8 plan, they were taken quite aback, and insisted that no decision be announced until after the Apollo 7 mission, which was to test the post-fire Apollo Command and Service Modules in Earth orbit. Although final approval of the preliminary decisions taken that day would be months in coming, it is remarkable that the basics of such a momentous choice could be put in place in just a few hours on one day, and then put in motion a few days later.

The Apollo 8 lunar orbit mission was designated C' (C Prime) because it was inserted in the previously planned Apollo mission sequence which included the following missions: C – test of the Apollo Command and Service module in low Earth orbit; D – test of the Apollo Command and Service and Lunar Modules in low Earth orbit; E – test of the Apollo Command and Service and Lunar Modules in a mission beyond Earth orbit, but not headed to the moon; F – test of all equipment in lunar orbit; and G – lunar landing mission.

There is no mention in any of these documents of any concern that the Soviet Union might be able to fly a cosmonaut crew around the Moon before the United States was able to send its astronauts to the lunar vicinity, even though intelligence estimates and several 1968 flights of the Soviet "Zond" spacecraft suggested that such a mission might be in preparation.

In addition to Low's notes, documents included here are reservations about the wisdom of undertaking the mission raised by Associate Administrator for Manned Space Flight George Mueller, Apollo Program Manager Lieutenant General Sam Phillip's memoranda making the changes in mission plans that would allow the circumlunar choice and formally recommending approval of the circumlunar Apollo 8 mission, and NASA Acting Administrator Thomas Paine's memorandum documenting his decision to approve that recommendation.

Document II-60

<u>SPECIAL NOTES FOR AUGUST 9, 1968, AND SUBSEQUENT</u>

<u>Background</u>:

<u>June, July 1968.</u> The current situation in Apollo was that LM - 3 had been delivered to KSC somewhat later than anticipated; and CSM 103 would be delivered to KSC in late July. Checkout of 101 at KSC was proceeding well, and a launch in the Fall of 1968 appeared to be assured. There was every reason to believe that 103 would also be a mature spacecraft but that for many reasons LM-3 might run into difficulties. Certification tests of LM were lagging; there were many open failures; and the number of changes and test failures at KSC was quite large.

It had been clear for some time that a lunar landing in this decade could be assured only if the AS 503/CSM 103/LM 3 mission could be flown before the end of 1968. During the June-July time period the projected launch date had slipped from November into December, and the December date was by no means assured. The over-all problem was compounded by the Pogo anomaly resulting from the Apollo 6 mission, and this remained a significant unknown.

In this time period also the possibility of a circumlunar or lunar orbit mission during 1968, using AS 503 and CSM 103, first occurred to me as a contingency mission to take a major step forward in the Apollo Program.

July 20 to August 5, 1968. By now the Pogo situation looked a lot more encouraging. MSFC had demonstrated analytically that a relatively simple launch vehicle fix was available to cure the problem. The results of many tests and analyses at MSC led to the general conclusion that the Spacecraft/LM Adapter problem would most likely be cured if the launch vehicle Pogo is cured.

In the same time period, work on CSM 103 continued to progress somewhat slower than expected but in a satisfactory manner. Delivery of the spacecraft to KSC during the second week of August was virtually assured. The spacecraft was extremely clean. LM-3, however, required much more work at KSC than anticipated. There was a significant number of changes in addition to test failures, requiring trouble-shooting, changeouts and retest, and a serious EMI problem that continued to persist. The [2] outlook for a 1968 launch, although mathematically still possible, appeared to be very dim.

August 6, 1968. Presented a long list of LM changes to the OMSF Management Council review in Houston. In collecting this information it had become more and more apparent that we still weren't quite on top of the situation and that the list of problems continued to grow instead of decreasing.

August 7, 1968. With the background of open work and continued problems on LM-3 and the real concern that the mission might not be able to fly until February or March, 1969, I asked Chris Kraft to look into the feasibility of a lunar orbit mission on AS 503 with CSM 103 and without a LM.

August 8, 1968. Spent the day at KSC, reviewing 503 open work and schedules with Debus, Petrone, Phillips, Hage, Bolender, and many others. The official KSC schedule showed an earliest possible launch date during the first week of January, 1969; however, the EMI problem was still open. KSC pointed out that the hardware changes were not the real cause of the problem. The many retest requirements and checkout problems caused real concern. There was little confidence in the assembled group that the early January launch date could be met. In fact, until the EMI problem was solved, things were essentially at a standstill.

Steps in Planning the Mission:

August 9, 1968. Met with Gilruth at 0845 and reported to him the detailed status of LM-3 and CSM 103 and informed him that I had been considering the possibility of an AS 503 lunar orbit mission. Gilruth was most enthusiastic and indicated that this would be a major step forward in the program.

Met with Chris Kraft at 0900, and he indicated that his preliminary studies had shown that the mission was technically feasible from the point of view of ground control and onboard computer software. (A step of major importance to make this possible had been taken several months ago when we had decided to use the Colossus onboard computer program for the 103 spacecraft.)

At 0930, I met with Gilruth, Kraft and Slayton. After considerable discussion, we agreed that this mission should certainly [3] be given serious consideration and that we saw no reason at the present time why it should not be done. We immediately decided that it was important to get both von Braun and Phillips on board in order to obtain their endorsement and enthusiastic support. Gilruth called von Braun, gave him the briefest description of our considerations, and asked whether we could meet with him in Huntsville that afternoon. I called Phillips at KSC and also informed him of our activities and asked whether he and Debus could join us in Huntsville that afternoon. Both von Braun and Phillips indicated their agreement in meeting with us, and we set up a session in Huntsville for 2:30 p.m.

August 9, 1968 - 2:30 p. m. Met in von Braun's office with von Braun, Rees, James and Richard from MSFC; Phillips and Hage from OMSF; Debus and Petrone from KSC; and Gilruth, Kraft, Slayton and Low from MSC. I described the background of the situation, indicating that LM-3 has seen serious delays and that presently we were one week down on the KSC schedule, indicating a 31 December launch. I went on to indicate that, under the best of circumstances, given a mature spacecraft, we might expect a launch at the end of January; however, with the present situation on LM, I would expect that the earliest possible D mission launch date would be during the middle of March. It therefore appeared that getting all of the benefits of the F (lunar orbit) mission before the D mission was both technically and programmatically advisable. Under this concept a lunar orbit mission, using AS 503 and CSM 103, could be flown in December, 1968. The most significant milestone in this plan would have to be an extremely successful C mission, using CSM 101. However, if 101 were not completely successful, an alternate to the proposed mission would be a CSM alone flight, still in December, using AS 503 and CSM 103 in an earth orbital flight rather than a lunar orbit flight. Under this plan the D mission would be flown on AS 504 with CSM 104 and LM-3, probably still in mid-March. In other words, we would get an extra mission in ahead of the D mission; would get the earliest possible Pogo flight; and would get much of the information needed from the F mission much earlier than we could otherwise. Chris Kraft made the strong point that, in order to gain the F mission flight benefits, the flight would have to be into lunar orbit as opposed to circumlunar flight.

During the remainder of the meeting in Huntsville, all present exhibited a great deal of interest and enthusiasm for this flight.

[4] Phillips outlined on the blackboard the actions that would have to be taken over the next several days. Generally, KSC indicated that they could support such a mission by December 1, 1968; MSFC could see no difficulties from their end;

MSC's main concern involved possible differences between CSM 103 and CSM 106, which was the first one that had been scheduled to leave earth orbit, and finding a substitute for the LM for this flight.

The Huntsville meeting ended at 5 p.m. with an agreement to get together in Washington on August 14, 1968. At that time the assembled group planned to make a decision as to whether to proceed with these plans or not. If the decision was affirmative, Phillips would immediately leave for Vienna to discuss the plans with Mueller and Webb, since it would be most important to move out as quickly as possible once the plan was adopted. It was also agreed to classify the planning stage of this activity secret, but it was proposed that, as soon as the Agency had adopted the plan, it should be fully disclosed to the public.

August 9, 1968 - 8:30 p.m. After returning to Houston, held a meeting with Kleinknecht, Bolender, Dale Myers of NR, and George Abbey. We agreed to move out as described earlier with a view toward identifying any difficulties over the weekend. Bolender immediately left for Bethpage to discuss the proposal with GAEC and to find the best possible LM substitute. Myers returned to Downey to work the problem from that end.

August 10, 1968. No difficulties identified as yet. Kleinknecht is defining detailed configuration differences between CSM 103 and 106, and the most outstanding difficulty will probably be in the area of the high gain antenna. (This was known at the time the plan was discussed on August 9.) Insofar as a LM substitute is concerned, it looks as though LM-2 might be able to support this flight. Kotanchik, however, made a strong point that we should not fly a LM but install a simple crossbeam instead. He indicated that if a residual Pogo problem remained, it would be best not to have a LM on this flight; and if Pogo were solved, the LM would not be necessary. I discussed this with Hage in Washington and Richard at MSFC. Both agreed that a high-fidelity LM would not be necessary but that a mass representation might be required to avoid Saturn V control systems dynamics problems.

I also discussed the proposed mission with Bill Bergen, who appeared less receptive than most of the people who had been exposed to this plan.

[5] August 12, 1968. Held many meetings and telephone conversations on the subject of the new mission during the day.

Kraft indicated that the biggest constraint was the launch window; a December 20 launch would be required if a daylight launch was desired. (All agreed that for the first manned Saturn V flight a daylight launch would be a requirement.) We thought it would be best to plan for a December 1 launch and build in a "hold" period until December 20 to give maximum assurance of meeting that date.

In the area of a LM substitute, LTA-B appeared on the scene. This test article had been through the dynamic test vehicle program at MSFC and was now stored at KSC, ready for an unmanned 503 launch. It has the proper mass distribution and is in a flight-ready condition. All except Kotanchik agreed that this would be a good choice. Kotanchik made a strong point that we should fly with a lightweight crossbeam in order to get a maximum possible safety factor in the SLA region. During several discussions with MSFC we determined that this was not possible for the previously stated reasons concerning the launch vehicle dynamics.

GAEC proposed that LM-2 should not be flown in order to save it for the drop test program. They suggested instead that we build an LTA-4, consisting of the LM-9 descent stage with LM-8 ascent stage. However, since this would take another flight LM out of the program, I concluded that LTA-B would be our best choice.

August 13, 1968. Continued working detailed problems in Houston. A thorough analysis of configuration differences between 103 and 106 identified the high gain antenna as the most critical item. However, Kraft indicated that it would be possible to fly the mission even if the high gain antenna should fail during the flight. There were no "show stoppers" in any of the spacecraft systems and, in fact, only minor changeouts would have to be made to bring the spacecraft into a position to fly the proposed mission.

Kraft had reviewed all of the operational elements and determined that there would be no insurmountable difficulties. The available launch window will be from December 20 to December 26 (with the exception of December 25). In early January a launch window with an Atlantic injection would become available, and toward the end of January another Pacific injection window would open up.

[6] Slayton had decided to assign the 104 crew to this mission (Borman, Lovell and Anders, backed up by Armstrong, Aldrin and Haise) in order to minimize possible effects on the D mission. Slayton had talked to Borman on Saturday and found him to be very much interested in making this flight.

August 14, 1968. Went to Washington with Gilruth, Kraft and Slayton to meet with Paine, Phillips, Hage, Schneider and Bowman from Headquarters; von Braun, James and Richard from Marshall; and Debus and Petrone from KSC. The meeting started with an MSC review of spacecraft, flight operations, and flight crew support for the proposed mission. I reviewed the Spacecraft 103 hardware configuration, the proposed LM substitute, consumable requirements, and the proposed alternate mission. Copies of the charts used in this review are attached. [not included]

Kraft indicated that there were no major problems with either the MSFN or the Mission Control Center and flight operations. He discussed the launch window constraints and indicated that NASA management would have to get with the Department of Defense in order to obtain recovery support. Our conclusion was that we should go for the December 20, 1968, launch window with a built-in two week hold prior to the launch. Then, if it is logistically possible to shift to the Atlantic insertion period, we should try for the January 3, 1969, launch window if we miss the December launch window. If this is not feasible, we would have to go from the December 20 date to the January 20 date.

MSFC indicated that there were no significant difficulties with the launch vehicle to support this mission. We agreed that LTA-B would be loaded for a total payload weight of 85,000 pounds. MSFC also agreed that they could provide telemetry for the LTA-B measurements.

Petrone outlined his plans for activities at KSC and indicated that the earliest possible launch date would be December 6, 1968. Other dates included the first manned altitude chamber run on September 14; the move to the VAB on September 28; and move to the pad on October 1.

We also discussed the mission sequence to be followed after the proposed mission and proposed that the best plan would be to fly the D mission next, followed by an F mission, which, in turn, would be followed by the first lunar landing mission. In other words, the [7] proposed mission would take the place of the E mission but would be flown before D. MSC also proposed that for internal planning purposes we should schedule the D mission for March 1, 1969; the F mission for May 15, 1969; and the G mission for July or August, 1969. However, dates two weeks later for D, one month later for F, and one month later for G should be our public commitment dates.

During the course of the meeting Phillips received a call from George Mueller in Vienna. Apparently Phillips had discussed the proposal with Mueller on the previous day, and after thinking it over, Mueller's reception was very cool. Mueller was concerned over stating the plan before the flight of Apollo 7 and was against announcing a plan as we might have to back away from it if 101 did not work. He also indicated that Phillips' arrival and departure in Vienna might create problems with the press and therefore urged Phillips not to come. Mueller's plans were to return to the country on August 21 for a speech in Detroit, and he would not be able to meet with us in Washington until August 22.

All present indicated that we would have to move out immediately in order to meet the December launch window and that a delay until August 22 or later would automatically mean the mission would have to slip until January. It was also hard for us to believe that Mueller was unwilling to accept the plan which was unanimously accepted by all Center Directors and Program Managers. We again urged Phillips to review our findings with Mueller and make a strong plea to visit Mueller in Vienna immediately, assuming, of course, that it was not possible for Mueller to return to this country. We also pointed out that if we were to implement our plan with any degree of confidence, so many people would have to become involved that it would be impossible to keep it quiet for very long.

Following the over-all discussions of the mission, Dr. Paine indicated that it had not been too long since we were uncertain as to whether the Apollo 503 mission should even be manned. Now we were proposing an extremely bold mission. Had we really considered all of the implications? He specifically wanted to know whether anyone present was against making this move. In going around the table, one by one, the following comments were made:

von Braun: Once a decision has been made to fly a man on 503, it doesn't matter to the launch vehicle how far we go. From the [8] program point of view, this mission appears to be simpler than the D mission. The mission should by all means be undertaken.

Hage: There are a number of way stations in the mission. Decision points can be made at each of these way stations, thereby minimizing the over-all risk. I am all for the mission.

Slayton: This is the only chance to get to the moon before the end of 1969. It is a natural thing to do in Apollo today. There are many positive factors and no negative ones.

Debus: I have no technical reservations; however, it will be necessary to educate the public, for if this is done wrong and we fail, Apollo will have a major setback. By all means fly the mission.

Petrone: I have no reservations.

Bowman: It is a shot in the arm for manned space flight.

James: Manned safety in this flight and in the following flights is enhanced. The over-all Apollo budget and schedule position is enhanced. An early go-ahead is needed.

Richard: The decision to fly manned has already been made for 503. Our lunar capability in Apollo is enhanced by flying this mission now.

Schneider: This has my whole-hearted endorsement. There are very valid reasons for pressing on.

Gilruth: Although this may not be the only way to make our goal, it certainly enhances our possibility. There is always risk in manned space flight, but this is a path of less risk. In fact, it has a minimum risk of all of our Apollo plans. If I had the key decision, I would make it in the affirmative.

Kraft: Probably the flight operations people have the most difficult job in this. We will need all kinds of priorities. It will not be easy to do, but I have every confidence in our doing it. However, it should be a lunar orbit mission and not a circumlunar mission.

[9]

Low: This is really the only thing to do technically in the current state of Apollo. Assuming a successful Apollo 7 mission, there is no other choice. The question is not whether we can afford to do it, it should be can we afford not to do it.

Following this set of comments, Paine congratulated the assembled group for not being prisoners of previous plans and indicated that he personally felt that this was the right thing for Apollo and that, of course, he would have to work with Mueller and Webb before it could be approved.

Phillips indicated that his conclusion was that this was a technically sound thing to do and does not represent a short cut introducing additional risks. Our plan would be to meet with Mueller on Thursday, August 22, in Washington. Phillips reiterated Mueller's reservations. These included reservations about program risks such as possible questions about irresponsible scheduling, possible program impact if the Apollo 7 mission should fail and we could not proceed with an announced major step forward, and the question concerning program impact of a catastrophic failure on this special mission.

At the conclusion of the meeting we agreed to move out on a limited basis. Since the day-by-day timing was critical, Phillips agreed that we should involve the next level of people required to carry forward with our plans, giving them, of course, proper instructions about the current security classification of the mission. At the conclusion of the meeting Phillips indicated the earliest possible decision would come in 7 to 10 days under the best of circumstances.

August 15, 1968. Received a call from Phillips while at Bethpage for a GAEC CCB. Phillips indicated "we broke the log jam" and that Mueller had agreed to our plan. However, he would prefer if publicly we did not commit to the total plan but indicate only that AS-503 mission would be flown without a LM; that we were reviewing many objectives for the actual mission; and that these objectives included plans for an earth orbital flight like the Apollo 7 mission and plans for a lunar orbital flight; the final mission decision would not be made until after the Apollo 7 flight. The internal program directive, however, would be that we should make our plans for the most difficult mission and that our planning should proceed for a lunar orbit mission in December.

[10]   Later in the day, Phillips and Paine discussed the plan with Webb, who apparently had not yet heard from Mueller. (Webb is in Vienna, too.) Webb wanted time to think about the plan and requested that information be sent to him via diplomatic courier. Paine and Phillips expected a call from Webb and Mueller on August 16, 1968.

I discussed our plans with Lew Evans at GAEC. He, of course, had previously been informed by Joe Gavin. Evans' reaction was very favorable, indicating that this was the best thing that Apollo could do at this time.

August 16, 1968. No news from Washington today. Apparently Phillips and Paine have been in meetings most of the day with some correspondence going back and forth to and from Vienna. Late in the day, Phillips called and indicated that he and Hage would come to Houston tomorrow to meet with Gilruth, Kraft, Slayton and Low to decide how to proceed within the constraints imposed by Mr. Webb.

In the meantime, we worked several of the detailed problems and moved out on many of the required spacecraft changes. Kleinknecht asked Arabian to be sure that we will have a high-gain antenna. We moved out on several other spacecraft changes, without divulging to the people involved why the changes are required. (Many of the changes we are authorizing today were firmly turned down in recent CCB's.)

August 17, 1968. Phillips and Hage came to Houston to meet with Gilruth, Kraft, Slayton and Low. Phillips indicated that we have clear-cut authority from Mr. Webb to prepare for a December 6 flight of 103/LTA-B/503; that this mission will be known as the C' [C prime] mission, designated as Apollo 8; that the E mission crew will fly this mission; that this will be an earth-orbital mission with basic objectives to mature the CSM and Saturn V systems; and that we may proceed with studies and plans to gain the maximum flexibility when the final C' mission objectives are defined after Apollo 7.

Webb also authorized preparation of 104/LM-3/504 for a February 20 flight of the D mission.

A copy of General Phillips' notes on this subject is attached. Also attached is a copy of a telegram from Mr. Webb to Dr. Paine. [not included]

Phillips indicated that the major problem expressed by Dr. Mueller was that we could not obtain clearance to proceed with a lunar orbit mission until after the results of Apollo 7 were available.

[11]

Phillips indicated that Webb's initial reaction (on August 15) was one of shock and that he was fairly negative to the proposed lunar orbital mission. Following this, Paine and Phillips sent a lengthy paper to Vienna, giving the rationale for the need to change the mission sequence and proposing that the full range of capabilities from earth orbital up to lunar orbit should be authorized and discussed publicly. However, for many reasons Webb was unwilling to permit a commitment at this time beyond an earth orbital mission. Phillips was convinced, however, that Webb would consider going all the way to a lunar orbital mission after Apollo 7, provided, of course, that Apollo 7 was a successful flight.

Our challenge, therefore, is to be prepared to carry out a full lunar orbit mission without committing the Agency to such a mission at this time. This had been our objective as well, even during our initial meeting in Huntsville, but at that time we saw no way to achieve this.

We discussed many alternatives, always keeping in mind that we had to be completely honest and forthright with Dr. Mueller, Dr. Paine and Mr. Webb, and be prepared to fly an earth orbital mission in December. However, we wanted to keep the door open to be able to fly a lunar orbital mission, should we be ready to do so after Apollo 7. At the same time we agreed that whatever we did, we would have to be perfectly honest within NASA and with the press in stating what we were doing and why we were doing it.

Our first consideration was to determine whether the C' mission as presently defined should be like a C mission (low earth orbit) or like an E mission (4000 miles apogee). My recommendation was to make it like E, because this would give a better public justification for selecting the Borman crew and because it would demonstrate a step forward, publicly, beyond the C mission. However, since Mr. Webb's main concern had been that we should not announce and implement plans from which we would later have to retrench, Phillips decided it would be best to define the C' mission to be like a C mission, with the Saturn V launch vehicle instead of the Saturn I-B.

After much discussion, we finally decided that the most important thing Apollo can achieve this year is a lunar capability in hardware, software, crew training, etc. This, we believe, is necessary whether the C' mission goes to the moon or not. We also agreed that the only [12] way to achieve this lunar capability is to plan the mission as though it were going to fly to the moon. By so doing, all involved would, without question, have to face the real issues and make the real decisions that would allow us to go to the moon. An earth orbital mission would, of course, be a natural fallout because such a mission would have to be an abort option for a lunar mission in the event that the S-IVB stage could not make its second burn. Therefore, by planning such a mission, we would have, in December, an earth orbital capability on the C' mission while at the same time having completed all the planning and preparation that would be necessary should conditions be such that we could go to the moon. We would not commit now, either within NASA or outside, to do any more than the earth orbital mission.

This plan was adopted, and the over-all program plan can best be summarized as follows:

a. AS-503, designated Apollo 8, will be prepared to be ready for launch on December 6, 1968. It will consist of CSM 103, LTA-B, and AS-503. The reasons for making the change from the previously defined mission are that this will give us the earliest possible Pogo checkout flight and that LM checkout delays have prevented us from making an early flight with the LM.

b. The mission will be designated as C'. It will be an earth orbital mission, including whatever elements of C need to be repeated and elements of D, E, F, and G that can be incorporated.

c. Final definition of the mission will not come after Apollo 7.

d. The crew will be the E mission crew so that the D mission crew can continue its active preparation for that mission.

e. We recognize that after the C' mission the remaining missions will be upon us and that it is essential to bring lunar capability into being while we are implementing the C' mission. This includes lunar capability in hardware, software, flight operations, and crew operations.

f. This capability can only be brought into being if we plan for it now. Therefore, we will do all of our planning for the C' mission as though it were a lunar orbit mission. This will give us maximum flexibility to fly the assigned earth orbital mission with whatever elements of all other missions, including the lunar landing mission, are best to put into that flight after the results of Apollo 7 are known.

[13]

August 19, 1968. Received a copy of the proposed press release and program directive sent from Phillips to Gilruth. (A copy is attached.) A supplement to the program directive, which will authorize the planning to obtain the capability for a lunar orbit mission, is still in work in Washington.

Held my regular ASPO staff meeting and summarized our proposed plans as outlined in the August 16 notes. Dr. Gilruth held a Senior Staff Meeting, informing other Center elements of this approach. Phillips held a press conference in Washington which, from all reports, also went according to plan. Our job now is to implement the C' mission and, as stated, bring along the lunar capability at the same time. These special notes will be discontinued and the effort in implementing the C' mission will be reported in my daily notes to Dr. Gilruth.

Document II-61

FOR OFFICIAL USE ONLY

NATIONAL AERONAUTICS AND SPACE ADMINISTRATION
WASHINGTON, D.C. 20546

19 August 1968

TO: Director
                     John F. Kennedy Space Center, NASA
                     Kennedy Space Center, Florida 32899

                     Director
                     George C. Marshall Space Flight Center, NASA
                     Marshall Space Flight Center, Alabama 35812

                     Director
                     Manned Spacecraft Center, NASA
                     Houston, Texas 77058

FROM: Apollo Program Director

SUBJECT: Apollo Mission Preparation Directive

The following changes will be made in planning and preparation for Apollo flight missions:

1.   Apollo-Saturn 503

      a.   Assignment of Saturn V 503, CSM 103 and LM-3 to Mission D is cancelled.

b. Saturn V 503 will be prepared to carry CSM 103 and LTA B on a manned CSM only mission to be designated the C prime mission.

c. The objectives and profile of the C prime mission will be developed to provide the maximum gain consistent with standing flight safety requirements in maturing of the Apollo-Saturn V space system in earth orbital operation. Studies will be carried out and plans prepared so as to provide reasonable flexibility in establishing final mission objectives after the flight of AS 205.

[2] d. All planning and preparations for the C prime mission will proceed toward a launch readiness date of 6 December 1968.

2. Apollo-Saturn 504

a. Saturn V 504, CSM 104, and LM-3 are reassigned to the D Mission.

b. The D Mission will be scheduled for launch readiness no earlier than 20 February 1969 with all mission and hardware preparations proceeding toward that goal.

3. Crew Assignment

a. The crews now assigned to the D Mission remain assigned to the D Mission. The crews currently assigned to the E Mission are reassigned to the newly defined C prime mission.

4. Crew Training and Equipping and Operational Preparations

b. Training and equipping of the D Mission crews and operational preparations will proceed as previously planned but to meet the newly established flight readiness date.

c. Training and equipping of the C prime crews and operational preparations will proceed as required to meet mission requirements and to meet the newly established flight readiness date.

/Signed/
Sam C. Phillips
Lt. General, USAF
Apollo Program Director

[3]

Proposed Press Release by NASA Headquarters

NASA Acting Administrator Thomas O. Paine announced that Lunar Module operations will be dropped from the first manned Apollo-Saturn V flight, Apollo

8. Dr. Paine also stated that the Office of Manned Space Flight will begin planning for an alternate manned Command and Service Module mission for launch in December.

Dr. Paine emphasized that no final decision will be made on the precise mission plan for the alternate flight until after the first manned Apollo flight (Apollo 7) this Fall. Apollo 7 is a mission of up to 10 days duration to complete flight qualification of the Command and Service Modules.

To assure greatest value from the mission, planning and training for Apollo 8 must begin in the period before the Apollo 7 mission is flown but the final content of the mission plan will be selected only after the Apollo 7 mission results are evaluated.

Lunar Module 3, which has been delayed in checkout, will be flown next year on the fourth Saturn V (AS 504) with Command and Service Modules No. 104. This decision is based on preliminary studies which indicate that many Apollo program objectives scheduled for later flights can be attained by utilizing the Apollo 8 Command Service Module mission.

[4]

2.

General Samuel Phillips, Apollo Program Director, said one very important advantage of flying Apollo 8 this year is the opportunity for earlier experience in the operation of the Saturn V and Command and Service Modules then can otherwise be obtained. Two problems previously experienced in the Saturn Apollo systems – vertical oscillation or "POGO effect" in the first stage of the Saturn V and the rupture of small propellant lines in the upper stages – have been corrected and the solutions verified in extensive ground tests.

Document II-62

NATIONAL AERONAUTICS AND SPACE ADMINISTRATION
Washington, D.C. 20546
November 4, 1968

Dr. Robert R. Gilruth, Director
Manned Spacecraft Center
National Aeronautics and Space Administration
Houston, Texas 77058

Dear Bob:

In inviting the Apollo Executives and their program managers to meet with us on November 10, it is with the deepest recognition that the Apollo 8 mission involves many issues in addition to the technical capabilities of the Apollo systems. Before reaching a decision of such importance to the total national space program, we must be sure that we have weighed all the considerations, and evaluated their advantages and disadvantages.

There are grave risks to the program as a whole, not just to the Apollo 8 mission, in embarking on a lunar orbit mission with the second manned flight of the CSM. We have to face the possibility that this type of mission could appear to the public, and to our peers in government, to be a precipitous, risky venture where the propaganda value is the only gain. In assessing the alternatives, I am concerned that I have seen no real criticism of a lunar orbit mission. The general reaction both inside and outside NASA has been one of enthusiasm and anticipation of a major feat. Yet, you and I know that if failure comes, the reaction will be that anyone should have known better than to undertake such a trip at this point in time. Considering the potential risks to the public acceptance of the program and the basic confidence in future manned space flight, the very vital issues are:

1. Does a C' mission move us measurably towards a lunar landing?

2. Does it enhance the probability of a safe landing in the future?

3. What do we gain in a technical sense from carrying out a C' mission?

4. What are the consequences of a failure?

[2]

On the pro side, it is quite clear that any vehicle can experience a failure; however, it is reasonable to believe that since the first manned CSM, Apollo 7, performed well for 10.8 days, the second manned CSM can be just as successful and as safe as the third, fourth, or fifth flight. Although different to some extent, each lunar-capable CSM is built and checked out to give a consistent performance. From the standpoint of the probability of reliable performance, unless basic design flaws are uncovered, each flight should be equally likely to succeed. The technical advantages of obtaining early information on communications, navigation, guidance and control, thermal conditions, and gravitational potential at lunar distances are clearly positive gains in increasing the safety and success of subsequent missions. Perhaps the greatest single advantage is the motivation that the alternate planning for a lunar orbital mission has given to the entire Apollo organization. Since the establishment of the lunar orbital mission as an Apollo 8 alternate three months ago, the Apollo Program has been meeting every one of its major milestones.

On the con side, a lunar orbital mission involves a very difficult decision in that we are dealing with a complex, new vehicle. The paradox between the 501 and 502 launch vehicle performances illustrates this point. In addition, there is the obvious risk of being three days instead of one hour away from land. I must say that as far as I can see, and depending on the detailed Apollo 7 results and Apollo 8 evaluations and reviews, the CSM should perform consistently, and the risks from a purely technical aspect are probably reasonable and acceptable. If such a mission failed, however, the risks to the program as a whole could be significant.

I would very much appreciate your thoughtful consideration of these aspects of the decision, as well as any other facets of the problem which we may not have considered, so that we may benefit from your views at the meeting on Sunday.

One technique that we have been using in our considerations of the risk involved is the Mission Risk Assessment Form. I am sending along a copy of the form and an explanation of its use. I have found it helpful in trying to arrive at an assessment of how to minimize the overall risk of a lunar landing. If you can find the time to complete the form and wish to provide me with a copy, I would be very grateful.

[3]

I am looking forward to our meeting on Sunday. Again, you have my personal thanks and appreciation for your willingness to give up so much of your time for the progress of the Apollo Program.

Sincerely,
[Signed George]

Associate Administrator
For Manned Space Flight

Document II-63

SPECIAL NOTES FOR NOVEMBER 10 AND 11, 1968

Introduction:

During the period of August 9 to August 19, 1968, I set forth in some special notes the activities that took place in that time period concerning the Apollo 8 lunar orbit flight.

In the intervening time since the middle of August, planning in the entire manned space flight organization has proceeded in accordance with the steps outlined in the earlier notes. Spacecraft checkout went extremely well, and a modification period to make those changes that were necessitated by the mission reassignment took place in good order. The spacecraft went through its unmanned and manned altitude chamber tests, was moved to the VAB, erected on AS 503, and moved to the launch pad several days prior to the Apollo 7 flight. In the same time period, all of the mission planning and flight crew training also focused on the planned circumlunar flight. No new factors came to light that weren't understood, at least in general terms, at the time of the mid-August decisions.

The Apollo 7 flight took place in the period from October 11 to October 22. All of the mission objectives were accomplished, and the spacecraft's performance far exceeded my expectations. There were, of course, some anomalies with the equipment; but, in general, these were explained either during the flight or shortly after the flight. There was no question in any of our minds after completion of the Apollo 7 flight that the Apollo 8 flight should perform the lunar orbit mission. During the flight, as well as after the flight, we had a series of reviews with Phillips, Mueller, and the Management Council, discussing the present status of the hardware, mission operations, and crew training, over and over again. If anything, the period was marked by so many reviews that many of us felt that we really didn't have the time to do the job at hand. The reviews culminated in two meetings in Washington on November 10 and 11, 1968, first with the Apollo Executives and then with NASA management. The details of these meetings are as follows:

Apollo Executives Meeting, November 10, 1968

This meeting started with an introduction by Phillips, giving the background of the Apollo 8 mission recommendation, the sequence of [2] of [*sic*] flight missions, and a summary of the present status. Following Phillips' introduction, Lee James reported on the launch vehicle status, its readiness for manned flight, and the results of all of the work in connection with POGO.

Following Lee James' briefing, it was my turn to discuss the spacecraft situation and our readiness to complete a lunar orbit flight. I indicated that the pertinent questions were:

a. Is the spacecraft design adequate?

b. Will the systems perform as designed?

c. Are the benefits worth the risks?

I felt that it was important to cast the issues in this light, since over the last several weeks we have been asked many questions that indicated that people really didn't understand that the mission we are about to fly is the design mission for the Apollo spacecraft. It is a mission that we would have had to face sooner or later anyway, and the risk involved in performing the mission now after a successful Apollo 7 flight is no greater than it would be a year from now. I went into considerable detail discussing the Apollo design redundancy in critical systems such as propulsion, power, environmental control, and communications. This was followed by a review of Apollo 7 anomalies and conclusions concerning the benefits and the risks of this flight. On the latter point, we indicated that the risks were no greater than those that are generally inherent in a progressive flight test program and that we believed that the probability of success of the ultimate lunar landing mission would be greatly enhanced.

My briefing was followed by a very clear discussion by Chris Kraft concerning the flight mission operations and a review by Deke Slayton of the flight plan, with emphasis on the lunar timeline. After Deke's briefing, Petrone reported on the checkout readiness status of the space vehicle, indicating that we would be ready to launch as early as December 10 or 12 and that he could foresee no problems with a launch on December 21 which is the day on which the lunar window opens. The work at KSC on AS 503 has been quite remarkable in that the very tight schedule which was laid down early in August was met in spite of a great deal of additional work.

[3] Phillips summed up at the conclusion of our meeting and repeated many of the thoughts expressed by all of us during the review. He indicated that he would make a firm recommendation on the next day to proceed with an Apollo 8 lunar orbit flight. Following Phillips' summation, Mueller asked the Apollo Executives for their personal views concerning this flight. The following is a brief summary of each of the Executive's opinions and views:

Walter Burke, McDonnell Douglas. The S-IVB is ready to do any of the missions listed; however, McDonnell Douglas feels that we ought to fly a circumlunar flight instead of a lunar orbit mission in order to minimize the risks.

Hilly Paige, GE. GE would like to go on record that we should go ahead with an Apollo 8 lunar orbit flight.

Paul Blasingame, AC Electronics. The G&N hardware is completely ready. Generalizing to the mission as a whole, when we risk the lives of people, we ought to get something for this risk. A lunar orbit flight looks like the right size of step to make.

Stark Draper, MIT. We should go ahead with the mission.

Bob Evans, IBM. The program is in good shape, and the instrumentation unit is ready to go.

George Bunker, Martin Marietta. The presentations made a persuasive case to fly a lunar orbit mission. The risk in lunar orbit is certainly greater than in earth orbit, but in assessing the risks for a lunar landing mission on a cumulative basis, it appears that the lunar orbit mission now will lessen the overall risk. I am for a lunar orbit mission.

Wilson, Boeing. There is every indication that the lunar orbit mission is the right thing to do.

Lee Atwood, North American Rockwell. As manufacturers of the spacecraft, our motivation to take chances is no higher than Frank Borman's, but we are ready to go.

Bob Hunter, Philco-Ford. I have no reservations in supporting the complete mission.

[4] Tom Morrow, Chrysler. We have no hardware on this mission, but we wish we had. We strongly feel that we ought to go for it. We must take steps like this one. We cannot move forward without progressing on each step. I vote yes.

Bill Gwinn, United Aircraft. It is difficult to quantify the risks. I am impressed by what I heard. The risks appear to be less than I thought before I came down here. George Low's recommendation not to change the fuel cells or the components is the right one.[1] I support the recommendation to proceed with the mission.

---

1 As a result of the condenser exit temperature problem on Apollo 7, Pratt & Whitney had first recommended that we should replace the fuel cells on Spacecraft 103, and on the morning of November 10, recommended that we should change-out the hydrogen pump motors in order to install the new higher temperature pinions. Myers and I held a meeting with Pratt & Whitney prior to the Executives meeting, and after discussing the whole situation in detail, decided that we should not replace these motors. The reasons for this decision were that: (a) The vibration flushing of the radiators decreased the probability of the problem's recurrence on Apollo 8; (b) Replacement of the pinion would only slightly increase the temperature margin, but would not really fix the problem; and (c) Detailed analyses have indicated that, even under the worst-case conditions of recurrence, there was no flight safety degradation, and it was unlikely that the mission would be degraded in any way. I reported the situation in detail during my briefing at the Executives review. Stu Conley, the Pratt & Whitney Program Manager, however, still felt that the motors should be replaced. This would have required breaking into systems that were already checked out, and KSC felt that they could not guarantee that the systems would not be degraded by so doing.

Joe Gavin, Grumman. Since we have no hardware on this flight, our interest is only with respect to the overall program. The mission makes a lot of sense. If we don't do it on this flight, we should do it anyway. I have no reservations.

Bill Bergen, Space Division, North American Rockwell. I agree that there are more risks in a lunar orbit mission than in an earth orbit mission. Also, it is unlikely that we will have as high performance of [5] our systems as we had on Apollo 7, but I am confident that our systems will perform satisfactorily. Although there would be less risks with a repeat flight, there are risks with no gain. We should make the lunar orbit flight.

George Stoner, Boeing. I endorse the recommendation without any reservations.

Gerry Smiley, GE. We have built up a head of steam in Apollo since we first started talking about C'. To do anything other than fly a lunar orbit mission now would set the program back.

The meeting was adjourned with the conclusion that a firm recommendation to fly the Apollo 8 mission to lunar orbit would be made the next day to the Acting Administrator.

NASA Management Meeting. November 11, 1968

On November 11, 1968, Dr. Mueller, the Center Directors, General Phillips, and the Center Program Managers met with Dr. Paine, Dr. Newell, Mr. Shapley, Mr. Finger, and a large number of staff members to discuss the Apollo 8 flight. The briefings were the same as those given to the Apollo Executives. The recommendations by Phillips and each of us were to firmly commit to a lunar orbit flight.

Following the briefings, Dr. Mueller indicated that this situation had been discussed with STAC, PSAC, DOD, and the Apollo Executives. He pointed out that STAC members had made a penetrating review of the flight and clearly understood the risks. Their reaction was a positive one, with the exception of

Gordon MacDonald who had reservations in that he believed the risks far outweighed the benefits.

PSAC was favorably disposed to support the mission, but had no firm recommendation. DOD also generally favors the mission. The Apollo Executives' reactions have already been reported in previous pages. Dr. Mueller also pointed out that Bellcomm had been quite negative. Bellcomm's reasoning was that the risk of a lunar orbit mission is considerably greater than that for an earth orbit mission. Bellcomm, therefore, believed that a lunar orbit mission should only be flown if this made it possible to reduce the total number of flights in the lunar landing program. If this were not possible, then Bellcomm believed the lunar orbit mission was not justified.

[6] Dr. Paine indicated that he had hoped that it would be possible to quantify the risks better than had been done in the course of the briefings. Dr. Mueller mentioned that we had tried to perform a numerical risk assessment, but that this had not turned out to be as positive as he had hoped it would be. However, in generalizing the results, he mentioned that the least cumulative risk in the lunar landing program resulted from making the minimum number of flights. Dr. Gilruth rebutted by stating that this is like saying that "the faster you drive your car, the safer you are because your exposure is less." Dr. Paine also felt that Dr. Mueller's statement was not valid since we will be in the flying business for a long time to come and we will fly on all Saturn V's, whether we use them in the lunar program or not. The general view expressed by many of us was that the highest probability of success for the lunar landing mission would come from a progressive buildup of flight experience. We felt that although there is risk in each manned flight, it was impossible to quantify this risk. Instead, the flight test program should be based on the best available judgment and experience and should, of course, be reviewed after each mission. Today's best indications are that the sequence of missions, C' (lunar orbit), D (earth orbit with LM), F (lunar orbit with LM), and G (lunar landing), would give us the best chance at a successful lunar landing in this decade.

At the conclusion of these discussions, Dr. Paine convened a smaller meeting, involving some of his immediate staff, Dr. Mueller, General Phillips, and the Center Directors. This was followed by a third meeting, involving only Paine, Newell, and Mueller. At the conclusion of these meetings, Dr. Paine announced that the Apollo 8 flight would be a lunar orbit mission. This was announced publicly in a press conference in Washington on Tuesday, November 12, 1968.

Document II-64

TO　　　　　　:　M/Associate Administrator for Manned Space Flight
　　　　　　　　　Date: 11 Nov, 1968

FROM　　　　　:　MA/Apollo Program Director

SUBJECT   :   Apollo 8 Mission Selection

The purpose of this memorandum is to obtain your approval to fly Apollo 8 on an open-ended lunar orbit mission in December 1968.

My recommendation is based on an exhaustive review of pertinent technical and operational factors and also on careful consideration of the impact that either a success or a failure in this mission will have on our ability to carry out the manned lunar landing in 1969.

THE APOLLO 8 C' LUNAR ORBIT MISSION:

Attachment I to this memorandum [not included] contains a detailed description of the Apollo 8 lunar orbit mission. Significant features of this mission plan are:

> Planned Schedule:
>
> Launch: 0750 EST, 21 December 1968
> Translunar Injection: 1040 EST, 21 December 1968
> Lunar Orbit Insertion:
> LOI_1 Initiate: (60X170 NM Orbit) 0457 EST, 24 December 1968
> LOI_2 Initiate: (60 NM Circular Orbit) 0921 EST, 24 December 1968
> Transearth Injection: 0105 EST, 25 December 1968
> Landing: 1053 EST, 27 December 1968
>
> Alternate Schedule:
>
> Monthly Launch Windows: 21-27 December 1968 or as soon thereafter as possible.
>
> Daily Launch Windows: Approximately 5 hours duration.
>
> Open-Ended Mission Concept:
>
> A large number of abort and alternate mission options are provided for in the Mission Plan and associate Mission Rules. Noteworthy examples of the way in which this open-ended concept could operate in this mission are the following:
>
>> A low earth orbital mission in the event of a "no go" in earth orbit prior to translunar injection.

[2]   Early return to earth in event of certain malfunction conditions during translunar coast.

> A circumlunar mission in event of a "no go" during checkout prior to the lunar orbit insertion burn.

APOLLO 8 MISSION SELECTION:

On August 19, 1968, we announced the decision to fly Apollo 8 as a Saturn V, CSM-only mission. The basic plan provided for Apollo 8 to fly a low earth orbital mission, but forward alternatives were to be considered up to and including a lunar orbital mission. Final decision was to be reserved pending completion of the Apollo 7 mission and a series of detailed reviews of all elements of the Apollo 8 mission including the space vehicle, launch complex, operational support system, and mission planning.

Apollo 7 Mission Results:

An important factor in the total decision process leading to my recommendation has been and continues to be the demonstrated performance of the Apollo 7 Command and Service Module (CSM) subsystems, and the compatibility of the CSM with crew functions, and the Manned Space Flight Network. Comprehensive understanding of all Apollo 7 flight anomalies and their impact on a lunar mission is fundamental to arriving at a proper decision. Attachment II to this memorandum [not included] provides a recap of the Apollo 7 flight anomalies, their disposition, and a statement of any known risk remaining on the proposed Apollo 8 mission together with the actions proposed.

Apollo 4 and Apollo 6 Results:

The results of the Apollo 4 and Apollo 6 missions, in which the performance of the 501 and 502 Saturn V launch vehicles was tested, have been carefully analyzed. All flight anomalies have been resolved. In particular, the two most significant problems encountered in Apollo 6– longitudinal oscillation or "POGO" effect in the first stage of the Saturn V and the rupture of small propellant lines in the upper stages–have been corrected and the solutions verified in extensive ground tests.

Meetings and Reviews:

The decision process, resulting in my recommendation, [sic] has included a comprehensive series of reviews conducted over the past several weeks to examine in detail all facets of the considerations involved in planning for and providing a capability to fly Apollo 8 on a lunar orbit mission. The calendar for and purpose of these meetings are presented in Attachment III. [not included] An important milestone [3] was achieved with successful completion of the Design Certification Review on November 7, 1968. A copy of the signed Design Certification is appended as Attachment IV. [not included]

Pros and Cons of a Lunar Orbital Flight:

My objective through this period has been to bring into meaningful perspective the trade–offs between total program risk and gain resulting from introduction of a CSM-only lunar orbit mission on Apollo 8 into the total mission sequence leading to the earliest possible successful Apollo lunar landing and return. As you know, this assessment process is inherently judgmental in nature. Many factors have been considered, the evaluation of which supports a recommendation to proceed forward with an Apollo 8 open-ended lunar orbit mission. These factors are:

PROS:

Mission Readiness:

- The CSM has been designed and developed to perform a lunar orbit mission and has performed very well on four unmanned and one manned flights (CSM's 009, 011,017,020, and 101).

- We have learned all that we need in earth orbital operation except repetition of performance already demonstrated.

- The extensive qualification and endurance-type subsystem ground testing conducted over the past 18 months on the CSM equipments has contributed to a high level of system maturity, as demonstrated by the Apollo 7 flight.

- Performance of Apollo 7 systems has been thoroughly reviewed, and no indication has been evidenced of design deficiency.

- Detailed analysis of Apollo 4 and Apollo 6 launch vehicle anomalies, followed by design modifications and rigorous ground testing gives us high confidence in successful performance of the Apollo 8 launch vehicle.

- By design all subsystems affecting crew survival (Environmental Control System, Electrical Power System, Reaction Control System, and Guidance and Navigation System) are redundant and can suffer significant degradation without crew or mission loss. The sole exceptions are the injector and thrust chamber of [4] the Service Propulsion System. These two engine components are of simple, rugged design, with high structural and thermal safety margins. (See Attachment V) [not included]

- Excellent consumables and performance margins exist for the first CSM lunar mission because of the reduction in performance requirements represented by omitting the weight of the lunar module. An example of the predicted spacecraft consumables usage is provided below to illustrate this point:

Consumable	Total Usable	Total Used	Reserve
Service Module Reaction Control System Propellant (Pounds)	1140	294.5	845.5
Command Module Reaction Control System Propellant (Pounds)	231.2	29.4	201.8
Service Propulsion System Propellant (Pounds)	40,013	28,987	11,026
Cryogenic Oxygen (Pounds)	640	410	230
Cryogenic Hydrogen (Pounds)	56	40	16

PROS:

Effect on Program Progress:

The lunar orbit mission will:

- Provide valuable operational experience on a lunar CSM mission for flight and ground and recovery crews. This will enhance probability of success on the subsequent more complex lunar missions by permitting training emphasis on phases of these missions as yet untried.

- Provide an opportunity to evaluate the quality of MSFN and on-board navigation in lunar orbit including the effects of local orbit perturbations. This will increase anticipated accuracy of rendezvous maneuvers and lunar touchdown on a lunar landing mission.

- Permit validation of Apollo CSM communications and navigation systems at lunar distance.

- [5] Serve to improve consumables requirements prediction techniques.

- Complete the final verification of the ground support elements and the onboard computer programs.

- Increase the depth of understanding of thermal conditions in deep space and lunar proximity.

- Confirm the astronauts' ability to see, use, and photograph landmarks during a lunar mission.

- Provide an early opportunity for additional photographs for operational and scientific uses such as augmenting Lunar Orbiter coverage and for obtaining data for training crewmen on terrain identification under different lighting conditions.

CONS:

Mission Readiness:

- Marginal design conditions in the Block II CSM may not have been uncovered with only one manned flight.

- The life of the crew depends on the successful operation of the Service Propulsion System during the Transearth Injection maneuver.

- The three days endurance level required of backup systems in the event of an abort from a lunar orbit mission is greater than from an earth orbit mission.

CONS:

Effect on Program Progress:

- Validation of Colossus spacecraft software program and Real Time Computer Complex ground software program could be accomplished in a high earth orbital mission.

- Only landmark sightings and lunar navigation require a lunar mission to validate.

Impact of Success or Failure on Accomplishing Lunar Landing in 1969:

A successful mission will:

- Represent a significant new international achievement in space.

- [6] Offer flexibility to capitalize on success and advance the progress of the total program towards a lunar landing without unreasonable risk.

- Provide a significant boost to the morale of the entire Apollo program, and an impetus which must, inevitably enhance our probability of successful lunar landing in 1969.

A mission failure will:

- Delay ultimate accomplishment of the lunar landing mission.

- Provide program critics an opportunity to denounce the Apollo 8 mission as precipitous and unconservative.

RECOMMENDATION:

In conclusion, but with the proviso that all open work against the Apollo 8 open-ended lunar orbit mission is completed and certified, I request your approval to proceed with the implementation plan required to support an earliest December 21, 1968, launch readiness date.

/Signed/
Sam C. Phillips
Lt. General, USAF

Document II-65

NATIONAL AERONAUTICS AND SPACE ADMINISTRATION

WASHINGTON, D.C. 20546

[stamped Nov 18, 1968]

MEMORANDUM to: Associate Administration for Manned Space Flight

FROM   :   Acting Administrator

REFERENCE   :   a. Memorandum for Acting Administrator from Associate for Manned Space Flight from Apollo Director, Subject: Apollo 8 Selection, dated November 11, 1968

b. Memorandum for Acting Administrator from Associate Administrator for Manned Space Flight, dated November 11, 1968

c. Memorandum to Acting Administrator from Associate Administrator for Manned Space Flight, Subject: Request for Approval to Man the Apollo Saturn V Launch Vehicle, dated November 5, 1968

Based on careful consideration and analyses of all of the information, comments, results of engineering tests and analysis, etc. provided to me, I approved on November 11 Lt. Gen. Samuel C. Phillips' recommendation (reference a), transmitted and agreed to by your memorandum to me (reference b), that the Apollo 8 mission be conducted as a manned lunar orbit mission with CSM 103 on Saturn 503 pending successful accomplishment of all necessary preparation and checkout activities for this mission. Included among the various inputs that I considered were:

1.   The recommendation of Lt. Gen. Samuel C. Phillips, Apollo Program Director, with the supporting reasoning attached to his memorandum to you dated November 11 (reference a);

2.   The presentations made to me on November 11 by Gen. Phillips, Mr. Lee James, Saturn V Program Manager-MSFC, Mr. George Low, Apollo Program Manager-MSC, Mr. Christopher C. Kraft, Director of Flight Operations-MSC, Mr. Rocco A. Petrone, Director of Launch Operations-KSC;

3.   The statements of Mr. Gerald M. Truszynski, Associate Administrator for Tracking and Data Acquisition, and Lt. Gen. Vincent Houston, USAF, indication the ability of their systems and forces to be ready for such a mission;

[2] 4. The statements supporting a manned lunar orbit mission by each of the following (in the separate meeting on November 11, following the formal presentation by Gen. Phillips and the Apollo Program Managers listed above):

Mr. Harold B. Finger, Associate Administrator for Organization and Management

Mr. Willis H. Shapley, Associate Deputy Administrator

Mr. Bob P. Helgeson, NASA Safety Director

Mr. Julian Scheer, Assistant Administrator for Public Affairs

Dr. Kurt H. Debus, Director KSC

Dr. Robert R. Gilruth, Director MSC

Dr. Wernher von Braun, Director MSFC

Dr. Floyd L. Thompson, Special Assistant to the Administrator

Mr. Eberhard F. M. Rees, Deputy Director-Technical, MSFC

5.   The information that you provided to me concerning the comments of the Science [and] Technology Advisory Committee (STAC), the reactions of PSAC, and the comments of the representatives of the industrial organizations responsible for various elements in the Apollo program;

6.   The separate statements that you and Dr. Newell, Associate Administrator, made also supporting this mission;

7.   The information provided to me in various briefings and in your memorandum of November 5 (reference c) to indicate that the problems or anomalies encountered in AS-502 have been solved and proven in analysis and tests;

8. My telephone conversation with Command Pilot Frank Borman who also supports this mission.

It should be made clear to all participating organizational elements throughout the Apollo program that any problem encountered during the preparation for this mission that may, in any way, increase the potential risk of the mission must be made known to all appropriate levels of NASA management as [3] soon as the problem is encountered. I will rely on you and those organizations to notify me as soon as such a problem is encountered, since my approval was based on consideration of the benefits to be derived from this mission and the risks involved in undertaking it.

/Signed/
T. O. Paine

Document II-66

Document Title: Memorandum from George M. Low, Manager of Apollo Spacecraft Program, "Program Plan revision," 20 August 1968.

Source: NASA Historical Reference Collection, History Division, NASA Headquarters, Washington, DC.

The tentative decision to transform the Apollo 8 mission into a flight into lunar orbit caused a significant revision to the previously planned Apollo flight schedule. In particular, for the first time the third mission after Apollo 8, i.e., Apollo 11, could, if all preceding missions went off without problems, become the first attempt at a lunar landing. The Apollo 8 lunar orbit mission was designated C' (C Prime) because it was inserted in the previously planned Apollo mission sequence which included the following missions: C—test of the Apollo Command and Service module in low Earth orbit; D—test of the Apollo Command and Service and Lunar Modules in low Earth orbit; E—test of the Apollo Command and Service and Lunar Modules in a mission beyond Earth orbit, but not headed to the moon; F—test of all equipment in lunar orbit; and G—lunar landing mission. With the success of Apollo 8, the E mission was dropped from NASA's planning. Apollo 9 flew the D mission, and Apollo 10 flew the F mission, clearing the way for Apollo 11 to fly the G mission, aimed at the first lunar landing.

[CONFIDENTIAL] [DECLASSIFIED]

DATE: [stamped AUG 20 1968]

TO   : See attached list

FROM   : PA/Manager, Apollo Spacecraft Program

SUBJECT   : Program Plan revision

The recent decision to fly a mission C ' (manned CSM on AS 503) prior to the first CSM/LM manned mission on AS 504 has resulted in significant program plan revisions.

Only the revised assignments, delivery, and launch schedules are provided to you at this time in order to expedite distribution of the revisions. I intend to provide you with a complete revised program plan during the first week in September.

The offices responsible for the timely completion of the Controlled Milestones are to notify Mr. C. L. Taylor, Assistant Chief, Program Control Division, immediately whenever a situation exists, or is anticipated to exist, that will impact or potentially impact these milestones.

                                        [Signed George M Low 8-20]

                                        George M. Low

Enclosure

PP3:GHJordan:jt 8-20-68

8-20-68 (Rev. 9)
Attachment A
Page 1 of 2

SPACECRAFT DELIVERY AND LAUNCH DATES Ø

MISSION DESIGNATION	MISSION TYPE	LAUNCH COMPLEX	LAUNCH VEHICLE	CSM	CSM DELIVERY	SLA	LM	LM DELIVERY	LAUNCH SCHEDULE
Apollo 4	A	39A	501	017	Dec 22, 1966A	8	10R	Sep 20, 1966A	Nov 9, 1967A
Apollo 5	B	37B	204	---	---	7	1	Jun 23, 1967A	Jan 22, 1968A
Apollo 6	A	39A	502	020	Nov 22, 1967A	9	2R	Feb 16, 1967A	Apr 4, 1968A
Apollo 7	C	34	205	101	May 29, 1968A	5	---	---	Oct 11, 1968
Apollo 8	C'	39A	503	103	Aug 12, 1968A	11	LTA B	Jan 9, 1968A	Dec 6, 1968
	D	39B	504	104	Sep 1, 1968	12	3	Jun 14, 1968A	Feb 20, 1969
	E	39A	505	106	Oct 27, 1968	13	4	Sep 15, 1968	May 1, 1969
	F or G	39A	506	107	Jan 6, 1969	14	5	Oct 5, 1968	Jul 10, 1969
	G	39A	507	108	Mar 6, 1969	15	6	Nov 27, 1968	Sep 18, 1969
	G	39A	508	109	May 7, 1969	16	7	Jan 15, 1969	Nov 27, 1969
	G	39A	509	110	Jul 8, 1969	17	8	Mar 15, 1969	Feb 5, 1970

A – Actual

[p. 2 not provided]

Document II-67

Document Title: Memorandum to George Mueller, NASA Associate Administrator for Manned Space Flight from Lt. General Sam C. Phillips, Apollo Program Director, "Extravehicular Activities for the First Lunar Landing Mission," 19 October 1968.

Source: NASA Historical Reference Collection, History Division, NASA Headquarters, Washington, DC.

As the Earth-orbiting Apollo 7 mission, launched on 11 October 1968, was underway, marking the return to flight of the redesigned Apollo spacecraft after the 27 January 1967 Apollo 1 fire, senior Apollo managers were deciding on the details of the first lunar mission. This memorandum lays out the somewhat conservative plans for what the astronauts would do as they became the first humans to step onto another celestial body. In fact, this original plan called for only one of the two astronauts who landed on the lunar surface to actually leave the lunar module, except in an emergency situation. There were a number of subsequent revisions to this original proposal as the first landing mission grew closer.

National Aeronautics and
Space Administration

DATE: Oct 19 1968

TO   : M/Dr. George E. Mueller

FROM   : MA/Lt. General Sam C. Phillips

SUBJECT   : Extravehicular Activities for the First Lunar Landing Mission

Since the inception of the Apollo Program the primary objective of the first lunar landing mission has been the safe manned lunar landing and return. The hardware has, however, been designed and procured to give us the capability to conduct significant scientific investigations in anticipation of a series of lunar missions. Our planning, testing and simulations to date have been such as to assure this capability.

In view of our current schedules and mission planning and crew training activities, I believe that it is now necessary to firmly commit to the scope of EVA activities for the first lunar landing mission. To this end this mission was reviewed in detail on August 26 and 27, 1968. Based on this review, a proposal was made for the EVA activities for the first mission:

1. Plan for one EVA of approximately two hours duration

2. Carry out this EVA with one crewman on the surface and the other in the spacecraft on the umbilicals but prepared to carry out rescue.

3. The EVA activity planning to provide for an early contingency sample, photography and Lunar Module inspection, and a more extensive second soil sample in that order of importance.

4. The EVA would not include the deployment of the erectable antenna, the Apollo Lunar Surface Experiments Package (ALSEP) or the Lunar Geology Investigation (LGI).

RATIONALE (PRO)

The rationale for this proposal is:

1. On the first lunar landing mission the LM descent, landing, surface activities and ascent will be accomplished for the first time under lunar conditions. As a result of these many new activities the timelines must be scheduled in a conservative manner. A comparison of scheduled times for one and two EVA plans is:

	Two EVA Plan	One EVA Plan
Awakening to touchdown	9	9
Touchdown to sleep	8:20	8:20
Total first day	17:20 hrs	17:20 hrs
Sleep	7	7
Awakening to ascent	10:30	3:30
Ascent to docking	4	4
Total	14:30 hrs	7:30 hrs

Under the two EVA plan the long first day, coupled with the tasks of deploying ALSEP and the LGI on the second EVA, could result in added risk in the rendezvous phase because of crew fatigue.

2. Safety is increased because of lower probability of random equipment failures as the LM is separated from the CSM for a shorter period of time. Although weight and consumables margins are not a motivating

factor, the proposal results in approximately 100 ft/sec increase in LM descent ΔV capability, which represents an increase of 30 per cent in the propellant budgeted for landing point redesignation and hover during descent. The consumables margins could also be increased because of the shorter separation time.

3. The first landing mission represents a large step from orbital operations. The descent, landing, EVA, and ascent are new operations in a new environment. From a training point of view the crew should concentrate on the crucial, necessary tasks to achieve a safe landing and return. By not including ALSEP (180 hours of training), the LGI and the erectable antenna on the first mission, additional training and concentration on the descent, landing and ascent phases can be accomplished.

4. Our Gemini EVA experience showed that a methodical increase in task complexity was necessary in order to understand the zero g environment. The 1/6 g lunar surface environment will be a new experience, one which cannot be simulated on earth. It seems prudent, therefore, to plan the lunar EVA sequence in a methodical fashion in increasing complexity. In this light, it appears that the deployment of ALSEP and the Lunar Geology Investigation should be deferred to the second mission. Planning to accomplish these tasks on the first mission and failing could result in a slower build-up of lunar exploration capability than if they were deferred to the second mission.

RATIONALE (CON)

Several arguments have been advanced against the proposal:

1. Scientific data from the moon will be lost. The significance of this loss can only be judged in the context of the magnitude of the follow-on lunar exploration program. If only two additional flights are authorized, then the loss would be most significant, as a viable seismic net could not be established. If there are ten additional flights, the loss may not be significant.

2. The reduction in scientific return will result in some adverse comments. The overall significance to manned space flight of these comments can only be assessed in terms of (1) above.

3. There are serious reservations that, if one one-man EVA is all that we can commit based on our current state of knowledge, the second flight will similarly be limited in scope of scientific investigation.

4. The proposed plan may be too conservative at this point in time. If the flight proves our pessimism was not warranted, then we could be criticized for not being in a position to capitalize on success.

DISCUSSION

The proposal and rationale were transmitted to the Science and Technology Advisory Committee, the Lunar and Planetary Missions Board, the Manned Spacecraft Center, Marshall Space Flight Center, Kennedy Space Center, and Headquarters offices for comment. The responses are in general agreement with the proposal, with some of the scientific community in opposition. Modifications to the proposal have been suggested:

1. TV on the first mission was accorded increased emphasis especially in the area of observing the initial EVA activities. To assure TV, either mission planning must be complex, hardware changes must be made, or the erectable antenna must be carried. Studies are in progress to more fully understand these alternatives. If a requirement for coverage of the first egress is generated, then the LM steerable antenna-Goldstone method is the only available path without hardware modifications. It was recommended that the erectable antenna be retained until the mission constraints on the use of Goldstone are more fully understood.

2. The Kennedy Space Center, the Manned Spacecraft Center, the Apollo Lunar Exploration Office and Bellcomm have suggested that if the second EVA period is eliminated, both crewmen should egress during the first period, either together or in sequence. LM failure modes should be examined to ascertain which would be safer. Other than the safety question, the psychological factor of going to the moon and not egressing must be considered. Further, the interaction of the two subjects with the lunar surface environment would give us twice the data upon which to plan the succeeding mission EVA, hence move the program more rapidly toward a scientific exploration capability.

3. Several comments have been made with respect to assuring that we are moving as rapidly as is prudent towards achieving a scientific exploration capability. It seems reasonable, therefore, that for the first mission a primary objective should be to obtain data on the capabilities and limitations of the astronaut plus Extravehicular Mobility Unit in the lunar surface environment. This specific data gathering should be well planned and covered as an approved experiment or Detailed Test Objective for the flight in order to assure that the full capabilities are achieved on the second mission.

4. Total EVA time is limited. We should, therefore, move as rapidly as possible to hardware modifications designed to free the crew from mechanical tasks (such as unstowing and transferring equipment from the descent stage to the ascent stage) and maximizing the time available for science.

5. It appears that a one-man two-hour EVA is the minimum-risk situation, but what is not clear is how the risk changes as the EVA activity is increased. It is also not clear as to the relative magnitude of the EVA risk to the total mission risk. Two 1.5 hour EVA's (separate astronauts)

may involve only a slight increase in total mission risk over one two-hour one-man EVA, yet the scientific return could be increased significantly. If it is planned to have both crewmen egress, it was suggested that it be in sequence with one in the LM at all times. This allows the status of the LM and the EVA crewman to be monitored at all times, one man is always on the LM life support system and the communications to earth (both voice and biomedical telemetry) are independent for the two crewmen.

6. If ALSEP and LGI are not carried, several suggestions were made for other scientific experiments. These included uprating the preliminary sample to be of greater scientific value, and to examine the possibility of including the laser ranging retroreflector, a Surveyor seismometer, and soil mechanics experiments.

OMSF ACTION

The proposal, comments, and recommendations of the Apollo Program Director were presented to the OMSF Management Council on September 11, 1968. The Council approved the following:

1. A single EVA period open-ended to three hours will be planned for the first mission. The surface traverse will be open-ended to a maximum of 300 feet from the LM. Training experience, simulations, timeline verification studies and failure mode analyses will be used as the basis for a decision between one-man and two-man EVA's and two one-man EVA's during the period.

2. The ALSEP and LGI will not be carried. A lunar soil sample will be collected in a manner which will maximize the scientific value, and other candidate scientific experiments will be identified and submitted for consideration by October 10, 1968.

3. TV will be carried. Planning will be such as to exploit both its operational and public information uses. The MSC will identify changes in mission planning and/or hardware necessary to utilize only the LM steerable antenna.

4. In order to maximize the scientific return from the second mission, a Primary Objective of the first mission will be to obtain data to assess the capabilities and limitations of the astronaut and his equipment in the lunar surface environment. The MSC will plan and implement Detailed Test Objectives and experiments for the first lunar landing mission to achieve this objective.

5. The MSC should study and schedule recommendations, including cost and schedules, to the Apollo Program Director for any changes in hardware for future lunar missions which would increase the percentage of EVA time available for scientific investigations.

[Signed Sam. C Phillips]

Sam C. Phillips
Lt. General, USAF
Apollo Program Director

Attachments 1-26

Cc: (w/o attachments)
CD/HKDeubs
DIR/WvonBraun
AA/RRGilruth
MA-A/GHHage
MA/WCSchneider
MAO/JKHolcomb
MAL/LRScherer

Document II-68

Document Title: Letter to George M. Low, Manager, Apollo Spacecraft Program, from Julian Scheer, Assistant Administrator for Public Affairs, 12 March 1969.

Source: Folder #148675, NASA Historical Reference Collection, History Division, NASA Headquarters, Washington, DC.

Document II-69

Document Title: Letter to Julian Scheer, Assistant Administrator for Public Affairs, from George M. Low, Manager, Apollo Spacecraft Program, 18 March 1969.

Source: Folder #18675, NASA Historical Reference Collection, History Division, NASA Headquarters, Washington, DC.

Julian Scheer was one of NASA Headquarter's "inner circle" during the Apollo program, in addition to his role as NASA top public spokesman. In this letter to NASA veteran manager of human space flight George Low, who assumed responsibility for the Apollo spacecraft project after the January 1967 Apollo 1 fire, Scheer suggested that it would be inappropriate to suggest to the Apollo 11 crewmembers what they might say as they reached the Moon. Low's reply indicates that he agreed with Scheer, and that there had been a misunderstanding of what actions Low had taken. The "Shapley Committee" was headed by senior NASA Headquarters staff member Willis Shapley, who was responsible for NASA's top-level political and budgetary strategy. Simon Bourgin was an employee of the U.S. Information Agency with a particular focus on the space program.

Document II-68

NATIONAL AERONAUTICS AND SPACE ADMINISTRATION

WASHINGTON. D.C 20546

March 12, 1969

Mr. George M. Low
Manager
Apollo Spacecraft Program
NASA Manned Spacecraft Center
Houston, Texas 77058

Dear George:

It has come to my attention that you have asked someone outside of NASA to advise you on what the manned lunar landing astronauts might say when they touch down on the Moon's surface. This disturbs me for several reasons.

The Agency has solicited from within NASA any suggestions on what materials and artifacts might be carried to the surface of the Moon on that historic first flight. But we have not solicited comment or suggestions on what the astronauts might say. Not only do I personally feel that we ought not to coach the astronauts, but I feel it would be damaging for the word to get out that we were soliciting comment. The ultimate decision on what the astronauts will carry is vested in a committee set up by the Administrator; the committee will not, nor will the Agency by any other means, suggest remarks by the astronauts.

Frank Borman solicited a suggestion from me on what would be appropriate for Christmas Eve. I felt–and my feeling still stands–that his reading from the Bible would be diminished in the eyes of the public if it were thought that NASA pre-planned such a thing. I declined both officially and personally to suggest words to him despite the fact that I had some ideas. I believed then [2] and I believe the same is true of the Apollo 11 crew that the truest emotion at the historic moment is what the explorer feels within himself not for the astronauts to be coached before they leave or to carry a prepared text in their hip pocket.

The Lunar Artifacts Committee, chaired by Willis Shapley, asked that all elements of NASA consider what might be carried on Apollo 11. I know that General Phillips has properly reiterated the request by asking all elements of Manned Flight to suggest things, but it was not the desire or intent of the committee to broaden the scope of the solicitation to verbal reactions.

There may be some who are concerned that some dramatic utterance may not be emitted by the first astronaut who touches the lunar surface. I don't share that concern. Others believe a poet ought to go to the Moon. Columbus wasn't a poet and he didn't have a prepared text, but his words were pretty dramatic to me. When he saw the Canary Islands he wrote, "I landed, and saw people running around naked, some very green trees, much water, and many fruits."

Two hundred years before Apollo 8, Captain James Cook recorded while watching the transit of Venus over the sun's disk, "We very distinctly saw an atmosphere or dusky shade around the body of the planet."

Meriwether Lewis, traveling with William Clark, recorded, "Great joy in camp. We are in view of the ocean, this great Pacific Ocean which we have been so long anxious to see, and the roreing [*sic*] or noise made by the waves braking [*sic*] on the rockey [*sic*] shore may be heard distinctly."

Peary was simply too tired to say anything in 1909 when he reached the North Pole. He went to sleep. The next day he recorded in a diary, "The pole at last. The [3] prize of three centuries [*sic*]. I cannot bring myself to realize it. It seems all so simple and commonplace."

The words of these great explorers tell us something of the men who explore and it is my hope that Neil Armstrong or Buzz Aldrin will tell us what they see and think and nothing that we feel they should say.

I have often been asked if NASA indeed plans to suggest comments to the astronauts. My answer on behalf of NASA is "no."

I'd appreciate your comments.

Regards,

[signed]
Julian Scheer
Assistant Administrator
for Public Affairs

Document II-69

March 18, 1969

Mr. Julian Scheer
Assistant Administrator
for Public Affairs
National Aeronautics and Space Administration
Washington, D. C. 20546

Dear Julian:

I have just received your letter of March 12, 1969, which apparently stemmed from a misunderstanding. Let me first point out that I completely agree with you that the words said by the astronauts on the lunar surface (or, for that matter, at any other time) must be their own. I have always felt that way and continue to do so.

I am, of course, aware of the Shapley Committee that was established by Dr. Paine, and have also received a copy of a telegram from General Phillips soliciting our comments on what should be carried to the lunar surface. I felt that in order to respond properly to General Phillips and to the Shapley Committee, I would like to seek the advice of Si Bourgin, whose judgment I respect a great deal in these matters. As you know, I met Si on our trip to South America and found that he offered excellent advice to all of us throughout our trip. I, therefore, called Si as soon as he returned from Europe and asked him whether he could offer any advice concerning what the astronauts should <u>do</u> (not <u>say</u>) when we have first landed on the moon. Si called me back [2] the night before the Apollo 9 launch, and we discussed his ideas at some length. We again agreed at that time that it is properly NASA's function to plan what artifacts should be left on the lunar surface or what should be brought back, but that the words that the astronauts should say must be entirely their own.

Since then, I have had a meeting with Neil Armstrong to discuss with him some of our ideas and suggestions, including those of Si Bourgin's, in order to solicit his views. Even though I had not yet received your letter at that time, we also discussed the point that whatever things are left on the lunar surface are things that he must be comfortable with, and whatever words are said must be his own words.

All of these activities—my discussions with Si, my discussions with Neil, and discussions with many others within and outside of NASA—are to gain the best possible advice that I can seek for what I consider to be a most important event. The result for of all of this will be my input to Dr. Gilruth so that he can forward it to the Shapley Committee, should he so desire.

I hope that this clarifies any misunderstanding that we might have had on this matter.

Sincerely yours,

/Signed/
George M. Low
Manager
Apollo Spacecraft Program

Document II-70

Document Title: Memorandum to Dr. [George] Mueller from Willis H. Shapley, Associate Deputy Administrator, "Symbolic Items for the First Lunar Landing," 19 April 1969.

Document Source: Folder #18675, NASA Historical Reference Collection, History Division, NASA Headquarters, Washington, DC.

Document II-71

Document Title: Memorandum to Dr. (George) Mueller from Willis Shapley, NASA Associate Deputy Administrator, "Symbolic Activities for Apollo 11," 2 July 1969.

Source: Folder #18675, NASA Historical Reference Collection, History Division, NASA Headquarters, Washington, DC.

As planning for the first lunar landing picked up in intensity, attention turned to the symbolic aspects of the mission. Willis Shapley, a veteran Washington bureaucrat who served as a policy advisor to the NASA Administrator, chaired a Symbolic Activities Committee that was set up to determine what items would be carried to the Moon, and what symbolic activities would be carried out on the lunar surface on the Apollo 11 mission. The final decisions on these matters were communicated to the Apollo program management just two weeks before the 16 July liftoff of the mission.

Document II-70

NATIONAL AERONAUTICS AND SPACE ADMINISTRATION
Washington, D.C. 20546

April 19, 1969

MEMORANDUM FOR: M/Dr. Mueller

Subject: Symbolic Items for the First Lunar Landing

This is to advise you, the Apollo Program Office, and MSC of the thinking that has emerged from discussions among members of the Symbolic Activities Committee to date on symbolic activities in connection with the first lunar landing, including articles to be left on the moon and articles to be taken to the moon and returned.

Further discussions will be necessary prior to the time we will make final recommendations for decision by the Administrator, and comments and suggestions from all members of the Committee and others are still in order. However, in view of the general agreement on approach that has been manifested so far and the tight deadlines for decisions on matters directly affecting preparations for the mission, the approach outlined below should be taken as the basis for further planning at this time.

1.   Symbolic activities must not, of course, jeopardize crew safety or unduly interfere with or degrade achievement of mission objectives. They should be simple, in good taste from a world-wide standpoint, and have no commercial implications or overtones.

2.   The intended overall impression of the symbolic activities and of the manner in which they are presented to the world should be to signalize the first lunar landing as an historic forward step of all mankind that has been accomplished by the United States of America.

3.   The "forward step of all mankind" aspect of the landing should be symbolized primarily by a suitable inscription to be left on the moon and by statements made on earth, and also perhaps by leaving on the moon miniature flags of all nations. The UN flag, flags of all other regional or international organizations, or other international or religious symbolism will not be used.

4.   The "accomplishment by the United States" aspect of the landing should be symbolized primarily by placing and leaving a U.S. flag on the moon in such a way as to make it clear that the flag symbolized the fact that an effort by American people reached the moon, not that the U.S. is [2] "taking possession" of the moon. The latter connotation is contrary to our national intent and would be inconsistent with the Treaty on Peaceful Uses of Outer Space.

5.   In implementing the approach outlined above, the following primary symbolic articles and actions or their equivalents should be considered for inclusion in the mission:

a.   A U.S. flag to be placed and left on the moon. The flag should be such that it can be clearly photographed and televised. If possible, the act of emplacing the flag by the astronaut, as well as the emplaced flag with an astronaut beside it, should be photographed and televised. Current thinking is that a recognizable traditional flag should be emplaced on the moon. The flag decal on the LM decent stage would not by itself suffice unless a flag proved to be clearly not feasible. Consideration of how best to emplace the flag should include but not be limited to the following suggestions:

(1) Cloth flag on vertically emplaced pole, with astronaut to hold flag in visible position for photographing.

(2) Cloth flag on pole emplaced at an angle so that flag is visible for photographing.

(3) An adaptation of the Solar Wind Experiment device in the form of a flag.

(4) Flag on a pole using the commemorative marker (item b below) as a base.

b.   A permanent commemorative marker, suitably inscribed, to be placed and left on the lunar surface, with photographic and television coverage as suggested above for the U.S. flag, if possible. Possibilities to be considered should include, but not be limited to:

(1) A thin-walled metal pyramid, with inscriptions on each of its three or four sides, which could also serve as a sealed repository for a set of miniature flags of all nations (item c below).

(2) A container of cylindrical or other more convenient shape to perform the same function as suggested in (1) above.

(3) A pyramid or other container, as above, which would also serve as the base for the U.S. flag to be emplaced on the moon.

[3]

c.   Miniature flags of all nations, one set to be left on the moon in a suitable container (see above), and a duplicate set to be returned to earth for possible presentation by the President to foreign Chiefs of State. If flag container is not feasible, the set of flags might be left on or in the LM decent stage.

d.   One or more U.S. flags to be presented to NASA prior to the mission by the President and/or other senior officials, taken to the moon and back, and then suitably displayed, perhaps with photographs of the astronauts on the moon, in suitable national locations such as the Capitol, White House, National Archives, Smithsonian Institution, Library of Congress, or elsewhere.

6.   The LM decent stage itself will be of prime symbolic significance since the descent stage will become a permanent monument on the surface of the moon. For this reason, the name given to the LM and any inscriptions to be placed on it must be consistent with the overall approach on symbolic articles and must be approved by the Administrator. The present thinking is that:

a.   The name of the vehicle should be dignified and hopefully convey the sense of "beginning" rather than "culmination" of man's exploration of other worlds.

b.   Assuming that a commemorative marker with inscription is carried, inscriptions on the LM should be limited to the present flag decal and words "United States."

7.   The principal secondary symbolic articles receiving favorable consideration so far include the following:

a.   A small <u>postage stamp</u> die to be taken to the moon and back from which commemorative stamps would be printed. Weight and dimensions alternatives are being investigated.

b.   A jeweler's die to be taken to the moon and back from which lapel type pins associated with the NASA special "Apollo Achievement Awards" now under consideration would be stamped out. Weight and size requirements are being investigated.

8.   It would be appreciated if any comments, further suggestions, or problems you or others receiving copies of this memorandum may have with respect to the foregoing tentative plans and conclusions are made known promptly to me and the Committee via the secretary, Mr. Daniels.

/Signed/
Willis H. Shapley
Associate Deputy Administrator

Document II-71

[stamped Jul 2 1969]

MEMORANDUM FOR: M/Dr. Mueller

Subject: Symbolic Activities for Apollo 11

As your office has previously been advised, the symbolic articles approved for the Apollo 11 mission as of this date are as follows:

A.   <u>Symbolic articles to be left on the moon</u>

1.   A <u>U.S. flag</u>, on a metal staff with an unfurling device, to be emplaced in the lunar soil by the astronauts. This will be the only flag emplanted [*sic*] or otherwise placed on the surface of the moon.

2.   A <u>commemorative plaque</u> affixed to the LM descent stage to be unveiled by the astronauts. The plaque will be inscribed with:

a.   A design showing the two hemispheres of the earth and the outlines of the continents, without national boundaries.

b.   The words: "Here men from the planet earth first set foot upon the moon. We came in peace for all mankind."

             c.   The date (month and year).

             d.   The signatures of the three astronauts and the President of the U.S.

     3.   A <u>microminiaturized photoprint</u> of letters of good will received from Chiefs of State or other representatives of foreign nations.

B.   <u>Symbolic articles to be taken to the moon and returned to earth</u>

     1.   <u>Miniature flags</u> (1 each) of all nations of the UN, and of the 50 states, District of Columbia, and U.S. territories—for subsequent presentation as determined by the President. "All nations" has been defined on the advice of the State Department to include "the members of the United Nations and the UN Specialized Agencies." These items will be stowed in the LM.

[2]

     2.   <u>Small U.S. flags</u>—for special presentation as determined by the President or the Administrator of NASA. These will also be stowed in the LM.

     3.   <u>Stamp die</u> from which Post Office Department will print special postage stamps commemorating the first lunar landing and a <u>stamped envelope</u> to be cancelled with the <u>cancellation stamping device</u>. Cancellation can be done as convenient during the mission in the CM. The stamp die will be stowed in the LM; the stamping device and envelope will be stowed in the CM. <u>These items will not be announced in advance.</u>

     4.   Two <u>full size U.S. flags</u>—which have been flown over the Capitol, the House and the Senate, to be carried in CM but will not be transferred to the LM.

C.   <u>Personal Articles</u>

    Personal articles of the astronauts' choosing under arrangements between Mr. Slayton and the flight crews.

With respect to all items under categories A and B above, it should be clearly understood that the articles are "owned" by the Government and that the disposition of the articles themselves or facsimiles thereof is to be determined by the Administrator or NASA. The articles returned from the mission should be turned over to a proper authority at MSC promptly upon return. In the case of Item B2, the Administrator has determined that a reasonable number of small U.S. flags will be made available to the flight crew for presentation as they see fit, subject to the avoidance of conflict with plans for presentation of these flags by the President or the Administrator.

With respect to articles in Category C above, Mr. Scheer should be notified in advance of the mission of any items which are or may appear to be duplicates of items the President or others might present to Governors, Heads of State, etc. The value of these "one-of-a-kind" presentations can be diminished if there is a proliferation of such items. Flags and patches particularly fall into this category.

Public announcement has or will be made of all items in Categories A and B in advance of the mission <u>except</u> for the items under B3, any release concerning which is subject to a separate decision.

[Signed Willis H. Shapley]

Willis H. Shapley
Associate Deputy Administrator

cc: A/Dr. Paine
       AA/Dr. Newell
       F/Mr. Scheer
       C/Mr. Allnutt
       I/Mr. Frutkin

Document II-72

Document Title: Letter from Frank Borman, NASA Astronaut, to Paul Feigert, 25 April 1969.

Source: Folder #18675, NASA Historical Reference Collection, History Division, NASA Headquarters, Washington, DC.

For the general public, the two highlights of the Apollo 8 mission at Christmas time 1968 were the photograph of Earth rising over the desolate lunar surface and the reading of the first 10 verses of Genesis from the Bible by the crew on Christmas Eve.

[stamped April 25, 1969]

Mr. Paul F. Feigert
1702 Terrace Drive
Lake Worth, Florida 33460

Dear Mr. Feigert:

Dr. Gilruth has asked me to answer your inquiry concerning the reading of the first 10 verses of Genesis.

    a.   Three small Bibles supplied by the Gideons did accompany us on the flight.

    b.   Because the Bibles were flammable, they were sealed in fireproof plastic and not opened during the flight.

    c.   The first 10 verses of Genesis were copied from the Bible and printed on the flame resistant paper of the fight plan.

Thank you for your interest in this matter.

Sincerely,

[Signed Frank Borman]

Frank Borman
Colonel, USAF
NASA Astronaut

Document II-73

Document Title: "General Declaration: Agriculture, Customs, Immigration, and Public Health," 24 July 1969.

Source: Folder #18675, NASA Historical Reference Collection, History Division, NASA Headquarters, Washington, DC.

Like all travelers who return to the United States from trips outside the country, the Apollo 11 crew had to file this declaration as the ship carrying them and their cargo reached their first port of entry, Honolulu, Hawaii, after their return from the Moon.

GENERAL DECLARATION

(Outward/Inward)

AGRICULTURE, CUSTOMS, IMMIGRATION, AND PUBLIC HEALTH

Owner or Operator .... NATIONAL AERONAUTICS AND SPACE ADMINISTRATION

Marks of Nationality and Registration ..... U.S.A. ..... Flight No. APOLLO 11 .... Date JULY 24, 1969

Departure from .... MOON .... Arrival at .... HONOLULU, HAWAII, U.S.A. ....
(Place and Country)   (Place and Country)

FLIGHT ROUTING

("Place" Column always to list origin, every en-route stop and destination)

PLACE	TOTAL NUMBER OF CREW	NUMBER OF PASSENGERS ON THIS STAGE	CARGO
CAPE KENNEDY	COMMANDER NEIL A. ARMSTRONG		
MOON		Departure Place:	MOON ROCK AND MOON DUST SAMPLES Cargo Manifests Attached
JULY 24, 1969 HONOLULU	COLONEL EDWIN E. ALDRIN, JR.	Embarking ..... NIL Through on same flight ..... NIL	
		Arrival Place:	
		Disembarking ..... NIL Through on same flight ..... NIL	
	LT. COLONEL MICHAEL COLLINS		

Declaration of Health

Persons on board known to be suffering from illness other than airsickness or the effects of accidents, as well as those cases of illness disembarked during the flight:

NONE

Any other condition on board which may lead to the spread of disease:

TO BE DETERMINED

Details of each disinsecting or sanitary treatment (place, date, time, method) during the flight. If no disinsecting has been carried out during the flight give details of most recent disinsecting:

Signed, if required ....
Crew Member Concerned

For official use only

HONOLULU AIRPORT

Honolulu, Hawaii

ENTERED

Customs Inspector

I declare that all statements and particulars contained in this General Declaration, and in any supplementary forms required to be presented with this General Declaration are complete, exact and true to the best of my knowledge and that all through passengers will continue/have continued on the flight.

Document II-74

Document Title: Memorandum to Captain Lee Scherer from Julian Scheer, NASA Assistant Administrator for Public Affairs, 24 July 1969.

Source: Folder #18675, NASA Historical Reference Collection, History Division, NASA Headquarters, Washington, DC.

The Apollo 11 crew brought back 44 pounds of lunar material. While most of this material was reserved for scientific investigations, a small amount was set aside for more public purposes. Lee Scherer was the Director of the Lunar Exploration Office at the Manned Spacecraft Center at the time of the Apollo 11 mission.

[ADMINISTRATIVE CONFIDENTIAL] [DECLASSIFIED]

NATIONAL AERONAUTICS AND SPACE ADMINISTRATION
Washington, D.C. 20546

Office of the Administrator

[stamped July 24, 1969]

MEMORANDUM to Captain Lee Scherer

Mittauer informed me of your preliminary plan of one percent lunar samples for "public affairs" purposes. This included suggestion of grains for Nixon to present heads of state, rotating exhibit, small rocks for Nixon, Agnew, others personally. We approve setting aside of this sample and wish it impounded immediately for purposes to be outlined only by Administrator. There should be no discussion of possible uses. Administrator emphatic that this sample and no others be used for this purpose and no other part or parts of sample be released to anyone for public or private giveaways. Suggest that egg-size samples be retained, since they can be used as large display or broken into grains, depending on Administrator's conclusion.

> Julian Scheer
> Assistant Administrator
> for Public Affairs

Document II-75

Document Title: Letter to Robert R. Gilruth, Director, Manned Spacecraft Center, from George E. Mueller, Associate Administrator for Manned Space Flight, 3 September 1969.

Source: Folder #18675, NASA Historical Reference Collection, History Division, NASA Headquarters, Washington, DC.

The managers of the Apollo program at the Manned Spacecraft Center in Houston were primarily from an engineering background, and tended to view the Apollo missions as engineering achievements rather than expeditions driven by scientific requirements. This led to continuing tensions between Houston and members of the scientific community interested in lunar science. This letter reflects such tensions. Ultimately, NASA decided to fly a scientist-astronaut, geologist Harrison "Jack" Schmitt, on the final Apollo mission, Apollo 17.

NATIONAL AERONAUTICS AND SPACE ADMINISTRATION

Sep 3, 1969

Dr. Robert R. Gilruth
Director
Manned Spacecraft Center
Houston, Texas 77058

Dear Bob,

To the public, the success of Apollo 11 is an historical fact. However, to your Center in particular, and to many of the rest of us, the mission is not yet completed and will not be for some time to come. As we have discussed informally, completion of data analysis, posturing solutions for the minor, yet important anomalies which occurred in flight, and, provision for adequate and continuing support of the science effort are items of priority. The latter item, science support, is of particular concern at this time.

Over the past couple of years we have taken steps both here at NASA Headquarters and at MSC to establish a science management, administration and support capability for the Apollo Program. This has been done with significant sacrifice to other program areas within a steadily reducing total Manned Flight and NASA personnel ceiling. During the pre-Apollo 11 time-period the workload of this group increased steadily and it was difficult to obtain a commensurate increase in the number and appropriate types of personnel to do the many jobs involved. Now, with operating experiments on the lunar surface returning data and the return of Apollo 11 lunar samples for analysis, the workload has increased many fold. The resulting increased interest and direct participation of the scientific community in Apollo is taxing our capability to the limit. Despite this, we will certainly detract measurably from the success of Apollo 11, and the missions yet to be flown, unless we meet the challenge. Therefore, we must provide the support required in the science area.

A problem of immediate concern is prompt and proper distribution of the lunar samples to the Principle [*sic*] Investigators through their home institutions. To protect the government and public interest in these materials, contractual coverage must be obtained. At the current rate of contract negotiation I am concerned that we will have clearance from the ICBC for sample release before all of the sample analysis contract processing is completed. I urge you to assign whatever resources are necessary to bring completion of contract processing into phase with sample release.

[2]The successful accomplishment of the initial Apollo lunar landing was necessarily the focus and emphasis in the program for many years. The operational complexity of the next few missions will also require concentration on that aspect. We will be increasing our capability to do more and more interesting science simultaneously. Still, some members of the scientific community are impatient and as you know, are willing to air their views without necessarily relating those views to what is practicable and possible.

Public discussion aside, it is our policy to do the maximum science possible in each Apollo mission and to provide adequate science support. For the long term we must assure ourselves and the world of science that we are making those adjustments which will provide steadily increasing and effective support for the science area. Good progress has been made to date, but we must do even better to meet the future challenge. I ask your personal involvement in this as well as in solving the immediate concerns relating to Apollo 11.

Sincerely yours,

/Signed/
George E. Mueller
Associate Administrator
for Manned Space Flight

Document II-76

Document Title: Space Science Board, National Academy of Sciences, "Report of Meeting on Review of Lunar Quarantine Program," 17 February 1970.

Source: Folder #18675, NASA Historical Reference Collection, History Division, NASA Headquarters, Washington, DC.

After two landings on the Moon, Apollo 11, and Apollo 12, and no sign of dangerous life forms being returned to Earth, NASA was contemplating the end of the quarantine that had been in place for those two missions. The National Academy of Sciences was asked to examine the question. This report contains the recommendation of the ad hoc committee set up to prepare the Academy's response. It also provides an overview of the testing for signs of life done on the returned lunar samples.

Report of

Meeting on

Review of Lunar Quarantine Program

February 17, 1970
At
The Manned Spacecraft Center
National Aeronautical and Space Administration
Houston, Texas

Space Science Board
National Academy of Sciences
2101 Constitution Avenue
Washington, D.C.

PREFACE

On December 24, 1969 the Administrator of the National Aeronautics and Space Administration requested the President of the National Academy of Sciences to form a committee to review the Academy's 1964 recommendations for a lunar quarantine program in light of information acquired from lunar flights, Apollos 11 and 12. The President referred the request to the Space Science Board where it was favorably considered at the Board's meeting on January 12-13, 1970. An ad hoc committee was authorized by the Space Science Board to consider new evidence accumulated about the earth's moon since its 1964 conference and to make recommendations pertinent to the continuation of the lunar quarantine program.

The ad hoc committee met at the National Aeronautics and Space Administration's Manned Spacecraft Center, Houston, Texas on February 17, 1970. A listing of the participants is shown.

The agenda for the meeting provided for an exchange of views and facts in geology, geochemistry, and biology, including microbiology and medicine. We believe the varied positions were thoroughly argued.

The recommendations of the ad hoc committee are summarized at the beginning of the report and the rationale in arriving at the recommendations follows in the body of the report. Minority views are attached in the appendix.

Participants in the Space Science Board's

Review of the National Academy of Sciences of Lunar

Quarantine Recommendations

Manned Spacecraft Center, Houston, Texas
February 17, 1970

Members

Martin Alexander	Cornell University
Klaus Biemann	Massachusetts Institute of Technology
Allan H. Brown, (Chairman)	University of Pennsylvania
Gustave J. Dammin	Harvard Medical School
Paul Gast	Columbia University
Lawrence B. Slobodki	University of New York at Stony Brook
John Spizizen	Scripps Clinic, La Jolla
Wolf Vishniac	University of Rochester
Frank G. Favorite	Space Science Board, National Academy of Sciences

Liaison Representatives and Other Participants

Earl H. Arnold	NASA Headquarters, Office of Manned Space Flight
Charles A. Berry	NASA, Manned Spacecraft Center
Howard H. Eckles	NASA, Manned Spacecraft Center
Lawrence B. Hall	NASA, Headquarters, Office of Space Science and Applications
Rufus R. Hessberg	NASA Headquarters, Office of Manned Space Flight
James W. Humphreys, Jr.	NASA Headquarters, Office of Manned Space Flight
R. E. Kallio	University of Illinois
Walter W. Kemmerer, Jr.	NASA, Manned Spacecraft Center

Adrian Mandel	NASA, Manned Spacecraft Center
John A. Mason	NASA, Manned Spacecraft Center
Carl Sagan	Cornell University
E. E. Salmon	U.S. Department of Agriculture
David J. Sencer	Department of Health, Education and Welfare, Communicable Disease Center Atlanta
Gerald R. Taylor	NASA, Manned Spacecraft Center
Bennic C. Wooley	NASA, Manned Spacecraft Center

Summary of Finding and Major Recommendations

Finding

In Apollo 13 the proposed highland landing site and core sample to a depth of 8 feet constitute a substantially new lunar environment in comparison to the landing sites and sampled areas of Apollo 11 and 12 missions.

Recommendations

Lunar Quarantine Program

A majority recommend continuance of the 3-week lunar quarantine period. A minority favor discontinuance of quarantine.

Lunar Samples

We recommend development of procedural changes in the handling of lunar samples to preclude alteration of the sample prior to analysis.

Biological Testing Program

We recommend the continued development of a research program within the LRL to develop greater confidence in the adequacy of the test program and the validity of both negative and positive findings.

Introduction

Our committee heard from representations of the National Aeronautics and Space Administration's Office of Manned Space Flight and Manned Spacecraft Center, from lunar sample experimenters and from persons responsible for operations in the Lunar Receiving Laboratory. Summaries were presented of medical tests on astronauts and other quarantined personnel and of the examination of lunar samples including tests for pathogenicity, or toxicity and for the presence of life forms.

We believe that the quarantine policy which has applied to lunar samples, spacecraft and astronauts was conscientiously implemented in Apollo 11 and 12 missions. It was noted that some procedures have been less than ideal. Nevertheless, a quarantine policy implementation, beset from the start with severe difficulties of interdisciplinary communication and inflexible schedules, was as successful as could have been expected.

It is noteworthy that the Interagency Committee on Back-Contamination (ICBC) was effective in formulating the policies and approving the operational procedures which guided the implementation of those policies by NASA. We feel credit is due, both to the ICBC and to NASA for meeting numerous challenges so successfully.

The committee agrees with the wisdom of lunar quarantine as a policy of caution, well justified at the time it was established by the potential hazard of back-contamination from what was a largely unknown environment. The possibility existed that Apollo astronauts, infected with a virulent, contagious, lunar, biological agent, would exhibit disease symptoms within the period of quarantine and thus alert attending physicians to the need for continued effective containment of the infectious agent. A small possibility of this still exists and views expressed by qualified persons and groups who have appraised the current status of the subject differ chiefly because everyone cannot agree on the magnitude of this possibility.

It is well recognized that quarantine at best is imperfect protection against diseases even of known etiology. Some members of our committee feel that close medical surveillance of the returned Apollo astronauts would be quite sufficient. However, the majority feel that astronaut quarantine, employing essentially the same procedures as were used on the Apollo 12 mission, ought to be in effect for any future missions which may be judged to involve a risk of back-contamination.

Discussion

It seems as it did prior to Apollo 11 that any change in the U.S. Quarantine Policy must be based on a revised or more confident assessment of the overall back-contamination hazard to man and his environment. Results from Apollo 11 and 12 missions have made available substantial new information about the moon, and some of this is directly relevant to the charge of our committee. Briefly stated, we view the evidence as follows:

Hazard to Human Beings

There have been no medical signs or symptoms of illness among lunar astronauts during or subsequent to quarantine which could reasonably be attributed to lunar pathogens. Moreover, no such indications of pathology have been reported among some 150 individuals who have had at least some contact either with Apollo astronauts or with lunar sample material, however, no formal medical surveillance of this group has been maintained. We consider these negative findings reassuring but not definitive. With the relatively short duration of exposure and the small number of astronauts involved, lack of observed infection is not equivalent to a confirmed absence of pathogens.

Hazards to Animals and Plants

The lunar sample material was not found to be pathogenic to any of a number of test species of plants and animals. Again it is our view that this evidence (which pertains to many species) is more reassuring than the absence of evident human pathology. Nevertheless, such negative evidence seems insufficient to warrant the conclusion that no pathogens exist on the moon. It has been noted that lunar material under some test conditions is capable of stimulating plant growth. It is not yet clear whether such effects are attributable to the direct biological action of lunar material or perhaps to nutritional stimulation which in this context would be trivial. Until these growth augmentation results, samples were biologically inert.

Evidence of Life Forms

No living organisms were detected in lunar sample material. We feel that this evidence by itself is inconclusive, partly because there may be some question that the biological assay was fully adequate to reveal exotic life forms, but chiefly because the material which has been tested represents a limited sampling of the lunar environment. What has been examined so far is essentially surface material from two mare sites, largely igneous in origin, predictably sterile, and not for certain representative of what may be found, for example, at several meters depth in the highland region which will be sampled during the Apollo 13 mission.

We find it exceptionally difficult to conceive of an ecological model whereby life forms could endure and maintain themselves even in the most favorable environment we can imagine which could be compatible with the analytical measurements of lunar samples from Apollo 11 and 12. It is this, perhaps, even more than the negative results from direct biological testing, which constitutes the more persuasive argument against lunar life existing in those particular mare sites.

Evidence of Water and Carbon

New chemical evidence of several kinds makes it seem improbable that indigenous life could ever have existed in the environment represented by the Apollo samples so far obtained. The salient evidence is first, the absence of any hydrous minerals in the samples examined (indicating that water was not an environmental constituent when crystallization took place, and the preservation of delicate glassy surfaces and finely divided particles of iron and iron sulfide indicate the samples have not been exposed to water, vapor or liquid since crystallization); and second, the extremely low content of organic carbon which characterizes the samples.

We recognize that if only a minute quantity of organic material is present, but that it includes some living organisms, it is quite reasonable to expect nearly all of the organic carbon to be contained in those organisms. The sensitivity of testing for such organisms by chemical or physical assays without the benefit of biological amplification (growth) is inadequate. The "noise level" of such test procedures would correspond to the carbon content of hundreds or thousands of microbial cells.

Lunar "Gardening"

Geological evidence of lunar surface turnover as this applies to Apollo 11 and 12 sites persuades us against the existence at these places of a protected region containing at least some water and organic matter, and therefore a possible abode for lunar organisms. Finally, mineralogical findings and evidence from isotope dating indicate a kind of sample heterogeneity which could best be explained by assuming transport of substantial amounts of material onto the mare, presumably from the neighboring highlands. It seems quite possible or even likely that in the Apollo 11 and 12 samples, several percent may represent highland material. Even so, it would hardly be permissible to generalize from knowledge of these two sites to the many particular local environments to be found on the moon. Much of the moon is as yet unknown and thus predictions of biological significance about the landing site of Apollo 13 may be in error.

Lunar Quarantine Program

We note that the Apollo 11 and 12 samples were in all likelihood from the upper surfaces of lava beds. It is therefore not surprising that the samples from both areas appear sterile. Any possible pre-existing life would have been destroyed by processes which created these formations, and the likelihood of reinoculation from other (highland?) areas might have been negligible. On or near the surface, radiation and temperature extremes probably preclude growth and perhaps even survival of live organisms. In any case, other Apollo landing sites are apt to have quite different and new chemical characteristics. Even the two mare sites, originally expected to be much the same, have turned out to be surprisingly different. It is surely unwise to generalize from this limited Apollo sampling and it seems to most of us that the new information gained from past Apollo missions is insufficient to justify a substantial change in lunar quarantine policy applicable to the Apollo 13 mission which is targeted for a highland landing site. We therefore endorse the policy established by the ICBC which asserts that each time a substantially new type of lunar environment is visited or sampled a maximum back-contamination hazard obtains and whatever quarantine measures have been agreed upon for that circumstances become fully applicable.

Quarantine of Lunar Samples

The overriding reason for continuing the quarantine of the astronauts and the lunar samples returned from the Apollo 13 mission is the possibility that materials that have not been exposed on the lunar surface for long periods will be returned in the lower portion of a drill core sample. An additional but secondary reason for continuing the quarantine is the planned return of materials that differ significantly in composition, age and origin from the Apollo 11 and 12 samples. In previous missions the sample chosen for the biological protocol was selected to be representative of all the returned rocks and soil. Detailed study of these rock and soil samples have not shown us that there is little variation among the rock types. The requirement of pooled test samples has resulted in severe time constraints in the preliminary examinations of the lunar samples. Handling lunar samples in the LRL under quarantine restrictions precludes some desirable operations, introduces chemical contamination of the samples, and is responsible for harming delicate surface features of the rock due to the awkward manipulations which are performed. We therefore recommend that Apollo 13 samples used for the biological protocol be restricted to a much smaller portion of the returned samples. An aliquot of the lower portion of the drill core and one soil sample might be adequate. As presently planned, both of these samples could come from the ALSRC containing the drill core section. The second ALSRC, and sample returned in other containers, need not be involved in the biological protocol.

Lunar-Planetary Quarantine Relationship

There are important long-range benefits to be gained from Apollo quarantine experience. Perhaps within two decades manned missions will explore Mars and perhaps other space objectives about which we have little biologically significant information. At this time it seems advisable for NASA to plan to establish and implement a quarantine policy applicable to those more ambitions missions on the assumption that the back-contamination risks, with respect to Mars at least, will continue to be much greater than was ever thought to be the case for the earth's moon. We believe that the recommendations of the 1964 Conference on Potential hazards of back-contamination from the Planets continue to apply to the planning for a manned Mars mission. In this connection it would be valuable for NASA to document its Apollo quarantine experience in such a manner that a future generation of planners can benefit maximally from what was learned during Apollo. Substantial savings in the cost of quarantine, avoidance of compromises and more effective communication between design engineers and those responsible for biomedical aspects of quarantine policy and procedures would be facilitated by an enlightened accounting of the many lessons which are being learned in the course of the lunar quarantine program.

---

*Space Science Board, National Academy of Sciences, 29-30 July 1964. Revised 19 February 1965. 15pp.

Biological Testing Program

In the course of our meeting we studied the design and results of biological tests performed with lunar samples and visited the biological laboratories of the Lunar Receiving Laboratory. Each specialist had reason to comment upon the design, conduct and results of these biological tests. An absence of direct testing methods such as microscopy scanning was noted. We found complete agreement with our views by resident scientists. The biological lunar testing program has raised many fundamental questions about the selection of host organisms and culture media, rout of inoculation with lunar material, incubation period and temperature, control samples and test procedures that we feel warrant immediate attention.

We recommend a continuing research effort at LRL to develop a wide based biological testing program, expanded to include other competent biological laboratories, with sufficient diversity not only to maximize the chance of positive findings but also to validate negative findings through adequate controls, particularly those inoculated with material known to be capable of infecting the host or culture.

Appendix

Minority Views

Dr Frank Favorite, Space Science Board

National Academy of Sciences 2101 Constitution Ave Was/DC

I disagree with continuation of lunar quarantine procedure. I recommend a post-flight isolation of one week for astronauts followed by surveillance of two months or longer. Samples should be contained in aseptic manner and released if biological testing proves negative after three weeks. Investigation should be made in depth, using expert consultants, of the plant stimulation and microbial toxicity test. Research on the survival of micro-organisms in lunar environment should be conducted as soon as possible. Better methods for detection of organisms should be investigated, especially direct methods with electron microscopy.

John Spizizen, Scripps Clinic

Harvard Medical School – Peter Bent Brigham Hospital

February 24, 1970

Memorandum for: Dr. Allan H. Brown, Chairman, ad hoc Committee of the Space Science Board in lunar quarantine, National Academy of Sciences

From: Gustave J. Dammin, M.D., member, ad hoc, Committee

1. The recommendations pertaining to "Lunar Samples" and "Biological Testing Program" contained in the "Summary of Finding and Major Recommendations" of the report of the ad hoc Committee, I concur in. However, I wish to dissent from the recommendation which calls for the continuance of the 3-week lunar quarantine procedure with reference to Apollo 13. The evidence gathered before, and the evidence presented at our meeting Feb. 17, was not sufficient in my evaluation to establish a basis for suspecting lunar samples might contain agents that would be inimical to man, animals or plants.

2. I would recommend, with reference to study of the astronauts, a period of isolation following return to earth during which specimens could be collected for such purposes as determining possible changes in flora, and the like. Conceivably no more than 3-4 days might be needed, depending upon the details of the protocol.

3. The experience gained with the 21-day quarantine procedure for Apollo 11 and 12 is indeed valuable. It will be helpful in planning the quarantine

protocol to be pursued with reference to the Mars exploration. Recording of the procedures employed in all their detail is essential since future teams of scientists concerned with the quarantine procedure may not include those who have profited from the recent Apollo experiences.

/Signed/
Gustave J. Dammin, M.D.

Document II-77

Document Title: George Low, Personal Notes No. 30, Interim Operating Budget and Apollo Decisions.

Source: Papers of George M. Low, Rensselaer Polytechnic Institute, Troy, New York.

Document II-78

Document Title: George M. Low, Acting Administrator, Letter to Edward E. David, Jr., Science Advisor to the President, "Apollo versus Skylab and Research Airplane Programs," 30 October 1970.

Source: Folder #18675, NASA Historical Reference Collection, History Division, NASA Headquarters, Washington, DC.

Document II-79

Document Title: James C. Fletcher, Administrator, Letter to Caspar W. Weinberger, Deputy Director, Office of Management and Budget, 3 November 1971.

Source: Papers of James C. Fletcher, University of Utah Library, Salt Lake City, UT.

George Low had become NASA Deputy Administrator in December 1969, and Dale Myers had replaced George Mueller as NASA Associate Administrator for Manned Space Flight in early 1970. During 1970, NASA was trying to gain White House approval to begin development of both an Earth-orbiting space station and a fully reusable space shuttle to service it, while the top White House priority was reducing the NASA budget. The future of the remaining Apollo missions and of the interim space station, Skylab, which was based on the conversion of the upper stage of an unneeded Saturn V booster, were caught up in the conflict between NASA's desire to get started on new development programs and the White House push for budget limitations.

NASA Administrator Thomas Paine had announced in early August 1970 his intention to resign on September 15: Low became Acting Administrator upon Paine's departure. Of the people mentioned in Low's note, Peter Flanigan was Special Assistant to President Nixon, with responsibility for space matters; Lee DuBridge was the President's Science Adviser and Russ Drew was his top staff person for space; Apollo 8 astronaut Bill Anders had become Executive Secretary of the National Aeronautics and Space Council; George Shultz was Director of the Office of Management and Budget.

NASA's decision to cancel two Apollo missions did not satisfy the White House; there was continuing pressure to either cancel additional Apollo missions and/or not fly the Skylab mission, planned for 1973. By the end of October 1970, Edward E. David, Jr. had replaced Lee Dubridge as Science Adviser, and he asked George Low to compare the priorities of additional Apollo missions and Skylab.

Even after the successful flights of Apollo 14 and Apollo 15, the White House gave serious consideration to canceling the last two Apollo missions, but ultimately NASA flew Apollo 16 and Apollo 17 in 1972 and launched the Skylab station on 14 May 1973. James Fletcher, former President of the University of Utah, became NASA Administrator in May 1971.

Document II-77

September 6, 1970

PERSONAL NOTES NO. 30

Interim Operating Budget and Apollo Decisions

        I spent most of the last two weeks in August on vacation, but did return to the office on August 24th for a meeting concerning whether we should have six additional Apollo flights as planned, or should reduce the number to four. At the meeting on August 24th, we heard Dale Myers' proposal to reduce the number of flights to four, with a saving of approximately $40 million in Fiscal Year 1971, but an overall saving over the next four or five years of approximately $800 million. Also, Drs. Findlay and Ruby reported the results of the Lunar and Planetary Missions Boards meetings and the Space Science Board meetings, looking into the question concerning the additional scientific aims that could be had by maintaining six Apollo flights. The scientists' view was that they strongly recommended flying out all six remaining missions; but that the loss of one mission (Apollo 15 with its lesser capability) would not be nearly as serious as the loss of both Apollo 15 and Apollo 19.

        In meetings later on August 24th and on August 31st, September 1st and September 2nd, we decided to delete two flights, Apollo 15 and Apollo 19. (The remaining Apollo flights would, of course, be redesignated Apollos 14, 15, 16, and 17.) At the same time, we developed an interim operating plan which we will use until we get a 1971 Appropriations Bill. This plan is based on the Appropriations Bill that was passed by the Congress but was subsequently vetoed by the

President. It is, therefore, at a level of approximately $64 million less than the 1971 President's budget.

In arriving at these decisions, we had invited Flanigan or his representative, DuBridge's representative, and Bill Anders to the August 24th meeting. Anders and Russ Drew, representing DuBridge, showed up, but Flanigan did not send anybody. Our intention had been to notify the White House and the Office of Management and Budget of our decision before making it public on September 1st or September 2nd. However, prior to our notification, we had a call from Shultz of OMB questioning the wisdom of making the decision at this time. His main concern was that we might lose additional funding in the Congress if we made the decision now. However, after a number of telephone conversations, we were allowed to move out with the decision. The interesting part was that the substance of the decision was not questioned, but merely our strategy relative to Congress. Shultz made the strong recommendation, however, that we do not mention the $800 million saving over a number of years. [2] We reluctantly agreed to this approach, which is probably the main reason why the publicity on the cancellation of Apollos 15 and 19 was not as good as it might have been. THE NEW YORK TIMES editorially stated that we were cancelling [*sic*] the potentially, scientifically, most fruitful missions for a relatively small amount of $40 million. Had we publicized the $800 million, I would guess that they could not have taken this stand.

Document II-78

OCT 30 1970 [stamped]
Honorable Edward E. David, Jr.
Science Advisor to the President
Executive Office Building
Washington, D.C. 20506

Dear Ed:

During our meeting last Monday, we promised to write to you on the following subjects: the relative priorities of Apollo and Skylab; and the requirement for the research airplane programs proposed in our FY 1972 budget.

Apollo Versus Skylab

Looking first to Apollo, we have already had a successful program that has met the fundamental objective laid down in 1961: to prove American technological superiority without military confrontation, to build a new level of national pride and prestige, and to create a base of science and technology for the future. The Apollo 11 and 12 missions have, in addition, opened a new field of lunar-related science with the return of samples, emplacement of seismic and other instruments on the surface, and erection of the laser reflector. These alone have already provided substantial scientific return on the nature of the moon and its environment, and will continue to do so for many years. Study of the data from these two missions should contribute a great deal to our understanding of the origin and evolution not only of the moon, but also of our own earth.

The remaining four Apollo missions will add incrementally to the science base as the radius of exploratory activity increases, as the diversity of sites visited enlarges, and as the sophistication of surface and orbital instrumentation grows with each flight. To reduce or constrain the scientific returns from Apollo by dropping one or more missions would involve very great losses. Moreover, any impression that each successive Apollo mission is constantly in jeopardy of being cancelled for budget reasons will have serious impact on the technical teams responsible for the safety of the flights, thereby adding to the existing dangers of the already difficult remaining missions. It would, of course, also reinforce the sentiment in the scientific community that the priority of science is decreasing on the national scene.

[2] Nevertheless, continuing Apollo missions through the next four flights, while significantly increasing our scientific understanding of the Earth-Moon system, would in another sense be dead-ended. No new capabilities or techniques would be explored that could be further exploited in the conduct of manned or unmanned programs; no major new opportunities for international leadership and prestige would likely accrue; and the potential of Apollo for international cooperation is limited.

A budgetary alternative to cutting back one or more Apollo missions would be the cancellation of Skylab. Here the situation differs, in that there has as yet been no return from the considerable investment to date; the basic objectives of Skylab are yet to be achieved. We simply have no data on man's ability to live and work in space for long periods of time. Our own 14-day and the U.S.S.R.'s 18-day manned mission experience is [sic] inadequate as a basis for future decisions. Our experience with man as a necessary contributor to science and applications tasks is severely limited. Our experience with long-duration habitable space systems is non-existent.

Although there are some who question the worth of space stations at this time, there is also a body of scientific and engineering opinion today that a space station will be an important and extremely valuable next step in man's exploration and utilization of space. (In fact, today's support, by scientists, for the space station appears to be greater than their support for Apollo as little as two years ago!) With Skylab, we can extend our experience from two weeks to two months; we can test realistically man's contribution to science, applications, and engineering functions; and we can develop an understanding of our future options early enough to permit the rational, deliberate evolution of our programs.

At the same time, Skylab-borne experiments are of unique scientific and technical value in themselves. The Apollo Telescope Mount (ATM) will, because of its capability to use film, have data acquisition rates a million times higher than that of the automated Orbiting Solar Observatory; the ATM is therefore ideally suited for the very high resolution study of rapidly varying solar phenomena. The earth resources survey package will give us the first meaningful intercomparison of photographic, infrared, and microwave remote sensors to correlate with aircraft ERTS experiments for determination of the next step in this exciting and relevant applications area. This package will also provide a special resolution far greater than the unmanned ERTS instruments.

[3] To forego Skylab would have a powerful negative impact on astronomy and earth resources surveys. It would leave the U.S. without the data base for any future manned mission decisions. It would surrender to the U.S.S.R. the option of having the first real space station in orbit. It would leave underdeveloped the desirable precedent of openly shared manned flight program scientific and technical results, a possibility currently underscored by the discussions in Moscow on the suggestion that the U.S. and U.S.S.R. use common docking hardware in their orbital spacecraft.

On balance, the weight of evidence seems to favor Skylab over Apollo if a choice must be made. The scientific returns from the single Skylab mission promise to be greater than those from a sixth Apollo lunar landing. We have already capitalized on our Apollo investment but not yet on that of Skylab; we will have more new options better developed stemming from Skylab than from Apollo; and, for this increased return, we risk less in earth orbit than at lunar distances.

[remainder of letter not included]

Sincerely yours,

/Signed/
George M. Low
Acting Administrator

Document II-79

PRIVILEGED INFORMATION [DECLASSIFIED]

NATIONAL AERONAUTICS AND SPACE ADMINISTRATION
WASHINGTON. D.C, 20546

November 3, 1971

Honorable Caspar W. Weinberger
Deputy Director
Office of Management and Budget
Executive Office of the President
Washington, D. C. 20503

Dear Cap:

In our conversation last week, you indicated that cancellation of Apollo 16 and 17 was being considered by the President and asked for my views on the actions that should be taken to offset or minimize the adverse consequences if such a decision is made.

From a scientific standpoint these final two missions are extremely important, especially Apollo 17 which will be the only flight carrying some of the most advanced experiments originally planned for Apollos 18 and 19, cancelled last year. With what we have learned from Apollo 15 and previous missions, we seem to be on the verge of discovering what the entire moon is like: its structure, its composition, its resources, and perhaps even its origin. If Apollo 16 and 17 lead to these discoveries, the Apollo program will go down in history not only as man's greatest adventure, but also as his greatest scientific achievement. Recognizing the great scientific potential and the relatively small saving ($133 million) compared to the investment already made in Apollo ($24 billion), I must as Administrator of NASA strongly recommend that the program be carried to completion as now planned.

If broader considerations, nevertheless, lead to a decision to cancel Apollo 16 and 17, the consequences would be much more serious than the loss of a major scientific opportunity. Unless compensatory actions are taken at the same time to offset and minimize the impact, this decision could be a blow from which the space program might not easily recover. As you requested, I will summarize the principal adverse consequences as I see them and then outline my recommendations on the compensatory actions necessary.

[2]   <u>PRINCIPAL ADVERSE CONSEQUENCES</u>

1.   <u>Negative Effect on Congressional and Public Support.</u>

Without strong compensatory actions, a decision to cancel Apollo 16 and 17 would undermine the support the space program now enjoys and jeopardize the continued support that is required over the years to sustain the nation's position in space. Even though enthusiasm for the space program has diminished since the first lunar landing, NASA has continued to receive better than 98 percent of its budget requests each year (99.94% in FY 1972) because a substantial majority has accepted the judgments of the Administration and NASA's leadership that the space program is vital to the United States and that the programs recommended each year are necessary to achieve our goals. Cancellation of Apollo 16 and 17 would undermine this support in two ways.

<u>First,</u> it would call into question our credibility on this and other major elements of the space program since it would be a sudden reversal of the position we have so recently strongly supported in defense of our FY 1972 budget.

<u>Second,</u> it would terminate our best known, most visible and most exciting program which, in the minds of many in Congress and the public, has been the symbol of the space program and its success.

These factors, unless offset by strong positive actions, could result in a loss of confidence and interest that would have a "domino" effect, causing us to lose support for the programs which are essential to the long-term future of the nation in space.

2.   Impact on Science and the Scientific Community

At this time, the entire cognizant scientific community is strongly in favor of Apollo 16 and 17. Cancellation would come as a shock and a surprise in view of the strong support these missions have received from the President's Science Adviser, all of NASA's science advisory groups, NASA management, and the Congress. There will be strong and vocal critical reaction.

[3] 3.   Impact on Industry.

Taken by itself, the direct impact of the cancellation of Apollo 16 and 17 would be further reductions in 1972 of over 6,000 aerospace jobs. The hardest hit areas would be Southern California, Long Island, Cape Kennedy, and Houston. Unless the decision is coupled with commitments and actions to proceed with and possibly expedite other programs, like the space shuttle, it will be a devastating blow, actually and psychologically, to an already hard-hit industry.

4.   Impact on NASA.

The impact on NASA will be felt most strongly at Houston and, to a lesser extent, at Huntsville and Cape Kennedy. A major problem will be to hold together for over a year the team we will need to rely on to conduct safely the Skylab missions in 1973. We will have to deal with the difficult and visible problem of the futures of the 16 astronauts now assigned to Apollo 16 and 17. The blow to morale throughout NASA will be serious unless, again, the decision is coupled with clear decisions and commitments on future programs.

5.   Impact on the Public.

The large segment of space enthusiasts in the population at large would be extremely disappointed by the proposed cancellation. Included in this group would be millions who have come to Cape Kennedy, often from very long distances, to witness Apollo launches, and the much larger numbers who follow each mission closely on TV. These groups may be a minority in the U.S. but they are quite vocal and certainly non-negligible in size.

6.   Impact Abroad

It is our understanding from USIA reports that the Apollo flights have been a major plus factor for the U.S. image abroad. The impact of cancelling [*sic*] Apollo 16 and 17 should be assessed in arriving at a decision.

[4]   RATIONALE AND ACTIONS REQUIRED

If a decision is made to cancel Apollo 16 and 17, it is essential to provide a clearly stated and defensible rationale and take constructive actions to minimize the adverse impacts of the cancellation on the space program, the Administration, and the individuals, "communities of interest," and organizations affected. The rationale and actions must make it clear that, in spite of the cancellation, the President continues to support a program involving man in space and with strong scientific content. Specifically:

1. The reason given for cancelling [*sic*] Apollo would be budgetary; there are no other limitations to carrying Apollo to completion.

2. The total space program recommended by the President must be one that does not put an end to manned space flight (or even portends to do so in the future) and must, therefore, include Skylab and a real commitment to the shuttle with a go-ahead in the spring of 1972, and some earth orbit Apollo ("gap-filler") missions between Skylab and the shuttle.

3. The scientific content of the space program should be enhanced to offset the science lost with Apollo 16 and 17.

4. The total NASA budget should not drop below the essentially constant level of FY 1971 and FY 1972 (about $3.3 billion in budget authority) to demonstrate the President's intent to maintain a strong space program.

Rationale

The rationale supporting this position would be as follows:

"Our space program has three basic purposes: exploration; the acquisition of scientific knowledge; and practical applications for man on earth. (See President's statement of March 7, 1970.) We must always strive to achieve the proper balance among these purposes.

"Today we must stress two aspects of our space program. We must give a top priority to practical applications now possible and press forward with [5] the development of earth oriented systems which will enable us to make wider and more effective practical uses of space in the future.

"The key to the future in space–in science and exploration as well as practical applications–is routine access to space. Space activities will be part of our lives for the rest of time. These activities cannot continue, for long, to be as complex, as demanding, or as costly as they are today. We must develop new, simpler, less expensive techniques to go to space and to return from space. This is the goal of the space shuttle program. The sooner we get on with this development, the sooner will we be able to turn our knowledge gained in space science and space exploration toward helping man on earth.

"To operate in space most effectively we must also learn more about how man can best live and work in space. So while we are developing the shuttle, we must conduct space operations over longer periods of time–with Skylab.

"But to do all these things within limited resources, we must give up something. And when all factors are considered, the best project to give up--most reluctantly–is the remainder of Apollo: Apollo 16 and 17. This will for a time curtail our program of <u>manned</u> exploration and science.

"But we will, of course, continue exploration deep into space with <u>unmanned</u> spacecraft, including a landing on Mars in July 1976 with Viking, and the exploration of all the outer planets, Jupiter and beyond, with the Grand Tour late in this decade. The unmanned science program, with its High Energy Astronomy Observatory and other spacecraft will also continue to expand our fundamental knowledge of the universe. It is only manned science, and manned exploration, that will be curtailed.

"The United States must continue to fly men in space. Man will fly in space, and we cannot forego our responsibility–to ourselves and to the free world–to take part in this great venture. But for a time man can devote his own efforts, from space, toward practical [6] needs here on earth, while leaving exploration beyond the earth to machines."

I believe that this is the best rationale that can be given, although it is admittedly somewhat complex, and neither it nor any other rationale will be accepted by the interested scientific community.

<u>Actions</u>

The actions required to offset the adverse impact of the cancellation of Apollo 16 and 17 should include:

     1.   A commitment to a strong manned space flight program including <u>Skylab</u> and a good start on the <u>space shuttle</u>.

     2.   The earth-oriented emphasis of manned space flight can be further amplified by flying "surplus" <u>Apollo spacecraft in earth orbit</u> in the 1974-76 time period, i.e., "gap-filler" missions after Skylab and before the shuttle, as proposed in the NASA FY 1973 budget submissions. These spacecraft can be equipped with sophisticated earth-oriented experiments as precursors to the type of operations to be carried out with the shuttle. At the same time, they could provide the means for a joint flight with the Soviet Union–a step that has already been hailed editorially as one in the right direction for the U.S. space program in that it could lead to an ultimate sharing of the expense of space among many nations.

     3.   <u>Science:</u> NASA needs the support of the "scientific community" to carry out its programs. And although the impact of this community on the Administration as a whole is small, it is important to minimize and divert the criticism the Administration will receive as a result of a decision to cancel Apollo 16 and 17.

     Nothing can be done to get general acceptance of a cancellation by the lunar scientists. However, the impact on them, as well as criticism by all scientists, can be minimized if the following steps are taken:

a. Announcement of a sound program for the continued analysis of lunar materials already obtained. Such a [7] program would have a great scientific value, and would also continue financial support to the scientists involved in lunar analysis, who would otherwise be out of a job.

b. Initiation of a small effort toward Jupiter orbiters and probes ("Pioneer" class spacecraft). One of the most important concerns of the National Academy of Sciences Space Science Board is that NASA's present plans for the Grand Tour missions to the outer planets do not include a parallel program for the detailed exploration of Jupiter. The inclusion of a continuing Pioneer program, in addition to the Grand Tour, would partially offset the negative impact of the cancellation of Apollo.

c. Reinstatement of the Orbiting Solar Observatories I, J, and K, proposed for deletion in NASA's FY 1973 budget proposal, and the full funding of the High Energy Astronomy Observatory, proposed for reduced funding in NASA's budget, would demonstrate the Administration's desire to support science to a large segment of the space science community.

4.   The effectiveness of the above compensatory actions will depend in large measure on the total budget level approved for NASA for FY 1973. Unless the NASA FY 1973 budget is essentially at or above the FY 1971 and FY 1972 budget authority level of about $3.3 billion, the decision to cancel Apollo 16 and 17 will be regarded by the Congress, the public, and the scientific community as a part of a general backing away from and downgrading of the space program.

Effect of Actions on Budget

The actions discussed above would result in a net reduction in NASA's FY 1973 minimum recommended budget estimates but would not take the total estimate for budget authority below $3.3 billion, as indicated below.

[8]

	(in millions)	
	Budget Authority	Budget Outlays
NASA FY 1973 Budget Submission—Minimum Recommended Program	$3,385	$3,225
Cancellation of Apollo 16 & 17	–133	–109
Start Space Shuttle	(no change)	(no change)
Reinstate OSO-I, J, K	+ 20	+ 15
Start Pioneer Orbiter/ Probes	+ 15	+ 5

Full Support for HEAO	+ 26	+ 20
1974-1976 Manned Orbital Flights	+ 38	+ 30
TOTAL	$3,351	$3, 186

Effect on Employment

If the actions previously discussed–the early go-ahead on the shuttle, the inclusion of the gap-filler missions, and the augmentation of science missions–are taken, then the negative impact on the industry, and on employment, will to some degree be alleviated. The effects of these actions on employment during calendar year 1972, in terms of changes in contractor employment projected under our FY 1973 budget recommendations, would be approximately as follows:

[9]

Employment End of 1972	Contractor
Estimated under NASA FY 1973 Budget Submission (Minimum Recommended Program)	109, 200
Cancellation of Apollo 16 & 17	- 6,200
Start Space Shuttle	(no change)
Reinstate OSO I, J, K	+ 700
Start Pioneer orbiter/probes	(no significant effect until 1973)
Full support for HEAO	+ 1,200
1974-1976 manned orbital flights	+ 1,900
TOTAL	106, 800

The net effect on employment will be downward since the decrease would be almost immediate but increases due to new programs obviously take a few months to materialize.

In a separate exercise, we have provided information to Fred Foy to show how employment on the shuttle could be increased above our FY 1973 budget recommendations.

CONCLUSIONS

I recommend against the cancellation of Apollo 16 and 17 because these flights are scientifically important, and because much of the overall support for NASA's space program depends on our actions with respect to these flights.

If, nevertheless, for reasons external to NASA, Apollo 16 and 17 must be cancelled, then it becomes necessary to:

      1.   Provide strong backing to the manned earth-oriented space program.

[10]   2.   Develop a rationale for the actions taken that is credible and supportable.

      3.   Take compensatory actions that will minimize the impact on the remaining NASA programs and their support.

The proposed rationale for the cancellation of Apollo 16 and 17 is that, in these times of pressing domestic needs, the manned space program should be earth-oriented instead of exploration and science-oriented.

The compensatory actions involve an early go-ahead for the space shuttle, the inclusion of "gap-filler" missions between Skylab and the shuttle, a number of augmented unmanned space science programs, and maintaining a total NASA budget at the FY 1971-1972 level of about $3.3 in budget authority.

I would be pleased to discuss these matters with you at your convenience.

Sincerely,

James C. Fletcher
Administrator

Document II-80

Document Title: Letter to Congressman G. P. Miller, Chairman of the House Committee on Science and Astronautics, from 39 Scientists, 10 September 1970.

Source: Folder #18675, NASA Historical Reference Collection, History Division, NASA Headquarters, Washington, DC.

NASA announced the cancellation of the Apollo 15 and Apollo 19 missions on 2 September 1970. There was an outcry from the media and many members of the scientific community, but the decision could not be reversed. This meant that the Apollo lunar landing program would end with the Apollo 17 mission in December 1972.

September 10, 1970

Congressman G. P. Miller
Chairman House Committee on Science
    and Astronautics
House Office Building
Washington, D. C. 20515

Dear Chairman Miller:

We, the undersigned scientists concerned with the space program, would like to express to you our deep misgivings about the NASA decision of cancelling [*sic*] two of the remaining lunar Apollo flights, resulting in a severe curtailment of the lunar exploration program. In particular, we would like to stress the following points:

1. The Apollo lunar program is intended to supply not merely information of interest to scientists, but to give us finally a clear understanding of the origin of the earth-moon system and with this, an understanding of the origin and mode of construction of our earth. The structure of the Apollo program is one of increasing capabilities, and the two cancelled missions represent much more than one third of the planned scientific program. With this curtailment, the program may fail in its chief purpose of reaching a new level of understanding.

2. The NASA policy leading to the cancellations appears to be one of favoring the early construction of large manned earth orbital systems following after Skylab A, and the effort and funds saved by the curtailment will probably go towards these. The merit of these programs for science or applications should be investigated, and the very important decision regarding their funding should in our view be made as a separate step. At present, it appears that the approved and scientifically most fruitful lunar program will suffer in favor of an as yet unapproved program for whose scientific value there is no consensus, and whose purpose is unclear.

3. The majority of the equipment saved by the proposed cancellations will in all probability be shelved indefinitely, since large funds would be required for its adaptation to other purposes or its rehabilitation at a later date for lunar flights, as well as for the re-creation of the Apollo launch capability.

We hope that these decisions are not yet final, and that the country will not give up a plan of very great significance when the preparation for it is so nearly complete.

Yours sincerely,

(See attached pages)

cc: Congressman Olin E. Teague
House Office Building
Washington, D.C. 20515

[Signed]   Dr. M. E. Langseth
Lamont-Doherty Geological Observatory –
Columbia University

[Signed]   Dr. William M. Kaula
Professor of Geophysics
University of California

[Signed]   Dr. Lincoln R. Page
U.S. Geological Survey

[Signed]   Dr. William R. Muehlberger
Professor of Geology
University of Texas

[Signed]   Dr. Rolf Meissner
Visiting Professor
University of Hawaii

[Signed]   Dr. T. W. Thompson
Jet Propulsion Laboratory

[Signed]   Dr. Brian H. Mason, Curator
Division of Meteorites
U.S. National Museum

[Signed]   Dr. Roman A. Schmitt
Radiation Center
Oregon State University

[Signed]   Dr. Ian D. MacGregor
Department of Geology
University of California

[Signed]   Professor Thomas Gold
Center for Radio Physics and Space Research –
Cornell University

[Signed]   Dr. William W. Ruby, Director
The Lunar Science Institute and
Prof. of Geology, the University
of California, Los Angeles

[Signed]   Dr. Leon T. Silver
Division of Geological Sciences
California Institute of Technology

[Signed]	S. O'Sullivan
[Signed]	Dr. Eugene Schoemaker [*sic*] Division of Geological Sciences California Institute of Technology
[Signed]	Dr. Jeffrey L. Warner Geology Branch Manned Spacecraft Center
[Signed]	Dr. Charles E. Helsley Acting Head, Geosciences University of Texas, Dallas
[Signed]	Dr. Warren G. Meinschein Department of Geology Indiana University
[Signed]	Dr. George Wetherill Department of Planetary and Space Science University of California
[Signed]	Dr. A. G. W. Cameron Goddard Space Flight Center Institute for Space Studies
[Signed]	Dr. John Wasson Department of Chemistry University of California
[Signed]	Dr. Bruce Doe NASA Headquarters Code MAL
[Signed]	Dr. John Wood Smithsonian Astrophysical Observatory
[Signed]	Manuel N. Bass
[Signed]	Dr. Harold C. Urey Department of Chemistry Revelle College University of California, San Diego
[Signed]	Dr. Robert A. Phinney Department of Geological and Geophysical Sciences Princeton University
[Signed]	Dr. Anthony W. England Code CB NASA Manned Spacecraft Center

[Signed]   Dr. Harold Masursky
                            Branch of Astrogeologic Studies
                            U.S. Geological Survey

[Signed]   Dr. Gerald Schubert
                            Dept. of Planetary and Space Sciences
                            University of California

[Signed]   Dr. Charles Sonnet
                            NASA Ames Research Center

[Signed]   Dr. Geoffrey Eglinton
                            Organic Geochemistry Unit
                            School of Chemistry
                            University of Bristol

[Signed]   Dr. N. U. Mayall, Director
                            Kitt Peak National Observatory

[Signed]   Dr. John B. Adams
                            Carribean Research Institute
                            College of the Virgin Islands

[Signed]   Dr. Thomas B. McCord
                            Assistant Professor, Planetary Physics
                            Dept. of Earth & Planetary Sciences
                            Massachusetts Institute of Technology

[Signed]   Dr. J. J. Papike
                            Dept. of Earth & Space Sciences
                            State University of New York

[Signed]   Mr. Ernest Schonfeld
                            Geology Branch
                            Manned Spacecraft Center

[Signed]   Dr. George E. Ulrich
                            U.S. Geological Survey

[Signed]   Mr. J. D. Strobell, Jr.
                            U.S. Geological Survey

[Signed]   Dr. David S. McKay

[Signed]   Dr. John Reynolds
                            Department of Physics
                            University of California, Berkeley

Document II-81

Document Title: Mission Evaluation Team, NASA Manned Spacecraft Center, "Apollo 11: Mission Report," 1971.

Source: Johnson Space Center Archives.

This report captures in flat prose what actually took place during the historic Apollo 11 lunar landing mission. Included here are a brief mission overview and the crew's report on mission activities.

Mission Report

PREPARED BY

MISSION EVALUATION TEAM

NASA MANNED SPACECRAFT CENTER

INDEX DATA
DATE OPR # T PGM SUBJECT SIGNATOR
00-00-71 NASA SP-238 2 GEN A MSC

2 copies

Scientific and Technical Information Office 1971
NATIONAL AERONAUTICS AND SPACE ADMINISTRATION
Washington, D.C.

[1-1]

1.0 SUMMARY

The purpose of the Apollo 11 mission was to land men on the lunar surface and to return them safely to earth. The crew were Neil A. Armstrong, Commander; Michael Collins, Command Module Pilot; and Edwin E. Aldrin, Jr., Lunar Module Pilot.

The space vehicle was launched from Kennedy Space Center, Florida, at 8:32:00 a.m., e.s.t., July 16, 1969. The activities during earth orbit checkout, translunar injection, transposition and docking, spacecraft ejection, and translunar coast were similar to those of Apollo 10. Only one midcourse correction, performed at about 27 hours elapsed time, was required during translunar coast.

The spacecraft was inserted into lunar orbit at about 76 hours, and the circularization maneuver was performed two revolutions later. Initial checkout of lunar module systems was satisfactory, and after a planned rest period, the Commander and Lunar Module Pilot entered the lunar module to prepare for descent.

The two spacecraft were undocked at about 100 hours, followed by separation of the command and service modules from the lunar module. Descent orbit insertion was performed at approximately 101-1/2 hours, and powered descent to the lunar surface began about 1 hour later. Operation of the guidance and descent propulsion systems was nominal. The lunar module was maneuvered manually approximately 1100 feet down range from the nominal landing point during the final 2-1/2 minutes of descent. The spacecraft landed in the Sea of Tranquility at 102:45:40. The landing coordinates were 0 degrees 41 minutes 15 seconds north latitude and 23 degrees 26 minutes east longitude reference to lunar map ORB-II-6(100), first edition, December 1967. During the first 2 hours on the lunar surface, the two crewmen performed a postlanding checkout of all lunar module systems. Afterward, they ate their first meal on the moon and elected to perform the surface operations earlier than planned.

Considerable time was deliberately devoted to checkout and donning of the back-mounted portable life support and oxygen purge systems. The Commander egressed through the forward hatch and deployed an equipment module in the descent stage. A camera in this module provided live television coverage of the Commander descending the ladder to the surface, with first contact made at 109:24:15 (9:56:15 p.m. e.s.t., July 20, 1969). The Lunar Module Pilot egressed soon thereafter, and both crewmen used the initial period on the surface to become acclimated to the reduced gravity and unfamiliar surface conditions. A contingency sample was taken from the surface, and the television camera was deployed so that most of the lunar module was included in its view field. The crew activated the scientific experiments, which included a solar wind detector, a passive [1-2] seismometer, and a laser retro-reflector. The Lunar Module Pilot evaluated his ability to operate and move about, and was able to translate rapidly and with confidence. Forty-seven pounds of lunar surface material were collected to be returned for analysis. The surface exploration was concluded in the allotted time of 2-1/2 hours, and the crew reentered the lunar module at 111-1/2 hours.

Ascent preparation was conducted efficiently, and the ascent stage lifted off the surface at 124-1/4 hours. A nominal firing of the ascent engine placed the vehicle into a 48- by 9-mile orbit. After a rendezvous sequence similar to that of Apollo 10, the two spacecraft were docked at 128 hours. Following transfer of the crew, the ascent stage was jettisoned, and the command and service modules were prepared for transearth injection.

The return flight started with a 150-second firing of the service propulsion engine during the 31st lunar revolution at 135-1/2 hours. As in the translunar flight, only one midcourse correction was required, and passive thermal control was exercised for most of transearth coast. Inclement weather necessitated moving the landing point 215 miles downrange. The entry phase was normal, and the command module landed in the Pacific Ocean at 195-1/4 hours. The landing coordinates, as determined from the onboard computer, were 13 degrees 19 minutes north latitude and 169 degrees 09 minutes west longitude.

After landing, the crew donned biological isolation garments. They were then retrieved by helicopter and taken to the primary recovery ship, USS Hornet. The crew and lunar material samples were placed in the Mobile Quarantine Facility for transport to the Lunar Receiving Laboratory in Houston. The command module was taken aboard the Hornet about 3 hours after landing.

With the completion of Apollo 11, the national objective of landing men on the moon and returning them safely to earth before the end of the decade had been accomplished.

[Sections 2 and 3 not included]

[4-1]

4.0 PILOTS' REPORT

4.1 PRELAUNCH ACTIVITIES

All prelaunch systems operations and checks were completed on time and without difficulty. The configuration of the environmental control system included operation of the secondary glycol loop and provided comfortable cockpit temperature conditions.

4.2 LAUNCH

Lift-off occurred precisely on time with ignition accompanied by a low rumbling noise and moderate vibration that increased significantly at the moment of hold-down release. The vibration magnitudes decreased appreciably at the time tower clearance was verified. The yaw, pitch, and roll guidance-program sequences occurred as expected. No unusual sounds or vibrations while passing through the region of maximum dynamic pressure and the angle of attack remained near zero. The S-IC/S-II staging sequence occurred smoothly and at the expected time.

The entire S-II stage flight was remarkably smooth and quiet, and the launch escape tower and boost protective cover were jettisoned normally. The mixture ratio shift of the was accompanied by a noticeable acceleration decrease. The S-II/S-IVB staging sequence occurred smoothly and approximately at the predicted time. The S-IVB insertion trajectory was completed without incident and the automatic guidance shutdown yielded an insertion-orbit ephemeris, from the command module computer, of 102.1 by 103.9 miles. Communications between the crewmembers and the Network were excellent throughout all stages of launch.

4.3 Earth Orbit Coast and Translunar Injection

The insertion checklist was completed, and a series of spacecraft systems checks disclosed no abnormalities. All tests of the navigation equipment, including alignments and drift checks, were satisfactory. The service module reaction control thrusters were fired in the minimum impulse mode and were verified by telemetry.

No abnormalities were noted during preparation for translunar injection. Initiation of translunar injection was accompanied by the proper onboard indications and the S-IVB propellant tanks were repressurized on schedule.

[4-2] The S-IVB stage reignited on time at 2:44:16 without ignition or guidance transients. An apparent 0.50- to 1.5- degree pitch-attitude error on the attitude indicators was not confirmed by the command module computer, which indicated that the attitude and the attitude rate duplicated the reference trajectory precisely (see section 8.6). The guided cutoff yielded a velocity very close to that expected, as indicated by the onboard computer. The entry monitor system further confirmed that the forward velocity error for the translunar injection maneuver was within 3.3 ft/sec.

4.4 Transposition and Docking

The digital autopilot was used for the transposition maneuver scheduled to begin 20 seconds after spacecraft separation from the S-IVB. The time delay was to allow the command and service modules to drift approximately 70 feet prior to thrusting back toward the S-IVB. The separation and the beginning of transposition were on time. In order to assure a pitch-up maneuver for better visibility through the hatch window, pitch axis control was retained in a manual mode until after a pitch-up rate of approximately 1 deg/sec was attained. Control was then given to the digital autopilot to continue the combined pitch/roll maneuver. However, the autopilot stopped pitching up at this point, and it was necessary to reestablish manual control (see section 8.6 for more discussion of this subject). This cycle was repeated several times before the autopilot continued the transposition maneuver. Consequently, additional time and reaction control fuel (18 pounds above preflight nominal) were required, and the spacecraft reached a maximum separation distance of at least 100 feet from the S-IVB.

The subsequent closing maneuvers were made normally under digital autopilot control, using a 2-deg/sec rate and 0.5-degree deadband control mode.

Contact was made at an estimated 0.1 ft/sec, without side velocity, but with a small roll misalignment. Subsequent tunnel inspection revealed a roll index angle of 2.0 degrees and a contact mark on the drogue 4 inches long. Lunar module extraction was normal.

4.5 Translunar Coast

The S-IVB was targeted to achieve a translunar injection cut-off velocity 6.5 ft/sec in excess of that required to place it on the desired free-return trajectory. This overspeed was then cancelled by a service propulsion correction of 20 ft/sec at 23 minutes after spacecraft ejection.

[4-3] Two periods of cislunar midcourse navigation, using the command module computer program (P23), were planned and executed. The first, at 6 hours, was primarily to establish the apparent horizon altitude for optical marks in the computer. The first determination was begun at a distance of approximately 30,000 miles, while the second determination, at 24 hours, was designed to accurately determine the optical bias errors. Excess time and fuel were expended during the first period because of difficulty in locating the substellar point of each star. Ground-supplied gimbal angles were used rather than those from the onboard computer. This technique was devised because computer solutions are unconstrained about the optics shaft axis; therefore, the computer is unable to predict if the lunar module structure might block the line of sight to the star. The ground-supplied angles prevented the lunar module structure from occulting the star, but were not accurate in locating the precise substellar point, as evidenced by the fact that the sextant reticle pattern was not parallel to the horizon. Additional maneuvers were required to achieve a parallel reticle pattern near the point of horizon-star superposition.

The second period of navigation measurements was less difficult, largely because the earth appeared much smaller and trim maneuvers to the substellar point could be made much more quickly and economically.

The digital autopilot was used to initiate the passive thermal control mode at a positive roll rate of 0.3 deg/sec, with the positive longitudinal axis of the spacecraft pointed toward the ecliptic North Pole during translunar coast (the ecliptic South Pole was the direction used during transearth coast). After the roll rate was established, thruster firing was prevented by turning off all 16 switches for the service module thrusters. In general, this method was highly successful in that it maintained a satisfactory spacecraft attitude for very long periods of time and allowed the crew to sleep without fear of either entering gimbal lock or encountering unacceptable thermal conditions. However, a refinement to the procedure in the form of a new computer routine is required to make it foolproof from an operator's viewpoint. [Editor's note: A new routine (routine 64) was available for Apollo 12.] On several occasions and for several different reasons, an incorrect computer-entry procedure was used, resulting in a slight waste of reaction control propellants. Satisfactory platform alignments (program P52, option 3) using the optics in the resolved mode and medium speed were possible while rotating at 0.3 deg/sec.

4.6 Lunar Orbit Insertion

The spacecraft was inserted into a 169.9- by 60.9-mile orbit based on the onboard computer with a 6-minute service propulsion maneuver. Procedurally, this firing was the same as all the other service propulsion [4-4] maneuvers, except that it was started by using the bank-B propellant valves instead of the bank-A valves. The steering of the docked spacecraft was exceptionally smooth, and the control of applied velocity change was extremely accurate, as evidenced by the fact that residuals were only 0.1 ft/sec in all axes.

The circularization maneuver was targeted for a 66- by 54-mile orbit, a change from the 60-mile circular orbit which had been executed in previous lunar flights. The firing was normally accomplished using bank-A propellant valves only, and the onboard solution of the orbit was 66.1 by 54.4 miles. The ellipticity of this orbit was supposed to slowly disappear because of irregularities in the lunar gravitational field, such that the command module would be in a 60-mile circular orbit at the time of rendezvous. However, the onboard estimate of the orbit during the rendezvous was 63.2 by 56.8 miles, indicating the ellipticity decay rate was less than expected. As a result the rendezvous maneuver solutions differed from the preflight estimates.

4.7 Lunar Module Checkout

Two entries were made into the lunar module prior to the final activation on the day of landing. The first entry was made at about 57 hours, on the day before lunar orbit insertion. Television and still cameras were used to document the hatch probe and drogue removal and the initial entry into the lunar module. The command module oxygen hoses were used to provide circulation in the lunar module cabin. A leisurely inspection period confirmed the proper positioning of all circuit breaker and switch settings and stowage items. All cameras were checked for proper operation.

4.8 Descent Preparation

4.8.1 Lunar Module

The crew was awakened according to the flight plan schedule. The liquid cooling garment and biomedical harnesses were donned. In anticipation, these items had been unstowed and prepositioned the evening before. Following a hearty breakfast, the Lunar Module Pilot transferred into the lunar module to accomplish initial activation before returning to the command module for suiting. This staggered suiting sequence served to expedite the final checkout and resulted in only two crewmembers being in the command module during each suiting operation.

[4-5] The sequence of activities was essentially the same as that developed for Apollo 10, with only minor refinements. Numerous Network simulations and training sessions, including suited operations of this mission phase, ensured the completion of this exercise within the allotted time. As in all previous entries into

the lunar module, the repressurization valve produced a loud "bang" whenever it was positioned to CLOSE or AUTO with the cabin regulator off. Transfer of power from the command module to the lunar module and then electrical power system activation were completed on schedule.

The primary glycol loop was activated about 30 minutes early, with a slow but immediate decrease in glycol temperature. The activation continued to progress smoothly 30 to 40 minutes ahead of schedule. With the Commander entering the lunar module early, the Lunar Module Pilot had more than twice the normally allotted time to don his pressure suit in the command module.

The early powerup of the lunar module computer and inertial measurement unit enabled the ground to calculate the fine gyro torquing angles for aligning the lunar module platform to the command module platform before the loss of communications on the lunar far side. This early alignment added more than an hour to the planned time available for analyzing the drift of the lunar module guidance system.

After suiting, the Lunar Module Pilot entered the lunar module, the drogue and probe were installed, and the hatch was closed. During the ascent-battery checkout, the variations in voltage produced a noticeable pitch and intensity variation in the already loud noise of the glycol pump. Suit-loop pressure integrity and cabin regulator repressurization checks were accomplished without difficulty. Activation of the abort guidance system produced only one minor anomaly. An illuminated portion of one of the data readout numerics failed, and this resulted in some ambiguity in data readout (see section 16.2.7).

Following command module landmark tracking, the vehicle was maneuvered to obtain steerable antenna acquisition and state vectors were uplinked into the primary guidance computer. The landing gear deployment was evidenced by a slight jolt to the vehicle. The reaction control system, the descent propulsion system, and the rendezvous radar system were activated and checked out. Each pressurization was confirmed both audibly and by instrument readout.

The abort guidance system calibration was accomplished at the preplanned vehicle attitude. As the command and service modules maneuvered both vehicles to the undocking attitude, a final switch and circuit breaker configuration check was accomplished, followed by donning of helmets and gloves.

[4-6]   4.8.2 Command Module

Activities after lunar orbit circularization were routine, with the time being used primarily for photographs of the lunar surface. The activation of the lunar module in preparation for descent was, from the viewpoint of the Command Module Pilot, a well organized and fairly leisurely period. During the abort guidance system calibration, the command module was maintained at a fixed attitude for several minutes without firing thrusters. It was easy to stabilize the spacecraft with minimum-impulse control prior to the required period so that thruster firings were needed for at least 10 minutes.

The probe, drogue, and hatch all functioned perfectly, and the operations of closing out the tunnel, preloading the probe, and cocking the latches were done routinely. Previous practice with installation and removal of the probe and drogue during translunar coast was most helpful.

Two periods of orbital navigation (P22) were scheduled with the lunar module attached. The first, at 83 hours, consisted of five marks on the Crater Kamp in the Foaming Sea. The technique used was to approach the target area in an inertial attitude hold mode, with the X-axis being roughly horizontal when the spacecraft reached an elevation angle of 35° from the target, at which point a pitch down of approximately 0.3 deg/sec was begun. This technique was necessary to assure a 2-1/2 minute mark period evenly distributed near the zenith, was performed without difficulty.

The second navigation exercise was performed on the following day shortly prior to separation from the lunar module. A series of five marks was taken on a small crater on the inner north wall of crater 130. The previously described technique was used, except that two forward-firing thrusters (one yaw and one pitch) were inhibited to preclude thrust impingement on the deployed rendezvous-radar and steerable antennas. The reduced pitch authority doubled the time required, to approximately 3 seconds when using acceleration command, to achieve a 0.3 deg/sec pitch-down rate. In both cases, the pitch rate was achieved without reference to any on board rate instrumentation by simply timing the duration of acceleration-command hand controller inputs, since the Command Module Pilot was in the lower equipment bay at the time.

To prevent the two vehicles from slipping and hence upsetting the docked lunar module platform alignment, roll thruster firings were inhibited after the probe preload until the tunnel had been vented to approximately 1 psi. Only single roll jet authority was used after the l-psi point was reached and until the tunnel pressure was zero.

[4-7]

4.9 UNDOCKING AND SEPARATION

Particular care was exercised in the operation of both vehicles throughout the undocking and separation sequences to ensure that the lunar module guidance computer maintained an accurate knowledge of position and velocity.

The undocking action imparted a velocity to the lunar module of 0.4 ft/sec, as measured by the lunar module primary guidance system. The abort guidance system disagreed with the primary system by approximately 0.2 ft/sec, which is well within the preflight limit. The velocity was nulled, assuming the primary system was assumed to be correct. The command module undocking velocity was maintained until reaching the desired inspection distance of 40 feet, where it was visually nulled with respect to the lunar module.

A visual inspection by the Command Module Pilot during a lunar module 360-degree yaw maneuver confirmed proper landing gear extension. The lunar module maintained position with respect to the command module at relative rates believed to be less than 0.1 ft/sec. The 2.5-ft/sec, radially downward separation maneuver was performed with the command and service modules at 100 hours to enter the planned equiperiod separation orbit.

4.10 LUNAR MODULE DESENT

The first optical alignment of the inertial platform in preparation for descent orbit insertion was accomplished shortly after entering darkness following separation. The torquing angles were approximately 0.3 degree, indicating an error in the docked alignment or platform drift. A rendezvous radar lock was achieved manually, and the radar boresight coincided with that of the crew optical sight. Radar range was substantiated by the VHD ranging in the command module.

4.10.1 Descent Orbit Insertion

The descent orbit insertion maneuver was performed with the descent engine in the manual throttle configuration. Ignition at the minimum throttle setting was smooth, with no noise or sensation of acceleration. After 15 seconds, the thrust level was advanced to 40 percent, as planned. Throttle response was smooth and free of oscillations. The guided cutoff left residuals of less than 1 ft/sec in each axis. The X- and Z-axis residuals were reduced to zero by using the reaction control system. The computer determined ephemeris was 9.1 by 57.2 miles, as compared with the [4-8] predicted value of 8.5 by 57.2 miles. The abort guidance system confirmed that the magnitude of the maneuver was correct. An additional evaluation was performed by using the rendezvous radar to check the relative velocity between the two spacecraft at 6 and 7 minutes subsequent to the maneuver. These values corresponded to the predicted data within 0.5 ft/sec.

4.10.2 Alignment and Navigation Checks

Just prior to powered descent, the angle between the line of sight to the sun and a selected axis of the inertial platform was compared with the onboard computer prediction of that angle and this provided a check on inertial platform drift. Three such measurements were all within the specified tolerance, but the 0.08-degree spread between them was somewhat larger than expected.

Visual checks of downrange and crossrange position indicated that ignition for the powered descent firing would occur at approximately the correct location over the lunar surface. Based on measurements of the line-of-sight rate of landmarks, the estimates of altitudes converged on a predicted altitude at ignition 52 000 feet above the surface. These measurements were slightly degraded because of a 10 - to 15-degree yaw bias maintained to improve communications margins.

4.10.3 Powered Descent

Ignition for powered descent occurred on time at the minimum thrust level, and the engine was automatically advanced to the fixed throttle point (maximum thrust) after 26 seconds. Visual position checks indicated the spacecraft was 2 or 3 seconds early over a known landmark, but with little cross-range error. A yaw maneuver to a face-up position was initiated at an altitude of about 45 900 feet approximately 4 minutes after ignition. The landing radar began receiving altitude data immediately. The altitude difference, as displayed from the radar and the computer, was approximately 2800 feet.

At 5 minutes 16 seconds after ignition, the first of a series of computer alarms indicated a computer overload condition. These alarms continued intermittently for more than 4 minutes, and although continuation of the trajectory was permissible, monitoring of the computer information display was occasionally precluded (see section 16.2.5).

Attitude-thruster firings were heard during each major attitude maneuver and intermittently at other times. Thrust reduction of the descent propulsion system occurred nearly on time (planned at 6 minutes 24 seconds after ignition), contributed to the prediction that the [4-9] landing would probably be down range of the intended point, inasmuch as the computer had not been corrected for the observed downrange error.

The transfer to the final-approach-phase program (P64) occurred at the predicted time. After the pitch maneuver and the radar antenna position change, the control system was transferred from the automatic to the attitude hold mode and control response checked in pitch and roll. Automatic control was restored after zeroing the pitch and yaw errors.

After it became clear that an automatic descent would terminate in a boulder field surrounding a large sharp-rimmed crater, manual control was again assumed, and the range was extended to avoid the unsatisfactory landing area. The rate-of-descent mode of throttle (program P66) was entered in the computer to reduce altitude rate so as to maintain sufficient height for landing-site surveillance.

Both the downrange and the crossrange positions were adjusted to permit final descent in a small, relatively level area bounded by a boulder field to the north and sizable craters to the east and south. Surface obscuration caused by blowing dust was apparent at 100 feet and became increasingly severe as the altitude decreased. Although visual determination of horizontal velocity, attitude, and altitude rate were degraded, cues for these variables were adequate for landing. Landing conditions are estimated to have been 1 or 2 ft/sec left, 0 ft/sec forward, and 1 ft/sec down; no evidence of vehicle instability at landing was observed.

4.11 COMMAND MODULE SOLO ACTIVITIES

The Command Module Pilot consolidated all known documentation requirements for a single volume, known as the Command Module Pilot Solo Book, which was very useful and took the place of a flight plan, a rendezvous book, an updates book, a contingency extravehicular checklist, and so forth. This book normally was anchored to the Command Module Pilot by a clip attached to the end of his helmet tie-down strap. The sleep period was timed to coincide with that of the lunar module crew so that radio silence could be observed. The Command Module Pilot had complete trust in the various systems experts on duty in the Mission Control Center and therefore was able to sleep soundly.

The method used for target acquisition (program P22) while the lunar module was on the surface varied considerably from the docked case. The optical alignment sight reticle was placed on the horizon image, and the resulting spacecraft attitude was maintained manually at the orbital rate in the minimum-impulse control mode. Once stabilized, the vehicle maintained this attitude long enough to allow the Command Module Pilot to [4-10] move to the lower equipment bay and take marks. He could also move from the equipment bay to the hatch window in a few seconds to cross-check the attitude. This method of operation in general was very satisfactory.

Despite the fact that the Command Module Pilot had several uninterrupted minutes each time he passed over the lunar module, he could never see the spacecraft on the surface. He was able to scan an area of approximately 1 square mile on each pass, and ground estimates of lunar module position varied by several miles from pass to pass. It is doubtful that the Command Module Pilot was ever looking precisely at the lunar module and more likely was observing an adjacent area. Although it was not possible to assess the ability to see the lunar module from 60 miles, it was apparent there were no flashes of specular light with which to attract his attention.

The visibility through the sextant was good enough to allow the Command Module Pilot to acquire the lunar module (in flight) at distances of over 100 miles. However, the lunar module was lost in the sextant field of view just prior to powered descent initiation (120-mile range) and was not regained until after ascent insertion (at an approximate range of 250 miles), when it appeared as a blinking light in the night sky.

In general, more than enough time was available to monitor systems and perform all necessary functions in a leisurely fashion, except during the rendezvous phase. During that 3-hour period when hundreds of computer entries, as well as numerous marks and other manual operations, were required, the Command Module Pilot had little time to devote to analyzing any off-nominal rendezvous trends as they developed or to cope with any systems malfunctions. Fortunately, no additional attention to these details was required.

4.12 LUNAR SURFACE OPERATIONS

4.12.1 Postlanding checkout

The postlanding checklist was completed as planned. Venting of the descent oxidizer tanks was begun almost immediately. When the oxidizer tank pressure was vented to between 40 and 50 psi, fuel was vented to the same pressure level. Apparently, the pressure indications received on the ground were somewhat higher and they increased with time (see section 16.2.2). At ground request, the valves were reopened and the tanks vented to 15 psi.

[4-11]

Platform alignment and preparation for early lift-off were completed on schedule without significant problems. The mission timer malfunctioned and displayed an impossible number that could not be correlated with any specific failure time. After several unsuccessful attempts to recycle this timer, it was turned off for 11 hours to cool. The timer was turned on for ascent, and it operated properly and performed satisfactorily for the remainder of the mission (see section 16.2.1).

4.12.2 Egress Preparation

The crew had given considerable thought to the advantage of beginning the extravehicular activity as soon as possible after landing instead of following the flight plan schedule of having the surface operations between two rest periods. The initial rest period was planned to allow flexibility in the event of unexpected difficulty with postlanding activities. These difficulties did not materialize, the crew were not overly tired, and no problem was experienced in adjusting to the 1/6-g environment. Based on these facts, the decision was made at 104:40:00 to proceed with the extravehicular activity prior to the first rest period.

Preparation for extravehicular activity began at 106:11:00. The estimate of the preparation time proved to be optimistic. In simulations, 2 hours had been found to be a reasonable allocation; however, everything had also been laid out in an orderly manner in the cockpit, and only those items involved in the extravehicular activity were present. In fact, items involved in the extravehicular activity were present. In fact, there were checklists, food packets, monoculars, and other miscellaneous items that interfered with an orderly preparation. All these items required some thought as to their possible interference or use in the extravehicular activity. This interference resulted in exceeding the time line estimate by a considerable amount. Preparation for egress was conducted slowly, carefully, and deliberately, and future missions should be planned and conducted with the same philosophy. The extravehicular activity preparation checklist was adequate and was closely followed. However, minor items that required a decision in real time or had not been considered before flight required more time than anticipated.

An electrical connector on the cable that connects the remote control unit to the portable life support system gave some trouble in mating (see section 16.3.2). This problem had been occasionally encountered with the same equipment before flight. At least 10 minutes were required in order to connect each unit, and at one point it was thought the connection would not be successfully completed.

Considerable difficulty was experienced with voice communications when the extravehicular transceivers were used inside the lunar module. At times communications were good, but at other times they were garbled on the [4-12] ground for no obvious reason. Outside the vehicle, there were no appreciable communication problems. Upon ingress from the surface, these difficulties recurred, but under different conditions. That is, the voice dropouts to the ground were not repeatable in the same manner.

Depressurization of the lunar module was one aspect of the mission that had never been completely performed on the ground. In the various altitude chamber tests of the spacecraft and the extravehicular mobility unit, a complete set of authentic conditions was never present. The depressurization of the lunar module through the bacteria filter took much longer than had been anticipated. The indicated cabin pressure did not go below 0.1 psi, and some concern was experienced in opening the forward hatch against this residual pressure. The hatch appeared to bend on initial opening, and small particles appeared to be blown out around the hatch when the seal was broken (see section 16.2.6).

4.12.3 Lunar Module Egress

Simulation work in both the water immersion facility and the 1/6-g environment in an airplane was reasonably accurate in preparing the crew for lunar module egress. Body positioning and arching-the-back techniques that were required in to exit the hatch were preformed, and no unexpected problems were experienced. The forward platform was more than adequate to allow changing the body position from that used in egressing the hatch to that required for getting on the ladder. The first ladder step was somewhat difficult to see and required caution and forethought. In general, the hatch, porch, and ladder operation were not particularly difficult and caused little concern. Operations on the platform could be performed without losing body balance, and there was adequate room for maneuvering.

The initial operation of the lunar equipment conveyor in lowering the camera was satisfactory, but after the straps had become covered with lunar surface material, a problem arose in transporting the equipment back into the lunar module. Dust from this equipment fell back onto the lower crewmember and into the cabin and seemed to bind the conveyor so as to require considerable force to operate it. Alternatives in transporting equipment into the lunar module had been suggested before flight, and although no opportunity was available to evaluate these techniques, it is believed they might be an improvement over the conveyor.

[4-13]

4.12.4 Surface Exploration

Work in the 1/6-g environment was a pleasant experience. Adaptation to movement was not difficult and movement seemed to be natural. Certain specific peculiarities, such as the effect of the mass versus the lack of traction, can be anticipated but complete familiarization need not be pursued.

The most effective means of walking seemed to be the lope that evolved naturally. The fact that both feet were occasionally off the ground at the same time, plus the fact that the feet did not return to the surface as rapidly as on earth, required some anticipation before attempting to stop. Although movement was not difficult, there was noticeable resistance provided by the suit.

On future flights, crewmembers may want to consider kneeling in order to work with their hands. Getting to and from the kneeling position would be no problem, and being able to do more work with the hands would increase productive capability.

Photography with the Hasselblad cameras on the remote control unit mounts produced no problems. The first panorama was taken while the camera was hand-held; however, it was much easier to operate on the mount. The handle on the camera was adequate, and very few pictures were triggered inadvertently.

The solar wind experiment was easily deployed. As with the other operations involving lunar surface penetration, it was only possible to penetrate the lunar surface material only about 4 or 5 inches. The experiment mount was not quite as stable as desired, but it stayed erect.

The television system presented no difficulty except that the cord was continually in the way. At first, the white cord showed up well, but it soon became covered with dust and was therefore more difficult to see. The cable had a "set" from being coiled around the reel and it would not lie completely flat on the surface. Even when it was flat, however, a foot could still slide under it, and the Commander became entangled several times (see section 16.3.1).

Collecting the bulk sample required more time than anticipated because the modular equipment stowage assembly table was in deep shadow, and collecting samples in that area was far less desirable than taking those in the sunlight. It was also desirable to take samples as far from the exhaust plume and propellant contamination as possible. An attempt was made to include a hard rock in each sample and approximately 20 trips were required to fill the box. As in simulations, the difficulty of scooping up the material without throwing it out as the scoop [4-14] became free created some problem. It was almost impossible to collect a full scoop of material, and the task required about double the planned time.

Several of the operations would have been easier in sunlight. Although it was possible to see in the shadows, time must be allowed for dark adaptation when walking from the sunlight into shadow. On future missions, it would advantageous to conduct a yaw maneuver just prior to landing so that the descent stage work area would be in sunlight.

The scientific experiment package was easy to deploy manually, and some time was saved here. The package was easy to manage, but finding a level area was quite difficult. A good horizon reference was not available, and in the 1/6-g environment, physical cues were not as effective as in a one-g. Therefore, the selection of a deployment site for the experiments caused some problems. The experiments were placed in an area between shallow craters in surface material of the same consistency as the surrounding area and which should be stable. Considerable effort was required to change the slope of one of the experiments. It was not possible to lower the equipment by merely forcing it down, and it was necessary to move the experiment back and forth to scrape away the excess surface material.

No abnormal conditions were noted during the lunar module inspection. The insulation on the secondary struts had been damaged from the heat, but the primary struts were only singed or covered with soot. There was much less damage than on the examples that had been seen before flight.

Obtaining the core tube sample presented some difficulty. It was impossible to force the tube more than 4 or 5 inches into the surface material, yet the material provided insufficient resistance to hold the extension handle in the upright position. Since the handle had to be held upright, this precluded using both hands on the hammer. In addition, the resistance of the suit made it difficult to steady the core tube and swing with any great force. The hammer actually missed several times. Sufficient force was obtained to make dents in the handle, but the tube could be driven only to a depth of about 6 inches. Extraction offered little or virtually no resistance. Two samples were taken.

Insufficient time remained to take the documented sample, although as wide a variety of rocks was selected as remaining time permitted.

The performance of the extravehicular mobility unit was excellent. Neither crewman felt any thermal discomfort. The Commander used the minimum cooling mode for most of the surface operation. The Lunar Module Pilot switched to the maximum diverter valve position immediately after [4-15] sublimator startup and operated at maximum position for 42 minutes before switching to the intermediate position. The switch remained in the intermediate position for the duration of the extravehicular activity. The thermal effect of shadowed areas in [sic] versus those areas in sunlight was not detectable inside the suit.

The crewmen were kept physically cool and comfortable, and the ease of performing in the 1/6-g environment indicate that tasks requiring greater physical exertion may be undertaken on future flights. The Commander experienced some physical exertion while transporting the sample return container to the lunar module, but his physical limit had not been approached.

4.12.5 Lunar Module Ingress

Ingress to the lunar module produced no problems. The capability to do a vertical jump was used to an advantage in making the first step up the ladder. By doing a deep knee bend, then springing up the ladder, the Commander was able to guide his feet to the third step. Movements in the 1/6-g environment were slow enough to allow deliberate foot placement after the jump. The ladder was a bit slippery from the powdery surface material, but not dangerously so.

As previously stated, mobility on the platform was adequate for developing alternate methods of transferring equipment from the surface. The hatch opened easily, and the ingress technique developed before flight was satisfactory. A concerted effort to arch the back was required when about half way through the hatch, to keep the forward end of the portable life support system low enough to clear the hatch. There was very little exertion associated with transition to a standing position.

Because of the bulk of the extravehicular mobility unit, caution had to be exercised to avoid bumping into switches, circuit breakers, and other controls while moving around the cockpit. One circuit breaker was in fact broken as a result of contact (see section 16.2.11).

Equipment jettison was performed as planned, and the time taken before flight in determining the items not required for lift-off was well spent. Considerable weight reduction and increase in space was realized. Discarding the equipment through the hatch was not difficult, and only one item remained on the platform. The post-ingress checklist procedures were performed without difficulty; the checklist was well planned and was followed precisely.

[4-16]

4.12.6 Lunar Rest Period

The rest period was almost a complete loss. The helmet and gloves were worn to relieve any subconscious anxiety about a loss of cabin pressure and presented no problem. But noise, lighting, and a lower-than-desired temperature were annoying. It was uncomfortably cool in the suits, even with the water-flow disconnected. Oxygen flow was finally cut off, and the helmets were removed, but the noise from the glycol pumps was then loud enough to interrupt sleep. The window shades did not completely block out light, and the cabin was illuminated by a combination of light through the shades, warning lights, and display lighting. The Commander rested on the ascent engine cover and was bothered by the light entering through the telescope. The Lunar Module Pilot estimated that he slept fitfully for perhaps 2 hours and the Commander did not sleep at all, even though body positioning was not a problem. Because of the reduced gravity, the positions on the floor and on the engine cover were both quite comfortable.

4.13 LAUNCH PREPERATION

Aligning the platform before lift-off was complicated by the limited number of stars available. Because of sun and earth interference, only two detents effectively remained from which to select stars. Accuracy is greater for stars close to the center of the field, but none were available at this location. A gravity/one-star alignment was successfully performed. A manual averaging technique was used to sample five successive cursor readings and then five spiral readings. The result was then entered into the computer. This technique appeared to be easier than taking and entering five separate readings. Torquing angles were close to 0.7° in all three axes and indicated that the platform drifted. (Editor's note: Platform drift was within specification limits.)

After the alignment, the navigation program was entered. It is recommended that future crews update the abort guidance system with the primary guidance state vector at this point and then use the abort guidance system to determine the command module location. The primary guidance system cannot be used to determine the command module range and range rate, and the radar will not lock on until the command module is within 400 miles range. The abort guidance system provides good data as this range is approached.

A cold-fire reaction control system check and an abort guidance system calibration were performed, and the ascent pad was taken. About 45 minutes prior to lift-off, another platform alignment was performed. The landing site alignment option at ignition was used for lift-off. The torquing angles for this alignment were approximately 0.09 degree.

[4-17]

In accordance with ground instructions, the rendezvous radar was placed in the antenna SLEW position with the circuit breakers off for ascent to avoid recurrence of the alarms experienced during a descent.

Both crewmembers had forgotten to watch for the small helium pressure decrease indication that the Apollo 10 crew experienced when the ascent tanks were pressurized, and the crew initially believed that only one tank had been pressurized. This oversight was temporary and delayed the crew verification of proper pressurization of both tanks.

4.14 ASCENT

The pyrotechnic noises at descent stage separation were quite loud, but ascent-engine ignition was inaudible. The yaw and pitch maneuvers were very smooth. The pitch- and roll-attitude limit cycles were as expected and were not accompanied by physiological difficulties. Both the primary and the abort guidance systems indicated the ascent to be a duplicate of the planned trajectory. The guided cutoff yielded residuals of less than 2 ft/sec; and the inplane components were nulled to within 0.1 ft/sec with the reaction control system. Throughout the trajectory, the ground track could be visually verified, although a pitch attitude confirmation by use of the horizon in the overhead window was found to be quite difficult because of the horizon lighting condition.

4.15 RENDEZVOUS

At orbital insertion, the primary guidance system showed an orbit of 47.3 by 9.5 miles, as compared to the abort guidance system solution of 46.6 by 9.5 miles. Since radar range-rate data were not available, the Network quickly confirmed that the orbital insertion was satisfactory.

In the preflight planning, stars had been chosen that would be in the field of view and that would require a minimum amount of maneuvering to get through alignment and back in plane. This maintenance of a nearly fixed attitude would permit the radar to be turned on and the acquisition conditions designated so that marks for a coelliptic sequence initiation solution would be immediately available. For some reason, during the simulations, these preselected stars had not been correctly located relative to the horizon, and some time and fuel were wasted in first maneuvering to these stars, then failing to mark on them, and then maneuvering to an alternate pair. Even with these problems, the alignment was finished about 28 minutes before coelliptic sequence initiation, and it was possible to proceed with radar lock-on.

[4-18]

All four sources for the coelliptic sequence initiation solution agreed to within 0.2 ft/sec, an accuracy that had never been observed before. The Commander elected to use the primary guidance solution without any out-of-plane thrusting.

The coelliptic sequence initiation maneuver was accomplished by using the plus Z thrusters, and the radar lock-on was maintained throughout the firing. Continued navigation tracking by both vehicles indicated a plane change maneuver of about 2-1/2 ft/sec, but the crew elected to defer this small correction until terminal phase initiation. The very small out-of-plane velocities that existed between the spacecraft orbits indicated a highly accurate lunar surface alignment. As a result of the higher-than-expected ellipticity of the command module orbit, backup chart solutions were not possible for the first two rendezvous maneuvers, and the constant differential height maneuver had a higher-than-expected vertical component. The computers in both spacecraft agreed closely on the maneuver values, and the lunar module primary guidance computer solution was executed, using the minus X thrusters.

During the coelliptic phase, radar tracking data were inserted into the abort guidance system to obtain an independent intercept guidance solution. The primary guidance solution was 6-1/2 minutes later than planned. However, the intercept trajectory was quite nominal, with only two small midcourse corrections of 1.0 and 1.5 ft/sec. The line-of-sight rates were low, and the planned braking schedule was used to reach a stationkeeping position.

In the process of maneuvering the lunar module to the docking attitude, while at the same time avoiding direct sunlight in the forward windows, the platform inadvertently reached gimbal lock. The docking was completed by using the abort guidance system for attitude control.

4.16 COMMAND MODULE DOCKING

Pre-docking activities in the command module were normal in all respects, as was docking up to the point of probe capture. After the Command Module Pilot ascertained that a successful capture had occurred, as indicated by "barberpole" indicators, the CMC-FREE switch position was used and one retract bottle fired. A right yaw excursion of approximately 15° immediately took place for 1 or 2 seconds. The Command Module Pilot went back to CMC-AUTO and made hand-controller inputs to reduce the angle between the two vehicles to zero. At docking thruster firings occurred unexpectedly in the lunar module when the retract mechanism was actuated, and attitude excursions of up to 15° were observed. The lunar module was manually realigned. While [4-19] this maneuver was in progress, all 12 docking latches fired, and docking was completed successfully. (See section 8.6.1 for further discussion.)

Following docking, the tunnel was cleared, and the probe and drogue were stowed in the lunar module. The items to be transferred to the command module were cleaned using a vacuum brush attached to the lunar module suit return hose. The suction was low and made the process rather tedious. The sample return containers and film magazines were placed in appropriate bags to complete the transfer, and the lunar module was configured for jettison according to the checklist procedure.

4.17 TRANSEARTH INJECTION

The time between docking and transearth injection was more than adequate to clean all equipment contaminated with lunar surface material and to return it to the command module for stowage so that the necessary preparations for transearth injection could be made. The transearth injection maneuver, the last service propulsion engine firing of the flight, was nominal. The only difference between it and previous firings was that without the docked lunar module the start transient was apparent.

4.18 TRANSEARTH COAST

During transearth coast, faint spots or scintillations of light were observed within the command module cabin. These phenomena became apparent to the Commander and Lunar Module Pilot after they became dark-adapted and relaxed. [Editor's note: The source or cause of the light scintillations is as yet unknown. One explanation involves primary cosmic rays with energies in the range of billions of electron volts, bombarding an object in outer space. The theory assumes that numerous heavy and high-energy cosmic particles penetrate the command module structure, causing heavy ionization inside the spacecraft. When liberated electrons recombine with ions, photons in the visible portion of the spectrum are emitted. If a sufficient number of photons are emitted, a dark-adapted observer can detect the photons as a small spot or a streak of light. Two simple laboratory experiments were conducted to substantiate the theory, but no positive results were obtained in a 5-psi pressure environment because a high enough energy source was not available to create the radiation at that pressure. This level of radiation does not present a crew hazard.]

[4-20] Only one midcourse correction, a reaction control system firing of 4.8 ft/sec, was required during transearth coast. In general, the transearth coast period was characterized by a general relaxation on the part of the crew, with plenty of time available to sample the excellent variety of food packets and to take photographs of the shrinking moon and the growing earth.

4.19 ENTRY

Because of the presence of thunderstorms in the primary recovery area (1285 miles downrange from the entry interface of 400 000 feet), the targeted landing point was moved to a range of 1500 miles from the entry interface. This change required the use of computer program P65 (skip-up control routine) in the computer, in addition to those programs used for the planned shorter range entry. This change caused the crew some apprehension, since such entries had rarely been practiced in preflight simulations. However, during the entry, these parameters remained within acceptable limits. The entry was guided automatically and was nominal in all respects. The first acceleration pulse reached approximately 6.5g and the second reached 6.0g.

4.20 RECOVERY

On the landing, the 18-knot surface wind filled the parachutes and immediately rotated the command module into the apex down (stable II) flotation position prior to parachute release. Moderate wave-induced oscillations accelerated the uprighting sequence, which was completed in less than 8 minutes. No difficulties were encountered in completing the postlanding checklist.

The biological isolation garments were donned inside the spacecraft. Crew transfer into the raft was followed by hatch closure and by decontamination of the spacecraft and crewmembers by germicidal scrubdown.

Helicopter pickup was performed as planned, but visibility was substantially degraded because of moisture condensation on the biological isolation garment faceplate. The helicopter transfer to the aircraft carrier was performed as quickly as could be expected, but the temperature increase inside the suit was uncomfortable. Transfer from the helicopter into the mobile quarantine facility completed the voyage of Apollo 11.

[remainder of report not included]

Biographical Appendix

A

William A. Anders (1933–) was a career United States Air Force officer, although a graduate of the U.S. Naval Academy. Chosen with the third group of astronauts in 1963, he was the backup pilot for Gemini 9 and lunar module pilot for Apollo 8. Anders resigned from NASA and the Air Force (active duty) in September 1969 and became Executive Secretary of the National Aeronautics and Space Council. He joined the Atomic Energy Commission in 1973, and became chairman of the Nuclear Regulatory Commission in 1974. He was named U.S. Ambassador to Norway in 1976. Later he worked as a Vice-President of General Electric and then as Senior Executive Vice President-Operations, Textron, Inc. Anders retired as Chief Executive Officer of General Dynamics in 1993, but remained Chairman of the Board. See "Anders, W. A.," biographical file 000082, NASA Historical Reference Collection, NASA History Division, NASA Headquarters, Washington, DC and (*http://www.jsc.nasa.gov/Bios/htmlbios/anders-wa.html*).

Neil A. Armstrong (1930–) was the first person to set foot on the Moon on 20 July, 1969. He became an astronaut in 1962 after having served as a test pilot with the National Advisory Committee for Aeronautics (1955-1958) and NASA (1958–1962). He flew as command pilot on Gemini 8 in March 1966 and commander of Apollo 11 in July 1969. In 1970 and 1971 he was Deputy Associate Administrator for the office of Advanced Research and Technology, NASA Headquarters. In 1971 he left NASA to become a professor of aerospace engineering at the University of Cincinnati and to work as a private consultant. See Neil A. Armstrong, *et al.*, *First on the Moon: A Voyage with Neil Armstrong, Michael Collins and Edwin E. Aldrin, Jr.* (Boston: Little, Brown, 1970); Neil A. Armstrong, *et al.*, *The First Lunar Landing: 20th Anniversary/as Told by the Astronauts, Neil Armstrong, Edwin Aldrin, Michael Collins* (Washington, DC: National Aeronautics and Space Administration EP-73, 1989); (*http://www.jsc.nasa.gov/Bios/htmlbios/armstrong-na.html*).

William S. Augerson (1929–) was assigned to the Human Factors Section of the NASA Space Task Group in 1958 where he worked on the development of Life Systems for Project Mercury. In 1945 he joined the U.S. Navy to serve as an electronics technician and the next year entered Bowdoin College where he majored in physics and English, graduating with honors in 1949. He continued his education at Cornell University where he earned his M.D. in 1955. Dr. Augerson then entered active duty in the U.S. Army, interning at Brooke Army Hospital, San Antonio, Texas. His other posts included Division Surgeon for the 4th Infantry in 1957–58 and Army Liaison Officer for Bioastronautics Research at the U.S. Air Force Aeromedical Laboratory at Wright-Patterson Air Force Base in 1958. He would eventually retire from the Army with the rank of general. See "Gen. Augerson, William S.," biographical file 000118, NASA Historical Reference Collection, History Division, NASA Headquarters, Washington, DC.

B

McGeorge Bundy (1919–1996) was a professor of government before serving as the national security adviser to Presidents Kennedy and Johnson from 1961-1966. See *Who's Who in America, 1996* (New Providence, NJ: Marquis Who's Who, 1995).

C

Richard L. Callaghan (1925–) served as NASA's Assistant Administrator for Legislative Affairs from 1963-1967. An Army veteran of the World War II European Theatre, he received a B.S. from Georgetown University Foreign Service School in 1950 and an LL.B. from the George Washington University Law School in 1957. While attending law school, Callaghan worked in various legislative offices in Washington, DC, including that of Montana Senator James E. Murray. He also served as the staff director of the Senate Committee on Interior and Insular Affairs from 1955 until he joined NASA in 1962 as Special Assistant to Administrator James E. Webb. In 1968 he received the NASA Exceptional Service Medal for his work in the organization. See "Callaghan, R. L.," biographical file 000279, NASA Historical Reference Collection, History Division, NASA Headquarters, Washington, DC.

M. Scott Carpenter (1925–) piloted the Mercury 7 mission in 1962, making him the second American to orbit Earth. He earned his bachelor's degree in aeronautical engineering from the University of Colorado in 1949, after which he was commissioned in the U.S. Navy. Carpenter served in the Korean War as a Naval aviator and then served as a test pilot for the Navy from 1954 to 1957. Two years later he was selected as one of the original seven astronauts to serve in the Mercury program. Upon completion of his mission, Carpenter took a leave of absence from NASA and participated in the Navy's SEALAB II program, thus making him the first person to hold both the titles of astronaut and aquanaut. After retiring from the Navy in 1969, he finished his distinguished career working in the private sector. Carpenter's awards include the Navy's Legion of Merit, the Distinguished Flying Cross, the NASA Distinguished Service Medal, and the Collier Trophy. (*http://www.jsc.nasa.gov/Bios/htmlbios/carpenter-ms.html*) accessed 27 September 2006.

Michael Collins (1930–) served as command module pilot on Apollo 11 in 1969, remaining in lunar orbit while Neil Armstrong and Buzz Aldrin became the first two people to walk on the Moon. Born in Rome, Italy, Collins graduated from high school in Washington, DC and went on to earn a bachelor of science degree from the United States Military Academy at West Point in 1952. Collins chose an Air Force career upon graduation from West Point and served as an experimental flight test officer at Edwards Air Force Base in California. He also piloted the Gemini 10 mission in 1966 during which he successfully rendezvoused and docked with separately launched target vehicles. His awards include the Presidential Medal for Freedom in 1969 as well as the NASA Exceptional Service medal. (*http://www.jsc.nasa.gov/Bios/htmlbios/collins-m.html*) accessed 2 October 2006.

L. Gordon Cooper Jr. (1927–2004) piloted the Mercury 9 mission in 1963, which concluded the operational phase of Project Mercury. He was commissioned into the Air Force after attending three years at the University of Hawaii. After serving four years in Munich, Germany, Carpenter came back to the U.S. and earned a bachelor of science in aeronautical engineering in 1956 from the Air Force Institute of Technology. He spent the next three years as a test pilot at Edwards Air Force Base and was then selected as one of the original seven Mercury astronauts. After Mercury, Carpenter also served as command pilot of the Gemini 5 mission, thus becoming the first person to make two orbital flights and in the process setting a new space endurance record. He retired from the Air Force and NASA in 1970 to finish his career working in private industry. His awards include the Air Force Legion of Merit, the Air Force Distinguished Flying Cross Cluster, the NASA Distinguished Service Medal, and the 1962 Collier Trophy for pioneering piloted spaceflight in the USA. See *"Cooper, L. Gordon, Jr.,"* biographical file 376, NASA Historical Reference Collection, NASA History Division, NASA Headquarters, Washington, DC and (*http://www.jsc.nasa.gov/Bios/ htmlbios/cooper-lg.html*) accessed 2 October 2006.

Walter Cunningham (1932–) was in the third group of astronauts selected by NASA in October 1963 and served as the lunar module pilot in the Apollo 7 mission, the first piloted flight test of the third generation United States spacecraft. After graduating from Venice High School in California, he joined the Navy in 1951 and began flight training the following year. In 1953, Cunningham joined a Marine squadron where he served on active duty until 1956. He then went on to earn both a bachelor's and a master's degree in physics at UCLA in 1960 and 1961, respectively. After receiving his master's, Cunningham was employed as a physicist by the Rand Corporation where he worked on problems with Earth's magnetosphere as well as projects for the Department of Defense. As an astronaut, he played a key role in all aspects of piloted space flight including training, planning, system design, public relations, and program management. Cunningham then completed the Advanced Management Program at Harvard Graduate School of Business in 1974 and attained senior executive positions in several highly successful businesses over the course of the following decades. (*http://www.jsc.nasa.gov/Bios/htmlbios/ cunningham-w.html*) accessed 2 October 2006.

D

Kurt H. Debus (1908–1983) earned a B.S. in mechanical engineering (1933), an M.S. (1935) and Ph.D. (1939) in electrical engineering, all from the Technical University of Darmstadt in Germany. He became an assistant professor at the university after receiving his degree. During the course of World War II he became an experimental engineer at the A-4 (V-2) test stand at Peenemünde (see entry for Wernher von Braun), rising to become superintendent of the test stand and test firing stand for the rocket. In 1945 he came to the United States with a group of engineers and scientists headed by von Braun. From 1945-1950 the group worked at Fort Bliss, Texas, and then moved to the Redstone Arsenal in Huntsville, Alabama. From 1952-1960 Debus was chief of the missile firing laboratory of the Army Ballistic Missile Agency. In this position, he was located at Cape Canaveral, Florida, where he supervised the launching of the first ballistic missile fired from there, an Army Redstone. When ABMA became part of NASA, Debus continued to supervise missile and space vehicle launchings, first as director of the Launch Operations Center and then of the Kennedy Space Center as it was renamed in December 1963. He retired from that position in 1974 See "Debus, Kurt H.," biographical file 000443, NASA Historical Reference Collection, History Division, NASA Headquarters, Washington, DC.

Charles J. Donlan (1916–) served the United States government for nearly 38 years in NACA and NASA. After graduating with a bachelor of science in aeronautical engineering from MIT in 1938, he joined the research staff of NACA's Langley Aeronautical Laboratory where he worked to improve aircraft design, stability, and control. In 1958 Donlan was appointed Associate Director of the NASA Space Task Group at Langley to conduct Project Mercury. Three years later he became Associate Director of Langley until 1967 when he was made Deputy Director of the facility. The following year he was transferred to NASA Headquarters to become the Deputy Associate Administrator for Manned Space Flight. In addition to this, he was Acting Director of the Space Shuttle Program from 1970 until 1973. Donlan retired from NASA in 1976 and then worked as a consultant for the Institute for Defense Analysis where he studied military uses for the Shuttle for the next twelve years. His awards include the NASA Distinguished Service Medal and the NASA Medal for Outstanding Leadership. See "Donlan, Charles J.," biographical file 000481, NASA Historical Reference Collection, NASA History Division, NASA Headquarters, Washington, DC.

Hugh L. Dryden (1898–1965) was a career civil servant and an aerodynamicist by discipline who had also begun life as something of a child prodigy. He graduated at age 14 from high school and went on to earn an A.B. in three years from Johns Hopkins (1916). Three further years later (1919) he earned his Ph.D. in physics and mathematics from the same institution even though he had been employed full-time in the National Bureau of Standards since June 1918. His career at the Bureau of Standards, which lasted until 1947, was devoted to studying airflow, turbulence, and particularly the problems of the boundary layer--the thin layer of air next to an airfoil that causes drag. In 1920 he became chief of the aerodynamics section in the bureau. His work in the 1920s on measuring turbulence in wind tunnels facilitated research in the NACA that produced the laminar flow wings used in the P-51 Mustang and other World War II aircraft. From the mid-1920s to 1947, his publications became essential reading for aerodynamicists around the world. During World War II, his work on a glide bomb named "the bat" won him a Presidential Certificate of Merit. He capped his career at the bureau by becoming its Assistant Director and then Associate Director during his final two years there. He then served as Director of the NACA from 1947-1958, after which he became Deputy Administrator of NASA under T. Keith Glennan and James E. Webb. See Richard K. Smith, *The Hugh L. Dryden Papers, 1898-1965* [Baltimore, MD: The Johns Hopkins University Library, 1974] and "Dr. Hugh L. Dryden" (*http://www.hq.nasa.gov/office/pao/History/Biographies/dryden.html*) accessed 23 October 2006.

E

Donn F. Eisele (1930–1987) served as the command module pilot during the Apollo 7 mission in 1968. He earned a bachelor of science degree from the United States Naval Academy in 1952 and a master of science degree in astronautics from the Air Force Institute of Technology in 1960. Prior to his selection as an Apollo astronaut, Eisele served as a project engineer and experimental test pilot at the Air Force Special Weapons Center at Kirtland Air Force Base, New Mexico. After he retired from both the Air Force and the space program in 1972 he became the Director of the U.S. Peace Corps in Thailand. Eisele finished his career working in private industry back in the United States. (*http://www.jsc.nasa.gov/Bios/htmlbios/eisele-df.html*) accessed 3 October 2006.

F

Maxime A. Faget (1921–2004) was an aeronautical engineer with a B.S. from LSU (1943), joined the staff at Langley Aeronautical Laboratory in 1946 and soon became head of the performance aerodynamics branch of the pilotless aircraft research division. There, he conducted research on the heat shield of the Mercury spacecraft. In 1958 he joined the Space Task Group in NASA, forerunner of the NASA Manned Spacecraft Center that became the Johnson Space Center, and he became its assistant director for engineering and development in 1962 and later its director. He contributed many of the original design concepts for Project Mercury's piloted spacecraft and played a major role in designing virtually every U.S.-crewed spacecraft since that time, including the Space Shuttle. He retired from NASA in 1981 and became an executive for Eagle Engineering, Inc. In 1982 he was one of the founders of Space Industries, Inc. and became its president and chief executive officer. See "Maxime A. Faget," biographical file 000602, NASA Historical Reference Collection, NASA History Division, NASA Headquarters, Washington, DC.

G

Yuri Gagarin (1934–1968) was the Soviet cosmonaut who became the first human in space with a one-orbit mission aboard the spacecraft *Vostok 1* on April 12, 1961. The great success of that feat made the gregarious Gagarin a global hero, and he was an effective spokesman for the Soviet Union until his death in an unfortunate aircraft accident. "Gagarin Vostok 1 (1961)," biographical file 745, NASA Historical Reference Collection, NASA History Division, NASA Headquarters, Washington, DC.

John H. Gibbons (1929–) headed the Office of Technology Assessment under Congress for fourteen years before becoming President Clinton's science advisor and head of the White House Office of Science and Technology Policy in 1993. He received a Ph.D. in physics from Duke University in 1954. See "Gibbons, John," biographical file 5237, NASA Historical Reference Collection, History Division, NASA Headquarters, Washington, DC.

Robert R. Gilruth (1913–2000) was a longtime NACA engineer working at the Langley Aeronautical Laboratory from 1937–1946, then as chief of the pilotless aircraft research division at Wallops Island from 1946–1952, who had been exploring the possibility of human spaceflight before the creation of NASA. He served as Assistant Director at Langley from 1952–1959 and as Assistant Director (piloted satellites) and head of Project Mercury from 1959–1961, technically assigned to the Goddard Spaceflight Center but physically located at Langley. In early 1961 Glennan established an independent Space Task Group (already the group's name as an independent subdivision of the Goddard Center) under Gilruth at Langley to supervise the Mercury program. This group moved to the Manned Spacecraft Center, Houston, Texas, in 1962. Gilruth was then director of the Houston operation from 1962–1972. See, Henry C. Dethloff, *"Suddenly Tomorrow Came . . .": A History of the Johnson Space Center* (Washington, DC: NASA SP-4307, 1993); James R. Hansen, *Engineer in Charge: A History of the Langley Aeronautical Laboratory, 1917-1958* (Washington, DC: NASA SP-4305, 1987), pp. 386-88.

John H. Glenn, Jr. (1921–) was chosen with the first group of astronauts in 1959. He was the pilot for the 20 February 1962 Mercury-Atlas 6 (*Friendship 7*) mission, the first American orbital flight. He made three orbits on this mission. He left the NASA astronaut corps in 1964 and later entered politics as a senator from Ohio. See Lloyd S. Swenson, Jr., James M. Grimwood, and Charles C. Alexander, *This New Ocean: A History of Project Mercury* (Washington, DC: NASA SP-4201, 1966) and (*http://www.jsc.nasa.gov/Bios/htmlbios/glenn-j.html*) accessed 23 October 2006.

Nicholas Golovin (1912–1969) served on the staff of the White House Office of Science and Technology from 1962 to 1968, during which time he played an antagonistic role towards NASA and the decision to use the lunar orbit rendezvous mode to achieve a piloted lunar landing. Born in Odessa, Russia, but educated in this country (Ph.D. in physics, George Washington University, 1955) he worked in various capacities for the government during and after World War II, including for the Naval Research Laboratory, 1946–1948. He held several administrative positions with the National Bureau of Standards from 1949 to 1958. In 1958 he was chief scientist for the White Sands Missile Range and then worked for the Advanced Research Projects Agency in 1959 as director of technical operations. He became a Deputy Associate Administrator of NASA in 1960. He joined private industry before becoming, in 1961, the director of the NASA-DOD large launch vehicle planning group. He joined the Office of Science and Technology at the White House in 1962 as a technical advisor for aviation and space and remained there until 1968 when he took a leave of absence as a research associate at Harvard and as a fellow at the Brookings Institution. Obituaries, *Washington Star*, 30 Apr. 1969, p. B-6, and *Washington Post*, 30 Apr. 1969, p. B14.

A. J. Goodpaster (1915–2005) was a career Army officer who served as defense liaison officer and secretary of the White House staff from 1954 to 1961, being promoted to brigadier general during that period. He later was deputy commander, U.S. forces in Vietnam, 1968–1969, and commander-in-chief, U.S. Forces in Europe, 1969–1974. He retired in 1974 as a four-star general but returned to active duty in 1977 and served as superintendent of the U.S. Military Academy, a post he held until his second retirement in 1981.

Edward Z. Gray (1915–) worked for Boeing Co. from 1943–1963 as a design engineer for the Boeing jet aircraft series as well as the DynaSoar and Minuteman programs. He held a number of positions in systems engineering management, the last one being as development program manager of advanced space systems. He served on numerous committees for the government and aerospace industry, including the NASA research advisory committee on structural loads in 1958–1959, of which he was chairman. In 1963 NASA appointed him to the directorship of its advanced piloted missions programs. He worked in that position through 1967, transferred to a position as assistant to the president of Grumman Aircraft Engineering Corp. from 1967–1973, and then returned to NASA as Assistant Administrator for industry affairs and technology utilization. By 1978 he had assumed a position as director of government/industry affairs. In 1979 he joined Bendix Corp.'s aerospace-electronics group as director of systems development. See "Edward Z. Gray," biographical file 000871, NASA Historical Reference Collection, NASA History Division, NASA Headquarters, Washington, DC.

Virgil I. "Gus" Grissom (1927–1967) was chosen with the first group of astronauts in 1959. He was the pilot for the 1961 Mercury-Redstone 4 (*Liberty Bell 7*) mission, a suborbital flight; command pilot for Gemini 3; backup command pilot for Gemini 6; and had been selected as commander of the first Apollo flight carrying three crew members at the time of his death in the Apollo 1 fire in January 1967. See Betty Grissom and Henry Still, *Starfall* (New York: Thomas Y. Crowell, 1974); The Astronauts Themselves, *We Seven* (New York: Simon and Schuster, 1962); (*http://www.jsc.nasa.gov/Bios/htmlbios/grissom-vi.html*) accessed 23 October 2006.

H

James C. Hagerty (1909–1981) was on the staff of the *New York Times* from 1934 to 1942, the last four years as legislative correspondent in the paper's Albany bureau. He served as executive assistant to New York Governor Thomas Dewey from 1943 to 1950 and then as Dewey's secretary for the next two years before becoming press secretary for President Eisenhower from 1953 to 1961.

D. Brainard Holmes (1921–) was involved in the management of high technology efforts in private industry and the federal government. He was on the staff of Bell Telephone Labs, 1945–1953, and at RCA, 1953–1961. He then became Associate Administrator for Manned Space Flight at NASA, 1961–1963. Thereafter he assumed a series of increasingly senior positions with Raytheon Corp., and since 1982 chairman of Beech Aircraft. See "D. Brainard Holmes" biographical file 001048, NASA Historical Reference Collection, History Division, NASA Headquarters, Washington, DC; "Holmes, D(yer) Brainerd," *Current Biography 1963*, pp. 191–92.

Donald F. Hornig (1920–), a chemist, was a research associate at the Woods Hole Oceanographic Lab, 1943–1944, and a scientist and group leader at the Los Alamos Scientific Laboratory, 1944–1946. He taught chemistry at Brown University starting in 1946, rising to the directorship of Metcalf Research Lab, 1949–1957, and also serving as associate dean and acting dean of the graduate school from 1952–1954. He was Donner Professor of Science at Princeton from 1957–1964 as well as chairman of the chemistry department from 1958–1964. He was a special assistant to the president of the U.S. for science and technology from 1964–1969 and president of Brown University from 1970–1976. See Gregg Herken, *Cardinal Choices: Science Advice to the President from Hiroshima to SDI* (New York: Oxford University Press, 1992).

John C. Houbolt (1919–) was an aeronautical engineer who helped conceptualize and was the primary advocate for the idea of lunar orbit rendezvous. He received both bachelor and master of science degrees in civil engineering from the University of Illinois in 1940 and 1942, and a doctorate in technical sciences from the Swiss Federal Institute of Technology in 1957. He first joined NACA as an aeronautical engineer in 1942 before serving in the Army Corps of Engineers from 1944 to 1946. In 1949, back at Langley, he was appointed Assistant Chief of the Dynamic Loads Division where he pursued research problems in aeroelasticity in application to aircraft and space vehicles. In 1961 Houbolt was named Chief of the Theoretical Mechanics Division at Langley where he successfully argued the case of lunar orbit rendezvous to the NASA Administration. He left NASA in 1963 to work as a senior vice president and consultant for a private research firm, but then returned to Langley in 1976 as Chief Aeronautical Scientist. Houbolt officially retired from NASA in 1985. See James R. Hansen, *Enchanted Rendezvous: John C. Houbolt and the Genesis of the Lunar-Orbit Rendezvous Concept* (Washington, DC: National Aeronautics and Space Administration Monographs in Aerospace History No. 4, 1995) and "Houbolt, John C.," biographical file 001100, NASA Historical Reference Collection, History Division, NASA Headquarters, Washington, DC.

J

Lyndon B. Johnson (1908–1973) was President of the United States from 1963–1969. Johnson was elected to the House of Representatives in 1937 and served until 1949. He was a senator from 1949-1961 and then Vice President of the U.S. from 1960–1963 under Kennedy. Best known for the social legislation he passed during his presidency and for his escalation of the war in Vietnam, he was also highly instrumental in revising and passing the legislation that created NASA and in supporting the U.S. space program as chairman of the Committee on Aeronautical and Space Sciences and of the preparedness subcommittee of the Senate Armed Services Committee, then later as chairman of the National Aeronautics and Space Council when he was vice president. (On his role in support of the space program, Robert A. Divine, "Lyndon B. Johnson and the Politics of Space," in *The Johnson Years: Vietnam, the Environment, and Science,* Robert A. Divine, ed. [Lawrence: University of Kansas Press, 1987], pp. 217-53; and Robert Dallek, "Johnson, Project Apollo, and the Politics of Space Program Planning," unpublished paper delivered at a symposium on "Presidential Leadership, Congress, and the U.S. Space Program," sponsored by NASA and American University, March 25, 1993.)

K

John F. Kennedy (1916–1963) was President of the United States, 1961–1963. In 1960 John F. Kennedy, a Senator from Massachusetts between 1953 and 1960, ran for president as the Democratic candidate with Lyndon B. Johnson as his running mate. Using the slogan, "Let's get this country moving again," Kennedy charged the Republican Eisenhower administration with doing nothing about the myriad social, economic, and international problems that festered in the 1950s. He was especially hard on Eisenhower's record in international relations, taking a cold warrior position on a supposed "missile gap" (which turned out not to be the case) wherein the United States lagged far behind the Soviet Union in ICBM technology. On 25 May, 1961, President Kennedy announced to the nation a goal of sending an American to the Moon before the end of the decade. The human spaceflight imperative was a direct outgrowth of it; Projects Mercury (at least in its latter stages), Gemini, and Apollo were each designed to execute it. On this subject see, Walter A. McDougall, . . . *The Heavens and the Earth: A Political History of the Space Age* (New York: Basic Books, 1985); John M. Logsdon, *The Decision to Go to the Moon: Project Apollo and the National Interest* (Cambridge, MA: MIT Press, 1970).

George Kistiakowsky (1900–1982) was a pioneering chemist at Harvard University, associated with the development of the atomic bomb, and later an advocate of banning nuclear weapons. He served as science advisor to President Eisenhower from July 1959 to the end of the Eisenhower administration. He later served on the advisory board to the United States Arms Control and Disarmament Agency from 1962 to 1969. See *New York Times*, December 9, 1982, p. B21 and "George B. Kistiakowsky," biographical file 001200, NASA Historical Reference Collection, History Division, NASA Headquarters, Washington, DC.

James R. Killian (1904–1988) was president of the Massachusetts Institute of Technology between 1949 and 1959, on leave between November 1957 and July 1959 when he served as the first presidential science advisor. President Dwight D. Eisenhower established the President's Science Advisory Committee (PSAC), which Killian chaired, following the Sputnik crisis. After leaving the White House staff in 1959, Killian continued his work at MIT but in 1965 began working with the Corporation for Public Broadcasting to develop public television. Killian described his experiences as a presidential advisor in *Sputnik, Scientists, and Eisenhower: A Memoir of the First Special Assistant to the President for Science and Technology* (Cambridge, MA: MIT Press, 1977). For a discussion of the PSAC see Gregg Herken, *Cardinal Choices: Science Advice to the President from Hiroshima to SDI* (New York: Oxford University Press, 1992).

Kenneth Kleinknecht started his career in 1942 at the Lewis Research Center after graduating from Purdue University with a B.S. in mechanical engineering. In 1951, Kleinknecht transferred to the Flight Research Center in Edwards, CA. After NASA formed, he then transferred to the Manned Spacecraft Center in Houston in 1959. Before being named the manager of the Mercury project, Kleinknecht was active in the National Air Races, served as supervisor for a number of avionics tests at Lewis, and was the head of the Project Engineering Station for the X-1E. Additionally, Kleinknecht served as the Advanced Projects Management Officer on the X-15 project and as the Technical Assistant to the Director of the Manned Spacecraft Center. Source: "Kenneth Kleinknecht" biographical file 001205, NASA Historical Reference Collection, NASA History Division, NASA Headquarters, Washington, DC.

Christopher C. Kraft, Jr. (1924–   ) was a long-standing official with NASA throughout the Apollo program. He received as B.S. in aeronautical engineering from Virginia Polytechnic Institute in 1944 and joined the Langley Aeronautical Laboratory of the National Advisory Committee for Aeronautics (NACA) the next year. In 1958, still at Langley, he became a member of the Space Task Group developing Project Mercury and moved with the Group to Houston in 1962. He was flight director for all of the Mercury and many of the Gemini missions and directed the design of Mission Control at the Manned Spacecraft Center (MSC), redesignated the Johnson Space Center in 1973. He was named the MSC Deputy Director in 1970 and its Director two years later, a position he held until his retirement in 1982. Since then he has remained active as an aerospace consultant. See "Kraft, Christopher C., Jr.," biographical file 001237, NASA Historical Reference Collection, NASA History Division, NASA Headquarters, Washington, DC.

Nikita Khrushchev (1894–1971) was premier of the USSR from 1958 to 1964 and first secretary of the Communist party from 1953 to 1964. He was noted for an astonishing speech in 1956 denouncing the crimes and blunders of Joseph Stalin and for gestures of reconciliation with the West in 1959–1960, ending with the breakdown of a Paris summit with President Eisenhower and the leaders of

France and Great Britain in the wake of Khrushchev's announcement that the Soviets had shot down an American U-2 reconnaissance aircraft over the Urals on 1 May 1960. Then in 1962 Khrushchev attempted to place Soviet medium range-missiles in Cuba. This led to an intense crisis in October, after which Khrushchev agreed to remove the missiles if the U.S. promised to make no more attempts to overthrow Cuba's Communist government. Although he could be charming at times, Khrushchev was also given to bluster (extending even to shoe-pounding at the U.N.) and was a tough negotiator, although he believed, unlike his predecessors, in the possibility of Communist victory over the West without war. See his *Khrushchev Remembers: The Last Testament* (Boston: Little, Brown, 1974); Edward Crankshaw, *Khrushchev: A Career* (New York: Viking, 1966); Michael R. Beschloss, *Mayday: Eisenhower, Khrushchev and The U-2 Affair* (New York: Harper and Row, 1986); and Robert A. Divine, *Eisenhower and the Cold War* (New York: Oxford University Press, 1981) for further information about him.

Joachin P. Kuettner (1909–) served as Chief of the Mercury-Redstone project at NASA's Marshall Space Flight Center. Born and raised in Germany, he earned a doctorate in law from the University of Breslau at the age of 21 and a doctorate in physics and meteorology from the University of Hamburg in 1939. During World War II, Dr. Kuettner served as a test pilot and later as the head of a flight test department for advanced airplanes such as the piloted version of the German V-1. He came to the United States in December 1948 and joined the Air Force Cambridge Research Center. Here he was in charge of geophysical flight research using jet aircraft and high-altitude sailplanes. He then worked for the Army Ballistic Missile Agency as Director of the agency's efforts in Project Mercury from 1958 until he transferred to NASA and Marshall Space Flight Center two years later. After Mercury-Redstone, he was put in charge of the Saturn-Apollo Systems Integration at Marshall. Over his long career, Dr. Kuettner published many papers in the fields of aeronautics, meteorology, and astronautics and holds numerous awards from several different countries.

L

James A. Lovell, Jr. (1928–) flew on four space flights and was a member of the first crew to circle the Moon. He was selected in the second group of astronauts in 1962 and flew in the Gemini 7, Gemini 12, Apollo 8, and Apollo 13 missions, thus making him the first person to fly twice to the Moon. Following his graduation with a bachelor of science degree from the U.S. Naval Academy in 1952, Lovell received his flight training and was later assigned as a test pilot at the Naval Air Test Center in Maryland. A graduate of the Aviation Safety School of the University of Southern California, he also served as a flight instructor and safety engineer with Fighter Squadron 101 at the Naval Air Station, Oceana, Virginia. In addition to the four missions in which Captain Lovell flew, he also served as backup pilot for Gemini 4, backup Commander for both Gemini 9 and Apollo 11. In 1971, he was named Deputy Director of Science and Applications at NASA's Manned Spacecraft Center in Houston. In addition to these duties, he was appointed by President Lyndon B. Johnson to serve as a consultant for Physical Fitness and Sports and was later made Chairman of the Council by President Nixon. Lovell retired from the Navy and NASA in1973 to accept a position as Senior Executive Vice President in the Bay Houston Towing Company. Among his many honors are the Presidential Medal for Freedom, the NASA Distinguished Service Medal, and two Navy Distinguished Flying Crosses. See "Lovell, James A., Jr. Apollo flights," biographical file 001350, NASA Historical Reference Collection, History Division, NASA Headquarters, Washington, DC and "James A. Lovell" (*http:// www.jsc.nasa.gov/Bios/htmlbios/lovell-ja.html*) accessed 31 October 2006.

George M. Low (1926–1984), a native of Vienna, Austria, came to the U.S. in 1940 and received an aeronautical engineering degree from Rensselaer Polytechnic Institute (RPI) in 1948 and an M.S. in the same field from the same school in 1950. He joined the NACA in 1949 and at Lewis Flight Propulsion Laboratory he specialized in experimental and theoretical research in several fields. He became chief of piloted space flight at NASA Headquarters in 1958. In 1960, he chaired a special committee that formulated the original plans for the Apollo lunar landings. In 1964 he became deputy director of the Manned Spacecraft Center in Houston, the forerunner of the Johnson Space Center. He became Deputy Administrator of NASA in 1969 and served as Acting Administrator in 1970–1971. He retired from NASA in 1976 to become president of RPI, a position he held until his death. In 1990 NASA renamed its quality and excellence award after him. See "Low, George M.," Deputy Administrator file 004133, NASA Historical Reference Collection, History Division, NASA Headquarters, Washington, DC and "George M. Low" (*http://www.hq.nasa.gov/office/pao/History/Biographies/low. html*) accessed 23 October 2006.

M

Charles W. Mathews (1921–2001) was NASA's Associate Administrator for Applications from 1971 until 1976. After earning a B.S. in aerospace engineering from Rensselaer Polytechnic Institute in 1943, he immediately joined the engineering staff at the National Advisory Committee for Aeronautics Langley Research Center. Here he conducted research on supersonic flight, automatic control devices and systems for use in the interception of enemy bombers, and piloted spacecraft studies. In 1958, Mathews became chief of the NASA Space Task Group Operations Division and was responsible for the overall operations of Project Mercury. Upon the successful completion of the Mercury program, he was named Gemini Program Manager at the Manned Spacecraft Center in 1963. Following Gemini's success, Mathews was made the Director of the Skylab Program in 1966 and moved to NASA Headquarters. Two years later he became the Deputy Associate Administrator for Manned Space Flight. He retired from the organization in 1976 after thirty-three years of government service. See "Mathews, Charles W.," biographical file 001443, NASA Historical Reference Collection, History Division, NASA Headquarters, Washington, DC.

Owen E. Maynard (1924–2000) was responsible for the conceptualization and design of the lunar module used in the Apollo program. After serving in the Royal Canadian Air Force in World War II, Maynard earned a degree in aeronautical engineering from the University of Toronto while working on and eventually designing aircraft at Avro Canada. He joined NASA in 1959 to work on the Mercury program and first became involved with Apollo the following year. Maynard was one of the early supporters of the lunar orbit rendezvous method and became the chief of engineering for the lunar module in 1963. He served as chief of the systems engineering division in the Apollo Spacecraft Program Office from 1964 to 1970, at which time he left NASA to work in the private sector for the remainder of his career. See (*http://history.nasa.gov/maynard.html*) accessed 27 September 2006.

James A. McDivitt (1929–) commanded the Gemini 4 and Apollo 9 missions and was the program manager for Apollo 12 through Apollo 16. He earned a bachelor of science in aeronautical engineering from the University of Michigan in 1959, graduating first in his class. Before he was selected by NASA as an astronaut in 1962, McDivitt served in the U.S. Air Force and flew 145 combat missions during the Korean War. He is a graduate of both the USAF Experimental Test Pilot School and the USAF Aerospace Pilot Research course, after which he served as an experimental test pilot at Edwards Air Force Base, California. He left NASA and retired from the Air Force with the rank of Brigadier General in 1972 to work in leading executive positions in various private firms. McDivitt's awards include two NASA Distinguished Service Medals, four Distinguished Flying Crosses, and four Honorary Doctorates in science and law. See (*http://www.jsc.nasa.gov/Bios/htmlbios/mcdivitt-ja.html*) accessed 2 October 2006.

George E. Mueller (1918–) was Associate Administrator for the Office of Manned Space Flight at NASA Headquarters, 1963–1969, where he was responsible for overseeing the completion of Project Apollo and of beginning the development of the Space Shuttle. He moved to the General Dynamics Corp., as senior vice president in 1969, and remained until 1971. He then became president of the Systems Development Corporation, 1971–1980, and its chairman and CEO, 1981-1983. He was for a number of years the President of the International Academy of Astronautics and a founder of Kistler Aerospace. See "Mueller, George E.," biographical file 001520, NASA Historical Reference Collection, History Division, NASA Headquarters, Washington, DC.

N

Homer Newell (1915–1983) earned his Ph.D. in mathematics at the University of Wisconsin in 1940 and served as a theoretical physicist and mathematician at the Naval Research Laboratory from 1944–1958. During part of that period, he was science program coordinator for Project Vanguard and was acting superintendent of the atmosphere and astrophysics division. In 1958 he transferred to NASA to assume responsibility for planning and development of the new Agency's space science program. He soon became deputy director of space flight programs. In 1961 he assumed directorship of the office of space sciences; in 1963, he became associate administrator for space science and applications. Over the course of his career, he became an internationally known authority in the field of atmospheric and space sciences as well as the author of numerous scientific articles and seven books, including *Beyond the Atmosphere: Early Years of Space Science* (Washington, DC: NASA SP-4211, 1980). He retired from NASA at the end of 1973. "Newell General," Deputy Administrator file 4493, NASA Historical Reference Collection, NASA History Division, NASA Headquarters, Washington, DC.

Richard M. Nixon (1913–1994) was president of the United States when the first man landed on the Moon, serving between January 1969 and August 1974. Early in his presidency, Nixon appointed a Space Task Group under the direction of Vice President Spiro T. Agnew to assess the future of spaceflight in the nation. Its report recommended a vigorous post-Apollo exploration program culminating in a human expedition to Mars. Nixon did not approve this plan, but did decide in favor of building one element of it, the Space Shuttle, which was approved on January 5, 1972. See Roger D. Launius, "NASA and the Decision to Build the Space Shuttle, 1969-72," *The Historian* 57 (Autumn 1994): 17–34.

Warren North (1922–) earned a B.S. from the University of Illinois in 1947. From then until 1955 he was an engineer and test pilot for the Lewis Laboratory. From 1956-1959 he served as assistant chief of the aerodynamics branch at Lewis. He then transferred to NASA Headquarters, where he took part in early planning for Project Mercury, including the selection and training of the seven Mercury astronauts. He moved in 1962 to the Manned Spacecraft Center (later the Johnson Space Center), where he headed the division responsible for training the astronauts for the Gemini rendezvous and docking operations and the Apollo lunar landings. He continued to work in the fields of astronaut selection and training until he retired in 1985 as special assistant to the director of flight operations in planning space shuttle crew training. ("Warren North," biographical file 001608, NASA Historical Reference Collection, NASA History Division, NASA Headquarters, Washington, DC.)

P

Thomas O. Paine (1921–1992) was appointed Deputy Administrator of NASA on January 31, 1968. Upon the retirement of James E. Webb on October 8, 1968, he was named Acting Administrator of NASA. He was nominated as NASA's third Administrator March 5, 1969, and confirmed by the Senate on March 20, 1969. During his leadership the first seven piloted Apollo missions were flown, in which 20 astronauts orbited Earth, 14 traveled to the Moon and four walked upon its surface. Paine resigned from NASA on September 15, 1970 to return to the General Electric Co. in New York City as Vice President and Group Executive, Power Generation Group, where he remained until 1976. In 1985 the White House chose Paine as chair of a National Commission on Space to prepare a report on the future of space exploration. Since leaving NASA fifteen years earlier, Paine had been a tireless spokesman for an expansive view of what should be done in space. The Paine Commission took most of a year to prepare its report, largely because it solicited public input in hearings throughout the United States. The Commission report, *Pioneering the Space Frontier,* was published in a lavishly illustrated, glossy format in May 1986. It espoused a "pioneering mission for 21st-century America"—"to lead the exploration and development of the space frontier, advancing science, technology, and enterprise, and building institutions and systems that make accessible vast new resources and support human settlements beyond Earth orbit, from the highlands of the Moon to the plains of Mars." The report also contained a "Declaration for Space" that included a rationale for exploring and settling the solar system and outlined a long-range space program for the United States. See Roger D. Launius, "NASA and the Decision to Build the Space Shuttle, 1969–72," *The Historian* 57 (Autumn 1994): 17-34 and "Thomas O. Paine" (*http://www.hq.nasa.gov/office/pao/History/Biographies/ paine.html*) accessed 23 October 2006.

Samuel C. Phillips (1921–1990), was trained as an electrical engineer at the University of Wyoming, but he also participated in the Civilian Pilot Training Program during World War II. Upon his graduation in 1942 Phillips entered the Army infantry but soon transferred to the air component. As a young pilot he served with distinction in the Eighth Air Force in England—earning two distinguished flying crosses, eight air medals, and the French croix de guerre—but he quickly became interested in aeronautical research and development. He became involved both in the development of the incredibly successful B-52 bomber in the early 1950s and headed the Minuteman intercontinental ballistic missile program in the latter part of the decade. In 1964 Phillips, by this time an Air Force general, was lent to NASA to head the Apollo Moon landing program. He went back to the Air Force in the 1970s and commanded Air Force Systems Command prior to this retirement in 1975. See "Gen. Samuel C. Phillips of Wyoming," *Congressional Record*, 3 August 1973, S-15689; Rep. John Wold, "Gen. Sarah H. Turner, "Sam Phillips: One Who Led Us to the Moon," *NASA Activities*, May/June 1990, pp. 18-19; obituary in *New York Times*, 1 February 1990, p. D1.

R

Milton Rosen (1915–), an electrical engineer by training, joined the staff of the Naval Research Laboratory in 1940, where he worked on guidance systems for missiles during World War II. From 1947 to 1955, he was in charge of Viking rocket development. He was technical director of Project Vanguard, the scientific earth satellite program, until he joined NASA in October 1958 as Director of Launch Vehicles and Propulsion in the Office of Manned Space Flight. In 1963 he became senior scientist in NASA's Office of the Deputy Associate Administrator for Defense Affairs. He later became Deputy Associate Administrator for Space Science (engineering). In 1974 he retired from NASA to become executive secretary of the National Academy of Science's Space Science Board. ("Milton W. Rosen," biographical file 001835, NASA Historical Reference Collection, NASA History Division, NASA Headquarters, Washington, DC; see also his *The Viking Rocket Story* [New York: Harper, 1955].)

S

Julian Scheer (1926–2001) served as NASA's Assistant Administrator for Public Affairs from 1963 until 1971. He began his career in 1939 as an apprentice for a chain of weekly newspapers in his native Richmond, VA and went on to serve in the Merchant Marines during World War II and later in the U.S. Naval Reserve. Scheer earned a bachelor's degree from the University of North Carolina in 1950 and worked as the university's Assistant Director of Sports Information until he joined NASA in 1962 as a consultant. As NASA's missions progressed in the 1960s they attracted unprecedented public and press attention, creating ever-increasing demands for instantaneous information in every form. Under Scheer's direction, NASA anticipated and planned for the press needs in connection with Apollo piloted flights, including a worldwide communications network for disseminating television pictures live from the Moon on Apollo 11. His Public Affairs program received several national awards, including the 1970 University of Missouri School of Journalism Special Achievement Award which cited the NASA program "for its outstanding, almost inconceivable, contributions to journalism technology." His personal awards include NASA's Exceptional Service Medal in 1968 and the Distinguished Service Medal in 1969. See "Scheer, Julian," biographical file 001902, NASA Historical Reference Collection, NASA History Division, NASA Headquarters, Washington, DC.

Walter M. Schirra, Jr. (1923– 2007) was one of the original seven astronauts chosen by NASA in 1959. He became the fifth American in space in 1963 when he piloted the Mercury 8 mission. Schirra earned a bachelor of science degree from the United States Naval Academy in 1945. As a Navy pilot he flew 90 combat missions over Korea and was awarded the Distinguished Flying Cross and two Air Medals for his service. He then attended the Naval Air Safety Officer School at the University of Southern California and completed test pilot training at the Naval Air Test center in 1958. Schirra was the only person to fly in America's first three space programs—Mercury, Gemini and Apollo—logging over 295 hours in space. In 1969 he was awarded three separate honorary doctorates in astronautical engineering, science, and astronautics. See "Schirra, Walter M. Mercury Flight," biographical file 001915, NASA Historical Reference Collection, History Division, NASA Headquarters, Washington, DC and (*http://www.jsc.nasa. gov/Bios/htmlbios/schirra-wm.html*) accessed 23 October 2006.

Harrison H. Schmitt (1935–) occupied the lunar module pilot seat as a scientist-astronaut on Apollo 17. Schmitt conducted the longest and most productive lunar exploration of the Apollo program during this mission, spending twenty-two hours exploring the surface of the Moon and bringing back the largest lunar sample to date. He earned a bachelor of science degree from the California Institute of Technology in 1957 and a doctorate in geology from Harvard in 1964. Before joining NASA in 1965, Schmitt worked with the U.S. Geological Survey's Astrogeology Center at Flagstaff, Arizona, where he was project chief for lunar field geological methods. While at this position, he was among the USGS astrogeologists that instructed NASA astronauts during their geological field trips. In 1974, after assuming additional duties as Chief of Scientist-Astronauts, he was appointed NASA Assistant Administrator for Energy Programs. Dr. Schmitt left NASA in 1975 to run for the United States Senate and subsequently served a six-year term in his home state of New Mexico. In 2005 he became chair of the NASA Advisory Council. See "Schmitt, Dr. Harrison (Jack) thru A-17," biographical file 001925, NASA Historical Reference Collection, NASA History Division, NASA Headquarters, Washington, DC and (*http://www.jsc.nasa.gov/Bios/ htmlbios/schmitt-hh.html*) accessed 3 October 2006.

William C. Schneider (1923–1999) joined NASA in June 1963 and was the Gemini mission director for seven of the ten piloted Gemini missions. From 1967 to 1968, he served as Apollo mission director and the Apollo program's deputy director for missions. He then served from 1968 to 1974 as the Skylab program's director. After that, he worked as the Deputy Associate Administrator for Space Transportation Systems for almost four years. From 1978 to 1980, he served as the Associate Administrator for Space Tracking and Data systems. He received a Ph.D. in engineering from Catholic University. See "Schneider, William C.," biographical file 001927, NASA Historical Reference Collection, History Division, NASA Headquarters, Washington, DC.

Russell L. Schweickart (1935–) served as lunar module pilot during the Apollo 9 mission in 1969, during which he tested the portable life support backpack which was subsequently used on the lunar surface explorations. He earned a bachelor of science degree from the Massachusetts Institute of Technology in 1956 and then served as a fighter pilot in the Massachusetts Air National Guard until 1963. He then returned to MIT as a graduate student and research scientist at the school's Experimental Astronomy Laboratory, earning a master of science degree in 1963. That same year, Schweickart was selected by NASA to be in the third group of astronauts and fly in the Apollo program. After Apollo he served as backup commander for the first Skylab mission in 1973 and assumed responsibility for the development of hardware and procedures associated with erecting the emergency solar shade and deployment of the jammed solar array wing following the loss of the Skylab vehicle's thermal shield. Schweickart finished his career at NASA serving as the Director of User Affairs in the Office of Applications in Washington, DC. (*http://www.jsc.nasa.gov/Bios/htmlbios/schweickart-rl.html*) accessed 3 October 2006.

David R. Scott (1932–) was selected as one of the third group of astronauts in 1963 and flew in the Gemini 8, Apollo 9, and Apollo 15 missions. He graduated near the top of his class at West Point with a bachelor of science degree and then chose to commission into the Air Force. He completed pilot training at Webb Air Force Base, Texas, in 1955 and was assigned to the 32d Tactical Fighter squadron stationed in Netherlands until 1960. Upon completing his tour of duty, Scott returned to the U.S. to study at MIT where he earned a master of science degree in aeronautics and astronautics as well as an engineering degree in aeronautics and astronautics, both in 1962. After leaving the astronaut corps in 1972, he was named Technical Assistant to the Apollo Program Manager at Johnson Space Center. He retired from the Air Force in March 1975 with the rank of Colonel and over 5600 hours of flying time. In that same year, Scott was appointed Director of Dryden Flight Research Center where he remained until he left NASA for private business ventures in 1977. Recently, Scott was the technical consultant to the 1998 HBO miniseries *From the Earth to the Moon*. See "Scott, David R. (Post – NASA)," biographical file 001958, NASA Historical Reference Collection, History Division, NASA Headquarters, Washington, DC and (*http://www.jsc.nasa.gov/Bios/htmlbios/scott-dr.html*) accessed October 3, 2006.

Robert C. Seamans, Jr. (1918–2008) was born on October 30, 1918, in Salem, Massachusetts. He attended Lenox School, Lenox, Massachusetts; earned a bachelor of science degree in engineering at Harvard University in 1939; a master of science degree in aeronautics at Massachusetts Institute of Technology (MIT) in 1942; and a doctor of science degree in instrumentation from MIT in 1951. Dr. Seamans also received the following honorary degrees: doctor of science from Rollins College (1962) and from New York University (1967); doctor of engineering from Norwich Academy (1971), from Notre Dame (1974), and from Rensselaer Polytechnic Institute (RPI) in 1974. In 1960, Dr. Seamans joined NASA as Associate Administrator. In 1965, he became Deputy Administrator, retaining many of the general management-type responsibilities of the Associate Administrator and also serving as Acting Administrator. During his years at NASA he worked closely with the Department of Defense in research and engineering programs and served as Co-chairman of the Astronautics Coordinating Board. Through these associations, NASA was kept aware of military developments and technical needs of the Department of Defense and Dr. Seamans was able to advise that agency of NASA activities which had application to national security. Seamans left NASA in late 1967; in 1969 President Nixon named him Secretary of the Air Force. He subsequently became the first Administrator of the Energy Research and Development Administration. For further information on Robert C. Seamans, Jr., see his autobiography, *Aiming at Targets* (Washington, DC: NASA

SP-4106, 1996), his monograph *Project Apollo: the Tough Decisions* (Washington, DC: NASA SP-2005-4536, 2005), and "Robert C. Seamans, Jr." (*http://www.hq.nasa.gov/office/pao/History/Biographies/seamans.html*) accessed 23 October 2006.

Joseph F. Shea (1926–1999) served NASA as Deputy Director of the Office of Manned Space Flight at Headquarters in Washington, DC, and as manager of the Apollo spacecraft program in Houston. He earned bachelor's degrees in both engineering and mathematics and a master's and doctorate degree in engineering mechanics, all at the University of Michigan. Shea worked in numerous positions in private companies, including Space Program Director at the Space Technology Laboratories in California, Advance Systems R & D Manager with General Motors, and Military Development Engineer with the Bell Telephone Laboratories. Shea officially retired from NASA in 1993 after his health began to fail him. He also was Senior Vice President for Engineering at Raytheon Co. from 1980 until his death in 1999. See "Shea, Joseph F.," biographical file 2007, NASA Historical Reference Collection, History Division, NASA Headquarters, Washington, DC.

Alan B. Shepard, Jr. (1923–1998) was a member of the first group of seven astronauts in 1959 chosen to participate in Project Mercury. He was the first American in space, piloting Mercury-Redstone 3 (*Freedom 7*), and was backup pilot for Mercury-Atlas 9. He was subsequently grounded due to an inner-ear ailment until May 7, 1969 (during which time he served as chief of the astronaut office). Upon returning to flight status Shepard commanded Apollo 14, and in June 1971 resumed duties as chief of the astronaut office. He retired from NASA and the U.S. Navy on August 1, 1974, to join the Marathon Construction Company of Houston, Texas, as partner and chairman. See Alan Shepard and Deke Slayton, *Moonshot: The Inside Story of America's Race to the Moon* (New York: Turner Publishing, Inc., 1994); The Astronauts Themselves, *We Seven* (New York: Simon and Schuster, 1962); (*http://www.jsc.nasa.gov/Bios/htmlbios/schirra-wm.html*) accessed 23 October 2006.

Hugh S. Sidey (1927–2005) was a top reporter for *Time* and *Life* magazines during the Kennedy Presidency. He graduated from Iowa State University with a bachelor's degree in 1950 and immediately began working with numerous publications such as the *Omaha World-Herald* and the *Free Press*. He would later author a biography of President Kennedy entitled *John F. Kennedy, President*. See *Who's Who in America, 1966-1967* (Chicago, IL: Marquis, 1966).

Abe Silverstein (1908–2001), who earned a B.S. in mechanical engineering (1929) and an M.E. (1934) from Rose Polytechnic Institute, was a longtime NACA manager. He had worked as an engineer at the Langley Aeronautical Laboratory between 1929 and 1943 and had moved to the Lewis Laboratory (later, Research Center) to a succession of management positions, the last (1961–1970) as director of the Center. Interestingly, in 1958 Case Institute of Technology had awarded him an honorary doctorate. When Glennan arrived at NASA, Silverstein was on a rotational assignment to the Washington headquarters as Director of the Office of Space Flight Development (later, Space Flight Programs) from the position of Associate Director at Lewis, which he had held since 1952. During his first tour at Lewis, he had directed investigations leading to significant improvements in reciprocating and early turbojet engines. At NASA Headquarters he helped create and direct the efforts leading to the space flights of Project Mercury and to establish the technical basis for the Apollo program. As Lewis's director, he oversaw a major expansion of the Center and the development of the Centaur launch vehicle. He retired from NASA in 1970 to take a position with Republic Steel Corp. On the career of Silverstein see, Virginia P. Dawson, *Engines and Innovation: Lewis Laboratory and American Propulsion Technology* (Washington, DC: NASA SP-4306, 1991), passim; "Silverstein, Abe," biographical file 002072, NASA Historical Reference Collection, History Division, NASA Headquarters, Washington, DC.

Donald K. Slayton (1924–1993) was named one of the original seven Mercury astronauts in 1959, but was relieved of this assignment following the discovery of a heart condition in August of that same year. Instead he assumed the role of Director of Flight Crew Operations in 1963, bringing upon himself the responsibilities of directing the activities of the astronaut office, the aircraft office, the flight crew integration division, the crew training and simulation division, and the crew procedures division. Born and raised in Sparta, Wisconsin, Slayton joined the Air Force after high school and earned his wings in 1943. As a B-25 pilot with the 340th and 319th Bombardment groups, he flew a total of 63 combat missions over Europe and Japan. Upon completion of his tour of duty he attended the University of Minnesota, earning a bachelor of science degree in aeronautical engineering in 1949. He then worked for two years as an aeronautical engineer with the Boeing Aircraft Corporation until he was recalled to active duty in 1951 with the Minnesota Air National Guard. After his second tour of duty, he attended the USAF Test Pilot School in 1955 at Edwards Air Force Base, California, where he subsequently served as a test pilot until 1959. Slayton resigned from the Air Force in 1963 to fully devote himself to his duties at NASA. In 1972, following a comprehensive review of his medical status, he was finally restored to full flight status and certified eligible for piloted space flight. Two years later he made his first space flight as Apollo docking module pilot of the Apollo-Soyuz Test Project, logging over 217 hours in space. Slayton retired from NASA in 1982 and founded a company to develop rockets for small commercial payloads. (*http://www.jsc.nasa.gov/Bios/htmlbios/slayton.html*) accessed 16 October 2006.

Charles P. Sonnett (1924–   ) served as chief of NASA's Lunar and Planetary Sciences from 1960–62. He earned a bachelor of arts degree in physics from the University of California at Berkeley in 1949 and a masters and Ph.D. both in Nuclear Physics from the University of California at Los Angeles in 1951 and 1954, respectively. From 1954 to 1960 he was the Senior Staff Head of the Space Physics Section of Space Technology Laboratories while at the same time lecturing in the U.C.L.A. department of engineering. In 1962 Dr. Sonnett became the head of the Space Sciences Division at Ames Research Center, where he oversaw research for the nation's space program in the areas of geophysics, interplanetary and planetary physics, planetary sciences, astronomy, and astrophysics. See "Sonnett, Dr. Charles P.," biographical file 002160, NASA Historical Reference Collection, History Division, NASA Headquarters, Washington, DC.

T

Edward Teller (1908–2003) was a naturalized American physicist born in Hungary who made important contributions to the development of both fission-

and fusion-type bombs. As a member of the advisory committee of the AEC, he advocated the hydrogen bomb as a U.S. tactical weapon, arousing a great deal of controversy. He also spoke publicly about Sputnik as showing that the Soviets were beginning to gain a lead on the U.S. in the fields of science and technology. Among other works on Teller, see the view of the insider, Herbert York, *The Advisors: Oppenheimer, Teller, and the Superbomb* (San Francisco: W. H. Freeman, 1976). For one perspective on Teller's more recent and still controversial activities in the world of science and defense technology, see William J. Broad, *Teller's War: The Top-Secret Story Behind the Star Wars Deception* (New York: Simon & Schuster, 1992).

Albert Thomas (1898–1966) (D-TX), a lawyer and World War I veteran, had first been elected to the House of Representatives in 1936 and served successively until 1962. In 1960–1962 he was chair of the independent offices subcommittee of the House Appropriations Committee and thus exercised considerable congressional power over NASA's funding. "Thomas, Albert," biographical file 002295, NASA Historical Reference Collection, History Division, NASA Headquarters, Washington, DC.

Howard W. Tindall (1925–1995) was an expert in orbital mechanics and a key figure in the development of rendezvous techniques for Gemini and lunar trajectories for Apollo. He was directly responsible for planning all ten of the Gemini missions at the Manned Spacecraft Center in Houston. Tindall received a bachelor of science degree in mechanical engineering from Brown University in 1948 and subsequently joined the National Advisory Committee for Aeronautics at Langley Research Center that same year. He moved to Houston in 1961 to assume mission planning responsibilities in the Flight Operations Directorate for Gemini. He gained popularity within the organization for his irreverently written "Tindallgrams" which captured the details of complicated aspects of key flight problems. In 1970, Tindall was appointed deputy director of Flight Operations, and in 1972, he became director. He retired from NASA in 1979 after thirty-one years of service. See "Tindall, Howard W., Jr.," biographical file 004812, NASA Historical Reference Collection, History Division, NASA Headquarters, Washington, DC.

V

Cyrus Vance (1917–2002) had a long career as a senior government official in various Democratic administrations. He had been general counsel for the Department of Defense during the Kennedy administration of the early 1960s, and as Secretary of the Army, 1962-1964. He was Deputy Secretary of Defense, 1964–1967. He served as Secretary of State for President Jimmy Carter in the latter 1970s. See "Vance, Cyrus R[oberts]," *Current Biography 1977*, pp. 408-11.

Robert B. Voas (1928–   ) was part of the first Space Task Group in 1958 and helped to conceptualize the criteria for the selection of astronauts. He earned a bachelor of arts, master of science and Ph.D. in psychology from the University of California in Los Angeles, as well as a bachelor of philosophy degree from the University of Chicago. Voas served in the United States Navy where he reached the rank of lieutenant and logged about three hundred hours in jet aircraft. After being assigned to NACA in 1958, Voas went on to serve as Training

Officer for project Mercury and later proposed the selection process for the Gemini astronauts. See "Voas, Robert B.: Biography," biographical file 002449, NASA Historical Reference Collection, History Division, NASA Headquarters, Washington, DC.

Wernher von Braun (1912–1977) was the leader of what has been called the "rocket team," which had developed the German V-2 ballistic missile in World War II. At the conclusion of the war, von Braun and some of his chief assistants—as part of a military operation called Project Paperclip—came to America and were installed at Fort Bliss in El Paso, Texas, to work on rocket development and use the V-2 for high altitude research. They used launch facilities at the nearby White Sands Proving Ground in New Mexico. Later, in 1950 von Braun's team moved to the Redstone Arsenal near Huntsville, Alabama, to concentrate on the development of a new missile for the Army. They built the Army's Jupiter ballistic missile, and before that the Redstone, used by NASA to launch the first Mercury capsules. The story of von Braun and the "rocket team" has been told many times. See, as examples, David H. DeVorkin, *Science With a Vengeance: How the Military Created the US Space Sciences After World War II* (New York: Springer-Verlag, 1992); Frederick I. Ordway III and Mitchell R. Sharpe, *The Rocket Team* (New York: Thomas Y. Crowell, 1979); Erik Bergaust, *Wernher von Braun* (Washington, DC: National Space Institute, 1976); "Wernher von Braun,"(*http://history.nasa.gov/sputnik/braun. html*) accessed 23 October 2006; "Marshall Space Flight Center (MSFC)," (*http:// history.nasa.gov/centerhistories/marshall.htm*) accessed 23 October 2006.

W

James E. Webb (1906–1992) was NASA Administrator between 1961 and 1968. Previously he had been an aide to a Congressman in New Deal Washington, an aide to Washington lawyer Max O. Gardner, and a business executive with the Sperry Corporation and the Kerr-McGee Oil Co. He had also been director of the Bureau of the Budget between 1946 and 1950 and Under Secretary of State, 1950–1952. See W. Henry Lambright, *Powering Apollo: James E. Webb of NASA* (Baltimore, MD: Johns Hopkins University Press, 1995) and "James E. Webb" (*http://www.hq.nasa.gov/office/pao/History/Biographies/webb.html*) accessed 23 October 2006.

Caspar W. Weinberger (1917–2006), longtime Republican government official, was a senior member of the Nixon, Ford, and Reagan administrations. For Nixon he was deputy director (1970–1972) and director (1972–1976) of the Office of Management and Budget. In this capacity he had a leading role in shaping the direction of NASA's major effort of the 1970s, the development of a reusable Space Shuttle. For Reagan he served as Secretary of Defense, where he also oversaw the use of the Shuttle in the early 1980s for the launching of classified Department of Defense payloads into orbit. See "Weinberger, Caspar W(illard)," *Current Biography 1973*, pp. 428-30.

Edward C. Welsh (1909–1990) had a long career in various private and public enterprises. He had served as legislative assistant to Senator Stuart Symington (D-MO), 1953–1961, and was the executive secretary of the National Aeronautics and Space Council through the 1960s. See "Welsh, Dr. Edward C.," biographical file 002546, NASA Historical Reference Collection, NASA History Division, NASA Headquarters, Washington, DC.

Jerome B. Wiesner (1915–1994) was Science Advisor to President John F. Kennedy. He had been a faculty member of the Massachusetts Institute of Technology, and had served on President Eisenhower's Science Advisory Committee. During the presidential campaign of 1960, Wiesner had advised Kennedy on science and technology issues and chaired a transition team report on the space program that questioned the value of human spaceflight. As Kennedy's Science Advisor he tussled with NASA over the lunar landing commitment and the method of conducting it. See Gregg Herken, *Cardinal Choices: Science Advice to the President from Hiroshima to SDI* (New York: Oxford University Press, 1992).

Edward H. White, Jr. (1930–1967) piloted the Gemini 4 mission during which he carried out the first extra vehicular activity. He graduated with a bachelor of science degree from the United States Military Academy in 1952 and then was commissioned into the Air Force. Following his flight training, he was stationed in Germany for three and a half years with a fighter squadron, flying F-86's and F-100's. White then returned to the United States and earned a master of science degree in aeronautical engineering from the University of Michigan in 1959. That same year he attended the Air Force Test Pilot School at Edwards Air Force Base, California, and was later reassigned to Wright-Patterson Air Force Base in Ohio as an experimental test pilot with the Aeronautical Systems Division. He was named a member of the second group of astronauts selected by NASA in 1962. After piloting Gemini 4 and serving as backup command pilot for Gemini 7, he was named as one of the pilots for the Apollo 1 mission. Lieutenant Colonel White died on January 27, 1967 in the Apollo spacecraft flash fire during a launch pad test at Kennedy Space Center, Florida, and was posthumously awarded the Congressional Space Medal of Honor. See "Edward H. White, II," (*http://www.jsc. nasa.gov/Bios/htmlbios/white-eh.html*) accessed 30 October 30, 2006.

Walter C. Williams (1919–1995) earned a B.S. in aerospace engineering from LSU in 1939 and went to work for the NACA in 1940, serving as a project engineer to improve the handling, maneuverability, and flight characteristics of World War II fighters. Following the war, he went to what became Edwards Air Force Base to set up flight tests for the X-1, including the first human supersonic flight by Capt. Charles E. Yeager in October 1947. He became the founding director of the organization that became Dryden Flight Research Facility. In September 1959 he assumed associate directorship of the new NASA Space Task Group at Langley, created to carry out Project Mercury. He later became director of operations for the project, then associate director of the NASA Manned Spacecraft Center in Houston, subsequently renamed the Johnson Space Center. In 1963 Williams moved to NASA Headquarters as Deputy Associate Administrator of the Office of Manned Space Flight. From 1964 to 1975, he was a vice president for Aerospace Corporation. Then from 1975-1982 he served as chief engineer of NASA, retiring in the latter year. See "Williams, W.C.," biographical file 002618, NASA Historical Reference Collection, History Division, NASA Headquarters, Washington, DC.

Z

Charles H. Zimmerman (1907–) was handpicked by Robert R. Gilruth to serve on the first Space Task Group in 1958 and served as Director of Aeronautical Research in NASA's Office of Advanced Research and Technology from 1962-1963. He received a B.S. in electrical engineering from the University of Kansas in 1929 and joined the staff of the National Advisory Committee for Aeronautics in that same year. He spent the next 33 years of his life in government and private industry developing and improving new aircraft. Zimmerman earned a master's degree in aeronautical engineering from the University of Virginia in 1954 and two years later was the recipient of both the Alexander Klemin Award of the American Helicopter Society and the Wright Brothers Medal of the Society of Automotive Engineers. See "Charles H. Zimmerman," biographical file 002882, NASA Historical Reference Collection, History Division, NASA Headquarters, Washington, DC.

Index

H

O

The NASA History Series

Reference Works, NASA SP-4000:

Grimwood, James M. *Project Mercury: A Chronology.* NASA SP-4001, 1963.

Grimwood, James M., and C. Barton Hacker, with Peter J. Vorzimmer. *Project Gemini Technology and Operations: A Chronology.* NASA SP-4002, 1969.

Link, Mae Mills. *Space Medicine in Project Mercury.* NASA SP-4003, 1965.

Astronautics and Aeronautics, 1963: Chronology of Science, Technology, and Policy. NASA SP-4004, 1964.

Astronautics and Aeronautics, 1964: Chronology of Science, Technology, and Policy. NASA SP-4005, 1965.

Astronautics and Aeronautics, 1965: Chronology of Science, Technology, and Policy. NASA SP-4006, 1966.

Astronautics and Aeronautics, 1966: Chronology of Science, Technology, and Policy. NASA SP-4007, 1967.

Astronautics and Aeronautics, 1967: Chronology of Science, Technology, and Policy. NASA SP-4008, 1968.

Ertel, Ivan D., and Mary Louise Morse. *The Apollo Spacecraft: A Chronology, Volume I, Through November 7, 1962.* NASA SP-4009, 1969.

Morse, Mary Louise, and Jean Kernahan Bays. *The Apollo Spacecraft: A Chronology, Volume II, November 8, 1962–September 30, 1964.* NASA SP-4009, 1973.

Brooks, Courtney G., and Ivan D. Ertel. *The Apollo Spacecraft: A Chronology, Volume III, October 1, 1964–January 20, 1966.* NASA SP-4009, 1973.

Ertel, Ivan D., and Roland W. Newkirk, with Courtney G. Brooks. *The Apollo Spacecraft: A Chronology, Volume IV, January 21, 1966–July 13, 1974.* NASA SP-4009, 1978.

Astronautics and Aeronautics, 1968: Chronology of Science, Technology, and Policy. NASA SP-4010, 1969.

Newkirk, Roland W., and Ivan D. Ertel, with Courtney G. Brooks. *Skylab: A Chronology.* NASA SP-4011, 1977.

Van Nimmen, Jane, and Leonard C. Bruno, with Robert L. Rosholt. *NASA Historical Data Book, Volume I: NASA Resources, 1958–1968.* NASA SP-4012, 1976, rep. ed. 1988.

Ezell, Linda Neuman. *NASA Historical Data Book, Volume II: Programs and Projects, 1958–1968.* NASA SP-4012, 1988.

Ezell, Linda Neuman. *NASA Historical Data Book, Volume III: Programs and Projects, 1969–1978.* NASA SP-4012, 1988.

Gawdiak, Ihor Y., with Helen Fedor, compilers. *NASA Historical Data Book, Volume IV: NASA Resources, 1969–1978.* NASA SP-4012, 1994.

Rumerman, Judy A., compiler. *NASA Historical Data Book, 1979–1988: Volume V, NASA Launch Systems, Space Transportation, Human Spaceflight, and Space Science.* NASA SP-4012, 1999.

Rumerman, Judy A., compiler. *NASA Historical Data Book, Volume VI: NASA Space Applications, Aeronautics and Space Research and Technology, Tracking and Data Acquisition/Space Operations, Commercial Programs, and Resources, 1979–1988.* NASA SP-2000-4012, 2000.

Astronautics and Aeronautics, 1969: Chronology of Science, Technology, and Policy. NASA SP-4014, 1970.

Astronautics and Aeronautics, 1970: Chronology of Science, Technology, and Policy. NASA SP-4015, 1972.

Astronautics and Aeronautics, 1971: Chronology of Science, Technology, and Policy. NASA SP-4016, 1972.

Astronautics and Aeronautics, 1972: Chronology of Science, Technology, and Policy. NASA SP-4017, 1974.

Astronautics and Aeronautics, 1973: Chronology of Science, Technology, and Policy. NASA SP-4018, 1975.

Astronautics and Aeronautics, 1974: Chronology of Science, Technology, and Policy. NASA SP-4019, 1977.

Astronautics and Aeronautics, 1975: Chronology of Science, Technology, and Policy. NASA SP-4020, 1979.

Astronautics and Aeronautics, 1976: Chronology of Science, Technology, and Policy. NASA SP-4021, 1984.

Astronautics and Aeronautics, 1977: Chronology of Science, Technology, and Policy. NASA SP-4022, 1986.

Astronautics and Aeronautics, 1978: Chronology of Science, Technology, and Policy. NASA SP-4023, 1986.

Astronautics and Aeronautics, 1979–1984: Chronology of Science, Technology, and Policy. NASA SP-4024, 1988.

Astronautics and Aeronautics, 1985: Chronology of Science, Technology, and Policy. NASA SP-4025, 1990.

Noordung, Hermann. *The Problem of Space Travel: The Rocket Motor.* Edited by Ernst Stuhlinger and J. D. Hunley, with Jennifer Garland. NASA SP-4026, 1995.

Astronautics and Aeronautics, 1986–1990: A Chronology. NASA SP-4027, 1997.

Astronautics and Aeronautics, 1990–1995: A Chronology. NASA SP-2000-4028, 2000.

Management Histories, NASA SP-4100:

Rosholt, Robert L. *An Administrative History of NASA, 1958–1963.* NASA SP-4101, 1966.

Levine, Arnold S. *Managing NASA in the Apollo Era.* NASA SP-4102, 1982.

Roland, Alex. *Model Research: The National Advisory Committee for Aeronautics, 1915–1958.* NASA SP-4103, 1985.

Fries, Sylvia D. *NASA Engineers and the Age of Apollo.* NASA SP-4104, 1992.

Glennan, T. Keith. *The Birth of NASA: The Diary of T. Keith Glennan. J. D. Hunley,* editor. NASA SP-4105, 1993.

Seamans, Robert C., Jr. *Aiming at Targets: The Autobiography of Robert C. Seamans, Jr.* NASA SP-4106, 1996.

Garber, Stephen J., editor. *Looking Backward, Looking Forward: Forty Years of U.S. Human Spaceflight Symposium.* NASA SP-2002-4107, 2002.

Mallick, Donald L. with Peter W. Merlin. *The Smell of Kerosene: A Test Pilot's Odyssey.* NASA SP-4108, 2003.

Iliff, Kenneth W. and Curtis L. Peebles. *From Runway to Orbit: Reflections of a NASA Engineer.* NASA SP-2004-4109, 2004.

Chertok, Boris. *Rockets and People, Volume 1.* NASA SP-2005-4110, 2005.

Laufer, Alexander, Todd Post, and Edward Hoffman. *Shared Voyage: Learning and Unlearning from Remarkable Projects.* NASA SP-2005-4111, 2005.

Dawson, Virginia P. and Mark D. Bowles. *Realizing the Dream of Flight: Biographical Essays in Honor of the Centennial of Flight, 1903-2003.* NASA SP-2005-4112, 2005.

Mudgway, Douglas J. *William H. Pickering: America's Deep Space Pioneer,* NASA SP-2007-4113, 2007.

Project Histories, NASA SP-4200:

Swenson, Loyd S., Jr., James M. Grimwood, and Charles C. Alexander. *This New Ocean: A History of Project Mercury.* NASA SP-4201, 1966; rep. ed. 1998.

Green, Constance McLaughlin, and Milton Lomask. *Vanguard: A History.* NASA SP-4202, 1970; rep. ed. Smithsonian Institution Press, 1971.

Hacker, Barton C., and James M. Grimwood. *On the Shoulders of Titans: A History of Project Gemini.* NASA SP-4203, 1977.

Benson, Charles D., and William Barnaby Faherty. *Moonport: A History of Apollo Launch Facilities and Operations.* NASA SP-4204, 1978.

Brooks, Courtney G., James M. Grimwood, and Loyd S. Swenson, Jr. *Chariots for Apollo: A History of Manned Lunar Spacecraft.* NASA SP-4205, 1979.

Bilstein, Roger E. *Stages to Saturn: A Technological History of the Apollo/Saturn Launch Vehicles.* NASA SP-4206, 1980, rep. ed. 1997.

SP-4207 not published.

Compton, W. David, and Charles D. Benson. *Living and Working in Space: A History of Skylab.* NASA SP-4208, 1983.

Ezell, Edward Clinton, and Linda Neuman Ezell. *The Partnership: A History of the Apollo-Soyuz Test Project.* NASA SP-4209, 1978.

Hall, R. Cargill. *Lunar Impact: A History of Project Ranger.* NASA SP-4210, 1977.

Newell, Homer E. B*eyond the Atmosphere: Early Years of Space Science.* NASA SP-4211, 1980.

Ezell, Edward Clinton, and Linda Neuman Ezell. *On Mars: Exploration of the Red Planet, 1958–1978.* NASA SP-4212, 1984.

Pitts, John A. *The Human Factor: Biomedicine in the Manned Space Program to 1980.* NASA SP-4213, 1985.

Compton, W. David. *Where No Man Has Gone Before: A History of Apollo Lunar Exploration Missions.* NASA SP-4214, 1989.

Naugle, John E. *First Among Equals: The Selection of NASA Space Science Experiments.* NASA SP-4215, 1991.

Wallace, Lane E. *Airborne Trailblazer: Two Decades with NASA Langley's Boeing 737 Flying Laboratory.* NASA SP-4216, 1994.

Butrica, Andrew J., editor. *Beyond the Ionosphere: Fifty Years of Satellite Communication.* NASA SP-4217, 1997.

Butrica, Andrew J. *To See the Unseen: A History of Planetary Radar Astronomy.* NASA SP-4218, 1996.

Mack, Pamela E., editor. *From Engineering Science to Big Science: The NACA and NASA Collier Trophy Research Project Winners.* NASA SP-4219, 1998.

Reed, R. Dale, with Darlene Lister. *Wingless Flight: The Lifting Body Story.* NASA SP-4220, 1997.

Heppenheimer, T. A. *The Space Shuttle Decision: NASA's Search for a Reusable Space Vehicle.* NASA SP-4221, 1999.

Hunley, J. D., editor. *Toward Mach 2: The Douglas D-558 Program.* NASA SP-4222, 1999.

Swanson, Glen E., editor. *"Before this Decade Is Out . . .": Personal Reflections on the Apollo Program.* NASA SP-4223, 1999.

Tomayko, James E. *Computers Take Flight: A History of NASA's Pioneering Digital Fly-by-Wire Project.* NASA SP-2000-4224, 2000.

Morgan, Clay. *Shuttle-Mir: The U.S. and Russia Share History's Highest Stage.* NASA SP-2001-4225, 2001.

Leary, William M. *"We Freeze to Please": A History of NASA's Icing Research Tunnel and the Quest for Flight Safety.* NASA SP-2002-4226, 2002.

Mudgway, Douglas J. *Uplink-Downlink: A History of the Deep Space Network 1957–1997.* NASA SP-2001-4227, 2001.

Dawson, Virginia P. and Mark D. Bowles. *Taming Liquid Hydrogen: The Centaur Upper Stage Rocket, 1958-2002.* NASA SP-2004-4230, 2004.

Meltzer, Michael. *Mission to Jupiter: A History of the Galileo Project.* NASA SP-2007-4231.

Heppenheimer, T. A. *Facing the Heat Barrier: A History of Hypersonics.* NASA SP-2007-4232, 2007.

Tsiao, Sunny. *Read You Loud and Clear.* NASA SP-2007-4233, 2007.

Center Histories, NASA SP-4300:

Rosenthal, Alfred. *Venture into Space: Early Years of Goddard Space Flight Center.* NASA SP-4301, 1985.

Hartman, Edwin P. *Adventures in Research: A History of Ames Research Center, 1940–1965.* NASA SP-4302, 1970.

Hallion, Richard P. *On the Frontier: Flight Research at Dryden, 1946–1981.* NASA SP-4303, 1984.

Muenger, Elizabeth A. *Searching the Horizon: A History of Ames Research Center, 1940–1976.* NASA SP-4304, 1985.

Hansen, James R. *Engineer in Charge: A History of the Langley Aeronautical Laboratory, 1917–1958.* NASA SP-4305, 1987.

Dawson, Virginia P. *Engines and Innovation: Lewis Laboratory and American Propulsion Technology.* NASA SP-4306, 1991.

Dethloff, Henry C. *"Suddenly Tomorrow Came . . .": A History of the Johnson Space Center.* NASA SP-4307, 1993.

Hansen, James R. *Spaceflight Revolution: NASA Langley Research Center from Sputnik to Apollo.* NASA SP-4308, 1995.

Wallace, Lane E. *Flights of Discovery: 50 Years at the NASA Dryden Flight Research Center.* NASA SP-4309, 1996.

Herring, Mack R. *Way Station to Space: A History of the John C. Stennis Space Center.* NASA SP-4310, 1997.

Wallace, Harold D., Jr. *Wallops Station and the Creation of the American Space Program.* NASA SP-4311, 1997.

Wallace, Lane E. *Dreams, Hopes, Realities: NASA's Goddard Space Flight Center, The First Forty Years.* NASA SP-4312, 1999.

Dunar, Andrew J., and Stephen P. Waring. *Power to Explore: A History of the Marshall Space Flight Center.* NASA SP-4313, 1999.

Bugos, Glenn E. *Atmosphere of Freedom: Sixty Years at the NASA Ames Research Center.* NASA SP-2000-4314, 2000.

Schultz, James. *Crafting Flight: Aircraft Pioneers and the Contributions of the Men and Women of NASA Langley Research Center.* NASA SP-2003-4316, 2003.

General Histories, NASA SP-4400:

Corliss, William R. *NASA Sounding Rockets, 1958–1968: A Historical Summary.* NASA SP-4401, 1971.

Wells, Helen T., Susan H. Whiteley, and Carrie Karegeannes. *Origins of NASA Names.* NASA SP-4402, 1976.

Anderson, Frank W., Jr. *Orders of Magnitude: A History of NACA and NASA, 1915–1980.* NASA SP-4403, 1981.

Sloop, John L. *Liquid Hydrogen as a Propulsion Fuel, 1945–1959.* NASA SP-4404, 1978.

Roland, Alex. *A Spacefaring People: Perspectives on Early Spaceflight.* NASA SP-4405, 1985.

Bilstein, Roger E. *Orders of Magnitude: A History of the NACA and NASA, 1915–1990.* NASA SP-4406, 1989.

Logsdon, John M., editor, with Linda J. Lear, Jannelle Warren-Findley, Ray A. Williamson, and Dwayne A. Day. *Exploring the Unknown: Selected Documents in the History of the U.S. Civil Space Program, Volume I, Organizing for Exploration.* NASA SP-4407, 1995.

Logsdon, John M., editor, with Dwayne A. Day and Roger D. Launius. *Exploring the Unknown: Selected Documents in the History of the U.S. Civil Space Program, Volume II, Relations with Other Organizations.* NASA SP-4407, 1996.

Logsdon, John M., editor, with Roger D. Launius, David H. Onkst, and Stephen J. Garber. *Exploring the Unknown: Selected Documents in the History of the U.S. Civil Space Program, Volume III, Using Space.* NASA SP-4407, 1998.

Logsdon, John M., general editor, with Ray A. Williamson, Roger D. Launius, Russell J. Acker, Stephen J. Garber, and Jonathan L. Friedman. *Exploring the Unknown: Selected Documents in the History of the U.S. Civil Space Program, Volume IV, Accessing Space.* NASA SP-4407, 1999.

Logsdon, John M., general editor, with Amy Paige Snyder, Roger D. Launius, Stephen J. Garber, and Regan Anne Newport. *Exploring the Unknown: Selected Documents in the History of the U.S. Civil Space Program, Volume V, Exploring the Cosmos.* NASA SP-2001-4407, 2001.

Siddiqi, Asif A. *Challenge to Apollo: The Soviet Union and the Space Race, 1945–1974.* NASA SP-2000-4408, 2000.

Hansen, James R., editor. *The Wind and Beyond: Journey into the History of Aerodynamics in America, Volume I, The Ascent of the Airplane.* NASA SP-2003-4409, 2003.

Hogan, Thor. *Mars Wars: The Rise and Fall of the Space Exploration Initiative.* NASA SP-2007-4410, 2007.

Hansen, James R., editor. *The Wind and Beyond: Journey into the History of Aerodynamics in America, Volume II, Reinventing the Airplane.* NASA SP-2007-4409, 2007.

Monographs in Aerospace History, NASA SP-4500:

Launius, Roger D. and Aaron K. Gillette, compilers, *Toward a History of the Space Shuttle: An Annotated Bibliography*. Monograph in Aerospace History, No. 1, 1992.

Launius, Roger D., and J. D. Hunley, compilers, *An Annotated Bibliography of the Apollo Program*. Monograph in Aerospace History, No. 2, 1994.

Launius, Roger D. *Apollo: A Retrospective Analysis*. Monograph in Aerospace History, No. 3, 1994.

Hansen, James R. *Enchanted Rendezvous: John C. Houbolt and the Genesis of the Lunar-Orbit Rendezvous Concept*. Monograph in Aerospace History, No. 4, 1995.

Gorn, Michael H. *Hugh L. Dryden's Career in Aviation and Space*. Monograph in Aerospace History, No. 5, 1996.

Powers, Sheryll Goecke. *Women in Flight Research at NASA Dryden Flight Research Center, from 1946 to 1995*. Monograph in Aerospace History, No. 6, 1997.

Portree, David S. F. and Robert C. Trevino. *Walking to Olympus: An EVA Chronology*. Monograph in Aerospace History, No. 7, 1997.

Logsdon, John M., moderator. *Legislative Origins of the National Aeronautics and Space Act of 1958: Proceedings of an Oral History Workshop*. Monograph in Aerospace History, No. 8, 1998.

Rumerman, Judy A., compiler, *U.S. Human Spaceflight, A Record of Achievement 1961–1998*. Monograph in Aerospace History, No. 9, 1998.

Portree, David S. F. *NASA's Origins and the Dawn of the Space Age*. Monograph in Aerospace History, No. 10, 1998.

Logsdon, John M. *Together in Orbit: The Origins of International Cooperation in the Space Station*. Monograph in Aerospace History, No. 11, 1998.

Phillips, W. Hewitt. *Journey in Aeronautical Research: A Career at NASA Langley Research Center*. Monograph in Aerospace History, No. 12, 1998.

Braslow, Albert L. *A History of Suction-Type Laminar-Flow Control with Emphasis on Flight Research*. Monograph in Aerospace History, No. 13, 1999.

Logsdon, John M., moderator. *Managing the Moon Program: Lessons Learned From Apollo*. Monograph in Aerospace History, No. 14, 1999.

Perminov, V. G. *The Difficult Road to Mars: A Brief History of Mars Exploration in the Soviet Union*. Monograph in Aerospace History, No. 15, 1999.

Tucker, Tom. *Touchdown: The Development of Propulsion Controlled Aircraft at NASA Dryden*. Monograph in Aerospace History, No. 16, 1999.

Maisel, Martin D., Demo J. Giulianetti, and Daniel C. Dugan. *The History of the XV-15 Tilt Rotor Research Aircraft: From Concept to Flight*. NASA SP-2000-4517, 2000.

Jenkins, Dennis R. *Hypersonics Before the Shuttle: A Concise History of the X-15 Research Airplane*. NASA SP-2000-4518, 2000.

Chambers, Joseph R. *Partners in Freedom: Contributions of the Langley Research Center to U.S. Military Aircraft in the 1990s.* NASA SP-2000-4519, 2000.

Waltman, Gene L. *Black Magic and Gremlins: Analog Flight Simulations at NASA's Flight Research Center.* NASA SP-2000-4520, 2000.

Portree, David S. F. *Humans to Mars: Fifty Years of Mission Planning, 1950–2000.* NASA SP-2001-4521, 2001.

Thompson, Milton O., with J. D. Hunley. *Flight Research: Problems Encountered and What They Should Teach Us.* NASA SP-2000-4522, 2000.

Tucker, Tom. *The Eclipse Project.* NASA SP-2000-4523, 2000.

Siddiqi, Asif A. *Deep Space Chronicle: A Chronology of Deep Space and Planetary Probes, 1958–2000.* NASA SP-2002-4524, 2002.

Merlin, Peter W. *Mach 3+: NASA/USAF YF-12 Flight Research, 1969–1979.* NASA SP-2001-4525, 2001.

Anderson, Seth B. *Memoirs of an Aeronautical Engineer—Flight Tests at Ames Research Center: 1940–1970.* NASA SP-2002-4526, 2002.

Renstrom, Arthur G. *Wilbur and Orville Wright: A Bibliography Commemorating the One-Hundredth Anniversary of the First Powered Flight on December 17, 1903.* NASA SP-2002-4527, 2002.

No monograph 28.

Chambers, Joseph R. *Concept to Reality: Contributions of the NASA Langley Research Center to U.S. Civil Aircraft of the 1990s.* SP-2003-4529, 2003.

Peebles, Curtis, editor. *The Spoken Word: Recollections of Dryden History, The Early Years.* SP-2003-4530, 2003.

Jenkins, Dennis R., Tony Landis, and Jay Miller. *American X-Vehicles: An Inventory-X-1 to X-50.* SP-2003-4531, 2003.

Renstrom, Arthur G. *Wilbur and Orville Wright: A Chronology Commemorating the One-Hundredth Anniversary of the First Powered Flight on December 17, 1903.* NASA SP-2003-4532, 2002.

Bowles, Mark D. and Robert S. Arrighi. *NASA's Nuclear Frontier: The Plum Brook Research Reactor.* SP-2004-4533, 2003.

Matranga, Gene J. and C. Wayne Ottinger, Calvin R. Jarvis with D. Christian Gelzer. *Unconventional, Contrary, and Ugly: The Lunar Landing Research Vehicle.* NASA SP-2006-4535.

McCurdy, Howard E. *Low Cost Innovation in Spaceflight: The History of the Near Earth Asteroid Rendezvous (NEAR) Mission.* NASA SP-2005-4536, 2005.

Seamans, Robert C. Jr. *Project Apollo: The Tough Decisions.* NASA SP-2005-4537, 2005.

Lambright, W. Henry. *NASA and the Environment: The Case of Ozone Depletion.* NASA SP-2005-4538, 2005.

Chambers, Joseph R. *Innovation in Flight: Research of the NASA Langley Research Center on Revolutionary Advanced Concepts for Aeronautics.* NASA SP-2005-4539, 2005.

Phillips, W. Hewitt. *Journey Into Space Research: Continuation of a Career at NASA Langley Research Center.* NASA SP-2005-4540, 2005.

Rumerman, Judith A., compiler, with Chris Gamble and Gabriel Okolski, *U.S. Human Spaceflight: A Record of Achievement, 1961-2006.* NASA SP-2007-4541, 2007.

Electronic Media, NASA SP-4600:

Remembering Apollo 11: The 30th Anniversary Data Archive CD-ROM. NASA SP-4601, 1999.

The Mission Transcript Collection: U.S. Human Spaceflight Missions from Mercury Redstone 3 to Apollo 17. NASA SP-2000-4602, 2001.

Shuttle-Mir: The United States and Russia Share History's Highest Stage. NASA SP-2001-4603, 2002.

U.S. Centennial of Flight Commission Presents Born of Dreams—Inspired by Freedom. NASA SP-2004-4604, 2004.

Of Ashes and Atoms: A Documentary on the NASA Plum Brook Reactor Facility. NASA SP-2005-4605, 2005.

Taming Liquid Hydrogen: The Centaur Upper Stage Rocket Interactive CD-ROM. NASA SP-2004-4606, 2004.

Fueling Space Exploration: The History of NASA's Rocket Engine Test Facility DVD. NASA SP-2005-4607, 2005.

Conference Proceedings, NASA SP-4700:

Dick, Steven J., and Keith L. Cowing, editors. *Risk and Exploration: Earth, Sea and the Stars.* NASA SP-2005-4701, 2005.

Dick, Steven J., and Roger D. Launius, editors. *Critical Issues in the History of Spaceflight.* NASA SP-2006-4702, 2006.

Societal Impact, NASA SP-4800:

Dick, Steven J., and Roger D. Launius, editors. *Societal Impact of Spaceflight.* NASA SP-2007-4801, 2007.